Logic, Argumentation & Reasoning

Interdisciplinary Perspectives from the Humanities and Social Sciences

Volume 34

Logic, Argumentation & Reasoning (LAR) explores links between the Humanities and Social Sciences, with theories (including decision and action theory) drawn from the cognitive sciences, economics, sociology, law, logic, and the philosophy of science.

Its main ambitions are to develop a theoretical framework that will encourage and enable interaction between disciplines, and to integrate the Humanities and Social Sciences around their main contributions to public life, using informed debate, lucid decision-making, and action based on reflection.

- Argumentation models and studies
- Communication, language and techniques of argumentation
- Reception of arguments, persuasion and the impact of power
- Diachronic transformations of argumentative practices

LAR is developed in partnership with the Maison Européenne des Sciences de l'Homme et de la Société (MESHS) at Nord - Pas de Calais and the UMR-STL: 8163 (CNRS).

This book series is indexed in SCOPUS.

Proposals should include:

- A short synopsis of the work, or the introduction chapter
- The proposed Table of Contents
- The CV of the lead author(s)
- If available: one sample chapter

We aim to make a first decision within 1 month of submission. In case of a positive first decision, the work will be provisionally contracted—the final decision about publication will depend upon the result of an anonymous peer review of the complete manuscript.

The complete work is usually peer-reviewed within 3 months of submission.

LAR discourages the submission of manuscripts containing reprints of previously published material, and/or manuscripts that are less than 150 pages / 85,000 words.

For inquiries and proposal submissions, authors may contact the editor-in-chief, Shahid Rahman at: shahid.rahman@univ-lille.fr, or the managing editor, Juan Redmond, at: juan.redmond@uv.cl

Teresa Lopez-Soto • Alvaro Garcia-Lopez
Francisco J. Salguero-Lamillar
Editors

The Theory of Mind Under Scrutiny

Psychopathology, Neuroscience, Philosophy of Mind and Artificial Intelligence

Springer

Editors
Teresa Lopez-Soto
Laboratory of Speech and Phonetic Sciences
University of Seville
Seville, Spain

Department of Forensic Medicine, Psychiatry and Pathology
Complutense University of Madrid
Madrid, Spain

Francisco J. Salguero-Lamillar
Department of Spanish Language, Linguistics and Theory of Literature
University of Seville
Seville, Spain

Alvaro Garcia-Lopez
Department of Applied Mathematics, Science and Technology of Materials and Electronic Engineering
University Rey Juan Carlos
Madrid, Spain

ISSN 2214-9120 ISSN 2214-9139 (electronic)
Logic, Argumentation & Reasoning
ISBN 978-3-031-46741-7 ISBN 978-3-031-46742-4 (eBook)
https://doi.org/10.1007/978-3-031-46742-4

This Springer imprint is published by the registered company Springer Nature Switzerland AG
The registered company address is: Gewerbestrasse 11, 6330 Cham, Switzerland

Paper in this product is recyclable.

Contents

Part I Fundamental Issues in ToM and New Perspectives

1 Interdisciplinary Debates and Approaches on TOM, AI, and Language . . . 3
Sergio Marin-Conejo

2 An Introduction to Theory of Mind: Fundamental Concepts and Issues . . . 11
Miriam Rivero-Contreras, David Saldaña, and Martina Micai

3 A Mental Files Theory of Mind: How Children Represent Belief and Its Aspectuality . . . 35
Michael Huemer

Part II Pathologies Associated with ToM

4 Theory of Mind and Reading . . . 73
Pablo Delgado and Isabel R. Rodríguez-Ortiz

5 Theory of Mind and Psychopathology: A Comprehensive Assessment and an Overview of Impairments in Neuropsychiatric Disorders . . . 103
Pilar de la Higuera-González, Alejandra Galvez-Merlin, Elisa Rodríguez-Toscano, Jorge Andreo-Jover, and Alejandro de la Torre-Luque

6 Theory of Mind in Autism: From a Primary Deficit to Just Mutual Misunderstanding? . . . 161
Gema Erena-Guardia, Mila Vulchanova, and David Saldaña

7 **Theory of Mind After Acquired Brain Injury: Basic Aspects, Evaluation and Intervention** 189
Inés Abalo-Rodríguez, Jesús Cabrera-Álvarez, Sandra Doval, Alberto Fernández Lucas, and Dolores Villalobos

8 **Theory of Mind in Children Who Are Deaf: The Importance of Early Language and Conversational Access** 243
Kimberly Peters and David B. Pisoni

9 **Analyzing the Dynamics Between Theory of Mind, Speech Disorders, and Brain Rewiring in Aphasia** 281
Teresa Lopez-Soto

10 **Sleep and Its Disorders: When ToM Is Not Awake** 327
Enrique Montes-Latorre and Paolo Porcacchia

11 **Hypotalamus-Pituitary-Adrenal (HPA) Axes and Their Relationship with Stress, Mood, Personality, and Neurocognitive Functioning** 341
Ana María Martínez Robayo

Part III ToM from the Perspective of Philosophy of Mind

12 **What We Are for Us, What We Are for Others: Consciousness and Identity** 369
Pilar López Segura and Tomás Ortiz Alonso

13 **The Mind-Body Problem: An Overview of Proposed Solutions** 435
Javier Alejandro Galadí

14 **Do I Really Believe That? A Mindreading Account of Belief Self-Ascription** 469
Sylvain Montalvo

15 **Study of the Theory of Mind (ToM) Through the Japanese Philosophy** 493
Diego López-García, Diego Lopez-Luque, Walter Federico Gadea-Aiello, and Emilio José Delgado-Algarra

16 **Content and Process in the Brain. Implications for Clinical and Educational Approaches** 527
Carlos M. Gómez, Brenda Y. Angulo-Ruiz, Elena I. Rodríguez-Martínez, Francisco J. Ruiz-Martínez, Eva María Padilla Muñoz, and María Dolores Lanzarote Fernández

Part IV ToM from the Perspective of Language

17 **Language, Mind and Thought: A General Overview** 561
Inmaculada Aguilar-Ponce

18 Relations Between Bilingualism and Theory of Mind, a Neurologic Challenge. From the *Bilingual Advantage* to a New Assessment of Conclusions 591
Priscila Sánchez Soriano

19 Processes of Early Language Acquisition and Its Implications for ToM in Autistic Children 613
Mayuresh Kumar

20 Gendered Theory of Mind: A Linguistic and Literary Approach 633
Sergio Marin-Conejo and Teresa Lopez-Soto

Part V ToM from the Perspective of Artificial Intelligence

21 Data-Driven Vs Model-Driven Approaches in Cognitive Speech Processing 649
Pedro Gómez-Vilda and Andrés Gómez-Rodellar

22 The Social Machine: Artificial Intelligence (AI) Approaches to Theory of Mind 681
Alberto Nebreda, Danylyna Shpakivska-Bilan, Carmen Camara, and Gianluca Susi

23 Theory of Mind in Artificial Intelligence Applications 723
Alvaro Garcia-Lopez

Part I
Fundamental Issues in ToM and New Perspectives

Chapter 1
Interdisciplinary Debates and Approaches on TOM, AI, and Language

Sergio Marin-Conejo

Abstract In this chapter, we embark on a factual exploration of the intricate threefold domain of the Theory of Mind (TOM), Artificial Intelligence and Language, recognizing its multifaceted pivotal role in shaping our comprehension of the array of debates and approaches that they trigger with the interrelated facets of social interactions, communication, empathy, moral reasoning, and the complex workings of the human brain and mind. In order to navigate the contents of this volume, we introduce first our esteemed editor, Professor Teresa Lopez-Soto, as she meticulously curated the selection of scholarly contributions from renowned experts in their fields. These are then presented organized. These are then organised into different transdisciplinary dimensions, each of which serves to contextualize and illuminate the contemporary landscape of the indispensable concept of ToM. Hence, as readers commence their journey through this compendium, this prefatory chapter provides a genuine overarching perspective on thematic threads that weave through the subsequent discourse, offering a glimpse into the thought-provoking insights presented within the pages of this volume.

Keywords Theory of Mind (TOM) · Artificial Intelligence · Cognitive development · Language and communication · Machine learning · Neuropsychiatric disorders · Philosophy · Social cognition

This volume has been possible thanks to the unwavering effort and remarkable perseverance of its esteemed editor, professor Teresa Lopez-Soto, hailing from the prestigious *University of Seville*. The professor's exceptional dedication to the project has been nothing short of inspiring, as she has demonstrated an unyielding commitment to the pursuit of knowledge and the advancement of academic excellence. With her profound expertise in the subject matter and her keen eye for detail, she has masterfully curated a collection that not only showcases the latest research and groundbreaking insights but also represents a testament to her profound passion

S. Marin-Conejo (✉)
University of Seville, Seville, Spain
e-mail: smarin@us.es

T. Lopez-Soto et al. (eds.), *The Theory of Mind Under Scrutiny*, Logic, Argumentation & Reasoning 34, https://doi.org/10.1007/978-3-031-46742-4_1

for scholarly exploration. Her tireless work ethic and visionary leadership have served as a guiding light, illuminating the path for all contributors involved in this monumental endeavour. Furthermore, Lopez-Soto's collaborative spirit and talented organizational skills have fostered a harmonious synergy among the various authors, enabling a seamless integration of diverse perspectives and facilitating the generation of innovative ideas. Her invaluable feedback and astute guidance have elevated the quality of each chapter, culminating in this volume pivoting in the Theory of Mind that promises to leave a lasting impact in its respective field of study.

The Theory of Mind (henceforth, TOM) is a fundamental cognitive ability that underlies our interactions, communication, empathy, and moral understanding. It plays a significant role in shaping how we perceive and interact with the world and the people in it, particularly in social contexts. It refers to the ability to attribute mental states to oneself and others, which includes beliefs, intentions, desires, emotions, and knowledge. This capacity allows individuals to make inferences about others' thoughts and feelings, predict their behaviour, and engage in complex social relationships. In general, children begin to develop ToM around the age of four or five. However, there is some variation in the timing of TOM development which is thought to be influenced by several factors, including genetics, environment, and experience. Beyond humans, TOM can also be considered towards other alive beings, our fellow animals, but also the coming state-of-the-art in machine learning and artificial intelligence, arising questions on this future already past.

This volume is therefore divided into four parts, each of which presents a current debate on TOM from four different perspectives, to adapt the body of knowledge to recognizable focal points. Part I is dedicated to the TOM from a clinical and psychopathological approach since this ongoing line of research suggests that TOM deficits are a common feature of a number of different mental situations: autism, brain injury, deafness, aphasia, among other. Regarding clinical practice, TOM assessments are used to evaluate social cognition in individuals with mental health issues showing that challenge in ToM understanding, disorders or impairments have been linked to various mental health conditions and, understanding TOM offers valuable insights into the underlying mechanisms and potential treatment approaches. Delving into the specific impairments in ToM, associated with different mental disorders, helps inform treatment strategies and interventions as well as provide ways in which our brains work. Part II is dedicated to TOM from a philosophical standpoint dedicated to the mind, covering identity, consciousness, and awareness. This complex cognitive ability is raising a large number of ontological questions: whether TOM is a form of knowledge or a form of inference, whether innate or learned, or a simple wondering if it is necessary. Part III is the glue among all four parts: language and communication with a language acquisition and bilingual approach. Language, verbal or not, is not merely a tool for communication; it also serves as a means for expressing and understanding the thoughts and emotions of oneself and others. Part IV takes us into the future. Beyond the literary and philosophical implications, the horizon seems to be painted with data used by self-learning machines able to create dystopias or utopias. This final part anticipates exponential growth in AI TOM in the coming years, driven by technological advancements and interdisciplinary collaborations between psychology, neuroscience, computer science, language, and communication engineering research fields.

Step by step, Part I gets underway with the international collaborative team made up of MIRIAM RIVERO-CONTRERAS and DAVID SALDAÑA, from the *Department of Developmental and Educational Psychology* at the *University of Seville*, along with MARTINA MICAI of the *Istituto Superiore di Sanità* in Rome. Their contribution on the 'fundamental concepts and issues' related to TOM explores the diverse approaches used to evaluate TOM in research, along with the methodological challenges associated with its measurement. They additionally examine its progression in children and adolescents, as well as factors influencing individual variations.

MICHAEL HUEMER, from the *Department of Psychology* at *Harvard University*, focuses on how children typically begin attributing beliefs to themselves and others around the age of 4 years, improving their ability on various tasks. The Mental Files Theory offers an explanation for this synchronized development and sheds light on why older children may struggle to grasp that beliefs about an object depend on their acquaintance with it (aspectuality). However, these difficulties tend to disappear once children master second-order belief tasks at around the age of six. Furthermore, the chapter addresses several challenges to this perspective that have been raised in the existing literature.

The next proposal, by PABLO DELGADO and ISABEL R. RODRÍGUEZ-ORTIZ, from the *Developmental and Educational Psychology* at *University of Seville*, reads the TOM reviewing recent empirical evidence that suggests that TOM appears to contribute to reading comprehension through its influence on listening comprehension. Reading comprehension is the result of a complex interplay of processes and components comprising a reader's skills and knowledge. Recently, TOM has been recognized as a crucial capacity in comprehending texts, particularly in narrative stories. This is because narrative comprehension involves understanding the emotions, mental states, and perspectives of characters, often requiring the inference of information not explicitly stated in the text. In this chapter, the authors examine relevant research that explores the association between TOM and reading comprehension in both typical and atypical populations such as individuals with autism spectrum disorders and those with deafness or reduced hearing compared to typically developing peers since these two atypical populations often experience challenges with TOM.

A clinical approach is provided by PILAR DE LA HIGUERA-GONZÁLEZ, ALEJANDRA GÁLVEZ-MERLÍN, ELISA RODRÍGUEZ-TOSCANO, JORGE ANDREO-JOVER, and ALEJANDRO DE LA TORRE-LUQUE from diverse hospitals and the *Complutense University of Madrid*. This chapter presents the characteristics of TOM assessment and provides a compilation of the main useful tools for this purpose. Additionally, it gathers specific TOM deficits observed in neuropsychiatric disorders. While ToM impairments have been extensively studied in conditions like autism spectrum disorders and schizophrenia, recent years have seen an extension of this research to many other clinical diagnoses and significant cross-cutting issues in mental health, such as suicidal behaviour. As a result, this chapter offers a comprehensive overview of TOM impairments across a broad range of neuropsychiatric disorders and discusses accurate assessment protocols. This information is essential for readers to understand TOM impairments in various neuropsychiatric conditions and to effectively employ assessment tools in clinical practice.

For nearly four decades, the Theory of Mind explanation of autism has been a central focus of cognitive research. Initially, it proposed that the varied symptoms observed in individuals with autism might stem from their difficulty in inferring the mental representations of others. In this chapter, GEMA ERENA-GUARDIA, *University of Seville*, MILA VULCHANOVA, *Norwegian University of Science and Technology* and DAVID SALDAÑA, *University of Seville*, delve into some of the key studies that have contributed significantly to this area of research, observing the two main challenges.

The interdisciplinary and interuniversity team formed by INÉS ABALO-RODRÍGUEZ, JESÚS CABRERA-ÁLVAREZ, SANDRA DOVAL, ALBERTO FERNÁNDEZ LUCAS, and DOLORES VILLALOBOS, sheds light to the specific intervention programs and techniques targeting general Social Cognition (SC) abilities, particularly TOM, in patients with *acquired brain injury* (ABI). While ToM rehabilitation has been more frequently implemented in other patient groups, such as those with autism and psychiatric disorders, newly designed programs for ABI patients show promising results. ABI implies a significant cause of global mortality and disability, leading to a wide range of symptoms affecting cognitive, behavioural, emotional, and social aspects. Their chapter examines TOM deficits in ABI individuals and provide relevant tools and considerations for clinicians.

KIMBERLY PETERS, professor at *Western Washington University*, and DAVID B. PISONI, at *Indiana University*, explore how ToM and language develop in the children who are deaf or hard of hearing. For them, delays in identifying their condition, receiving appropriate hearing technology and language interventions late, and not developing strong early language and conversational skills can result in delayed TOM development during preschool and early elementary school years. Conversely, children who are deaf, identified early, receive timely treatment, and acquire conversational competence within the typical timeframe, are more likely to demonstrate ToM development similar to their peers with typical hearing.

The following chapter provides a review of essential aspects of TOM in patients with aphasia and schizophrenia, focusing on their unique challenges and implications for understanding TOM in the context of language-related disorders. TERESA LOPEZ-SOTO, from *University of Seville*, explores the relationship between TOM and Language, focusing on the biological foundations of TOM, including anatomical and neurochemical aspects. It investigates the evolution of TOM and neuroplasticity throughout our lifespan, highlighting the dynamic nature of our brain's relationship to maintain health and safety. This understanding has significant implications for the ways language pathologies and cognitive decline are approached that may impede TOM.

Eight hours a day. Throughout history, many theories have attempted to explain the ultimate purpose of sleep, from various philosophers, poets, and artists to Hans Berger's introduction of electroencephalography in 1929. It allowed a new approach to studying the sleeping brain, marking the inception of the field of somnology, which is less than a century old. ENRIQUE MONTES-LATORRE, *department of Clinical Neurophysiology* at *University Hospital Virgen del Rocío* in Seville, and PAOLO PORCACCHIA, from the *Sleep Pathology Unit* at the same Hospital start with the liminal hypothesis of considering TOM in the context of the Science of Sleep, since consciousness and its limits come to play an important role both in daytime functioning and health consequences. "When ToM is not awake" sheds light on sleep disorders which encompass a diverse range of phenomena that not only affect sleep itself but also impact daytime functioning, and some of these conditions may lead to injuries or serious health consequences. This chapter reviews some of these sleep-related issues and their implications.

The neuropsychologist and neurocognitive independent researcher ANA MARÍA MARTÍNEZ ROBAYO provides an overview of the brain structures and neurohormonal agents involved in the hypothalamic-pituitary-adrenal (HPA) axis, including the limbic system, hypothalamus, regulation of pituitary secretion, and their connections to stress, mood, personality, and neurocognition. The components of the HPA axis encompass the hypothalamus, limbic system, pituitary gland, and adrenal glands, working in tandem to maintain hormone balance and regulate the stress response. Research has extensively shown that disruptions in the HPA axis and changes in neurocognitive and psychological states lead to elevated stress levels in individuals.

Opening Part II, dealing with the philosophical approach on mind and ToM, PILAR LÓPEZ SEGURA and TOMÁS ORTIZ ALONSO from the department of Forensic Medicine, Psychiatry and Pathology at *Universidad Complutense de Madrid*, discuss on consciousness briefly exploring the mechanisms underlying meta-cognition and our sense of identity as individual or socially constructed. Finally, the chapter addresses the neuropsychological and philosophical aspects of how lying can affect our reality, identity, and relationships with others.

From a neuro-philosophical approach too, JAVIER ALEJANDRO GALADÍ, researcher at the *Center for Brain and Cognition, Universitat Pompeu Fabra*, delves on the question of how physical-biological systems can have states like thoughts, fears, and hopes the relationship between the mind and the body, or more broadly, between mental and physical phenomena. Known as the mind-body problem, this has been a fundamental concern in Philosophy, especially disseminated by René Descartes and his critics four centuries ago. Physical properties are observable by everyone, while mental properties are private and can only be directly experienced by the individual. Conscious mental events are exclusive to the subject, and no one else has privileged access to them as they do to physical phenomena. Nonetheless, this chapter clarifies that conscious experiences do not seem to arise from nothing, but rather emerge from physical-biological processes within the body, especially neural processes in the brain.

SYLVAIN MONTALVO suggests "that the cognitive mechanism responsible for self-ascription of belief is nothing but our mindreading ability, applied to ourselves". As a result, TOM is applied to the self as an evolutive process in interaction and mindreading bringing us to the social dimension of belief self-attribution. Montalvo concludes stating that echo chambers can be a breeding ground for self-fulfilling endorsement of beliefs since they provide a safe space for people to validate their own beliefs and to reject dissenting ideas.

The right time for DIEGO LÓPEZ-GARCÍA, DIEGO LÓPEZ-LUQUE, WALTER FEDERICO GADEA-AIELLO, AND EMILIO JOSÉ DELGADO-ALGARRA, from both the *University of Huelva* and *University of Seville*, to take us out from the Eurocentric Western mindset establishing a connection between the TOM and nipponese philosophy. In Japan, TOM is intertwined with the concepts of *mushin*, also known as "no mind" and *Ikigai*, which relates to the health and well-being of individuals. To comprehend the 心の理論 /ko'koro no riro:n/(TOM in Japanese), it is essential to grasp the underlying philosophy in the Asian island, as it profoundly influences the behaviour and way of life of the Japanese people, thus providing insights into their mental processes: Taoism, Buddhism, Confucianism, on the one hand, and on the other, Shintoism and Zen offers a fresh perspective in understanding the nature of the mind, awareness, empathy, meditation, and compassion in Japanese culture. This chapter directly connect us to the next.

Part III focuses on the TOM from a linguistic, language acquisition and bilingual approach. Let us explore how our brains deal with content and processes and its implications with the chapter by CARLOS M. GÓMEZ, BRENDA Y. ANGULO-RUIZ, ELENA I. RODRIGUEZ-MARTÍNEZ, FRANCISCO J. RUIZ-MARTÍNEZ, EVA MARÍA PADILLA MUÑOZ, MARÍA DOLORES LANZAROTE FERNÁNDEZ from the *University of Seville*. It raises the question of whether process and content are governed by distinct neural mechanisms and its significant clinical applications, as conditions like depression, psychosis, and anxiety are experienced as specific contents or states, but their underlying causes may be related to abnormal information processing as well as from an educational perspective, this topic relates to biased educational strategies that prioritize memory (content) versus those that emphasize problem-solving approaches (processes).

"Language, mind and thought" would have been a perfect title for this Part III. Professor INMACULADA AGUILAR-PONCE, from the *department of Clinical Neurophysiology*, at *Universidad de Seville*, came to us with this 'general overview', given that TOM is associated with thought and language as they play a vital role in acquiring it. Neuroscience experts have been curious about the point at which language begins to interconnect with TOM, questioning whether language is a prerequisite for the development of TOM as well as if TOM can emerge independently, without relying on language.

Professor PRISCILA SÁNCHEZ SORIANO, from *Universidad Pablo de Olavide, de Seville*, dedicates her chapter on the studies on bilingualism and TOM, revealing notable differences among various forms of bilingualism. This includes understanding the specific brain regions engaged in performing tasks related to bilingualism, as well as examining key factors that relate to TOM, such as sociolinguistics, executive functions, and metalinguistic awareness.

MAYURESH KUMAR's proposal, from *Aligarh Muslim University* in India. The chapter examines various studies that delve into the processes of early language acquisition in autistic children, scrutinizing the methodologies employed to draw specific conclusions. Understanding early language acquisition in autistic children holds significant implications for TOM, professor KUMAR identifies the constraints on language acquisition as a crucial step forward.

We finish this Part III with a chapter by SERGIO MARÍN CONEJO and TERESA LOPEZ SOTO, from *Universidad de Seville*. In this piece, we propose an original and innovative correlation between TOM and the sex/gender system, as little has been written, through the lens of linguistic insights exemplified with literary excerpts. These resources serve as valuable tools to examine how individuals connect, mainly through language and narratives and discourse. Our contention is that TOM is not a singular skill; instead, it encompasses an intricate array of mechanisms that are influenced by gender within a historical and culturally patriarchal context. We reach the conclusion that it is imperative to trace the roots of the sex/gender paradigm, as it fundamentally shapes individuals and groups to grasp an essential contrast in vitality, animosity,and agency both in linguistic expression and communication.

Part IV, TOM and Artificial Intelligence, starts as PEDRO GÓMEZ-VILDA from *Universidad Politécnica de Madrid*, and ANDRÉS GÓMEZ-RODELLAR, from the *Faculty of Medicine* at *University of Edinburgh*, foresee the link of machine learning platforms and TOM establishing a cooperative relationship between both approaches to produce explainable representations that can be valuable for monitoring pathological speech. Their research investigates how the collaboration of these approaches can be fruitful and lead to improved outcomes in the context of speech analysis and monitoring. The abundance of data generated from brain functional activity analysis presents a significant challenge for machine learning systems. These systems are not only expected to perform traditional tasks like prediction, regression, clustering, classification, and characterizing specific task-related behavioural communication skills (e.g., speech and phonation), but they also need to differentiate between pathological and normal brain activity, providing cause-and-effect insights. These requirements are especially critical in the biomedical data processing domain, where proposed methods aim to assist clinicians in tasks such as diagnosis, prognosis, treatment, and rehabilitation.

ALBERTO NEBREDA, DANYLYNA SHPAKIVSKA-BILAN, and GIANLUCA SUSI, from the *Centre for Cognitive and Computational Neuroscience* at *Complutense University of Madrid,* and CARMEN CAMARA, from the *department of Computer Science* at *Carlos III University of Madrid*, outline promising applications of AI TOM models, which span across various domains such as industry (e.g., autonomous vehicles), speech processing, biometry, healthcare, education, marketing, and more. However, it is essential to acknowledge that the concept of an Artificial TOM is still in its early stages, with existing algorithms being domain-specific and limited to certain environments and tasks. The authors aim to explore the current understanding of TOM in the context of AI computational models.

ALVARO GARCIA-LOPEZ from the *department of Electronic Technology* at University Carlos III in Madrid, makes clear that the combination of TOM and AI has the potential to significantly improve mental health therapy and increase access to care using smartphones. Personalized emotional support and guided conversations through mental health apps can aid in the development of TOM abilities, leading to improved communication and better interpersonal relationships.

This 23-chapter book explores TOM's application in diverse disciplines such as brain studies, mental health, philosophy, language and communication, and AI. Emphasizing these interdisciplinary debates, it offers an adventure into emerging frontiers, revealing the profound impact of ToM on interconnected fields. A captivating journey into the world of mind and cognition awaits.

Chapter 2
An Introduction to Theory of Mind: Fundamental Concepts and Issues

Miriam Rivero-Contreras, David Saldaña, and Martina Micai

Abstract Theory of Mind (ToM) is a construct that reflects people's ability to understand the mental states of others. In our daily lives, we continuously infer what others are thinking, understanding, feeling, and interpreting, especially with respect to us and other persons. The present chapter is focused on different definitions and concepts of ToM, its development in children and adolescents, and factors related to individual differences. We also analyse how ToM has been assessed in different studies and the methodological issues surrounding its measurement.

Keywords Theory of mind · Mindreading · Development · Methods of investigation · Executive function · Metacognition social · Skills · Culture · Social context

2.1 Theory of Mind: Definition and Concept

Theory of Mind (ToM) can be defined as the cognitive representation of our own mental states and those of others (Adolphs 2009; Kanske 2018; Keysers and Gazzola 2009; Premack and Woodruff 1978; Schurz et al. 2021), which allows prediction of behavior in social contexts (Baron-Cohen 1999). This includes what other people may perceive of a given situation or reality, their intentions in that context, or their motivations to act in a particular way. Every day, we put into place this ability to infer what others are thinking, understanding, feeling, and interpreting. It is a key social cognitive skill for a person's functioning, social competence, and acceptance.

M. Rivero-Contreras (✉)
Department of Psychology and Anthropology, University of Extremadura, Seville, Spain
e-mail: mrivero5@us.es

D. Saldaña
Department of Evolutionary Psychology and Education, University of Seville, Seville, Spain
e-mail: dsaldana@us.es

M. Micai
Research Coordination and Support Service, Istituto Superiore di Sanità, Roma, Italy
e-mail: martina.micai@iss.it

T. Lopez-Soto et al. (eds.), *The Theory of Mind Under Scrutiny*, Logic, Argumentation & Reasoning 34, https://doi.org/10.1007/978-3-031-46742-4_2

Even now, in an apparently less social situation, while we are writing we think about how the reader might interpret, understand our words, and try to make them as clear and interesting as possible.

It was Premack and Woodruff (1978) who coined the expression ToM, the most used term to describe this ability. However, researchers have also called it *mindreading* (Byrne and Whiten 1991), *mentalizing* (Frith et al. 1991), *folk psychology* (Wellman 1990), and *intentional stance* (Dennett 1987). In this chapter, we refer to ToM, keeping in mind that other synonyms may be applicable.

Traditionally, ToM had been understood mainly as related to an *explicit* understanding of the beliefs and desires of others. We see explicit ToM at work when someone can describe what others know or believe in a specific situation, for example (e.g., why did Julia look under the couch for her ball, when everyone – except for her– saw that her mother had put it in her bedroom) (Sodian et al. 2020). This form of ToM has been considered to develop between the ages of 3 and 5 years, depending on linguistic and cognitive individual differences (Wimmer and Perner 1983). At this age, we can observe children distinguishing between what people *think* about reality and reality itself (Wellman et al. 2001). But findings from the past 10–15 years suggest that a form of ToM may emerge before children can describe the representations that guide others' behaviors. Infants appear to predict the actions of people with whom they interact based on their intentions long before they are able to talk about them, using some form of *implicit* ToM (Kulke et al. 2018). How common the processes involved in explicit and implicit ToM might be is a matter of debate, which we shall return to in the next section.

2.2 The Development of Theory in Mind in Children

ToM development in children traverses several stages, beginning at 1 year of age and ending up around 9–11 years (Brüne and Brüne-Cohrs 2006), with substantial improvements between the ages of 2 and 5 years in children's understanding of mental states (Harris 2006). Explicit ToM begins to develop fully about 2 years. Around 4 years of age, children start to understand explicitly that people's beliefs can be different from reality (Wellman et al. 2001).

But, as we have pointed out above, implicit ToM signals can be observed in the first and second years of life (Baillargeon et al. 2016). Researchers have increasingly used nonverbal and simplified tests (e.g., observation of imitation behaviors, violation of expectation paradigms, eye tracking; Beaudoin et al. 2020) and measures for preverbal children that have allowed them to show that some ToM abilities may already be present in infancy (Slaughter et al. 2015). At 5 months, infants preferentially encode the action goal over the spatiotemporal properties of the reaching and grasping action (Woodward 1998). At 6 months, infants integrate information on agent perception with goal representation by, for example, observing whether the agent who grasped one of two objects had explored both before choosing (Luo and Johnson 2009). At 9 months of age, infants try to align their own goals and attention with the interlocutor (ToMasello 2018). Around 12–18 months of age, infants demonstrate knowledge of intentions (Kristen et al. 2011), at 18 months of age others' desires (Repacholi and Gopnik 1997; Poulin-Dubois et al. 2007), and some comprehension of incorrect representations of reality in others (*false beliefs*) at 15 months (Onishi and Baillargeon 2005; Southgate et al. 2007; Senju 2012). When they reach 2 years of age, children can attempt to help an adult reach a goal, mimic a failed attempt (Meltzoff 1995; Warneken et al. 2012), and take the false beliefs of the interlocutors into account when building action expectations (Baillargeon et al. 2010; Sodian 2016).

As Carlson et al. (2013) reviewed, by 2 years of age, children have a basic understanding of emotion, intention, and desire (Wellman 2002). However, at this stage, it is still difficult for children to understand the beliefs and knowledge states of others (Wimmer and Perner 1983). At 3 years of age, children still have difficulty recognizing that reality differs from the apparencies (Flavell et al. 1983) and that people can have different visual perspectives on the same input (Flavell et al. 1981). It is not until the age of five that it is possible for them to use a variety of more complex explicit ToM tasks (Beaudoin et al. 2020). At 4–5 years of age, children begin to present a more adult-like understanding of ToM (Harris 2006). For example, they improve their ability to understand second-order false belief tasks (in which they must determine another person's representation about a third person's representation, not just about reality) around 5–6 years of age. Comprehension of sarcasm and white lies is not really evident until adolescence (Miller 2009).

Later development of ToM skills could reflect a lack of sensitivity of measurement instruments and tasks to ToM processes in younger children, rather than limitations in early ToM skills. But even if our tasks reliably measure ToM throughout development, it seems to be clear that long before children can explain the representations of other persons, they are already showing indications of taking them into account when interacting with them. These prior skills could be developmental precursors to later more complex skills, in a conceptually continuous developmental pathway (Sodian et al. 2020).

An alternative view proposes that early and later ToM skills belong to two relatively independent mindreading systems, implicit and explicit, based on different neurocognitive mechanisms. Implicit false-belief processing would be fast and automatic, but inflexible, while explicit processing is more flexible, but slower and effortful, and dependent on language (Sodian et al. 2020). Explicit ToM is characterized by the attribution of beliefs and desires to agents. Implicit ToM mechanisms occur in infants and adults in spontaneous, automatic, and unconscious ways (for a review, see Schneider et al. 2017, and Sodian et al. for alternative theories). However, the presence of these two systems is far from established: A recent systematic replication study did not show evidence for implicit ToM as a robust phenomenon, because anticipatory looking false-belief paradigms seem less reliable and valid than previously assumed (Kulke et al. 2018).

2.3 Individual Differences in ToM

ToM abilities appear to vary not only with age, but also between different individuals (but not everyone agrees, e.g., Gordon et al. 2014; Sodian and Frith 1992; Sai et al. 2021), with varied impact on social skills and interactions. For example, some studies find that children with better ToM are more likely to lie and maintain their lies (e.g., Fu et al. 2018; Ma et al. 2015; Talwar and Lee 2008). In adolescence, well-developed perspective-taking abilities improve social relationships (Van der Graaff et al. 2018; Derksen et al. 2018), and are associated with greater quality of friendship and lower levels of loneliness (Flannery and Smith 2017). Impaired perspective taking abilities have been associated with internalizing and externalizing problems (Halfon et al. 2020), adolescent-onset schizophrenia (Li et al. 2017), and a wide range of mental disorders in adults (Wang et al. 2018).

Higher levels of perspective taking during adolescence have been found to be associated with relatively more mature cortical structures in specific regions of the brain, compared to development in those with lower levels of perspective taking (Tamnes et al. 2018). Some other studies found that females outperform males on emotion recognition tasks (Baron-Cohen et al. 1997), understanding of faux pas (Baron-Cohen et al. 1999), and empathy measures (Baron-Cohen and Wheelwright 2004). In addition, individual differences in ToM are determined by family context. In fact, adolescents with higher quality relationships with their parents showed higher levels of perspective taking (Silke et al. 2018). For a review of factors that can promote the development of ToM, see the dedicated paragraph below.

2.4 Theory of Mind in Non-Human Animals

TOM is one of the most advanced cognitive skills of human beings and has been extensively studied in psychology and neuroscience. However, recent research has shown that it may not be unique to humans, and that some other animals may also possess this ability. Studies on ToM in nonhuman animals are important because they can answer several open research questions regarding the origin of ToM and the factors influencing its development. Many suggest that ToM is partly driven by culture and socialization practices (e.g., Heyes and Frith 2014; Liu et al. 2008). However, studies from nonhuman animals appear to exclude the role and input of language and culture in the development and functioning of ToM. Furthermore, the role of memory and inhibitory control in ToM is still unknown (Carlson et al. 2002; Powell and Carey 2017). Studying ToM in species with more limited inhibitory control or memory than humans would provide some clues towards understanding the minimum requirements for ToM to emerge (Krupenye and Call 2019).

To study ToM in animals, researchers use a variety of scientific methods that are like those used to study human cognition. These methods include observational studies, behavioral experiments, and neuroimaging techniques. Observational studies involve examining the natural behavior of animals in their environment. For example, researchers may observe how animals interact with each other in social situations and look for evidence of behaviors that suggest an understanding of mental states, such as deception or cooperation. Behavioral experiments involve designing tasks that test an animal's ability to understand mental states. For example, researchers may use a false-belief task to see if an animal can anticipate the actions of another based on their false beliefs. Other tasks may involve understanding the intentions of others, recognizing oneself in a mirror, or recognizing emotions in others. Neuroimaging techniques, such as functional magnetic resonance imaging (fMRI), can be used to study the neural basis of TOM in animals. For example, researchers may use fMRI to observe which areas of an animal's brain are active when they are engaging in tasks that require an understanding of mental states. Also, the application of eye tracking and looking-time analyses are useful tools for exploring ToM in nonhuman animals, for example, to measure their predictions about how agents will behave, use anticipatory-looking false belief test, and capture the attention (Kano et al. 2017). Overall, the methods used to study TOM in animals are varied and depend on the specific research question being asked. The extensive research on ToM in nonhuman animals such as corvids, primates, lemurs, and dogs shows that some concepts of ToM are not unique to our species. One of the most studied animals in relation to ToM is the chimpanzee. In one study, researchers used a classic false-belief task (Call and ToMasello 2008).. In this task, a person hides an object in one location, and then leaves the room. While the person is away, someone else moves the object to a different location. When the original person returns, the researchers observe whether the chimpanzee anticipates that the person will look for the object in its original location or the new location. The results showed that chimpanzees can anticipate the person's false belief, suggesting that they have some understanding of mental states.

Other animals that have been studied in relation to TOM include dolphins, elephants, and ravens. In one study with dolphins, researchers used a mirror test to see if dolphins could recognize themselves, which is an indication of self-awareness (Reiss and Marino 2001). Similarly, mirror-self recognition has been observed in elephants showing to have some understanding of others' mental states (Plotnik et al. 2006). Finally, ravens have been observed to engage in deceptive behaviors, such as hiding food from others and then later retrieving it when the others are not around. This suggests that ravens may be able to understand that others have different mental states and beliefs, and that they can use this knowledge to their advantage (Bugnyar and Heinrich 2005). As Krupenye and Call (2019) described in their review, Eurasian jays showed understanding of subjective desires, and the great apes responded appropriately to false beliefs.

While the research on ToM in animals is still relatively new, these studies suggest that some animals may possess a rudimentary understanding of mental states. This has important implications for our understanding of animal cognition and raises ethical questions about the treatment of animals. Further research is needed to fully understand the extent and nature of ToM in animals, and to explore the cognitive and neural mechanisms underlying this ability.

2.5 Cognitive Processes Related to Theory of Mind

ToM is very closely related to other cognitive domains and skills (Apperly 2012). For example, empathy, as the affective route for understanding others (Gallese 2003) has been considered close to *affective ToM* (the ability to understand the desires and emotions of others, not only the cognitive mental states, such as beliefs and thoughts) (e.g., Schlaffke et al. 2015; Sebastian et al. 2012). Although these socio-affective and sociocognitive routes are served by independent brain networks, they are jointly required in many complex social situations (Preckel et al. 2018). Furthermore, Leslie et al. (2004) described that metarepresentation works for the processing of other mental states, regardless of cognitive or affective content. Meta-representations – the representation of a representation – are at the core of ToM.

Metacognition is defined as the knowledge and regulation of cognition (Schraw and Dennison 1994). In metacognitive research, we can find metacognitive knowledge or declarative metacognition, a component related to factual knowledge about cognition, and procedural metacognition, an executive component devoted to the regulation and monitoring of cognitive processes (Haberkorn et al. 2014). Factual knowledge about the cognition of others or yourself is close to the construct of ToM.

A striking example of the application of metacognition to daily life can be seen during reading. Metacognition in reading is conscious awareness and knowledge of cognition, cognitive processes, and strategies used during reading (Jacobs and Paris 1987). Metacognition is a very important component of reading because it allows the adaptation of behavior according to task demands (Woolley et al. 2010). To successfully comprehend a text, readers need to think about the process of reading, learn from the monitoring of the reading, set goals, activate strategies, and assess goal progress and outcomes (Zimmerman 2002). Skilled readers can predict the content of the text, activate memories, and test themselves to see if they have sufficient knowledge about the reading material (Glazer and Burke 1994). The ability to predict the content of the text promotes the understanding and commitment of a story and helps in verifying the comprehension of the text (Duke and Pearson 2002). Skilled readers generally verify their predictions by monitoring meaning and using repair strategies (e.g., reading back when they find difficulties in the comprehension of the text or reading on when their predictions failed) (Zinar 2000). Metacognitive comprehension strategies are used more often by skilled readers compared to less skilled readers (e.g., Myers and Paris 1978) and deficits in metacognition are correlated with reading comprehension difficulties (Garner 1987). Furthermore, poor readers showed less than adequate knowledge about reading (Myers and Paris 1978) and about cognitive processes (Papetti et al. 1992). Metacognition evolves with age (e.g., Flavell et al. 1970) and increases with training. Guthrie et al. (2004) explained that children become more cognitively engaged when they are taught to use metacognitive skills (e.g., asking themselves questions and monitoring their own response for understanding). This behavior leads individuals to become more goal directed and active in constructing meaning while reading (Afflerbach et al. 2008). Metacognitive skills can be improved with a positive impact on reading comprehension (e.g., Borduin et al. 1994; Pressley 1976). How metacognition and ToM interact during reading is the specific focus of another chapter in this book.

Another important component of metacognition in reading is the relationship between the ToM and story processing. On several occasions, it has been proposed that reading fiction improves social cognition in general, and ToM in particular. Mumper and Gerrig (2017) conducted a meta-analysis with the aim of investigating the relationship between reading habits and measures of empathy or ToM. Specifically, they found a positive correlation of .22 between fiction reading habits and ToM versus a positive correlation of .09 between non-fiction reading habits and ToM. Similarly, Dodell-Feder and Tamir (2018), in another meta-analysis of 14 research studies on fiction reading involving 1.615 fiction readers, aged 18–37, found slightly higher results on social cognition tasks, with an overall effect size of 0.15. However, despite the overall positive results found in the meta-analyses, we should take a closer look at this research. The experiments conducted so far, and their replications, show contradictory results that reading literary fiction improves TOM. Kidd and Castano (2013) conducted five experiments which found that reading literary fiction would lead to better performance in ToM. However, several subsequent studies have not found the same results, or the results could not be replicated (e.g., Camerer et al. 2018; De Mulder et al. 2017; Panero et al. 2016; Samur et al. 2018). Eekhof et al. (2022), after reviewing these contradictory effects, propose that part of the explanation for these conflicting results could be because all research has collapsed various types of texts, readers, and sociocognitive processes, tacitly assuming that any (literary) narrative will affect all readers in the same way. Therefore, they suggest that future research should focus on (1) the specific text characteristics that drive the social-cognitive potential of narratives, (2) individual differences between readers with respect to their sensitivity to this potential, and (3) the various aspects of social cognition that are potentially affected by reading narratives.

ToM has been extensively explored in relation to *executive function*, with often conflicting results. Executive function is an umbrella term that includes initiation, sustaining, shifting, and inhibition (Denckla 1996; Diamond 2013) and refers to the neurocognitive processes that control and coordinate cognition and guide goal-directed behavior (Barkley 1997; Garner 2009; Salthouse 2005). Cognitive complexity and control theory (Frye et al. 1995; Zelazo and Frye 1997) indicates that "reflective awareness of one's knowledge is required to use that knowledge to guide behavior under conditions of interference" (Zelazo and Frye 1997). It predicts that ToM and executive function are strongly intertwined because both involve higher-order rules.

There is ample evidence to support the relationship between both constructs. Joseph and Tager-Flusberg (2004) showed that, in a sample of autistic people, ToM performance was significantly and positively related to working memory and inhibition, while Fahie and Symons (2003), in children with attentional and behavioral problems, found ToM and working memory to be significantly and positively related. Fisher and Happé (2005) showed that performance in executive function skills was positively associated with performance in ToM in an autistic group. ToM and executive function difficulties have been associated with children (Mary et al. 2016) and adults (Tatar and Cansız 2022) with attention deficit hyperactivity disorder (ADHD) and hard-to-manage preschoolers (Hughes et al. 1998). The executive functions most correlated with ToM are inhibitory control, working memory, cognitive flexibility, and attention (Pineda-Alhucema et al. 2018).

The directionality of the relationship between executive function and ToM has been the subject of considerable debate. Perner and Lang (1999) proposed that a better understanding of one's own mind provides better insight into how to exert self-control and improves the exercise of self-control itself. But those supporting it is executive function which mainly influences ToM (and not the other way around) argue that a sense of personal agency and top-down self-control allows children to understand the mental states of other people (Carlson et al. 2013). The development of ToM appears to be supported by several components of executive function, such as analysis, inference, deduction, and estimation (Singh et al. 2021). For instance, Carlson et al. (2004) observed that executive function skills measured at 24 months predicted later ToM comprehension at 39 months. The relationship between executive function and ToM appears to remain unchanged throughout development during childhood and adolescence (Austin et al. 2014). However, the cognitive process that best predicts ToM is still unknown since the different studies report contradictory findings. Some studies find that working memory is the only skill that predicts ToM performance (Lecce and Bianco 2018; Lecce et al. 2017), while others reveal that it is predicted by inhibitory control (Bock et al. 2015; Cassetta et al. 2018; Vetter et al. 2013). Weimer et al. (2021) proposed that these differences could be due to the measures themselves, as these would require different subcomponents of executive function. However, in studies with participants with ADHD, where a poorer ToM would be predicted because of limitations in executive function, the degree of prediction and predictability of executive function are extremely heterogeneous between different studies (Pineda-Alhucema et al. 2018).

To complicate things further, ToM and executive function recruit common neural regions (Perner and Lang 1999). Brain regions typically involved in executive function also appear to be active during mental state reasoning, but there are also other neurobiological mechanisms underlying ToM and executive functions (Wade et al. 2018).

Emotional self-regulation is another cognitive process related to ToM. It manages and changes how/when/whether one experiences emotions and physiological states (Eisenberg et al. 2007). This skill would also be related to ToM, as it is necessary for behaviors aimed at achieving goals (Gross 2015). Children who experience greater physiological reactivity (indicative of emotional reactivity) have been found to show better ToM (Lane et al. 2013). Schurz et al. (2021) propose that understanding mental states of other people can be described through a hierarchical structure model of several levels that integrates ToM and empathy (understanding someone's emotions): a higher level with broader and abstract functioning, and a lower level with functions in specific contexts given by particular stimuli and task formats. This higher level would consist of three groups of neurocognitive processes:

(a) predominantly cognitive processes, which are activated when mentalizing requires self-generated cognition linked to the physical world; (b) more affective processes, which are activated when we witness emotions in other people, based on shared emotional, motor, and somatosensory representations; and (c) combined processes, which concurrently involve cognitive and affective functions. However, there is not as much research on emotional self-regulation and ToM as on executive function. This could be because emotional self-regulation is related to executive function, specifically to inhibitory control (Evans and Rothbart 2007), and thus not enough attention has been paid to it separately.

2.6 Other Factors Related to the Development of ToM

There are other factors that appear to interact with the development of good ToM, such as family relationships, peer relationships, cultural context, language skills, and reading (Fig. 2.1).

Regarding family relationships, greater parent-child conversational elaboration about mental states appears to be a significant predictor of ToM development in children (Brown et al. 1996; Ontai and Thompson 2008). For example, Lenna and Ross (2008) analyzed the relationships between attachment, mother-child discourse, and ToM in 76 children aged 4 years. These authors found that maternal conversational elaboration was a significant predictor of ToM development. Sabbagh and Callanan (1998) evaluated conversations between parents and children in 36 children aged three to five to explore two aspects related to the development of meta-representation: the contrast between two different mental states and the responses of the parents to the answers "I don't know" and the contradictions of the children, to determine whether the parents used these opportunities to highlight the representational nature of mental states. The authors observed that the children regularly caused mentalist responses in their parents, and in some cases, these parents' responses were positively related to those of the children. Therefore, it seems that including children in conversational interactions with adults makes the former see the latter as intentional and mental agents, and not as merely animated agents.

Furthermore, children who have safe attachment show better performance in ToM tasks (Arranz et al. 2002; Steele et al. 1999). For instance, Arranz et al. (2002) analyzed the influence of different variables on the development of ToM in a group of 114 pre-school children. This study evaluated family interaction quality, attachment quality, and child performance against a progressive series of ToM tasks, ranging from simple desire situations to a false belief task. The percentage of children classified with safe attachment who responded correctly to the false-belief task was significant. Even parenting style is an important variable, as Kuntoro et al. (2013) found that authoritarian parenting style was related to lower rates of ToM development.

Regarding peer relationships, different studies associate ToM with the popularity effect, prosocial behavior, and cooperation in play contexts. Regarding the

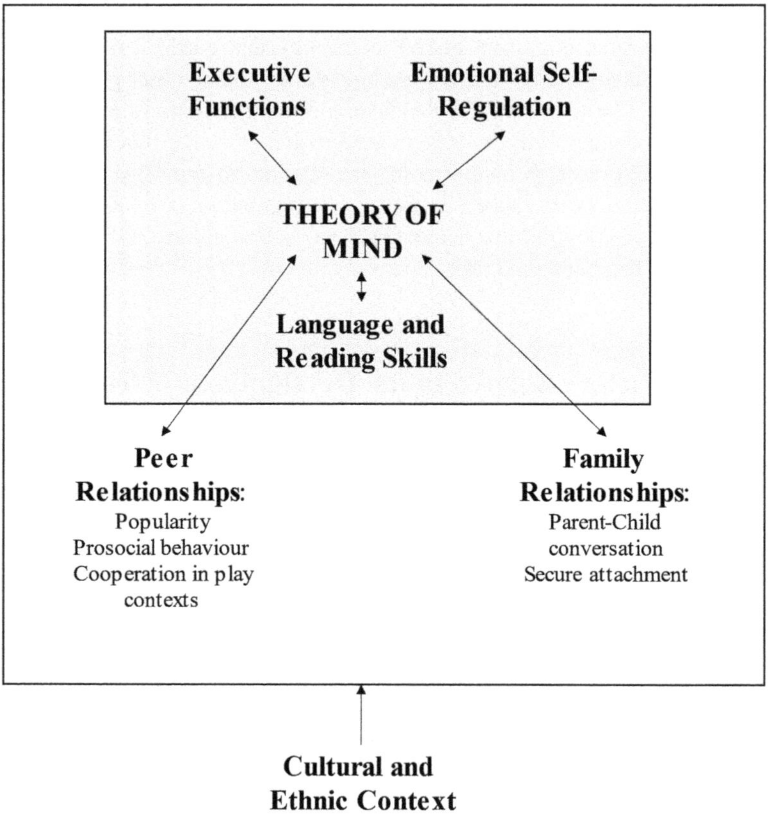

Fig. 2.1 Representation of the factors that promote ToM development

popularity effect, Slaughter et al. (2015) performed a meta-analysis of 20 studies that included 2096 children. This meta-analysis revealed a general significant association ($r = 0.19$), indicating that children with higher ToM scores were also more popular in their group of peers. Furthermore, these authors did not find any effect of age, although they found a weaker effect for boys ($r = 0.12$) compared to girls ($r = 0.30$). Regarding prosocial behavior, Imuta et al. (2016) also conducted a meta-analysis that included 76 studies with 6432 children aged 2–12 years. These authors evaluated the relationship between ToM and prosocial behavior, finding a significant, although weak, association between these two constructs ($r = 0.19$). This indicates that children with higher ToM scores also received higher scores on simultaneous measures of prosocial behavior. Regarding cooperation in play contexts, Etel and Slaughter (2019) evaluated its association with ToM in preschool-age children, finding that those children who showed better performance on the ToM task were related to cooperation among peers in the context of independent play, but not in the context of group play.

In general, all this suggests that children in preschool and school age with a lower understanding of mental states, compared to their peers, are at increased risk of developing behavioral and relational problems (Weimer et al. 2021). Similarly, higher rates of aggression among children and adolescents appear to be associated with poorer ToM skills (Gomez-Garibello and Talwar 2015; Weimer et al. 2017).

The cultural context is another factor that must be considered in the development of ToM. It seems that Western preschool children follow similar trajectories in the development of ToM, while Asian children show an alternative developmental sequence (Shahaeian et al. 2011; Zhang et al. 2016). For example, Kuntoro et al. (2017) evaluated the attitudes of mothers toward collectivism versus individualism, and authoritarian versus authoritative parenting, through self-reports, and ToM in their children. Indonesian mothers preferred collectivism over individualism and authoritativeness over authoritarianism. Moreover, the authors found that performance in the ToM Scale of the children was negatively correlated with authoritativeness, but not with other attitudes of the parents. In addition, differences were found between the two ethnic groups, one of which coincided with the sequence of the Western ToM scale, while the other coincided with the Asian sequence. The authors hypothesized that this difference could be due to differences in parenting attitudes and values of the two ethnic groups.

Regarding linguistic skills, the results indicate that verbal communication, especially speaking and elaborating on mental entities, promotes the development of ToM (De Villiers and Pyers 2002; Ebert 2020; Ebert et al. 2017). For example, Ebert (2020) found a bidirectional relationship between early and late measures of language skills of children and ToM. Specifically, the changes in ToM were predicted by language skills, especially by receptive grammar/sentence comprehension, and the changes in the receptive vocabulary of the children were predicted by early ToM.

Furthermore, other studies have found a relationship between ToM and reading comprehension (Atkinson et al. 2017; Boerma et al. 2017; Guajardo and Cartwright 2016; Kim 2017). This could be the result of the shared relationships of reading and ToM with EF skills (Weimer et al. 2021). Atkinson et al. (2017) explored the relationship between ToM and emergent reading comprehension in 80 children with typical development through a 2-year longitudinal study. At time point 1 (at the age of 3 years and 1 month), the authors evaluated ToM, decoding, linguistic skills, and executive functions; then, at time point 2 (at the age of 6 years and 3 months), the authors tested the efficiency of word reading, linguistic skills, and reading comprehension. The results showed that ToM, at timepoint 1, indirectly predicted reading comprehension at timepoint 2 through linguistic skills after controlling for age, nonverbal skills, decoding, EF, and previous linguistic skills.

2.7 How Has Theory of Mind Been Tested?

Since this construct of social cognition began to be investigated, different experimental paradigms have been used:

- Recognition of target words with a subsequent action word (Mason et al. 2004). Participants must indicate whether the action word (specific and nonspecific for humans or dogs) can be used to describe the target word. In this task, only actions associated with humans are expected to evoke attribution to mental states.
- Listening to two-word lists (Baron-Cohen et al. 1994). Participants must listen to two-word lists (one with terms related to mental state and the other with terms related to the body) and decide whether each word is consistent with the topic of one of the lists.
- Responding to questions about three categories of passages in prose (ToM, physical causality, and detached sentences) (Fletcher et al. 1995; Happé 1994).
- Sequence of stories, presented as comic strips or representing puppets, that would indicate true and false beliefs (Wimmer and Perner 1983; Baron-Cohen et al. 1985; Sommer et al. 2007). In typical first-order false-belief tasks, a change is made in the location of the object with or without the presence of the critical protagonist (true or false belief, respectively). In an example of one of these tasks, a character leaves a room, and another changes the location of an object while they are absent. The participant is then asked where the character who left will look for the object, whether in its original location or in the new one. Responding to the original one is an indication of having acquired a certain level of ToM, since it requires understanding the representation that character has of the situation (a representation of where the object was when they left the room), different from reality. In second-order belief tasks, participants are asked where a third character thinks a second person believes to object may be, for example.
- The use of comic strips (Brunet et al. 2000; Ciaramidaro et al. 2007; Walter et al. 2004). The participants observe a short sequence of events and must choose the end among three slide conditions: one condition evokes the attribution of intentions to people, and another two conditions represent sequences of physical causality (with and without characters).
- The use of animations (German et al. 2004; Castelli et al. 2000; Iacoboni et al. 2005; Mosconi et al. 2005), including videos of animated characters or actors, silent animations with interacting triangles, or videoclips.

Despite such a broad availability of tasks, Quesque and Rossetti (2020) argue that most do not require the participant to represent the mental state of another person, or any mental state at all, but are measuring lower-level processes. They recommend applying two criteria to determine whether a ToM measure is valid: the criterion of no fusion and the mentalizing criterion. The former refers to the distinction between the mental state of the other person and one's own mental state, whereas the latter indicates that the lower-level processes (e.g., orientation to attention and associative learning) must not explain the successful performance of any ToM task. These authors found that of 23 experimental tests analyzed, 15 did not meet one or both criteria. Thus, these tests could be measuring lower-level processes. Following these criteria, Quesque and Rossetti identify the following as potentially poor measures of ToM:

- Detection of devious intentions from kinematics (Sebanz and Shiffrar 2009).
- Recognition of emotions from images (Ekman and Friesen 1971).
- Recognition of emotions from voices (Golan et al. 2007).
- Attribution of the intention of the film (Premack and Woodruff 1978).
- Access-to-knowledge task (Povinelli et al. 1990).
- Level-1 representation of the visual experience of another person (Samson et al. 2010).
- Attribution of the mental state from animated shapes (Heider and Simmel 1944).
- Attribution of the mental state from face pictures (Baron-Cohen et al. 2001).
- Attribution of motor intention to a previous rational action (Brunet et al. 2000).
- Attribution of social intention to kinematics (Lewkowicz et al. 2015).
- Judgements of visual accessibility (Masangkay et al. 1974).
- Description of the scene (Quesque et al. 2018).
- Social spatial compatibility (Freundlieb et al. 2016).
- Spontaneous influence of the beliefs of a spectator on the decision-making process (e.g., Kovács et al. 2010).
- Spontaneous influence of the beliefs of a character on anticipatory looking behavior (Surian and Geraci 2012).

Another potential issue with many ToM measures is the fact that most of them are presented from the perspective of a third person, with the participant of the experimental task being a mere observer. It is known that social cognition is different depending on this perspective (third or first person) (Gallotti and Frith 2013) and this has probably been controlled for in very few studies.

Methodological limitations could affect the validity of the ToM construct itself. If tasks designed to measure ToM do not measure it adequately, there would be a lack of construct validity. A recent psychometric assessment of the structure of ToM (the only one conducted to date in this regard) shows that tasks evaluate more than one ToM construct (Navarro 2022). In this evaluation, they aimed to (1) understand whether ToM should be considered a monolithic skill and (2) explore whether ToM tasks adequately evaluate ToM, beyond general cognitive capacity (crystallized intelligence (Gc) and fluid intelligence (Gf)). To this end, they performed confirmatory factor analyses (CFAs), exploratory factor analyses (EFA), and exploratory network meta-analyses (NMA). Results indicate that the ToM tasks did not merely evaluate the cognitive capacity. In other words, ToM is not simply a subproduct of general intelligence. However, they also did not purely evaluate a single ToM construct. Therefore, it seems that ToM should be explored and measured as multiple domains and dimensions that represent different aspects of social cognition: some of lower order (kinematic processing, social attention, and emotion recognition, among others) and others of higher order (inhibiting one's own perspective, creating alternative emotional responses, and updating one's own knowledge, among others).

2.8 Conclusions

ToM is defined as the cognitive representations of the mental states of others. Thus, it allows us to infer what others are thinking, understanding, feeling, and interpreting about our behavior. Furthermore, this capacity allows us to predict the behaviour of other individuals.

Although many have proposed that ToM is partially driven by culture and socialization practices, some ToM concepts are not exclusive to our species. ToM is closely related to other cognitive domains and abilities, such as executive function, empathy, and meta-representation. Family and peer relationships, cultural context, language skills, and reading are also closely related to the development of ToM. ToM has been measured in many ways. Research in this field reveals that despite this diversity, many measures appear to lack specificity. These measures should systematically meet two main criteria: the no-fusion criterion and the mentalization criterion. Many of the tests do not. However, ToM does not seem to be a monolithic skill, and thus must be explored as multiple domains or dimensions that represent different aspects of social cognition.

References

Adolphs, R. 2009. The social brain: Neural basis of social knowledge. *Annual Review of Psychology* 60 (1): 693–716. https://doi.org/10.1146/annurev.psych.60.110707.163514.

Afflerbach, P., P.D. Pearson, and S.G. Paris. 2008. Clarifying the differences between reading skills and reading strategies. *The Reading Teacher* 61 (5): 364–373. https://doi.org/10.1598/RT.61.5.1.

Apperly, I.A. 2012. What is "theory of mind"? Concepts, cognitive processes and individual differences. *Quarterly Journal of Experimental Psychology* 65 (5): 825–839. https://doi.org/10.1080/17470218.2012.676055.

Arranz, E., J. Artamendi, F. Olabarrieta, and J. Martín. 2002. Family context and theory of mind development. *Early Child Development and Care* 172 (1): 9–22. https://doi.org/10.1080/03004430210880.

Atkinson, L., L. Slade, D. Powell, and J.P. Levy. 2017. Theory of mind in emerging reading comprehension: A longitudinal study of early indirect and direct effects. *Journal of Experimental Child Psychology* 164: 225–238. https://doi.org/10.1016/j.jecp.2017.04.007.

Austin, G., K. Groppe, and B. Elsner. 2014. The reciprocal relationship between executive function and theory of mind in middle childhood: A 1-year longitudinal perspective. *Frontiers in Psychology* 5 (JUN): 1–11. https://doi.org/10.3389/fpsyg.2014.00655.

Baillargeon, R., R.M. Scott, and L. Bian. 2016. Psychological reasoning in infancy. *Annual Review of Psychology* 67 (1): 159–186. https://doi.org/10.1146/annurev-psych-010213-115033.

Baillargeon, R., R.M. Scott, and Z. He. 2010. False-belief understanding in infants. *Trends in Cognitive Sciences* 14 (3): 110–118. https://doi.org/10.1016/j.tics.2009.12.006.

Barkley, R.A. 1997. Behavioral inhibition, sustained attention, and executive functions: Constructing a unifying theory of ADHD. *Psychological Bulletin* 121 (1): 65–94.

Baron-Cohen, S. 1999. The evolution of a theory of mind. In *The descent of mind: Psychological perspectives on hominid evolution*, ed. M. Corballis and S.E.G. Lea, 261–277. Oxford University Press.

Baron-Cohen, S., A.M. Leslie, and U. Frith. 1985. Does the autistic child have a "theory of mind"? *Cognition* 21 (1): 37–46. https://doi.org/10.1016/0010-0277(85)90022-8.

Baron-Cohen, S., M. O'Riordan, V. Stone, et al. 1999. Recognition of faux pas by normally developing children and children with Asperger syndrome or high-functioning autism. *Journal of Autism and Developmental Disorders* 29: 407–418.

Baron-Cohen, S., H. Ring, J. Moriarty, B. Schmitz, D. Costa, and P. Ell. 1994. Recognition of mental state terms. *Clinical findings in children with autism and a functional neuroimaging study of normal adults. British Journal of Psychiatry* 165 (NOV.): 640–649. https://doi.org/10.1192/bjp.165.5.640.

Baron-Cohen, S., S. Wheelwright, J. Hill, Y. Raste, and I. Plumb. 2001. The "Reading the mind in the eyes" test revised version: A study with normal adults, and adults with Asperger syndrome or high-functioning autism. *Journal of Child Psychology and Psychiatry and Allied Disciplines* 42 (2): 241–251. https://doi.org/10.1017/S0021963001006643.

Baron-Cohen, S., T. Jolliffe, C. Mortimore, and M. Robertson. 1997. Another advanced test of theory of mind: Evidence from very high functioning adults with autism or Asperger syndrome. *Journal of Child Psychology and Psychiatry* 38 (7): 813–822. https://doi.org/10.1111/j.1469-7610.1997.tb01599.x.

Baron-Cohen, S., and S. Wheelwright. 2004. The empathy quotient: An investigation of adults with Asperger syndrome or high functioning autism, and normal sex differences. *Journal of Autism and Developmental Disorders* 34 (2): 163–175. https://doi.org/10.1023/B:JADD.0000022607.19833.00.

Beaudoin, C., É. Leblanc, C. Gagner, and M.H. Beauchamp. 2020. Systematic review and inventory of theory of mind measures for young children. *Frontiers in Psychology* 10: 2905. https://doi.org/10.3389/fpsyg.2019.02905.

Bock, A.M., K.C. Gallaway, and A.M. Hund. 2015. Specifying links between executive functioning and theory of mind during middle childhood: Cognitive flexibility predicts social understanding. *Journal of Cognition and Development* 16 (3): 509–521. https://doi.org/10.1080/15248372.2014.888350.

Boerma, I.E., S.E. Mol, and J. Jolles. 2017. The role of home literacy environment, mentalizing, expressive verbal ability, and print exposure in third and fourth graders' reading comprehension. *Scientific Studies of Reading* 21 (3): 179–193. https://doi.org/10.1080/10888438.2016.1277727.

Borduin, B.J., C.M. Borduin, and C.M. Manley. 1994. The use of imagery training to improve reading comprehension of second graders. *The Journal of Genetic Psychology* 155 (1): 115–118. https://doi.org/10.1080/00221325.1994.9914764.

Brown, J.R., N. Donelan-McCall, and J. Dunn. 1996. Why talk about mental states? The significance of children's conversations with friends, siblings, and mothers. *Child Development* 67: 836–849. https://psycnet.apa.org/doi/10.2307/1131864.

Brüne, M., and U. Brüne-Cohrs. 2006. Theory of mind – Evolution, ontogeny, brain mechanisms and psychopathology. *Neuroscience and Biobehavioral Reviews* 30 (4): 437–455. https://doi.org/10.1016/j.neubiorev.2005.08.001.

———. 2000. A PET investigation of the attribution of intentions with a nonverbal task. *NeuroImage* 11 (2): 157–166. https://doi.org/10.1006/nimg.1999.0525.

Byrne, R., and A. Whiten. 1991. Natural theories of mind. In *Computation and mindreading in primate tactical deception*, ed. A. Whiten. London: Basil Blackwell.

Bugnyar, T., and B. Heinrich. 2005. Ravens, Corvus corax, differentiate between knowledgeable and ignorant competitors. *Proceedings of the Royal Society B: Biological Sciences* 272 (1573): 1641–1646.

Call, J., and M. ToMasello. 2008. Does the chimpanzee have a theory of mind? 30 years later. *Trends in Cognitive Sciences* 12 (5): 187–192. https://doi.org/10.1016/j.tics.2008.02.010.

Camerer, C.F., A. Dreber, F. Holzmeister, T.-H. Ho, J. Huber, M. Johannesson, M. Kirchler, G. Nave, B.A. Nosek, T. Pfeiffer, A. Altmejd, N. Buttrick, T. Chan, Y. Chen, E. Forsell, A. Gampa, E. Heikensten, L. Hummer, T. Imai, et al. 2018. Evaluating the replicability of social science experiments in nature and science between 2010 and 2015. *Nature Human Behaviour* 2 (9): 637–644. https://doi.org/10.1038/s41562-018-0399-z.

Carlson, S.M., M.A. Koenig, and M.B. Harms. 2013. Theory of mind. *Wiley Interdisciplinary Reviews: Cognitive Science* 4 (4): 391–402. https://doi.org/10.1002/wcs.1232.

Carlson, S.M., D.J. Mandell, and L. Williams. 2004. Executive function and theory of mind: Stability and prediction from ages 2 to 3. *Developmental Psychology* 40 (6): 1105–1122. https://doi.org/10.1037/0012-1649.40.6.1105.

Carlson, S.M., L.J. Moses, and C. Breton. 2002. How specific is the relation between executive function and theory of mind? Contributions of inhibitory control and working memory. *Infant and Child Development* 11 (2): 73–92. https://doi.org/10.1002/icd.298.

Cassetta, B.D., P.M. Pexman, and V.M. Goghari. 2018. Cognitive and affective theory of mind and relations with executive functioning in middle childhood. *Merrill-Palmer Quarterly* 64 (4): 514–538. https://muse.jhu.edu/article/711731.

Castelli, F., F. Happé, U. Frith, and C. Frith. 2000. Movement and mind: A functional imaging study of perception and interpretation of complex intentional movement patterns. *NeuroImage* 12 (3): 314–325. https://doi.org/10.1006/nimg.2000.0612.

Ciaramidaro, A., M. Adenzato, I. Enrici, S. Erk, L. Pia, B.G. Bara, and H. Walter. 2007. The intentional network: How the brain reads varieties of intentions. *Neuropsychologia* 45 (13): 3105–3113. https://doi.org/10.1016/j.neuropsychologia.2007.05.011.

De Mulder, H.N.M., F. Hakemulder, R. van den Berghe, F. Klaassen, and J.J.A. van Berkum. 2017. Effects of exposure to literary narrative fiction. *Scientific Study of Literature* 7 (1): 129–169. https://doi.org/10.1075/ssol.7.1.06dem.

De Villiers, J.G., and J.E. Pyers. 2002. Complements to cognition: A longitudinal study of the relationship between complex synthax and false-belief-understanding. *Cognitive Development* 17 (1): 1037–1060. https://doi.org/10.1016/S0885-2014(02)00073-4.

Denckla, M.B. 1996. A theory and model of executive function: A neuropsychological perspective. In *Attention, memory, and executive function*, ed. G.R. Lyon and N.A. Krasnegor, 263–277. Baltimore: Paul H. Brookes.

Dennett, D. 1987. *The intentional stance*. Cambridge, MA: MIT Press/Bradford Books.

Derksen, D.G., M.C. Hunsche, M.E. Giroux, D.A. Connolly, and D.M. Bernstein. 2018. A systematic review of theory of mind's precursors and functions. *Zeitschrift für Psychologie* 226 (2): 87–97. https://doi.org/10.1027/2151-2604/a000325.

Diamond, A. 2013. Executive functions. *Annual Review of Psychology* 64: 135–168. https://doi.org/10.1146/annurev-psych-113011-143750.

Dodell-Feder, D., and D.I. Tamir. 2018. Fiction reading has a small positive impact on social cognition: A meta-analysis. *Journal of Experimental Psychology: General* 147 (11): 1713–1727. https://doi.org/10.1037/xge0000395.

Duke, N., and P.D. Pearson. 2002. Effective practices for developing reading comprehension. In *What research has to say about reading instruction*, ed. A.E. Farstrup and S.J. Samuels, 205–242. Newark: International Reading Association.

Ebert, S. 2020. Theory of mind, language, and reading: Developmental relations from early childhood to early adolescence. *Journal of Experimental Child Psychology* 191: 104739. https://doi.org/10.1016/j.jecp.2019.104739.

Ebert, S., C. Peterson, V. Slaughter, and S. Weinert. 2017. Links among parents' mental state language, family socioeconomic status, and preschoolers' theory of mind development. *Cognitive Development* 44 (August): 32–48. https://doi.org/10.1016/j.cogdev.2017.08.005.

Eekhof, L.S., K. van Krieken, and R.M. Willems. 2022. Reading about minds: The social-cognitive potential of narratives. *Psychonomic Bulletin and Review* 29 (5): 1703–1718. https://doi.org/10.3758/s13423-022-02079-z.

Eisenberg, N., C. Hofer, and J. Vaughan. 2007. Effortful control and its socioemotional consequences. In *Handbook of emotion regulation*, ed. J.J. Gross, 287–306. New York: Guilford.

Ekman, P., and W.V. Friesen. 1971. Consants across cultures in the face and emotion. *Journal of Personality and Social Psychology* 17 (2): 124–129. https://doi.org/10.1037/h0030377.

Etel, E., and V. Slaughter. 2019. Theory of mind and peer cooperation in two play contexts. *Journal of Applied Developmental Psychology* 60 (October 2017): 87–95. https://doi.org/10.1016/j.appdev.2018.11.004.

Evans, D.E., and M.K. Rothbart. 2007. Developing a model for adult temperament. *Journal of Research in Personality* 41 (4): 868–888. https://doi.org/10.1016/j.jrp.2006.11.002.

Fahie, C.M., and D.K. Symons. 2003. Executive functioning and theory of mind in children clinically referred for attention and behavior problems. *Journal of Applied Developmental Psychology* 24 (1): 51–73. https://doi.org/10.1016/S0193-3973(03)00024-8.

Fisher, N., and F. Happé. 2005. A training study of theory of mind and executive function in children with autistic spectrum disorders. *Journal of Autism and Developmental Disorders* 35 (6): 757–771. https://doi.org/10.1007/s10803-005-0022-9.

Flannery, K.M., and R.L. Smith. 2017. The effects of age, gender, and gender role ideology on adolescents' social perspective-taking ability and tendency in friendships. *Journal of Social and Personal Relationships* 34 (5): 617–635. https://doi.org/10.1177/0265407516650942.

Flavell, J.H., B.A. Everett, K. Croft, and E.R. Flavell. 1981. Young children's knowledge about visual perception: Further evidence for the Level 1–Level 2 distinction. *Developmental Psychology* 17 (1): 99–103. https://doi.org/10.1037/0012-1649.17.1.99.

Flavell, J.H., E.R. Flavell, and F.L. Green. 1983. Development of the appearance-reality distinction. *Cognitive Psychology* 15 (1): 95–120. https://doi.org/10.1016/0010-0285(83)90005-1.

Flavell, J.H., A.G. Friedrichs, and J.D. Hoyt. 1970. Developmental changes in memorization processes. *Cognitive Psychology* 1 (4): 324–340. https://doi.org/10.1016/0010-0285(70)90019-8.

Fletcher, P.C., F. Happé, U. Frith, S.C. Baker, R.J. Dolan, R.S.J. Frackowiak, and C.D. Frith. 1995. Other minds in the brain: A functional imaging study of "theory of mind" in story comprehension. *Cognition* 57 (2): 109–128. https://doi.org/10.1016/0010-0277(95)00692-R.

Freundlieb, M., Á.M. Kovács, and N. Sebanz. 2016. When do humans spontaneously adopt another's visuospatial perspective? *Journal of Experimental Psychology: Human Perception and Performance* 42 (3): 401–412. https://doi.org/10.1037/xhp0000153.

Frith, U., J. Morton, and A.M. Leslie. 1991. The cognitive basis of a biological disorder: Autism. *Trends in Neurosciences* 14: 433–438. https://doi.org/10.1016/0166-2236(91)90041-R

Frye, D., P.D. Zelazo, and T. Palfai. 1995. Theory of mind and rule-based reasoning. *Cognitive Development* 10 (4): 483–527. https://doi.org/10.1016/0885-2014(95)90024-1.

Fu, G., L. Sai, F. Yuan, and K. Lee. 2018. Young children's self-benefiting lies and their relation to executive functioning and theory of mind. *Infant and Child Development* 27 (1): e2051. https://doi.org/10.1002/icd.2051.

Gallese, V. 2003. The roots of empathy: The shared manifold hypothesis and the neural basis of intersubjectivity. *Psychopathology* 36 (4): 171–180. https://doi.org/10.1159/000072786.

Gallotti, M., and C.D. Frith. 2013. Social cognition in the we-mode. *Trends in Cognitive Sciences* 17 (4): 160–165. https://doi.org/10.1016/j.tics.2013.02.002.

Garner, J.K. 2009. Conceptualizing the relations between executive functions and self-regulated learning. *The Journal of Psychology* 143 (4): 405–426. https://doi.org/10.3200/JRLP.143.4.405-426.

Garner, R. 1987. *Metacognition and reading comprehension*. Norwood: Ablex Publishing.

German, T.P., J.L. Niehaus, M.P. Roarty, B. Giesbrecht, and M.B. Miller. 2004. Neural correlates of detecting pretense: Automatic engagement of the intentional stance under covert conditions. *Journal of Cognitive Neuroscience* 16 (10): 1805–1817. https://doi.org/10.1162/0898929042947892.

Glazer, S.M., and E.M. Burke. 1994. *An integrated approach to early literacy: Literature to language*. Boston: Allyn and Bacon, Order Processing.

Golan, O., S. Baron-Cohen, J.J. Hill, and M.D. Rutherford. 2007. The "Reading the mind in the voice" test-revised: A study of complex emotion recognition in adults with and without autism spectrum conditions. *Journal of Autism and Developmental Disorders* 37 (6): 1096–1106. https://doi.org/10.1007/s10803-006-0252-5.

Gomez-Garibello, C., and V. Talwar. 2015. Can you read my mind? Age as a moderator in the relationship between theory of mind and relational aggression. *International Journal of Behavioral Development* 39 (6): 552–559. https://doi.org/10.1177/0165025415580805.

Gordon, H.M., T.D. Lyon, and K. Lee. 2014. Social and cognitive factors associated with Children's secret-keeping for a parent. *Child Development* 85 (6): 2374–2388. https://doi.org/10.1111/cdev.12301.

Gross, J.J. 2015. Emotion regulation: Current status and future prospects. *Psychological Inquiry* 26 (1): 1–26. https://doi.org/10.1080/1047840X.2014.940781.

Guajardo, N.R., and K.B. Cartwright. 2016. The contribution of theory of mind, counterfactual reasoning, and executive function to pre-readers' language comprehension and later reading awareness and comprehension in elementary school. *Journal of Experimental Child Psychology* 144: 27–45. https://doi.org/10.1016/j.jecp.2015.11.004.

Guthrie, J.T., A. Wigfield, P. Barbosa, K.C. Perencevich, A. Taboada, M.H. Davis, N.T. Scafiddi, and S. Tonks. 2004. Increasing reading comprehension and engagement through concept-oriented reading instruction. *Journal of Educational Psychology* 96 (3): 403–423. https://doi.org/10.1037/0022-0663.96.3.403.

Haberkorn, K., K. Lockl, S. Pohl, S. Ebert, and S. Weinert. 2014. Metacognitive knowledge in children at early elementary school. *Metacognition and Learning* 9 (3): 239–263. https://doi.org/10.1007/s11409-014-9115-1.

Halfon, S., A. Coşkun, Ö. Bekar, and H. Steele. 2020. Imbalances and impairments in mental state talk of children with internalizing and externalizing problems. *Journal of Infant, Child, and Adolescent Psychotherapy* 19 (3): 283–302. https://doi.org/10.1080/15289168.2020.1771531.

Happé, F. 1994. An advanced test of theory of mind: Understanding of story characters' thoughts and feelings by able autistic, mentally handicapped, and normal children and adults. *Journal of Autism and Developmental Disorders* 24 (2): 129–154.

Harris, P.L. 2006. Social Cognition. In *Handbook of child psychology: Cognition, perception, and language*, ed. D. Kuhn, R.S. Siegler, W. Damon, and R.M. Lerner, 811–858. Hoboken: Wiley.

Heider, F., and M. Simmel. 1944. An experimental study of apparent behavior. *The American Journal of Psychology* 57: 243–259. https://doi.org/10.2307/1416950.

Heyes, C.M., and C.D. Frith. 2014. The cultural evolution of mind reading. *Science* 344 (6190): 1243091. https://doi.org/10.1126/science.1243091.

Hughes, C., J. Dunn, and A. White. 1998. Trick or treat?: Uneven understanding of mind and emotion and executive dysfunction in "hard-to-manage" preschoolers. *Journal of Child Psychology and Psychiatry* 39 (7): 981–994. https://doi.org/10.1111/1469-7610.00401.

Iacoboni, M., I. Molnar-Szakacs, V. Gallese, G. Buccino, and J.C. Mazziotta. 2005. Grasping the intentions of others with one's own mirror neuron system. *PLoS Biology* 3 (3): 0529–0535. https://doi.org/10.1371/journal.pbio.0030079.

Imuta, K., J.D. Henry, V. Slaughter, and T. Ruffman. 2016. Supplemental material for theory of mind and prosocial behavior in childhood: A meta-analytic review. *Developmental Psychology* 52 (8): 1192–1205. https://doi.org/10.1037/dev0000140.supp.

Jacobs, J.E., and S.G. Paris. 1987. Children's metacognition about reading: Issues in definition, measurement, and instruction. *Educational Psychologist* 22 (3–4): 255–278. https://doi.org/10.1080/00461520.1987.9653052.

Joseph, R.M., and H. Tager-Flusberg. 2004. The relationship of theory of mind and executive functions to sympToM type and severity in children with autism. *Development and Psychopathology* 16 (1): 137–155. https://doi.org/10.1017/S095457940404444X.

Kano, F., C. Krupenye, S. Hirata, and J. Call. 2017. Eye tracking uncovered great apes' ability to anticipate that other individuals will act according to false beliefs. *Communicative and Integrative Biology* 10 (2): e1299836. https://doi.org/10.1080/19420889.2017.1299836.

Kanske, P. 2018. The social mind: Disentangling affective and cognitive routes to understanding others. *Interdisciplinary Science Reviews* 43 (2): 115–124. https://doi.org/10.1080/03080188.2018.1453243.

Keysers, C., and V. Gazzola. 2009. Expanding the mirror: Vicarious activity for actions, emotions, and sensations. *Current Opinion in Neurobiology* 19 (6): 666–671. https://doi.org/10.1016/j.conb.2009.10.006.

Kidd, D.C., and E. Castano. 2013. Reading literary fiction improves theory of mind. *Science* 342 (6156): 377–380. https://doi.org/10.1126/science.1239918.

Kim, Y.S.G. 2017. Why the simple view of reading is not simplistic: Unpacking component skills of reading using a direct and indirect effect model of reading (DIER). *Scientific Studies of Reading* 21 (4): 310–333. https://doi.org/10.1080/10888438.2017.1291643.

Kovács, Á.M., E. Téglás, and A.D. Endress. 2010. The social sense: Susceptibility to others' beliefs in human infants and adults. *Science* 330 (6012): 1830–1834. https://doi.org/10.1126/science.1190792.

Kristen, S., B. Sodian, C. Thoermer, and H. Perst. 2011. Infants' joint attention skills predict toddlers' emerging mental state language. *Developmental Psychology* 47 (5): 1207–1219. https://doi.org/10.1037/a0024808.

Krupenye, C., and J. Call. 2019. Theory of mind in animals: Current and future directions. *WIREs Cognitive Science* 10 (6): 1–25. https://doi.org/10.1002/wcs.1503.

Kulke, L., B. von Duhn, D. Schneider, and H. Rakoczy. 2018. Is implicit theory of mind a real and robust phenomenon? Results from a systematic replication study. *Psychological Science* 29 (6): 888–900. https://doi.org/10.1177/0956797617747090.

Kuntoro, I.A., C.C. Peterson, and V. Slaughter. 2017. Culture, parenting, and children's theory of mind development in Indonesia. *Journal of Cross-Cultural Psychology* 48 (9): 1389–1409. https://doi.org/10.1177/0022022117725404.

Kuntoro, I.A., L. Saraswati, C. Peterson, and V. Slaughter. 2013. Micro-cultural influences on theory of mind development: A comparative study of middle-class and pemulung children in Jakarta, Indonesia. *International Journal of Behavioral Development* 37 (3): 266–273. https://doi.org/10.1177/0165025413478258.

Lane, J.D., H.M. Wellman, S.L. Olson, A.L. Miller, L. Wang, and T. Tardif. 2013. Relations between temperament and theory of mind development in the United States and China: Biological and behavioral correlates of preschoolers' false-belief understanding. *Developmental Psychology* 49 (5): 825–836. https://doi.org/10.1037/a0028825.

Lecce, S., and F. Bianco. 2018. Working memory predicts changes in children's theory of mind during middle childhood: A training study. *Cognitive Development* 47 (May): 71–81. https://doi.org/10.1016/j.cogdev.2018.04.002.

Lecce, S., F. Bianco, R.T. Devine, and C. Hughes. 2017. Relations between theory of mind and executive function in middle childhood: A short-term longitudinal study. *Journal of Experimental Child Psychology* 163: 69–86. https://doi.org/10.1016/j.jecp.2017.06.011.

Leslie, A.M., O. Friedman, and T.P. German. 2004. Core mechanisms in 'theory of mind'. *Trends in Cognitive Sciences* 8 (12): 528–533. https://doi.org/10.1016/j.tics.2004.10.001.

Lewkowicz, D., F. Quesque, Y. Coello, and Y.N. Delevoye-Turrell. 2015. Individual differences in reading social intentions from motor deviants. *Frontiers in Psychology* 6 (August): 1–12. https://doi.org/10.3389/fpsyg.2015.01175.

Li, D., X. Li, F. Yu, X. Chen, L. Zhang, D. Li, Q. Wei, Q. Zhang, C. Zhu, and K. Wang. 2017. Comparing the ability of cognitive and affective theory of mind in adolescent onset schizophrenia. *Neuropsychiatric Disease and Treatment* 13: 937–945. https://doi.org/10.2147/NDT.S128116.

Liu, D., H.M. Wellman, T. Tardif, and M.A. Sabbagh. 2008. Theory of mind development in Chinese children: A meta-analysis of false-belief understanding across cultures and languages. *Developmental Psychology* 44 (2): 523–531. https://doi.org/10.1037/0012-1649.44.2.523.

Luo, Y., and S.C. Johnson. 2009. Recognizing the role of perception in action at 6 months. *Developmental Science* 12 (1): 142–149. https://doi.org/10.1111/j.1467-7687.2008.00741.x.

Ma, F., A.D. Evans, Y. Liu, X. Luo, and F. Xu. 2015. To lie or not to lie? The influence of parenting and theory-of-mind understanding on three-year-old children's honesty. *Journal of Moral Education* 44 (2): 198–212. https://doi.org/10.1080/03057240.2015.1023182.

Mary, A., H. Slama, P. Mousty, I. Massat, T. Capiau, V. Drabs, and P. Peigneux. 2016. Executive and attentional contributions to theory of mind deficit in attention deficit/hyperactivity disorder (ADHD). *Child Neuropsychology* 22 (3): 345–365. https://doi.org/10.1080/09297049.2015.1012491.

Masangkay, Z.S., K.A. McCluskey, C.W. McIntyre, J. Sims-Knight, B.E. Vaughn, and J.H. Flavell. 1974. The early development of inferences about the visual percepts of others. *Child Development* 45 (2): 357–366.

Mason, M.F., J.F. Banfield, and C.N. Macrae. 2004. Thinking about actions: The neural substrates of person knowledge. *Cerebral Cortex* 14 (2): 209–214. https://doi.org/10.1093/cercor/bhg120.

Meltzoff, A.N. 1995. Understanding the intentions of others: Re-enactment of intended acts by 18-month-old children. *Developmental Psychology* 31 (5): 838–850. https://doi.org/10.1037/0012-1649.31.5.838.

Miller, S.A. 2009. Children's understanding of second-order mental states. *Psychological Bulletin* 135 (5): 749–773. https://doi.org/10.1037/a0016854.

Mosconi, M.W., P.B. Mack, G. McCarthy, and K.A. Pelphrey. 2005. Taking an "intentional stance" on eye-gaze shifts: A functional neuroimaging study of social perception in children. *NeuroImage* 27 (1): 247–252. https://doi.org/10.1016/j.neuroimage.2005.03.027.

Mumper, M.L., and R.J. Gerrig. 2017. Leisure reading and social cognition: A meta-analysis. *Psychology of Aesthetics, Creativity, and the Arts* 11 (1): 109–120. https://doi.org/10.1037/aca0000089.

Myers, M., and S.G. Paris. 1978. Children's metacognitive knowledge about reading. *Journal of Educational Psychology* 70 (5): 680–690. https://doi.org/10.1037/0022-0663.70.5.680.

Navarro, E. 2022. What is theory of mind? A psychometric study of theory of mind and intelligence. *Cognitive Psychology* 136 (June): 101495. https://doi.org/10.1016/j.cogpsych.2022.101495.

Onishi, K.H., and R. Baillargeon. 2005. Do 15-month-old infants understand false beliefs? *Science* 308 (5719): 255–258. https://doi.org/10.1126/science.1107621.

Ontai, L.L., and R.A. Thompson. 2008. Attachment, parent-child discourse and theory-of-mind development. *Social Development* 17 (1): 47–60. https://doi.org/10.1111/j.1467-9507.2007.00414.x.

Panero, M.E., D.S. Weisberg, J. Black, T.R. Goldstein, J.L. Barnes, H. Brownell, and E. Winner. 2016. Does reading a single passage of literary fiction really improve theory of mind? An attempt at replication. *Journal of Personality and Social Psychology* 111 (5): e46–e54. https://doi.org/10.1037/pspa0000064.

Papetti, O., C. Cornoldi, A. Pettavino, G. Mazzoni, and J. Borkowsky. 1992. Memory judgments and allocation of study times in good and poor comprehenders. *Advances in Learning and Behavioral Disabilities* 7: 3–33.

Perner, J., and B. Lang. 1999. Development of theory of mind and executive control. *Trends in Cognitive Sciences* 3 (9): 337–344. https://doi.org/10.1016/S1364-6613(99)01362-5.

Pineda-Alhucema, W., E. Aristizabal, J. Escudero-Cabarcas, J.E. Acosta-López, and J.I. Vélez. 2018. Executive function and theory of mind in children with ADHD: A systematic review. *Neuropsychology Review* 28 (3): 341–358. https://doi.org/10.1007/s11065-018-9381-9.

Plotnik, J.M., F.B. De Waal, and D. Reiss. 2006. Self-recognition in an Asian elephant. *Proceedings of the National Academy of Sciences* 103 (45): 17053–17057.

Poulin-Dubois, D., B. Sodian, U. Metz, J. Tilden, and B. Schoeppner. 2007. Out of sight is not out of mind: Developmental changes in infants' understanding of visual perception during the second year. *Journal of Cognition and Development* 8 (4): 401–425. https://doi.org/10.1080/15248370701612951.

Povinelli, D.J., K.E. Nelson, and S.T. Boysen. 1990. Inferences about guessing and knowing by chimpanzees (pan troglodytes). *Journal of Comparative Psychology* 104 (3): 203–210. https://doi.org/10.1037//0735-7036.104.3.203.

Powell, L.J., and S. Carey. 2017. Executive function depletion in children and its impact on theory of mind. *Cognition* 164: 150–162. https://doi.org/10.1016/j.cognition.2017.03.022.

Preckel, K., P. Kanske, and T. Singer. 2018. On the interaction of social affect and cognition: Empathy, compassion and theory of mind. *Current Opinion in Behavioral Sciences* 19: 1–6. https://doi.org/10.1016/j.cobeha.2017.07.010.

Premack, D., and G. Woodruff. 1978. Does the chimpanzee have a theory of mind? *Behavioral and Brain Sciences* 1 (4): 515–526. https://doi.org/10.1017/S0140525X00076512.

Pressley, G.M. 1976. Mental imagery helps eight-year-olds remember what they read. *Journal of Educational Psychology* 68 (3): 355–359. https://doi.org/10.1037/0022-0663.68.3.355.

Quesque, F., E. Chabanat, and Y. Rossetti. 2018. Taking the point of view of the blind: Spontaneous level-2 perspective-taking in irrelevant conditions. *Journal of Experimental Social Psychology* 79 (January): 356–364. https://doi.org/10.1016/j.jesp.2018.08.015.

Quesque, F., and Y. Rossetti. 2020. What do theory-of-mind tasks actually measure? Theory and practice. *Perspectives on Psychological Science* 15 (2): 384–396. https://doi.org/10.1177/1745691619896607.

Reiss, D., and L. Marino. 2001. Mirror self-recognition in the bottlenose dolphin: A case of cognitive convergence. *Proceedings of the National Academy of Sciences* 98 (10): 5937–5942. https://doi.org/10.1073/pnas.101086398.

Repacholi, B.M., and A. Gopnik. 1997. Early reasoning about desires: Evidence from 14- and 18-month-olds. *Developmental Psychology* 33 (1): 12–21. https://doi.org/10.1037/0012-1649.33.1.12.

Sabbagh, M.A., and M.A. Callanan. 1998. Metarepresentation in action: 3-, 4-, and 5-year-olds' developing theories of mind in parent–child conversations. *Developmental Psychology* 34 (3): 491–502. https://doi.org/10.1037/0012-1649.34.3.491.

Sai, L., S. Shang, C. Tay, X. Liu, T. Sheng, G. Fu, X.P. Ding, and K. Lee. 2021. Theory of mind, executive function, and lying in children: A meta-analysis. *Developmental Science* 24 (5): e13096. https://doi.org/10.1111/desc.13096.

Salthouse, T.A. 2005. Relations between cognitive abilities and measures of executive functioning. *Neuropsychology* 19 (4): 532–545. https://doi.org/10.1037/0894-4105.19.4.532.

Samson, D., I.A. Apperly, J.J. Braithwaite, B.J. Andrews, and S.E. Bodley Scott. 2010. Seeing it their way: Evidence for rapid and involuntary computation of what other people see. *Journal of Experimental Psychology: Human Perception and Performance* 36 (5): 1255–1266. https://doi.org/10.1037/a0018729.

Samur, D., M. Tops, and S.L. Koole. 2018. Does a single session of reading literary fiction prime enhanced mentalising performance? Four replication experiments of Kidd and Castano (2013). *Cognition and Emotion* 32 (1): 130–144. https://doi.org/10.1080/02699931.2017.1279591.

Schlaffke, L., S. Lissek, M. Lenz, G. Juckel, T. Schultz, M. Tegenthoff, T. Schmidt-Wilcke, and M. Brüne. 2015. Shared and nonshared neural networks of cognitive and affective theory-of-mind: A neuroimaging study using cartoon picture stories. *Human Brain Mapping* 36 (1): 29–39. https://doi.org/10.1002/hbm.22610.

Schneider, D., V.P. Slaughter, and P.E. Dux. 2017. Current evidence for automatic theory of mind processing in adults. *Cognition* 162: 27–31. https://doi.org/10.1016/j.cognition.2017.01.018.

Schraw, G., and R.S. Dennison. 1994. Assessing metacognitive awareness. *Contemporary Educational Psychology* 19 (4): 460–475. https://doi.org/10.1006/ceps.1994.1033.

Schurz, M., J. Radua, M.G. Tholen, L. Maliske, D.S. Margulies, R.B. Mars, J. Sallet, and P. Kanske. 2021. Toward a hierarchical model of social cognition: A neuroimaging meta-analysis and integrative review of empathy and theory of mind. *Psychological Bulletin* 147 (3): 293–327. https://doi.org/10.1037/bul0000303.

Sebanz, N., and M. Shiffrar. 2009. Detecting deception in a bluffing body: The role of expertise. *Psychonomic Bulletin and Review* 16 (1): 170–175. https://doi.org/10.3758/PBR.16.1.170.

Sebastian, C.L., N.M.G. Fontaine, G. Bird, S.-J. Blakemore, S.A. De Brito, E.J.P. McCrory, and E. Viding. 2012. Neural processing associated with cognitive and affective theory of mind in adolescents and adults. *Social Cognitive and Affective Neuroscience* 7 (1): 53–63. https://doi.org/10.1093/scan/nsr023.

Senju, A. 2012. Spontaneous theory of mind and its absence in autism spectrum disorders. *The Neuroscientist* 18 (2): 108–113. https://doi.org/10.1177/1073858410397208.
Shahaeian, A., C.C. Peterson, V. Slaughter, and H.M. Wellman. 2011. Culture and the sequence of steps in theory of mind development. *Developmental Psychology* 47 (5): 1239–1247. https://doi.org/10.1037/a0023899.
Silke, C., B. Brady, C. Boylan, and P. Dolan. 2018. Factors influencing the development of empathy and pro-social behaviour among adolescents: A systematic review. *Children and Youth Services Review* 94: 421–436. https://doi.org/10.1016/j.childyouth.2018.07.027.
Singh, J., P. Arun, and M.K. Bajaj. 2021. Theory of mind and executive functions in children with attention deficit hyperactivity disorder and specific learning disorder. *Indian Journal of Psychological Medicine* 43 (5): 392–398. https://doi.org/10.1177/0253717621999807.
Slaughter, V., K. Imuta, C.C. Peterson, and J.D. Henry. 2015. Meta-analysis of theory of mind and peer popularity in the preschool and early school years. *Child Development* 86 (4): 1159–1174. https://doi.org/10.1111/cdev.12372.
Sodian, B. 2016. Is false belief understanding continuous from infancy to preschool age? In *Core knowledge and conceptual change*, ed. D. Barner and A.S. Baron. Oxford University Press. https://doi.org/10.1093/acprof:oso/9780190467630.003.0017.
Sodian, B., and U. Frith. 1992. Deception and sabotage in autistic, retarded and normal children. *Journal of Child Psychology and Psychiatry* 33 (3): 591–605. https://doi.org/10.1111/j.1469-7610.1992.tb00893.x.
Sodian, B., S. Kristen-Antonow, and D. Kloo. 2020. How does children's theory of mind become explicit? A review of longitudinal findings. *Child Development Perspectives* 14 (3): 171–177. https://doi.org/10.1111/cdep.12381.
Sommer, M., K. Döhnel, B. Sodian, J. Meinhardt, C. Thoermer, and G. Hajak. 2007. Neural correlates of true and false belief reasoning. *NeuroImage* 35 (3): 1378–1384. https://doi.org/10.1016/j.neuroimage.2007.01.042.
Southgate, V., A. Senju, and G. Csibra. 2007. Action anticipation through attribution of false belief by 2-year-olds. *Psychological Science* 18 (7): 587–592. https://doi.org/10.1111/j.1467-9280.2007.01944.x.
Steele, H., M. Steele, C. Croft, and P. Fonagy. 1999. Infant-mother attachment at one year predicts children's understanding of mixed emotions at six years. *Social Development* 8 (2): 161–178. https://doi.org/10.1111/1467-9507.00089.
Surian, L., and A. Geraci. 2012. Where will the triangle look for it? Attributing false beliefs to a geometric shape at 17 months. *British Journal of Developmental Psychology* 30 (1): 30–44. https://doi.org/10.1111/j.2044-835X.2011.02046.x.
Talwar, V., and K. Lee. 2008. Social and cognitive correlates of childrens lying behavior. *Child Development* 79 (4): 866–881. https://doi.org/10.1111/j.1467-8624.2008.01164.x.
Tamnes, C.K., K. Overbye, L. Ferschmann, A.M. Fjell, K.B. Walhovd, S.-J. Blakemore, and I. Dumontheil. 2018. Social perspective taking is associated with self-reported prosocial behavior and regional cortical thickness across adolescence. *Developmental Psychology* 54 (9): 1745–1757. https://doi.org/10.1037/dev0000541.
Tatar, Z.B., and A. Cansız. 2022. Executive function deficits contribute to poor theory of mind abilities in adults with ADHD. *Applied Neuropsychology: Adult* 29 (2): 244–251. https://doi.org/10.1080/23279095.2020.1736074.
ToMasello, M. 2018. How children come to understand false beliefs: A shared intentionality account. *Proceedings of the National Academy of Sciences* 115 (34): 8491–8498. https://doi.org/10.1073/pnas.1804761115.
Van der Graaff, J., G. Carlo, E. Crocetti, H.M. Koot, and S. Branje. 2018. Prosocial behavior in adolescence: Gender differences in development and links with empathy. *Journal of Youth and Adolescence* 47 (5): 1086–1099. https://doi.org/10.1007/s10964-017-0786-1.
Vetter, N.C., M. Altgassen, L. Phillips, C.E.V. Mahy, and M. Kliegel. 2013. Development of affective theory of mind across adolescence: Disentangling the role of executive functions. *Developmental Neuropsychology* 38 (2): 114–125. https://doi.org/10.1080/87565641.2012.733786.

Wade, M., H. Prime, J.M. Jenkins, K.O. Yeates, T. Williams, and K. Lee. 2018. On the relation between theory of mind and executive functioning: A developmental cognitive neuroscience perspective. *Psychonomic Bulletin and Review* 25 (6): 2119–2140. https://doi.org/10.3758/s13423-018-1459-0.

Walter, H., M. Adenzato, A. Ciaramidaro, I. Enrici, L. Pia, and B.G. Bara. 2004. Understanding intentions in social interaction: The role of the anterior paracingulate cortex. *Journal of Cognitive Neuroscience* 16 (10): 1854–1863. https://doi.org/10.1162/0898929042947838.

Wang, Y., Y. Wang, Y. Zou, K. Ni, X. Tian, H. Sun, S.S.Y. Lui, E.F.C. Cheung, J. Suckling, and R.C.K. Chan. 2018. Theory of mind impairment and its clinical correlates in patients with schizophrenia, major depressive disorder and bipolar disorder. *Schizophrenia Research* 197: 349–356. https://doi.org/10.1016/j.schres.2017.11.003.

Warneken, F., M. Gräfenhain, and M. ToMasello. 2012. Collaborative partner or social tool? New evidence for young children's understanding of joint intentions in collaborative activities. *Developmental Science* 15 (1): 54–61. https://doi.org/10.1111/j.1467-7687.2011.01107.x.

Weimer, A.A., S.J. Parault Dowds, W.V. Fabricius, P.J. Schwanenflugel, and G.W. Suh. 2017. Development of constructivist theory of mind from middle childhood to early adulthood and its relation to social cognition and behavior. *Journal of Experimental Child Psychology* 154: 28–45. https://doi.org/10.1016/j.jecp.2016.10.002.

Weimer, A.A., K.R. Warnell, I. Ettekal, K.B. Cartwright, N.R. Guajardo, and J. Liew. 2021. Correlates and antecedents of theory of mind development during middle childhood and adolescence: An integrated model. *Developmental Review* 59 (December 2020): 100945. https://doi.org/10.1016/j.dr.2020.100945.

Wellman, H. 1990. *Children's theories of mind*. Cambridge, MA: MIT Press.

Wellman, H.M. 2002. Understanding the psychological world: Developing a theory of mind. In *Blackwell handbook of childhood cognitive development*, ed. U. Goswami, 167–187. Malden: Blackwell. https://doi.org/10.1002/9780470996652.ch8.

Wellman, H.M., D. Cross, and J. Watson. 2001. Meta-analysis of theory-of-mind development: The truth about false belief. *Child Development* 72 (3): 655–684. https://doi.org/10.1111/1467-8624.00304.

Wimmer, H., and J. Perner. 1983. Beliefs about beliefs: Representation and constraining function of wrong beliefs in young children's understanding of deception. *Cognition* 13 (1): 103–128. https://doi.org/10.1016/0010-0277(83)90004-5.

Woodward, A. 1998. Infants selectively encode the goal object of an actor's reach. *Cognition* 69 (1): 1–34. https://doi.org/10.1016/S0010-0277(98)00058-4.

Woolley, A. W., C.F. Chabris, A. Pentland, N. Hashmi, and T.W. Malone. 2010. Evidence for a collective intelligence factor in the performance of human groups. *Science* 330 (6004): 686–688. https://doi.org/10.1126/science.1193147.

Zelazo, P.D., and D. Frye. 1997. Cognitive complexity and control: A theory of the development of deliberate reasoning and intentional action. In *Language structure, discourse, and the access to consciousness*, ed. M.I. Stamenov. Amsterdam: John Benjamins.

Zhang, T., Z. Shao, and Y. Zhang. 2016. Developmental steps in theory of mind of typical Chinese children and Chinese children with autism spectrum disorder. *Research in Autism Spectrum Disorders* 23 (1): 210–220. https://doi.org/10.1016/j.rasd.2015.10.005.

Zimmerman, B.J. 2002. Becoming a self-regulated learner: An overview. *Theory Into Practice* 41 (2): 64–70. https://doi.org/10.1207/s15430421tip4102_2.

Zinar, S. 2000. The relative contributions of word identification skill and comprehension-monitoring behavior to reading comprehension ability. *Contemporary Educational Psychology* 25 (4): 363–377. https://doi.org/10.1006/ceps.2000.1024.

Chapter 3
A Mental Files Theory of Mind: How Children Represent Belief and Its Aspectuality

Michael Huemer

Abstract The standard view on explicit theory of mind development holds that children around the age of 4 years start to ascribe beliefs to themselves and others. At this age they begin to master FB tasks in which they have to ascribe a mistaken belief to someone else. The emerging competence in FB tasks goes hand in hand with the developing ability to master various tasks that also require the understanding of different perspectives, like the alternative naming game, false sign or identity tasks. Mental Files Theory allows to explain this developmental synchrony. It also helps to explain why older children struggle to understand that beliefs about an object depend on how one is acquainted with it (aspectuality), and why these difficulties disappear once children master second-order belief tasks at age 6. In this chapter, I focus on children's developing mental file management, and how this accounts for the developmental synchrony of tasks that require taking into account different perspectives. In addition, I address several challenges for this view from the literature.

Keywords Mental files theory · Theory of mind · False belief · Aspectuality of belief · Knowledge ("true belief") error

3.1 Introduction

Humans show a remarkable proficiency in social cognition which depends on our folk psychological competence to predict, explain, interpret, and influence the behavior of people based on reasoning about their minds (Davies and Stone 1995; Frith and Frith 2005; Ho et al. 2022; Saxe and Kanwisher 2003; Wellman 2014). The ability of ascribing mental states – like intentions, desires, beliefs, and knowledge – to oneself and others, and to understand that the mental states of others can be different to one's own mental states is termed 'theory of mind' (Premack and Woodruff 1978). Since having a theory of mind is a crucial part of human's social

M. Huemer (✉)
Laboratory for Developmental Studies, Department of Psychology, Harvard University, Cambridge, MA, USA
e-mail: michaelhuemer@fas.harvard.edu

T. Lopez-Soto et al. (eds.), *The Theory of Mind Under Scrutiny*, Logic, Argumentation & Reasoning 34, https://doi.org/10.1007/978-3-031-46742-4_3

cognitive abilities it plays an apparently important role in the cohesion of human societies (Tomasello 2014). Intensive and interdisciplinary efforts in this area of research are made in developmental psychology, anthropology, ethology, neuroscience, and philosophy of mind. Yet, there is still a lack of understanding of the cognitive representations and processes underlying this important ability. Mental Files Theory enables to overcome this lack of understanding. In this chapter, I provide a cognitive analysis of children's developing understanding of belief and its aspectuality using Mental Files Theory.

The standard view on explicit theory of mind development holds that children around the age of 4 years start to ascribe beliefs to themselves and others. At this age they begin to master false belief (FB) tasks in which they have to ascribe a mistaken belief to someone else (Wellman et al. 2001). In the original version of the FB task, Maxi puts his chocolate into the green cupboard and leaves for the playground. In his absence his mother moves the chocolate to the blue cupboard, and leaves. Maxi returns, and the child is asked where Maxi will look for his chocolate (Wimmer and Perner 1983). The emerging competence in FB tasks goes hand in hand with the developing ability to master various tasks that also require the understanding of different perspectives, like visual perspective taking level-2 (Hamilton et al. 2009; Masangkay et al. 1974), false sign (Parkin 1994; Schuster et al. 2021), alternative naming (Doherty and Perner 1998; Perner et al. 2002) or identity tasks (Perner et al. 2011; Weingartner and Haring 2020; for an overview see Doherty and Perner 2020; Perner and Leahy 2016; Perner and Roessler 2012). This synchronous development suggests the onset of robust, explicit metarepresentational thinking, allowing children to grasp the notion of propositional attitudes (Perner 1991; Rakoczy 2017, 2022; Wellman 2011). Although children at age 4 are able to ascribe mental states to themselves and others, and subsequently pass standard FB tasks, it seems they still do not understand all features of mental states. Empirical evidence suggests that children this age still lack competence in taking into account that beliefs and other mental states about an object depend on the aspect under which the object is represented (Apperly and Robinson 1998, 2001, 2003; Huemer et al. 2018; Kamawar and Olson 1999, 2009, 2011; Perner et al. 2015; Russell 1987; Schünemann et al. 2022; Sprung et al. 2007) and the recursive nature of higher-order mental states (e.g., "John thinks that Mary thinks that...", Perner and Wimmer 1985). Children overcome their problems with aspectuality of belief when they start passing second-order belief tasks about 2 years later, at around 6 years of age (Huemer et al. 2018; Perner et al. 2015; Sprung et al. 2007).

When we think or talk about the mind, we do this by the use of propositional attitudes, e.g., "Mary believes that the ball is in the box". A property of propositional attitudes is aspectuality, representing objects and situations only under certain aspects, with intensionality (in linguistic contexts) being a consequence of it (McKay and Nelson 2014; Searle 1983). In intensional (opaque) contexts there is limited substitutability of coreferential terms, as opposed to extensional contexts where coreferential terms can be substituted *salva veritate* (without changing the truth value). For example, given that:

(1) The *red ball* is *Mary's birthday present*

is true, then the term "the red ball" can be substituted by the coreferential term "her birthday present" without affecting the truth in the following sentence,

(2) Mary finds the *red ball* in the box
(3) Mary finds *her birthday present* in the box

If (2) is true, then (given premise (1) is also true) (3) also has to be true. Contrary to this, in intensional contexts, for instance when dealing with propositional attitude reports, the substitution of coreferential terms may lead to change of the truth value. For example, even if the following sentence,

(4) Mary believes the *red ball* is in the box

is true, then by substituting "the red ball" with the coreferential term "her birthday present" we get the sentence.

(5) Mary believes *her birthday present* is in the box

which may well be false. Here, the truth of (5) does not follow from the truth of (4) and the premise (1). Mary may simply be unaware that the red ball is her birthday present. Beliefs about an object may be only under one but not another aspect.

In Section (1), I describe the empirical problems to which I apply Mental Files Theory. In Section (2), I introduce Mental Files Theory. In Section (3), based on the mental files framework, I present provide a cognitive analysis of the existing empirical evidence, integrating it into a coherent theory. In Section (4), I address two challenges for Mental Files Theory arising from the so-called knowledge and true belief error. Section (5) discusses the findings and concludes.

3.2 Children's Developing Understanding of the Aspectuality of Belief

Understanding the aspectuality of belief is difficult for children in the preschool age. The first study on this subject was done by Russell (1987), where he examined children's explicit understanding of talking about beliefs. Children learned about a boy named George and his watch was stolen, and – unbeknownst to George – that the thief has curly red hair. Children were then asked, "Can we say George is thinking 'I must find the man with curly red hair who stole my watch'?". Even 6- to 8-year-olds struggle with such explicit judgements (see also Kamawar and Olson 1999, 2009, 2011). In a subsequent study (Hulme et al. 2003), linguistic demands were reduced by showing children some pictures depicting a stereotypically looking thief, the actual thief, etc., from which they had to choose one that best fits what George thinks the thief looks like. About half of the 6-year-olds chose the picture of the actual thief even though almost all of them correctly stated that George is not aware what the actual thief looks like.

Apperly and Robinson (1998, 2001) introduced a new task in which children were told about Heinz and a ball, and Heinz being not aware that the ball is his birthday present. Heinz can see the ball being put into a box. The children were then asked, "Does Heinz know that the ball is a present?" (Correct answer: "no"), and "Does Heinz know there's a present in the box?" (Correct answer: "no"). While the majority of 5-year-old children competently answered the first question (69% correct), they had significant problems with answering the second (15% correct). The level of correct responses to a FB task (76%) was comparable to that to the first question. In contrast 4-year-olds had remarkable problems with this question (34% correct) and the FB task (38% correct) and showing even more problems with the second question (11%). A similar gap was also found in the study conducted by Kamawar and Olson (1999).

Together with the aforementioned task, Apperly and Robinson (1998, 2001) also introduced an action-prediction task where no metalinguistic judgements were required. In their Heinz scenario (see Fig. 3.1), 4- to 6-year-old children were

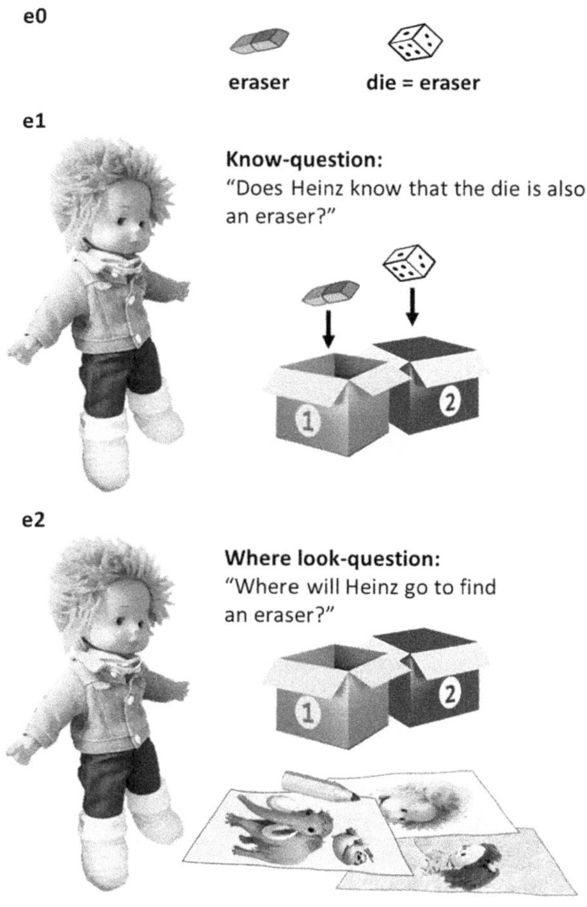

Fig. 3.1 Apperly and Robinson's (1998, 2001) Heinz task

confronted with a standard eraser and a dual function object with an obvious function, being a die, while the other function, being an eraser, could only be detected via close inspection. Children were made familiar with the dual nature of the die/eraser, while Heinz was absent and, therefore, ignorant about this fact. The child and Heinz both observed that the die/eraser was put into one box and the standard eraser into another box. When the children were asked "Does Heinz know that the die is an eraser?" a majority correctly denied that Heinz knew about the rubber die's dual nature, but when asked "Where will Heinz go to find an eraser?" both boxes were equally often indicated as the location where Heinz will search. This incoherent pattern of responses occurs in a developmental window between passing first-order FB tasks and passing second-order belief tasks (e.g., "Heinz thinks he knows..."; Perner and Howes 1992) about 2 years later (Huemer et al. 2018; Sprung et al. 2007). Hereafter, I refer to children who pass both first-order and second-order verbal false belief tasks as [++]; children who pass first-order but fail second-order verbal false belief tasks as [+−]; and children who fail both as [−].

Apperly and Robinson (2003) argued that children this age have difficulties with Heinz' partial knowledge about the object, knowing it only under one but not the other description.

To explain why children, master the Heinz task at the same time as second-order belief tasks, Sprung et al. (2007) argued that both tasks require higher-order perspective taking. The first level of perspective taking is due to adopting the perspective of Heinz. The second level of perspective taking arises as different conceptions of objects are given, for example the die/eraser can be conceived of as "die" or as "eraser". As different labels for an object create different perspectives (Clark 1997; Tomasello 1999), the different conceptions referring to the same object create a conceptual perspective problem. Therefore, Sprung and his colleagues (Sprung et al. 2007) argued that the Heinz task requires children to understand embedded perspectives: Heinz has come to know the object only as a die, so from his perspective only the die-perspective is available but not the eraser-perspective. In two experiments they could demonstrate for the Heinz story (action-prediction version) that children who passed second-order belief tasks predicted that Heinz would look for an eraser at the location of the standard eraser but not at the location of the die/rubber.

In a more recent attempt to investigate children's reasoning about the aspectuality of belief, Rakoczy et al. (2015) found that 4- to 6-year-olds' difficulties all but vanish when reducing task complexity by using only a single object. In the crucial task (Experiment 3; see Fig. 3.2), they used an object with two aspects, e.g., a ball that is a rattle. Then Susi appeared and both the child and Susi observed the ball being put into a box (box 1). Susi left and in her absence the object was then taken out of the box and children were shown that the ball is a rattle. The object was then put back into box 1. Susi returned and in her presence the object was moved *as rattle* (hidden in the experimenter's hands) from box 1 to box 2. Children correctly denied that Susi knew that the ball is the rattle, and when asked where Susi will look for the ball, they correctly responded with box 1. Passing this task requires that children take into account that Susi does not know that the ball is also the rattle, and subsequently must suppress their own knowledge. Rakoczy et al. (2015) argue that their task tests the

Fig. 3.2 Rakoczy et al.'s (2015) Susi task

same abilities as does the action-prediction task of Apperly and Robinson (1998, 2001)) and Sprung et al. (2007). They suggest that children's difficulties with aspectuality tasks used in earlier studies can be explained in terms of excessive linguistic and other demands like resolution of ambiguous reference. Once these factors are removed, as in the Susi task, aspectuality tasks are no more difficult than standard FB tasks. Then Rakoczy et al. (2015); see also Rakoczy 2017) arrive at the conclusion that by age 4 children acquire a fully-fledged concept of propositional attitudes, including their aspectuality.

However, not all existing data can be explained by task complexity, as suggested by Rakoczy et al. (2015). Apperly and Robinson (2003) compared a double identity (aspectuality) condition with a FB false belief condition, where both conditions contained only one critical object and identical test questions and found that the former was more difficult than the latter. Moreover, the complexity explanation does not account for the fact that tasks in which information about the object was given in a predicative manner (e.g., the ball rattles) were significantly easier than tasks in which information about the object was given in an individuating manner (e.g., the ball is the rattle) (Perner et al. 2015; Sprung et al. 2007). Conversely, neither Apperly and Robinson's (2003) partial knowledge explanation nor the embedded perspective account proposed by Sprung et al. (2007) can explain Rakoczy et al.'s (2015) data. Because Susi has only partial knowledge of the object, knowing it only as a ball but not as a rattle, Apperly and Robinson would incorrectly predict that children struggle with this task. Likewise, an incorrect prediction would result from the embedded perspectives account (Sprung et al. 2007). To pass this task, children must assume that from Susi's perspective, the rattle-perspective on the object is not available. Therefore, children should not succeed until they can also master second-order belief. However, Rakoczy et al.'s (2015) tasks were much easier, being not more difficult than FB tasks.

None of these accounts can explain all the data. In what follows, I will introduce Mental Files Theory, and provide a cognitive analysis of the existing empirical evidence, integrating it into a coherent theory.

3.3 Mental Files

Mental Files Theory plays an important role in philosophy, where it is used to address issues about Frege's (1892) fundamental problem of identity and the distinction between sense and reference (Perry 2002) and about Russell's (1910) problem of acquaintance (Recanati 2012). In linguistics mental files play a role in discourse representation (Heim 2002) where they are related to the notion of discourse referents (Kamp and Reyle 1993; Karttunen 1976). In psychology mental files were used as object files (Pylyshyn 2007; Treisman and Gelade 1980; Xu and Carey 1996) and in Theory of Mind research they were used to analyze cognitive development of how beliefs are represented (Doherty and Perner 2020; Huemer et al. 2018; Perner and Brandl 2005; Perner et al. 2003, 2015; Perner and Leahy 2016; Perner et al. 2007; Wolf 2021). Mental files capture how one represents the things in the world he thinks about. Mental Files Theory can account for different perspectives on the same thing, what characterizes the mental domain. It paves the way for considering perspective – visual perspective, conceptual perspective generated by different names for an object, mental perspective generated by differing beliefs – as a unified phenomenon (Perner and Leahy 2016). This feature of Mental Files Theory gives rise to an explanation of the developmental correlation of belief understanding and a number of tasks that require some awareness of perspective differences (identity statements, alternative naming, false sign, visual perspective; for a review see Perner and Roessler 2012), which children start to master around the same age (around 4 years) as the false belief task (Doherty and Perner 2020; Perner and Leahy 2016).

A mental file represents a particular object, its referent, *as* something. The function of mental file is to track its referent over time and cluster information about it (Millikan 2000; Recanati 2012). An important feature of mental files is that they capture the structure of language and thought: the distinction between what one is talking or thinking of (the subject or topic; what the file is tracking) and what one says or thinks about it (the information about the subject; what is noted on the file). A mental file accumulates information about its referent from a particular perspective, which can be different *conceptual* perspectives – in what way an object is individuated – and different *mental* perspectives – what different people believe or think about the object. The different files provide the different perspectives, the information on the files provide the content of each perspective. When one uses a new label for an object this object is individuated in a new way, a new *conceptual* perspective is put on that object and therefore a new file with this label is opened. This file contains all predicative information one has about this object from this conceptual perspective (Doherty and Perner 2020; Perner et al. 2015; Perner and Leahy 2016). Figure 3.3 depicts the essential features of a mental file.

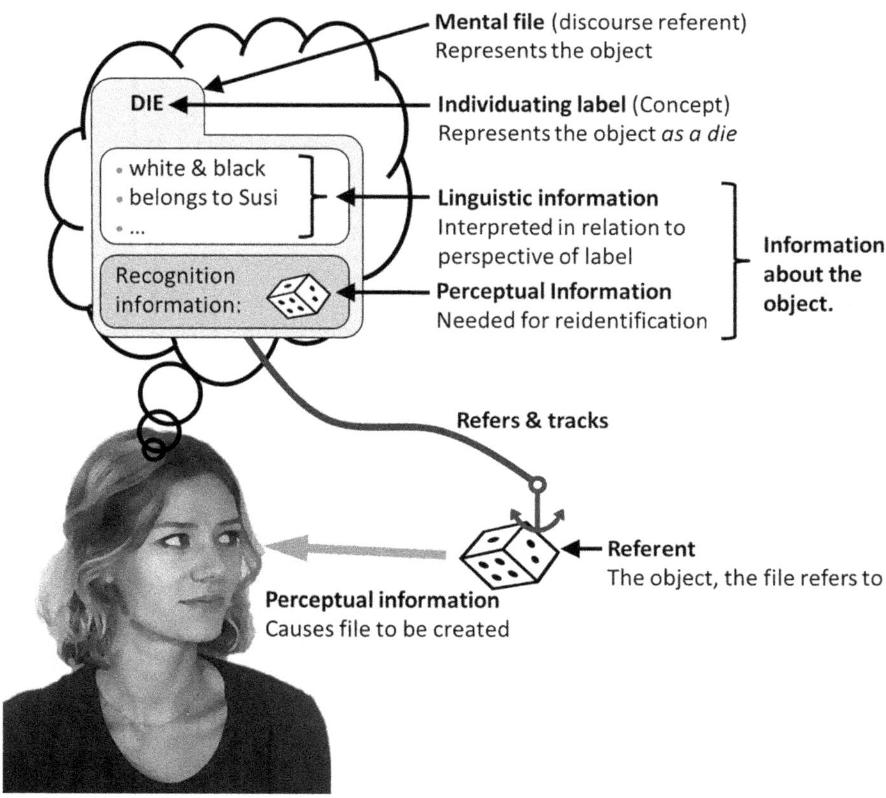

Fig. 3.3 Essential features of a mental file (Doherty and Perner 2020)

When one individuates an object with different names it gets different "identities" (Markman 1989; Flavell 1988) and different perspectives on that object are created (Clark 1997; Tomasello 1999). Thinking of an object in multiple ways, is putting multiple *conceptual* perspectives on that object. Each file anchored to the same object conceptualises it under a different perspective. If one has more than one file that refers to the same object but is not aware that these files corefer, then one also thinks of as many objects as there are files. The files just happen to have the same referent. When one becomes aware that two (or more) files corefer, for instance by learning that the die is the eraser, then these files are connected by a *horizontal* link. This link enables access to information stored on one file from the other file (Perry 2002; see Fig. 3.4a). Different people can hold different beliefs about the state of the world, and subsequently they can have different *mental* perspectives on the same object. One's own thoughts about an object are represented by *regular* files, while *vicarious* files (indexed to the other agent) capture one's representation of the other's beliefs about the object. To indicate sameness of referent, those vicarious files and regular files are linked. As the information stored in the vicarious file needs to be kept in quarantine (Leslie 1987) to avoid confusion and incoherence, vicarious files are linked to regular files via a v*ertical* link (Recanati 2012) which constrains the information flow between them (see Fig. 3.4b).

The developing mental file management in childhood departs in some respects from the standard of the mature system. Children's emerging competence in understanding identity problems, and thereby making sense of different conceptual perspectives, coincides with their arising ability to handle different mental perspectives as shown by their mastery of false belief tasks (Huemer et al. 2019; Perner et al. 2011; Weingartner and Haring 2020). Mental Files Theory can account for this evidence by making the claim that the underlying common ability for a successful performance in these tasks is the competence of linking files – horizontally as well as vertically – which children gain around the age of 4 (Doherty and Perner 2020; Perner and Leahy 2016).

One of the stories in the identity tasks of Perner et al. (2011) was about helping Max, the zookeeper, to find the right keys for the snake cage, the lion cage and the food storage from a bowl of unmarked keys (see Fig. 3.5a). First, the experimenter picked a key from a bowl with several keys, and then tried whether this key would open the food storage, which it did not. The experimenter then checked if this key would open the lion's cage. The key fit and was marked with a yellow sticker. In the *identity condition*, the key was then returned to the bowl. Next, another key (supposedly a new one, in fact the key just marked yellow, with the still unmarked up) was taken out of the bowl. The experimenter then checked if this key would open the snake's cage. The key fit and was marked with a green sticker. While putting the key back, the experimenter pretended to notice with surprise that the key with the green marking also had the yellow marking on the other side. The key was then put with its green marked side up in front of the child who was then asked the three test questions: (1) "Does this [green] key open the [yellow] lion's cage?" [yes], (2) "Does it open the food storage?" [no], and (3) "Does it open the [green] snake's cage?" [yes]. The first of these questions tested whether children understood that the green

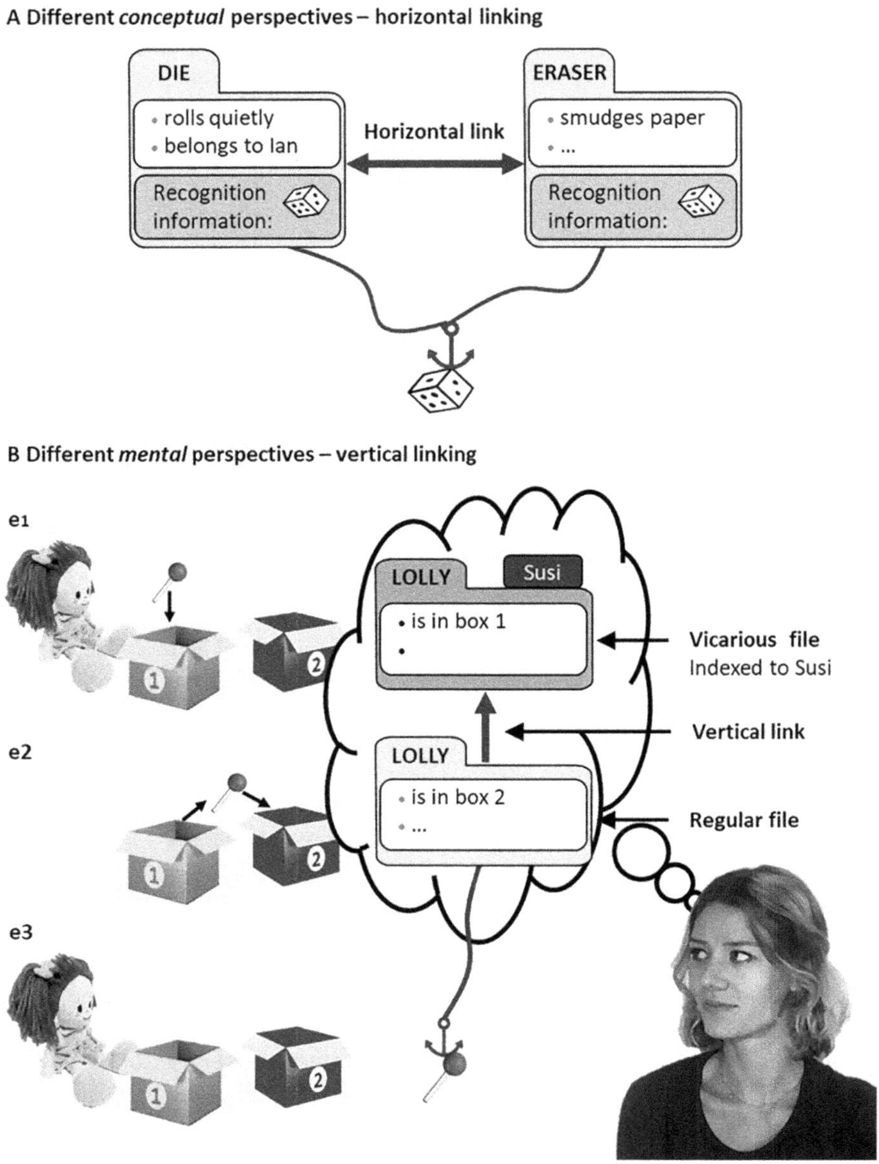

Fig. 3.4 (**a**) Horizontal linking. Two mental files are anchored to the same external referent and are horizontally linked. Each conceptual perspective on the external referent is captured by the respective file. (**b**) Vertical linking. The girl's own thoughts about the lolly are represented by the regular file where the information about the lolly – as she knows it – is stored, while in the vicarious file indexed to Susi the information about the lolly what the girl thinks that Susi believes is stored. Both files are linked by a vertical link which constrains the flow of information between the two files

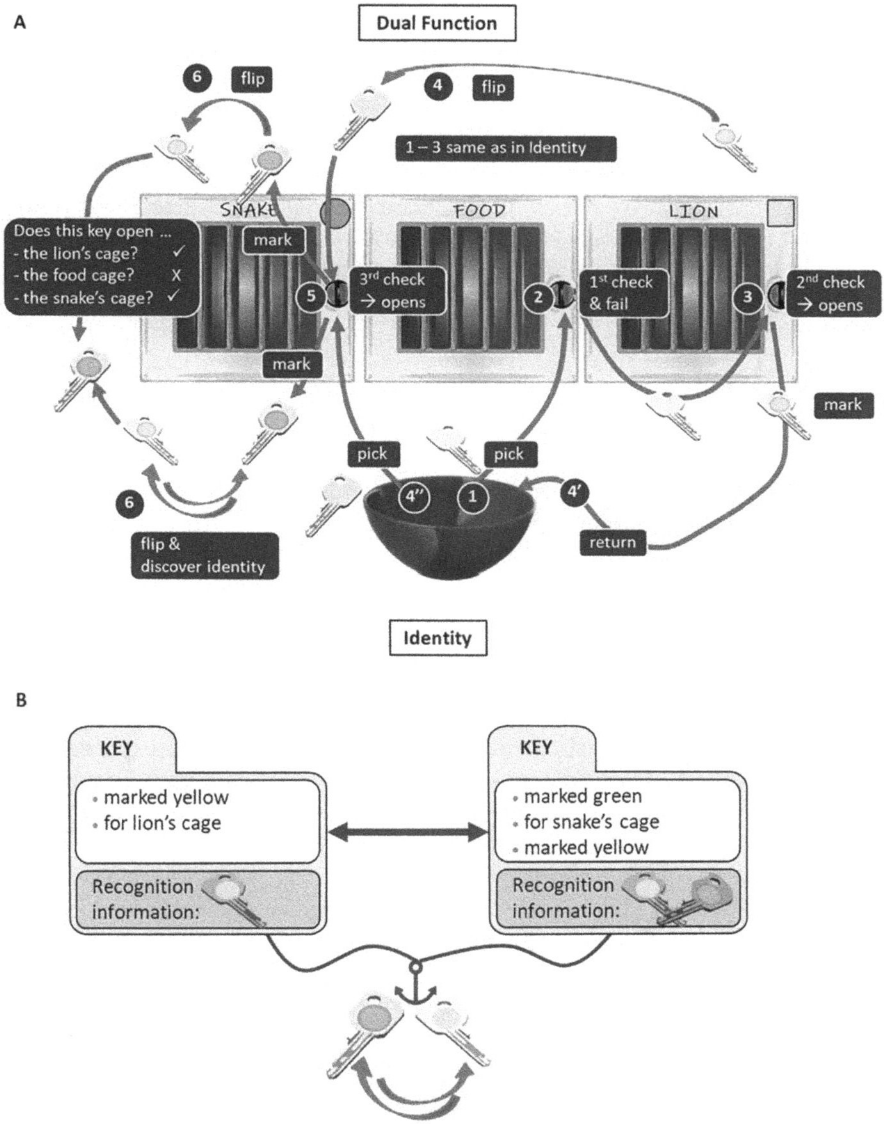

Fig. 3.5 Perner et al.'s (2011) identity task. (**a**) Procedure of the identity condition and the dual function control condition. (**b**) Mental files analysis: The link between the two files allows to access the information "for lion's cage" from the conceptual perspective of the green marked key. Without that link, this information is not accessible

marked key is (identical with) the yellow marked key and, hence, must also open the lion's cage. The identity task and the FB task were equally hard for children and performance in the two tasks was correlated (even when controlling for age). Children's difficulties were specific to the identity condition, they had no problems with the *dual function control condition*, which was the same as the identity condition except that the key was not put back in the bowl after it was marked green, instead it was just flipped over and then the experimenter checked whether it opens the lion' cage. Figure 3.5b shows that the relevant information to answer the crucial test question (whether the key opens the lion's cage) is only accessible when the two corefering files are linked.

But the management of mental files in children this age may still lack at least two aspects of the mature system: (1) when to assign a vicarious file, and (2) when should vicarious files be horizontally linked to indicate that the person to whom the files are indexed is aware of the sameness of referent. Perner et al. (2015) argued that [+−] children do not consider which conceptual perspective is available from the agent's perspective. Therefore, they (1) deploy too many vicarious files they shouldn't deploy, and (2) do not consider horizontal linking between vicarious files, subsequently information transfer between these files is unfeasible. Children's difficulties disappear when they develop an understanding of embedded perspectives. This helps them both to appreciate which conceptual perspective is available from the agent's perspective and to master second-order beliefs.

3.4 A Mental Files Account of Children's Developing Understanding of Aspectuality of Belief

Following the argumentation from the previous section, it can now be explained why the critical Susi task of Rakoczy et al. (2015) is easy for [+−] children (Perner et al. 2015; see Fig. 3.6, absent condition). Children are acquainted with the object both under the ball- and the rattle-aspect. They thus form a regular ball-file and a regular rattle-file. These two files are horizontally linked, indicating the child's awareness of the identity. But [+−] children do not consider horizontal links between vicarious files. In the beginning, the child sees how Susi observes the object referred to as "ball" being put in box 1, and the location information "in box 1" is noted on the child's regular ball-file and on the vicarious ball-file. Later, when the object is transferred as "rattle" in both the child's and Susi's presence, the location information "in box 2" is noted on both the regular and the vicarious rattle-file. Since the two vicarious files are not horizontally linked the now outdated location information "in box 1" is not updated on the vicarious ball-file. In contrast, the regular files are linked horizontally, and therefore the information about the transfer is also available in the regular ball-file. Hence, the child represents that the ball is actually in box 2, but that Susi falsely believes that the ball is in box 1. When asked where Susi will look for the ball they correctly respond with "box 1".

Perner et al. (2015) replicated this finding and then empirically tested the assumption that children do not consider horizontal links between vicarious files. Based on this assumption, they made a surprising prediction: When Susi knows that the ball is the rattle, [+−] children will erroneously predict that Susi will look for the

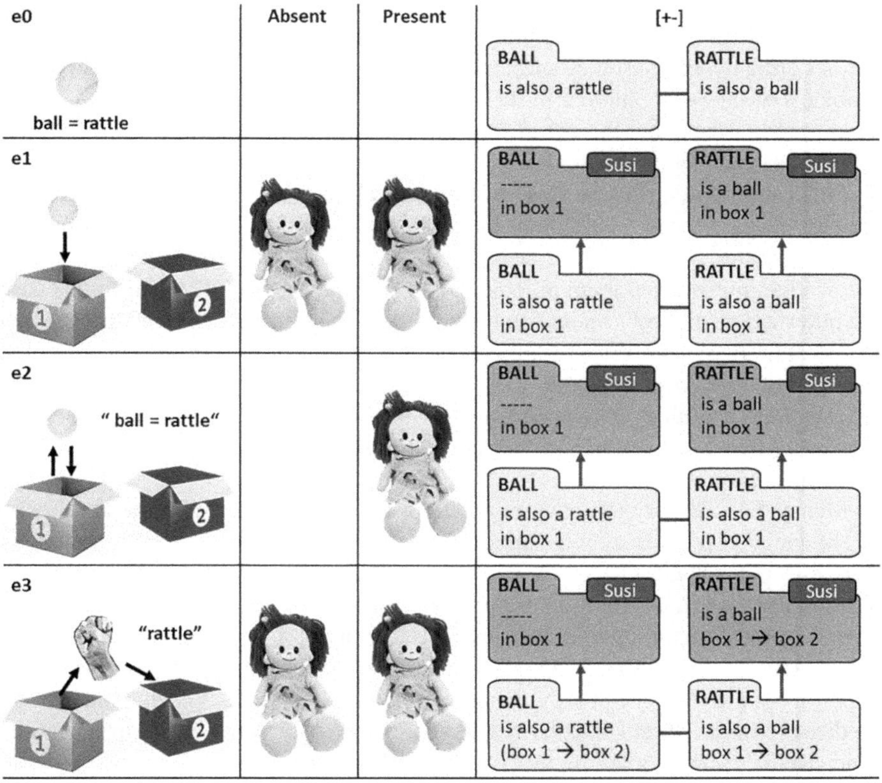

Fig. 3.6 Mental Files analysis of Perner et al.'s (2015) experiment. At the preparatory stage e_0, when the child is shown that the ball is also a rattle, a regular ball- and a regular rattle-file are deployed. In e_1, when the child sees how puppet Susi observes the object being put in the box, the child opens a vicarious ball- and a vicarious rattle-file, indexed to Susi. Then, in e_2, Susi is either present or absent, depending on the condition, when it is again shown that the ball is also a rattle. As a result, Susi either knows or is ignorant about the fact that the ball is also the rattle. If, from the child's point of view, Susi knows that the ball is the rattle, then the two vicarious files should be linked. If, on the contrary, she is ignorant about this fact, then that vicarious files should not be linked. According to assumption (2), [+−] children do not consider horizontal linking between vicarious files. Therefore, in both conditions, regardless of whether or not Susi knows that the ball is the rattle, they do not horizontally link the two vicarious files. This is an error in the present condition. When, in e_3, the object was moved as rattle (hidden in the experimenter's hands) from box 1 to box 2, this information is noted on both the regular and the vicarious rattle-file. As the vicarious rattle-file is not horizontally linked to the vicarious ball-file, the outdated location-information "in box 1" on the ball-file will not be updated. Due to the horizontal link between the regular files the information about the move is also available in the regular ball-file. This analysis shows that [+−] children's failure to not consider horizontal linking between vicarious files leads to correct responses in the absent condition but wrong responses in the present condition

ball in box 1. A correct prediction of where Susi will look for the ball would require linking the vicarious ball- and rattle-files. Only in the rattle-file is the current location (box 2) of the object available, while in the ball-file only the outdated location (box

1) is registered (see Fig. 3.6). The observed responses of the [+−] children showed their difficulties with this condition. Only when children become able to master second-order belief tasks, they succeed in this version of the Susi task. Moreover, the problems of the [+−] children in this condition vanished (as in Sprung et al. 2007) when information about the object was given in a predicative manner, like "the ball rattles", instead of individuating it as "a ball" and as "a rattle".

Predication tasks, such as those of Perner et al. (2015) and Sprung et al. (2007), are easy to pass for [+−] children. Individuating information about an object, like "this is a rattle", causes that a new file is created, but this is not the case when predicative information about an object, like "the ball rattles" is given. In this case the information "rattles" is noted on the ball-file. When Susi is informed that the ball is the rattle, two vicarious files are opened, but when she gets information that the ball rattles, only one vicarious file, for the ball-perspective, is opened and the information "it rattles" is recorded on that file. In this scenario, however, it does not matter that [+−] children do not consider linking of vicarious files, as there is only one vicarious file.

Mental Files Theory (Huemer et al. 2018; Perner et al. 2015; Perner and Leahy 2016) can also provide an explanation for 4- to 6-year-olds' incoherent pattern of responses in the Heinz task. Children represent the dual nature of the rubber die with two regular files, a die- and an eraser-file. When Heinz has informational contact with the object vicarious files indexed to Heinz are assigned, representing Heinz' acquaintance with the object. But [+−] children are still naïve about which vicarious files to assign. They correctly assign a vicarious die-file, which contains all predicative information Heinz has gained about the object from the die perspective. Since Heinz cannot not tell by looking that the rubber die is an eraser, this information is not noted on this file. When asked if Heinz knows whether the die is also an eraser, they properly attest Heinz' lack of knowledge of this fact. But they also – erroneously – assign a vicarious eraser-file even though Heinz does not conceive of the rubber die as an eraser. Considering that another person conceives of an object in a different way as oneself requires to understand which conceptual perspectives on the object are available from the other person's perspective. Mental Files Theory argues that children start to manage this challenge with the emergence of the competence to handle higher order problems (Huemer et al. 2018; Perner et al. 2015). The assumption of excessive assignment of vicarious files allows to explain why [+−] children randomly choose between the location of the die/eraser and the location of the standard eraser (see Fig. 3.7).

The mental files analysis of the Heinz scenario leads to a fundamental question about the development of file management, "When do [+-] children assign vicarious files?". Children's assignment of vicarious files could depend on when (*timing*) and how (*modality*) the other agent learns about the object.

Timing should affect which vicarious files are assigned. In the Heinz scenario, Heinz encounters the die/eraser after the child learns about the object and its dual nature. The child entertains regular files representing each of the object's two aspects; when Heinz then gets acquainted with the die/eraser the child assigns vicarious files for the die- and the eraser-aspect. If the child and Heinz both see

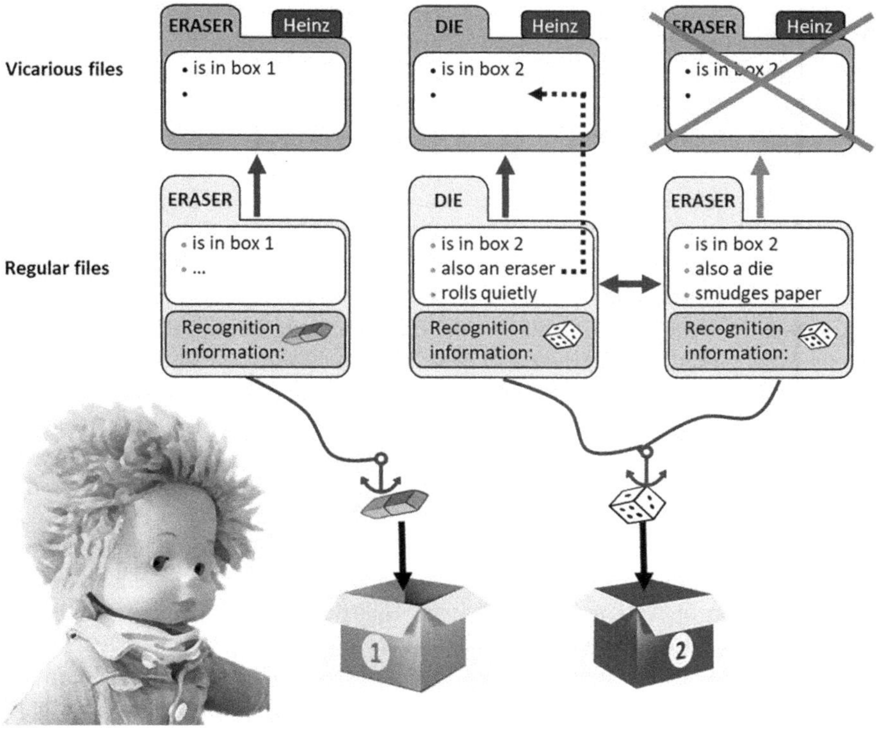

Fig. 3.7 Mental files analysis of Apperly and Robinson's (Apperly 1998) Heinz scenario for [+−] children (Huemer et al. 2018). The child, seeing the scenario, has a regular file for the standard eraser and a regular die- and eraser-files for the die/eraser. For Heinz, seeing the objects, the child deploys a vicarious file (indexed to Heinz) for each of her own regular files. However, the vicarious eraser-file for the die/eraser should not be deployed, as Heinz cannot tell by seeing that the object is an eraser. To indicate this error, this file is crossed out (red X) in the figure. The child has registered the information "also an eraser" on her regular die-file, and correctly does not transfer this information to the vicarious die-file (that the child correctly does not execute this transfer is indicated by the broken arrow). The anchors indicate the respective referents of the files

the die/eraser being put in the box, but do not know about its dual nature, then the child only entertains a regular die-file and a vicarious die-file. When learning in a later episode in Heinz' absence that the die is also an eraser the child also opens a regular eraser-file. Since Heinz has no informational contact to the object at this time, they do not assign a vicarious eraser-file. When asked where Heinz will look for an eraser, children should respond that Heinz will exclusively look for the eraser where he knows the standard eraser is. Hence, children's problems with this question should vanish when Heinz learns the die's location before children learn that it is also an eraser.

The modality of the agent's information is another determining factor: Does Heinz see the die/eraser or is he only told about the die? Semantic and pragmatic theories like file change semantics and discourse representation theory (Heim 2002;

Karttunen 1976; Kamp and Reyle 1993) use structures similar to mental files. Novel overt referring expressions like “a die” prompt the construction of a new file to store information about the expression’s referent. When the child notices that Heinz is told about “a die”, but does not see it, a vicarious die-file is opened. Since Heinz has no information about the eraser – neither verbally nor perceptually – no vicarious eraser-file is created. In contrast, when the child observes that Heinz sees the die/eraser a vicarious die-file and a vicarious eraser-file are assigned, as Heinz non-epistemically sees (Dretske 1969) the die and the eraser. Children do not distinguish *as what* Heinz perceives the die/eraser and, hence, they indifferently assign both vicarious files. Therefore, in a setting where Heinz is told about the die but does not see the object, children should not be prone to indicate the location of the die when asked where Heinz will search for an eraser.

Huemer et al. (2018) tested 35- to 86-month-olds with a modified version of Apperly and Robinson’s task. Each child faced four tasks resulting from two experimental factors, *timing* and *mode of information*. *Timing*: Heinz learns the die’s location either *before* or *after* children learn that the die is an eraser. *Mode of information*: Heinz either *sees* or is *told* about where the die is. When Heinz is told in which box the die is put, he never sees the die at all (Fig. 3.8). They found that Apperly and Robinson’s problem occurs only in the seen-after condition (the original version of Apperly and Robinson’ Heinz task), where Heinz sees the die after children had learnt that it was also an eraser. It vanishes when Heinz learns where the die is before children learn that it is also an eraser. The problem also vanishes when Heinz learns where the die is purely verbally (e.g., “The die is in the red box”) and never sees it (Fig. 3.9). The results of this study help identify important features of children’s mental file management, extending the scope of the existing framework.

Only the Mental Files Theory can account for all the data observed in the different versions of both the Heinz and Susi task. None of the other theories presented – partial knowledge (Apperly and Robinson 2003), task complexity (Rakoczy et al. 2015), and embedded perspectives (Sprung et al. 2007) – can do so.

3.5 A Challenge for Mental Files Theory

A challenge for the mental files account of children’s developing understanding of belief and its aspectuality arises from 4- to 7-year-olds surprising difficulties with knowledge and true belief (TB) tasks.[1] Knowledge tasks have been the focus of interest in almost all relevant studies (for an overview, see Fabricius et al. 2021). In these tasks Maxi watches how the chocolate is moved and thus knows where it is

[1] In the literature, knowledge tasks are often referred to as “true belief (TB) tasks”. I will keep these two terms separate, as they refer to different types of tasks and one of the accounts presented to explain the phenomenon – pragmatics account – is only applicable to knowledge tasks.

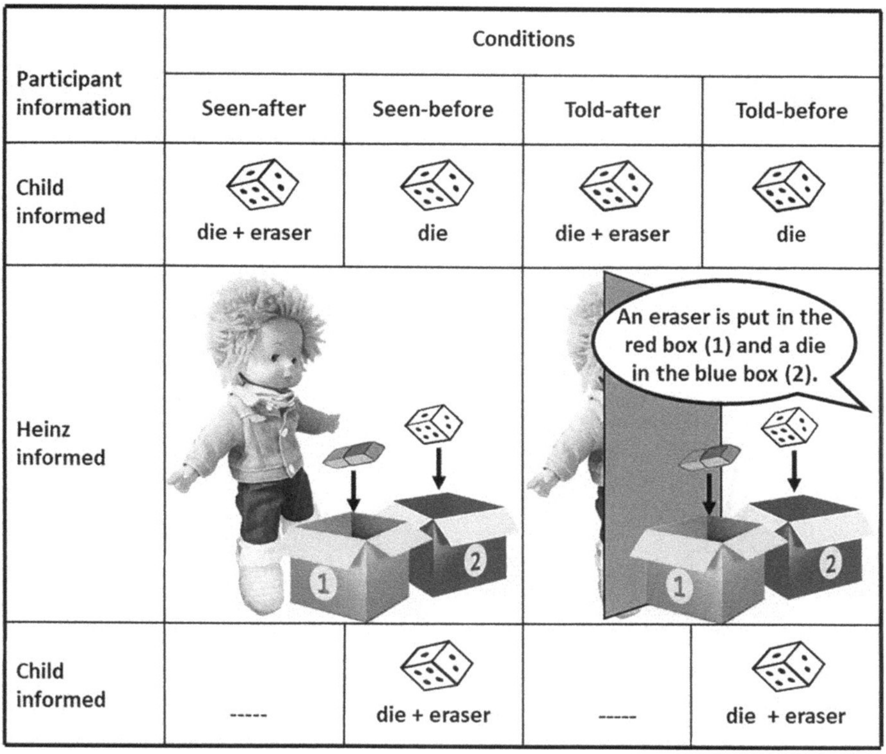

Fig. 3.8 Illustration of experimental conditions with factors timing and mode of information (Huemer et al. 2018). *Timing*: Children see Heinz learn where the die/ eraser is either *before* or *after* they learn about the dual nature of this object. *Mode of information*: Heinz learns where the die/eraser is either *perceptually (seen)* or *linguistically (told)*

now. To keep knowledge tasks structurally as similar as possible to FB tasks, Maxi is absent either before or after the teddy is moved (see Fig. 3.10, top and middle row). In TB tasks, in Maxi's absence, the chocolate is taken out of its original place and then put back in the same place (see Fig. 3.10, bottom row). Maxi has a justified true belief about the chocolate's but does not have proper knowledge of it. Scenarios like this resemble "Gettier cases" introduced in epistemology to demonstrate that knowledge cannot be equated with justified true belief (Gettier 1963).

Recent evidence shows that younger children typically fail FB tasks but have no problems with knowledge and TB tasks. However, once children start to pass FB tasks, they begin to fail knowledge and TB tasks (e.g., Fabricius et al. 2010, 2021; Friedman et al. 2003; Oktay-Gür and Rakoczy 2017; Rakoczy and Oktay-Gür 2020; Schidelko et al. 2021, 2022a, b). It is only from around the age of 10 that children's difficulties with these tasks reliably disappear (Oktay-Gür and Rakoczy 2017; Schidelko et al. 2022b). For an overview, see the recent meta-analysis on children's performance in knowledge and TB tasks provided by Fabricius et al. (2021).

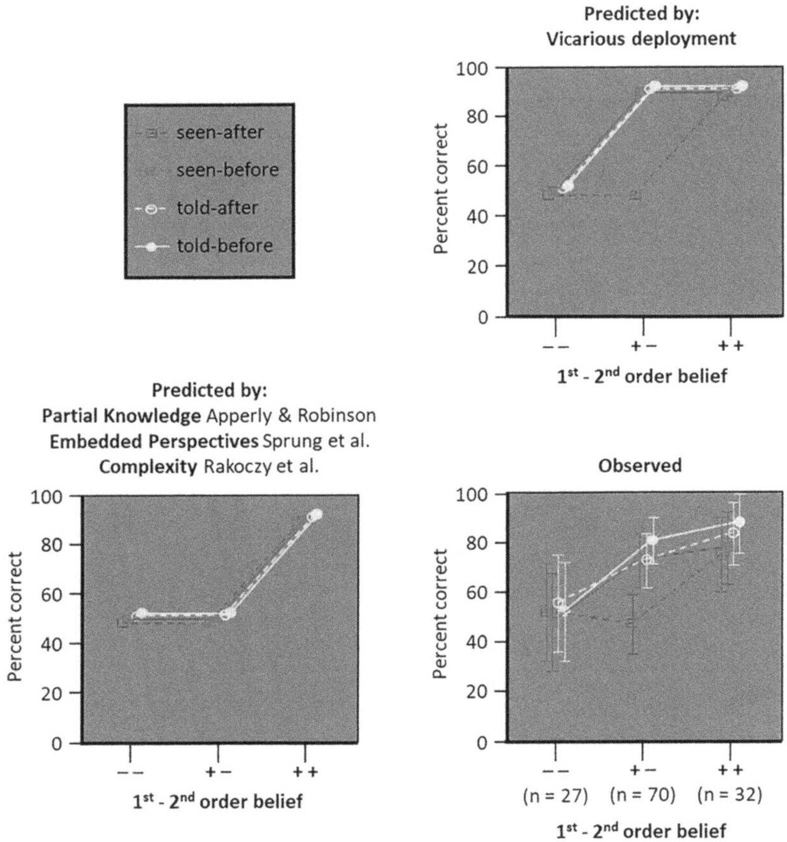

Fig. 3.9 Predicted and observed percentage of correct responses to the where-look question (Huemer et al. 2018). The x-axis in each panel shows the three categories of children according to their failing or passing the first-order and second-order false belief test (pass criterion: At least one of 2 questions correct). The four lines correspond to the four conditions. Error bars for the observed data show ± CI. Note: The three alternative explanations for the findings in Apperly and Robinson's (1998, 2001) original Heinz scenario – partial knowledge, embedded perspectives and task complexity – apply in the same way to all four conditions, and, hence, they predict that [+−] children will find them equally hard

There are two principal ways to explain this paradoxical phenomenon – *competence* vs. *performance* limitations. Proponents of competence limitations argue that children do not have a fully developed concept of belief by age 4. For instance, the perceptual access reasoning (PAR) account posits that 4- and 5-year-old children reason that someone without perceptual access lacks knowledge and will therefore act in the wrong way ("doesn't see ⇒ doesn't know ⇒ gets it wrong"). When Maxi leaves the scene after the chocolate was transferred and comes back later, then upon his return he cannot see the chocolate. Using PAR, children reason from his lack of perceptual access that he lacks knowledge and will act incorrectly

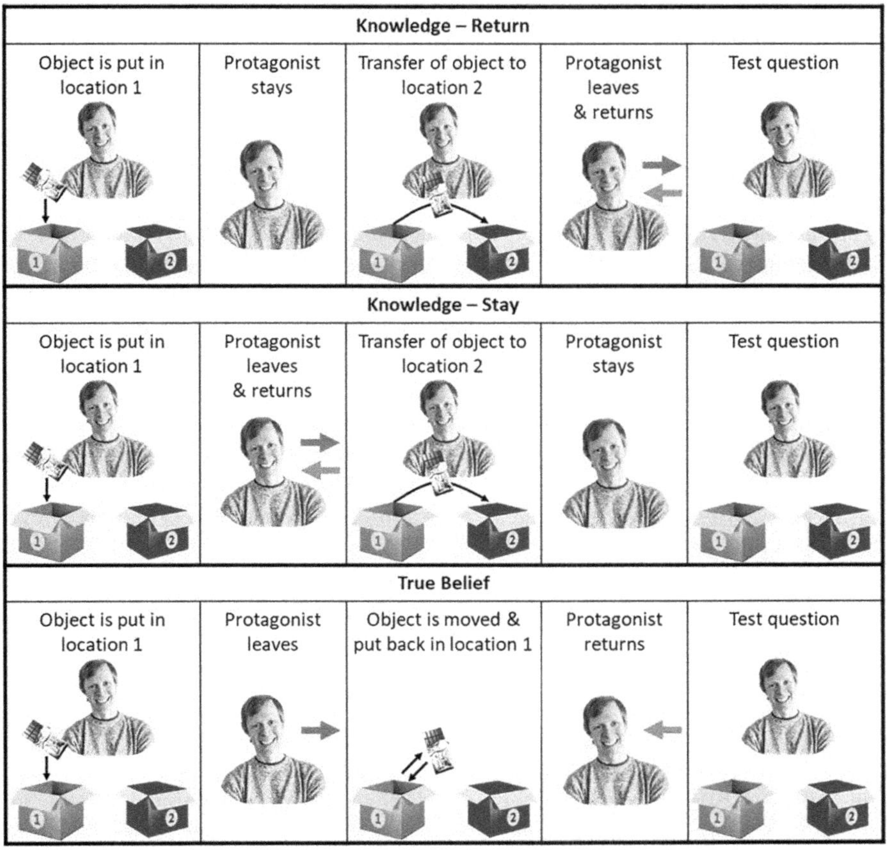

Fig. 3.10 Knowledge and true belief (TB) tasks

(i.e., go to the empty box). Only later, from 6 years on, children use belief reasoning and therefore master both FB and knowledge tasks (Fabricius et al. 2010; Fabricius and Imbens-Bailey 2000; Fabricius and Khalil 2003; Fabricius et al. 2021, 2023; Hedger 2016; Hedger and Fabricius 2011). In contrast, performance limitation explanations hold that 4-year-old children have the conceptual prerequisites to master belief tasks but fail because of extraneous performance factors in knowledge tasks. The pragmatic account (Oktay-Gür and Rakoczy 2017; Rakoczy and Oktay-Gür 2020; Schidelko et al. 2022a, b) attributes children's problems with knowledge tasks to a combination of several factors which jointly make the test question pragmatically confusing and thus challenging. They are academic test questions that are trivial, asking where Maxi thinks his chocolate is, or to predict where he will look for it, in a situation where he shares common ground and is not mistaken in any way.

Children's errors in knowledge and TB tasks and two of the proposed explanations for their failures challenge the mental files account of children's developing understanding of beliefs in two ways. (1) According to the PAR approach, children are not capable of belief reasoning at all until they are about 6 years old. This poses a challenge not only to Mental Files Theory and its account of children's developing understanding of belief as presented here, but to the standard view of the Theory of Mind development in general. (2) Oktay-Gür and Rakoczy (2017) and Rakoczy and Oktay-Gür 2020; see also Rakoczy (2017) applied their pragmatic account of the knowledge error to the Susi task as well. They argue that children from about 4 years of age acquire a fully-fledged concept of beliefs including aspectuality. According to their account, children's correct responses in the condition of the Susi task where Susi is ignorant about the identity of the pen and the rattle reflect their competence, while children's difficulties with the condition, where Susi knows that the pen is the rattle, are due to the fact that children this age have problems with the pragmatics posed by knowledge tasks. This explanation of children's responses in the Susi task is in direct contrast to the mental files account of Perner et al. (2015), who argued that [+−] children do not consider linking vicarious files, resulting in incorrect responses in the present condition (where Susi learns that the ball is the rattle) and fortuitously correct responses in the absence condition (where Susi is ignorant about the fact that the ball is the rattle).

3.5.1 PAR Account

According to the *PAR account* (Fabricius et al. 2010; Fabricius and Imbens-Bailey 2000; Fabricius and Khalil 2003; Fabricius et al. 2021, 2023; Hedger 2016; Hedger and Fabricius 2011), children's development of reasoning in belief and knowledge tasks is characterised by three stages. In the *first* stage, until around age 4, children base their responses to the test on their own knowledge about reality. This results in incorrect responses in FB tasks and correct responses in knowledge and TB tasks. In the *second* stage, children of age 4 and 5 base their reasoning in these tasks on agents' perceptual access (an idea originally put forward by Ruffman 1996). PAR is composed of two rules:

(1) perceptual access ⇒ knowing, lack of perceptual ⇒ not knowing.
(2) knowing ⇒ acting correctly, not knowing ⇒ acting incorrectly.

Children using PAR assume that agents without perceptual access lack knowledge and will therefore act in the wrong way (e.g., Maxi goes to empty location), while agents with perceptual access possess knowledge, which enables them to get things right (e.g., Maxi goes go to location with chocolate). No ascription of (false) beliefs is involved in this way of reasoning. In the *third* stage, from 6 years on, children engage in belief reasoning leading to correct answers in FB, knowledge, and TB tasks.

In the PAR account (see, e.g., Fabricius et al. 2023), there is a level-1 conception of perceptual access that allows one to determine *what* someone perceives (e.g., whether an object is in someone's visual field or obscured by another object), but not *how* something appears from the other's perspective (e.g., a picture of turtle can be seen by one as right-side up, but by someone looking at it from the other side as upside down). This ability begins to develop after the age of 18 months (Moll and Tomasello 2006) and is well established in cognitive development a year later (Flavell et al. 1981; Lempers et al. 1977; Masangkay et al. 1974). The PAR account posits that knowledge ascription is based on a non-representational, level-1 conception of knowing (Fabricius et al. 2023). There is no presumption of children having a mature concept of knowing as a representational mental state that one maintains over time. Thus, children's reasoning whether someone knows or not is constrained to an agents' perceptual access in the current situation. When the situation changes (e.g., when Maxi leaves the scene and returns later), PAR is applied anew based on the new situation (Fabricius et al. 2010; Hedger and Fabricius 2011). No memory is attributed to others when they lose perceptual contact with a situation (Fabricius et al. 2021).

This can account for why 4- and 5-year-olds respond correctly in FB tasks without deploying belief reasoning. Maxi sees and therefore knows in the beginning where the chocolate is put, but children do not ascribe any memory to him about this. Instead, upon his return (after the chocolate was transferred in his absence), they see him as being in a new situation and apply PAR anew. When he returns, they assess whether he can see the chocolate. As he cannot, they infer from his lack of perceptual access that he does not know and will therefore act incorrectly (i.e., going to the now empty location). Children engage in the same kind of reasoning in TB and knowledge tasks where Maxi leaves and returns after the chocolate was moved. The use of PAR results in these tasks in incorrect predictions by the children about where Maxi will be looking for the chocolate. In contrast, in stay conditions of the knowledge stories where Maxi does not leave after the chocolate was moved, children will not perceive him as being in a new situation and, hence, they use their initial analysis that Maxi had perceptual access, knows about the location of the chocolate, and thus acts correctly (i.e., going to the location where the chocolate currently is). Similarly, when Maxi does not return after he left following the transfer, it is less likely that children perceive him as being in a new situation and, hence, they will tend to stay with their initial assessment that he had perceptual access, knows, and therefore will get it right. Therefore, children are expected to provide correct predictions about Maxi's behavior in both stay and no-return conditions. (e.g., Fabricius et al. 2010, 2021; Hedger and Fabricius 2011).

In their meta-analysis, Fabricius et al. (2021, p. 45f) compared, among others, conditions where Maxi stays or leaves without returning after the chocolate was transferred with conditions where Maxi leaves and returns after transfer. They report that they found no difference between stay and no-return conditions. Thus, for the subsequent analysis they collapsed the two conditions; children's performance (estimated at 54 months) in the 24 stay/no-return conditions were higher (94% correct) than in the five return conditions (78% correct). This difference is predicted by the PAR account. However, it is worth noting, that in the five return conditions included in the meta-analysis, the children did not make a lot of errors, instead they responded correctly (78% correct) more often than not. To see whether this is consistent with the PAR account, one would need to know how these children perform in a FB task, their correct answers in the knowledge tasks could also reflect reality-based thinking.

The predictions of the PAR account on differences between conditions where Maxi (after the transfer) stays, leaves and does not return, or leaves and returns have been directly investigated in two studies. Friedman et al. (2003, Experiment 2) found that a return condition (82% incorrect responses) was more difficult than a no-return condition (53% incorrect responses) for 4.5-year-old children. Huemer et al. (2023, Experiments 1 and 2) compared stay and return conditions. In Experiment 1, only a few FB+ children predicted that Maxi would look for the chocolate in its current location in both the stay condition (36%) and the return condition (25%). Interestingly, some of the children asked whether Maxi had seen the transfer. Questioning this seems not implausible as in the animation shown to the children, Maxi is standing in the middle of the scene without showing any reaction while his sister takes the chocolate out and puts it in its new place. Therefore, in Experiment 2, it was introduced that Maxi followed his sister while she was moving the chocolate, and children were then asked if Maxi could see the transfer. Now, FB+ children predicted in both conditions that Maxi would look for the chocolate in the current location (stay condition: 81%, return condition: 88%). In both experiments no difference between the conditions was found. In Experiment 3, Huemer et al. directly contrasted a condition where Maxi was passively present (as in Experiment 1) with an active watching condition where he accompanies his sister during the transfer (as in Experiment 2), both in a return version. They found that FB+ children predicted in the passive presence condition that Maxi would look the chocolate where it was initially, whereas in the active watching condition these children indicated that Maxi would look for the chocolate in its current location. The results from the latter two experiments cannot be explained with the PAR account. According to the PAR account, FB+ children should (contrary to the results of Experiment 2) respond incorrectly in the return condition but correctly in the stay condition; and they should (contrary to the results of Experiment 3) respond incorrectly in return conditions regardless of what Maxi sees before he leaves the scene. However, an important caveat when interpreting this study is that in the online study by Schidelko et al. (2022a) return conditions were used with very similar stories to Huemer et al. (2023), and there the children committed a knowledge error.

A further challenge for the PAR account arises from the developmental synchrony of mastering FB, false sign (Parkin 1994; Schidelko et al. 2022a; Schuster et al. 2021) and identity tasks (Perner et al. 2011; Weingartner and Haring 2020; for an overview see Perner and Roessler 2012). It is difficult to see how the PAR account could explain this pattern. There seems to be no story that would show how PAR could be applied in false sign or identity tasks. For example, if we look at the identity story described above with the yellow and green marked key, this becomes immediately obvious. Perceptual access, knowledge and acting right or wrong do not play a role in this task. A relation to the FB task, if it is claimed to be mastered through the use of PAR, would then be challenging to explain. Even more so as the correlation of this identity task with FB is also present when controlling for age and the control task (dual function), which makes an explanation via general cognitive development seem rather unlikely. This stands in stark contrast to Mental Files Theory, which, as discussed earlier, can offer a straightforward explanation for this relation.

At the same time, the evidence that there is a relation between reasoning about second-order beliefs and aspectuality of belief and both are mastered from about 6 years of age is consistent with PAR, as, according to the PAR account, the transition from PAR to belief reasoning occurs at about 6 years of age and the ability for belief reasoning is required for both reasoning about second-order beliefs and aspectuality of belief (see Fabricius et al. 2021). However, the PAR account struggles to explain the specific response patterns of [+−] children in both the Susi and Heinz task.

In the Susi task, [+−] children predict in both individuation conditions (e.g., the ball is the rattle), regardless of whether Susi knows that the ball is the rattle or whether she is ignorant about this fact, that she will look for the ball in the now empty box 1. According to the PAR explanation (as offered by Fabricius et al. 2021, p. 37f), when the die/eraser was transferred – hidden in the experimenter's hands and referred to as "rattle" – children reason that Susi lacks perceptual access to the ball, so she does not know where it is and therefore acts incorrectly (i.e., going to the now empty box). This yields the observed results. Children should reason in the same way in the predication conditions (e.g., the ball rattles), where the ball/rattle was referred to as "thing that rattles" during the transfer, and, thus, should make the same predictions about where Susi will look for the ball as in the individuation conditions. But the observed pattern of results was different: while children predict that Susi will look for the ball in its previous location when she was ignorant about the fact that the ball rattles (in accordance with PAR), they predict that Susi will look for the ball in its current location when she was informed that the ball rattles (in contrast to PAR).

In the Heinz task, when Heinz is verbally informed that an eraser (i.e., the standard eraser) is put in the red bx and a die (i.e., the die/eraser) is put in the blue box (see Fig. 3.8), the PAR account should predict that children expect Heinz to go to the red box when he is in need of an eraser. Heinz was given information about that eraser; thus he knows where it is and will act correctly. This is in line with the data. Likewise, when Heinz sees how the standard eraser and the die/eraser are put in the boxes after the child was informed about the dual nature of the die/eraser, children would reason that he had perceptual access to both, thus knows about both, and consequently will act correctly. Acting correctly, I think, would mean that he could go to any of the two boxes, which is consistent with the findings of children indicating both locations equally often. However, when Heinz sees the two objects being put in the boxes before the child was informed in his absence that the die is an eraser, upon his return, children should reason that he cannot see the two objects, hence does not know, and therefore will act incorrectly. It is hard to see how applying PAR should lead to the observed responses that children have a clear tendency to predict that Heinz will go to the box that contains the standard eraser.

3.5.2 Pragmatic Account

The *pragmatic account* (Oktay-Gür and Rakoczy 2017; Rakoczy and Oktay-Gür 2020; Schidelko et al. 2022a, b) attributes children's difficulties with knowledge tasks to performance rather than competence limitations. A combination of three factors jointly makes the test question pragmatically confusing and thus challenging. *First*, the critical test questions in these tasks are academic test questions. They make perfect sense to us who are familiar with test questions; questions that are not asked to learn the answer but are asked to find out whether the listener knows the answer. Children in the relevant age range are likely do not be able to make sense of test questions because embedded mental states need to be attributed (Perner et al. 1984), for example, "the speaker *wants to know* whether I *know* what Maxi will do". Children until 6 years or older tend to fail second-order belief questions (Perner and Wimmer 1985) and have problems with test questions (Siegal 1999). If the point of this kind of question escapes young children, then the test question is understood as a sincere question. *Second*, the test questions are highly trivial. If the question is understood as a sincere question, then the answer is unnaturally obvious. This may lead children to ask themselves whether the experimenter could really have meant that or think they must have missed something. Likewise, if the question is understood as a test question, children might wonder why the experimenter would ask such a trivial question to which the answer is so obvious. *Third*, children are asked where Lisa thinks her teddy is, or to predict Lisa's action in a situation where she shares common ground and is not subject to any error. But, as Oktay-Gür and Rakoczy (2017) and Rakoczy and Oktay-Gür (2020) point out, typically, we only wonder about an agent's subjective perspective when there is the possibility that the other is misinformed and thus has diverging beliefs from oneself. There is no reason to talk or think about Maxi's belief when it is clear that everyone present has seen the same events and, therefore, Maxi's belief does not differ from anyone else's beliefs. Usually, we only talk about Maxi's belief if we think his belief might be false (Papafragou et al. 2007). Thus, if a question is asked about what someone thinks who is in a situation where he shares common ground, without the possibility of erring, this may be confusing. Likewise, it can be confusing when we are asked to predict the action of someone who cannot be mistaken. Since the answer to the test questions in knowledge tasks seems too obvious, especially because it asks about behavior or beliefs which could hardly be otherwise, children may ask themselves whether this is what the experimenter could really have meant, or they may think they must have missed something, and respond with the answer different from the obvious one.

These three factors together can make the test questions in knowledge tasks confusing for 4- to 7-year-old children with immature pragmatic skills. At this age (and contrary to 3-year-olds), children are already developing metarepresentational thinking (Perner 1991), that lays the foundation for their developing theory of mind competence, as reflected by increased mastery of the FB task, and their emerging sensitivity to pragmatics (e.g., Fernández 2013; Happé 1993; Winner and Gardner

1993). These children in particular should be affected by the pragmatic confusion caused by the test questions. This would lead to a negative correlation between FB and knowledge tasks and has been supported empirically by a number of studies (Oktay-Gür and Rakoczy 2017; Rakoczy and Oktay-Gür 2020). In contrast, 3-year-old children are not subject to such pragmatic confusion. Neither have they yet acquired a concept of belief, nor have their pragmatic abilities developed sufficiently; they respond to the test question purely reality based. Older children, from around 8 to 10 years of age, can resolve the pragmatic confusion of knowledge tasks when their recursive theory of mind competencies reach a higher-level beyond second-order (Schidelko et al. 2022b).

Several studies provide empirical evidence in favor of the importance of the three proposed factors. When tested with a nonverbal knowledge and FB task (modelled after Call and Tomasello's (Call and Tomasello 1999) task[2]) with no test question at all, there was no evidence of a U-shaped pattern in performance on the knowledge task as found in standard tasks (such that both younger and older children pass, while children in between fail), and once children's performance as a group was significantly above chance levels on the FB task, performance on FB and knowledge tasks were positively correlated (Rakoczy and Oktay-Gür 2020, Experiment 1). Likewise, when the triviality of the test question in a standard knowledge task was made explicit by telling children that they would be presented with some tasks that were for younger children and would be easy for them, 4- to 6-year-olds had no difficulties with the knowledge task (Rakoczy and Oktay-Gür 2020, Experiment 5, TB first condition).

Addressing the third factor, Rakoczy and Oktay-Gür (2020, Experiment 2) used "false" and "true" photo (FP/TP) tasks (based on Zaitchik 1990; e.g., in FP task, a photo was taken in situation$_1$ when a book was in the living room, which now lays in the bedroom). These tasks are structurally similar to FB and knowledge tasks but with the crucial difference that the test question does not involve the subjective perspective of an agent but veridical and outdated photos. Children exhibited comparable performance on FB and FP tasks, but worse performance on knowledge than on TP tasks. This indicates the importance of whether trivial test questions refer to an agent's subjective perspective (rather than to non-mental representations). However, the FP task does not pose a perspective problem in the same way as the FB task: Since the photo does not refer to the present situation but to a previous situation, it is not false but only outdated (Perner and Leekam 2008). In a subsequent study, Schidelko et al. (2022a) used false and true sign (FS/TS) tasks as FS tasks involve a perspective problem with falsehood (e.g., a sign showing that the ice-cream van is at the church when it is actually at the school), but not an agent's subjective perspective (Perner and Leekam 2008). They found similar results to those of Rakoczy and Oktay-Gür (2020). The performance of 3- to 6-year-old children in FB and FS tasks developed in parallel and was correlated, while the

[2]Although Call and Tomasello's (1999) task seems to be generally more difficult than standard tasks (Schröder et al. 2021), this does not affect the conclusion drawn here.

developmental trajectory in TS and Knowledge tasks was remarkably different. Whereas performance on the TS task was close to the ceiling for all age groups, performance in the knowledge task declines with age (at a generally lower level), and both tasks were not correlated. In line with the findings of Rakoczy and Oktay-Gür (2020), this pattern of results supports the hypothesis that a crucial part of children's difficulties with the test questions in knowledge tasks is that these questions are about an agent's actions and, thereby, evoke thinking about the subjective perspective of someone who cannot be mistaken.

While these findings of pragmatic difficulties with test questions in standard knowledge tasks are in line with the mental files framework, Oktay-Gür and Rakoczy (2017) and Rakoczy and Oktay-Gür (2020); see also Rakoczy (2017)) applied their pragmatic account of the knowledge error to the Susi task as well, which in turn is in contrast to the mental files explanation of children's developing understanding of the aspectuality of belief. According to their account, children's correct responses in the condition of the Susi task where Susi is ignorant about the identity of the pen and the rattle reflect their already developed competence to reason about beliefs and its aspectuality, while children's difficulties with the condition, where Susi knows that the pen is the rattle, can be explained by children's problems with the pragmatic demands posed by knowledge tasks. A serious problem for Rakoczy and Oktay-Gür's argument, however, is that Perner et al. (2015) demonstrated that children's problems with the Susi task were limited to the individuation condition (e.g., the ball is the rattle), but did not occur in the predication condition (e.g., the ball rattles); a pattern that was also found for the Heinz task (Sprung et al. 2007). While Mental Files Theory can provide a natural interpretation of this pattern, it is difficult to see what explanation the pragmatics account could possibly offer. Another hint that children's respective problems with standard knowledge tasks and with tasks that require their appreciation of aspectuality might be of different nature comes from an experiment by Rakoczy and Oktay-Gür (2020, Experiment 3a). They tested 4- to 6-year-olds with a version of both standard and aspectuality (Susi) tasks, where the test question was asked by a second experimenter who was not present when the first experimenter presented the story to the child. This procedure was introduced to turn the test questions into sincere questions (where information is genuinely asked for) as they were asked by a naïve person. Interestingly, children had much less problems in the standard knowledge task (around 65% correct responses) than in the aspectuality task (less than 30%).

Performance limitation is also suggested as an explanation for children's problems with the Heinz task by Rakoczy et al. (2015). According to them, children's difficulties with the Heinz task (and other aspectuality tasks used in earlier studies) can be explained in terms of excessive linguistic and other demands like resolution of ambiguous reference. This explanation, however, cannot account for the findings of Huemer et al. (2018) that timing (Heinz learns the die's location either before or after children learn that the die is an eraser) and mode of information (Heinz either sees or is told about where the die is) moderate children's performance on the Heinz task. The alleged pragmatic challenges that children face in the Heinz task are present in all four conditions, and should make those conditions equally difficult, but they do not.

3.5.3 Discussion of the Knowledge and TB Error

The knowledge and TB error poses two challenges to the mental files framework as presented here. *First*, proponents of the PAR account claim that children's emerging mastery of FB tasks at around age 4 does not reflect competence in belief reasoning. Instead, these children apply PAR until they begin to engage in belief reasoning about two years later, at age 6. This view poses a challenge not only to Mental Files Theory in particular, but to the standard view of explicit theory development in general, according to which children are undergoing a conceptual shift to metarepresentational thinking as a basis for thinking about beliefs and other mental states (Perner 1991; Rakoczy 2017, 2022; Wellman 2011). In contrast, according to the pragmatic account, children's problems with knowledge tasks are not due to a conceptual deficit, but to a performance limitation caused by features of the tasks used.

Evaluating these two accounts for explaining the knowledge and TB error, one can say that the empirical evidence is not entirely conclusive. Unfortunately, the meta-analysis of Fabricius et al. (2021) does not allow us to decide between the two accounts under discussion as long as we do not know how FB+ children performed in both the return and the stay conditions. In more recent studies, not included in the meta-analysis, Rakoczy and Oktay-Gür (2020, Experiment 2 and Experiment 4) as well as Schidelko et al. (2021, Experiment 2) found that 4- to 6-year-old children showed the knowledge error in scenarios where Maxi stays after the transfer; contrary to what the PAR account predicts. The direct comparison of return, no return and stay conditions provides a thorough test of the two approaches. Whereas the PAR account predicts incorrect responses in return conditions, correct responses in stay conditions and a tendency of giving more correct responses in return than no return conditions, the pragmatic accounts does not predict such differences. To my knowledge, only two studies have directly tested two of these conditions against each other. Friedman et al. (2003, Experiment 2) found that 4.5-year-old children had more difficulties with a return condition than a no-return condition, which is in line with the PAR account but not with the pragmatic account. Huemer et al. (2018) reported that FB+ children responded correctly in both a return and a stay condition (but bear in mind the caveat mentioned earlier), which is inconsistent with the PAR pragmatic account. Regarding the pragmatic account, the picture is more complex. This account would predict that FB+ children would struggle in both conditions. However, as Huemer et al. (2023, p. 8) pointed out, a possibility to reconcile their findings with the pragmatic account is that their manipulation of Maxi actively watching the transfer and asking children whether Maxi saw the transfer "may have alerted children that the protagonist's perspective is of relevance, and that h[is] perspective is part of the discourse in this situation". Introducing talk about perspectives in the discourse during the task might help eliminating children's confusion with the test question.

Two further pieces of empirical evidence are of interest when evaluating the PAR and the pragmatic account. *First*, children's errors in Gettier-type TB tasks (as used in Fabricius et al. 2010, Experiment 1; see Fig. 3.10, bottom row), it seems, can only be explained by the PAR account. It is not clear how the pragmatic account could explain these results. The TB task differs from an FB task only in that in Maxi's absence the chocolate, after being taken out, is put back in the same place instead of being moved to a different place. Thus, The TB task does not appear to be more trivial than the FB task. Moreover, the third factor, that we don't talk about beliefs when no one could possibly err, shouldn't apply either, since in Maxi's absence someone else moved the chocolate, which he is not aware of. Also, Huemer et al.'s (2023) suggestion that the knowledge error might be due to procedural features of the material – children may interpret the scenarios differently than intended and think that Maxi did not notice the chocolate being transferred – is not applicable to explain children's failure of TB tasks. However, I think it is important to mention that the data base for TB tasks is much scarcer than for knowledge tasks. As far as I could ascertain, it was only implemented in the study by Fabricius et al. (2010, Experiment 1). Second, the PAR account fails to explain the developmental synchrony of success in FB tasks and various tasks that also require the understanding of different perspectives, like false sign, alternative naming, or identity tasks, as repeatedly demonstrated in numerous studies (for an overview, see Perner and Roessler 2012). The PAR account's level-1 conception of both perceptual access and knowing does not allow to explain these findings, as children are not yet able to represent mental and other perspectives.

The knowledge and TB error poses a challenge not only to Mental Files Theory but also, more generally, to the standard view of the theory of mind development. Neither the PAR account nor the pragmatic account capable of explaining all the data. Nor can the overall pattern of findings be explained by children interpreting the scenarios in the knowledge tasks differently than intended because of certain task characteristics, and actually giving reasonable answers on that basis. In order to further evaluate the different accounts from the literature, future research should put these accounts to a direct test by experimentally contrasting the relevant factors like, e.g., stay, no-return and return scenarios in knowledge tasks. It seems particularly important to me that Gettier-type TB tasks should also be investigated to a greater extent. The investigation of the knowledge and TB error is not only interesting with regard to Mental Files Theory but also for theory of mind research in general.

Second, with regard to the alternative explanation proposed by Rakoczy and Oktay-Gür that children's difficulties with aspectuality tasks, which they do not overcome until around 6 years of age, can be attributed to performance limitations, I have argued that this is not compatible with the available evidence from the different conditions of the Heinz and Susi tasks. Furthermore, data from one of their own experiments (Rakoczy and Oktay-Gür (2020, Experiment 3a) suggest that children's respective problems with standard knowledge tasks and with tasks that require them to reason about aspectuality may be different in nature. Both these aspects help to surmount this challenge for the Mental Files Theory.

3.6 General Discussion

In this section, I reflect on how the presented work contributes to the extension of Mental Files Theory and theory of mind research. I will also provide a brief glance at ongoing research and an outlook on future projects.

The presented work help identify important features of children's mental file management, extending the scope of the existing framework by showing how children represent belief and its aspectuality. Up to now, Mental Files Theory has successfully been used to offer an explanation for why children around age 4 begin to ascribe (false) beliefs at the same time they start to understand identity statements (Perner et al. 2011; Weingartner and Haring 2020) and the alternative naming game (Doherty and Perner 2020; Perner and Leahy 2016). In addition to explaining the developmental synchrony of these tasks, mental files allows us to integrate that children as young as 4 can attribute (false) beliefs to others, but it is not until 6 years old that they understand that beliefs about an object depend on the aspect under which that object is represented into a coherent account (Perner et al. 2015). Children begin to appreciate the aspectuality of belief when they start passing second-order belief tasks (Huemer et al. 2018; Perner et al. 2015; Sprung et al. 2007).

The management of mental files in children between the ages of 4 and 6 may lack at least two aspects of the mature system: (1) when to assign a vicarious file, and (2) when should vicarious files be horizontally linked to indicate that the person to whom the files are indexed is aware of the sameness of referent. Perner et al. (2015) argued that [+−] children do not consider which conceptual perspective is available from the agent's perspective. Therefore, they (1) deploy too vicarious files excessively, and (2) do not consider horizontal linking between vicarious files.

Huemer et al.'s (2018) study addresses a fundamental question about the development of file management that emerges from the mental files analysis of the Heinz task, as to when [+−] children assign vicarious files. They tested the hypothesis that children's deployment of vicarious files may depend on when (*timing*) and how (*modality*) the other agent learns about the object by manipulating these two factors. Each child was presented with four tasks based on these two experimental factors. *Timing*: Heinz was informed in which box the die is put either *before* or *after* the children learned that the die is an eraser. *Modality*: Heinz *sees* where the die is placed or is informed only *verbally*. [+−] children showed difficulties only in the original version of the Heinz story introduced by Apperly and Robinson (their seen-after condition). These difficulties disappeared when Heinz was informed only verbally where the die is placed, and they also disappeared when the children learned that the die is an eraser only after Heinz encountered the die. This supports the suggested account of vicarious deployment of mental files.

Building on the theoretical considerations and empirical evidence of this study, the following picture of the developmental course of mental file management can be drawn:

[−−] children base their predictions of other's behavior on their own regular files.

[+−] children use vicarious files to predict the behavior of others. However, their file management is still immature. When the agent sees the object, they deploy vicarious files for all the conceptions they have of that object. This leads to excess deployment of vicarious files which results in their erroneous responses in the Heinz task in the see-after condition. When the information is provided only verbally, then only a vicarious file is deployed with the label used. Hence, there is no excess deployment of vicarious files in this case. They assign vicarious files for an object only at the time when the agent encounters the object and receives information about it.

[++] children's use of vicarious files is more sophisticated. They only deploy vicarious files for conceptions of the object available to the agent. To avoid this excessive deployment, one must understand what conceptual perspectives are available from the agent's perspective. This is a second-order perspective problem: The conceptual perspective is embedded in a mental perspective. Therefore, children do not master the Heinz task until they master second-order problems such as the second-order belief task.

The vicarious deployment account predicts that the responses of the [+−] children in the Heinz scenario should vary in certain ways depending on modality and timing. This prediction is supported by the observed data. Conversely, these data cannot be explained by the alternative theories presented unless the sensitivity to modality and timing can be incorporated into these theories. The findings of this study provide an answer to the question of when vicarious files are assigned and helps us to understand children's developing file management.

The second still immature aspect of [+−] children's file management is that they do not consider linking vicarious files. In the Susi task, this yields wrong answers in the condition where Susi knows that the ball is the rattle. After the child and Susi witnessed how the object was transferred as rattle to the second location, children would have to link the vicarious ball- and rattle-files to correctly predict that Susi would look for the ball at the same location she knows the rattle is now. Children start to succeed in this condition with the mastery of second-order belief reasoning.

The findings from the Susi task (see also Sprung et al. 2007, for findings along the same line) also provide empirical support for one of the essential features of mental files, that they capture the structure of language and thought: The distinction between what one is talking or thinking of (the subject or topic; what the file is tracking) and what one says or thinks about it (the information about the subject; what is noted on the file). While [+−] children fail when the ball/rattle is individuated both as "a ball" and as "a rattle" and Susi knows about that fact (as they do not consider linking between vicarious files), they are passing a version of this task when information about the object was given in a predicative manner, like "the ball rattles". Individuating information about an object, like "this is a rattle", causes that a new file is created, but this is not the case when predicative information about an object, like "the ball rattles" is given. In this case the information "rattles" is noted on the ball-file. Thus, when Susi is informed that the ball is the rattle, two vicarious files are opened, but when she gets information that the ball rattles, only one vicarious file, for the ball-perspective, is opened and the information "it rattles" is recorded on that file. In this scenario, however, it does not matter that [+−] children do not consider linking of vicarious files, as there is only one vicarious file. This observation reflects the distinction between the way an object is conceived under a particular concept from the information one has about the object.

Only the Mental Files Theory can account for all the data observed in the different versions of both the Heinz and Susi task. None of the other theories presented – partial knowledge (Apperly and Robinson 2003), task complexity (Rakoczy et al. 2015), and embedded perspectives (Sprung et al. 2007) – is able to do so.

Mental Files Theory provides the analytical resources for a theoretical integration of various phenomena relating to the general capacity of representing conflicting states as investigated in conjunction with belief understanding for, for example, visual perspective-taking, identity understanding and alternative naming. In ongoing and future studies, we aim to further extend the mental files framework by, for instance, exploring whether understanding possibility also falls into this category. Recent studies showed that young children often fail to consider alternative possibilities. In one of these studies, for instance, children see three cups, arranged as a singleton and a pair of cups, and a coin is hidden in the singleton and another coin is hidden in the pair. Children are then told that they have one choice to choose a cup and receive its content. The wise choice is to opt for the singleton, as it must hold a coin whereas each member of the pair might be empty. Three-year-olds choose a member of the pair almost half the time. They are not guessing at random (chance is .33), yet they are far from maximizing expected reward. (e.g., Leahy et al. 2022; Mody and Carey 2016; for an overview and discussion, see Leahy and Carey 2020). It was hypothesized that representing mutual exclusive possibilities, as, e.g., in the described cups task, requires "the recursive capacity for metarepresentation" (Redshaw and Suddendorf 2020, p. 57). Representing multiple, incompatible possible states of the world may also be related to a more general capacity of representing conflicting states as, for example, in visual perspective-taking, appearance-reality, alternative naming (Goddu et al. 2021) and false belief. Mental Files Theory seems to be able to provide the analytical resources for a theoretical integration of these phenomena, and helps to understand them within a unified theoretical framework.

To sum up, the present chapter provides a cognitive analysis of children's developing understanding of belief and its aspectuality based on Mental Files Theory. It reviews previous work by revealing important features of children's file management and its development. Children's file management is still premature and lacks important aspects until they start passing second-order belief tasks at around age 6. The findings presented here rule out alternative accounts for the development of understanding aspectuality and provides new insights in the nature of the knowledge and TB error.

Acknowledgements The author received financial support from Carey Fund, Harvard University.

References

Apperly, I.A., and E.J. Robinson. 1998. Children's mental representation of referential relations. *Cognition* 67: 287–309. https://doi.org/10.1016/S0010-0277(98)00030-4.

———. 2001. Children's difficulties handling dual identity. *Journal of Experimental Child Psychology* 78 (4): 374–397. https://doi.org/10.1006/jecp.2000.2571.

———. 2003. When can children handle referential opacity? Evidence for systematic variation in 5-and 6-year-old children's reasoning about beliefs and belief reports. *Journal of Experimental Child Psychology* 85: 297–311. https://doi.org/10.1016/s0022-0965(03)00099-7.

Call, J., and M. Tomasello. 1999. A nonverbal false belief task: The performance of children and great apes. *Child Development* 70 (2): 381–395. https://doi.org/10.1111/1467-8624.00028.

Clark, E.V. 1997. Conceptual perspective and lexical choice in acquisition. *Cognition* 64: 1–37. https://doi.org/10.1016/S0010-0277(97)00010-3.

Davies, M., and T. Stone. 1995. Introduction. In *Folk psychology: The theory of mind debate*, ed. M. Davies and T. Stone, 1–48. Oxford: Blackwell.

Doherty, M.J., and J. Perner. 1998. Metalinguistic awareness and theory of mind: Just two words for the same thing? *Cognitive Development* 13: 279–305. https://doi.org/10.1016/S0885-2014(98)90012-0.

———. 2020. Mental files: Developmental integration of dual naming and theory of mind. *Developmental Review* 56: 100909. https://doi.org/10.1016/j.dr.2020.100909.

Dretske, F.I. 1969. *Seeing and knowing*. University of Chicago Press.

Fabricius, W.V., and A.L. Imbens-Bailey. 2000. False beliefs about false beliefs. In *Children's reasoning about the mind*, ed. P. Mitchell and K. Riggs, 267–280. Psychology Press.

Fabricius, W.V., and S.L. Khalil. 2003. False beliefs or false positives? Limits on children's understanding of mental representation. *Journal of Cognition and Development* 4 (3): 239–262. https://doi.org/10.1207/s15327647jcd0403_01.

Fabricius, W.V., T.W. Boyer, A.A. Weimer, and K. Carroll. 2010. True or false: Do 5-year-olds understand belief? *Developmental Psychology* 46 (6): 1402–1416. https://doi.org/10.1037/a0017648.

Fabricius, W.V., C.R. Gonzales, A. Pesch, A.A. Weimer, J. Pugliese, K. Carroll, et al. 2021. Perceptual access reasoning (PAR) in developing a representational theory of mind. *Monographs of the Society for Research in Child Development* 86 (3): 7–154. https://doi.org/10.1111/mono.12432.

Fabricius, W.V., C.R. Gonzales, A. Pesch, and A.A. Weimer. 2023. Perceptual access reasoning: What are the alternatives? *Cognitive Development* 66: 101306. https://doi.org/10.1016/j.cogdev.2023.101306.

Fernández, C. 2013. Mindful storytellers: Emerging pragmatics and theory of mind development. *First Language* 33 (1): 20–46. https://doi.org/10.1177/0142723711422633.

Flavell, J.H. 1988. The development of children's knowledge about the mind: From cognitive connections to mental representations. In *Developing theories of mind*, ed. J.W. Astington, P.L. Harris, and D.R. Olson, 244–267. Cambridge University Press.

Flavell, J.H., B.A. Everett, K. Croft, and E.R. Flavell. 1981. Young children's knowledge about visual perception: Further evidence for the level 1 – level 2 distinction. *Developmental Psychology* 17: 99–103. https://doi.org/10.1037/0012-1649.17.1.99.

Frege, G. 1892. Über Sinn und Bedeutung. *Zeitschrift für Philosophie und philosophische Kritik* 100: 25–50.

Friedman, O., R. Griffin, H. Brownell, and E. Winner. 2003. Problems with the seeing = knowing rule. *Developmental Science* 6 (5): 505–513. https://doi.org/10.1111/1467-7687.00308.

Frith, C., and U. Frith. 2005. Theory of mind. *Current Biology* 15 (17): R644–R645.

Gettier, E.L. 1963. Is justified true belief knowledge? *Analysis 23* (6): 121–123. https://doi.org/10.1093/analys/23.6.121.

Goddu, M.K., J.N. Sullivan, and C.M. Walker. 2021. Toddlers learn and flexibly apply multiple possibilities. *Child Development* 92 (6): 2244–2251. https://doi.org/10.1111/cdev.13668.

Hamilton, A.F., R. Brindley, and U. Frith. 2009. Visual perspective taking impairment in children with autistic spectrum disorder. *Cognition* 113 (1): 37–44. https://doi.org/10.1016/j.cognition.2009.07.007.

Happé, F.G.E. 1993. Communicative competence and theory of mind in autism: A test of relevance theory. *Cognition* 48 (2): 101–119. https://doi.org/10.1016/0010-0277(93)90026-R.

Hedger, J.A. 2016. Perceptual access reasoning: Developmental stage or system 1 heuristic? *Phenomenology and the Cognitive Sciences* 15 (2): 207–226. https://doi.org/10.1007/s11097-015-9412-4.

Hedger, J.A., and W.V. Fabricius. 2011. True belief belies false belief: Recent findings of competence in infants and limitations in 5-year-olds, and implications for theory of mind development. *Review of Philosophy and Psychology* 2 (3): 429–447. https://doi.org/10.1007/s13164-011-0069-9.

Heim, I. 2002. File change semantics and the familiarity theory of definiteness. In *Formal semantics: The essential readings*, ed. P. Portner and B.H. Partee, 223–248. Blackwell Publishing.

Ho, M.K., R. Saxe, and F. Cushman. 2022. Planning with theory of mind. *Trends in Cognitive Sciences.* https://doi.org/10.1016/j.tics.2022.08.003.

Huemer, M., J. Perner, and B. Leahy. 2018. Mental files theory of mind: When do children consider agents acquainted with different object identities? *Cognition* 171: 122–129. https://doi.org/10.1016/j.cognition.2017.10.011.

Huemer, M., T. Haslehner, and J. Perner 2019, March 21–23. *The development of conceptual perspective taking* [Poster presentation]. Poster presented at the Biennial Meeting of SRCD, 21 March 2019, Baltimore. https://www.srcd.org/sites/default/files/file-attachments/srcd19programguide_web_1.pdf.

Huemer, M., L.M. Schröder, S.J. Leikard, S. Gruber, A. Mangstl, and J. Perner. 2023. The knowledge ("true belief") error in 4-to 6-year-old children: When are agents aware of what they have in view? *Cognition* 230: 105255. https://doi.org/10.1016/j.cognition.2022.105255.

Schröder, L.M., S.J. Leikard, M. Huemer, and J. Perner. 2021. The 'false' False Belief Task: A replication and reanalysis of Call and Tomasello's (1999) nonverbal false belief task [Poster presentation]. Budapest CEU Conference on Cognitive Development, January 4 – 8, Budapest, Hungary. https://cdc.ceu.edu/bcccd21.

Hulme, S., P. Mitchell, and D. Wood. 2003. Six-year-olds' difficulties handling intensional contexts. *Cognition* 87: 73–99. https://doi.org/10.1016/s0010-0277(02)00199-3.

Kamawar, D., and D.R. Olson. 1999. Children's representational theory of language: The problem of opaque contexts. *Cognitive Development* 14: 531–548. https://doi.org/10.1016/S0885-2014(99)00018-0.

———. 2009. Children's understanding of referentially opaque contexts: The role of metarepresentational and metalinguistic ability. *Journal of Cognition and Development* 10 (4): 285–305. https://doi.org/10.1080/15248370903389499.

———. 2011. Thinking about representations: The case of opaque contexts. *Journal of Experimental Child Psychology* 108 (4): 734–746. https://doi.org/10.1016/j.jecp.2010.10.005.

Kamp, H., and Reyle, U. (1993). From discourse to logic. Introduction to modeltheoretic sematics of natural language, formal logic and discourse representation theory. Kluwer Academic Publishers.

Karttunen, L. 1976. Discourse referents. In *Notes from the linguistic underground (syntax and semantics)*, ed. J. McCawley, vol. 7, 363–385. Academic Press.

Leahy, B.P., and S.E. Carey. 2020. The acquisition of modal concepts. *Trends in Cognitive Sciences* 24 (1): 65–78. https://doi.org/10.1016/j.tics.2019.11.004.

Leahy, B., M. Huemer, M. Steele, S. Alderete, and S. Carey. 2022. Minimal representations of possibility at age 3. *Proceedings of the National Academy of Sciences* 119 (52): e2207499119. https://doi.org/10.1073/pnas.2207499119.

Lempers, J.D., E.R. Flavell, and J.H. Flavell. 1977. The development in very young children of tacit knowledge concerning visual perception. *Genetic Psychology Monograph* 95: 3–53.

Leslie, A.M. 1987. Pretense and representation: The origins of 'theory of mind'. *Psychological Review* 94 (4): 412–426. https://doi.org/10.1037/0033-295X.94.4.412.

Markman, E. M. (1989). Categorization and naming in children: Problems of induction. A Bradford Book.

Masangkay, Z.S., K.A. McCluskey, C.W. McIntyre, J. Sims-Knight, B.E. Vaughn, and J.H. Flavell. 1974. The early development of inferences about the visual percept of others. *Child Development* 45: 357–366. https://doi.org/10.2307/1127956.

McKay, T., and M. Nelson. 2014. Propositional attitude reports (spring 2014 edition). In *The Stanford encyclopedia of philosophy*, ed. E.N. Zalta. Retrieved from http://www.plato.stanford.edu/archives/spr2014/entries/prop-attitude-reports/.

Millikan, R.G. 2000. *On clear and confused ideas. An essay about substance concepts*. Cambridge University Press.

Mody, S., & Carey, S. 2016. The emergence of reasoning by the disjunctive syllogism in early childhood. *Cognition* 154: 40–48. https://doi.org/10.1016/j.cognition.2016.05.012

Moll, H., and M. Tomasello. 2006. Level 1 perspective-taking at 24 months of age. *British Journal of Developmental Psychology* 24 (3): 603–613. https://doi.org/10.1348/026151005X55370.

Oktay-Gür, N., and H. Rakoczy. 2017. Children's difficulty with true belief tasks: Competence deficit or performance problem? *Cognition* 166: 28–41. https://doi.org/10.1016/j.cognition.2017.05.002.

Papafragou, A., K. Cassidy, and L. Gleitman. 2007. When we think about thinking: The acquisition of belief verbs. *Cognition* 105 (1): 125–165. https://doi.org/10.1016/j.cognition.2006.09.008.

Parkin, L.J. 1994. *Children's understanding of misrepresentation*. Unpublished D. Phil Thesis, University of Sussex.

Perner, J. 1991. *Understanding the representational mind*. MIT Press.

Perner, J., and J. Brandl. 2005. File change semantics for preschoolers: Alternative naming and belief understanding. *Interaction Studies* 6: 483–501. https://doi.org/10.1075/is.6.3.11per.

Perner, J., and D. Howes. 1992. 'He thinks he knows'; and more developmental evidence against the simulation (role-taking) theory. *Mind and Language* 7: 72–86. https://doi.org/10.1111/j.1468-0017.1992.tb00197.x.

Perner, J., and B. Leahy. 2016. Mental files in development: Dual naming, false belief, identity and intensionality. *Review of Philosophy and Psychology* 7: 491–508. https://doi.org/10.1007/s13164-015-0235-6.

Perner, J., and S. Leekam. 2008. The curious incident of the photo that was accused of being false: Issues of domain specificity in development, autism, and brain imaging. *The Quarterly Journal of Experimental Psychology* 61: 76–89. https://doi.org/10.1080/17470210701508756.

Perner, J., and J. Roessler. 2012. From infants to children's appreciation of belief. *Trends in Cognitive Sciences* 16: 519–525. https://doi.org/10.1016/j.tics.2012.08.004.

Perner, J., and H. Wimmer. 1985. "John thinks that Mary thinks that . . .": Attribution of second-order beliefs by 5-to 10-year-old children. *Journal of Experimental Child Psychology* 39: 437–471. https://doi.org/10.1016/0022-0965(85)90051-7.

Perner, J., S.R. Leekam, and H. Wimmer. 1984. *The insincerity of conservation questions: Children's growing sensitivity to experimenter's epistemic intentions*. Unpublished manuscript, Laboratory of Experimental Psychology, University of Sussex. Presented as poster at the APA (American Psychological Association) convention, August 1984, and at the BPS-Developmental Section Meeting in Lancaster, September 1984.

Perner, J., S. Stummer, M. Sprung, and M. Doherty. 2002. Theory of mind finds its Piagetian perspective: Why alternative naming comes with understanding belief. *Cognitive Development* 17 (3–4): 1451–1472.

Perner, J., J.L. Brandl, and A. Garnham. 2003. What is a perspective problem? Developmental issues in belief ascription and dual identity. *Facta Philosophica* 5: 355–378.

Perner, J., B. Rendl, and A. Garnham. 2007. Objects of desire, thought, and reality: Problems of anchoring discourse referents in development. *Mind and Language* 22: 475–513. https://doi.org/10.1111/j.1468-0017.2007.00317.x.

Perner, J., M.C. Mauer, and M. Hildenbrand. 2011. Identity: Key to children's understanding of belief. *Science* 333 (6041): 474–477. https://doi.org/10.1126/science.1201216.

Perner, J., M. Huemer, and B. Leahy. 2015. Mental files and belief: A cognitive theory of how children represent belief and its intensionality. *Cognition* 145: 77–88. https://doi.org/10.1016/j.cognition.2015.08.006.

Perry, J. 2002. *Identity, personal identity and the self.* Hackett Publishing.

Premack, D.G., and G. Woodruff. 1978. Does the chimpanzee have a theory of mind? *Behavioral and Brain Sciences* 1 (4): 515–526. https://doi.org/10.1017/S0140525X00076512.

Pylyshyn, Z.W. 2007. *Things and places: How the mind connects with the world.* MIT Press.

Rakoczy, H. 2017. In defense of a developmental dogma: Children acquire propositional attitude folk psychology around age 4. *Synthese* 194: 689–707. https://doi.org/10.1007/s11229-015-0860-8.

———. 2022. Foundations of theory of mind and its development in early childhood. *Nature Reviews Psychology* 1 (4): 223–235. https://doi.org/10.1038/s44159-022-00037-z.

Rakoczy, H., and N. Oktay-Gür. 2020. Why do young children look so smart and older children look so dumb on true belief control tasks? An investigation of pragmatic performance factors. *Journal of Cognition and Development* 21 (2): 213–239. https://doi.org/10.1080/15248372.2019.1709467.

Rakoczy, H., D. Bergfeld, I. Schwarz, and E. Fizke. 2015. Explicit theory of mind is even more unified than previously assumed: Belief ascription and understanding aspectuality emerge together in development. *Child Development* 86 (2): 486–502. https://doi.org/10.1111/cdev.12311.

Recanati, F. 2012. *Mental files.* Oxford University Press.

Redshaw, J., and T. Suddendorf. 2020. Temporal junctures in the mind. *Trends in Cognitive Sciences* 24 (1): 52–64. https://doi.org/10.1016/j.tics.2019.10.009.

Ruffman, T. 1996. Do children understand the mind by means of simulation or a theory? Evidence from their understanding of inference. *Mind and Language* 11 (4): 388–414. https://doi.org/10.1111/j.1468-0017.1996.tb00053.x.

Russell, B. 1910. Knowledge by acquaintance and knowledge by description. *Proceedings of the Aristotelian Society* 11: 108–128.

Russell, J. (1987). "Can we say ...?" Children's understanding of intensionality. Cognition, 25, 289–308. https://doi.org/10.1016/S0010-0277(87)80007-0.

Saxe, R., and N. Kanwisher. 2003. People thinking about thinking people: The role of the temporo-parietal junction in 'theory of mind'. *NeuroImage* 19 (4): 1835–1842.

Schidelko, L.P., B. Schünemann, H. Rakoczy, and M. Proft. 2021. Online testing yields the same results as lab testing: A validation study with the false belief task. *Frontiers in Psychology* 12: 703238. https://doi.org/10.3389/fpsyg.2021.703238.

Schidelko, L.P., M. Huemer, L.M. Schröder, A.S. Lueb, J. Perner, and H. Rakoczy. 2022a. Why do children who solve false belief tasks begin to find true belief control tasks difficult? A test of pragmatic performance factors in theory of mind tasks. *Frontiers in Psychology* 12: 6505. https://doi.org/10.3389/fpsyg.2021.797246.

Schidelko, L.P., M. Proft, and H. Rakoczy. 2022b. How do children overcome their pragmatic performance problems in the true belief task? The role of advanced pragmatics and higher-order theory of mind. *PLoS One* 17 (4): e0266959. https://doi.org/10.1371/journal.pone.0266959.

Schünemann, B., M. Proft, and H. Rakoczy. 2022. Children's developing understanding of the subjectivity of intentions – A case of "advanced theory of mind". *Journal of Cognition and Development* 23 (2): 231–253. https://doi.org/10.1080/15248372.2021.2003366.

Schuster, B.A., M. Huemer, B. Priewasser, and J. Perner. 2021, January 4–8. Preschoolers' understanding of misrepresentation: Why the false sign test overestimates performance in German-speaking children [Poster Presentation]. Budapest CEU conference on cognitive development, Budapest, Hungary. https://cdc.ceu.edu/bcccd21.

Searle, J. 1983. *Intentionality*. Cambridge: Cambridge University Press.

Siegal, M. 1999. Language and thought: The fundamental significance of conversational awareness for cognitive development. *Developmental Science* 2 (1): 1–14. https://doi.org/10.1111/1467-7687.00048

Sprung, M., J. Perner, and P. Mitchell. 2007. Opacity and embedded perspectives: Object identity and object properties. *Mind and Language* 22: 215–245. https://doi.org/10.1111/j.1468-0017.2007.00307.x.

Tomasello, M. 1999. *The cultural origins of human cognition*. Harvard University Press.

———. 2014. *A natural history of human thinking*. Harvard University Press.

Treisman, A.M., and G. Gelade. 1980. A feature-integration theory of attention. *Cognitive Psychology* 12 (1): 97–136. https://doi.org/10.1016/0010-0285(80)90005-5.

Weingartner, P., and S. Haring. 2020. Children's understanding of identity – The kind of refer-ent that children have in mind when understanding identity arguments. *Humanities and Social Sciences* 8 (2): 51–62.

Wellman, H.M. 2011. Developing a theory of mind. In *The Wiley Blackwell handbook of childhood cognitive development*, ed. U. Goswami, 2nd ed., 258–284. Wiley-Blackwell.

———. 2014. *Making minds: How theory of mind develops*. Oxford: Oxford University Press.

Wellman, H., D. Cross, and J. Watson. 2001. Meta-analysis of theory-of-mind development: The truth about false belief. *Child Development* 72 (3): 655–684. https://doi.org/10.1111/1467-8624.00304.

Wimmer, H., and J. Perner. 1983. Beliefs about beliefs: Representation and constraining function of wrong beliefs in young children's understanding of deception. *Cognition* 13 (1): 103–128. https://doi.org/10.1016/0010-0277(83)90004-5.

Winner, E., and H. Gardner. 1993. Metaphor and irony: Two levels of understanding. In *Metaphor and thought*, ed. A. Ortony, 425–443. Cambridge University Press. https://doi.org/10.1017/CBO9781139173865.021.

Wolf, J. 2021. *The paradox of false belief understanding: The role of cognitive and situational factors for the development of social cognition*. Vol. 50. Walter de Gruyter GmbH and Co KG.

Xu, F., and S. Carey. 1996. Infants' metaphysics: The case of numerical identity. *Cognitive Psychology* 30 (2): 111–153. https://doi.org/10.1006/cogp.1996.0005.

Zaitchik, D. 1990. When representations conflict with reality: The preschooler's problem with false beliefs and "false" photographs. *Cognition* 35: 41–68. https://doi.org/10.1016/0010-0277(90)90036-J.

Part II
Pathologies Associated with ToM

Chapter 4
Theory of Mind and Reading

Pablo Delgado and Isabel R. Rodríguez-Ortiz

Abstract Reading comprehension is the product of a complex set of processes and components (i.e., reader's skills and knowledge). Among them, Theory of Mind (ToM) has recently been pointed out as a capacity that plays a crucial role in comprehending texts. This has been especially highlighted in the case of narrative stories because it requires understanding of emotions, mental states, and perspectives of the characters, which often demands inferring information that is not explicitly stated in the text. In the present chapter, we review some of the most relevant research work that has examined the association between ToM and reading comprehension in both the typical and atypical population. To that end, we first introduce the basic assumptions of some of the more relevant theoretical models of reading comprehension. We then review recent empirical evidence showing that ToM seems to contribute to reading comprehension via listening comprehension, as well as that it is related to low-level linguistic skills, such as vocabulary or grammatical knowledge, and to high-level linguistic skills, such as inference-making or comprehension monitoring. In addition, as we show below, the association between ToM and reading comprehension appears to be stronger in people with autism spectrum disorders and in individuals with deafness or reduced hearing than in their typically developing peers, two atypical populations in which difficulties in ToM are usual. Lastly, we conclude the chapter by outlining the main limitations, research gaps, and future directions in this field.

Keywords Theory of mind · Reading comprehension · Components of reading · Autism spectrum disorder · Deafness

P. Delgado · I. R. Rodríguez-Ortiz (✉)
Universidad de Seville, Seville, Spain
e-mail: pdelgado@us.es; ireyes@us.es

T. Lopez-Soto et al. (eds.), *The Theory of Mind Under Scrutiny*, Logic, Argumentation & Reasoning 34, https://doi.org/10.1007/978-3-031-46742-4_4

4.1 Reading Comprehension and Theory of Mind

A wide variety of cognitive and linguistic skills underlie the apparent simplicity of reading, and they are responsible for a series of processes through which the reader simultaneously extracts and builds meaning from his/her interaction with the text (Snow 2002). This complexity is demonstrated by the existence of numerous theoretical models that aim to explain reading comprehension (see McNamara and Magliano 2009). In general, these models place different emphasis on the cognitive processes responsible for reading comprehension or on the cognitive and linguistic components (i.e., skills and knowledge) required to successfully execute such processes (see Butterfuss et al. 2020; Kendeou et al. 2016). To set the explanatory basis of the probable role of ToM as one of such components, we will focus on some of the most relevant theoretical contributions.

One of the most ambitious explanations of the cognitive processes responsible for discourse comprehension, either written or oral, is the one proposed by Kintsch's Construction-Integration (CI) model (Kintsch 1988, 1998). Its complex view of these processes ultimately aims to explain human cognition (Kintsch 1998; Wharton and Kintsch 1991). Regarding reading comprehension, the CI model is based on the idea that the reader produces three levels of mental representation of the content of the text: (1) the surface structure, (2) the textbase, and (3) the situation model. From the surface structure, which includes the recognition of the words and their order within the sentence, the reader generates the representation of the textbase, which is the first result of the semantic and syntactic processing of the discourse. It consists of a corpus of propositions that represent the minimum processing units as complete ideas or units of meaning, which gather the semantic content of the text, dispensing with the exact words of the surface structure. Figure 4.1 presents an example, adapted from Kintsch (1998), of the propositions that can be extracted from a sentence. As can be observed, the sentence is represented by a main proposition, constituted by the predicate and its arguments (in this case, agent, object, and objective). Moreover, the propositional

Sentence (*surface structure*): Yesterday, Mary gave Fred the old book in the library

Propositional representation (*textbase*):

- Main proposition: Give (Mary [agent], Book [object], Fred [goal])
- Subordinate propositions:
  - Old (book)
  - Yesterday (time)
  - Library (place)

Fig. 4.1 Example of the propositional structure that the reader extracts from the surface structure of a sentence

structure consists of three subordinate propositions. One of them modifies one of the arguments (*old*), whereas the other two contain information of the circumstances of the main proposition (*yesterday*, *library*).

All propositions of the example in Fig. 4.1 are based on information that is explicitly found in the text. However, in many cases, especially in narrative literature, texts do not explicitly present all the relevant information to understand their content. Thus, to establish the necessary coherence for comprehension, the reader must make inferences, that is, construct information that solves the cohesion ruptures of the text. In the following example: "*John, feeling guilty, confessed to Susan: Your bird escaped through the window; it was very hot*", John's confession contains two explicit clauses between which there is a certain lack of coherence. Without the capacity to make inferences, the fact that the bird escaped through the window and the fact that it was very hot do not seem to be related. Nevertheless, the skilled reader uses his/her prior knowledge to infer an implicit fact that solves the incoherence: John opened the window because he was hot, and that is how the bird could escape. When the reader starts making inferences based on his/her prior knowledge to transcend the explicit information of the text, he/she is constructing the third representational level of its content: the situation model, according to Kintsch's CI model (Kintsch 1988, 1998).

The situation model represents the most complex level of comprehension. It is the result of refining the propositions of the textbase and combining them with the reader's prior knowledge, generating a complete and coherent representation of the content. Furthermore, it goes far beyond the inferences that solve the lack of local coherence (i.e., the one between two close ideas in the text) and contains a whole set of inferences that favor the global coherence, that is, the general structure of the content. In this sense, in the previous example, John's feeling of guilt and the fact that he was not as careful as he should have could serve to understand other aspects of his relationship with Susan during the story between these two characters. As was stated by Van Dijk et al. (1983), "a situational model is the cognitive representation of the events, actions, persons, and in general the situation that a text is about" (p. 11).

Therefore, the more coherent, complete, adequate, and accurate the situation model constructed by the reader, the higher his/her level of comprehension of the content of the text. Moreover, this representational level is fundamental for learning, since the construction of inferences necessarily connects the information of the text with the prior knowledge of the reader, thus favoring retention in long-term memory (Kintsch 1998).

4.1.1 Inference-Making: A Core Process for Reading Comprehension

In some cases, although certain implicit contents of the text may be extracted directly from the meaning of the words and their syntactic order (for a discussion about this topic, see Perfetti and Stafura 2015), inference making during reading is considered necessary and fundamental for the accurate and full comprehension of a text (Cain et al. 2001; Elleman 2017; Oakhill and Cain 2012). Therefore, the study of the different types of inferences that can be generated by the reader has been the focus of the different theoretical contributions that aim to explain the processes that underlie reading comprehension. This chapter is focused on the inferences required by narrative texts due to their clear relation to ToM as an underlying cognitive skill (for a view that integrates the necessary inferences to understand narrative and expository texts, see van den Broek et al. 1993).

Among the theoretical contributions about the inferential comprehension of narrative texts, it is worth highlighting the Constructionist model (Graesser et al. 1994; Singer et al. 1994), which considers inference making as the result of processes that are mainly driven by the reader –as opposed to other views that contemplate such activation as automatic and triggered by the content of the text (see Long and Lea 2005). This model is founded on three main assumptions of the principle of *search after meaning* (Stein and Trabasso 1985): the reader goal assumption, the coherent assumption, and the explanation assumption. According to Graesser et al. (1994), these three assumptions jointly state that "readers attempt to construct a meaningful referential situation model that addresses the readers' goals, that is coherent, and that explains why actions, events, and states are mentioned in the text" (p. 372). Based on this, the main objective of the Constructionist model is to describe and predict how the readers generate, while they read, three specific types of inferences: (1) *the superordinate goals,* which represent the reasons of the actions of the characters; (2) *the causal antecedents*, which explain why the different actions, events, or physical and mental states are mentioned in the text; and (3) *global thematic inferences*, which construct the essence of the text and establish relationships between the main ideas, facts, or circumstances and reader's prior knowledge.

Thus, inference making is the product of complex and high-level comprehension processes (see O'Brien et al. 2015) that require the coordination and combination of different cognitive skills (mainly those among the executive and cognitive functions, such as comprehension monitoring), linguistic skills, such as vocabulary or grammatical knowledge, and the prior knowledge of the reader, including both world knowledge and topic-specific knowledge (Cain et al. 2001; Kendeou et al. 2016; Potocki et al. 2017; van den Broek et al. 2013; Yang et al. 2022). These skills and knowledge, which are thus necessary for successful comprehension, are the focus of the comprehension models that revolve around the so-called components of reading. Their study is necessary not only to know how we read, but also to identify those elements that must be addressed in education to develop reading comprehension, especially to intervene in students who present reading difficulties.

Considering the abovementioned types of inferences required to understand narrative texts proposed by the Constructionist model (including those aimed at understanding the reasons for the actions of the characters and why such actions, as well as other events or states, appear in the story), ToM presents as a skill that could play a fundamental role, since it is, precisely, the capacity of people to attribute mental states –thoughts, desires, intentions or beliefs– both to oneself and to others, thereby allowing the reader to infer causes and motivations of the behavior, as well as to predict the latter (Flavell 1999; Wellman 2002; see Chap. 2 in this volume). Therefore, this skill is fundamental for understanding the characters of any story as actors with motivations, desires, intentions, and other mental states. In turn, understanding such mental states or the perspective of the characters is related to a better understanding of the causal structure of the narrative story –either written or oral– (e.g., Pelletier and Astington 2004; Ronfard and Harris 2014; Shanahan and Shanahan 1997; Strasser et al. 2013). For example, the results of Strasser et al. (2013, Study 2), in a study with pre-school children, revealed that an intervention based on asking questions to the children about the reasons, objectives, and thoughts of the characters during the dialogical reading with an adult improved the comprehension of the story compared to other questions focused on descriptions, predictions, or establishing associations between the content of the text and the prior knowledge of the children.

However, as was pointed out by Dore et al. (2018), the role of ToM as a component of reading comprehension has gained little attention from researchers, who have mostly focused on ToM as a product from the study of children's cognitive development. Consequently, during the last decade, researchers have proposed some models of components and theoretical approaches to reading comprehension that include this skill as one of the several skills that unfold during reading (e.g., Dore et al. 2018; Duke and Cartwright 2018; Kim 2017). Next, we will review the empirical evidence available to this respect, starting with the results of studies conducted in typically developing students, and then we will go over the results of studies carried out in atypical populations, specifically in people with autism spectrum disorder (see Chap. 7 in this volume) and people with deafness or reduced hearing (see Chap. 9 in this volume), two populations that usually present difficulties related to ToM and to reading comprehension.

4.2 Theory of Mind as a Component of Reading Comprehension: Studies in Typical Population

It is currently assumed that the determinants of reading comprehension are not only related to individual factors of the readers, as these interact with other aspects, such as the characteristics of the texts, the type of reading task, the immediate context, and even the belonging to sociocultural groups (e.g., Duke and Cartwright 2018; Rouet et al. 2017; Snow 2002). However, since ToM belongs to the range of skills and knowledge of the reader (although, of course, its development also depends on contextual and sociocultural factors, e.g., Lane et al. 2012), we will begin this section with one of the theoretical models which usually provides the basis for research about the individual components of reading comprehension: the Simple View of Reading (SVR) model.

The SVR model considers that reading comprehension is the product (not the sum) of two fundamental skills: word recognition (decoding) and listening comprehension (Hoover and Gough 1990). While the decoding skill is the basis for the construction of the representation of the surface structure of the text in terms of Kintsch's (1988, 1998) CI model, linguistic comprehension enables the representations of the textbase and the situation model. Nevertheless, a wide range of sub-components underlie this apparent simplicity, which influences the success of the execution of decoding and linguistic comprehension processes (e.g., Kim 2017). To make it even more complex, the roles of these components seem to be related to each other through three types of interactions: hierarchical, bidirectional, and dynamic. Regarding their hierarchical relationships, the available evidence indicates that there are low-level and high-level components, in a structure in which the low-level components determine the reading comprehension by indirect influences through their contribution in the high-level components. Furthermore, among the components of the same hierarchical level, there are bidirectional relations, that is, they are skills or knowledge that seem to influence each other reciprocally. Lastly, it is concluded that their relationships are also dynamic, since the influence level of the different components varies throughout the different developmental moments of the learning of reading comprehension (e.g., Kim 2017, 2020a, b). This dynamic relationship is especially evident in the highest hierarchical level, that is, the one proposed by the SVR model (i.e., decoding and listening comprehension skills). During the first stages of such development, the decoding skill is the one that best explains the inter-individual variability in reading comprehension. However, as the students automatize this skill and, therefore, require less cognitive effort to perform it, the individual differences in listening comprehension become the ones that best distinguish skilled readers from those with poor reading comprehension (Lonigan et al. 2018; van den Broek and Espin 2012).

Among the subcomponents of listening comprehension, ToM is considered a high-level linguistic skill that ranks hierarchically above other cognitive and linguistic components, such as executive functions (e.g., working memory), vocabulary, or syntactic processing (Devine and Hughes 2014; Dore et al. 2018; Duke and Cartwright 2018; Kim 2017, 2020a). Thus, the components of lower hierarchical level contribute to linguistic comprehension (and, consequently, to reading comprehension) through ToM and other high-level skills to which ToM seems to be bidirectionally and dynamically related, such us inference making or metacomprehension.

The most thorough and robust work conducted to date to determine the role of ToM and its relationships with other components of discourse comprehension (either written or oral) is that of Young-Suk G. Kim, throughout a series of studies performed with South Korean children (Kim 2015a, b, 2016) and American children (Kim 2017, 2020a, b; Kim and Petscher 2021; Kim and Phillips 2014). One of the greatest strengths of this series of studies was the use of *structural equation modelling* to analyze the contribution of the different components of discourse comprehension. This type of analysis, which is more sophisticated than the usual correlation analyses or hierarchical regression analyses, allows exploring the direct and indirect relationship between the different components, as well as between these and the main variable that is intended to be explained (in this case, discourse comprehension).

The results in Kim's research showed that ToM, measured in all cases through first- and second-order false-belief tasks (see Chap. 10 in this volume for a description of these tasks), contributes to reading comprehension always through listening comprehension, of which it explains a unique variance in pre-school children (mean age = 6.08; Kim 2015b), Grade-3 children (Kim 2017, 2020b) and Grade-5 children (Kim 2020a). In Kim (2015a), although ToM explained a unique variance of reading comprehension in pre-school children (mean age = 6 years and 1 month), this study did not include listening comprehension in the model as an intermediate component between ToM and reading comprehension. Moreover, in those studies that only analyzed the contribution of ToM to the listening comprehension of texts, ToM also explained a unique variance of this skill in Grade-2 children (Kim 2016, 2020b; Kim and Phillips 2014; Kim and Petscher 2021), which indirectly also provides evidence of the contribution of ToM to reading comprehension.

It is important to highlight that the unique contribution of ToM to the listening comprehension of texts (and thus indirectly to reading comprehension) in the studies of Kim appeared even when other cognitive and linguistic components were included in the model. Additionally, the results reveal that ToM is a skill that mediates the contribution of some of these components to listening comprehension and is bidirectionally related to other variables of its hierarchical level. For example, in one of these studies, in which the author tested the Direct and Indirect Effects Model of Reading (DIER; Kim 2017), working memory contributed to explain the vocabulary level and grammatical knowledge of Grade-3 students, and, in turn, these two variables contributed to ToM. Lastly, ToM explained a unique variance of listening comprehension, which, in turn, explained reading comprehension. These results indicate that ToM, in addition to directly contributing to listening comprehension, is a skill through which working memory, vocabulary, and grammatical knowledge explain such comprehension. Furthermore, ToM seems to be bidirectionally related to the general skill of inference-making and to the ability of the students to monitor their own comprehension (see Fig. 4.2).

At this point, it is necessary to point out that the relationships shown in Fig. 4.2 have a correlational nature, thus, as was stated by the author, the data do not allow inferring causal relations. Moreover, the different hierarchical levels cannot be statistically established; therefore, they were set a priori based on theoretical decisions. We will come back to these circumstances at the end of the chapter when we discuss the limitations of studying the relationship between ToM and reading comprehension.

Similar results arose in the rest of Kim's studies, which also showed ToM as a mediator of the influence of the attention span on oral comprehension in a sample of Grade-3 and Grade-5 children (Kim 2020a, b), and of syntactic knowledge in pre-school children (Kim 2015b) and Grade-2 children (Kim 2016); these studies also showed that ToM is bidirectionally related to inhibitory control (Kim and Phillips 2014). In addition, in a recent study in which generalized linear mixed models were used instead of structural equation modelling as the analysis method, Kim and Petscher (2021) found that ToM significantly predicted the variability in oral comprehension in a sample of Grade-3 students after including in the model: the

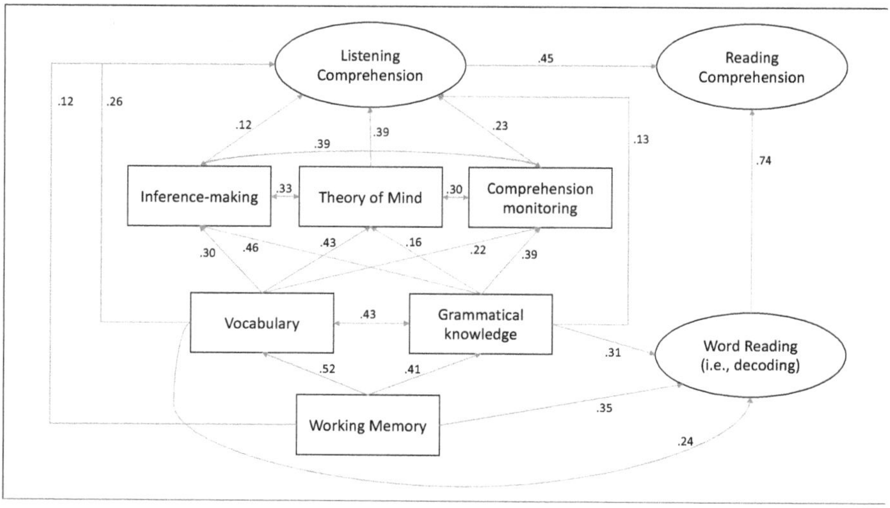

Fig. 4.2 Direct and indirect contributions (i.e., standardized structural regression weights) of the components of the DIER model on word decoding, listening comprehension, and reading comprehension. Adapted from Kim (2017)

components that are presented in Fig. 4.2, the reading comprehension skill of the students, the type of text that was read to the students (narrative vs. expository), and the types of listening comprehension questions that they had to answer (literal vs. inferential). Furthermore, the results in this set of studies showed that the predictive capacity of ToM did not interact with any of these last three variables, that is, the contribution of ToM to listening comprehension was similar regardless of reading comprehension skill, the type of comprehension questions, and whether the text genre was narrative or expository.

Other authors have also recently explored ToM as a component of reading comprehension (Atkinson et al. 2017; Boerma et al. 2017; Ebert 2020a, b; Guajardo and Cartwright 2016; Lockl et al. 2017; Pelletier and Beatty 2015; Strasser and Río 2014). Among these authors, Boerma et al. (2017) performed *path analyses* to examine the association of children's reading comprehension skill with children's home literacy environment (a variable that included parental print exposure, number of parents' books at home, and number of children's books at home), children's print exposure, expressive verbal ability, and ToM (measured through the Strange Stories test; Happé 1994; see Chap. 2 in this volume) in a sample of Grade-4 and Grade-5 students. According to their findings, ToM played a determining role in reading comprehension above the verbal expression skill, partially mediating the contribution of home literacy environment to reading comprehension (Boerma et al. 2017). Similar results have also been reported by Pelletier and Beatty (2015) regarding the listening comprehension of tales in pre-school children, although, in this case, only ToM measured through second-order (vs. first-order) false-belief tasks predicted listening comprehension after controlling for children's vocabulary level and early linguistic skills.

An exception to the results reviewed so far are the findings of Strasser and Río (2014) in pre-school children. In this study, ToM did not predict the comprehension of a story in images (no text) or the oral comprehension of a narrative text after controlling for the contribution of children's breadth and depth of vocabulary, executive functions, inference-making ability, and metacognitive monitoring of comprehension. The authors suggest that ToM is highly confounded with other linguistic variables that, when controlled, make ToM lose its capacity to predict the comprehension of stories. However, many of these linguistic variables have been introduced in the explanatory models of the different studies conducted in this field without the relationship between ToM and discourse comprehension disappearing (see, for example, Kim's studies reviewed earlier).

The limitations of correlational analyses when inferring causal relationships can be partially overcome through longitudinal studies, although such studies are still scarce in the literature about the relationship between ToM and reading comprehension. Among these works, there are two studies carried out by Ebert (2020a, b) in which the DIET model of Kim (2017) was also tested. In Ebert (2020a), the author explored the relationships between vocabulary level, sentence comprehension, working memory, and non-verbal reasoning, measured at the age of 3 years; ToM, measured through first- and second-order false-belief tasks at the age of 5 years; the language used to express mental states and metacognitive knowledge, both measured at 9 years; and, lastly, oral and reading comprehension at 13 years.

Among the relationships found, the early linguistic skills predicted ToM measured 2 years later. Moreover, and more relevantly for the present chapter, ToM measured at 5 years of age explained a unique variance of the level of listening comprehension 8 years later (i.e., at the age of 13 years), although it did not reach statistical significance (but $p < .10$). That is, in agreement with the results of the set of studies by Kim, ToM contributed to reading comprehension indirectly through oral comprehension in Ebert (2020a), although, in this case, the findings show that this relationship, despite a smaller regression weight and the fact that it did not reach statistical significance, was maintained even when comprehension was measured 8 years later. Additionally, in Ebert (2020b), the author performed path analyses of the same sample and variables from the previous study (Ebert 2020a), although also including a measurement of advanced ToM skills (the Strange Stories test; Happé 1994) and of children's vocabulary level, both recorded 1 year before (at 12 years of age) the reading comprehension measurement. Among the results of this second study of Ebert (2020b), the author highlights that early ToM predicted the vocabulary level 7 years after, whereas advanced ToM indirectly predicted reading comprehension 1 year after through oral comprehension, although, again, without reaching statistical significance (but $p < .10$).

In another longitudinal study, Atkinson et al. (2017) evaluated ToM, non-verbal intelligence, executive functions, decoding skill, receptive vocabulary, and expressive and receptive language skills in British pre-school children aged 4 years (assessment time 1; T1) to analyze their capacity to predict reading comprehension measured 2 years later, at an average age of slightly over 6 years (T2). Among other results, ToM predicted reading comprehension after controlling for the rest of the evaluated skills. Furthermore, the results of a mediation analysis that only included ToM (T1), listening comprehension (T2), and reading comprehension (T2) indicated that ToM predicted reading comprehension 2 years after, both indirectly, through listening comprehension, and directly. It is noteworthy that this direct association between ToM and reading comprehension did not appear in the results of the reviewed studies of Kim and Ebert. A possible explanation for such a discrepancy is that the analyses performed by these two authors, in addition to being more sophisticated, included a larger number of components in their explanatory models of reading comprehension. Thus, the direct association found in the mediation analysis performed by Atkinson et al. (2017) would disappear in Kim's and in Ebert's studies due to the presence of other components in the same hierarchical level as ToM, such as inference-making or comprehension monitoring, with which it seems to relate bidirectionally.

To sum up, almost all the studies that included ToM as an explanatory component of reading comprehension found that its contribution is indirect through listening comprehension. These results would explain the absence of a direct relationship between ToM and reading comprehension in other longitudinal studies carried out by Guajardo and Cartwright (2016) and Lockl et al. (2017), since none of them explored its indirect contribution. Lockl et al. (2017) evaluated ToM in a sample of 4-year-old children through first-order false-belief tasks, which did not significantly predict the reading comprehension of the sample in two time points, i.e., 2 and 3 years later. Although the authors argued that this absence of relationship could have been since the listening comprehension skills are not very relevant in the first stages of the development of reading comprehension, their results are not in line with those found in the previously mentioned studies, some of which were conducted with children of similar age. However, Lockl et al. (2017) did not evaluate the indirect contribution of ToM to reading comprehension through listening comprehension, but only its direct contribution. Furthermore, such direct contribution was explored by statistically controlling, through multiple regression analysis, the contribution of linguistic listening comprehension skills. Therefore, the absence of relationship between ToM and reading comprehension in this study seems to be explained by these circumstances. Moreover, ToM was only measured through first-order false-belief tests, which require ToM skills that are less sophisticated than those measured through second-order false-belief tests, or other more advanced ToM skills, such as the ones measured by the Strange Stories test (Happé 1994). This could have also contributed to the absence of relationship between ToM and reading comprehension measured 2 and 3 years later. As we mentioned above, Pelletier and Beatty (2015) found that only the results of second-order false-belief comprehension were related to tale comprehension in children aged 4–5 years. The comprehension of narrative texts usually requires the understanding of the perspectives and mental states of multiple characters, as well as the understanding of the perspectives of each character toward the rest of characters. Therefore, the first-order false-belief recognition skills seems to be not enough for a full and accurate comprehension of narrative texts.

On their part, in the study of Guajardo and Cartwright (2016), the results of the hierarchical regression analysis showed that ToM evaluated in children aged 3–5 years (also through first-order false-belief tasks) predicted oral comprehension at the same measurement time point. Moreover, in line with the results of Lockl et al. (2017), neither this nor the other second-order false-belief measure evaluated approximately 4 years later predicted reading comprehension at this same second measurement time point, again, after controlling for the contribution of listening comprehension. Thus, they also did not analyze the indirect contribution of ToM through it. Additionally, it is interesting that the second-order measurements of ToM did predict the results of the sample in a measurement of reading awareness. This measure evaluated certain metacognitive skills related to reading through a set of questions whose answers required the children to reflect on their reading habits and strategies. However, since Guajardo and Cartwright (2015) performed regression analyses including up to 12 predicting variables in a sample of only 31 children, their results must be considered cautiously, due to the lack of statistical power.

To conclude this section, we would like to briefly highlight that, although the relationship between ToM and reading comprehension is more evident with respect to narrative text comprehension due to the high load of mental states of the characters in the comprehension of the causal relationships in the story (the most studied relationship in the literature), certain evidence indicate that ToM can also play a relevant role in the comprehension of expository texts. Firstly, some of the studies reviewed here include both narrative and expository texts as a measure of listening and reading comprehension (e.g., Atkinson et al. 2017; Kim 2020b). Secondly, Kim and Petscher (2021) not only included both types of texts, as we mentioned previously, but they also analyzed the possible interaction between the contribution of ToM to comprehension and both text genres. Since they did not find any significant interaction, the results indicated that the predictive capacity of ToM is also relevant in the case of the expository texts.

Lastly, different studies have found that ToM is related to other skills that are, in turn, associated with the capacity to understand texts regardless of the text genre. These skills include the development of metacognitive skills (Flavell et al. 2000; Kuhn 2000), counterfactual thinking (Cartwright 2015), and pragmatic language (Russell and Grizzle 2008). Furthermore, ToM presents bidirectional relationships with inferential comprehension (e.g., Kim 2017; Lee et al. 2022), facilitates the comprehension of the perspective of the authors when reading multiple texts about the same topic (Florit et al. 2020), and it even plays a fundamental role in learning situations in which the relationship between the learner, the teacher, and the rest of the classmates is essential for sharing ideas and reviewing beliefs or concepts during academic learning (Astington and Pelletier 1998). Therefore, it is plausible to consider that ToM is not only fundamental for the comprehension of narrative texts, but it also seems to contribute to the comprehension of expository texts.

4.3 Theory of Mind and Reading Comprehension in Atypical Population

As was pointed out in the previous section, ToM plays an important role in reading comprehension, mainly influencing the comprehension skills of oral language. In addition, the studies conducted in this specific field have reported that there are, at least, two populations characterized by their difficulties in reading comprehension and which, in turn, usually present delays or deficits in the development of ToM: people with autism spectrum disorder (ASD), and people with deafness or reduced hearing (DRH). In this section, we will explore the studies that relate these two competences (reading comprehension and ToM) in these populations.

4.3.1 Theory of Mind and Reading Comprehension in Population with ASD

ASD refers to a set of neurodevelopmental disorders characterized by salient difficulties in social interaction and communication, as well as by the presence of repetitive and restricted patterns of behaviors, interests, or activities (American Psychiatric Association 2013). The chapter dedicated to ToM and ASD population (see Chap. 7 in this volume) describes the difficulties of this population in attributing mental states and interpreting other people's behaviors from this attribution. For this reason, and due to the association observed between ToM and reading comprehension in typically developing children (e.g., Kim 2017), it has also gained special interest in the last years whether deficits in ToM could explain usual reading comprehension difficulties in the population with ASD, especially since it was observed that greater difficulties in reading comprehension appear when the deficits in social interaction and communication in this population are more pronounced (Jones et al. 2009).

The reader profile of the population with ASD has been traditionally characterized by lower performance in listening/reading comprehension than in word reading, which was called *hyperlexia* by some researchers (Nation 1999; Snowling and Frith 1986). However, due to the different interpretation of what the mastery of word reading means and when it must appear (Fleury et al. 2014), the term *poor comprehender* is preferred to refer to a person with good word reading and poor listening/reading comprehension (Davidson 2021). This profile is the one which people with ASD are most frequently associated with (Brown et al. 2013; Jones et al. 2009; Tong et al. 2020), although different studies have concluded that this is not the only reader profile that can be found in this population. For example, Williamson et al. (2012), in a study with high-functioning adolescents with ASD, identified three profiles of reading comprehension: *text bound comprehenders*, who were attached to the text, unable to interpret it; *imaginative comprehenders*, who understood the text better if it was accompanied by images and were able to make subjective interpretations of it; and *strategic comprehenders*, who used reading strategies like skilled readers, such as connecting what is read with their prior knowledge, although they had difficulties in making predictions. Nevertheless, the participants of the study of Williamson et al. (2012) only represented a part of the population with ASD. A larger population was contemplated in the classification of Davidson (2021). This author, based on the results in Davidson and Ellis Weismer (2014), McIntyre et al. (2017), and Solari et al. (2019), proposed five reader profiles in children with ASD: *Typical readers* with ASD, who represent a small group of readers that had no difficulties in word reading or in listening/reading comprehension; *discrepant poor comprehenders*, who showed an average or above-average word reading and lower comprehension skills; *below-average poor comprehenders*, who showed a borderline-average to average word reading and below-average comprehension (this group is the equivalent of typical poor comprehenders); *mixed-deficit profile*, which includes children with ASD characterized by word reading and comprehension skills below one standard deviation (*SD*) of the mean; *severe mixed-deficit profile*, which includes those with word reading and comprehension below 1.5 *SD* or lower with respect to the mean. Therefore, four of the five profiles identified by Davidson (2021) presented difficulties in comprehension.

Regarding the development of ToM in children with ASD, it has been demonstrated that, although they do not have difficulties with the basic ToM skills (for example, first-order false-belief tasks), they usually have difficulties with more advanced ToM skills (for example, second-order false-belief tasks) (Frith 2012; Tong et al. 2020; and see Chap. 7 in this volume); however, despite the relationship between ToM and reading comprehension (e.g., Boerma et al. 2017; Dore et al. 2018; Tong et al. 2020), few studies have been conducted in populations with ASD (Davidson 2021; McIntyre et al. 2018; Wang et al. 2022).

The idea that the differences in ToM may be associated with reading comprehension in this population is demonstrated in the meta-analysis carried out by Brown et al. (2013). In this meta-analysis, the authors explored 36 studies that compared the reading comprehension of a group of participants with ASD and a group of participants with typical development, with the aim of identifying factors that could predict the strengths and weaknesses of the reading comprehension of the population with ASD. Among their results, they found that most of the readers with ASD presented the profile of word decoding skills equivalent to that of their peers with typical development, but also greater heterogeneity and difficulties in reading comprehension. Moreover, in general, the individuals with ASD had difficulties in comprehending texts that required a good understanding of the social world; however, they tended to show less deficits in reading comprehension compared to the individuals with typical development when they read texts that required lower social knowledge. The reading comprehension of the individuals with ASD used to be 3 *SD* below that of their peers with typical development when reading texts that required high social knowledge.

In line with the findings of the abovementioned meta-analysis, White et al. (2009) analyzed the possibility that the difficulties in reading comprehension of children with ASD were associated with a low social knowledge. To this end, these authors gave the participants a series of stories that required thinking about mental states, and another series of stories that only required integrating the information of the text to reach its comprehension. The results showed that the children with ASD who had a limited ToM had more difficulties to understand both types of texts compared to their peers with typical development, although the children with ASD with higher deficits in ToM had a worse comprehension than those with lower deficits in ToM when reading mentalist texts (which required thinking about the mental states of others) and showed better performance when the text only required integrating the information. The authors concluded that the comprehension difficulties of the participants with ASD were rather related to a limited knowledge of mental states (i.e., to ToM) than to comprehension problems.

Nevertheless, in an attempt to separate the deficits in ToM from language disorders that are also frequent in this population (American Psychiatric Association 2013), some authors have separately analyzed the relationship between reading comprehension and ToM evaluated with verbal and non-verbal tasks, also controlling for the oral language level. Among them, Ricketts et al. (2013) observed that ToM, measured with verbal tasks (the Strange Stories test; Happé 1994) and non-verbal tasks (the Frith-Happé animations, which require the attribution of mental states to two interacting triangles; Abell et al. 2000), contributed to the reading comprehension of adolescents with ASD after controlling for word reading and oral language skill. In their study with a sample of 100 adolescents with ASD (mean age: 15 years and 6 months), both social behavior and comprehension of mental states were associated with reading comprehension, which led the authors to conclude that the difficulties to understand the social world could contribute to the difficulties in inference making shown by the readers with ASD (Ricketts et al. 2013), and these difficulties could even explain the heterogeneity in the reading performance of this population (Lee et al. 2022).

McIntyre et al. (2018), with the aim of replicating the findings of Ricketts et al. (2013), also used verbal tasks to explore ToM (Strange Stories test; Happé 1994) and non-verbal tasks (the Silent Films Task; Devine and Hughes 2013) analogous to the former. In this case, they evaluated a sample of participants with a wider age range than the sample in Ricketts et al. (2013) and they also analyzed the relationship between reading comprehension, word reading, oral language, and ToM in a sample of 70 children and adolescents with high-functioning ASD and 40 peers with typical development. The results showed that ToM contributed to reading comprehension after controlling for intelligence quotient, word recognition, and oral language, although only in the sample of children and adolescents with ASD. The predictors of reading comprehension were different between the two groups, and, in the case of the people with ASD, social cognition (i.e., ToM) appeared as a very relevant factor for the interventions to improve reading comprehension in this population. In this study, the pattern of the influence of ToM on reading comprehension was maintained when ToM was evaluated with more verbal tasks (Strange Stories) and with less verbal tasks (Silent Films).

The relationship between ToM and reading comprehension has also been observed in children with ASD in non-alphabetical languages, such as Chinese.

For example, Tong et al. (2020) demonstrated that the adequate development of reading comprehension is compromised by the deficits in ToM in school children with ASD (7–9 years old). Although matched in age, non-verbal intelligence, verbal working memory, and word reading skills with a group of school children with typical development, reading comprehension in the children with ASD was lower than that of their typically developing peers, in both the literal and non-literal reading comprehension (which implied inferences, evaluation, and mentalization) of the texts. Furthermore, the children with ASD presented skills in basic ToM tasks that were like those of their peers with typical development, but their performance in advanced ToM tasks was lower. In this case, the results showed that ToM was associated with both literal and non-literal reading comprehension only in the group with ASD.

Lee et al. (2022) also found that specific difficulties in executive functions and in ToM could be associated with the different difficulties in reading comprehension that appeared in the children (8–10 years old) with ASD who use non-alphabetic Chinese logographic script. Specifically, Lee et al. evaluated the influence of deficits in executive functions, ToM, and language abilities on literal and inferential reading comprehension, separately, in a sample of children with ASD and children with typical development of the same age. In their evaluation of ToM, they used a battery of tests that included six basic ToM tasks and five advanced ToM tasks (Tong et al. 2020). The children with ASD scored significantly lower than their peers with typical development in six ToM tasks (five advanced ToM tasks and one basic ToM task, which involved inference making) and in questions of both literal and inferential reading comprehension. The most interesting finding of the study was that, although literal and inferential comprehension required cognitive and linguistic skills, they were associated with different ToM skills. Specifically, two advanced ToM skills that required perspective-taking of a character's emotional state predicted literal reading comprehension, whereas inferential comprehension was more dependent on some perspective-taking ToM skills. From these results, the authors highlight the role of the affective components of ToM in finding information during the reading of a text, in the case of literal comprehension; thus, the difficulty in understanding the emotional state of others or in showing empathy could explain the difficulties that these children encounter in the processing of explicit information of written texts. Regarding inferential comprehension, the relationship that appears in the results with the perceptual role-taking skill is justified, according to Lee et al. (2022), by the association between this ToM skill and joint attention. The latter would help children to understand the context information and would thereby facilitate inference making.

To sum up, people with ASD present large heterogeneity in terms of reading comprehension performance, which could indicate the influence of different factors beyond the diagnosis of ASD (Brown et al. 2013). Among these factors, the reviewed studies reveal that it is necessary to consider the specific ToM skills.

4.3.2 Theory of Mind and Reading Comprehension in Population with DRH

In the population with DRH, difficulties in reading comprehension have also been observed (Kyle and Cain 2015; Moreno-Pérez et al. 2015; see Harris 2015 for a review), as well as in ToM (Marschark et al. 2019; Peterson 2020; Torres and Rodríguez-Ortiz 2011; Tucci and Easterbrooks 2020). As was already described in this chapter, according to the DIER model (Kim 2017), ToM would influence reading comprehension through its contribution to listening comprehension, although, in turn, it would be influenced by vocabulary and grammatical skills. These two aspects of the mastery of oral language have been reported as deficient in the population with DRH (Bishop 1983; Kelly 1996; Kyle and Harris 2010; Lillo-Martin et al. 1992; Musselman 2000; Paul 1996; and see Lederberg et al. 2013, for a review), thus poor oral language is considered one of the factors responsible for ToM delay in this population (Mancini et al. 2016; Meristo et al. 2012), although the latter has also been attributed to poor pragmatic skills, educational strategies, and the lack of social relations (Peterson 2020; Stanzione and Schick 2014; Tucci and Easterbrooks 2020). Moreover, executive functions have also been explored as determinants of ToM delay, especially the role of working memory (Holmer et al. 2016b; Liu et al. 2018; Meristo and Hjelmquist 2009).

Even though children with DRH usually obtain lower scores in ToM than their hearing peers, it has been observed that both groups follow the same developmental stages in this skill (Holmer et al. 2016b; Ketelaar et al. 2012; Peterson 2016; Peterson et al. 2005, 2012). This conclusion indicates that the population with DRH presents a delayed –not impaired– ToM development. As was mentioned in a previous paragraph, ToM delays in the population with DRH have frequently been attributed to difficulties in oral language and, specifically, to the imbalance between the limited linguistic skills of the child, difficulties derived from auditory deprivation, and the language of the people around the child, who mostly communicate with him/her using oral language (Lederberg et al. 2013; Peterson and Peterson 2009; Sundqvist and Heimann 2014). This imbalance places children with DRH in a situation in which they do not have enough understanding of the social context, they do not enjoy an adequate linguistic development to establish rich and varied social relations, and they are not exposed to mentalist language. In contrast with this situation, it has been shown that native signing children with DRH, who are exposed to sign language since birth, present a ToM development that is similar to that of their hearing peers (Lederberg et al. 2013), thus reinforcing the role of language in the development of ToM. However, some authors consider that poor linguistic skills of children with DRH are also associated with poor representations of mental states, which hinders the processing of these mental representations in working memory (Ma et al. 2014). Therefore, ToM delay in children with DRH could be attributed to both poor linguistic development and difficulties in working memory. In fact, it has been found that there is a correlation between ToM and working memory in children with DRH (Meristo and Hjelmquist 2009).

Numerous studies have been focused on the role of oral language in the ToM deficits of the population with DRH (e.g., Mancini et al. 2016; Meristo et al. 2012). Among them, the study of ToM in the case of children with a cochlear implant has gained special interest due to the early access to oral language granted by this hearing aid, which is inserted in the cochlea and allows reducing the gap between chronological age and auditory age caused by early hearing losses (Geers and Nicholas 2013). Nevertheless, several studies have observed that children with cochlear implant continue to present difficulties in understanding ToM, in its affective and cognitive version (Ketelaar et al. 2012; Peterson 2016; Peterson et al. 2016; Wiefferink et al. 2013; see Chap. 2 in this volume for the definition of these two types of ToM). In other cases, the implantation age seems to mark some difference regarding the development of ToM. Thus, Sundqvist et al. (2014) compared affective and cognitive ToM performance between a group of children with early cochlear implant, a group of children with late cochlear implant, and a group of hearing peers, all of them aged 4–9 years. Of the three groups, only the late implant group showed delayed affective ToM, although both groups of children with DRH obtained scores below those of the hearing children in the cognitive ToM tasks. This result suggests that the development of some aspects of ToM could require the involvement of some cognitive processes beyond language development.

Regarding the studies about the relationship between ToM and reading comprehension in the population with DRH, these are scarce and very heterogeneous in terms of the characteristics of the participants. However, the view of this relationship in these studies is the same as that in the population with ASD: ToM is related to reading comprehension in the population with DRH. One of the studies that shows such relationship is the one conducted by Holmer et al. (2016a). These authors explored ToM, visuo-spatial working memory, and reading comprehension in 13 children aged 10.1 years (range: 7.3–14.5) who were Swedish sign language users. ToM was evaluated using a Swedish version of the Wellman and Liu (2004) five-step ToM scale adapted to this sign language. The children showed delayed ToM with respect to their peers, although also a developmental progression similar to that of their hearing peers (Henning et al. 2011; Peterson et al. 2005, 2012; Wellman and Liu 2004; Wu and Su 2014) and that of signing children with DRH in other studies (Peterson et al. 2005, 2012). The reading comprehension of these children was also significantly lower than that of their hearing peers (Holmer et al. 2016b), but they presented a similar working memory capacity. The results also showed a significant correlation between ToM and reading comprehension ($r = .69$), as well as between ToM and non-verbal working memory ($r = .61$). However, in contrast with the predictions of the authors, ToM was not correlated to sign language comprehension, which, in turn, was not correlated to reading comprehension. The latter results led Holmer and colleagues to consider that a signing context and an adequate signing competence could be necessary, but not sufficient, for a typical development of ToM in signing children with DRH. Nevertheless, they also highlighted that only two parents of the participants were native signers, thus the lack of correlation between ToM and sign language comprehension should be addressed with caution. Moreover, after controlling for sign language comprehension, the correlation between ToM and reading comprehension remained unaltered ($r = .63$), which is in line with the results obtained in the typical population (e.g., Kim 2015a) and in children with ASD (Ricketts et al. 2013) This would thereby lead to the conclusion that the relationship between ToM and reading comprehension could not be fully explained by the mediation of the general language skills. Thus, some authors have moved to explore other cognitive processes involved in this relationship, such as inference-making capacity.

The capacity to understand metaphors and sarcasm could reflect the ability to infer the mental state of the speaker; therefore, these aspects could be directly related to ToM or part of it (Happé 1993; Mo et al. 2008; Norbury 2005; O'Reilly et al. 2014; Peterson et al. 2012; Wellman and Liu 2004). Edwards et al. (2021) evaluated the understanding of metaphors and sarcasm and the capacity to make different types of inferences from texts in two groups of university students with DRH: 32 cochlear implant users and 42 cochlear implant non-users, with a control group of 38 hearing university students. In this study, the authors examined the capacity to make two types of inferences from the texts: inferences that required general knowledge of the world and inferences of emotions in social scenarios. Regarding sarcasm comprehension, the students with DRH without cochlear implant obtained a score below that of their hearing peers, whereas the students with DRH who did have a cochlear implant obtained similar scores compared to their hearing peers. Regarding comprehension of metaphors, the use of the cochlear implant did not pose an advantage, and the hearing students scored higher than both groups with DRH. Accordingly, reduced hearing appeared to be associated with delays in advanced ToM skills, which is in line with the results of other studies (Orlando and Shulman 1989; Peterson 2020). Similar differences between groups were found when evaluating the ability to make two types of inferences (general and emotional). However, unlike other studies that analyzed ToM as a predictor of reading comprehension, the study by Edwards et al. (2021) aimed to determine which cognitive or linguistic factors contributed to the difficulties that university students with DRH encountered in understanding sarcasm and metaphors. Specifically, regarding sarcasm comprehension, it was found that 40% of the variance was explained by the capacity to infer others' emotions in social contexts, 10% was negatively predicted by the students' self-rated receptive sign language skill (lower comprehension of sarcasm was associated with higher receptive sign language skill), and, lastly, the capacity to make inferences that required general knowledge of the world, and working memory capacity explained another 10% of the variance. With respect to metaphor comprehension, the capacity to infer others' emotions was another major predictor (26% of the variance), followed by the best way of communicating (spoken over sign language), working memory capacity, and the capacity to make non-mental state inferences (general knowledge), which, together, represented 17.6% of the variance. Thus, this study showed that understanding sarcasm and metaphors is strongly related to the capacity to make inferences during reading comprehension, especially when inferring the mental state of the writer or the characters of the text. Consequently, the relationship between ToM and reading comprehension can be a bidirectional relationship.

Despite the access to oral language facilitated by the cochlear implant, children and adolescents with this device are not exempt from presenting difficulties in terms of reading comprehension (Geers and Hayes 2011). They especially appear when it comes to answering questions that require inference-making (Kyle and Cain 2015), which is an ability that, as has been demonstrated throughout this chapter, is related to ToM. Nevertheless, few studies have analyzed the correlation between ToM and reading comprehension in people with DRH, in general, and in people with cochlear implant. One of these few studies is the one conducted by Roh and Yim (2013), who explored reading comprehension and mind-reading in 16 children with cochlear implant and 16 hearing children of the same age (9–12 years old), with the aim of comparing their performance in both skills and examining the variables that predicted reading comprehension in each group. Both groups provided better answers to questions about literal text comprehension than to those about inferential comprehension, and the group of children with cochlear implant showed lower ToM compared to the group of hearing children. However, ToM was a better predictor of inferential comprehension in the case of hearing children, whereas, in the children with cochlear implant, the best predictor of both types of comprehension (literal and inferential) was expressive vocabulary.

Figueroa et al. (2020) evaluated the relationship between reading comprehension and ToM in 36 adolescents with cochlear implant (received at an average age of 2.6 years), comparing this relationship with a group of 54 hearing adolescents and using a false-belief task, in order to evaluate cognitive ToM and two stories from the Faux Pas Recognition test (Stone et al. 1998) for the analysis of affective ToM. In their results, these authors observed that the hearing adolescents obtained better scores than their peers with DRH in reading competence (in both expository and narrative comprehension) and cognitive ToM, although those with early cochlear implant or binaural audition (two cochlear implants or one implant and a hearing aid device) obtained similar scores as the hearing adolescents. Regarding the correlation between ToM and reading comprehension, a linear regression analysis revealed that in the group with cochlear implant the predictors of reading comprehension were cognitive ToM and syntactic knowledge. On the other hand, in the group of hearing adolescents, only the linguistic factors (lexical and syntactic indices) predicted reading comprehension. The latter result contradicts the finding of Roh and Yim (2013) and that of other studies that have linked cognitive ToM tasks to reading or oral comprehension in hearing children (e.g., Kim 2015a; Kim 2017), although Figueroa et al. (2020) highlighted the influence of linguistic factors in such comprehension in both groups. In addition, we also argue that given that linear regression analysis cannot capture indirect relationships, it could be the case that ToM also contributes to reading comprehension in hearing children via these linguistic factors, as shown by the studies previously reviewed in this chapter. If so, results in Figueroa et al. (2020) still suggest that the association between ToM and reading comprehension is stronger (and above linguistic factors) in adolescents with DRH than in their hearing peers, as it remains significant even after controlling for linguistic factors.

Regarding the adolescents with cochlear implant in Figueroa et al. (2020), the results also showed that bilateral devices and early cochlear implantation could place them in an advantageous position in terms of incidental learning, the early development of oral language, social relations and, as a consequence of all that, reading comprehension and ToM performance. Even with these results, it is necessary to carry out further studies that clarify the relationship between reading comprehension and ToM, not only in the population with cochlear implant, but also in the population with DRH in general, exploring the role of executive functions in this relationship.

For more details on this topic, read Chap. 9 in this same volume.

4.4 Conclusions, Limitations, Research Gaps, and Future Directions

Throughout the present chapter, we have shown that ToM is related to reading comprehension, particularly indirectly through its contribution to listening comprehension. Thus, the recent theoretical approaches to the components of reading comprehension include ToM as one of the underlying factors, that is, as one of the reader's abilities that play a determining role in text comprehension, especially narrative texts (Dore et al. 2018; Duke and Cartwright 2018; Kim 2017). In addition to its relationship with general abilities of inference making and metacomprehension, as is shown by the bidirectional relationships found in previous studies, as well as its role as a mediator of other skills of lower hierarchical level such as early cognitive and linguistic skills (e.g., Kim 2017, 2020a, b), ToM seems to contribute to explaining a unique variance of listening comprehension. Importantly, the association between ToM and reading comprehension appears both in the population with typical development and in the populations with ASD and with DRH, although seems to be stronger in these atypical populations, probably due to an increased heterogeneity in ToM skills. Additionally, although the contribution of ToM to reading comprehension seems to be more evident with respect to narrative texts, some evidence suggests that it could also play a certain role in the comprehension of expository texts.

However, the research on the contribution of ToM to reading comprehension still presents important limitations and other research gaps. Firstly, in addition to the fact that research in this field is still scarce, the correlational nature of the studies does not allow establishing causal relationships between these two skills, especially when the results indicate that language development and ToM present bidirectional associations (e.g., Miller 2006; Milligan et al. 2007). In fact, the directions of the correlations proposed in the studies that used structural equation modelling (e.g., Kim 2017, see Fig. 4.2) or path analyses are based on a priori theoretical decisions. Therefore, given that narrative reading seems to increase people's knowledge of the social world (e.g., Doise et al. 2013; Lysaker et al. 2011; Mar and Oatley 2008), the relationship between reading comprehension and ToM likely is also bidirectional. For instance, in a set of experiments, Kidd and Castano (2013) showed that the reading of narrative texts led to an improvement in the development of ToM even in adult readers. In this respect, it is necessary to carry out further longitudinal studies that analyze the evolution of ToM development and reading comprehension, as well as rigorous intervention studies that explore improvement in ToM and the possible gains in reading comprehension, and vice versa. As was suggested by Dore et al. (2018), ToM improvement should increase the comprehension of narrative texts, which could be established as a line of educational intervention to enhance the performance of students with difficulties in reading comprehension, which would be more effective in early learning stages.

Secondly, it is important to highlight that almost all the reviewed studies used ToM measures based on false-belief tasks or other tasks that require the oral comprehension of stories (e.g., Strange Stories test; Happé 1994). The linguistic load of such tests may influence the relationships found between ToM, other components of reading comprehension and, especially, reading comprehension itself. Future studies that use evaluation tools which do not require receptive language skills (e.g., Frith-Happé animations; Abell et al. 2000) could solve this methodological limitation.

Thirdly, as was pointed out also by Dore et al. (2018) and in the present chapter, the more advanced and sophisticated ToM skills seems to be more strongly related to reading comprehension. Thus, ToM could be more strongly associated to the comprehension of contents specifically related to the perspective of the characters and their mental states. However, most of the studies conducted to date have used tests that evaluate low-level ToM skills, such as false-belief tasks, and global reading comprehension questions. Further research should include measures of advanced ToM skills and evaluate comprehension focusing on the inferences related to the social content of the stories rather than on their global comprehension.

Fourthly, it is important to highlight that, although some findings allow suggesting that ToM is also positively related to expository text comprehension (e.g., Kim and Petscher 2021) and to other components of reading comprehension (e.g., Kuhn 2000), the studies in this regard are also still scarce and methodologically limited in terms of the variety of texts, types of questions, and text topics. Moreover, some studies show that the time spent reading expository texts is negatively related to the social skills (Mar et al. 2006), and other studies demonstrate that people with developmental disorders who usually show ToM deficiencies (e.g., ASD) have less difficulties to understand texts with low social content (Brown et al. 2013). Thus, the contribution of ToM to the understanding of this text genre is still unclear. To fill this gap, we suggest that, in addition to further studies that analyze the contribution of ToM to the comprehension of expository texts, such research must be rooted on models of reading comprehension processes that include the different inferences required for the comprehension of this text genre (e.g., the Landscape model; van den Broek et al. 1993, 1999). In addition, further research is needed to explore to what extent ToM contributes to the understanding of authors' perspective and intentions when the reader faces multiple expository texts on a single topic (Florit et al. 2020), which is especially crucial to comprehend and integrate inter-text conflicting information (Braasch and Bråten 2017).

Lastly, continuing to explore the contribution of ToM to reading comprehension in atypical populations could not only help to clarify the relationship between both abilities, but it would also increase the knowledge of how the variability in ToM in these populations (such as the populations with ASD and with DRH reviewed in this chapter) could explain their usual deficits in reading comprehension and facilitate specific intervention lines. Given that the difficulties in ToM are regarded as characteristic deficits of the developmental disorder in the population with ASD, whereas in the population with DRH they seem to be associated to a linguistic delay, knowing the particularities regarding their relationship with reading comprehension would allow developing a more personalized educational intervention.

Acknowledgments This work was supported by a grant Juan de la Cierva-Formación (grant number: FJC2020-044486-I) awarded to the first author by the Spanish Ministry of Science and Innovation (Spain); and by two grants awarded to the second author by the Junta de Andalucía (Spain), via its I+D+i FEDER Andalucía 2014-2020 Projects (grant number: US-1264792), and by the Spanish Ministry of Science and Innovation (Spain) (grant number: PGC2018-096094-B-I00), respectively.

References

Abell, F., F. Happé, and U. Frith. 2000. Do triangles play tricks? Attribution of mental states to animated shapes in normal and abnormal development. *Cognitive Development* 15: 1–16. https://doi.org/10.1016/S0885-2014(00)00014-9.

American Psychiatric Association. 2013. *Diagnostic and statistical manual of mental disorders*. https://doi.org/10.1176/APPI.BOOKS.9780890425596.

Astington, J.W., and J. Pelletier. 1998. The language of mind: Its role in teaching and learning. In *The handbook of education and human development: New models of learning, teaching and schooling*, ed. D.R. Olson and N. Torrance, 569–593. Blackwell Publishers. https://doi.org/10.1111/b.9780631211860.1998.00027.x.

Atkinson, L., L. Slade, D. Powell, and J.P. Levy. 2017. Theory of mind in emerging reading comprehension: A longitudinal study of early indirect and direct effects. *Journal of Experimental Child Psychology* 164: 225–238. https://doi.org/10.1016/j.jecp.2017.04.007.

Bishop, D.V.M. 1983. Comprehension of English syntax by profoundly deaf children. *Journal of Child Psychology and Psychiatry* 24: 415–434. https://doi.org/10.1111/J.1469-7610.1983.TB00118.X.

Boerma, I.E., S.E. Mol, and J. Jolles. 2017. The role of home literacy environment, Mentalizing, expressive verbal ability, and print exposure in third and fourth graders. *Reading Comprehension 21*: 179–193. https://doi.org/10.1080/10888438.2016.1277727.

Braasch, J.L., and I. Bråten. 2017. The discrepancy-induced source comprehension (D-ISC) model: Basic assumptions and preliminary evidence. *Educational Psychologist* 52: 167–181. https://doi.org/10.1080/00461520.2017.1323219.

Brown, H.M., J. Oram-Cardy, and A. Johnson. 2013. A meta-analysis of the reading comprehension skills of individuals on the autism spectrum. *Journal of Autism and Developmental Disorders* 43: 932–955. https://doi.org/10.1007/s10803-012-1638-1.

Butterfuss, R., J. Kim, and P. Kendeou. 2020. Reading comprehension. *Oxford Research Encyclopedia of Education.* https://doi.org/10.1093/acrefore/9780190264093.013.865.

Cain, K., J.V. Oakhill, M.A. Barnes, and P.E. Bryant. 2001. Comprehension skill, inference making ability, and the relation to knowledge. *Memory and Cognition* 29: 850–859. https://doi.org/10.3758/BF03196414.

Cartwright, K.B. 2015. *Executive skills and reading comprehension: A guide for educators*. Guilford Publications.

Davidson, M.M. 2021. Reading comprehension in school-age children with autism Spectrum disorder: Examining the many components that may contribute. *Language, Speech, and Hearing Services in Schools* 52: 181–196. https://doi.org/10.1044/2020_LSHSS-20-00010.

Davidson, M.M., and S. Ellis Weismer. 2014. Characterization and prediction of early reading abilities in children on the autism Spectrum. *Journal of Autism and Developmental Disorders* 44: 828–845. https://doi.org/10.1007/S10803-013-1936-2.

Devine, R.T., and C. Hughes. 2013. Silent films and strange stories: Theory of mind, gender, and social experiences in middle childhood. *Child Development* 84: 989–1003. https://doi.org/10.1111/CDEV.12017.

———. 2014. Relations between false belief understanding and executive function in early childhood: A meta-analysis. *Child Development* 85: 1777–1794. https://doi.org/10.1111/cdev.12237.

Doise, W., G. Mugny, A.S. James, N. Emler, and D. Mackie. 2013. *The social development of the intellect*. Vol. 10. Elsevier.

Dore, R.A., S.J. Amendum, R.M. Golinkoff, and K. Hirsh-Pasek. 2018. Theory of mind: A hidden factor in reading comprehension? *Educational Psychology Review* 30: 1067–1089. https://doi.org/10.1007/s10648-018-9443-9.

Duke, N.K., and K.B. Cartwright. 2018. The DRIVE model of reading: Deploying reading in varied environments. In *Theoretical models and processes of literacy*, ed. D.E. Alvermann, N.J. Unrau, M. Sailors, and R.B. Ruddell, 118–135. New York: Routledge.

Ebert, S. 2020a. Early language competencies and advanced measures of mental state understanding are differently related to listening and reading comprehension in early adolescence. *Frontiers in Psychology* 11: 952. https://doi.org/10.3389/fpsyg.2020.00952.

———. 2020b. Theory of mind, language, and reading: Developmental relations from early childhood to early adolescence. *Journal of Experimental Child Psychology* 191: 104739. https://doi.org/10.1016/j.jecp.2019.104739.

Edwards, L., M. Marschark, W.G. Kronenberger, K. Crowe, and D. Walton. 2021. Inferencing abilities of deaf college students: Foundations and implications for metaphor comprehension and theory of mind. *Journal of Developmental and Physical Disabilities* 33: 233–258. https://doi.org/10.1007/s10882-020-09746-w.

Elleman, A.M. 2017. Examining the impact of inference instruction on the literal and inferential comprehension of skilled and less skilled readers: A meta-analytic review. *Journal of Educational Psychology* 109: 761–781. https://doi.org/10.1037/edu0000180.

Figueroa, M., S. Darbra, and N. Silvestre. 2020. Reading and theory of mind in adolescents with Cochlear implant. *Journal of Deaf Studies and Deaf Education* 25: 212–223. https://doi.org/10.1093/deafed/enz046.

Flavell, J.H. 1999. Cognitive development: Children's knowledge about the mind. *Annual Review of Psychology* 50: 21–45. https://doi.org/10.1146/annurev.psych.50.1.21.

Flavell, J.H., F.L. Green, and E.R. Flavell. 2000. Development of children's awareness of their own thoughts. *Journal of Cognition and Development* 1: 97–112. https://doi.org/10.1207/S15327647JCD0101N_10.

Fleury, V.P., S. Hedges, K. Hume, D.M. Browder, J.L. Thompson, K. Fallin, F. El Zein, C.K. Reutebuch, and S. Vaughn. 2014. Addressing the academic needs of adolescents with autism Spectrum disorder in secondary education. *Remedial and Special Education* 35: 68–79. https://doi.org/10.1177/0741932513518823.

Florit, E., K. Cain, and L. Mason. 2020. Going beyond children's single-text comprehension: The role of fundamental and higher-level skills in 4th graders' multiple-document comprehension. *British Journal of Educational Psychology* 90: 449–472. https://doi.org/10.1111/bjep.12288.

Frith, U. 2012. The 38th sir Frederick Bartlett lecture why we need cognitive explanations of autism. *Quarterly Journal of Experimental Psychology* 65: 2073–2092. https://doi.org/10.1080/17470218.2012.697178.

Geers, A.E., and H. Hayes. 2011. Reading, writing, and phonological processing skills of adolescents with 10 or more years of cochlear implant experience. *Ear and Hearing* 32 (1 Suppl): 49S–59S. https://doi.org/10.1097/AUD.0B013E3181FA41FA.

Geers, A.E., and J.G. Nicholas. 2013. Enduring advantages of early Cochlear implantation for spoken language development. *Journal of Speech, Language, and Hearing Research* 56: 643–653. https://doi.org/10.1044/1092-4388(2012/11-0347).

Graesser, A.C., M. Singer, and T. Trabasso. 1994. Constructing inferences during narrative text comprehension. *Psychological Review* 101: 371–395. https://doi.org/10.1037/0033-295X.101.3.371.

Guajardo, N.R., and K.B. Cartwright. 2016. The contribution of theory of mind, counterfactual reasoning, and executive function to pre-readers' language comprehension and later reading awareness and comprehension in elementary school. *Journal of Experimental Child Psychology* 144: 27–45. https://doi.org/10.1016/j.jecp.2015.11.004.

Happé, F.G.E. 1993. Communicative competence and theory of mind in autism: A test of relevance theory. *Cognition* 48: 101–119. https://doi.org/10.1016/0010-0277(93)90026-R.

———. 1994. An advanced test of theory of mind: Understanding of story characters' thoughts and feelings by able autistic, mentally handicapped, and normal children and adults. *Journal of Autism and Developmental Disorders* 24: 129–154. https://doi.org/10.1007/BF02172093.

Harris, M. 2015. The impact of new technologies on the literacy attainment of deaf children. *Top Lang Disorders* 35: 120–132. https://doi.org/10.1097/TLD.0000000000000052.

Henning, A., F.M. Spinath, and G. Aschersleben. 2011. The link between preschoolers' executive function and theory of mind and the role of epistemic states. *Journal of Experimental Child Psychology* 108: 513–531. https://doi.org/10.1016/J.JECP.2010.10.006.

Holmer, E., M. Heimann, and M. Rudner. 2016a. Theory of mind and Reading comprehension in deaf and hard-of-hearing signing children. *Frontiers in Psychology* 7: 854. https://doi.org/10.3389/fpsyg.2016.00854.

———. 2016b. Imitation, sign language skill and the developmental ease of language understanding (d-elu) model. *Frontiers in Psychology* 7: 107. https://doi.org/10.3389/FPSYG.2016.00107/BIBTEX.

Hoover, W.A., and P.B. Gough. 1990. The simple view of reading. *Reading and Writing* 2: 127–160. https://doi.org/10.1007/BF00401799.

Jones, C.R.G., F. Happé, H. Golden, A.J.S. Marsden, J. Tregay, E. Simonoff, A. Pickles, G. Baird, and T. Charman. 2009. Reading and arithmetic in adolescents with autism spectrum disorders: Peaks and dips in attainment. *Neuropsychology* 23: 718–728. https://doi.org/10.1037/A0016360.

Kelly, L. 1996. The interaction of syntactic competence and vocabulary during reading by deaf students. *Journal of Deaf Studies and Deaf Education* 1 (1): 75–90. https://doi.org/10.1093/OXFORDJOURNALS.DEAFED.A014283.

Kendeou, P., K.L. McMaster, and T.J. Christ. 2016. Reading comprehension: Core components and processes. *Policy Insights From the Behavioral and Brain Sciences* 3: 62–69. https://doi.org/10.1177/2372732215624707.

Ketelaar, L., C. Rieffe, C.H. Wiefferink, and J.H.M. Frijns. 2012. Does hearing lead to understanding? Theory of mind in toddlers and preschoolers with cochlear implants. *Journal of Pediatric Psychology* 37: 1041–1050. https://doi.org/10.1093/JPEPSY/JSS086.

Kidd, D.C., and E. Castano. 2013. Reading literary fiction improves theory of mind. *Science* 342 (6156): 377–380. https://doi.org/10.1126/science.1239918.

Kim, Y.S.G. 2015a. Developmental, component-based model of reading fluency: An investigation of predictors of word-reading fluency, text-reading fluency, and reading comprehension. *Reading Research Quarterly* 50: 459–481. https://doi.org/10.1002/rrq.107.

———. 2015b. Language and cognitive predictors of text comprehension: Evidence from multivariate analysis. *Child Development* 86: 128–144. https://doi.org/10.1111/cdev.12293.

———. 2016. Direct and mediated effects of language and cognitive skills on comprehension of oral narrative texts (listening comprehension) for children. *Journal of Experimental Child Psychology* 141: 101–120. https://doi.org/10.1016/j.jecp.2015.08.003.
———. 2017. Why the simple view of Reading is not simplistic: Unpacking component skills of Reading using a Direct and Indirect Effect Model of Reading (DIER). *Scientific Studies of Reading* 21: 310–333. https://doi.org/10.1080/10888438.2017.1291643.
———. 2020a. Hierarchical and dynamic relations of language and cognitive skills to reading comprehension: Testing the direct and indirect effects model of reading (DIER). *Journal of Educational Psychology* 112: 667–684. https://doi.org/10.1037/edu0000407.
———. 2020b. Theory of mind mediates the relations of language and domain-general cognitions to discourse comprehension. *Journal of Experimental Child Psychology* 194: 104813.
Kim, Y.S.G., and Y. Petscher. 2021. Influences of individual, text, and assessment factors on text/discourse comprehension in oral language (listening comprehension). *Annals of Dyslexia* 71: 218–237. https://doi.org/10.1007/s11881-020-00208-8.
Kim, Y.S., and B. Phillips. 2014. Cognitive correlates of listening comprehension. *Reading Research Quarterly* 49: 269–281. https://doi.org/10.1002/rrq.74.
Kintsch, W. 1988. The role of knowledge in discourse comprehension: A construction-integration model. *Psychological Review* 95: 163–182. https://doi.org/10.1037/0033-295X.95.2.163.
———. 1998. *Comprehension. A paradigm for cognition*. Cambridge, MA: Cambridge University Press.
Kuhn, D. 2000. Metacognitive development. *Current Directions in Psychological Science* 9 (5): 178–181. https://doi.org/10.1111/1467-8721.00088.
Kyle, F.E., and K. Cain. 2015. A comparison of deaf and hearing Children's Reading comprehension profiles. *Topics in Language Disorders* 35: 144–156. https://doi.org/10.1097/TLD.0000000000000053.
Kyle, F.E., and M. Harris. 2010. Predictors of reading development in deaf children: A 3-year longitudinal study. *Journal of Experimental Child Psychology* 107: 229–243. https://doi.org/10.1016/J.JECP.2010.04.011.
Lane, J.D., H.M. Wellman, and E.M. Evans. 2012. Sociocultural input facilitates children's developing understanding of extraordinary minds. *Child Development* 83: 1007–1021. https://doi.org/10.1111/j.1467-8624.2012.01741.x.
Lederberg, A.R., B. Schick, and P.E. Spencer. 2013. Language and literacy development of deaf and hard-of-hearing children: Successes and challenges. *Developmental Psychology* 49: 15–30. https://doi.org/10.1037/A0029558.
Lee, H.K., W.S. Chan, and S.X. Tong. 2022. The heterogeneity and interrelationships among theory of mind, executive function, and reading comprehension deficits in Hong Kong Chinese children with autism. *Reading and Writing.* 36 (6): 1359–1368. https://doi.org/10.1007/s11145-022-10298-y.
Lillo-Martin, D.C., V.L. Hanson, and S.T. Smith. 1992. Deaf readers' comprehension of relative clause structures. *Applied PsychoLinguistics* 13: 13–30. https://doi.org/10.1017/S0142716400005403.
Liu, M., L. Wu, W. Wu, G. Li, T. Cai, and J. Liu. 2018. The relationships among verbal ability, executive function, and theory of mind in young children with cochlear implants. *International Journal of Audiology 57*: 875–882. https://doi.org/10.1080/14992027.2018.1498982.
Lockl, K., S. Ebert, and S. Weinert. 2017. Predicting school achievement from early theory of mind: Differential effects on achievement tests and teacher ratings. *Learning and Individual Differences* 53: 93–102. https://doi.org/10.1016/j.lindif.2016.11.007.
Long, D.L., and R.B. Lea. 2005. Have we been searching for meaning in all the wrong places: Defining the "search after meaning" principle in comprehension. *Discourse Processes* 39: 279–298. https://doi.org/10.1080/0163853X.2005.9651684.
Lonigan, C.J., S.R. Burgess, and C. Schatschneider. 2018. Examining the simple view of reading with elementary school children: Still simple after all these years. *Remedial and Special Education* 39: 260–273. https://doi.org/10.1177/0741932518764833.

Lysaker, J.T., C. Tonge, D. Gauson, and A. Miller. 2011. Reading and social imagination: What relationally oriented reading instruction can do for children. *Reading Psychology* 32: 520–566. https://doi.org/10.1080/02702711.2010.507589.

Ma, W.J., M. Husain, and P.M. Bays. 2014. Changing concepts of working memory. *Nature Neuroscience* 17: 347–356. https://doi.org/10.1038/NN.3655.

Mancini, P., I. Giallini, L. Prosperini, H.D. D'alessandro, L. Guerzoni, A. Murri, D. Cuda, G. Ruoppolo, M. De Vincentiis, and M. Nicastri. 2016. Level of emotion comprehension in children with mid to long term cochlear implant use: How basic and more complex emotion recognition relates to language and age at implantation. *International Journal of Pediatric Otorhinolaryngology 87*: 219–232. https://doi.org/10.1016/J.IJPORL.2016.06.033.

Mar, R.A., and K. Oatley. 2008. The function of fiction is the abstraction and simulation of social experience. *Perspectives on Psychological Science* 3: 173–192. https://doi.org/10.1111/j.1745-6924.2008.00073.x.

Mar, R.A., K. Oatley, J. Hirsh, J. Dela Paz, and J.B. Peterson. 2006. Bookworms versus nerds: Exposure to fiction versus non-fiction, divergent associations with social ability, and the simulation of fictional social worlds. *Journal of Research in Personality* 40: 694–712. https://doi.org/10.1016/j.jrp.2005.08.002.

Marschark, M., L. Edwards, C. Peterson, K. Crowe, and D. Walton. 2019. Understanding theory of mind in deaf and hearing college students. *The Journal of Deaf Studies and Deaf Education* 24: 104–118. https://doi.org/10.1093/DEAFED/ENY039.

McIntyre, N.S., E.J. Solari, R.P. Grimm, L.E. Lerro, J.E. Gonzales, and P.C. Mundy. 2017. A comprehensive examination of reading heterogeneity in students with high functioning autism: Distinct reading profiles and their relation to autism symptom severity. *Journal of Autism and Developmental Disorders* 47: 1086–1101. https://doi.org/10.1007/S10803-017-3029-0.

McIntyre, N.S., T.M. Oswald, E.J. Solari, M.C. Zajic, L.E. Lerro, C. Hughes, R.T. Devine, and P.C. Mundy. 2018. Social cognition and Reading comprehension in children and adolescents with autism spectrum disorders or typical development. *Research in Autism Spectrum Disorders* 54: 9–20. https://doi.org/10.1016/j.rasd.2018.06.004.

McNamara, D.S., and J. Magliano. 2009. Toward a comprehensive model of comprehension. In *The psychology of learning and motivation, Vol. 51*, ed. B.H. Ross, 297–384. Burlington: Elsevier Science. https://doi.org/10.1016/S0079-7421(09)51009-2.

Meristo, M., and E. Hjelmquist. 2009. Executive functions and theory-of-mind among deaf children: Different routes to understanding other minds? *Journal of Cognition and Development* 10: 67–91. https://doi.org/10.1080/15248370902966552.

Meristo, M., G. Morgan, A. Geraci, L. Iozzi, E. Hjelmquist, L. Surian, and M. Siegal. 2012. Belief attribution in deaf and hearing infants. *Developmental Science* 15: 633–640. https://doi.org/10.1111/J.1467-7687.2012.01155.X.

Miller, C.A. 2006. Developmental relationships between language and theory of mind. *American Journal of Speech-Language Pathology* 15: 142–154. https://doi.org/10.1044/1058-0360(2006/014).

Milligan, K., J.W. Astington, and L.A. Dack. 2007. Language and theory of mind: Meta-analysis of the relation between language ability and false-belief understanding. *Child Development* 78: 622–646. https://doi.org/10.1111/j.1467-8624.2007.01018.x.

Mo, S., Y. Su, R.C.K. Chan, and J. Liu. 2008. Comprehension of metaphor and irony in schizophrenia during remission: The role of theory of mind and IQ. *Psychiatry Research* 157: 21–29. https://doi.org/10.1016/J.PSYCHRES.2006.04.002.

Moreno-Pérez, F.J., D. Saldaña, and I.R. Rodríguez-Ortiz. 2015. Reading efficiency of deaf and hearing people in Spanish. *Journal of Deaf Studies and Deaf Education* 20: 374–384. https://doi.org/10.1093/deafed/env030.

Musselman, C. 2000. How do children who can't hear learn to read an alphabetic script? A review of the literature on reading and deafness. *The Journal of Deaf Studies and Deaf Education* 5: 9–31. https://doi.org/10.1093/DEAFED/5.1.9.

Nation, K. 1999. Reading skills in hyperlexia: A developmental perspective. *Psychological Bulletin* 125: 338–355. https://doi.org/10.1037/0033-2909.125.3.338.

Norbury, C.F. 2005. The relationship between theory of mind and metaphor: Evidence from children with language impairment and autistic spectrum disorder. *British Journal of Developmental Psychology* 23: 383–399. https://doi.org/10.1348/026151005X26732.

O'Reilly, K., C.C. Peterson, and H.M. Wellman. 2014. Sarcasm and advanced theory of mind understanding in children and adults with prelingual deafness. *Developmental Psychology* 50: 1862–1877. https://doi.org/10.1037/A0036654.

Oakhill, J.V., and K. Cain. 2012. The precursors of reading ability in young readers: Evidence from a four-year longitudinal study. *Scientific Studies of Reading* 16: 91–121. https://doi.org/10.1080/10888438.2010.529219.

O'Brien, E.J., A.E. Cook, and R.F. Lorch Jr. 2015. *Inferences during reading*. Cambridge University Press.

Orlando, A.-M., and B.B. Shulman. 1989. Severe-to-profound hearing-impaired children's comprehension of figurative language. *Journal of Childhood Communication Disorders* 12: 157–165.

Paul, P.V. 1996. Reading vocabulary knowledge and deafness. *Journal of Deaf Studies and Deaf Education* 1: 3–15. https://doi.org/10.1093/OXFORDJOURNALS.DEAFED.A014279.

Pelletier, J., and J.W. Astington. 2004. Action, consciousness and theory of mind: Children's ability to coordinate story characters' actions and thoughts. *Early Education and Development* 15: 5–22. https://doi.org/10.1207/s15566935eed1501_1.

Pelletier, J., and R. Beatty. 2015. Children's understanding of Aesop's fables: Relations to reading comprehension and theory of mind. *Frontiers in Psychology* 6: 1448. https://doi.org/10.3389/fpsyg.2015.01448.

Perfetti, C.A., and J.Z. Stafura. 2015. Comprehending implicit meanings in text without making inferences. In *Inferences during reading*, ed. E.J. O'Brien, A.E. Cook, and R.F. Lorch Jr., 1–18. Cambridge, MA: Cambridge University Press.

Peterson, C.C. 2016. Empathy and theory of mind in deaf and hearing children. *The Journal of Deaf Studies and Deaf Education* 21: 141–147. https://doi.org/10.1093/DEAFED/ENV058.

———. 2020. Theory of mind and conversation in deaf and hearing children. In *The Oxford handbook of deaf studies in learning and cognition*, ed. M. Marschark and H. Knoors, 212–231. Oxford University Press. https://doi.org/10.1093/OXFORDHB/9780190054045.013.13.

Peterson, C.C., and C.C. Peterson. 2009. Development of social-cognitive and communication skills in children born deaf. *Scandinavian Journal of Psychology* 50: 475–483. https://doi.org/10.1111/J.1467-9450.2009.00750.X.

Peterson, C.C., H.M. Wellman, and D. Liu. 2005. Steps in theory-of-mind development for children with deafness or autism. *Child Development* 76: 502–517. https://doi.org/10.1111/J.1467-8624.2005.00859.X.

Peterson, C.C., H.M. Wellman, and V. Slaughter. 2012. The mind behind the message: Advancing theory-of-mind scales for typically developing children, and those with deafness, autism, or Asperger syndrome. *Child Development* 83: 469–485. https://doi.org/10.1111/J.1467-8624.2011.01728.X.

Peterson, C.C., K. O'Reilly, and H.M. Wellman. 2016. Deaf and hearing children's development of theory of mind, peer popularity, and leadership during middle childhood. *Journal of Experimental Child Psychology* 149: 146–158. https://doi.org/10.1016/J.JECP.2015.11.008.

Potocki, A., M. Sanchez, J. Ecalle, and A. Magnan. 2017. Linguistic and cognitive profiles of 8-to 15-year-old children with specific reading comprehension difficulties: The role of executive functions. *Journal of Learning Disabilities* 50: 128–142. https://doi.org/10.1177/0022219415613080.

Ricketts, J., C.R.G. Jones, F. Happe, and T. Charman. 2013. Reading comprehension in autism Spectrum disorders: The role of Oral language and social functioning. *Journal of Autism and Developmental Disorders* 43: 807–816. https://doi.org/10.1007/s10803-012-1619-4.

Roh, J., and D. Yim. 2013. Relationships between reading comprehension and mind-reading in children with Cochlear implants from fourth through sixth grades. *Communication Sciences and Disorders-CSD* 18: 183–193. https://doi.org/10.12963/csd.13018.

Ronfard, S., and P.L. Harris. 2014. When will little red riding Hood become scared? Children's attribution of mental states to a story character. *Developmental Psychology* 50: 283–292. https://doi.org/10.1037/a0032970.

Rouet, J.F., M.A. Britt, and A.M. Durik. 2017. RESOLV: Readers' representation of reading contexts and tasks. *Educational Psychologist* 52: 200–215. https://doi.org/10.1080/00461520.2017.1329015.

Russell, R.L., and K.L. Grizzle. 2008. Assessing child and adolescent pragmatic language competencies: Toward evidence-based assessments. *Clinical Child and Family Psychology Review*, 11: 59–73. https://doi.org/10.1007/s10567-008-0032-1.

Shanahan, T., and S. Shanahan. 1997. Character perspective charting: Helping children to develop a more complete conception of story. *The Reading Teacher 50*: 668–677. https://www.jstor.org/stable/20201844.

Singer, M., A.C. Graesser, and T. Trabasso. 1994. Minimal or global inference during reading. *Journal of Memory and Language* 33: 421–441. https://doi.org/10.1006/jmla.1994.1020.

Snow, C.E. 2002. *Reading for understanding: Toward a research and development program in reading comprehension.* RAND.

Snowling, M., and U. Frith. 1986. Comprehension in "hyperlexic" readers. *Journal of Experimental Child Psychology* 42: 392–415. https://doi.org/10.1016/0022-0965(86)90033-0.

Solari, E.J., R.P. Grimm, N.S. McIntyre, M. Zajic, and P.C. Mundy. 2019. Longitudinal stability of reading profiles in individuals with higher functioning autism. *Autism* 23: 1911–1926. https://doi.org/10.1177/1362361318812423.

Stanzione, C., and B. Schick. 2014. Environmental language factors in theory of mind development: Evidence from children who are deaf/hard-of-hearing or who have specific language impairment. *Topics in Language Disorders* 34: 296–312. https://doi.org/10.1097/TLD.0000000000000038.

Stein, N.L., and T. Trabasso. 1985. The search after meaning: Comprehension and comprehension monitoring. In *Applied developmental psychology, Vol. 2*, ed. F.J. Morrison, C. Lord, and D. Keating, 33–58. San Diego: Academic.

Stone, V.E., S. Baron-Cohen, and R.T. Knight. 1998. Frontal lobe contributions to theory of mind. *Journal of Cognitive Neuroscience* 10: 640–656. https://doi.org/10.1162/089892998562942.

Strasser, K., and F.D. Río. 2014. The role of comprehension monitoring, theory of mind, and vocabulary depth in predicting story comprehension and recall of kindergarten children. *Reading Research Quarterly* 49: 169–187. https://doi.org/10.1002/rrq.68.

Strasser, K., A. Larraín, and M.R. Lissi. 2013. Effects of storybook reading style on comprehension: The role of word elaboration and coherence questions. *Early Education and Development* 24: 616–639. https://doi.org/10.1080/10409289.2012.715570.

Sundqvist, A., and M. Heimann. 2014. The development of theory of mind – Considerations for deaf children with cochlear implants. *Otorinolaringologia* 64 (4): 179–189. https://www.minervamedica.it/en/journals/Otorhinolaryngology/article.php?cod=R27Y2014N04A0179.

Sundqvist, A., B. Lyxell, R. Jönsson, and M. Heimann. 2014. Understanding minds: Early cochlear implantation and the development of theory of mind in children with profound hearing impairment. *International Journal of Pediatric Otorhinolaryngology* 78: 538–544. https://doi.org/10.1016/J.IJPORL.2013.12.039.

Tong, S.X., R.W.Y. Wong, J.L.Y. Kwan, and J. Arciuli. 2020. Theory of mind as a mediator of Reading comprehension differences between Chinese school-age children with autism and typically developing peers. *Scientific Studies of Reading* 24: 292–306. https://doi.org/10.1080/10888438.2019.1666133.

Torres, J., and I.R. Rodríguez-Ortiz. 2011. La comprensión de falsa creencia en niños y adolescentes sordos: Tareas gráficas versus clásicas. *Infancia y Aprendizaje* 34: 31–47. https://doi.org/10.1174/021037011794390157.

Tucci, S.L., and S.R. Easterbrooks. 2020. Theory of mind: Implications for cognition and learning in deaf and hard-of-hearing learners. In *The Oxford handbook of deaf studies in learning and cognition*, ed. M. Marschark and H. Knoors, 232–247. New York: Oxford University Press. https://doi.org/10.1093/OXFORDHB/9780190054045.013.24.

van den Broek, P., and C.A. Espin. 2012. Connecting cognitive theory and assessment: Measuring individual differences in reading comprehension. *School Psychology Review* 41: 315–325. https://doi.org/10.1080/02796015.2012.12087512.

van den Broek, P., C.R. Fletcher, and K. Risden. 1993. Investigations of inferential processes in reading: A theoretical and methodological integration. *Discourse Processes* 16: 169–189. https://doi.org/10.1080/01638539309544835.

van den Broek, P., M. Young, Y. Tzeng, and T. Linderholm. 1999. The landscape model of reading: Inferences and the online construction of a memory representation. In *The construction of mental representations during reading*, ed. H. van Oostendorp and S.R. Goldman, 71–98. Mahwah: Erlbaum.

van den Broek, P., A. Helder, and L. Van Leijenhorst. 2013. Sensitivity to structural centrality: Developmental and individual differences in reading comprehension skills. In *Reading: From words to multiple texts*, ed. M.A. Britt, S.R. Goldman, and J.F. Rouet, 132–146. New York: Routledge.

Van Dijk, T.A., W. Kintsch, and T.A. Van Dijk. 1983. *Strategies of discourse comprehension.* Academic.

Wang, Y., Z. Lan, I. Duan, P. Peng, W. Wang, and T. Wang. 2022. A meta-analysis on the cognitive and linguistic correlates of reading skills among children with ASD. *Reading and Writing* 36 (4): 1487–1514. https://doi.org/10.1007/s11145-022-10338-7.

Wellman, H.M. 2002. Understanding the psychological world: Developing a theory of mind. In *Blackwell handbook of childhood cognitive development*, ed. U. Goswami, 167–187. Malden: Blackwell. https://doi.org/10.1002/9780470996652.ch8.

Wellman, H.M., and D. Liu. 2004. Scaling of theory-of-mind tasks. *Child Development* 75: 523–541. https://doi.org/10.1111/J.1467-8624.2004.00691.X.

Wharton, C., and W. Kintsch. 1991. An overview of construction-integration model: A theory of comprehension as a foundation for a new cognitive architecture. *ACM SIGART Bulletin* 2 (4): 169–173. https://doi.org/10.1145/122344.122379.

White, S., E. Hill, F. Happé, and U. Frith. 2009. Revisiting the strange stories: Revealing Mentalizing impairments in autism. *Child Development* 80: 1097–1117. https://doi.org/10.1111/J.1467-8624.2009.01319.X.

Wiefferink, C.H., C. Rieffe, L. Ketelaar, L. De Raeve, and J.H.M. Frijns. 2013. Emotion understanding in deaf children with a Cochlear implant. *The Journal of Deaf Studies and Deaf Education* 18: 175–186. https://doi.org/10.1093/DEAFED/ENS042.

Williamson, P., C.R. Carnahan, and J.A. Jacobs. 2012. Reading comprehension profiles of high-functioning students on the autism spectrum: A grounded theory. *Exceptional Children* 78: 449–469.

Wu, Z., and Y. Su. 2014. How do preschoolers' sharing behaviors relate to their theory of mind understanding? *Journal of Experimental Child Psychology* 120: 73–86. https://doi.org/10.1016/J.JECP.2013.11.007.

Yang, C., W. Zhao, B. Yuan, L. Luo, and D.R. Shanks. 2022. Mind the gap between comprehension and metacomprehension: Meta-analysis of metacomprehension accuracy and intervention effectiveness. *Review of Educational Research* 93 (2): 003465432210940. https://doi.org/10.3102/00346543221094083.

Chapter 5
Theory of Mind and Psychopathology: A Comprehensive Assessment and an Overview of Impairments in Neuropsychiatric Disorders

Pilar de la Higuera-González, Alejandra Galvez-Merlin, Elisa Rodríguez-Toscano, Jorge Andreo-Jover, and Alejandro de la Torre-Luque

Abstract Theory of Mind (ToM) is a cognitive construct that involves different levels of complexity. Development of the different ToM skills follows an evolutionary course in line with the increasing requirements from environmental demands through childhood and adolescence to adulthood. Accordingly, ToM assessment should comprise the heterogeneity of the components – affective or cognitive – and the rising complexity of tasks according to the developmental course – first-order, second-order and higher-order beliefs. In this chapter, features of ToM assessment are presented with a compilation of the main useful tools. Furthermore, specific ToM deficits in neuropsychiatric disorders are collected. ToM impairments have been extensively studied in autism spectrum disorders and schizophrenia, but in recent years their study has been extended to many other clinical diagnoses and significant

P. de la Higuera-González · A. Galvez-Merlin (✉)
Health Research Instituto, Hospital Clínico San Carlos (IdISSC), Madrid, Spain

Universidad Complutense de Madrid, Madrid, Spain
e-mail: alegalve@ucm.es

E. Rodríguez-Toscano
Universidad Complutense de Madrid, Madrid, Spain

Department of Child and Adolescent Psychiatry, Institute of Psychiatry and Mental Health, Hospital General Universitario Gregorio Marañón, IiSGM, School of Medicine, Universidad Complutense de Madrid, Madrid, Spain

J. Andreo-Jover
Universidad Autónoma de Madrid, Madrid, Spain

Instituto de Investigación del Hospital Universitario La Paz (IdiPAZ), Madrid, Spain

A. de la Torre-Luque
Universidad Complutense de Madrid, Madrid, Spain

Biomedical Research Networking Consortium for Mental Health (CIBERSAM ISCII), Madrid, Spain

T. Lopez-Soto et al. (eds.), *The Theory of Mind Under Scrutiny*, Logic, Argumentation & Reasoning 34, https://doi.org/10.1007/978-3-031-46742-4_5

transversal problems in mental health such as suicidal behavior. As a result, this chapter outlines a comprehensive picture of ToM impairments in a broad range of neuropsychiatric disorders and the accurate assessment of the different deficits, to provide the reader an essential knowledge for understanding ToM impairments across neuropsychiatric disorders and assessment protocols in clinical practice.

Keywords Theory of mind · Cognitive theory of mind · Affective theory of mind · Assessment · Psychopathology · Neuropsychiatric disorder suicidal behavior

5.1 Theory of Mind: Conceptual Background

Theory of mind (ToM) refers to the capacity to understand other people by ascribing mental states to them, inferring what is happening in their minds.

The first mention of ToM was done by Premack and Woodruff in 1978, in their study about ToM in chimpanzees. These authors established that having a Theory of the Mind implies the capacity of imputing mental states to oneself and others. It implies the assumption that others have their own beliefs, emotions, knowledge and purposes. On the one hand, a feature of ToM is that the attribution of these mental states is based on non-directly observable inferences. On the other hand, these inferences can be used to predict the behaviors of others. Consequently, inferring mental states makes up a theory, a Theory of the Mind (Premack and Woodruff 1978), which is an essential capacity to manage oneself in a social environment (Green et al. 2015).

The study of ToM in humans began with the aim for detecting the "first manifestations" of ToM in children. Within these first manifestations, it is possible to distinguish two types of mental representations: first-order and second-order beliefs. First-order false belief refers to the ability to understand that a person's mental representation (thought) is false in relation to the real situation (Wimmer and Perner 1983). On the other hand, second-order beliefs refer to the ability to understand the mental representations of others (e.g., "Mary thinks her mother thinks. . .") and it involves higher levels of information processing. This ability also requires considering what people think of other people's thoughts, which is essential in social interactions and needs more complex tasks for its assessment (Perner and Wimmer 1985). The increasing complexity of the mental representations considered in each category follows a maturation trajectory: children with normal development acquire the ability to detect first-order false beliefs at the age of 4–5 years and second-order false beliefs at the age of 6–7 years (Wimmer and Perner 1983; Perner and Wimmer 1985). Subsequently, the study of ToM was extended from childhood to adult population with increasingly complex tasks involving the understanding of social situations.

Traditionally, ToM has been considered a unitary construct, and studies for its assessment were focused on the related cognitive skills of mentalizing: thoughts, intentions and beliefs of others. However, advances in neuroscience show a neuroanatomical differentiation of different components of ToM. It may allow for a ToM differentiation regarding two components: affective and cognitive ToM (Shamay-Tsoory and Aharon-Peretz 2007). On the one hand, cognitive ToM refers to the ability to infer thoughts, beliefs, and intentions in others, as well as the ability to reflect on one's own thoughts, beliefs, and intentions. On the other hand, affective ToM involves thinking about the emotions of others and from oneself. Shamay-Tsoory and Aharon-Peretz (2007) pointed out that although cognitive and affective mentalizing abilities are interrelated, they are dissociable and depend on distinct, yet overlapping, neuroanatomical substrates. In their study, they assessed performance on mentalization tasks in patients with acquired brain damage. The results showed that patients with damage to the ventromedial prefrontal cortex (VM) showed significantly lower scores on the affective ToM, while patients with extensive damage to the prefrontal cortex (PF) and those with damage to the VM extending to the temporal poles showed significantly lower scores on the cognitive ToM. In short, both cognitive and affective ToM rely on an intact functioning of the prefrontal cortex. However, VM would play a fundamental role in affective ToM, whereas PF is more related to cognitive ToM.

In conclusion, the complexity of ToM implies the creation of different and very heterogeneous tasks for its proper assessment. In the present chapter, assessment tools and protocols to measure ToM when processing cognitive information (i.e., objects' state or people's thoughts) will be first presented, followed by the instruments to evaluate ToM ability when it refers to processing affective information (i.e., emotions or feelings). This part will be followed by the description of the main neuropsychiatric disorders and relations between their symptomatology and ToM deficits. Strong evidence comes from populations with an autism spectrum disorder (ASD) and schizophrenia (SCZ). However, the interest in studying ToM profile in other neuropsychiatric disorders, has been raised in last years, amplifying the disorders studied to other disorders like bipolar disorder (BD), major depression disorder (MDD), social anxiety disorder (SAD), eating disorders (ED) a borderline personality disorder (BPD), and, also in Suicide Behavior Disorder (SBD), a possible transdiagnostic between disorders.

5.2 Assessment of Theory of Mind

5.2.1 Assessment of Theory of Mind Early Manifestations

5.2.1.1 Assessment of First-Order Beliefs

Starting with classic studies on ToM assessment, we should mention the *Test of Understanding False Beliefs* or *Maxi Task Test*. It was developed by Wimmer and Perner (1983) to assess the presence of early manifestations of ToM in children: first-order false beliefs. Subsequently, this task was versioned by Baron-Cohen et al. (1985) resulting in the famous *Sally-Anne Test* which is the most frequently replicated tool for the first-order false beliefs' assessment in children. Both the *Maxi Task* and the *Sally-Anne Test* consist of sequences of vignettes or representations in which two characters appear. Both tests include tasks in which one of the characters keeps an object in one place and leaves the scene. Subsequently, the second character picks up the object and changes its location. Once the first character comes back to the scene, the participant is asked to indicate where the first character will be looking for the object. Figure 5.1, below, shows the representations of the *Sally-Anne Test.*

In the first scene, the characters are introduced: Sally (on the left) and Anne (on the right). In the second scene, Sally puts a marble in a basket. In the third scene, Sally leaves the room. In the fourth scene, Anne takes the marble out of the basket and moves it into a box. In the fifth and final scene, Sally comes back, and the participants being assessed are asked: "*Where will Sally look for her marble*?" A person with normative ToM skills would answer that Sally will look for the marble in the basket, whereas a person with a deficit in mentalizing would answer that Sally would look for the marble into the box. The explanation of these different responses relies on the different ToM skills: a person with normative ToM skills understands that Sally cannot know that Anne changed the place where she kept the marble

Fig. 5.1 Representations of the Sally-Anne Test. (Reproduced from Frith 1989, 2002. Figure copyright of the artist Axel Scheffler)

because she was not present to see it. On the contrary, a person with ToM impairments may not be able to make the proper inference considering the situation exposed: people with ToM impairments may not understand that others have their own mental states and that they may be wrong, different from reality and differ from their own. Baron-Cohen et al. (1985) used the *Sally-Anne Test* to show that difficulties in understanding the mental state of the other are present in the ASD, and that this deficit is independent of the intelligence quotient (IQ).

5.2.1.2 Assessment of Second-Order Beliefs

To assess second-order beliefs, Perner and Wimmer (1985) designed the *Ice Cream Story*. It consists of a story in which a character holds a false belief because of their lack of awareness of changes in another person's beliefs. In contrast to the tests that measure first-order ToM, in this case, the false belief is not about the state of the real situation, but it is related to the mental state of one of the characters in the story: *"What does he think she thinks?"* (See Perner and Wimmer (1985) for a detailed description of the task). As this test measures more complex mental functions involving other cognitive abilities (e.g., working memory), it has control questions to ensure that the person under assessment understands the story.

According to Perner and Wimmer (1985), second-order epistemic intentions may be crucial for the understanding of ambiguous pragmatic situations, such as those that occur in non-literal senses situations (joking, lying, irony, misunderstanding, etc.). Therefore, this expanded the study of ToM to a new line of research into more advanced aspects of ToM. Second-order false beliefs are essential to understanding deception. When a situation of deception occurs, information is manipulated with the goal of creating a false belief about reality in another person. Due to that, understanding deception is crucial to understand social situations and to be able to manage them in a useful way to oneself.

5.2.2 Assessment of Higher-Order beliefs: Advanced Tasks for Theory of Mind

There was a growing interest in the study of mentalization in clinical settings after the previous mentioned pioneering lines of research. New assessment tools were developed to study mentalization in adolescent and adult populations thereafter. These new assessment instruments were characterized by the analysis of social situations in which misunderstandings take place. Owing to that, considering social variables was required. As a common feature, the instruments were developed since ambiguous social interactions in diverse situations.

These instruments aimed at overcoming one of the main criticisms of the pioneering ToM instruments: their poor ecological validity. The lack of generalization was due to the high artificiality of most of their items because they described scenes too distant from everyday life situations. To overcome these limitations, the *Strange Stories Test* was developed by Happé in 1994: a figurative-language understanding test, whose items were designed to cover real-life situations. It evaluates skills on mental state reasoning and emotion recognition across the lifespan. The instrument comprises 30 items: 24 stories involving cognitive mentalizing skills (mentalistic stories) and six control "physical stories". Mentalistic stories include situations involving lying, white lies, jokes, fiction, misunderstanding, persuasion, appearance/reality, metaphor, sarcasm, forgetfulness, double bluffing[1] and contrary emotions. The control physical stories did not involve neither mental state ascertainment nor social interaction scenes. However, its resolution implies extracting proper global inferences. Table 5.1 presents two examples of mentalistic story items and two examples of physical stories from the *Strange Stories Test* (Happé 1994).

Happé (1994) used the instrument to assess ToM skills in adults and adolescents with ASD, finding that these individuals tended to make significantly more errors in attributing mental states, regardless of their IQ. Afterwards, Jolliffe and Baron-Cohen (1999) replicated the study differentiating two groups of people with high functioning ASD: those with an Asperger syndrome,[2] and those with another disorder. The study revealed that both groups performed worse than controls on the mentalistic story items, but not on the physical story ones. However, it is noteworthy mentioning that, although both clinical groups had difficulties on the mentalistic stories, the deficit was weaker in the Asperger syndrome group.

Even the *Strange Stories Test* (Happé 1994) is a robust well-established test to assess ToM. Its writing format just allows a visual evaluation remote to realistic social stimulus. To overcome this issue, a test which assessed mental state attribution ToM and emotional processing in a more naturalistic way was developed by Murray et al. (2017): the *Strange Stories Film Task*. It is based on the original *Strange Stories Test* (Happé 1994). It includes filmed scenes with similar situations than the original test to evaluate the ability of attributing the speaker's intention: lie, irony, double bluff, pretense, joke, appearance/reality, white-lie, persuasion, misunderstanding, forgetting, contrary emotions and idioms. The test also includes control questions which require logical reasoning instead of attributing mental states. After each clip three questions are done: intention ("*Why did X say that?*"), interaction (assessing the ability to generate a response to the inferred mental state, e.g.: "*If you were in Y's* [other character i.e., not X] *situation, what would you say next?*") and

---

[1] Double bluffing refers to trying to deceive a person by telling the truth, when the other person believes you are lying.

[2] Since DSM-5 -Diagnostic and Statistical Manual of Mental Disorders Fifth Edition Revised- (APA 2013) autism spectrum disorders (ASD) are a unified category which encomprises previous diagnoses as autism, Asperger syndrome and Global Developmental Disorders into Pervasive Developmental Disorders (APA 2000). Currently DSM-5-TR (APA 2022) establishes that ASD is a unique category which reflects the dimensionality of the deficits and the difficulties (see Sec. 5.3.1. for a detailed description of the ASD diagnosis).

Table 5.1 Examples of mentalistic stories ("Banana" and "Picnic") and examples of physical stories ("Army" and "Glasses") of the Strange Stories Test

Examples of mentalistic stories
Banana
Katie and Emma are playing in the house. Emma picks up a banana from the fruit bowl and holds it up to her ear. She says to Katie, "Look! This banana is a telephone!" Is it true what Emma says? Question: Why does Emma say this?
Picnic
Sarah and Tom are going on a picnic. It is Tom's idea; he says it is going to be a lovely sunny day for a picnic. But just as they are unpacking the food, it starts to rain, and soon they are both soaked to the skin. Sarah is cross. She says, "Oh yes, a lovely day for a picnic alright!" Is it true, what Sarah says? Question: Why does she say this?
Examples of physical stories
Army
Two enemy powers have been at war for a very long time. Each army has won several battles, but now the outcome could go either way. The forces are equally matched. However, the Blue army is stronger than the Yellow army in foot soldiers and artillery. But the Yellow army is stronger than the Blue army in air power. On the day of the final battle, which will decide the outcome of the war, there is a heavy fog over the mountains where the fighting is about to occur. Low-lying clouds hang above the soldiers. By the end of the day the Blue army had won. Question: Why did the Blue army win?
Glasses
Sarah is very long-sighted. She has only one pair of glasses, which she keeps losing. Today she has lost her glasses again and she needs to find them. She had them yesterday evening when she looked up the television programs. She must have left them somewhere that she has been today. She asks Ted to find her glasses. She tells him that today she went to her regular early morning keep fit class, then to the post office, and last to the flower shop. Ted goes straight to the post office. Question: Why is the post office the most likely place to look?

Reproduced from (Happé 1994)

memory question (controlling attention and memory). The *Strange Stories Film Task* has been demonstrated as an effective test to discriminate against adults with and without ASD, even superior to other well-evidenced measures of social cognition (Murray et al. 2017).

Based on Happé's *Strange Stories Test*, Kaland et al. (2002) developed the *Stories from Everyday Life* test. This test comprises 26 stories and assesses the ability to infer physical and mental states in children and adolescents with Asperger syndrome. The first part of each story describes either a physical or a mechanical event, while the second part presents a situation in which inferring the mental state of the character from contextual cues is required. The mentalistic content of the story may include lie, white lie, metaphor, misunderstanding, double bluff, irony, persuasion, contrary emotions, forgetfulness, jealousy, intentions, empathic requirements and blunder. For the score of each story, physical inference questions, mental inference questions, and control questions are asked to ensure understanding of the content. Kaland et al. (2002) found that people with Asperger syndrome showed significant impairments extracting inferences about mental states in comparison with the neurotypical group: they showed a tendency to interpret behavior and statements literally, independently of the context. Furthermore, they would tend to choose a physical explanation instead of a mentalization one, despite the correct answer being a mental state.

Within these advanced aspects of mentalization, the ability to attribute intentions to others is noteworthy. This refers to the cognitive ability to understand what the others think and desire, which allows for adapting to others and recognizing their intentions behind expressions, going beyond the literal meaning.

Despite the study of ToM has mainly focused on patients with ASD, the ability to attribute intentions to others has been extensively studied in patients with schizophrenia.[3] In this regard, the first and most famous instrument developed is the Corcoran et al. (1995)'s *Hint Task.* This task aims at measuring the ability to infer "true intentions" behind indirect speech. It was created to assess mentalizing skills in patients with SCZ). Corcoran et al. (1995) found serious deficits in SCZ patients in comparison with people without neuropsychiatric disorders using their *Hint Task.* The instrument consists of ten situations in which a social interaction between two persons is described. Each situation was featured by the introduction of a hint. The individual must deduce the true meaning of the hint. If the test taker answers incorrectly, a "hint" with an easier question is provided. On the contrary, if the tested person answers the first question correctly, a score of two points is provided. If only the second question (the easiest one) is answered correctly, one point is scored. If both questions are answered incorrectly, the score is zero. The maximum total score is 20 points.

A couple of years later, Sarfati et al. (1997) were also interested in assessing intention decoding skills in patients with SCZ by means of nonverbal performance tasks. For this purpose, they developed the *Intention Inference Task*, with 30 comic strips. Each strip comprises a scenario. In each scenario a character performs a very simple action, showing the intention behind it. After seeing the story, the individual was asked to choose which of the three response cards was the most appropriate to complete the comic sequence. Only one of the cards represented the appropriate intention. To perform the task, subjects had to infer the character's intention within its context.

In the study of Sarfati et al. (1997), only patients with SCZ who presented language deficits did not adequately solve the *Intention Inference Task.* However, those patients without language and communication difficulties did not show any impairment in solving the task. Therefore, these results allow authors to relate the thought and speech disorganization, typical symptoms in SCZ and an underlying distorted attribution of mental states.

Among others who extended the study of ToM beyond autism are Adachi et al. (2004), who developed a test to detect differences in social cognitive abilities

[3] This volume actually shows applications of ToM to a varied number of situations (note by the Editors).

between people with high functioning pervasive developmental disorders and attention deficit and hyperactivity disorders. For this purpose, the authors developed the *Metaphor and Sarcasm Scenario Test.* This test comprises five metaphorical situations and five sarcastic situations. The results by the Adachi et al. (2004)'s study on Japanese school population showed that both children with ADHD and children with high functioning pervasive developmental disorders showed difficulties in understanding metaphors. However, the inability to understand a sarcastic situation was specific to children with a High-Functioning Pervasive Developmental Disorder (HFPDD). In conclusion, the *Metaphor and Sarcasm Scenario Test* may be able to distinguish between HFPDD-related and ADHD-related deficits in children. Some explanations regarding ADHD ToM deficits will be discussed in the following sections.

5.2.3 Assessment of Affective Theory of Mind

Within the affective-emotional component of ToM, basic aspects -such as emotional recognition through facial expressions, tone of voice, prosody-, and complex ones -such as thinking about one's own and others' emotions- can be described.

A pioneering instrument to evaluate emotional recognition skills is the *Pictures of Facial Affect Test (POFA)*, by Ekman and Friesen (1976). This test is based on Ekman's (1984) theory of basic emotions. According to this theory, beyond cultural differences and over and above possible cognitive interpretations, there are six basic and universal emotions: anger, disgust, joy, fear, surprise and sadness. Each basic emotion is associated with a distinctive facial expression. The test consists of 110 photographs of facial expressions that have been widely used in cross-cultural studies, and more recently, in neuropsychological research. The *POFA* assesses the subject's competence to recognize the six basic emotions through facial expressions. The original study by Ekman and Friesen (1976) was conducted on people without neuropsychiatric disorders, and it found that disgust, sadness, fear and joy seem to be the four most easily recognized facial emotions, in comparison with anger and surprise. Furthermore, visual recognition of disgust and fear appears to be independent of ageing.

Inspired by the Ekman and Friesen test, Baron-Cohen et al. (1997) developed the *Reading the Mind in the Eyes Test (Eyes Test).* This test does not only measure emotional decoding, but also evaluates the ability to understand mental states from other people through the gaze. In other words, it is not limited to emotional recognition, since it includes other's mental states (e.g., abstracted, playful, bored, fantasy, distracted...) which implies more complex decoding skills (inferring the beliefs or intentions of the person from her eyes). The objective of the test is to determine whether the participant can put herself in the other person's perspective and "tune into" her mental state. However, in the first version (Baron-Cohen et al. 1997) the *Eyes Test* not assessed ten complex mental states, but also ten basic emotions, with two possible answers for each item. But then, in the revised version

(Baron-Cohen et al. 2001), the *Eyes Test* just assessed complex mental states. It consists of 36 images of facial expressions that reflect emotional states through the gaze (including eyebrows), using a four-point scale in which the correct answer should resemble what the person in the image is feeling or thinking. A summary with glossary for the terms is also given to the participants. Table 5.2 below shows two examples of items.

Subsequently, Rutherford et al. (2002) published the *Reading the Mind in the Voice Test*. It is a similar test as the *Eyes Test* but developed using auditory rather than visual stimuli. The task consists of listening to spoken conversations over audio and choosing the adjective that best describes the state of mind of the speaker. Table 5.3 below shows two examples of the items that make up the test.

Table 5.2 Two examples of items of the *Reading the Mind in the Eyes Test – Test Revised Version*

For each pair of eyes, choose and circle the word that best describes what the person in the picture is thinking or feeling.

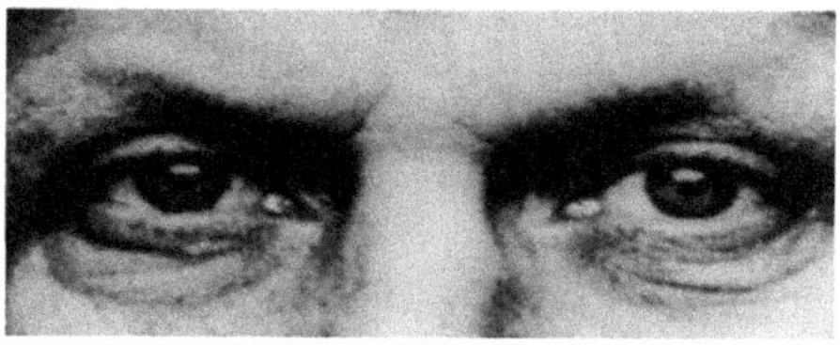

Response options:
(a) Serious
(b) Ashamed
(c) Bewildered
(d) Alarmed

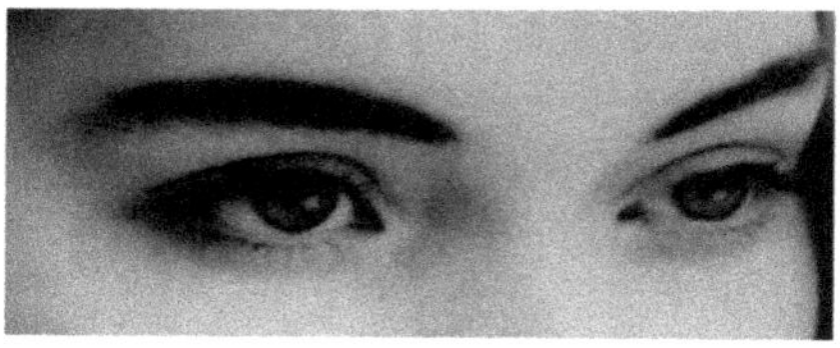

Response options:
(a) Impatient
(b) Aghast
(c) Irritated
(d) Reflective

Reproduced from (Baron-Cohen et al. 2001)

Table 5.3 Two examples of items of the *Reading the Mind in the Voice Test*

Spoken phrase	Response options			
"Where did you get them?"	Frustrated	Suspicious	Intrigued	Doubtful
"Keep the damn thing!"	Bossy	Irritated	Cruel	Surprised

Reproduced from (Rutherford et al. 2002)

The results of both original studies, *Reading the Mind in the Eyes Test* (Baron-Cohen et al. 2001) and the *Reading the Mind in the Voice Test* (Rutherford et al. 2002), reflected a deficit in the interpretation of mental states in patients with ASD.

For the assessment of more complex affective mentalistic concepts, the *Faux Pas Test* by Stone et al. (1998) should be mentioned. This goes a step forward in complex affective mental states recognition. Its aim is to assess the degree of understanding of social situations with emotional content. It consists of 10 stories in which a character makes a "gaffe" or a "faux pas" (an unfortunate mistake in a social interaction situation). The test also contains 10 other stories in which no faux pas appears, which are used as control questions. For each story there are eight possible questions. The first question is used to distinguish when a faux pas has occurred. The second question pretends to identify the character who commits a faux pas. The third question asks why the comment realized is inappropriate (why it is a faux pas). The fourth question inquires why the person made the inappropriate comment, the faux pas. The fifth question asks about the person's awareness of the faux pas committed and if she realized about the situation/context in which it was said. The sixth question inquiries about the way the person affected has been feeling. Finally, questions 7 and 8 are control questions about the content of the story. A control question is used to check whether the individual correctly understood the story. Thanks to control questions, incorrect answers may be linked with mentalization skill deficits, but not other deficits in more general cognition skills (e.g., working memory deficits). Table 5.4 below shows an example of a faux pas story and an example of a control story from the *Faux Pas Test.*

As can be seen in the example, for each story there is a first question that is mandatory: "*Did anyone say something they shouldn't have said or something awkward?*". If the answer to this question is "*Yes*", the person continues with the rest of questions 2, 3, 4, 5 and 6. If the answer is "*No*", control questions are directly made, because it is assumed that the person does not distinguish a faux pas (a complex mentalizing content), so the rest of mentalizing answers make no sense. Thus, question 1, and control questions 7 and 8 are always mandatory to be delivered.

Baron-Cohen et al. (1999) administered the *Faux Pas Test* to a sample of neurotypical children and another sample of children with Asperger syndrome or high functioning autism. The authors concluded that the latter showed higher difficulty in task performance relative to typically developing children. In addition, their results revealed that the ability to adequately identify a faux pas is a skill under development across childhood, observing an earlier consolidation in girls than in boys. These findings were in line with Werner's previously mentioned observations (Wimmer and Perner 1983; Perner and Wimmer 1985).

Table 5.4 Example of a faux pas story and an example of a story without a faux pas from the *Faux Pas Test*

Example of faux pas story
Jean West, a manager in Abco Software Design, called a meeting for all the staff. "I have something to tell you" she said. "John Morehouse, one of our accountants, is very sick with cancer and he's in hospital." Everyone was quiet, absorbing the news, when Robert, a software engineer, arrived late. "Hey, I heard this great joke last night!" Robert said. "What did the terminally ill patient say to his doctor?" Jean said, "Okay, let's get down to business in the meeting."
Question 1: Did anyone say something they shouldn't have said or something awkward?
If the answer is yes, then ask:
Question 2: Who said something they shouldn't have said or something awkward?
Question 3: Why shouldn't he/she have said it or why was it awkward?
Question 4: Why do you think he/she said it?
Question 5: When he came in, did Robert know that the accountant was sick with cancer?
Question 6: How do you think Jean, the manager, felt?
Control questions:
Question 7: In the story, what did Jean, the manager, tell the people in the meeting?
Question 8: Who arrived late to the meeting?
Example of a story without a faux pas
Jim was shopping for a shirt to match his suit. The salesman showed him several shirts. Jim looked at them and finally found one that was the right color. But when he went to the fitting room and tried it on, it didn't fit. "I'm afraid it's too small," he said to the salesman. "Not to worry," the salesman said. "We'll get some in the next week in a larger size." "Great. I'll just come back then," Jim said.
Question 1: Did anyone say something they shouldn't have said or something awkward?
If the answer is yes, then ask:
Question 2: Who said something they shouldn't have said or something awkward?
Question 3: Why shouldn't he/she have said it or why was it awkward?
Question 4: Why do you think he/she said it?
Question 5: When he tried on the shirt, did Jim know they didn't have it in his size?
Question 6: How do you think Jim felt?
Control questions:
Question 7: In the story, what was Jim shopping for?
Question 8: Why was he going to come back next week?

Reproduced from (Stone et al. 1998. Copyright VE Stone and S Baron-Cohen 1997, used with permission, from https://www.autismresearchcentre.com/tests/faux-pas-test-adult/)

5.2.4 *Assessment of Theory of Mind and other related constructs*

As we have elaborated throughout this chapter, the study of ToM has developed an increasingly complex conception of it, joined to more complex and heterogeneous assessment instruments. At this point we should mention a concept that arises precisely from this enlargement in the study of ToM Social Cognition (SC). SC is a diverse defined cognitive domain. It is understood as a psychological process which participates in perception, encoding, storage, retrieval, and regulation of intrapersonal and interpersonal information (Green et al. 2015). As an effort to uniform definitions of social cognitions, experts came to terms at the National

Institute of Mental Health in 2006, in the context of the "Measurement and Treatment Research to Improve Cognition in Schizophrenia" (MATRICS) Initiative (Green et al. 2008). Consequently, five areas of SC were delineated: Theory of the Mind, social perception, social knowledge, attributional bias, and emotional processing (Green et al. 2008).

This higher cognitive domain is strongly related to perspective taking (i.e., understanding the situation from the other's point of view), emotion recognition and empathy. SC plays a fundamental role in social interactions and is linked to the acquisition of ToM (Patin and Hurlemann 2015). In addition, research also suggests that there is a distinction between social, emotional, and neurocognitive cognition, in the same lines as ToM (Etchepare and Prouteau 2018). SC is a highly complex concept, and there are diverse positions about what does or does not constitute SC. Tomás-Labbé et al. (2019) propose a model of SC understanding it as an integration of the processes by which subjects perceive social signals (social perception), and from them, infer mental states of other people (Theory of Mind), to finally generate emotional responses that motivate and modulate behavior (empathy). In short, the concept of ToM would be framed within the complexity of SC, being the most important component of it. However, some authors use both concepts (ToM and SC) as equivalent terms (see, e.g., Dziobek et al. 2006).

The same applies to the term Reflective Function (RF) is presented as a crucial component of ToM, but it is used synonymously with the term mentalization. It is defined by Fonagy and Target (1997) as the capacity to understand ourselves and others in terms of intentional mental states. RF represents the process prior to interpreting and forecasting behavior. The concept of mentalization arose in an attempt to operationalize the ToM and it involves awareness of the mental state of oneself and others, especially when behavior intentions are involved (Valdivieso-Jiménez 2020). Fonagy and Target (1997) defined two subtypes of RF impairments: hypomentalization, an extreme difficulty in developing complex models of the mind of oneself and others; and hypermentalization, opposite to hypomentalization, defined as a tendency to develop highly complex models of mind that does not correspond to observable evidence. Therefore, the RF has a crucial role in the acquisition and development of ToM, mainly across childhood. In the last decade, several studies have addressed mentalizing skills in the parental educational context, or also called Parental Reflective Functioning (PRF), defined as the parental ability to understand their child's mental states, in order to secure children's mental health and development (Camoirano 2017). PRF has been associated with an accelerated or an inhibited ToM development in children at early age (Nijssens et al. 2021).

Most of the previously mentioned instruments focus on just one of the multiple dimensions of ToM (e.g., facial emotion recognition, false belief understanding...). To overcome these shortcomings, Dziobek et al. (2006) developed the *Movie for the Assessment of Social Cognition* (*MASC*). Although the name of the test refers to SC, the authors use the concepts of SC and ToM interchangeably. This instrument consists of a 15-min video in which a social interaction between four characters takes place. The assessed subject has to make inferences about the mental states of the characters. The *MASC* presents a high ecological validity, with the aim of integrating different components such as facial expression, gaze, body language, understanding of pragmatic aspects of language (such as irony or sarcasm) and interpretation of contextual information into the assessment. The test consists of 48 questions: 42 of them with mentalistic content, and six control questions. Control questions are used to ensure that mentalization errors are not due to impairments in other cognitive abilities such as memory. This instrument allows differentiation between three different types of mentalization errors: (1) an excessive ToM error or hypermentalization (a mental state being attributed when there is no mental explanation for the situation); (2) a reduced ToM error or hypomentalization (when a present mental state is wrongly attributed); (3) a total absence of mental inference, no mentalization (i.e. making physical causality attributions to social situations and mental states). Therefore, for each mentalistic content question, there are four possible outcomes: correct mentalization, hypermentalization, hypomentalization and no mentalization. For the control questions, there are only two possible outcomes: correct or incorrect answers. The aim of this instrument was widening the study of mentalization beyond the simpler tests designed for the study of mentalization in children with ASD, assessing mentalization by an ecological tool. Although Dziobek et al. (2006) originally applied this instrument in adults with Asperger syndrome, its study has currently been extended to other mental health conditions such as SCZ (Montag et al. 2011), Bipolar Disorder (BD) (Montag et al. 2010), stress (Smeets et al. 2009), depression (Wolkenstein et al. 2011) or personality disorder (Preißler et al. 2010).

Finally, RF assessment is described in the original RF manual (Fonagy et al. 1998) and the most complete and accurate mentalizing evaluation is the Reflective Functioning Scale (RFS), an interview-based measure applied during adult or parent development interview that provides a global RF score. After the interview, the examiner can rate the patient on an 11-point scale from antireflective to exceptional reflective. The main disadvantage of this tool is its duration, from 1 to 3 h to administer, and therefore the professional cost. Due to that, authors recommend the Reflective Functioning Questionnaire (RFQ), a self-report questionnaire with eight items on a seven-point Likert scale (1 = strongly disagree, 7 = strongly agree) measuring the ability to understand self and others mental states (Fonagy et al. 2016). This instrument has two subscales: uncertainty (hypomentalization) and certainty (hypermentalization) subscales. Uncertainty reflects an extreme lack of understanding about mental states, so higher scores on this subscale indicate worse mentalizing capacity. Certainty points to an excessive certainty about mental states, so then, lower scores indicate a worse mentalizing capacity (Spitzer et al. 2021). In case of parental RF measure, the Parental RFQ (PRFQ) is composed by 18 self-reported items, and it includes three subscales: pre-mentalizing subscale, or the inability to understand the subjective experiences of the child, the certainty in mental states subscale, the capacity to identify mental states, and the interest and curiosity in mental states subscale (Luyten et al. 2017).

5.3 Theory of Mind in Neuropsychiatric Disorders

ToM deficits are proposed as a nuclear cognitive phenotype in developmental and neuropsychiatric disorders (Ballespí et al. 2018; Cotter et al. 2018), placing ToM as a transdiagnostic clinical marker (Cotter et al. 2018; Ibáñez et al. 2018; Porcelli et al. 2019). Thus, impairments in ToM functioning are greatly relevant in both clinical and functional facets thorough disorders (Cotter et al. 2018).

With regards to clinical aspects, ToM is strongly related with progressive worsening in psychopathology (Porcelli et al. 2019). In addition, it can be used as a longitudinal predictor of clinical outcomes and as a marker for disease onset and disease progression between diagnoses (Cotter et al. 2018; Porcelli et al. 2019).

Considering functional aspects, ToM is strongly linked to social functioning across disorders (Braak et al. 2022), being associating through studies to a progressive deterioration of the general functioning, the patient's quality of life and social withdrawal (Ibáñez et al. 2018; Porcelli et al. 2019).

For the above reasons, deficits in social processes -like ToM- are strongly linked to a potential development of intrapersonal and interpersonal problems (Ballespí et al. 2018; Porcelli et al. 2019). Due to that, it is necessary to examine mechanism links to this diminished social function which could be detrimental to patients' daily actions and to their level of psychopathology (Cacioppo et al. 2014; Ibáñez et al. 2018).

In this context, the exhaustive study of ToM deficits in neuropsychiatric populations is just one main part of the whole complexity of mental health (Cacioppo et al. 2014). However, studying them in a deep way will allow clinicians to set the needed basis to understand neuropsychiatric disorders in a comprehensive way.

Concerning ToM, it may be an indicative biomarker through neuropsychiatric conditions. Distinguishing between them in terms of deficits in ToM according to recent meta-analysis results (Cotter et al. 2018), people with developmental and neurological disorders display medium to large deficits, whereas psychiatric disorders show wide variations of these deficits (possibly because of a bigger variance in clinical heterogeneity and severity of psychiatric disorders).

Even though Neuropsychiatry comprises a wide range of disorders as the above categories mentioned, neuropsychiatry gathers – in vast terms – neurology and psychiatry, two different fields (Northoff 2008). Several cognitive and behavioral domains are impaired in typical neurological and psychiatric disorders, as well as psychopathology (Ibáñez et al. 2018). However, both areas of knowledge have studied deficits from different points of view: neurology has been focusing on brain anatomic dysfunctions and psychiatry has been investigating functional alterations of the mind. These divergent specializations have contributed to a paucity in integration studies (Ibáñez et al. 2018). Although this is currently changing and similar problems are examined from different areas of specialization to enrich the complexity of human problems (Cacioppo et al. 2014), there are scarce studies which investigated the same deficits in both typical neurological and psychiatric conditions (a worth mention example of this needed integration effort is Braak et al. 2022): studies are usually conducted in ToM deficits in neurologic disorders (e.g., Poletti et al. 2012; Sandoz et al. 2014) or in psychiatric entities (e.g., Buhlmann et al. 2015; Plana et al. 2013; Preti et al. 2022; Seitz et al. 2022; Vaskinn et al. 2013; Washburn et al. 2015).

Due to this current research reality, this section will just be focused on psychiatric conditions, also including a developmental disorder (ASD), but its aim is not to examine neurological conditions. Moreover, it includes suicidality, because of its seriousness in mental health. This aims to gather uniform results and uniform ways of studying ToM deficits and implications. To achieve this purpose, in this section ToM deficits are described following the Diagnostic and Statistical Manual of Mental Disorders Fifth Edition Revised -DSM-5-TR- (APA 2022), for diagnostic criteria and for the diagnosis organization.

5.3.1 Autism Spectrum Disorder

Autism spectrum disorders (ASD) are an assortment of neurodevelopmental disorders featured by persistent deficits in social communication and social interaction across multiple contexts. Deficits may be found in social domains, such as social reciprocity, nonverbal communicative behaviors, and abilities to initiate, to hold and to understand interpersonal relationships. According to the American Psychiatric Association (APA 2022), children with an ASD may present critical difficulties to respond to gestures and glances from those around them, do not seek contact with others, and prefer to play alone. They may not have an intelligible language either echolalia (repeating the words they hear), alter the order of words, or use them with a particular meaning. Besides social communication deficits, the endorsement of ASD requires the presence of restricted and repetitive patterns of behavior, interests, or activities. For example, they may have excessive and inflexible interests towards some types of objects, animals, etc. (types of cars, types of birds…). They may also perform repetitive rituals or routines without any function (unlike obsessive-compulsive disorder, in which rituals are performed to avoid discomfort) and sometimes perform stereotyped movements (e.g., walking on tiptoe, moving hands in a certain way…). Moreover, they have hyper or hypo reactivity to sensory inputs or they are unusually interested in some aspects of the environment,

Symptoms are usually identified in the early stages. However, ASD may go unnoticed for years until social demands overcome limited socialization skills. Symptoms may also be masked by later learned strategies, and vary over the years, but significant impairment in quality of life is permanent (APA 2022).

As was forementioned, Asperger syndrome was previously considered a separate condition from autism, but in present-day classification systems both are included in the so-called ASD (APA 2022). People with Asperger syndrome may also manifest difficulties in social communication and thought flexibility. Even though, fluent language and intellectual performance) are often preserved in people with Asperger syndrome (the intellectual performance scores may be even higher in people with Asperger syndrome than the population average). Therefore, ASD may not inherently be linked with language impairment. In addition, they may also be associated with a known medical/genetic or environmental/acquired condition, or other neurodevelopmental, mental, or behavioral disorder (APA 2022).

The current lifetime prevalence of ASD in developed countries is estimated to be at least 1.5%, although age standardized prevalence is 0.364 % (GBD Mental Disorders Collaborators 2022). Prevalence rates considerably increase when persons without comorbid intellectual disability are included as falling within the spectrum (Baxter et al. 2015; Lyall et al. 2017).

A very striking feature of ASD that has been extensively studied is the absence of symbolic play. Symbolic or pretend play is a skill that typically develops between 18 and 24 months of life. It is characterized by using objects with a different function than the inherently attributed to the object, making use of the imagination. Through symbolic play, children imitate everyday situations (Leslie 1987; Thiemann-Bourque et al. 2012). Leslie (1987) distinguished between three ways of symbolic play: (1) using objects as if they were other objects (i.e., using a banana as if it were a telephone), (2) attributing false features to an object (e.g., wiping a doll's face as it would be dirty when it is actually clean), and (3) referring to an object as if it was present (i.e., pretending to drink tea from a cup that is in fact empty).

Baron-Cohen (1987) concluded that symbolic play difficulties in children with an ASD may be highly influenced by related cognitive ToM deficits. According to this, ASD would be the consequence of a deficit in the ability to attribute mental states (desires, beliefs, intentions, etc.) to others. Deficits in symbolic play, stereotyped behavior, language disturbances, and difficulties in the attribution of mental states would be the consequence of a dysfunction in the metarepresentational capacity (Ibáñez-Brassi and M. 2005).

ToM studies by Baron-Cohen fostered the study of basic mentalization processes in children with ASD. In a seminal study, Baron-Cohen et al. (1985) applied their *Sally-Anne Test* to assess first-order false belief in children with ASD, neurotypical children, and children with Down syndrome. As a result, they found that neurotypical children and children with Down syndrome performed adequately on the test from the age of 4 years. However, 80% of the age-matched children with an ASD showed a poorer performance in the test. The results support that the difficulties in mental state recognition should be considered a distinctive difficulty in people with an ASD, regardless of intelligence quotient (IQ). This study has been extensively replicated and has marked a milestone in the study of ToM in ASD.

Since then, research on ToM deficits in individuals with ASD has not ceased. These deficits are considered as nuclear cognitive characteristics of ASD, emphasizing global impairments in different mentalization tasks in comparison with typically developed children (Baron-Cohen et al. 2001).

In this line, studies are consistent in supporting that children with an ASD may show an atypical development of ToM abilities. A more robust finding is that children with an ASD may show systematically worse performance on false belief tasks before the age of 11. While typically developing 4-year-olds perform false belief tasks correctly, 4-year-olds with ASD may not be able to correctly predict the behavior of others based on their mental states. This is because they do not consider the other person's mental states (i.e., intentions, knowledge, and beliefs) (Happé 1994). It is also studied that ASD children with intellectual disabilities may show wider deficits in ToM skills (even basic ToM skills, such as false beliefs task), whereas high functioning adults with ASD show difficulties in complex ToM tasks too (Pino et al. 2017; Ponnet et al. 2004; Rosello et al. 2020; Velikonja et al. 2019). Regarding emotional decoding, persons with ASD and low cognitive functioning show severe impairment, while high functioning ASD persons show results comparable to participants without a neuropsychiatric disorder (Rojahn et al. 1995). Similarly, it appears that primary emotions are correctly decoding by persons with high and low functioning ASD. However, persons with low functioning ASD show more difficulties with more secondary emotions (Balconi et al. 2012). Neuroimaging evidence indicates that ASD children do activate different neural patterns, confirming autistic individuals rely on different strategies to process information of faces (Amenta et al. 2014).

Another line of research on ToM in ASD patients aimed to make distinctions between two types of mentalization: "implicit" and "explicit" mentalization. "Explicit" mentalization refers to a more conscious process, where language is involved (e.g., tasks such as the *Faux Pas Test*). In implicit mentalization tasks, no instructions are given to attribute a character's mental state. On the contrary, a researcher may measure spontaneous behavior during the tasks as an indirect outcome of mentalization (e.g., gazing and eye movements that reflect participants' expectations about the actor's beliefs). From this domain-based perspective of mentalization, Schneider et al. (2013) and Callenmark et al. (2014) found no differences in explicit mentalizing tasks between neurotypical people and people with ASD. However, neurotypical people showed higher performance in implicit mentalization tasks in comparison to people with ASD, regardless of IQ. Some individuals with an ASD may use some "compensation" strategies that may lead to preserved mentalization skills, with a subsequent better adjustment to social environment. Compensation is the process of generating new behaviors to avoid negative consequences. It is based on past experiences of socialization, so it is an adaptive process. Due to that, it may improve with learning and people can practice applying them to everyday situations to progress in their execution. Examples of this compensation include suppressing atypical behaviors, planning and training conversations, learning rules about verbal and non-verbal behaviors, following social norms, maintaining eye contact, asking others questions about themselves, etc. (Livingston et al. 2019b). The use of compensation strategies was found to be associated with a higher IQ and a better performance in executive function tasks. In turn, these compensatory mechanisms have been found to be related to higher levels of self-reported anxiety (Livingston et al. 2019a).

Considering neurobiological aspects, theories of deficits in "broken mirror neurons" underlying ToM deficits have been proposed (Ramachandran and Oberman 2006). Mirror neurons are found in the inferior frontal cortex, the premotor cortex, the supplementary motor area, the primary somatosensory cortex, and the inferior parietal cortex (Molenberghs et al. 2009). They are activated both during the performance of an action, as well as during the observation of the performed action (Rizzolatti and Craighero 2004).

An aberrant pattern has been found in the normal activity of the neural circuitry involved in the efficient perception of the motion activity (Kana et al. 2015; Ramachandran and Oberman 2006). However, the recent review by Andreou and Skrimpa (2020) concludes that it is necessary to shed more light on the mechanisms underlying the connection(s) between ToM and neurophysiological operations. Moreover, functional connectivity magnetic resonance imaging (fcMRI) studies, which assesses the correlation of the fMRI blood-oxygen-level dependent (BOLD) time-series has been used by some studies to characterize the neural circuitry underlying ToM in ASD. In these studies, children and adolescents with ASD showed atypical neurophysiological response on the mentalizing network -including medial prefrontal cortex, posterior cingulate and lateral temporal cortices- during mentalizing task (White et al. 2014). However, there were no differences between neurotypical participants and adults diagnosed with ASD in brain activation during mentalizing tasks (Dufour et al. 2013). In conclusion, it is possible that heterogeneity in ASD and methodological issues -such as the wide range of ToM assessment instruments- had contributed to those results and further investigation is required.

5.3.2 *Schizophrenia and Other Psychotic Disorders*

Psychotic disorders are an ensemble of disorders which can be placed into a gradient of psychopathology severity. However, all of them are defined by abnormalities in terms of psychotic symptoms: positive symptoms -delusions, hallucinations, disorganized thinking, disorganized or abnormal motor behavior- and negative symptoms (APA 2022).

Delusions are inflexible beliefs which are not permeable to change instead of contrary evidence to their veracity. Depending on the features of their content, they can be persecutory delusions (someone is going to harm oneself), referential delusions (environmental cues or interactional content is referred to oneself), grandiose delusions (beliefs about having exceptional features or conditions), erotomaniac delusions (the person wrongly belief that another one had fall in love with her), nihilistic delusions (beliefs about a catastrophe is going to take place) and somatic delusions (worry beliefs about health and organ functions). Hallucinations are perception-like experiences which do not happen with an external stimulus, but which are experienced vividly and clearly without voluntary control. Hallucinations can occur in any sensory modality. Disorganized thinking is inferred by an impaired speech which blocks functional communication. It is characterized by a loose of associations between topics (derailment), unrelated answers (tangentiality), severe disorganized speech which is incomprehensible (incoherence). Grossly disorganized or abnormal motor behavior is a goal-directed behavior which varies from childlike "silliness", agitation, and catatonia (a severe reactivity decreases to the environment) (APA 2022).

Negative symptoms are unmet in SCZ, and they strongly contribute to the long-term disability and impaired functionality in SCZ (Correll and Schooler 2020). They may be grouped in diminished expression symptoms and avolition or apathy symptoms (Correll and Schooler 2020). Blunted affect (diminished emotional expression) and alogia (diminished speech output) would form the diminished expression dimension; whereas avolition (decreased motivated self-initiated behavior), anhedonia (decreased ability to experience pleasure) and asociality (apparent lack of interest in social interactions) would form the avolition or apathy dimension (APA 2022; Correll and Schooler 2020).

Within psychotic disorders, SCZ is characterized by the presence of delusions, hallucinations or disorganized speech and disorganized behavior or negative symptoms. These symptoms must be present for at least 1 month for the diagnosis. During this period, the personal level of functioning has severely diminished. In addition, continuous symptoms persist for 6 months, with periods of active symptoms and other periods of residual symptoms (negative symptoms or other two main psychotic symptoms) (APA 2022). Age standardized prevalence for SCZ is 0.2874% (GBD Mental Disorders Collaborators 2022).

The current bulk of scientific knowledge shows a marked impairment in ToM in people with SCZ (Bora et al. 2009; Savla et al. 2013; Sprong et al. 2007). SCZ patients may show wider ToM impairments in comparison with patients with other neuropsychiatric disorders (except for ASD), regardless of executive function, IQ, memory, or general psychopathology. It would suggest that the ToM deficit is nuclear in SCZ. (Harrington et al. 2005).

In SCZ, SC has been linked to functional capacity -in terms of psychosocial function-, showing stronger associations than other neurocognitive variables (Addington et al. 2010; Cowman et al. 2021; Fett et al. 2011; Halverson et al. 2019).

Negative symptoms, and deficits in neurocognition (verbal memory and learning, working memory and problem solving) and ToM skills have been proposed to be significant predictors of social competence in SCZ (Couture et al. 2011; Halverson et al. 2019). More concretely, ToM skills have shown the strongest correlations with functioning in social contexts (Fett et al. 2011; Thibaudeau et al. 2021). In fact, ToM skills may work as robust mediators between wider cognitive skills and real-world functioning (Couture et al. 2011; Fett et al. 2011). Specifically, cognitive aspects of ToM (inferring implicit messages) have shown strong relationships with functioning, partly dependent on processing speed skills (Braak et al. 2022). Moreover, mental state decoding predicts quality of life of people with SCZ regarding the internal aspects, such as motivation, sense of purpose, curiosity, aimless inactivity, empathy, and emotional interaction (Tas et al. 2013).

Regarding neurocognition, although general neurocognitive impairments may contribute to ToM impairments (Thibaudeau et al. 2021), evidence supports the independence between these higher order cognitive domains -neurocognition and ToM- (van Hooren et al. 2008): whereas a non-impaired ToM functioning is not related to neurocognitive deficits, an impaired ToM performance is related to a poorer neurocognitive performance (Fanning et al. 2012; Zhang et al. 2018). It suggests that neurocognition is necessary but not sufficient to ToM, implying the specificity of ToM because they cannot be fully explained by neurocognition (Fanning et al. 2012).

However, both domains are strongly related and ToM deficits have been moderately associated with overall neurocognition and all neurocognitive domains, such as attention, working memory, episodic memory, speed of processing, language, visual memory, executive functions (plan/organize, inhibition, flexibility, abstraction and fluency), early processing and perception and autobiographical memory (Ayesa-Arriola et al. 2016; Braak et al. 2022; Catalan et al. 2018; Fanning et al. 2012; Thibaudeau et al. 2020).

It should be considered that the performance in a ToM task is related with its complexity -neurocognitive requirements- (Thibaudeau et al. 2021), and performance becomes easier to SCZ patients if the task joins visual and verbal contents than using solely tasks in which items are read, because it needs a large working memory load to retain the heard information, increasing the difficulty of the ToM task (Brüne 2005).

Focusing on ToM impairments, deficits in ToM have been robustly described across the different phases of psychotic diseases (Bora and Pantelis 2013): at ultra-high risk psychosis (UHR), at first-episode psychosis (FEP), in chronic diagnosis and in remitted phases of the illness.

UHR comprises a phase of imminent risk of psychosis, in a state of incipient psychotic disorder or prodromal (Schultze-Lutter et al. 2010; Yung and McGorry 2007). In this early stage of a psychosis disease, moderate ToM deficits have been found (Bora and Pantelis 2013; Lee et al. 2015; van Donkersgoed et al. 2015; Zhang et al. 2018). Impairments in both verbal and visual ToM tasks have been found in UHR (Bora and Pantelis 2013; van Donkersgoed et al. 2015). However, in the meta-analysis of van Donkersgoed et al. (2015), deficits in visual ToM, even moderate, were not significant. This result was explained by authors referring to the neurocognitive complexity -forementioned before- which implies a verbal task: if verbal tasks require bigger neurocognitive efforts and neurocognition it is also impaired in UHR, it is likely to find bigger deficits in verbal tasks (van Donkersgoed et al. 2015).

Moreover, it is worth noting that, within the UHR population, ToM deficits were stronger in those who later developed a psychotic disorder (Lee et al. 2015; Zhang et al. 2018), suggesting a specific association between ToM and psychosis.

Since the first-episodes of a psychotic disorder, substantial ToM deficits have been also described from the beginning of the psychotic illness, independently of the psychotic diagnosis (Bora et al. 2009; Bora and Pantelis 2013; Healey et al. 2016). Deficits have been consistently described even in verbal and visual ToM tasks (Bora and Pantelis 2013; Healey et al. 2016), however verbal deficits are more robust than visual ones due to the high methodological quality of the studies which assess ToM through verbal tasks: they control for the effect of neurocognition (Healey et al. 2016; van Neerven et al. 2021). A plausible explanation of that is related to verbal neurocognitive deficits, which have been larger than other neurocognitive domains in FEP populations (Mesholam-Gately et al. 2009). Considering ToM deficits, deficits in hypomentalization, hypermentalization and no mentalization have been found in FEP (Catalan et al. 2018).

In chronic stages of a psychotic diagnosis -SCZ-, highly ToM impairments have also been found (Brüne 2005; Bora et al. 2009; Bora and Pantelis 2013; Green et al. 2015; Healey et al. 2016; van Neerven et al. 2021). ToM deficits in SCZ are large, and they are comparable to such in FEP (Bora et al. 2009; Healey et al. 2016; van Neerven et al. 2021). As we have mentioned before, this large impairment in both stages of the illness -FEP, at the start of the psychotic onset and SCZ, with larger development of the psychosis- suggest that ToM deficits are not explained by effects of illness deterioration, chronicity, and longer exposure to pharmacotherapy (Bora and Pantelis 2013). Regarding specific ToM deficits, in SCZ a bigger impairment in second-order tasks (Bora et al. 2009; Harrington et al. 2005; van Neerven et al. 2021). Considering other psychotic disorders in addition to SCZ, deficits have also been found, both in cognitive and affective ToM (Dorn et al. 2021; Montag et al. 2011).

Finally, in remitted phases of the illness, moderate ToM deficits are also present (Bora et al. 2009; Harrington et al. 2005; van Neerven et al. 2021). They are more prominent in irony and faux pas tasks (Bora et al. 2009).

Considering the whole results about ToM deficits in psychosis, impairments in ToM are point out as a stable feature of psychosis and it has been proposed as a trait marker for psychotic disorders (Ayesa-Arriola et al. 2016; Bora et al. 2009; Bora and Pantelis 2013; Brüne 2005; Lee et al. 2015; van Neerven et al. 2021). Deficits cannot be explained by sociodemographic or clinical variables (Ayesa-Arriola et al. 2016; Savla et al. 2013), but it should be borne in mind that in SCZ high levels of social withdrawal -social disengagement- have been associated with negative emotional decoding (de la Torre-Luque et al. 2022b): given that emotional decoding is needed in affective ToM and that it is impaired in SCZ (Ayesa-Arriola et al. 2016), the increased sensitivity found in isolated SCZ patients linked to a worse emotional decoding may also be considered in future studies of ToM. ToM deficits can neither be explained by duration of illness, disease decline or pharmacological effects (Ayesa-Arriola et al. 2016; Bora and Pantelis 2013).

Considering previous meta-analytic and experimental results, it may be concluded that ToM deficits are more severe in acute stages of psychotic disease (Bora et al. 2008, 2009). Considering specifically psychotic symptoms, different results have been delineated, but severity of symptoms have been related with severity of ToM deficits (Bora et al. 2008; Harrington et al. 2005).

Regarding SCZ symptoms, Frith (1992) postulated the role of mentalization skills on illness symptoms development. Following this theory, regarding positive symptoms, control thoughts and delusions (e.g., believing that their body or mind is being managed by someone else) would reflect deficits in the SCZ patient's ability to represent her own intentions to act. On the other hand, delusions of reference (e.g. mistakenly believing that comments, situations, statements or facts are made about herself, or that they have special meaning, as for example, believing that television news sends special messages to her) and persecution would reflect a difficulty in representing other people's mental states. Due to that SCZ patients with positive symptoms would understand that others have mental states -they have a Theory of the Mind (i.e., they explain the behaviors of others referring to their intentions)-, but this capacity would be impaired: they would infer wrong intentions to others based on the wrong beliefs, hypermentalization (Frith 2004). By contrast, SCZ patients with negative symptoms would show no mentalizing abilities -no having a Theory of Mind-, because they do not consider beliefs, desires, and intentions of others to understand their behavior: they do not predict behavior considering beliefs but considering just actual aspects of the world (Frith 2004).

Meta-analytic results show that deficits in ToM are strongly associated with psychotic symptoms, mainly with negative ones (Bora et al. 2009; Healey et al. 2016). Specifically, negative symptoms have been related to affective ToM impairments (Dorn et al. 2021). Moreover, negative symptoms have been associated with no mentalizing impairments (Montag et al. 2011).

In relation to positive symptoms, delusions have been related with cognitive ToM (Dorn et al. 2021), suggesting ToM deficits as a mechanism under delusion formation (Catalan et al. 2018). Positive symptoms have been related to hypermentalizing mistakes (Fretland et al. 2015; Montag et al. 2011).

Needless to say, ToM impairments have not only strong implications for the patient's everyday life functioning, it also has heavy implications in psychosis treatment. One of the aspects involved in psychosis treatment is the common weak insight or awareness of the illness. Problems in being conscious of the disorder is a predictor of nonadherence to treatment, higher relapse rates, more involuntary treatments, poorer psychosocial functioning, aggression, and a poorer course of illness (APA 2022).

Low illness insight (i.e., consciousness of the illness and its consequences) features SCZ during the whole entire course of the illness, with clear implications during the disorder (David 1990; Rebolleda 2017; APA 2022). In terms of ToM skills, literature supports that the higher illness insight, the better the mentalizing performance and the perspective taking and considering the other's point of view about oneself, could be essential factors to check self-experience and disorder symptoms (Vaskinn et al. 2013; Konstantakopoulos et al. 2014; Ng et al. 2015).

Insight requires the capacity of being able to adopt the perspective of others to be aware of the disorder and the symptoms (Konstantakopoulos et al. 2014; Ng et al. 2015). The associations between ToM and insight remain significant regardless of neurocognitive and clinical profile (Ng et al. 2015). Bora (2017) also found modest correlations between ToM performance and clinical insight (covering symptoms, illness consciousness and necessity for treatment), regardless of clinical profile. These associations point to the involvement of perspective taking skills in clinical insight. Therefore, deficits in mental state attribution could contribute to a worse clinical insight and a worse clinical prognosis (Bora 2017; Konstantakopoulos et al. 2014; Ng et al. 2015; Vaskinn et al. 2013).

5.3.3 Bipolar Disorders

Bipolar disorders (BD) are characterized by a well-established period with abnormally euphoric, expansive, or irritable mood and an increase of energy or activity. During this period, different symptoms can be present, entailing a great change for the habitual personal behavior: grandiosity, decreased need for sleep, increased talkativeness, flight of ideas, distractibility, increased goal-directed or non-goal directed activity and excessive participation in potentially risked activities. This ensemble of mood symptoms can constitute a manic episode or a hypomanic one. In a manic episode these mood symptoms are present during 1 week, they can present psychotic features and they are severe enough to cause a great impairment in personal functioning or the need of hospitalization. In a hypomanic episode, mood symptoms are present for at least four days, they do not present psychotic features and they are not severe enough to cause a great impairment in personal functioning or the need of hospitalization (APA 2022).

Considering these mood disturbances, two BD syndromes must be differentiated: Bipolar I disorder and Bipolar II disorder. Bipolar I disorder is diagnosed if a manic episode has taken place, but hypomanic and depressive episodes could also appear after it. Bipolar II disorder is diagnosed when a hypomanic episode and a Major Depressive Disorder (MDD) had been present but not a manic one (APA 2022). BD has an age standardized prevalence of 0.4898% (GBD Mental Disorders Collaborators 2022).

Bipolar disease may show a medium-size impairment in affective and cognitive ToM tasks and in verbal and visual tasks (Bora et al. 2016). Stronger deficits in patients in acute phase, compared to patients under remission or with a subsyndromal condition -which did not show differences between them- have been found (Bora et al. 2016; Hawken et al. 2016).

BD patients in manic phases showed more deficits decoding mental states more than depressed or euthymic BD patients (Hawken et al. 2016). It must be considered that these deficits were related to psychotic symptoms (disorganized speech, language/thought disorder, and delusions). They suggest that manic phases add ToM impairments to the general BD deficits (Hawken et al. 2016) and, specifically, psychotic symptoms appear to have a specific role in ToM deficit (Bora et al. 2016). However, a study comparing performance between BD patients with or without psychotic symptoms does not find differences (Lahera et al. 2007). All these deficits suggest ToM as a risk endophenotype (Bora et al. 2016), making possible that ToM could be a trait marker to BD (Santos et al. 2017).

In BD patients, longer disease duration (stronger predictor), higher number of depressive relapses and cognitive deficit, have been suggested to have a negative effect in ToM performance (Santos et al. 2017; Wolf et al. 2010).

Neurocognitive deficits have been pointed out as possible contributors of ToM deficits in Bipolar disease (Bora et al. 2016). Low scores in *MASC* were described also in control items, highlighting a possible influence of executive function and memory domains, because these items require a high abstract reasoning (Montag et al. 2010). In the same vein, deficits in sustained attention have been involved in ToM deficits of BD (Lahera et al. 2007). Verbal processing and processing speed have been also related to BD deficits in ToM tasks (Navarra-Ventura et al. 2022). However, ToM deficits have remained significant after controlling for neurocognitive domains as IQ, executive function, memory, attention, and verbal learning (Montag et al. 2010; Wolf et al. 2010). They suggest deficits in ToM are specific and they could not be explained only by neurocognitive performance although it has an influence in this domain (Lahera et al. 2007; Montag et al. 2010; Santos et al. 2017).

Deficits in ToM have been found in cognitive but not emotional ToM (Barrera et al. 2012; Lahera et al. 2007; Montag et al. 2010; Santos et al. 2017; Wang et al. 2018). Deficits in cognitive ToM may be related to the higher requirements in those tasks to interpret non-literal questions, which could underlie the deficits (Montag et al. 2010).

Considering specific deficits, BD patients showed more hypomentalizing mistakes, and they have been related to the number of hypomanic or manic phases. *MASC* emotional mental state recognition and the hypomentalizing scores showed a significant negative correlation with the number of manic episodes (Montag et al. 2010). Moreover, a worse cognitive ToM impairment in euthymic phase was linked to an increased number of depressive episodes during the whole illness (Barrera et al. 2012). In addition, depression has been linked to ToM deficits (Wang et al. 2018). It must be considered that studies do not differentiate between Bipolar I disorder and Bipolar II (van Neerven et al. 2021), even the symptomatic profile is different regarding the manic episode and the depressive one (whereas in Bipolar I disorder manic phase is compulsory and MDD depressive phase is frequent but not necessary for a diagnostic, in Bipolar II disorder MDD is required for the diagnosis). Because of that, different contributions of the manic or depressive symptoms according to the type of BD included in the study, may influence results, and differentiate between types of BD must be considered in further investigations.

Congruently with these results, normative performance was found in affective ToM by the *Eyes Test* in BD in remission (Barrera et al. 2012) and in affective ToM tasks in the *MASC* (Montag et al. 2010). No differences were neither found in affective ToM within BD remitted patients (Purcell et al. 2013).

5.3.4 Depressive Disorders

Depressive disorders are highly common (with a lifetime prevalence ranging between 5-15%) affective disorders, featured by a predominance of negative mood (mainly governed by sadness and anger states), anhedonia (i.e., inability to feel pleasure) and a wide assortment of other somatic and cognitive symptoms (recurrent guilty feelings, concentration difficulty, agitation/psychomotor retard, etc.). A depressive disorder may lead to elevated levels of functioning impairment and disability (de la Torre-Luque and Ayuso-Mateos 2020; Ferrari et al. 2013; GBD Mental Disorders Collaborators 2022; Ojagbemi et al. 2022; Reddy 2010).

Depressive disorders is a diagnostic category which encompasses diverse disorders into it. Depressive disorders can be distinguished considering duration of the disease, timing, and aetiology. Depressive symptoms can be present for weeks or years, they can constitute an unique episode or a recurrent one, they can entail different levels of severity (mild, moderate and severe) and also they can present diverse symptomatology: mood symptoms can solely be present at the disorder, or joined with anxiety symptoms and even psychotic ones (APA 2022).

Within depressive disorders, MDD remains the most prevalent depressive disorder worldwide, with increasing prevalence trends in recent decades (United Nations 2022). MDD endorsement necessarily involves the presence of a depressive episode, in which typical symptoms are persistent depressive mood, lack of pleasure and alterations in feeding, sleep, psychomotor behavior and energy consumption patterns. It is also featured by excessive or inappropriate feelings of guilty or incompetence, disable ability to think and suicidal thoughts (APA 2022).

Considering depressive symptomatology, decreased social functioning and interpersonal problems must be examined. In relation to social functioning, SC is essential to understand proposed deficits in depressive disorders. In fact, although general cognition is nuclear in social functioning, SC impairments, and, specifically, ToM deficits, have been suggested to be clinically significant to MDD (Weightman et al. 2019).

ToM and social functioning have been linked suggesting that interpersonal troubles for MDD patients may be related to their impairment in understanding other's beliefs, intentions, and emotions (Cusi et al. 2013): difficulties in identifying other's mental states may contribute to conflicts and distress in social relations, provoking withdrawal from social stimuli and isolation (Berecz et al. 2016). In addition, negative bias typical of depressed patients may influence wrong inferences about other's mental states (i.e., a hyperfocus on negative characteristics of stimuli) and exacerbate maladaptive cognitive processes affecting mood and triggering depressive symptoms (Berecz et al. 2016; Weightman et al. 2014). All in all, deficits in ToM in MDD patients may be linked to a diminished motivation to communicate and to establish relations with others, reducing rewards from social interactions and reinforcing isolation (Weightman et al. 2014).

Specifically, deficits in ToM, such as identifying emotions and understanding emotional and cognitive mental states have been proposed as essential factors involved in MDD due to an adequate ToM allowing proper social reactions. Taking it into account, deficits in ToM may contribute to withdrawal in depressive patients because a wrong understanding of others (Cusi et al. 2013; Kupferberg et al. 2016; Maleki et al. 2020; Weightman et al. 2014; Wolkenstein et al. 2011).

Deficits in ToM in depressive disorders have been found in review and meta-analytic studies (Bora and Berk 2015; Nestor et al. 2022; Weightman et al. 2014; van Neerven et al. 2021). They have been described both in cognitive and affective ToM and in verbal and visual tasks (Bora and Berk 2015; Nestor et al. 2022; Wang et al. 2018). However, it is worth noting that Weightman et al. (2014) stress that deficits in affective ToM are more robust than in cognitive ToM owing to less studies assessing cognitive ToM in depressed patients.

Regarding ToM deficits, the severity of the depressive symptomatology must be considered. As aforementioned, altered ToM has been found in depressive disorders but clinical heterogeneity between disorders must be considered in those results due to a great deal of diversity and severity within depressive disorders (APA 2022; Berecz et al. 2016).

Despite a growing interest is observed among researchers in the study of ToM in depressive disorders, unfortunately, mixed evidence has been found so far. On that subject, it must be borne in mind that different assessment tools have been used across studies, and differences in complexity may influence the different results found (Berecz et al. 2016; Bora and Berk 2015) as was also mentioned in a previous section. At this point, neurocognition should be mentioned.

Neurocognitive measures are not consistently measured in ToM studies in patients with a depressive disorder (Weightman et al. 2014). A systematic review highlights a relationship between deficits in ToM and executive functions (Pagnoni et al. 2022). Regarding specific executive functions, ToM deficits and memory in chronic depressed patients (Zobel et al. 2010) and between cognitive flexibility and ToM during a current MDD episode (Förster et al. 2018). However, other studies pointed out that differences in ToM deficits have been independently found from neurocognitive measures in depressive patients (Bora and Berk 2015; Inoue et al. 2004; Ladegaard et al. 2014; Wang et al. 2008).

Focusing on ToM, a recent meta-analysis which distinguishes decoding and reasoning ToM abilities in depressive patients has found robust deficits in depressive patients in both abilities (Nestor et al. 2022). However, experimental results from single studies have obtained different results.

Regarding the *Eyes Test*, no differences in decoding ability were found between patients with depressive disorders and healthy participants, but patients with depressive disorders were more accurate decoding negative stimuli (Wolkenstein et al. 2011). This negative interpretation bias has been suggested in review studies as such from Weightman et al. (2014) and in meta-analysis (Nestor et al. 2022), considering it congruent with Beck's cognitive paradigm of depression. It may reflect a hypervigilance to negative emotional stimuli which may interfere in the depressed patients' social behavior.

In relation to that, Wolkenstein et al. (2011) found that an older age in the depressive disorder onset was related to less accuracy decoding negative stimuli: this age-dependent effect may reflect a link into an early onset of the depressive disorder and high accuracy to decode negative stimuli. This explanation could reflect a selection bias to prioritize negative social stimulus processing possible caused by a paucity of pleasant interpersonal relations in patients with an early onset of the disorder that could lead to better negative stimuli's decoding and also that this hypersensitivity could be a risk factor for early depressive disorder's onset.

On the contrary, other studies assessing mental decoding through the *Eyes Test* have found deficits in MDD patients decoding mental states (Lee et al. 2005; Nestor et al. 2022; Maleki et al. 2020; Wang et al. 2008). Moreover, severe MDD patients were significatively less accurate than healthy participants in *Eyes Test*, and moderate MDD patients just show impairments in a trend level (Lee et al. 2005). In addition, great deficits in decoding mental states in severe MDD patients were found in the study of Wang et al. (2008) also considering patients with MDD and psychotic symptoms, who obtained worse results than MDD patients without psychotic symptoms. These deficits were not found in neurocognitive tasks, reflecting a specific role of MDD symptomatology in ToM impairment.

About reasoning about mental states (i.e., (comprehension that the other has a different mind and making predictions considering others mental states), patients with depressive disorders have shown a hypomentalizing trend both in reasoning (i.e., deficits in deducing mental states from non-direct cues) about cognitive and affective mental states through the *MASC* (Wolkenstein et al. 2011). This deficit in reasoning was not attributable to deficits in attention, short term memory, verbal intelligence, or low educational level, and it reflects no deficits in inferring mental states from observable stimuli but in deducing mental states from contextual cues.

In the same line, a specific reduced ability inferring and reasoning about others' complex intentions through an ecological assessment was found in MDD. However, literal tasks in this assessment did not show any difference between healthy participants and MDD patients. Reduced appreciation of mental states and difficulties in social reasoning has been found in MDD patients (Ladegaard et al. 2014).

However, contradictory results have been found. A study from Wilbertz et al. (2009) found no differences between patients with a depressive disorder and "healthy participants", but the study reflects a trend into a better affective ToM compared to cognitive ToM through the *MASC*. However, these results should be interpreted cautiously due to "healthy participants" being not assessed regarding its psychopathology level. In addition, Seitz et al. (2022) did not find impairments in early onset MDD patients through the *MASC*. These patients had moderate levels of trauma, and Seitz et al. (2022) suggested that results may be due to clinical characteristics of their sample and also that patients with MDD and a history of emotional abuse are more sensitive to interpersonal cues and they are better at reasoning about other's mental states.

Considering complexity of ToM tasks, deficits in second-order tasks have been found in MDD (Cusi et al. 2013; Inoue et al. 2004; Maleki et al. 2020; Wang et al. 2018). Specifically, deficits in affective ToM through second-order false belief task was associated to depressive symptomatology and a worse executive function (Wang et al. 2018). These deficits were found in patients with mild severe MDD symptomatology (Cusi et al. 2013) and even when depressive symptomatology was remitted (Inoue et al. 2004).

Regarding clinical prognosis, ToM deficits have been placed also as essential in depressive disorders: impairments in ToM may affect the prognosis since deficits in ToM have also been related to increased probability to relapse. This relationship has been described after relapse from a depressive episode (Inoue et al. 2006) and during the remission state of a MDD episode (Yamada et al. 2015). Inoue et al. (2006) found a relationship to deficits in second-order beliefs after recovery from a depressive episode and a depressive relapse after 1 year of the assessment. Furthermore, deficits in second-order beliefs were linked to poor social functioning (Inoue et al. 2006; Yamada et al. 2015). It may reflect that a deficit in ToM, even after recovery from the affective disorder, may be a risk factor for poorer outcome and patients with ToM deficit may present problems to properly adapt to daily routine and social relations.

Regarding RF skills, the most described allusion in literature is referred to the depression treatment thought RF, Mentalization-Based Treatment (MBT) may contribute to moderate depression symptoms: patients with higher mentalizing abilities would benefit from MBT while patients with poorer RF capacity might have a resistance treatment (Halstensen et al. 2021). Also, it has been described the therapeutic effect of the MBT in depression symptoms, such as alexithymia (Bressi et al. 2017). Other studies described RF skills as a protective factor, to reduce the effect of adverse childhood experiences in the subsequent depression in mid adult life (Li et al. 2020).

5.3.5 *Social Anxiety Disorder*

Anxiety disorders are characterized by an intense fear -emotional response to an imminent threat real or imagined, and anxiety, an emotional response to a future threat, real or imagined, with behavior alterations associated, as avoiding behaviors. Different constellations of anxiety-related symptoms may be seen according to the type of stimuli or situations which may trigger negative emotional responses (APA 2022).

Specifically, social anxiety disease (SAD) is an anxiety disorder characterized by an excessive strong fear or anxiety in almost all social situations. Characteristic of these social situations is that the person could be evaluated in them and the emotional response is produced by the fear of a negative appraisal or to be embarrassed in those social situations. Due to that, social situations are usually avoided or lived with intense fear or anxiety. It has important consequences as patients tend to avoid social situations, limiting an usual life (APA 2022). Whereas the age standardized prevalence for anxiety disorders is 3.7795 % (GBD Mental Disorders Collaborators 2022) the life prevalence for SAD is 4% (Stein et al. 2017).

In SAD, ToM skills may be compromised. In this regard, higher levels of anxiety have been related with decreased ToM abilities (Alvi et al. 2022). Also, subclinical symptoms of social anxiety were found as being unrelated to ToM performance (An and Kochanska 2021). In general, a worse performance was found for SAD patients (Washburn et al. 2015), being stronger these deficits when a negative valence stimulus is present.

In SAD in children, relations between worse ToM abilities referred from parents, and children anxiety were found. In addition, a worse ToM was related to interpretation of neutral stimuli as threatening, and to a high anxiety symptomatology (Moldovan and Visu-Petra 2022).

Behaviorally-inhibited children and teenagers have shown an inhibited behavior, in terms of avoiding expressing fear and shyness emotions or thoughts; this may univocally facilitate the development of a SAD (An and Kochanska 2021). In a longitudinal study, ToM was pointed as a mediator through inhibited behaviors and SAD: inhibited behaviors were related with SAD for those children with low ToM but this relationship was not found for those with good ToM abilities. Consequently, ToM was related to the development of SAD just in children with low ToM skills (An and Kochanska 2021).

Moreover, early manifestations of ToM in children were negatively associated with expression of shyness in a negative way (e.g., silence, frown, angry, upset) and with social anxiety (Colonnesi et al. 2016). It could reflect that difficulties in understanding social interactions could be related to the development and generalization of an avoidant behavior, typical of anxiety. It would entail restricted adaptative reactions in the social interactions, having less positive social interactions and a significant loss of social rewards. However, advanced ToM tasks were positively associated with positive expressions of shyness (e.g., silence but smiling), suggesting that a tendency of expressing shyness in a positive way, would be related with higher social levels which could stimulate contact and social experiences, facilitating the development of advanced levels of social understanding. This good ToM development in childhood could facilitate social compression and promote positive social experience and high self-confidence (Colonnesi et al. 2016).

Adverse social experiences, inhibited behavioral pattern and a deficitarian ToM, could contribute to biased interpretation of others' mental states, facilitating an engagement into social avoidance and social isolation. Same experiences for good interpretations of others' mental states, could be useful for children to face these situations, even if they could have difficulties related to the social world, but these do not facilitate social avoidance and SAD development (An and Kochanska 2021).

Specifically, in adults, SAD patients make more ToM mistakes than healthy participants when inference about mental states must be derived from negative valence stimuli in the *Eyes Test* (Hezel and McNally 2014).

SAD participants showed an impaired performance on the *MASC*, with more hypermentalizing mistakes. It reflects a wrong interpretation of others' mental states, but not a wider bias leading to an overall negative interpretation, regardless of stimulus: deficits were specific to social information, without showing this deficit pattern for control questions (Hezel and McNally 2014). In comparison with other disorders or mental conditions, such as body dysmorphic disorder and obsessive–compulsive disorder, SAD may show a worse ToM performance, even when non-social stimuli are present (Buhlmann et al. 2015). It is worth considering that it is more likely that SAD patients show impairments in cognitive ToM than in affective ToM (Buhlmann et al. 2015).

Regarding this hypermentalizing pattern, there is a tendency in SAD children and adults to overthink others' mental states of close relations (Pequet and Warnell 2021). Authors specified that it could be due to a higher amount of familiar information from close people, which may facilitate this overthinking bias. In addition, close relations could have more impact on SAD patients than others and, owing to its closeness, their potential evaluation could be more considered. It could trigger a distinct mentalizing pattern in which SAD patients are more hypervigilant and they hypermentalize mental states from its close relations: their mental states would have a great importance to SAD patients and because of that SAD patients would need to exactly know their close relations' mental states (Pequet and Warnell 2021).

It is essential to consider that SAD patients, fearing others' evaluation, could understand the study as an evaluative context, because they would know that their results in tasks would be evaluated by researchers. It may influence their performance in tasks (Hezel and McNally 2014).

In addition, SAD symptoms may influence ToM performance in other ways. Social avoidance could dramatically interrupt the normative development of other's mental state decoding. For example, considering the *Eyes Test*, in which SAD patients were impaired decoding mental states from negative valence stimuli (Hezel and McNally 2014), negative valence eyes may evoke fear or anxiety to SAD patients. If SAD patients feel fear or anxiety, they may react by avoiding eye contact and answering very quickly without a proper analysis of the mental state. In the study from Washburn et al. (2015), SAD and comorbidities SAD and MDD people were studied, and SAD showed this unique impaired pattern.

Regarding the *MASC*, its format may stimulate a bias towards an answer pattern in which the participant feels as being part of the situation which is viewing. It may facilitate the trigger of SAD symptoms, such as fear or anxiety, due to the social situations that are taking place in the *MASC*'s film. It could lead to an anxiety state in the participant while is answering the questions. It could interfere with performance (e.g., not reading the question properly, answering quickly...), diminishing the correct answers and achieving worse results on the task (Buhlmann et al. 2015).

5.3.6 Eating Disorders

Eating disorders (ED) is a diagnostic category of disorders characterized by a pervasive alteration of eating behavior which provokes an altered consumption of food, impairing health, and psychosocial functioning (APA 2022). The prevalence of ED is 0.7173% (Santomauro et al. 2021) and the age standardized prevalence is 0.174 % (GBD Mental Disorders Collaborators 2022). Within ED, anorexia nervosa (AN), bulimia nervosa (BN) and binge-eating disorders (BED) will be addressed (APA 2022).

AN is characterized by the restriction of the intake in relation to the individual necessities leading to a significantly low weight, an intense fear to gain weight and alteration of the body image, a self-evaluation based on body image or weight and a lack of recognition of the severe risk of the low body weight. Two presentation forms may be adopted by an AN: restricting (AN-R) or binge-eating/purging

(AN-P). In AN-R, in the last 3 months no episodes of binge eating[4] or purging have been present and weight loss has been reached by dieting, fasting or excessive exercise. In AN-P, in the last 3 months recurrent episodes of binge eating or purging behavior (i.e., vomiting, using laxatives or diuretics) have been present (APA 2022). Its age-standardized rate prevalence is 0.094 (van Hoeken and Hoek 2020).

BN is characterized by recurrent binge eating episodes and inappropriate compensatory behaviors to prevent weight gain, which take place at least once a week, and a self-evaluation centered on the body shape and weight (APA 2022). Its age-standardized rate prevalence is 0.338 (van Hoeken and Hoek 2020).

BED is characterized by recurrent binge eating episodes associated with fast ingest, experiencing being full uncomfortably, isolated binge eating episodes (to avoid embarrassment from the quantity of the food eaten) and feeling guilty or depressed after the binge eating episode. Binge eating episodes are experienced with distress and they take place at least once a week (APA 2022). Its prevalence is 0.7173 (Santomauro et al. 2021).

Significative impairments on the mentalization ability of ED patients have been found, displaying a trend into mentalization deficits. Nevertheless different mentalization profiles according to ED diagnosis have to be considered (Simonsen et al. 2020).

In general, ED patients performed worse in *Eyes Test*, showing a worse affective ToM, with significant impairments in total accuracy score and differentiating by emotional valence, ED patients show deficits in the positive and neutral mental states but not in negative emotions. It could reflect a trouble in understanding neutral mental states, with a bias into a search for an underlying emotion, falling into mentalization mistakes with a negative trend (Medina-Pradas et al. 2012).

Systematic results of deficits in ToM in AN patients have been described, specially on cognitive ToM, without finding consistently affective ToM impairments in the literature (Tauro et al. 2022).

One of the pioneering study of ToM in AN showed no specific cognitive ToM deficits in AN patients (Tchanturia et al. 2004). Instead, patients did show impaired performance in mentalizing condition and even control condition, and no differences were observed between AN-R and AN-BP subtypes. However, post hoc analysis highlighted a worse performance in the mentalizing items than in control ones.

This trend into a worse cognitive ToM was again observed in the study by Russell et al. (2008), where AN patients showed impairments on mentalizing condition and control condition, indicating non-specific deficits in ToM in AN, at the same time as showing difficulties in AN patients to take advantage from the mentalization content to solve the tasks (i.e., to use mental states to predict behavior). In this study, however, body mass index and years of illness showed associations with performance in cognitive ToM, opening a debate on whether the deficits are associated

---

[4] A binge eating episode is characterised by eating in a concrete period of time a greater food than most individuals would eat in similar circumstances in that period of time. A binge eating is also featured by a sense of lack of control during the episode.

with either the AN symptomatology or with developmental and metabolic factors. In this regard, a robust meta-analysis demonstrated that AN patients may show evident deficits in reasoning about mental states and these deficits may be present in acute phases of the illness and in even in recovered ones (Bora and Köse 2016).

Konstantakopoulos et al. (2020) also showed that AN patients had significantly poorer cognitive ToM performance. This poorer performance was independent of other neurocognitive deficits. It could influence self-evaluation, which precises of mentalizing about our-selves as the same way as we mentalize about how others are looking at us, showing a dysfunctional self-reflection functioning.

Regarding affective ToM, some studies found impairments in affective ToM in AN. In the study by Russell et al. (2008), AN patients showed an impaired performance in the *Eyes Test*, specifically regarding female eye stimuli. Authors suggest that the sexualized nature of these eyes could be a handicap for AN patients, a clinical group in which problems with sexuality, and facial and body image have been widely described. In the same line, Harrison et al. (2009) found less correct answers in AN patients in the *Eyes Test*, and the ToM results were related with problems of accepting and regulating emotions.

A tendency toward lower scores of acute AN participants in comparison to healthy participants, with moderate effect of the ToM deficits, was found in *Eyes Test* in a study by Oldershaw et al. (2010). In the *Reading the Mind in the Eyes Test* -the *Eyes Test*- and the *Reading the Mind in the Voice Test* AN participants performed moderately lower in comparison with healthy participants and patients under remission, specially in terms of decoding negative emotions (differences in neutral or cognitive mental states were not found). Indeed, regarding positive emotions, recovered AN participants did not show differences to AN in the acute phase in the *Eyes Test*, stressing a deficitarian ToM profile in both groups. On the contrary, in the *Reading the Mind in the Voice Test*, recovered AN patients performed much better than AN inferring positive emotions from auditory stimuli (voices). Tapajóz Pereira de Sampaio et al. (2013) found that the AN group showed impairments in total score in -the *Eyes Test* and in decoding negative emotions, indicating troubles in the emotion decoding and mentalizing, showing a specific deficit. Unlikely, in this study the worse performance was found reading the mental states from male eyes.

On the contrary, other studies have not found impairments in affective ToM (Adenzato et al. 2012; Medina-Pradas et al. 2012). Adenzato et al. (2012), studying ToM performance in AN patients with the shorter chronicity in comparison with other studies, through -the *Eyes Test*, no deficits were found in AN. Adenzato et al. (2012) suggest that a normal body-index mass in their AN patients and the relative short duration of the illness may influence their results.

In relation to the clinical profile, contradictory results have been found in the literature. In this regard, no relationship with AN severity, anxiety, depression, illness chronicity and age of onset have been found across the different experimental studies which have been described in this ED section. Although worse deficits have been found in acute phases of AN, in meta-analysis studies (Preti et al. 2022) and in experimental ones (Oldershaw et al. 2010), no correlations with ToM deficits have been found between them.

Conversely, two key exceptions to this lack of relationship between the clinical status and ToM deficits were, on the one hand, the study of Harrison et al. (2010). In it, the severity of the ED symptomatology was related with a worse affective ToM performance and they did not find differences between healthy participants and ED participants (AN and BN) when controlling for antidepressant medication prescription. On the other hand, in the study of Russell et al. (2008), positive relations between the *Eyes Test* scores and functionality were found, connecting them with AN social problems. However, implications of clinical characteristics in ToM impairments have been found in meta-analytic results, where depressive symptomatology appears to moderate ToM capacity (Bora and Köse 2016). ToM deficit severity has been related to symptom chronicity and body mass index (which could reflect starvation consequences) have been highlighted (Preti et al. 2022). In addition, ToM abilities could have implications regarding the insight of the illness, because they could also contribute to resistance to the treatment, common on acute phases of AN, impeding recovering (Bora and Köse 2016).

At this point it should be mentioned that mentalization in AN is determined by the loss of social-cognitive capacity based on external features of others. It is an overestimation of the identifying of specific maladaptive processes that requires an understanding of the other´s interior mental world, which supposes a tendency to hypermentalization (Cortés-García et al. 2021).

Regarding BN, BN patients show deficits similar to AN patients under remission. These deficits are circumscribed to make inferences about mental states and not in decoding them (Bora and Köse 2016; Tapajóz Pereira de Sampaio et al. 2013). The BN mentalization profile seems to be milder than in AN, because in some studies, BN may score in a similar way than people without an ED, a non impaired ToM (Simonsen et al. 2020).

However, emotion decoding skills in BN shows a tendency into a better recognition of complex negative emotions, suggesting a negative bias. Kenyon et al. (2012) did not find signs of a worse performance in the *Reading the Mind in the Eyes Test* -the *Eyes Test*- and in the *Reading the Mind in the Voice Test*. However, BN patients were superior to healthy participants in negative emotional decoding. It could indicate a greater sensitivity to evaluations from others. It may interact with the mentalization of others in a negative way, provoking a low self-esteem and isolation, typically seen in BN. BN patients may exhibit deficits in decoding positive emotions and cognitive and neutral states. Difficulties in inferring positive emotions may possibly be linked with a smaller ability to personally experience positive emotional states (Medina-Pradas et al. 2012).

Considering together both main ED diagnoses, there is a paucity of studies comparing AN and BN. A mixed sample of ED patients, comprising both AN and BN patients, detected fewer correct emotions than healthy participants, showing that affective ToM measured by emotion recognition into the *Eyes Test*, was not specific to AN. BN patients scored significantly more correct answers than those with AN. Larger differences were found between healthy participants and AN-R participants. AN-P patients did not show different performance than healthy participants. Differences in ToM between AN and BN point out that abnormalities in inferring other's mental state are linked specifically to AN, and they are not associated with ED core symptoms, such as the worries about weight and image (Bora and Köse 2016). Meta-analysis results highlight an AN worse mentalization than BN (Simonsen et al. 2020) and experimental studies such as that from Tapajóz Pereira de Sampaio et al. (2013) concluded that AN patients may be less accurate in affective and cognitive ToM tasks, than BN patients.

Regarding BED, Aloi et al. (2017), studied obese patients with and without BED, obtained that there were no differences in affective ToM between them through the *Eyes Test*. However, these authors pointed out that BED scores were lower in comparison with others ED -which results have been outlined in this section-. It is essential to consider that obese participants with BED displayed a deficit in decoding their own emotions. However, deficits to decode affective mental states from others were not found, indicating a possible first-person ToM deficit which could have relations with depression levels, but further research is needed in this regard.

Turan et al. (2019) comparing adolescent with normal weight, obese and obese with BED though the *Eyes Test* and *Faux Pas Test* – and other ToM tasks -, emphasized a trend in their results in which obese BED adolescent had not better results than obese non BED ones, and worse than normal weight participants without BED. These results correlated negatively with depression levels and, again, suggest that obese people could have problems with the recognition of their own emotions, independently of the presence of an ED.

Tonelli and de Siqueira Rotenberg (2021) found that deficits in ToM tasks in obese people could be underlying emotional and social troubles which could be strongly linked to BED: they have shown impairments in the ability to identify and manage emotions and mental states of others and of themselves, independently of their clinical profile. Consequently, worse scores could be found through these domains in comparison with non-obese people of all ages.

In this field, a great heterogeneity into results and scarce investigation has a great impact on the current knowledge: results are very dissimilar between studies (Simonsen et al. 2020), and diagnostic criteria are also a point of divergence.

Studies with patients diagnosed by DSM-IV or DSM-IV-TR are more severe than those diagnosed by DSM-5, due to different symptoms criteria demand (Preti et al. 2022). In addition, BED is a category since the DSM-5 and was forming part of eating disorders not otherwise specified before it. Due to that, it is not possible nowadays to affirm these current results into a robust way and further research is highly required.

5.3.7 Borderline Personality Disorder

Borderline personality disorder (BPD) is defined as a stable pattern of behavior characterized by instability in interpersonal relationships, self-image and emotions, high levels of impulsivity and limited impulse control. Other problems that may be present in BPD patients are: fear of abandonment, chronic feelings of emptiness, self-injurious behavior (non-suicidal self-harm and suicide attempt), antisocial behavior, and transient paranoid ideas related to stress or severe dissociative symptoms. Sometimes high impulsivity and lack of emotional regulation end up leading to varying neuropsychiatric problems, such as addiction problems (e.g., drugs, alcohol, sex, gambling) and disordered eating problems (e.g., binge eating, vomiting) (APA 2022). BPD is also featured by significant impairment and profound distress at individual and community levels. BPD may coexist with other medical and psychiatric pathologies like mood disorders, anxiety, substance abuse and eating disorders (Shah and Zanarini 2018).

Epidemiological studies indicate a BPD lifetime prevalence rate of 1.6% in the general population. Figures in clinical populations are quite impressive. In this regard, studies have shown that one in five psychiatric patients may show a BPD diagnosis (Ellison et al. 2018). For all these reasons, BPD is a problem of important relevance in the psychiatric clinic.

It has been suggested that the recurrent interpersonal difficulties in interpersonal relationships typical of BPD could be related to a possible alteration in mentalizing processes, due to its involvement in social interactions. Numerous studies have investigated ToM in people with BPD, but the meta-analysis of Németh et al. (2018) concludes that results appear to be heterogeneous due to low sample sizes, variability in the ToM processes and components assessed, and the high comorbidities with other psychiatric diagnoses.

Some studies have detected deficits in both cognitive and affective ToM skills, with BPD patients showing more errors than the healthy participants in tests of false beliefs, faux pas, and greater difficulties in attributing affective mental states (Hillmann et al. 2021; Pourmohammad et al. 2021). Rogoff et al. (2021) assessed mentalizing skills towards the self and the others. In this study, BPD patients showed a worse performance on both types of mentalization tasks. However, it appears that on tests of affective mental state recognition (affective items of the *Eye Test*), people with BPD respond better than healthy participants – more hits and shorter response time – (Fertuck et al. 2009). Fertuck et al. (2009) found that the better performance on affective emotional state recognition tasks observed in BPD patients compared to healthy participants was associated with an increased activation of the amygdala and other cortical areas involved in overall emotion processing (e.g., medial frontal gyrus, left temporal pole and medial temporal gyrus). Moreover, RF impairments in BPD, described as ineffective mentalization capacity which resulted in a deficit though interpersonal relations, is derived from an insecure attachment style (Valdivieso-Jiménez 2020). In fact, due to the high impulsivity, BPD is associated with a strong tendency to hypermentalize (Sharp and Vanwoerden 2015). Interventions designed to improve RF, also called Mentalization-Based Treatment (MBT), have been associated with a therapeutic management of patients with this personality disorder (De Oliveira et al. 2017).

Considering all these findings, they may suggest that BPD patients show some kind of selection bias to prioritize social emotional stimuli processing. These results may also go in line with the evident link between attachment and mentalization processes. According to Fonagy and Bateman (2008), an individual with BPD who grows up in a non-reflective, non-validating and often abusive family environment, may develop an increased emotional hypervigilance to social stimuli, particularly to those stimuli with negative emotional content. ToM skill deficits in BPD patients may be independent from overall social cognition performance, as some studies point that recognition of isolated facial emotions (e.g., typical in experimental emotion facial recognition paradigms) and prosodic emotions remain preserved in BPD patients (Lynch et al. 2006; Wagner and Linehan 1999). BPD patients show impairments when emotion recognition should be unfolded in complex stimulus contexts (e.g., those where information from facial/prosodic stimuli should be integrated), as well as in the discrimination of non-emotional facial features. This higher-order impairment may be associated with interpersonal antagonism (e.g. hostility, suspiciousness, aggressive behavior), specifically in distrust and aggression contexts (Minzenberg et al. 2006). Consistent evidence on specific mental state attribution and cognitive ToM deficits of BPD patients was provided by Németh et al. (2018) in their robust meta-analysis. First, the meta-analysis supports the lack of a poorer performance of BPD patients in mental state decoding tasks (using the *Eyes Test*). Interestingly, the authors concluded that performance in complex ToM tasks with higher contextual demands seem to be impaired in BPD patients (Németh et al. 2018). In this regard, the *MASC* task may be an optimal tool to assess complex ToM skill deficits. The study by Ortega-Díaz et al. (2021) highlighted that both the BPD patients and their healthy first-degree relatives display significant deficits in ToM (measured by the *MASC* task), suggesting a genetic predisposition to the disorder, and proposing ToM as a possible endophenotypic marker of BPD and a potential marker for early detection of the disorder and intervention for BPD.

5.3.8 *Theory of Mind in Other Neuropsychiatric Disorders*

Nowadays the studying of ToM is widespread to other psychiatric disorders in which social relations are used to be impaired. Promising results are being obtained and they will allow a better understanding of ToM impairments in neuropsychiatric disorders to develop suitable clinical treatments.

5.3.8.1 Theory of Mind in Posttraumatic Stress Disorder

In Posttraumatic Stress Disorder (PTSD) large impairments in ToM ability and emotional recognition have been found (Plana et al. 2013). Deficits in both affective and cognitive have been described (Seitz et al. 2022). Regarding cognitive ToM a trend into a worse cognitive ToM and hypermentalizing mistakes have been described (Seitz et al. 2022). Considering affective ToM, Couette et al. (2020)

pointed out that deficits in PTSD are circumscribed to affective ToM and they may be related with the emotional impact of the trauma, showing an improvement of affective ToM abilities as time goes by. Mentalization deficits could increase consequences of trauma, leading to wrong inferences about others mental states and emotional mental states, contributing to PTSD symptomatology.

5.3.8.2 Theory of Mind in Attention Deficit and Hyperactivity Disorder

Regarding Attention Deficit and Hyperactivity Disorder (ADHD), impairments in ToM have been described in elemental ToM abilities (e.g., understanding of basic emotions), first-order and second-order false beliefs (Berenguer-Forner et al. 2017). However, they are comparable to other neurocognitive deficits such as verbal memory, inhibition, and attention (Bora and Pantelis 2016), congruent with the inherent executive dysfunction of the disorder. Owing to that, the study of ToM impairments in ADHD implies discriminating if deficits are primary or secondary to executive deficits: it is possible that executive dysfunction may alter the irrelevant stimulus suppression necessary to appropriately consider other's mental states. Following this argument, compromised executive functions, as characterized ADHD, influences ToM performance in children and adults (Berenguer-Forner et al. 2017; Hutchins et al. 2016; Tatar and Cansız 2022).

ToM deficits have been link with low inhibitory control, which is highly involved in the management of cognitive and emotional processes in social situations, influencing the ability to perceive the other's point of view and to express emotions in a suitable way (Pineda-Alhucema et al. 2018). In a similar way, inhibition of self-perspective has been found to be impaired in ADHD children, a deficit which strongly disrupts a good performance of a ToM task: it complicates the coordination of different perspectives such as the own-self and the other's, even if the ADHD patient recognize the other's mental state (Fizke et al. 2014).

Also, impairments in ToM have been positively related to working memory capacity, because it is essential to store and manipulate social information allowing a proper mental state inference (Imanipour et al. 2021; Papadopoulos et al. 2005).

In relation to attention, a worse attentional capacity has been linked to worse ToM abilities in demanding tasks like second-order false beliefs task and faux pas (Papadopoulos et al. 2005; Mary et al. 2015). Inattention symptoms could hinder ADHD patients to attend to relevant social clues, blocking the correct interpretation of the social situation.

5.3.8.3 Theory of Mind in Disruptive, Impulse-Control and Conduct Disorders

Considering Disruptive, Impulse-control and Conduct disorders -Oppositional Defiant Disorder (ODD) and Conduct Disorder (CD)- deficits in ToM have been reported. Delays in the emergence of ToM abilities are linked to the development of chronic externalizing problems with an early onset (before 3 years) as longitudinal studies suggest: comparing trajectories of children with high, low, or decreasing externalizing problems, those with high conduct problems show the more impaired ToM skills. It suggests that ToM abilities may play a nuclear role to face high intense emotional situations in a functional way without incur in disrupted behaviors (Olson et al. 2017).

In addition, a distorted mentalization (i.e., an overly positive bias into their own-self mentalization) has been independently associated with disrupted behaviors or conduct problems, predicting significant conduct problems: if individuals with disrupted behaviors or conduct problems inflate the image of themselves, they could feel questioned when the feedback of others is contrary, making them act out. It may reflect a deficit to understand that others have their own mental states, and to understand that they could be different to theirs (Ha et al. 2011).

Moreover, relations between ToM and aggression have been found. Firstly, regarding indirect aggression, is important to consider the valence that the children ascribe to it due to a possible impaired ToM in children who frequently received aggression from others: these negative experiences with others and an impaired ToM may lead to interpret the situation as threatening and to initiate aggression behaviors of others, developing an hostile attribution bias (Renouf et al. 2010a). Contrary, ToM has been also positively related to proactive aggression in bullied children: preserved ToM abilities could facilitate considering the risk and benefits of aggression and anticipating others' reaction and use aggression to its own interest, like gain social recognition (Renouf et al. 2010b).

Focusing specifically on ODD, moderate negative relations were found between first-order beliefs, as well as second-order beliefs, and severity of ODD symptoms. This relationship was much stronger for second-order tasks and *Faux Pas Test* performance (Pal et al. 2021). ToM performance was positively related with headstrong dimension of ODD (defying rules, annoying and blaming or arguing) mostly in affective ToM tasks and it was negatively associated with reaction time to answer the tasks. Contrary, the irritable dimension (being angry easily with others, being easily annoyed and losing one's temper) was linked to slower reactions to social stimuli, which was also related with higher levels of anxiety. These results from the study of de la Osa et al. (2016) may reflect that deficits in recognition of mental states lead to a pattern of irritable behavior, whereas a preserved understanding of other's mental states could be used in a non adaptive way to defy authority (headstrong dimension).

Referring to CD, results about cognitive ToM have shown that it is not impaired in children with CD (Jones et al. 2010). Classical studies such as that from Happé and Frith (1996) described that children with CD may show the largest impairments in behaviors which need a functional ToM. CD patients may have a functional, intact ToM, but it could be influenced by adverse contextual contexts: adverse environments could be linked to deficits in everyday situations which require mentalization, showing a tendency to wrongly ascribe a mental state in terms of threat and aggressive intentions despite a conserved ToM (Happé and Frith 1996). However, a current study found that the ability of recognizing others' intentions and emotions in CD predicts significatively conduct problems. This deficient ToM pattern could support the hypothesis that difficulties in the interpretation of the intentions and emotions of others may lead to an inappropriate behavior as observed in CD (Wells et al. 2019).

All in all, the improvement of the disrupted symptomatology, with the development of CD at 10 years after presenting ODD at 7 years, was moderated by the capacity to understand subjective states of others. As these results highlight, children with ODD who later develop CD could have worse ToM than those ODD children who do not develop CD (Mandy et al. 2013). Taking it into consideration Mandy et al. (2013) proposed two explanations: ToM may be a mediator between the deterioration of disrupted symptoms or ToM deficits could predispose high difficulties to understand the world and disrupted symptoms could be a way to face them, even in a non adaptive way.

5.3.9 Theory of Mind in Suicide Behavior Disorder

Suicide is one of the leading causes of preventable death in the world, responsible for more than 700,000 deaths a year, a 1.3% of all the deaths in 2019 (World Health Organization 2021). It´s rates have been raising in the last years, in 2020 there was 8,2 suicides per 100,000 inhabitants (de la Torre-Luque et al. 2022a). Additionally, experts estimate for every consummate suicide almost 20 suicide attempters in Spain, even data show an increase trend emergency department visits due to suicidality (Hernández-Calle et al. 2022).

Suicide behavior disorder (SBD) was introduced in DSM-5 as a disorder for further consideration and potential acceptance into the diagnostic system (Fehling and Selby 2021). All following criteria must be present for diagnostic: within the last 24 months, the individual has made a suicide attempt, the act does not meet criteria for non-suicidal self-injury (NSSI), diagnosis is not applied to suicidal ideation or to preparatory acts, the act was not initiated during a state of delirium or confusion and the act was not undertaken solely for a political or religious objective. Prevalence rate of SBD with other disorders such as bipolar disorder is elevated (42%), and with depression (34%) as well (Baldessarini and Tondo 2020), in this last especially in adolescent population (Liu et al. 2022); even though its diagnostic is independent from other disorders and its validity is growing with evidence (Fehling and Selby 2021).

The literature also reveals a variety of long-term environmental factors on risk for suicide behavior, such as adverse childhood experiences or short-term social stressors, for example breaking-up of intimate relationships. ToM is a social-cognitive ability that allows individuals to understand the feelings and thoughts of others, what is crucial for a healthy development of social interactions. A recent meta-analysis relates deficits in ToM with suicidality, describing a negative relationship between ToM abilities and suicide behavior (Nestor and Sutherland 2022). The lack of ToM, which is named hypomentalization, involves limited constructive social interactions, derived from a poor sense of self, that can contribute to increasing feelings of social withdrawal which in turn, can increase the risk of pathologies, such as SDB (Stagaki et al. 2022). Bearing in mind the key corollaries from the varying models of RF, deficits in mentalization skills may disturb normative emotional regulation processes, the capacity to build networks of social support and an effective social learning. Note that self-destructive behavior (in its main forms, non-suicidal self-harm and suicidal behavior) may be considered a maladaptive coping strategies to deal with stress (Klonsky et al. 2016; Martin et al. 2017; Hatkevich et al. 2019a; Berardelli et al. 2022). In addition, results of French RFQ-validation study, one of the most commonly used RF evaluation test, also described an association between hypomentalization and the apparition of non-suicidal self-harm in adults (Badoud et al. 2015).

On the other hand, other studies show how excessive ToM abilities, or over-attributing the mental/emotional states of others (hypermentalization), has been positively related with recent suicidal ideation and behavior as well (Hatkevich et al. 2019b). As previously mentioned, hypermentalization has been strongly described in BPD, one of the most suicidality related disorders (Bo et al. 2017; McLaren et al. 2022). It is characterized by making excessively inferences based on social cues of the others, i.e., a tendency to go beyond the evidence provided making inaccurate assumptions. Interventions designed to improve the ineffective mentalization tendencies, also called Mentalization-Based Treatment (MBT), have been related to the achievement even superior compared with other treatment in the reduce of BPD symptoms such as non-suicidal self-harm (Vogt and Norman 2019). Even, a randomized controlled trial with BDP has been described a reduction in the rate of suicide attempts after 18 months of MBT (Carlyle et al. 2020). In the same line, MBT has been compared with the treatment as usual in the management of depressive and suicidality symptoms during 1-year follow-up, founding MBT was more effective in the reducing self-harm (Rossouw and Fonagy 2012).

As can be seen, RF studies shed light into the influence of the mentalization on the self-harm and suicidality; i.e., an effective mentalizing could reduce the risk of the apparition of suicide ideation.

5.4 Conclusions

Theory of Mind is a cognitive construct that involves different levels of complexity. It refers to the capacity to understand other people by ascribing mental states to them, inferring what is happening in their minds. Advances in neuroscience show two dissociable but related components: cognitive ToM and affective ToM. Cognitive ToM assessment includes simpler tasks such as understanding false beliefs, as well higher-order beliefs such as lie, sarcasm, metaphor, or attribution of intentions. On the other hand, Affective ToM assessment involves facial recognition of emotions and more complex task of understanding social situations with emotional content. Assessment tools have evolved to become more complex in assessing the ability to abstract affective and cognitive mental states, as well as other related constructs such as social cognition and reflective function.

Regarding psychopathology, ToM deficits appear to be very close to clinical symptomatology in all disorders studied, with relations with an impaired social functioning and with symptoms through diagnoses.

In addition, even ToM depends on neurocognitive skills, deficits in ToM -with the exception of ADHD- appears to be specific and independent from them. This positioned ToM as a specific target in the whole diagnoses revised and further investigation will be needed to fulfil research gaps.

Research in ToM is making essential progress, but unifying ToM constructs, considering all ToM dimensions and using suitable standardized tools to assess them is required to improve current knowledge.

Studies are consistent in supporting that children with an ASD may show an atypical development of ToM abilities, and mentalization deficits are considered as nuclear cognitive characteristics of ASD. The results in the studies on ASD patients are very heterogeneous and dependents of the IQ. Children with ASD and intellectual disability show wider deficits in basic ToM skills (like false belief tasks) and emotional decoding. On the other hand, persons with high and low functioning ASD present impairments in more complex tasks.

In SCZ, substantial ToM deficits are present in different degrees but in the whole psychotic spectrum and during all the phases of a psychotic illness. An impaired ToM may be related with the development of psychotic symptomatology: a wrong interpretation or even an interpretation of the personal context without considering other's mental states, may lead to wrong misattributions and a general misunderstanding of the world, promoting a distance from reality in based on these lacked inferences.

ToM deficits in BD are found specifically in cognitive ToM, but heterogeneous results have been found. As in SCZ, an impaired ToM is present during the whole illness. However, deficits are milder than in SCZ and they are just found in cognitive ToM. Because BD is formed by two main diagnoses, studies must address their differences and consider them to address more robust conclusions about the role of ToM deficits and depressive and manic symptomatology. Despite that, ToM deficits are found in cognitive ToM, hypomentalization, rather than inferring emotional mental states.

Deficits in ToM in MDD, a wrong interpretation of other's mental states is made based on an over attribution of negative mental states, which promotes interpersonal problems and enhances isolation, diminishing the possibility of social rewards. This hypervigilance to negative stimuli, this impaired ToM, may be related to the high rates of MDD relapses into people with worse ToM.

SAD appears to have deficits in ToM, especially hypermentalizing deficits about mental states, possible to diminish the fear evaluation. However, the proper research situation may be an evaluated one for SAD people, interfering in results obtained.

In ED, deficits appear to be focused on AN, which show deficits in cognitive ToM. Deficits in ToM in AN may contribute a wrong inference of mental states about themselves, maintaining AN symptomatology to overcome this wrong inferences about evaluation. In BN, although no significant ToM deficits have been reported, a tendency to fix on negative stimuli may contribute to infer a negative evaluation from other's, maintaining BN symptomatology. ToM deficits in BED are just starting to be studied, so it is difficult to drive robust conclusions, but deficits appear to be more related to the recognition of emotions. However, the disparity of research efforts in BN and, mostly, in BED is unmet for future research.

For people with BDP, it has been suggested that the characteristic interpersonal dysfunction of the disorder could be related to a possible alteration in mentalizing processes. Some studies have detected deficits in both cognitive and affective ToM skills. However, the impairments may be related to processing and state attribution in complex stimulus contexts from social world. More basic ToM skills (e.g., isolated recognition of emotional states) seem to be preserved. Hypermentalizing mistakes are common in BPD patients, and have been associated with high impulsivity.

Considering PTSD, ADHD and ODD, promising results have been obtained about how ToM deficits may influence symptomatology in PTSD and ODD and how symptomatology -executive dysfunction- affected social understanding in ADHD. However, research must continue.

Ineffective mentalizing has been associated with SBD. On the one hand, the deficits in ToM abilities have been negative related with suicide behavior. The tendency to hypomentalization, due to a poor sense of self, can affect to the emotional regulation and the social learning, which is the base of external supporting, increasing the risk of suicide behavior. On the other hand, hypermentalization has been positively associated with the suicidality, and the more often excessive inferences of the mental states of others are made, the risk of suicide behavior is increased. Considering exposed results, considering ToM in clinical practice through a transdiagnostic approach may lead to better results in treatment of neuropsychiatric disorders (Ibáñez et al. 2018), due to its relevance to be conscious about the consequences of a neuropsychiatric disorder (e.g., insight in SCZ, BD and AN) and for its contribution to disease prognosis (MDD). Because of that, considering ToM in treatment is a promising therapeutic approach. Specifically, MBT studies shed light into the influence of the mentalization in the reduction of the self-harm and suicidality; i.e., an effective mentalizing could reduce the risk of suicide attempt.

Having a preserved ToM does not prevent from psychopathology, but ToM skills may improve facing and managing symptoms of neuropsychiatric disorders, contributing to a better social functioning (Ballespí et al. 2018) and a better quality of life.

Acknowledgements This study was supported by the Instituto de Salud Carlos III-FIS research grant PI20/00229, co-funded by the European Regional Development Fund (ERDF) "A Way to Build Europe"; by the Instituto de Salud Carlos III grant number: PI19/00941 SURVIVE, co-funded by the European Union (grants numbers: COV20/00988, PI17/00768); by the European Union's Horizon 2020 research and innovation programme Societal Challenges (grant number: 101016127); by the "Fundación Española de Psiquiatría y Salud Mental" and by a

Complutense University of Madrid predoctoral grant (ref. FPU21/04434). Authors express their gratitude to professors Uta Frith, Francesca Happé, Simon Baron-Cohen, Mel Rutherford, and Doctor Valerie Stone, who kindly permit to reproduce pieces of their test on this chapter. The authors declare that they do not have any conflict of interest to disclose.

References

Adachi, T., T. Koeda, S. Hirabayashi, Y. Maeoka, M. Shiota, E.C. Wright, and A. Wada. 2004. The metaphor and sarcasm scenario test: A new instrument to help differentiate high functioning pervasive developmental disorder from attention deficit/hyperactivity disorder. *Brain & Development* 26 (5): 301–306. https://doi.org/10.1016/S0387-7604(03)00170-0.

Addington, J., T.A. Girard, B.K. Christensen, and D. Addington. 2010. Social cognition mediates illness-related and cognitive influences on social function in patients with schizophrenia-spectrum disorders. *Journal of Psychiatry & Neuroscience* 35 (1): 49–54. https://doi.org/10.1503/jpn.080039.

Adenzato, M., P. Todisco, and R.B. Ardito. 2012. Social cognition in anorexia nervosa: Evidence of preserved theory of mind and impaired emotional functioning. *PloS One* 7 (8): e44414. https://doi.org/10.1371/journal.pone.0044414.

Aloi, M., M. Rania, M. Caroleo, P. De Fazio, and C. Segura-García. 2017. Social cognition and emotional functioning in patients with binge eating disorder. *European Eating Disorders Review* 25 (3): 172–178. https://doi.org/10.1002/erv.2504.

Alvi, T., D. Kumar, and B.A. Tabak. 2022. Social anxiety and behavioral assessments of social cognition: A systematic review. *Journal of Affective Disorders* 15 (311): 17–30. https://doi.org/10.1016/j.jad.2022.04.130.

Amenta, S., C. Ferrari, and M. Balconi. 2014. Facial expression decoding in autistic and Asperger children. In *Comprehensive guide to Autism*, ed. V. Patel, V. Preedy, and C. Martin. New York, NY: Springer. https://doi.org/10.1007/978-1-4614-4788-7_114.

American Psychiatric Association. 2000. *Diagnostic and statistical manual of mental disorders (4ª ed. Revised text)*. Washington, DC: APA.

———. 2013. *Diagnostic and statistical manual of mental disorders*. 5ª ed. Washington, DC: APA.

———. 2022. *Diagnostic and statistical manual of mental disorders (Fifth Edition, Revised Text)*. Washington, DC: American Psychiatric Association.

An, D., and G. Kochanska. 2021. Theory of mind as a mechanism that accounts for the continuity or discontinuity of behavioral inhibition: A developmentally informed model of risk for social anxiety. *Journal of Abnormal Child Psychology* 49 (10): 1333–1344. https://doi.org/10.1007/s10802-021-00831-y.

Andreou, M., and V. Skrimpa. 2020. Theory of mind deficits and neurophysiological operations in autism spectrum disorders: A review. *Brain Sciences* 10 (6): 393. https://doi.org/10.3390/brainsci10060393.

Ayesa-Arriola, R., E. Setién-Suero, K.D. Neergaard, A. Ferro, M. Fatjó-Vilas, M. Ríos-Lago, S. Otero, J.M. Rodríguez-Sánchez, and B. Crespo-Facorro. 2016. Evidence for trait related theory of mind impairment in first episode psychosis patients and its relationship with processing speed: A 3 year follow-up study. *Frontiers in Psychology* 4 (7): 592. https://doi.org/10.3389/fpsyg.2016.00592.

Badoud, D., P. Luyten, E. Fonseca-Pedrero, S. Eliez, P. Fonagy, and M. Debbané. 2015. The french version of the reflective functioning questionnaire: Validity data for adolescents and adults and its association with non-suicidal self-injury. *PloS One* 10 (12): e0145892. https://doi.org/10.1371/journal.pone.0145892.

Balconi, M., S. Amenta, and C. Ferrari. 2012. Emotional decoding in facial expression, scripts and videos: A comparison between normal, autistic and Asperger children. *Research in Autism Spectrum Disorders* 6 (1): 193–203. https://doi.org/10.1016/j.rasd.2011.04.005.

Baldessarini, R.J., and L. Tondo. 2020. Suicidal risks in 12 DSM-5 psychiatric disorders. *Journal of Affective Disorders* 271: 66–73. https://doi.org/10.1016/j.jad.2020.03.083.

Ballespí, S., J. Vives, M. Debbané, C. Sharp, and N. Barrantes-Vidal. 2018. Beyond diagnosis: Mentalization and mental health from a transdiagnostic point of view in adolescents from non-clinical population. *Psychiatry Research* 270: 755–763. https://doi.org/10.1016/j.psychres.2018.10.048.

Baron-Cohen, S. 1987. Autism and symbolic play. *British Journal of Developmental Psychology* 5 (2): 139–148. https://doi.org/10.1111/j.2044-835X.1987.tb01049.x.

Baron-Cohen, S., A.M. Leslie, and U. Frith. 1985. Does the autistic child have a "theory of mind"? *Cognition* 21 (1): 37–46. https://doi.org/10.1016/0010-0277(85)90022-8.

Baron-Cohen, S., T. Jolliffe, C. Mortimore, and M. Robertson. 1997. Another advanced test of theory of mind: Evidence from very high functioning adults with autism or Asperger syndrome. *Journal of Child Psychology and Psychiatry, and Allied Disciplines* 38 (7): 813–822. https://doi.org/10.1111/j.1469-7610.1997.tb01599.x.

Baron-Cohen, S., M. O'Riordan, V. Stone, R. Jones, and K. Plaisted. 1999. Recognition of faux pas by normally developing children and children with Asperger syndrome or high-functioning autism. *Journal of Autism and Developmental Disorders* 29 (5): 407–418. https://doi.org/10.1023/A:1023035012436.

Baron-Cohen, S., S. Wheelwright, J. Hill, Y. Raste, and I. Plumb. 2001. The "Reading the Mind in the Eyes" test revised version: A study with normal adults, and adults with Asperger syndrome or high-functioning autism. *Journal of Child Psychology and Psychiatry, and Allied Disciplines* 42 (2): 241–251.

Barrera, Á., G. Vázquez, L. Tannenhaus, M. Lolich, and L. Herbst. 2012. Theory of mind and functionality in bipolar patients with symptomatic remission. *Revista De Psiquiatría Y Salud Mental* 6 (2): 67–74. https://doi.org/10.1016/j.rpsmen.2012.07.003.

Baxter, A.J., T.S. Brugha, H.E. Erskine, R.W. Scheurer, T. Vos, and J.G. Scott. 2015. The epidemiology and global burden of autism spectrum disorders. *Psychological Medicine* 45 (3): 601–613. https://doi.org/10.1017/S003329171400172X.

Berardelli, I., S. Sarubbi, E. Rogante, D. Erbuto, C. Giuliani, D.A. Lamis, M. Innamorati, and M. Pompili. 2022. Association between childhood maltreatment and suicidal ideation: A path analysis study. *Journal of Clinical Medicine* 11 (8): 2179. https://doi.org/10.3390/jcm11082179.

Berecz, H., T. Tényi, and R. Herold. 2016. Theory of mind in depressive disorders: A review of the literature. *Psychopathology* 49 (3): 125–134. https://doi.org/10.1159/000446707.

Berenguer-Forner, C., R. Miranda, I.B. Fortea, R. García Castellar, C.C. Diago, and A.M. Casas. 2017. ADHD and peer problems. *Psicothema* 29 (4): 514–519. https://doi.org/10.7334/psicothema2016.376.

Bo, S., C. Sharp, P. Fonagy, and M. Kongerslev. 2017. Hypermentalizing, attachment, and epistemic trust in adolescent BPD: Clinical illustrations. *Personality Disorders: Theory, Research, and Treatment* 8 (2): 172. https://doi.org/10.1037/per0000161.

Bora, E. 2017. Relationship between insight and theory of mind in schizophrenia: A meta-analysis. *Schizophrenia Research* 190: 11–17. https://doi.org/10.1016/j.schres.2017.03.029.

Bora, E., and M. Berk. 2015. Theory of mind in major depressive disorder: A meta-analysis. *Journal of Affective Disorders* 191: 49–55. https://doi.org/10.1016/j.jad.2015.11.023.

Bora, E., and S. Köse. 2016. Meta-analysis of theory of mind in anorexia nervosa and bulimia nervosa: A specific impairment of cognitive perspective taking in anorexia nervosa? *International Journal of Eating Disorders* 49 (8): 739–740. https://doi.org/10.1002/eat.22572.

Bora, E., and C. Pantelis. 2013. Theory of mind impairments in first-episode psychosis, individuals at ultra-high risk for psychosis and in first-degree relatives of schizophrenia: Systematic review and meta-analysis. *Schizophrenia Research* 144 (1): 31–36. https://doi.org/10.1016/j.schres.2012.12.013.

———. 2016. Meta-analysis of social cognition in attention-deficit/hyperactivity disorder (ADHD): Comparison with healthy controls and autistic spectrum disorder. *Psychological Medicine* 46 (4): 699–716. https://doi.org/10.1017/S0033291715002573.

Bora, E., S. Gökçen, B. Kayahan, and B. Veznedaroglu. 2008. Deficits of social-cognitive and social-perceptual aspects of theory of mind in remitted patients with schizophrenia: Effect of residual symptoms. *The Journal of Nervous and Mental Disease* 196 (2): 95–99. https://doi.org/10.1097/NMD.0b013e318162a9e1.

Bora, E., M. Yucel, and C. Pantelis. 2009. Theory of mind impairment in schizophrenia: Meta-analysis. *Schizophrenia Research 109* (1-3): 1–9. https://doi.org/10.1016/j.schres.2008.12.020.

Bora, E., C. Bartholomeusz, and C. Pantelis. 2016. Meta-analysis of theory of mind (ToM) impairment in bipolar disorder. *Psychological Medicine* 46 (2): 253–264. https://doi.org/10.1017/S0033291715001993.

Braak, S., T. Su, W. Krudop, Y.A.L. Pijnenburg, L.M. Reus, N. van der Wee, A.C. Bilderbeck, G.R. Dawson, I.W. van Rossum, A.V. Campos, C. Arango, I.M.J. Saris, M.J. Kas, and B.W.J.H. Penninx. 2022. Theory of mind and social functioning among neuropsychiatric disorders: A transdiagnostic study. *European Neuropsychopharmacology* 64: 1–11. https://doi.org/10.1016/j.euroneuro.2022.08.005.

Bressi, C., S. Fronza, E. Minacapelli, E.P. Nocito, E. Dipasquale, L. Magri, F. Lionetti, and L. Barone. 2017. Short-term psychodynamic psychotherapy with mentalization-based techniques in major depressive disorder patients: relationship among alexithymia, reflective functioning, and outcome variables – A pilot study. *Psychology and Psychotherapy* 90 (3): 299–313. https://doi.org/10.1111/papt.12110.

Brüne, M. 2005. Theory of mind in schizophrenia: A review of the literature. *Schizophrenia Bulletin* 31 (1): 21–42. https://doi.org/10.1093/schbul/sbi002.

Buhlmann, U., R. Wacker, and I. Dziobek. 2015. Inferring other people's states of mind: Comparison across social anxiety, body dysmorphic, and obsessive–compulsive disorders. *Journal of Anxiety Disorders* 34: 107–113. https://doi.org/10.1016/j.janxdis.2015.06.003.

Cacioppo, J.T., S. Cacioppo, S. Dulawa, and A.A. Palmer. 2014. Social neuroscience and its potential contribution to psychiatry. *World psychiatry: Official journal of the World Psychiatric Association (WPA)* 13 (2): 131–139. https://doi.org/10.1002/wps.20118.

Callenmark, B., L. Kjellin, L. Rönnqvist, and S. Bölte. 2014. Explicit versus implicit social cognition testing in autism spectrum disorder. *Autism: The International Journal of Research and Practice* 18 (6): 684–693. https://doi.org/10.1177/1362361313492393

Camoirano, A. 2017. Mentalizing makes parenting work: A review about parental reflective functioning and clinical interventions to improve it. *Frontiers in Psychology* 8: 14. https://doi.org/10.3389/fpsyg.2017.00014.

Carlyle, D., R. Green, M. Inder, R. Porter, M. Crowe, R. Mulder, and C. Frampton. 2020. A randomized-controlled trial of mentalization-based treatment compared with structured case management for borderline personality disorder in a mainstream public health service. *Frontiers in Psychiatry* 11: 561916. https://doi.org/10.3389/fpsyt.2020.561916.

Catalan, A., V. Angosto, A. Díaz, N. Martínez, D. Guede, M. Pereda, A. Madrazo, S. Bustamante, A. Bilbao, L. Osa, L. Inchausti, and M.A. Gonzalez-Torres. 2018. The relationship between theory of mind deficits and neurocognition in first episode-psychosis. *Psychiatry Research* 268: 361–367. https://doi.org/10.1016/j.psychres.2018.06.066.

Colonnesi, C., M. Nikolić, W. De Vente, and S.M. Bögels. 2016. Social anxiety symptoms in young children: Investigating the interplay of theory of mind and expressions of shyness. *Journal of Abnormal Child Psychology* 45: 997–1011. https://doi.org/10.1007/s10802-016-0206-0.

Corcoran, R., G. Mercer, and C.D. Frith. 1995. Schizophrenia, symptomatology and social inference: Investigating "theory of mind" in people with schizophrenia. *Schizophrenia Research* 17 (1): 5–13. https://doi.org/10.1016/0920-9964(95)00024-G.

Correll, C.U., and N.R. Schooler. 2020. Negative symptoms in schizophrenia: A review and clinical guide for recognition, assessment, and treatment. *Neuropsychiatric Disease and Treatment* 16: 519–534. https://doi.org/10.2147/NDT.S225643.

Cortés-García, L., Ö.F. Akça, K. Wall, and C. Sharp. 2021. Exploring mentalizing in adolescents with anorexia nervosa and borderline personality disorder: A comparative study of psychiatric inpatients and healthy controls. *Scandinavian Journal of Child and Adolescent Psychiatry and Psychology* 9: 16–26. https://doi.org/10.21307/sjcapp-2021-003.

Cotter, J., K. Granger, R. Backx, M. Hobbs, C.Y. Looi, and J.H. Barnett. 2018. Social cognitive dysfunction as a clinical marker: A systematic review of meta-analyses across 30 clinical conditions. *Neuroscience and Biobehavioral Reviews* 84: 92–99. https://doi.org/10.1016/j.neubiorev.2017.11.014.

Couette, M., S. Mouchabac, A. Bourla, P. Nuss, and F. Ferreri. 2020. Social cognition in post-traumatic stress disorder: A systematic review. *British Journal of Clinical Psychology* 59 (2): 117–138. https://doi.org/10.1111/bjc.12238.

Couture, S.M., E.L. Granholm, and S.C. Fish. 2011. A path model investigation of neurocognition, theory of mind, social competence, negative symptoms and real-world functioning in schizophrenia. *Schizophrenia Research* 125 (2-3): 152–160. https://doi.org/10.1016/j.schres.2010.09.020.

Cowman, M., L. Holleran, E. Lonergan, K. O'Connor, M. Birchwood, and G. Donohoe. 2021. Cognitive predictors of social and occupational functioning in early psychosis: A systematic review and meta-analysis of cross-sectional and longitudinal data. *Schizophrenia Bulletin* 47 (5): 1243–1253. https://doi.org/10.1093/schbul/sbab033.

Cusi, A.M., A. Nazarov, G.M. Macqueen, and M.C. McKinnon. 2013. Theory of mind deficits in patients with mild symptoms of major depressive disorder. *Psychiatry Research* 210 (2): 672–674. https://doi.org/10.1016/j.psychres.2013.06.018.

David, A.S. 1990. Insight and psychosis. *The British Journal of Psychiatry: The Journal of Mental Science* 156 (6): 798–808. https://doi.org/10.1192/bjp.156.6.798.

de la Osa, N., R. Granero, J.M. Domenech, S. Shamay-Tsoory, and L. Ezpeleta. 2016. Cognitive and affective components of theory of mind in preschoolers with oppositional defiance disorder: Clinical evidence. *Psychiatry Research* 241: 128–134. https://doi.org/10.1016/j.psychres.2016.04.082.

de la Torre-Luque, A., and J.L. Ayuso-Mateos. 2020. The course of depression in late life: A longitudinal perspective. *Epidemiology and Psychiatric Sciences* 29: e147. https://doi.org/10.1017/S204579602000058X.

de la Torre-Luque, A., A. Pemau, V. Perez-Sola, and J.L. Ayuso-Mateos. 2022a. Suicide mortality in Spain in 2020: The impact of the COVID-19 pandemic. *Revista De Psiquiatria Y Salud Mental.* https://doi.org/10.1016/j.rpsm.2022.01.003.

de la Torre-Luque, A., A. Viera-Campos, A.C. Bilderbeck, M.T. Carreras, J. Vivancos, C.M. Diaz-Caneja, M. Aghajani, I.M.J. Saris, A. Raslescu, A. Malik, J. Clark, B.W.J.H. Penninx, N. van der Wee, I.W. Rossum, B. Sommer, H. Marston, G.R. Dawson, M.J. Kas, J.L. Ayuso-Mateos, and C. Arango. 2022b. Relationships between social withdrawal and facial emotion recognition in neuropsychiatric disorders. *Progress in Neuro-Psychopharmacology & Biological Psychiatry* 113: 110463. https://doi.org/10.1016/j.pnpbp.2021.110463.

De Oliveira, C., H. Rahioui, M. Smadja, M.A. Gorsane, and F. Louppe. 2017. Thérapie basée sur la mentalization et le trouble de personnalité limite. *L'Encephale* 43 (4): 340–345. https://doi.org/10.1016/j.encep.2016.02.020.

de Sampaio, Tapajóz Pereira, S. Fernanda Soneira, A. Aulicino, and R.F. Allegri. 2013. Theory of mind in eating disorders and their relationship to clinical profile. *European Eating Disorders Review* 21 (6): 479–487. https://doi.org/10.1002/erv.2247.

Dorn, N.L., N. Struck, F. Bitsch, I. Falkenberg, T. Kircher, W. Rief, and S. Mehl. 2021. The relationship between different aspects of theory of mind and symptom clusters in psychotic disorders: Deconstructing theory of mind into cognitive, affective, and hyper theory of mind. *Frontiers in Psychiatry* 9 (12): 607154. https://doi.org/10.3389/fpsyt.2021.607154.

Dufour, N., E. Redcay, L. Young, P.L. Mavros, J.M. Moran, C. Triantafyllou, J.D. Gabrieli, and R. Saxe. 2013. Similar brain activation during false belief tasks in a large sample of adults with and without autism. *PloS One* 8 (9): e75468. https://doi.org/10.1371/journal.pone.0075468.

Dziobek, I., S. Fleck, E. Kalbe, K. Rogers, J. Hassenstab, M. Brand, J. Kessler, J.K. Woike, O.T. Wolf, and A. Convit. 2006. Introducing MASC: A movie for the assessment of social cognition. *Journal of Autism and Developmental Disorders* 36 (5): 623–636. https://doi.org/10.1007/s10803-006-0107-0.

Ekman, P. 1984. Expression and the nature of emotion. In *Approaches to emotions*, ed. K.R. Scherer and P. Ekman. Hillsdale: Erlbaum.

Ekman, P., and W.V. Friesen. 1976. Measuring facial movement. *Environmental Psychology & Nonverbal Behavior* 1 (1): 56–75. https://doi.org/10.1007/BF01115465.

Ellison, W.D., L.K. Rosenstein, T.A. Morgan, and M. Zimmerman. 2018. Community and clinical epidemiology of borderline personality disorder. *The Psychiatric Clinics of North America* 41 (4): 561–573. https://doi.org/10.1016/j.psc.2018.07.008.

Etchepare, A., and A. Prouteau. 2018. Toward a two-dimensional model of social cognition in clinical neuropsychology: A systematic review of factor structure studies. *Journal of the International Neuropsychological Society: JINS* 24 (4): 391–404. https://doi.org/10.1017/S1355617717001163.

Fanning, J.R., M.D. Bell, and J.M. Fiszdon. 2012. Is it possible to have impaired neurocognition but good social cognition in schizophrenia? *Schizophrenia research* 135 (1–3): 68–71. https://doi.org/10.1016/j.schres.2011.12.009.

Fehling, K.B., and E.A. Selby. 2021. Suicide in DSM-5: Current evidence for the proposed suicide behavior disorder and other possible improvements. *Frontiers in Psychiatry* 11: 499980. https://doi.org/10.3389/fpsyt.2020.499980.

Ferrari, A.J., A.J. Somerville, A.J. Baxter, R. Norman, S.B. Patten, T. Vos, and H.A. Whiteford. 2013. Global variation in the prevalence and incidence of major depressive disorder: A systematic review of the epidemiological literature. *Psychological Medicine* 43 (3): 471–481. https://doi.org/10.1017/S0033291712001511.

Fertuck, E.A., A. Jekal, I. Song, B. Wyman, M.C. Morris, S.T. Wilson, B.S. Brodsky, and B. Stanley. 2009. Enhanced 'Reading the Mind in the Eyes' in borderline personality disorder compared to healthy controls. *Psychological Medicine* 39 (12): 1979–1988. https://doi.org/10.1017/S003329170900600X.

Fett, A.K., W. Viechtbauer, M.D. Dominguez, D.L. Penn, J. van Os, and L. Krabbendam. 2011. The relationship between neurocognition and social cognition with functional outcomes in schizophrenia: A meta-analysis. *Neuroscience and biobehavioral reviews* 35 (3): 573–588. https://doi.org/10.1016/j.neubiorev.2010.07.001.

Fizke, E., D. Barthel, T. Peters, and H. Rakoczy. 2014. Executive function plays a role in coordinating different perspectives, particularly when one's own perspective is involved. *Cognition* 130 (3): 315–334. https://doi.org/10.1016/j.cognition.2013.11.017.

Fonagy, P., and M. Target. 1997. Attachment and reflective function: Their role in self-organization. *Development and Psychopathology* 9 (4): 679–700. https://doi.org/10.1017/S0954579497001399.

Fonagy, P., and A. Bateman. 2008. The development of borderline personality disorder – A mentalizing model. *Journal of Personality Disorders* 22 (1): 4–21. https://doi.org/10.1521/pedi.2008.22.1.4.

Fonagy, P., Target, M., Steele, H., and Steele, M. 1998. *Reflective-functioning manual Version 5 for application to adult attachment interviews*. https://discovery.ucl.ac.uk/id/eprint/1461016/.

Fonagy, P., P. Luyten, A. Moulton-Perkins, Y.-W. Lee, F. Warren, S. Howard, R. Ghinai, P. Fearon, and B. Lowyck. 2016. Development and validation of a self-report measure of mentalizing: The reflective functioning questionnaire. *Plos One* 11 (7): e0158678. https://doi.org/10.1371/journal.pone.0158678.

Förster, K., S. Jörgens, T.M. Air, C. Bürger, V. Enneking, R. Redlich, D. Zaremba, D. Grotegerd, K. Dohm, S. Meinert, E.J. Leehr, J. Böhnlein, J. Repple, N. Opel, E. Kavakbasi, V. Arolt,

P. Zwitserlood, U. Dannlowski, and B.T. Baune. 2018. The relationship between social cognition and executive function in Major Depressive Disorder in high-functioning adolescents and young adults. *Psychiatry Research* 263: 139–146. https://doi.org/10.1016/j.psychres.2018.02.046.

Fretland, R.A., S. Andersson, K. Sundet, O.A. Andreassen, I. Melle, and A. Vaskinn. 2015. Theory of mind in schizophrenia: Error types and associations with symptoms. *Schizophrenia Research* 162 (1): 42–46. https://doi.org/10.1016/j.schres.2015.01.024.

Frith, U. 1989/2002. *Autism: Explaining the enigma* (1st ed. 1989; 2nd ed. 2002). Blackwell.

Frith, C.D. 1992. *The cognitive neuropsychology of schizophrenia*. Hove: Lawrence Erlbaum Associates, Inc.

———. 2004. Schizophrenia and theory of mind. *Psychological Medicine* 34 (3): 385–389. https://doi.org/10.1017/S0033291703001326.

GBD Mental Disorder Collaborators. 2022. Global, regional, and national burden of 12 mental disorders in 204 countries and territories, 1990–2019: A systematic analysis for the Global Burden of Disease Study 2019. *Lancet Psychiatry* 9: 137–150. https://doi.org/10.1016/S2215-0366(21)00395-3.

Green, M.F., D.L. Penn, R. Bentall, W.T. Carpenter, W. Gaebel, R.C. Gur, A.M. Kring, S. Park, S.M. Silverstein, and R. Heinssen. 2008. Social cognition in schizophrenia: A NIMH workshop on definitions, assessment, and research opportunities. *Schizophrenia Bulletin* 34 (6): 1211–1220. https://doi.org/10.1093/schbul/sbm145.

Green, M.F., W.P. Horan, and J. Lee. 2015. Social cognition in schizophrenia. *Nature reviews. Neuroscience* 16 (10): 620–631. https://doi.org/10.1038/nrn4005.

Ha, C., C. Sharp, and I. Goodyer. 2011. The role of child and parental mentalizing for the development of conduct problems over time. *European Child & Adolescent Psychiatry* 20 (6): 291–300. https://doi.org/10.1007/s00787-011-0174-4.

Halstensen, K., R. Gjestad, P. Luyten, B. Wampold, P. Granqvist, G. Stålsett, and S.U. Johnson. 2021. Depression and mentalizing: A psychodynamic therapy process study. *Journal of Counseling Psychology* 68 (6): 705–718. https://doi.org/10.1037/cou0000544.

Halverson, T.F., M. Orleans-Pobee, C. Merritt, P. Sheeran, A.K. Fett, and D.L. Penn. 2019. Pathways to functional outcomes in schizophrenia spectrum disorders: Meta-analysis of social cognitive and neurocognitive predictors. *Neuroscience and Biobehavioral Reviews* 105: 212–219. https://doi.org/10.1016/j.neubiorev.2019.07.020.

Happé, F.G.E. 1994. An advanced test of theory of mind: Understanding of story characters' thoughts and feelings by able autistic, mentally handicapped, and normal children and adults. *Journal of Autism and Developmental Disorders* 24 (2): 129–154. https://doi.org/10.1007/BF02172093.

Happé, F., and U. Frith. 1996. Theory of mind and social impairment in children with conduct disorder. *British Journal of Developmental Psychology* 14 (4): 385–398. https://doi.org/10.1111/j.2044-835X.1996.tb00713.x.

Harrington, L., R. Siegert, and J. McClure. 2005. Theory of mind in schizophrenia: A critical review. *Cognitive Neuropsychiatry* 10 (4): 249–286. https://doi.org/10.1080/13546800444000056.

Harrison, A., S. Sullivan, K. Tchanturia, and J. Treasure. 2009. Emotion recognition and regulation in anorexia nervosa. *Clinical Psychology and Psychotherapy* 16 (4): 348–356. https://doi.org/10.1002/cpp.628.

———. 2010. Emotional functioning in eating disorders: Attentional bias, emotion recognition and emotion regulation. *Psychological Medicine* 40 (11): 1887–1897. https://doi.org/10.1017/S0033291710000036.

Hatkevich, C., F. Penner, and C. Sharp. 2019a. Difficulties in emotion regulation and suicide ideation and attempt in adolescent inpatients. *Psychiatry Research* 271: 230–238. https://doi.org/10.1016/j.psychres.2018.11.038.

Hatkevich, C., A. Venta, and C. Sharp. 2019b. Theory of mind and suicide ideation and attempt in adolescent inpatients. *Journal of Affective Disorders* 256: 17–25. https://doi.org/10.1016/j.jad.2019.05.051.

Hawken, E.R., K.L. Harkness, L.K. Lazowski, D. Summers, N. Khoja, J.G. Gregory, and R. Milev. 2016. The manic phase of bipolar disorder significantly impairs theory of mind decoding. *Psychiatry Research* 239: 275–280. https://doi.org/10.1016/j.psychres.2016.03.043.

Healey, K.M., C.F. Bartholomeusz, and D.L. Penn. 2016. Deficits in social cognition in first episode psychosis: A review of the literature. *Clinical Psychology Review* 50: 108–137. https://doi.org/10.1016/j.cpr.2016.10.001.

Hernández-Calle, D., J. Andreo-Jover, J. Curto-Ramos, D.G. Martínez, L.V. Valor, G. Juárez, M. Alcamí, A. Ortiz, N. Iglesias, M.F. Bravo-Ortiz, B.R. Vega, and G. Martínez-Alés. 2022. Pediatric mental health emergency visits during the COVID-19 pandemic. *Scandinavian Journal of Child and Adolescent Psychiatry and Psychology* 10 (1): 53–57. https://doi.org/10.2478/sjcapp-2022-0005.

Hezel, D.M., and R.J. McNally. 2014. Theory of mind impairments in social anxiety disorder. *Behavior Therapy* 45 (4): 530–540. https://doi.org/10.1016/j.beth.2014.02.010.

Hillmann, K., C. Neukel, M. Krauch, A. Spohn, K. Schnell, S.C. Herpertz, and K. Bertsch. 2021. Cognitive and affective theory of mind in female patients with borderline personality disorder. *Journal of Personality Disorders* 35 (5): 672–690. https://doi.org/10.1016/j.psc.2018.07.008.

Hutchins, T.L., P.A. Prelock, H. Morris, J. Benner, T. LaVigne, and B. Hoza. 2016. Explicit vs. applied theory of mind competence: A comparison of typically developing males, males with ASD, and males with ADHD. *Research in Autism Spectrum Disorders* 21: 94–108. https://doi.org/10.1016/j.rasd.2015.10.004.

Ibáñez, A., A.M. García, S. Esteves, A. Yoris, E. Muñoz, L. Reynaldo, M.L. Pietto, F. Adolfi, and F. Manes. 2018. Social neuroscience: Undoing the schism between neurology and psychiatry. *Social Neuroscience* 13 (1): 1–39. https://doi.org/10.1080/17470919.2016.1245214.

Ibáñez-Brassi, A., and M. 2005. Autismo, funciones ejecutivas y mentalismo: Reconsiderando la heurística de descomposición modular. *Revista Argentina de Neuropsicología* 6: 25–49.

Imanipour, S., M. Sheikh, M. Shayestefar, and T. Baloochnejad. 2021. Deficits in working memory and theory of mind may underlie difficulties in social perception of children with ADHD. *Neurology Research International, Volume 2021*: 3793750. https://doi.org/10.1155/2021/3793750.

Inoue, Y., Y. Tonooka, K. Yamada, and S. Kanba. 2004. Deficiency of theory of mind in patients with remitted mood disorder. *Journal of Affective Disorders* 82 (3): 403–409. https://doi.org/10.1016/j.jad.2004.04.004.

Inoue, Y., K. Yamada, and S. Kanba. 2006. Deficit in theory of mind is a risk for relapse of major depression. *Journal of Affective Disorders* 95 (1): 125–127. https://doi.org/10.1016/j.jad.2006.04.018.

Jolliffe, T., and S. Baron-Cohen. 1999. A test of central coherence theory: Linguistic processing in high-functioning adults with autism or Asperger syndrome: Is local coherence impaired? *Cognition* 71 (2): 149–185. https://doi.org/10.1016/s0010-0277(99)00022-0.

Jones, A.P., F.G.E. Happé, F. Gilbert, S. Burnett, and E. Viding. 2010. Feeling, caring, knowing: different types of empathy deficit in boys with psychopathic tendencies and autism spectrum disorder. *Journal of Child Psychology and Psychiatry, and Allied Disciplines* 51 (11): 1188–1197. https://doi.org/10.1111/j.1469-7610.2010.02280.x.

Kaland, N., A. Møller-Nielsen, K. Callesen, E.L. Mortensen, D. Gottlieb, and L. Smith. 2002. A new 'advanced' test of theory of mind: Evidence from children and adolescents with Asperger syndrome. *Journal of Child Psychology and Psychiatry, and Allied Disciplines* 43 (4): 517–528. https://doi.org/10.1111/1469-7610.00042.

Kana, R.K., J.O. Maximo, D.L. Williams, T.A. Keller, S.E. Schipul, V.L. Cherkassky, N.J. Minshew, and M.A. Just. 2015. Aberrant functioning of the theory-of-mind network in children and adolescents with autism. *Molecular Autism* 6: 59. https://doi.org/10.1186/s13229-015-0052-x.

Kenyon, M., N. Samarawickrema, H. Dejong, F. Van den Eynde, H. Startup, A. Lavender, E. Goodman-Smith, and U. Schmidt. 2012. Theory of mind in bulimia nervosa. *The International Journal of Eating Disorders* 45 (3): 377–384. https://doi.org/10.1002/eat.20967.

Klonsky, E.D., A.M. May, and B.Y. Saffer. 2016. Suicide, suicide attempts, and suicidal ideation. *Annual Review of Clinical Psychology* 12: 307–330. https://doi.org/10.1146/annurev-clinpsy-021815-093204.

Konstantakopoulos, G., D. Ploumpidis, P. Oulis, P. Patrikelis, S. Nikitopoulou, G.N. Papadimitriou, and A.S. David. 2014. The relationship between insight and theory of mind in schizophrenia. *Schizophrenia Research* 152 (1): 217–222. https://doi.org/10.1016/j.schres.2013.11.022.

Konstantakopoulos, G., N. Ioannidi, P. Patrikelis, and F. Gonidakis. 2020. Empathy, mentalizing, and cognitive functioning in anorexia nervosa and bulimia nervosa. *Social Neuroscience* 15 (4): 477–488. https://doi.org/10.1080/17470919.2020.1760131.

Kupferberg, A., L. Bicks, and G. Hasler. 2016. Social functioning in major depressive disorder. *Neuroscience and Biobehavioral Reviews* 69: 313–332. https://doi.org/10.1016/j.neubiorev.2016.07.002.

Ladegaard, N., E.R. Larsen, P. Videbech, and P.H. Lysaker. 2014. Higher-order social cognition in first-episode major depression. *Psychiatry Research* 216 (1): 37–43. https://doi.org/10.1016/j.psychres.2013.12.010.

Lahera, G., J.M. Montes, A. Benito, M. Valdivia, E. Medina, I. Mirapeix, and J. Sáiz-Ruiz. 2007. Theory of mind deficit in bipolar disorder: Is it related to aprevious history of psychotic symptoms? *Psychiatry Research* 161 (3): 309–317. https://doi.org/10.1016/j.psychres.2007.08.009.

Lee, L., K.L. Harkness, M.A. Sabbagh, and J.A. Jacobson. 2005. Mental state decoding abilities in clinical depression. *Journal of Affective Disorders* 86 (2): 247–258. https://doi.org/10.1016/j.jad.2005.02.007.

Lee, T.Y., S.B. Hong, N.Y. Shin, and J.S. Kwon. 2015. Social cognitive functioning in prodromal psychosis: A meta-analysis. *Schizophrenia Research* 164 (1–3): 28–34. https://doi.org/10.1016/j.schres.2015.02.008.

Leslie, A.M. 1987. Pretense and representation: The origins of "theory of mind.". *Psychological Review* 94 (4): 412–426. https://doi.org/10.1037/0033-295X.94.4.412.

Li, E.T., E. Carracher, and T. Bird. 2020. Linking childhood emotional abuse and adult depressive symptoms: The role of mentalizing incapacity. *Child Abuse & Neglect* 99: 104253. https://doi.org/10.1016/j.chiabu.2019.104253.

Liu, R.T., R.F.L. Walsh, A.E. Sheehan, S.M. Cheek, and C.M. Sanzari. 2022. Prevalence and correlates of suicide and nonsuicidal self-injury in children: A systematic review and meta-analysis. *JAMA Psychiatry* 79 (7): 718–726. https://doi.org/10.1001/jamapsychiatry.2022.1256.

Livingston, L.A., E. Colvert, Social Relationships Study Team, P. Bolton, and F. Happé. 2019a. Good social skills despite poor theory of mind: Exploring compensation in autism spectrum disorder. *Journal of Child Psychology and Psychiatry, and Allied Disciplines* 60 (1): 102–110. https://doi.org/10.1111/jcpp.12886.

Livingston, L.A., P. Shah, and F. Happé. 2019b. Compensatory strategies below the behavioral surface in autism: A qualitative study. *The lancet. Psychiatry* 6 (9): 766–777. https://doi.org/10.1016/S2215-0366(19)30224-X.

Luyten, P., L. Nijssens, P. Fonagy, and L.C. Mayes. 2017. Parental reflective functioning: Theory, research, and clinical applications. *The Psychoanalytic Study of the Child* 70 (1): 174–199. https://doi.org/10.1080/00797308.2016.1277901.

Lyall, K., L. Croen, J. Daniels, M.D. Fallin, C. Ladd-Acosta, B.K. Lee, B.Y. Park, N.W. Snyder, D. Schendel, H. Volk, G.C. Windham, and C. Newschaffer. 2017. The changing epidemiology of autism spectrum disorders. *Annual Review of Public Health* 38: 81–102. https://doi.org/10.1146/annurev-publhealth-031816-04431.

Lynch, T.R., M.Z. Rosenthal, D.S. Kosson, J.S. Cheavens, C.W. Lejuez, and R.J.R. Blair. 2006. Heightened sensitivity to facial expressions of emotion in borderline personality disorder. *Emotion (Washington, D.C.)* 6 (4): 647–655. https://doi.org/10.1037/1528-3542.6.4.647.

Lysaker, P.H., G. Salvatore, M.L. Grant, M. Procacci, K.L. Olesek, K.D. Buck, G. Nicolò, and G. Dimaggio. 2010. Deficits in theory of mind and social anxiety as independent paths to paranoid features in schizophrenia. *Schizophrenia Research* 124 (1–3): 81–85. https://doi.org/10.1016/j.schres.2010.06.019.

Maleki, G., A. Zabihzadeh, M.J. Richman, Z. Demetrovics, and F. Mohammadnejad. 2020. Decoding and reasoning mental states in major depression and social anxiety disorder. *BMC Psychiatry* 20 (1): 463. https://doi.org/10.1186/s12888-020-02873-w.

Mandy, W., D. Skuse, C. Steer, B. St Pourcain, and B.R. Oliver. 2013. Oppositionality and socioemotional competence: Interacting risk factors in the development of childhood conduct disorder symptoms. *Journal of the American Academy of Child and Adolescent Psychiatry* 52 (7): 718–727. https://doi.org/10.1016/j.jaac.2013.04.011.

Martin, J., K.L. Raby, M.H. Labella, and G.I. Roisman. 2017. Childhood abuse and neglect, attachment states of mind, and non-suicidal self-injury. *Attachment & Human Development* 19 (5): 425–446. https://doi.org/10.1080/14616734.2017.1330832.

Mary, A., H. Slama, P. Mousty, I. Massat, T. Capiau, V. Drabs, and P. Peigneux. 2015. Executive and attentional contributions to theory of mind deficit in attention deficit/hyperactivity disorder (ADHD). *Child Neuropsychology: A Journal on Normal and Abnormal Development in Childhood and Adolescence* 22 (3): 345–365. https://doi.org/10.1080/09297049.2015.1012491.

McLaren, V., M. Gallagher, C.J. Hopwood, and C. Sharp. 2022. Hypermentalizing and Borderline Personality Disorder: A Meta-Analytic Review. *American Journal of Psychotherapy* 75 (1): 21–31. https://doi.org/10.1176/appi.psychotherapy.20210018.

Medina-Pradas, C., J.B. Navarro, E.M. Álvarez-Moya, A. Grau, and J.E. Obiols. 2012. Emotional theory of mind in eating disorders. *International Journal of Clinical and Health Psychology* 12 (2): 189–202.

Mesholam-Gately, R.I., A.J. Giuliano, K.P. Goff, S.V. Faraone, and L.J. Seidman. 2009. Neurocognition in first-episode schizophrenia: A meta-analytic review. *Neuropsychology* 23 (3): 315–336. https://doi.org/10.1037/a0014708.

Minzenberg, M.J., J.H. Poole, and S. Vinogradov. 2006. Adult social attachment disturbance is related to childhood maltreatment and current symptoms in borderline personality disorder. *The Journal of Nervous and Mental Disease* 194 (5): 341–348. https://doi.org/10.1097/01.nmd.0000218341.54333.4e.

Moldovan, M., and L. Visu-Petra. 2022. Theory of mind, anxiety, and interpretive bias during middle childhood. *Journal of Child and Family Studies* 31 (1): 99–113. https://doi.org/10.1007/s10826-021-02023-0.

Molenberghs, P., R. Cunnington, and J.B. Mattingley. 2009. Is the mirror neuron system involved in imitation? A short review and meta-analysis. *Neuroscience and Biobehavioral Reviews* 33 (7): 975–980. https://doi.org/10.1016/j.neubiorev.2009.03.010.

Montag, C., A. Ehrlich, K. Neuhaus, I. Dziobek, H.R. Heekeren, A. Heinz, and J. Gallinat. 2010. Theory of mind impairments in euthymic bipolar patients. *Journal of Affective Disorders* 123 (1-3): 264–269. https://doi.org/10.1016/j.jad.2009.08.017.

Montag, C., I. Dziobek, I.S. Richter, K. Neuhaus, A. Lehmann, R. Sylla, H.R. Heekeren, A. Heinz, and J. Gallinat. 2011. Different aspects of theory of mind in paranoid schizophrenia: Evidence from a video-based assessment. *Psychiatry Research* 186 (2-3): 203–209. https://doi.org/10.1016/j.psychres.2010.09.006.

Murray, K., K. Johnston, H. Cunnane, C. Kerr, D. Spain, N. Gillan, N. Hammond, D. Murphy, and F. Happé. 2017. A new test of advanced theory of mind: The "Strange Stories Film Task" captures social processing differences in adults with autism spectrum disorders. *Autism Research: Official Journal of the International Society for Autism Research* 10 (6): 1120–1132. https://doi.org/10.1002/aur.1744.

Navarra-Ventura, G., M. Vicent-Gil, M. Serra-Blasco, J. Cobo, S. Fernández-Gonzalo, X. Goldberg, M. Jodar, J.M. Crosas, D. Palao, G. Lahera, E. Vieta, and N. Cardoner. 2022. Higher order theory of mind in patients with bipolar disorder and schizophrenia/schizoaffective disorder. *European Archives of Psychiatry and Clinical Neuroscience* 272 (3): 497–507. https://doi.org/10.1007/s00406-021-01265-9.

Németh, N., P. Mátrai, P. Hegyi, B. Czéh, L. Czopf, A. Hussain, J. Pammer, I. Szabó, M. Solymár, L. Kiss, P. Hartmann, Á.L. Szilágyi, Z. Kiss, and M. Simon. 2018. Theory of mind disturbances in borderline personality disorder: A meta-analysis. *Psychiatry Research* 270: 143–153. https://doi.org/10.1016/j.psychres.2018.08.049.

Nestor, B.A., and S. Sutherland. 2022. Theory of mind and suicidality: A meta-analysis. *Archives of Suicide Research* 26 (4): 1666–1687. https://doi.org/10.1080/13811118.2021.1939209.

Nestor, B.A., S. Sutherland, and J. Garber. 2022. Theory of mind performance in depression: A meta-analysis. *Journal of Affective Disorders* 303: 233–244. https://doi.org/10.1016/j.jad.2022.02.028.

Ng, R., S. Fish, and E. Granholm. 2015. Insight and theory of mind in schizophrenia. *Psychiatry Research* 225 (1-2): 169–174. https://doi.org/10.1016/j.psychres.2014.11.010.

Nijssens, L., P. Luyten, S. Malcorps, N. Vliegen, and L. Mayes. 2021. Parental reflective functioning and theory of mind acquisition: A developmental perspective. *The British Journal of Developmental Psychology* 39 (4): 584–602. https://doi.org/10.1111/bjdp.12383.

Northoff, G. 2008. Neuropsychiatry. An old discipline in a new gestalt bridging biological psychiatry, neuropsychology, and cognitive neurology. *European Archives of Psychiatry and Clinical Neuroscience* 258 (4): 226–238. https://doi.org/10.1007/s00406-007-0783-6.

Ojagbemi, A., E. Estrada, A. de la Torre-Luque, D. Moreno-Agostino, E. Lara, F.F. Caballero, T. Bello, B. Olaya, J.M. Haro, O. Gureje, and J.L. Ayuso-Mateos. 2022. Late-life disability trajectories in Yoruba Nigerians and the Spanish population: A state space model in continuous time. *Aging & Mental Health* 26 (12): 2447–2453. https://doi.org/10.1080/13607863.2021.2008307.

Oldershaw, A., D. Hambrook, K. Tchanturia, J. Treasure, and U. Schmidt. 2010. Emotional theory of mind and emotional awareness in recovered anorexia nervosa patients. *Psychosomatic Medicine* 72 (1): 73–79. https://doi.org/10.1097/PSY.0b013e3181c6c7ca.

Olson, S.L., D.E. Choe, and A.J. Sameroff. 2017. Trajectories of child externalizing problems between ages 3 and 10 years: Contributions of children's early effortful control, theory of mind, and parenting experiences. *Development and Psychopathology* 29 (4): 1333–1351. https://doi.org/10.1017/S095457941700030X.

Ortega-Díaz, E., J. García-Campos, A. Moya-Martínez, C. Ramírez-Cremades, J.M. Rico-Gomis, C. Cuesta-Moreno, A. Palazón-Bru, G. Estan-Cerezo, J.A. Piqueras, and J. Rodríguez-Marín. 2021. Theory of mind in borderline personality disorder: A possible endophenotypic factor? *International Journal of Environmental Research and Public Health* 18 (6): 3193. https://doi.org/10.3390/ijerph18063193.

Pagnoni, I., E. Gobbi, C. Alaimo, E. Campana, R. Rossi, R. Manenti, M. Balconi, and M. Cotelli. 2022. The relationship between theory of mind and executive functions in major depressive disorders: A review. *Frontiers in Psychiatry* 13: 980392. https://doi.org/10.3389/fpsyt.2022.980392.

Pal, P., P. Mahour, A. Arya, and V. Agrawal. 2021. A study on theory of mind among adolescents with oppositional defiant disorder. *Journal of Indian Association for Child and Adolescent Mental Health* 17 (4): 115–132. https://doi.org/10.1177/0973134220210408.

Papadopoulos, T.C., G. Panayiotou, G. Spanoudis, and D. Natsopoulos. 2005. Evidence of poor planning in children with attention deficits. *Journal of Abnormal Child Psychology* 33 (5): 611–623. https://doi.org/10.1007/s10802-005-6741-8.

Patin, A., and R. Hurlemann. 2015. Social cognition. *Handbook of Experimental Pharmacology* 228: 271–303. https://doi.org/10.1007/978-3-319-16522-6_10.

Pequet, A., and K.R. Warnell. 2021. Thinking of you: Relations between mind-mindedness, theory of mind, and social anxiety traits in middle childhood and adulthood. *Social Development* 30 (1): 95–112. https://doi.org/10.1111/sode.12468.

Perner, J., and H. Wimmer. 1985. "John thinks that Mary thinks that…": Attribution of second-order beliefs by 5- to 10-year-old children. *Journal of Experimental Child Psychology* 39 (3): 437–471. https://doi.org/10.1016/0022-0965(85)90051-7.

Pineda-Alhucema, W., E. Aristizabal, J. Escudero-Cabarcas, J.E. Acosta-López, and J.I. Vélez. 2018. Executive function and theory of mind in children with ADHD: A systematic review. *Neuropsychology Review* 28 (3): 341–358. https://doi.org/10.1007/s11065-018-9381-9.

Pino, M.C., M. Mazza, M. Mariano, S. Peretti, D. Dimitriou, F. Masedu, M. Valenti, and F. Franco. 2017. Simple mindreading abilities predict complex theory of mind: Developmental delay in autism spectrum disorders. *Journal of Autism and Developmental Disorders* 47 (9): 2743–2756. https://doi.org/10.1007/s10803-017-3194-1.

Plana, I., M. Lavoie, M. Battaglia, and A.M. Achim. 2013. A meta-analysis and scoping review of social cognition performance in social phobia, posttraumatic stress disorder and other anxiety disorders. *Journal of Anxiety Disorders* 28 (2): 169–177. https://doi.org/10.1016/j.janxdis.2013.09.005.

Poletti, M., I. Enrici, and M. Adenzato. 2012. Cognitive and affective theory of mind in neurodegenerative diseases: Neuropsychological, neuroanatomical and neurochemical levels. *Neuroscience and Biobehavioral reviews* 36 (9): 2147–2164. https://doi.org/10.1016/j.neubiorev.2012.07.004.

Ponnet, K.S., H. Roeyers, A. Buysse, A. De Clercq, and E. Van der Heyden. 2004. Advanced mind-reading in adults with Asperger syndrome. *Autism: The International Journal of Research and Practice* 8 (3): 249–266. https://doi.org/10.1177/1362361304045214.

Porcelli, S., N. Van Der Wee, S. van der Werff, M. Aghajani, J.C. Glennon, S. van Heukelum, F. Mogavero, A. Lobo, F.J. Olivera, E. Lobo, M. Posadas, J. Dukart, R. Kozak, E. Arce, A. Ikram, J. Vorstman, A. Bilderbeck, I. Saris, M.J. Kas, and A. Serretti. 2019. Social brain, social dysfunction and social withdrawal. *Neuroscience and Biobehavioral Reviews* 97: 10–33. https://doi.org/10.1016/j.neubiorev.2018.09.012.

Pourmohammad, P., M. Imani, M.A. Goodarzi, and M.R. Sarafraz. 2021. Impaired complex theory of mind and low emotional self-awareness in outpatients with borderline personality disorder compared to healthy controls: A cross-sectional study. *Journal of Psychiatric Research* 143: 445–450. https://doi.org/10.1016/j.jpsychires.2021.10.009.

Preißler, S., I. Dziobek, K. Ritter, H.R. Heekeren, and S. Roepke. 2010. Social cognition in borderline personality disorder: Evidence for disturbed recognition of the emotions, thoughts, and intentions of others. *Frontiers in Behavioral Neuroscience* 4: 182. https://doi.org/10.3389/fnbeh.2010.00182.

Premack, D., and G. Woodruff. 1978. Does the chimpanzee have a theory of mind? *Behavioral and Brain Sciences* 1 (4): 515–526. https://doi.org/10.1017/S0140525X00076512.

Preti, A., S. Siddi, E. Marzola, and G. Abbate Daga. 2022. Affective cognition in eating disorders: A systematic review and meta-analysis of the performance on the reading the mind in the eyes test. *Eating and Weight Disorders* 27 (7): 2291–2307. https://doi.org/10.1007/s40519-022-01393-8.

Purcell, A.L., M. Phillips, and J. Gruber. 2013. In your eyes: Does theory of mind predict impaired life functioning in bipolar disorder? *Journal of Affective Disorders* 151 (3): 1113–1119. https://doi.org/10.1016/j.jad.2013.06.051.

Ramachandran, V.S., and L.M. Oberman. 2006. Broken mirrors: A theory of autism. *Scientific American* 295 (5): 62–69. https://doi.org/10.1038/scientificamerican1106-62.

Rebolleda, C. 2017. Cognición social en la Esquizofrenia. *Informaciones psiquiátricas* 227: 10–23.

Reddy, M.S. 2010. Depression: The disorder and the burden. *Indian Journal of Psychological Medicine* 32 (1): 1–2. https://doi.org/10.4103/0253-7176.70510.

Renouf, A., M. Brendgen, S. Parent, F. Vitaro, P.D. Zelazo, M. Boivin, G. Dionne, R.E. Tremblay, D. Pérusse, and J.R. Séguin. 2010a. Relations between theory of mind and indirect and physical aggression in kindergarten: Evidence of the moderating role of prosocial behaviors. *Social Development* 19 (3): 535–555. https://doi.org/10.1111/j.1467-9507.2009.00552.x.

Renouf, A., M. Brendgen, J.R. Séguin, F. Vitaro, M. Boivin, G. Dionne, R.E. Tremblay, and D. Pérusse. 2010b. Interactive links between theory of mind, peer victimization, and reactive and proactive aggression. *Journal of Abnormal Child Psychology* 38 (8): 1109–1123. https://doi.org/10.1007/s10802-010-9432-z.

Rizzolatti, G., and L. Craighero. 2004. The mirror-neuron system. *Annual Review of Neuroscience* 27: 169–192. https://doi.org/10.1146/annurev.neuro.27.070203.144230.

Rogoff, S., A. Moulton-Perkins, F. Warren, T. Nolte, and P. Fonagy. 2021. 'Rich' and 'poor' in mentalizing: Do expert mentalizers exist? *PloS one* 16 (10): e0259030. https://doi.org/10.1371/journal.pone.0259030.

Rojahn, J., M. Lederer, and M.J. Tassé. 1995. Facial emotion recognition by persons with mental retardation: A review of the experimental literature. *Research in Developmental Disabilities* 16 (5): 393–414.

Rosello, B., C. Berenguer, I. Baixauli, R. García, and A. Miranda. 2020. Theory of mind profiles in children with autism spectrum disorder: Adaptive/social skills and pragmatic competence. *Frontiers in Psychology* 11: 567401. https://doi.org/10.3389/fpsyg.2020.567401.

Rossouw, T.I., and P. Fonagy. 2012. Mentalization-based treatment for self-harm in adolescents: A randomized controlled trial. *Journal of the American Academy of Child and Adolescent Psychiatry* 51 (12): 1304–1313.e3. https://doi.org/10.1016/j.jaac.2012.09.018.

Russell, T.A., U. Schmidt, L. Doherty, V. Young, and K. Tchanturia. 2008. Aspects of social cognition in anorexia nervosa: Affective and cognitive theory of mind. *Psychiatry Research* 168 (3): 181–185. https://doi.org/10.1016/j.psychres.2008.10.028.

Rutherford, M.D., S. Baron-Cohen, and S. Wheelwright. 2002. Reading the mind in the voice: A study with normal adults and adults with Asperger syndrome and high functioning autism. *Journal of Autism and Developmental Disorders* 32 (3): 189–194. https://doi.org/10.1023/a:1015497629971.

Sandoz, M., J.F. Démonet, and M. Fossard. 2014. Theory of mind and cognitive processes in aging and Alzheimer type dementia: A systematic review. *Aging & Mental Health* 18 (7): 815–827. https://doi.org/10.1080/13607863.2014.899974.

Santomauro, D.F., S. Melen, D. Mitchison, T. Vos, H. Whiteford, and A.J. Ferrari. 2021. The hidden burden of eating disorders: An extension of estimates from the Global Burden of Disease Study 2019. *The Lancet. Psychiatry* 8 (4): 320–328. https://doi.org/10.1016/S2215-0366(21)00040-7.

Santos, J., E. Pousa, E. Soto, A. Comes, P. Roura, F. Arrufat, and J. Obiols. 2017. Theory of mind in euthymic bipolar patients and first-degree relatives. *The Journal of Nervous and Mental Disease* 205 (3): 207–212. https://doi.org/10.1097/NMD.0000000000000595.

Sarfati, Y., M.C. Hardy-Baylé, C. Besche, and D. Widlöcher. 1997. Attribution of intentions to others in people with schizophrenia: A non-verbal exploration with comic strips. *Schizophrenia Research* 25 (3): 199–209. https://doi.org/10.1016/S0920-9964(97)00025-X.

Savla, G.N., L. Vella, C.C. Armstrong, D.L. Penn, and E.W. Twamley. 2013. Deficits in domains of social cognition in schizophrenia: A meta-analysis of the empirical evidence. *Schizophrenia Bulletin* 39 (5): 979–992. https://doi.org/10.1093/schbul/sbs080.

Schneider, D., V.P. Slaughter, A.P. Bayliss, and P.E. Dux. 2013. A temporally sustained implicit theory of mind deficit in autism spectrum disorders. *Cognition* 129 (2): 410–417. https://doi.org/10.1016/j.cognition.2013.08.004.

Schultze-Lutter, F., S. Ruhrmann, J. Berning, W. Maier, and J. Klosterkötter. 2010. Basic symptoms and ultrahigh risk criteria: Symptom development in the initial prodromal state. *Schizophrenia Bulletin* 36 (1): 182–191. https://doi.org/10.1093/schbul/sbn072.

Seitz, K.I., N. Ehler, M. Schmitz, S.E. Schmitz, I. Dziobek, S.C. Herpertz, and K. Bertsch. 2022. Affective and cognitive theory of mind in posttraumatic stress, major depressive, and somatic symptom disorders: Association with childhood trauma. *The British Journal of Clinical Psychology* 61 (3): 680–700. https://doi.org/10.1111/bjc.12357.

Shah, R., and M.C. Zanarini. 2018. Comorbidity of borderline personality disorder: Current status and future directions. *The Psychiatric clinics of North America* 41 (4): 583–593. https://doi.org/10.1016/j.psc.2018.07.009.

Shamay-Tsoory, S.G., and J. Aharon-Peretz. 2007. Dissociable prefrontal networks for cognitive and affective theory of mind: A lesion study. *Neuropsychologia* 45 (13): 3054–3067. https://doi.org/10.1016/j.neuropsychologia.2007.05.021.

Sharp, C., and S. Vanwoerden. 2015. Hypermentalizing in borderline personality disorder: A model and data. *Journal of Infant, Child, and Adolescent Psychotherapy* 14: 33–45. https://doi.org/10.1080/15289168.2015.1004890.

Simonsen, C.B., A.G. Jakobsen, S. Grøntved, and G. Kjaersdam Telléus. 2020. The mentalization profile in patients with eating disorders: A systematic review and meta-analysis. *Nordic Journal of Psychiatry* 74 (5): 311–322. https://doi.org/10.1080/08039488.2019.1707869.

Smeets, T., I. Dziobek, and O.T. Wolf. 2009. Social cognition under stress: Differential effects of stress-induced cortisol elevations in healthy young men and women. *Hormones and Behavior* 55 (4): 507–513. https://doi.org/10.1016/j.yhbeh.2009.01.011.

Spitzer, C., J. Zimmermann, E. Brähler, S. Euler, L. Wendt, and S. Müller. 2021. Die deutsche Version des Reflective Functioning Questionnaire (RFQ): Eine teststatistische Überprüfung in der Allgemeinbevölkerung. *PPmP – Psychotherapie · Psychosomatik · Medizinische Psychologie* 71 (03-04): 124–131. https://doi.org/10.1055/a-1234-6317.

Sprong, M., P. Schothorst, E. Vos, J. Hox, and H. van Engeland. 2007. Theory of mind in schizophrenia: Meta-analysis. *The British Journal of Psychiatry: The Journal of Mental Science* 191: 5–13. https://doi.org/10.1192/bjp.bp.107.035899.

Stagaki, M., T. Nolte, J. Feigenbaum, B. King-Casas, T. Lohrenz, P. Fonagy, and P.R. Montague. 2022. The mediating role of attachment and mentalizing in the relationship between childhood maltreatment, self-harm and suicidality. *Child Abuse & Neglect* 128: 105576. https://doi.org/10.1016/j.chiabu.2022.105576.

Stein, D.J., C.C.W. Lim, A.M. Roest, P. de Jonge, S. Aguilar-Gaxiola, A. Al-Hamzawi, J. Alonso, C. Benjet, E.J. Bromet, R. Bruffaerts, G. de Girolamo, S. Florescu, O. Gureje, J.M. Haro, M.G. Harris, Y. He, H. Hinkov, I. Horiguchi, C. Hu, A. Karam, et al. 2017. The cross-national epidemiology of social anxiety disorder: Data from the world mental health survey initiative. *BMC Medicine* 15 (1): 143. https://doi.org/10.1186/s12916-017-0889-2.

Stone, V.E., S. Baron-Cohen, and R.T. Knight. 1998. Frontal lobe contributions to theory of mind. *Journal of Cognitive Neuroscience* 10 (5): 640–656. https://doi.org/10.1162/089892998562942.

Tas, C., E. Brown, Z. Cubukcuoglu, O. Aydemir, A.E. Danaci, and M. Brüne. 2013. Towards an integrative approach to understanding quality of life in schizophrenia: The role of neurocognition, social cognition, and psychopathology. *Comprehensive Psychiatry* 54 (3): 262–268. https://doi.org/10.1016/j.comppsych.2012.08.001.

Tatar, Z.B., and A. Cansız. 2022. Executive function deficits contribute to poor theory of mind abilities in adults with ADHD. *Applied Neuropsychology: Adult* 29 (2): 244–251. https://doi.org/10.1080/23279095.2020.1736074.

Tauro, J.L., T.A. Wearne, B. Belevski, M. Filipčíková, and H.M. Francis. 2022. Social cognition in female adults with anorexia nervosa: A systematic review. *Neuroscience and Biobehavioral Reviews* 132: 197–210. https://doi.org/10.1016/j.neubiorev.2021.11.035.

Tchanturia, K., F. Happé, J. Godley, J. Treasure, N. Bara-Carril, and U. Schmidt. 2004. 'Theory of mind' in anorexia nervosa. *European Eating Disorders Review* 12 (6): 361–366. https://doi.org/10.1002/erv.608.

Thibaudeau, É., A.M. Achim, C. Parent, M. Turcotte, and C. Cellard. 2020. A meta-analysis of the associations between theory of mind and neurocognition in schizophrenia. *Schizophrenia Research* 216: 118–128. https://doi.org/10.1016/j.schres.2019.12.017.

Thibaudeau, É., C. Cellard, M. Turcotte, and A.M. Achim. 2021. Functional impairments and theory of mind deficits in schizophrenia: A meta-analysis of the associations. *Schizophrenia Bulletin* 47 (3): 695–711. https://doi.org/10.1093/schbul/sbaa182.

Thiemann-Bourque, K.S., N.C. Brady, and K.K. Fleming. 2012. Symbolic play of preschoolers with severe communication impairments with autism and other developmental delays: More similarities than differences. *Journal of Autism and Developmental Disorders* 42 (5): 863–873. https://doi.org/10.1007/s10803-011-1317-7.

Tomás-Labbé, T., E. Ciampi Diaz, J. Venegas Bustos, R. Uribe San Martín, and C. Cárcamo Rodríguez. 2019. Cognición social: Conceptos y Bases Neurales. *Revista chilena de neuro-psiquiatría* 57 (4): 365–376. https://doi.org/10.4067/S0717-92272019000400365.

Tonelli, H., and L. de Siqueira Rotenberg. 2021. Emotion perception and theory of mind in obesity: A systematic review on the impact of social cognitive deficits on dysfunctional eating behaviors. *Surgery for Obesity and Related Diseases* 17 (3): 618–629. https://doi.org/10.1016/j.soard.2020.10.016.

Turan, S., G. Özyurt, G. Çatlı, Y. Öztürk, A. Abacı, and A.P. Akay. 2019. Social cognition and emotion regulation may be impaired in adolescents with obesity independent of the presence of binge eating disorder: A two-center study. *Psychiatry and Clinical Psychopharmacology* 29 (4): 887–894. https://doi.org/10.1080/24750573.2019.1693727.

United Nations. 2022. *The sustainable development goals report.* https://unstats.un.org/sdgs/report/2022/The-Sustainable-Development-Goals-Report-2022.pdf

Valdivieso-Jiménez, G. 2020. *Papel de la mentalización en la Personalidad y sus Trastornos*. 1 (1): 16–27. https://www.revistalatinoamericanadelapersonalidad.org/wp-content/uploads/2020/07/Revista-1-1.pdf

van Donkersgoed, R.J., L. Wunderink, R. Nieboer, A. Aleman, and G.H. Pijnenborg. 2015. Social cognition in individuals at ultra-high risk for psychosis: A meta-analysis. *PloS one* 10 (10): e0141075. https://doi.org/10.1371/journal.pone.0141075.

van Hoeken, D., and H.W. Hoek. 2020. Review of the burden of eating disorders: mortality, disability, costs, quality of life, and family burden. *Current Opinion in Psychiatry* 33 (6): 521–527. https://doi.org/10.1097/YCO.0000000000000641.

van Hooren, S., D. Versmissen, I. Janssen, I. Myin-Germeys, J.À. Campo, R. Mengelers, J. van Os, and L. Krabbendam. 2008. Social cognition and neurocognition as independent domains in psychosis. *Schizophrenia Research* 103 (1-3): 257–265. https://doi.org/10.1016/j.schres.2008.02.022.

van Neerven, T., D.J. Bos, and N.E. van Haren. 2021. Deficiencies in Theory of Mind in patients with schizophrenia, bipolar disorder, and major depressive disorder: A systematic review of secondary literature. *Neuroscience and Biobehavioral Reviews* 120: 249–261. https://doi.org/10.1016/j.neubiorev.2020.11.011.

Vaskinn, A., K. Sundet, T. Ueland, I. Agartz, I. Melle, and O.A. Andreassen. 2013. Social cognition and clinical insight in schizophrenia and bipolar disorder. *The Journal of Nervous and Mental Disease* 201 (6): 445–451. https://doi.org/10.1097/NMD.0b013e31829480c8.

Velikonja, T., A.K. Fett, and E. Velthorst. 2019. Patterns of nonsocial and social cognitive functioning in adults with autism spectrum disorder: A systematic review and meta-analysis. *JAMA Psychiatry* 76 (2): 135–151. https://doi.org/10.1001/jamapsychiatry.2018.364.

Vogt, K.S., and P. Norman. 2019. Is mentalization-based therapy effective in treating the symptoms of borderline personality disorder? A systematic review. *Psychology and Psychotherapy* 92 (4): 441–464. https://doi.org/10.1111/papt.12194.

Wagner, A.W., and M.M. Linehan. 1999. Facial expression recognition ability among women with borderline personality disorder: Implications for emotion regulation? *Journal of Personality Disorders* 13 (4): 329–344. https://doi.org/10.1521/pedi.1999.13.4.329.

Wang, Y., Y. Wang, S. Chen, C. Zhu, and K. Wang. 2008. Theory of mind disability in major depression with or without psychotic symptoms: A componential view. *Psychiatry Research* 161 (2): 153–161. https://doi.org/10.1016/j.psychres.2007.07.018.

Wang, Y.Y., Y. Wang, Y.M. Zou, K. Ni, X. Tian, H.W. Sun, S.S.Y. Lui, E.F.C. Cheung, J. Suckling, and R.C.K. Chan. 2018. Theory of mind impairment and its clinical correlates in patients with schizophrenia, major depressive disorder and bipolar disorder. *Schizophrenia Research* 197: 349–356. https://doi.org/10.1016/j.schres.2017.11.003.

Washburn, D., G. Wilson, M. Roes, K. Rnic, and K.L. Harkness. 2015. Theory of mind in social anxiety disorder, depression, and comorbid conditions. *Journal of Anxiety Disorders* 37: 71–77. https://doi.org/10.1016/j.janxdis.2015.11.004.

Weightman, M.J., T.M. Air, and B.T. Baune. 2014. A review of the role of social cognition in major depressive disorder. *Frontiers in Psychiatry* 5: 179. https://doi.org/10.3389/fpsyt.2014.00179.

Weightman, M.J., M.J. Knight, and B.T. Baune. 2019. A systematic review of the impact of social cognitive deficits on psychosocial functioning in major depressive disorder and opportunities for therapeutic intervention. *Psychiatry Research* 274: 195–212. https://doi.org/10.1016/j.psychres.2019.02.035.

Wells, A.E., L.M. Hunnikin, D.P. Ash, Van Goozen, and H.M. Stephanie. 2019. Children with behavioral problems misinterpret the emotions and intentions of others. *Journal of Abnormal Child Psychology* 48 (2): 213–221. https://doi.org/10.1007/s10802-019-00594-7.

White, S.J., U. Frith, J. Rellecke, Z. Al-Noor, and S.J. Gilbert. 2014. Autistic adolescents show atypical activation of the brain's mentalizing system even without a prior history of mentalizing problems. *Neuropsychologia* 56: 17–25. https://doi.org/10.1016/j.neuropsychologia.2013.12.013.

Wilbertz, G., E. Brakemeier, I. Zobel, M. Härter, and E. Schramm. 2009. Exploring preoperational features in chronic depression. *Journal of Affective Disorders* 124 (3): 262–269. https://doi.org/10.1016/j.jad.2009.11.021.

Wimmer, H., and J. Perner. 1983. Beliefs about beliefs: Representation and constraining function of wrong beliefs in young children's understanding of deception. *Cognition* 13 (1): 103–128. https://doi.org/10.1016/0010-0277(83)90004-5.

Wolf, F., M. Brüne, and H.J. Assion. 2010. Theory of mind and neurocognitive functioning in patients with bipolar disorder. *Bipolar Disorders* 12 (6): 657–666. https://doi.org/10.1111/j.1399-5618.2010.00854.x.

Wolkenstein, L., M. Schönenberg, E. Schirm, and M. Hautzinger. 2011. I can see what you feel, but I can't deal with it: Impaired theory of mind in depression. *Journal of Affective Disorders* 132 (1-2): 104–111. https://doi.org/10.1016/j.jad.2011.02.010.

World Health Organization. 2021. *Suicide worldwide in 2019: Global health estimates.* https://www.who.int/publications/i/item/9789240026643

Yamada, K., Y. Inoue, and S. Kanba. 2015. Theory of mind ability predicts prognosis of outpatients with major depressive disorder. *Psychiatry Research* 230 (2): 604–608. https://doi.org/10.1016/j.psychres.2015.10.011.

Yung, A.R., and P.D. McGorry. 2007. Prediction of psychosis: Setting the stage. *The British Journal of Psychiatry. Supplement* 51: s1–s8. https://doi.org/10.1192/bjp.191.51.s1.

Zhang, T., H. Cui, Y. Wei, Y. Tang, L. Xu, X. Tang, Y. Zhu, L. Jiang, B. Zhang, Z. Qian, A. Chow, X. Liu, C. Li, Z. Xiao, and J. Wang. 2018. Progressive decline of cognition during the conversion from prodrome to psychosis with a characteristic pattern of the theory of mind compensated by neurocognition. *Schizophrenia Research* 195: 554–559. https://doi.org/10.1016/j.schres.2017.08.020.

Zobel, I., D. Werden, H. Linster, P. Dykierek, T. Drieling, M. Berger, and E. Schramm. 2010. Theory of mind deficits in chronically depressed patients. *Depression and Anxiety* 27 (9): 821–828. https://doi.org/10.1002/da.20713.

Chapter 6
Theory of Mind in Autism: From a Primary Deficit to Just Mutual Misunderstanding?

Gema Erena-Guardia, Mila Vulchanova, and David Saldaña

Abstract The theory-of-mind account of autism has been central to cognitive research in the field for nearly 40 years. It initially proposed that the diversity of symptoms in autism could derive from a deficit in the ability of autistic people to infer other individuals' mental representations. An extraordinary amount of research has been carried out within or questioning this perspective, including numerous and different tasks that assess first-order, second-order, and advanced theory of mind. We review and describe in detail some of the more prominent studies in the field. We also explore two of the main challenges to this theory, namely the research highlighting the role of executive function and language development as alternative explanations of autistics' performance on experimental theory-of-mind tasks. But all these accounts tend to focus on the limitations of the person with autism, ignoring the way they respond to and adapt to a world organized around the viewpoints of people without autism, and how that context functions in response to them. Recent studies into camouflaging and compensation, and hypotheses like the double-deficit model, further expand the field in this direction.

Keywords Theory of mind · Autism · Executive function · Language · Camouflaging

6.1 What Is Autism Spectrum Disorder?

Autism spectrum disorder (ASD) is described in diagnostic systems as a neurodevelopmental disorder characterised by the presence of difficulties in communication and social interaction and a range of restricted and repetitive patterns of

G. Erena-Guardia · D. Saldaña (✉)
Universidad de Seville, Seville, Spain
e-mail: gerena@us.es; dsaldana@us.es

M. Vulchanova
NTNU, Trondheim, Norway
e-mail: mila.vulchanova@ntnu.no

T. Lopez-Soto et al. (eds.), *The Theory of Mind Under Scrutiny*, Logic, Argumentation & Reasoning 34, https://doi.org/10.1007/978-3-031-46742-4_6

behaviour and interests (American Psychiatric Association 2013; World Health Organization 2018).[1] Throughout development, autistic children show atypical and often reduced social interactions with peers. Other characteristic social behaviours (which are not necessarily observed in all individuals) include displaying little or unusual eye contact, a limited range of facial expressions, and reduced sharing of emotions with others. Social interactions are sometimes developmentally or contextually inappropriate. Imagine, for example, 10-year-old Julia, who is autistic. One day she is walking down the street with her mother. Mom sees a friend, and both stop to talk for a moment. Suddenly, Julia makes the following comment out loud: "*Mom, this woman has very ugly hair*". By standard social norms, Julia's comment will be seen as inappropriate and out of place.

Together with atypical social development, reduced cognitive flexibility and unusually focused and intense patterns of behaviour and interests are part of the diagnostic criteria and features of autism. John, a 13-year-old autistic boy, loves cars. When he goes out for recess, he starts telling a classmate that a new model of a Japanese brand has just been launched. He then goes on to talk about the features of all Japanese cars, such as the type of engine, the power, and the price. He also compares them with the Spanish car brands. In addition, he keeps asking him what car he has, what brands he likes, if he prefers five-door or three-door cars, or what are the factory colours of certain brands. Furthermore, it is possible that his peer may get bored or feel angry because he asks him so many questions about the same topic. This intense interest may not be shared, and John may find it hard to understand why the conversation comes to an end or what he should do in that situation. John's peer may also find it difficult to comprehend this degree of interest in a single topic, which is not necessarily shared in his context.

The most recent review of epidemiological studies reports a prevalence of 1 in 100 individuals with ASD worldwide over the past decade (Zeidan et al. 2022). Behaviours characteristic of autism-wide form a continuum both from a quantitative and a qualitative perspective. Thus, focused interests can range from specific objects to political themes. Support needs are also very different: autism can be accompanied with different degrees of language impairment (or not) and by intellectual disabilities, as well as other cooccurring conditions.

ASD appears early in development and can be identified around 18–24 months of age. It is in this period that many children begin to show clear signs in

[1] Although it is included as such in mainstream diagnostic systems, many argue that autism should not be considered a disorder and object to the use of the term autism spectrum disorder. The neurodiversity movement, for example, maintains that autistic behaviour should not be construed as *less* adaptive than typical development, but as a reflection of different developmental trajectories which are to be valued as such (Silberman 2016). We address these issues in the final sections of the chapter. We have attempted to make the language used throughout the chapter respectful to the different existing perspectives (Dwyer 2022), while at the same time reflecting the various theories as they were conceived at the time of their development. Where the authors have used other terms such as "Asperger syndrome", we have included their terminology. Also, we shall often be using the word autism as an umbrella term for all the spectrum.

communication and social interaction that distinguish them from children with other neurodevelopmental conditions or from typically developing infants. As they grow older, the different patterns of social interaction often lead to rejection or exclusion by others. This, in turn, provides them with fewer opportunities to relate to peers and create bonds, and they are more likely to feel isolated in multiple contexts (Umagami et al. 2022).

6.2 A Cognitive Explanation for Autism

As we have described above, and like many other conditions, there is a great variability in the actual behavioural manifestations of autism. There could even be some doubt as to whether it is a single entity or whether there are actually various 'autisms' (we shall return to this later). To support it as a single unified category, there have been numerous efforts in the fields of biological and psychological research. In the latter, cognitive models of autism have been enormously influential in setting the mainstream psychological perspectives (Rajendran and Mitchell 2007). For many years, in parallel to efforts to determine genetic and environmental etiological factors causing the condition, the search for a *primary cognitive deficit* that could explain the behaviours and heterogeneous forms we observe in autism was a main driver for autism research.

This search for a cognitive explanation for autism has been tremendously fruitful for the field. First, cognitive models have made an important contribution to bridging the levels of analysis of the brain and genetics, on the one hand, and behaviour, on the other (Frith 2012). They have provided units of analysis and hypotheses that have helped guide researchers looking at brain functioning in autism in the search for biological markers and endophenotypes. Although these models were initially intended to discover one single cognitive factor which could explain all autism, they have also provided an avenue for finding different subcategories or groups of individuals with autism. They have contributed to moving our perspective of autism away from the initial views of individuals with autism as less socially motivated and capable. At the same time, although with important limitations as we shall see, cognitive models have been useful in orienting the practice and models of intervention supporting clinicians and educators across the world.

One of the most influential of these cognitive models is the *theory-of-mind account* of autism (Baron-Cohen et al. 1985). The initial formulation of this model was based on the idea that autistic individuals are impaired in understanding the minds of others. Successful social interactions require understanding other people's intentions, interpreting nonverbal signals, anticipating what they may need or what they could feel, and what they might do. In short, it requires theory of mind (ToM) to take into account another person's feelings, the listener's interest in our conversation, to detect the figurative meaning of sentences, to pick up on deception or irony, to take into account what the other person knows, so they could understand what we are talking about, how our behaviour can influence the other person's response, and to understand unwritten rules or conventions. Therefore, ToM was a strong candidate for a primary cognitive factor to explain many of the atypical behaviours we see in autism.

Since the first studies in the field, this perspective has generated an enormous body of research and debate from the most varied of approaches and methods. We shall review the initial studies providing data which would make this a potentially fruitful hypothesis, then attempts to overcome the first limitations of the theory, and we shall finalize with some more recent critical perspectives.

6.3 The Beginnings of the Theory-of-Mind Account of Autism

6.3.1 Sally and Anne: The First Test

Baron-Cohen et al. (1985) believed that the ability to make inferences about what other people think was a crucial component of social skills. The hypothesis was that this would be a specific component of autistic cognition, not just the result of the general developmental delay observed in most children with autism at the time. To test this hypothesis, they evaluated 20 autistic children with typical and borderline IQ and an average age of approximately 12 years, 14 children with Down syndrome of approximately the same age, and 27 typically developing children of 4 and a half years, the age around when they should be able to solve the ToM tasks they were to be presented. They adapted Wimmer and Perner's (1983) false-belief task. In this kind of task, there is a difference between reality and the representation of a task by a given individual. If both coincide (a "true belief"), we cannot really assess ToM: the participant's own belief or their knowledge of the true state of reality will be sufficient to respond correctly to the task. However, if a character in a story has an incorrect representation of reality, the only way of realizing that this is the case is to infer his or her representation of that reality. In other words, to apply ToM (Dennett 1978).

In their specific version, a story was told with two dolls named Sally and Anne. The story and the scene with the dolls went like this:

> *There are two girls called Sally and Anne. Sally has a basket and Anne has a box. Sally picks up a marble and puts it in her basket. Then Sally leaves the room. While Sally is outside, Anne takes the marble out of Sally's basket and puts it in her box. Then Sally comes back into the room.*

Throughout the story, the researchers asked the participants different questions, one at the beginning and three at the end of the story.

1. Naming Question: "Who is Sally? Who is Anne?" This question is necessary to ensure that the participant knew which doll was which.

2. Control questions. One about the current location of the marble (Reality question: "Where is the marble really?") and another about the previous location of this object (Memory question: "Where was the marble at the beginning?").
3. Belief question: "Where will Sally look for her marble?" In this case, the answer informed the researchers whether the participant correctly attributed the belief. If participants pointed to the initial location of the marble (which was the basket), it was understood that they realized that Sally had a 'false-belief' about it. However, if the participant pointed to the current location, it was considered that they had failed the task.

All participants gave correct answers to the naming, reality, and memory questions, and 85% of the typically developing children and 86% of those with Down syndrome passed the belief question. But 80% of the children with ASD appeared to not understand Sally's representation of where the marble was. These results strongly supported the idea that autistic children differed from another neurodevelopmental disorder in ToM, and thus also the idea that ToM could be a possible specific primary cognitive factor in autism.

6.3.2 *Is It Specific to Mental States?*

Baron-Cohen et al. (1986) further studied the issue by exploring the comprehension of visual stories of different nature: they combined stories in which *intention* was central (i.e., ToM stories), with others in which it was not. Most of the participants were the same as in the previous study.

They tested participants on a picture sequencing task about different events. Researchers gave the participant the first card from the sequence and told him: "This is the first picture. Look at the other pictures and see if you can make a story with them". Then, when the children completed the sequencing task, they also had to tell the story. Each story was divided into four cards and corresponded to one of three conditions (*mechanical*, *behavioural*, and *intentional*):

1. Mechanical situations. In this case, there was a person and objects interacting. For example, the content for one situation divided into four cards was: (1) Man with rock, (2) Man pushes rock, (3) Rock rolls downhill, (4) Rock falls in water.
2. Behavioural situations. In this case, two people appeared doing a routine activity, but it did not require attributing mental states. For example: a boy is sitting and eating an ice cream. A girl appears and sits next to him: (1) Boy eats ice cream, (2) Girl sits down, (3) Girl takes away ice cream, (4) Girl eats it.
3. Intentional situations. In this sequence, the attribution of mental states of the protagonist was like a false-belief task. For example: (1) Girls puts teddy down, (2) Turns to pick up flower, (3) Boy takes teddy, (4) Girl sees teddy gone.

The researchers hypothesized that the participants with autism would show a good understanding of mechanical and behavioural situations and not so much of intentional ones. In contrast, the Down syndrome group and typically developing children may not have difficulties in the intentional situation. The results of this study showed that, in general, participants with Down syndrome had consistently poor performance in the three conditions (mechanical, behavioural, and intentional). But, more importantly, the group of children with autism performed much worse than the Down group in the intentional situation. In contrast, performance in mechanical conditions was reversed, with twice as many correct responses in the autism group than in the Down group.

Narratives were also coded according to the presence or absence of three types of verbal expression: causality, description, and use of expressions that manifest mental states. Here are some examples:

1. Causality. It was applied when a causal verb phrase appeared in the statement (e.g., "The man rolled the rock into the water"), the causal agent verb-object interaction was mentioned, or the adverb "because" was used (e.g., "The rock fell into the water because the man kicked it").
2. Mental states. It was applied when there were verbal expressions of mental state (e.g., want, believe, know, pretend, wish...) or if the participant marked the statement with a special intonation (e.g., "She is saying 'Where is my teddy bear!? !'")
3. Description. All examples that were not clear were considered as descriptions (e.g., "The man hit the stone") or those in which the narrative made no reference to causation or mental states.

No participants used expressions related to mental states in the mechanical sequences. However, children with autism used more causality terms with this type of situation. This makes sense because to perform well on the task, one needs to understand the physical causality of the situation. However, in behavioural situations, all participants used more descriptive expressions, except the typically developing group, who started to use mental state expressions. The most relevant finding in this experiment is related to the intentional situations. Here, the group of children with autism used less than half as many mental state expressions as the typically developing and Down group.

These results expanded the evidence that autistic children find it difficult to specifically attribute mental states and understand situations in which false expectations play a role.

6.3.3 From False Belief to 'True' Belief, or Knowledge and Ignorance

Hogrefe et al. (1986) had found that the concepts of knowing and ignoring are not present in most children until almost 4 years of age. Leslie and Frith (1988) aimed to confirm whether the limitations observed in solving false-belief tasks also extended to understanding whether someone has knowledge about a certain fact. Both share a meta-representational nature, but simply understanding that someone in a given situation "knows" something (for example, do they know where an object is if they have not seen it placed there?) is less complex than guessing correctly at that person's specific false representation.

In this study, 18 children with a diagnosis of autism (with typical or borderline IQ, mental age over 4 years old, and an average chronological age of 13 years) and 12 children with developmental language impairment (with an average age of almost 9 years) participated, matched on verbal comprehension. Children with autism responded to a limited knowledge task or true belief task and an adaptation of the false-belief task to a real-life situation. In contrast, children with language impairment only completed the false-belief task.

In the limited knowledge or true-belief task, the experimenters presented the child with a square yellow box, a red bag, and a round brown container. One experimenter also presented a red token to the child and hid it under the yellow box. He asked the child and another experimenter (Experimenter 2) "Can you see me hide this token under the box?" to make sure the child could see what was happening. Experimenter 1 then asked Experimenter 2 to leave the room. While outside, Experimenter 1 drew the child's attention to the fact that Experimenter 2 was not in the room and could not see what he/she was doing. Then he showed another similar token and asked the child "Can you put the token somewhere else?" (Control Question 1). Then the experimenter asked the child: "Where did Experimenter 2 see me hide the token?" (Control Question 2). These two answers had to be correct to continue with the task. After the correct answers, Experimenter 1 pointed to the object where the child had hidden the second token and asked: "Does Experimenter 2 know that there is a token under here" (Knowing question). Then he asked again: "When Experimenter 2 returns, where will he look for the token" (Prediction question). It was only considered correct if the participant pointed to the box where the token had first been hidden. In this task, the criterion of success was correct responses in both the knowledge and prediction questions.

In the false-belief task in a real-life scenario, the experimenters again presented the child with a yellow box, a red bag, and a basket with toys covered with a handkerchief. Experimenter 1 showed a coin and gave it to Experimenter 2. Experimenter 2 hid the coin in the basket, making sure the child saw it. Experimenter 2 left the room again. Then Experimenter 1 called the child's attention to ensure that the child knew that he was not seeing what was going on inside, and he/she asked: "Where did Experimenter 2 hide the coin?" (Control Question 1. It had to be a correct answer). The experimenter then took the coin out of the basket and hid it in the red bag. Next, Experimenter 1 asked the child several questions: "Does Experimenter 2 know that the coin is in the red bag?" (Knowledge question); "When he comes back, where will he look for the coin?" (Prediction question); "Where did you put the coin at the beginning?" (Control Question 1); "Where is the coin now?" (Control Question 2). If both answers were correct, they were finally asked: "Where does the Experimenter 2 believe the coin is?" (Reflection question).

Five of the 18 autistic participants passed the false-belief task and eight the knowledge task, whereas all participants with language disorder passed the knowledge tasks and all but one the false-belief task. Moreover, there appeared to be a clear interaction between performance on the limited knowledge and false-belief tests in autistic children. If children failed on the limited knowledge task, they also fail on false-belief task.

Consistent with previous research results, this study found that children with autism have clear difficulty in attribution of knowledge and belief states of mind. This was evident because the group of autistic children performed significantly lower on both tasks compared to the language-impaired group. A relevant observation was that this last group followed the false-belief scenario with enthusiasm and even found the idea of playing a prank on Experimenter 2 amusing. But this emotional state was not found in children with autism. Therefore, in addition to replicating the findings of Baron-Cohen et al. (1985) and confirming that they extend to a knowledge task, Leslie and Frith (1988) concluded that it is not exactly the language problems that are causing the performance on false-belief tasks.

6.3.4 Second-Order False-Belief Tasks

In the first study by Baron-Cohen et al. (1985), eighty percent of autistic participants performed poorly on the false-belief task. Leslie and Frith (1988) replicated these same results. Children with autism also found it difficult to solve a picture sequencing task in those situations where mental state attributions were required (Baron-Cohen et al. 1986). From these results, it appeared at the time that there could be a specific cognitive difference affecting ToM. However, in all these studies, approximately 20 to 30% of the autistic participants correctly performed the tasks. When researchers conducted a more specific analysis of those 20% who passed the false-belief task, they realized that they were the oldest participants. From these findings Baron-Cohen (1989) considered the possibility that there was a delay in the development of ToM and not so much a specific impairment. He hypothesized that if there was a true difficulty in the development of this capacity in the population with autism, it should also manifest in the older subgroup if asked to use a more advanced ToM. Perner and Wimmer (1985) had shown that typically developing children could make *second-order belief attributions,* solving what are known as *second-order false-belief tasks*. The main difference between the classical false-belief task and the second-order task lies in the following (Flavell et al. 1968): in a first-order task, we have the ability to think about a person's thoughts about an objective event; in the second-order task, we need to think about what another person thinks about the thoughts of a third person, about an objective event.

Perner and Wimmer (1985) had devised "The story of the ice-cream man". A small toy village was placed in front of the children: a house for Mary, a house for John, a church, a park, a road, and an ice-cream van, and also a row of trees, so that the characters in the story could not "see" the church or John's house from the park.

In the village, this story took place (notice there are questions posed by the experimenter at different intervals):

> *This is John and this is Mary. They live in this village* (Naming question). *They are in the park now. Along comes the ice-cream man. John would like to buy an ice cream, but he has left the money at home. He is sad. The ice-cream man tells him: "Don't worry, you can go home and get some money and buy some ice cream later. I will be in the park all afternoon". John replied: "Oh good, I'll be back in the afternoon to buy an ice cream"* (Control question 1). *So John goes home to get some money. He lives in this house.*
>
> *Meanwhile, the ice-cream man says, "I am going to drive my van to the church, maybe there I can sell some ice cream outside there"* (Control questions 2 and 3). *The ice-cream man drives over the church. On his way, he passes by John's house. John sees him and asks: "Where are you going?" The ice-cream man replies: "I'm going to sell some ice cream outside the church". And the ice-cream man continues his way.* (Prompt Questions 4 and 5)
>
> *Now, Mary goes home (she was in the park). She lives in this house. Then she decides to go to John's house. She knocks on the door and asks for her friend John, but his mother says: "He's just gone out to buy an ice cream"*(Belief question, Justification and Control questions)

The specific questions were as follows (included in the corresponding sections as indicated above):

- Naming question: "Who is Mary? Who is John?" This question is necessary to ensure that the participant knew which doll was which.
- Prompt Questions 1 to 5:

  1. "Where did the ice-cream man tell John that he would be all afternoon?"
  2. "Where did the ice-cream man say he was going?"
  3. "Did John hear that?"
  4. "Where did the ice-cream man tell John that he was going?"
  5. "Does Mary know that the ice-cream man has spoken to John?"

- Second-order belief question: "Where does Mary think John has gone to buy an ice cream?" The correct answer to this question had to be "the park". The child may use the following reasoning: Mary thinks (John thinks (the ice-cream van is in the park)), John wants (to buy an ice-cream), hence: John has gone to the park. As we described above, what makes this a second-order question is the fact that we need to think about what another person thinks about a third person's thoughts about an objective event to respond correctly.
- Justification question: "Why?". The reason for doing this was to check on what level the participants were making attributions (second-order, first-order, or zero-order attribution) according to whether the participant took account of: (a) John and Mary's beliefs (second order); (b) Only John or Mary's beliefs (first order); or (c) neither of their beliefs (zero order).
- Control questions. Reality question: "Where did John really go to buy his ice cream?" and a memory question: "Where was the ice-cream man at the beginning?" with the same aim as in the Sally and Anne task.

On this occasion, Baron-Cohen (1989) tested three different groups with ten participants in each one. The first group was composed of children with autism with typical and borderline IQ and chronological age of 15 years. The second group was made up of children with Down syndrome with an average age of 14. The third group were typically developing children aged 7 years. Typically, developing children of this age were chosen, as this is when second-order attribution ability seems to emerge (Perner and Wimmer 1985). All the participants had to be able to perform tasks that only involved first-order attributions (e.g., Sally and Anne's task).

The results of the three groups were as expected. Ninety percent of typically developing children and 60% of children with Down syndrome answered the belief question correctly. They were able to make attributions at a more advanced level. However, none of the children in the autism group managed to pass this question. All subjects who passed the belief question justified their answer with narratives such as "Mary doesn't know the ice-cream man talked to him" or "Mary thinks he doesn't know the ice-cream man is in the church". In both, it was clear that they were making second-order attributions. However, participants who failed the belief question showed that they misused first-order attribution, because they justified their answer with "He knows that the ice-cream man is in the church". Five participants with autism used a zero-order strategy, such as "The van is in the church".

These results supported the prediction that children with autism, despite being older, had a specific delay in the acquisition of a more complex mental-state attribution skill. This is not to say that these children are not able to make complex attributions which do not involve ToM, because most participants demonstrated the ability from the quick help and justification questions. Again, it seemed that this profile of sociocognitive processes was specific to the autistic population since typically developing children with an average age of 7 years did not respond in this way, nor did the Down syndrome group, albeit their lower IQ.

6.4 Advanced Theory of Mind Tasks

Research reviewed so far suggested that the atypical social interaction observed in autistic people could be related to atypical development of ToM. However, some studies questioned the idea that ToM deficits in autism were universal. Both Bowler (1992) and Ozonoff et al. (1991) found that at least some adults with Asperger syndrome or autism without intellectual disability did pass second-order false-belief tasks. But these tasks often show ceiling effects for children under the age of 6 years. First- and second-order false-belief tasks were designed to assess ToM skills for 4- to 6-year-old children (Baron-Cohen et al. 1997). This led to the development of more 'advanced' tasks, which could be considered appropriate for ToM testing at the adult level.

6.4.1 The "Strange Stories" Task

To explore ToM beyond the levels of the first- and second-order false-belief tasks, Happé (1994) conducted a study with autistic participants of typical and borderline verbal IQ, six of whom had responded incorrectly to first-order false-belief tasks, six who succeeded only on this task, and six who also passed the second-order attribution tasks. She also included ten neurotypical adults, 26 typically developing children, and 13 participants with intellectual disabilities, of whom 11 had passed the false-belief tasks.

The experimental task, which was called the *strange stories task*, consisted of 24 short vignettes related to 12 types of stories. They reflected natural social situations which require mentalizing[2] to be fully understood: lies, white lies, jokes, deceit, misunderstanding, persuasion, appearance/reality, figure of speech, sarcasm, forgetting, double bluff, and contrary emotions. Participants also had to respond to control stories that did not involve mental states but physical causal reasoning (e.g., a power cut at home that prevented food from being cooked properly). In each strange story, there was an image and two test questions (comprehension and justification questions). The comprehension questions — "Was what X (some character in the story) said true?" — and the justification questions — "Why did X say that?" — were the same across all stories. The justification questions were scored in two different ways. First, the experimenters scored correct or incorrect responses. Second, they registered whether the response was related to mental or physical states. For example, in a joke story, a boy calls a dog an elephant. A physical justification would be: "The dog is as big as an elephant." On the other hand, a justification that contains a mental state would be "the boy is only kidding". All correct mental state responses included expressions such as *like*, *want*, *happy*, *angry*, *scared*, *know*, *think*, *joke, pretend, lie*, or *fool someone*. It should be noted here, however, that there is no clear-cut distinction between the physical and the mental state justification. The so-called physical justification is more complex than just referring to a physical property in that it presupposes establishing a metaphorical relationship between the two referents, the dog and the elephant, thus increasing the cognitive load of the response.

Happé predicted that participants with autism would have more difficulty with these stories than the other control participants and that performance would be related to their performance on a false-belief ToM battery. In this way, information about the cognitive processes that underlie success or failure in mental state attribution tasks could be revealed.

The results appeared to support the validity of the stories. The lowest scores in the strange stories corresponded to the group that did not pass the false-belief tasks, and the group with the highest score was the one who passed all of them. Also, both the group that did not pass any false-belief tasks and the one who only did so for the

[2] Although in some cases distinctions are drawn between *mentalizing* and ToM itself, here we shall be using both terms indistinctively (see for example, Kliemann and Adolphs 2018).

first-order tests gave less accurate and mentalistic responses to the justification questions. Participants' responses suggested that the variety of materials, with different levels of complexity, requires the same ToM capacity as the classical false-belief tasks. In turn, they also supported the validity of the false-belief and deception tasks, since correctly responding to these tasks predicted subsequent understanding in the different natural situations of these strange stories, such as deception, joke, persuasion, etc.

Furthermore, the differences in performance between the autistic participants who did not respond correctly to false-belief tasks and the group with intellectual disabilities, matched on IQ, suggested that performing well on these tasks did not require verbal intellectual abilities.

Moreover, participants in the group who did not pass the false-belief tasks tended to give similar justifications about a mental state repeatedly. For example, one participant gave the same response "He is making a joke" for 15 of the 24 stories. While this answer was correct for the joke stories, it was incorrect for the rest of them. Happé (1994) proposed that these repeated justifications about a single mental state could mean that this population has one or two explanations stored for why people say disconcerting things in social situations. Furthermore, it was striking that the autism group with lower ToM, but with good cognitive ability, performed worse on strange stories than nonautistic participants with lower IQ. They also tended to give incorrect justifications regarding the mental state.

Since the strange stories are tasks that are closer to the natural context than the rest of the ToM task, they could reflect the real barriers that the population with autism could encounter in everyday life. To verify this, the most important comparison comes from the differences in the performance of the group that passed the second-order task with typically developing adults. Again, these participants with autism gave more atypical responses related to mental state than these adults and the much lower IQ group. Happé concluded that the strange stories can inform more than other traditional tests about the real-life difficulties many people with autism appear to encounter.

6.4.2 "Reading the Mind in the Eyes" Task

Although the strange stories tasks are widely used and have provided some interesting results, they are still highly verbal. Baron-Cohen developed another advanced ToM task, which aimed to avoid this limitation (Baron-Cohen et al. 1997; Baron-Cohen et al. 2001a).

The task consists in the presentation of a series of items which each include the presentation of a picture of the eye region of male and female faces representing different emotions. The total number of faces was 25, taken from black and white pictures from magazines. After seeing the pictures for 3 s, participants were asked "which word best describes what this person is feeling or thinking?"

In a first study (Baron-Cohen et al. 1997), participants had to choose between two options, which reflected the target emotion and the opposite valence emotion. Examples are "concerned-unconcerned", "decisive-undecisive", and "noticing you versus not noticing you". The items included basic and more complex emotions. Three groups of participants took part in this study: high-functioning autistic or Asperger syndrome, typically developing age-matched controls, and adults with Tourette syndrome. Tourette syndrome participants show similar frontal lobe impairments and social and academic disruptions in everyday life. Typically developing and Tourette participants correctly responded to 20 of 25 items approximately, as opposed to 16 approximately in the case of autistic participants. Furthermore, only 50% of these scored above chance, while 100% of the participants in the other groups did.

However, the task in this study had some limitations (Baron-Cohen et al. 2001a, b). Since there were only two options of response, and these were of opposite valence, the chance level for a correct response was very low, and there was a very small range between this level and the maximum score. This also contributed to ceiling effects. In addition, another study showed that parents of persons with Asperger syndrome, without a clinical diagnosis of autism, also had poorer scores on this task than nonrelated adults (Baron-Cohen and Hammer 1997). This indicates a poor discrimination of the task. A revised version excluded basic emotions (which do not strictly require ToM), those that could be simply responded by attending to gaze direction (e.g., "noticing"), and included a total of four response options, all of the same valence. This new version found significant differences between a group of autistic adults and typically developing IQ-matched adults, with approximate means scores of 21 and 26. More significantly, scores on the test were negatively correlated with a measure of autistic symptomatology, the Autism Quotient (Baron-Cohen et al. 2001a, b) ($r = -.53$).

The test is useful but does have some problems. The authors themselves recognize that it only actually involves the first phase of the ToM processing. It does evaluate the attribution of mental states, but it does not address the *content* of that attribution (why is the individual concerned, for example) (Baron-Cohen et al. 2001a, b).

6.4.3 *The Animations Task*

Both the strange stories and the eyes test contributed to the field by increasing the level of difficulty of the ToM tasks presented to individuals with autism and avoiding some of the measurement issues (e.g. ceiling effects) of previous studies. But both clearly load heavily on either verbal skills or emotional processing (Abell et al. 2000). More importantly, both assess explicit responses to ToM questions. A more recent group of studies focused on a task without these limitations and on the tendency to spontaneously attribute mental states, in what the authors see as a test more oriented toward the measurement of implicit ToM.

In the *animated shapes* task, participants are presented with one large red triangle and a smaller sized blue triangle which move around a white background screen (Abell et al. 2000). Ten different items are distributed into three conditions: random (two scenes), goal-directed (four), and ToM (four). The random scenes consist of the triangles moving around the screen without specific trajectories, i.e., bouncing off the sides or appearing to float in space. The goal-directed items involved a triangle responding to the behaviour of the other, e.g., following its trajectory or moving in parallel.

ToM items imply in some way that a triangle responded to the *mental state* of the other. For example, in one, the larger triangle is trying to coax the smaller one out of a rectangle. Another difference among conditions was the way the triangles were identified: in the first they were presented as triangles, in the second as animals (two deer fighting, mother duck and duckling following, two cats chasing, and two ponies dancing), and in the third as people (grandmother and grandson surprising, mother and child coaxing, teacher and boy mocking, and girl prisoner and guard seducing or escaping).

They are then scored as 0, 1 or 2, according to the degree in which they correspond to the actual intended meaning of the presentation. They are also scored according to whether they describe random action (a simple action statement with no reference to interaction between triangles or mental state or psychological language), interaction (reference to interaction between the triangles, *without* indicating mental language) or mental state attribution.

In an initial study, differences were found between groups of autistic children (with typical and borderline IQ and mild cognitive impairment), typically developing children, and children with mild cognitive impairment, in the ToM items. Both the cognitive impairment and autism group used less mentalizing descriptions than the typically developing children. Also, the autism group differed from the preestablished definitions for the different ToM scenarios in 36% of the cases, as opposed to 3% in the cognitive impairment group, 7% in the typically developing children, and 2% in the adult group.

In additional studies, the scoring system and instructions were slightly modified to assess intentionality (ranging from 0 to 5), appropriateness (0 to 3), and length (number of clauses, up to 4) (Castelli et al. 2000, 2002), with similar results. A more recent version has replaced open-ended questions with a multiple-choice format suitable for online testing (White et al. 2011).

A recent review and meta-analysis shows that this task, like those mentioned above, has been widely used in the field (Wilson 2021). The meta-analysis with 33 articles found that, after controlling for any differences between groups in verbal ability, the absolute effect sizes [95% confidence intervals (CI)] were small for random animations, $g = -0.35$ [−0.51, −0.19] and goal-directed movement, $g = -0.35$ [−0.48, −0.22], and there was a medium effect size for mentalizing animations, $g = -0.62$ [−0.74, −0.50], all $p < 0.001$. These effect sizes were overall smaller than those of the initial studies, but none-the-less significant. The moderator analysis also demonstrated a trend-level link between administration of the multiple-choice (rather than verbal) task format and slightly smaller group differences between autistic and nonautistic samples. There were no differences in effect size when comparing children and adults.

In addition to the different performance of the autism and typically developing groups, there is certain support within the social neuroscience literature for the validity of the animations as a test of mentalizing. Studies show that the mentalizing condition reliably activates social-cognitive networks in the brain that partially overlap with activation patterns observed for other mentalising tasks (Schurz et al. 2021).

However, the animations correlate very little, if at all, with other mentalizing tasks (Wilson 2021). Also, autistic individuals tend to achieve lower scores than typically developing participants in the control items. Both of these results would appear to indicate that the test is not a pure measure of ToM and could include other processing that accounts for individual variance.

6.5 What If It Is Not the Theory of Mind But the Executive Function?

The theory-of-mind account of autism is not without its critics. In the previous section, we have revised the limitations in interpreting many of the studies in the field and in the tasks that have been used. Here we turn to a competing model of autism which, to different degrees, calls into question the primary nature of ToM as a cognitive explanation for autism: the executive-function account.

Executive function (EF) is an umbrella term which includes different processes involved in the control and regulation of goal-directed behaviour (Hill 2004). Closely related to the prefrontal cortex, it includes inhibition, working memory, planning, and flexibility, among others. In typical development, both EF and ToM experience important developments between the ages of 4 and 6 years, and therefore the relationship between them has been the object of considerable interest (Wade et al. 2018). The nature of the relationship between the two constructs could come under a series of different possibilities. There are various options: EF develops as a consequence of ToM; ToM relies on the prior development of EF; neither is exact, but the tasks with which we measure ToM do require a certain degree of development of EF; or both EF and ToM tasks tap a common underlying process, which might also be influencing them both (Perner and Lang 1999; Wade et al. 2018). Perner et al. argued that for a full development of EF, it is necessary for the individual to understand the representational nature of the mind, i.e., ToM is a prerequisite for EF. However, others support the idea certain development of EF is necessary for ToM to take off (Ozonoff et al. 1991).

In the specific case of autism, a series of researchers soon observed that some of the criteria for autism are like those found in patients with executive function deficits (Ozonoff et al. 1991): need for invariability, difficulties in switching attention, tendency to perseverate. These are exactly the behaviours that are harder to explain from a ToM perspective (as opposed to behaviours related to social interaction, for example). The study of EF as a potential primary cognitive deficit for autism was soon to come (Rajendran and Mitchell 2007).

In one of the first studies to explore this issue, Ozonoff et al. (1991) included autistic participants with typical and borderline full IQ and typically developing individuals, aged 8–20 years, matched in age, sex, verbal, nonverbal, and full IQ, socioeconomic status, race, and handedness. They expected to find all autistic participants performing below the typically developing group on the ToM tasks, but only a few did so on the EF tasks, consistent with a ToM account of autism. They did find that the scores of the autistic group were lower than those of the typically developing individuals on all composite scores used in the study (first-order ToM, second-order ToM, and EF). But, looked at individually, while 96% of autistics scored below the control-group mean on EF, only 52% did so on the first-order composite and 87% on the second-order composite. EF and second-order theory-of-mind composites best predicted group membership in a discriminant analysis. But again, second-order ToM limitations were less specific to the autism group: 56% of the autistic group failed all tasks, but so did 15% of controls.

Russell and his team took a different approach, attempting to distinguish the role of ToM and EF in the same task (Russell et al. 2003). They produced a series of experiments in which they combined the need for understanding mental states, while requiring participants to inhibit a prepotent response and hold an arbitrary rule in memory. They set up a contraption with two containers above two corresponding buttons, with a central outlet that delivered sweets when the appropriate button was pressed. The way it worked was that the child had to press the button under an empty container to obtain a sweet, which was shown to be in another adjacent container. In one of their studies, they used three conditions: no-opponent, opponent-nondeceptive, opponent-deceptive. In the first they obtained sweets from the machine; in the second when they got a sweet, another player did not, and vice versa; and in the third they had to indicate the button they wanted the opponent to press (for them to get the sweet). Russell and colleagues interpret that the errors in this task are more related to the difficulty to inhibit the tendency to press the button under the sweet (inhibition) than in understanding the position of another player.

Another example of research that examines the role of EF with respect to ToM can be found in the studies by Pellicano (2007, 2010). She found that ToM and EF were closely correlated in a sample of 30 four- to seven-year-old participants with and without autism (and typical IQ) (Pellicano 2007). In the autism group especially, ToM and set-shifting EF measures were significantly related, even after taking verbal and nonverbal ability and age into consideration. Although in many cases, impairments in both EF and ToM were evident, when it was not, it was more often that EF developed typically and ToM did not, than the contrary (although the tasks could not be considered fully equivalent in difficulty). She further investigated this issue in a longitudinal study with 45 children aged 4–7 years with ASD at Time 1, of which 37 were followed up 3 years later. She found that EF skills were predictive of ToM skills, but not the opposite (Pellicano 2010).

Hill's (2004) extensive review appeared to corroborate the differences in development in EF between autistic and typically developing individuals. However, although the results we have presented would seem to point in the direction of a primacy of EF, several limitations of the theory have been pointed out. Few studies other than these have carefully looked at the issue of universality of EF differences with respect to typically developing participants, so it is not clear that this is a marker of autism (Rajendran and Mitchell 2007). Also, even accepting that EF plays a role in false-belief tasks and in everyday social interaction, it is difficult to see why domain-general processes in EF should differentially affect social interaction and problem solving in such a specific way as we can appreciate in autism (Fletcher-Watson and Happé 2019). Finally, differences in EF are also seen in other diagnostic groups. Although it has been suggested that the autism EF profile affects cognitive flexibility specifically and not inhibition, studies trying to typify autism based on EF profiles have produced mixed results (Hill 2004; Rajendran and Mitchell 2007).

6.6 What If It Is Language?

The language profiles of individuals on the autism spectrum are highly heterogeneous, characterised by varying degrees of structural language competence. While some autistic children may produce little language and may have substantial language difficulties, others acquire age-adequate structural language skills and can use language appropriately (Tager-Flusberg 2000). Still, despite preserved structural language skills and, sometimes, strengths in grammar, even highly verbal individuals with autism are faced with problems in certain domains of language, such as figurative language comprehension, and display a delayed developmental trajectory in that domain (Chahboun et al. 2016; Vulchanova et al. 2015). Furthermore, weaknesses have been attested in the domain of discourse comprehension and ability to establish relations of coherence in text (Joliffe and Baron-Cohen 1999; Micai et al. 2017). Given that language is a central part of the understanding of the verbal ToM tasks, an important question is whether language skills might be causing the variation observed in performance on both first- and second-order ToM tasks. Indeed, there is evidence of a strong relationship between structural language skills (e.g., vocabulary and grammar) and success at ToM tasks in participants with autism (Fisher et al. 2005; Tager-Flusberg and Joseph 2005). Furthermore, a body of research has provided evidence of the impact of specific language skills which might be playing a role in successful performance on ToM scenarios (Durrleman et al. 2017; Durrleman and Franck 2015; Lind and Bowler 2009; Tager-Flusberg 2000; Tager-Flusberg and Joseph 2005). In addition, given the variability in autistic language competence, a useful comparison will be with children with developmental language disorder.

Lind and Bowler (2009) investigated the hypothesis that autistic children use their syntactic competence in the domain of clausal complementation to "hack out" solutions to false-belief tasks. Thus, the idea is that these children rely on complement syntax skills to resolve ToM tasks in the absence of representational ToM. This is based on extant evidence from typical development (De Villiers 1995; de Villiers and Pyers 2002) and a suggestion by Tager-Flusberg (2000) that a specific type of syntactic competence, the syntax of complementation, may facilitate performance on ToM tasks in children with autism. Tager-Flusberg established that competence in the domain of complementation after verbs of communication was a significant predictor of performance on change of location false-belief tasks (e.g., Sally-Anne task) in children and adolescents with autism. Verbs of communication (*say, tell*) and mental verbs (*believe, know, think*) provide an interesting context where the truth-values of the two clauses, the root clause and the complement clause, can be independently assessed. Thus, in a sentence like "*She believed that John was a spy, but it turned out he wasn't*", the fact that John actually wasn't a spy does not affect the (true) fact that she believed so. In such a way, developing an understanding of the underlying logic of such linguistic structures may pave the way for resolving ToM/false-belief scenarios, as presented in typical tasks.

In their study, Lind and Bowler (2009) tested 48 children with autism in comparison to 48 children matched on verbal age on the Sally-Anne task, and in second sample, 53 children with autism and a comparison group of 53 children matched on verbal age, on the Smarties false-belief task. In this task, children are shown a box of chocolates (Smarties) which, in fact contains a pencil. When they are shown the content, they are then asked what a person who had not seen the content would think was in the box. Both comparison groups consisted of children with learning disability in order to match the children with autism who had a learning disability and typically developing children (for those children with ASD who did not have a learning disability). Both samples were also tested on the BPVS to assess their verbal age and on a memory for complements task based on the design by De Villiers and Pyers (2002). In this type of task, the participant hears a sentence like "She said *she found a monster under her chair,* but it was really the neighbour's dog" and is asked "What did she say?". Thus, if the child understands the relationship between the two clauses, they will correctly respond what she actually said, rather than what the complement clause implies. Consistent with the hypothesis of the study, the authors found a stronger relationship between the scores on syntax complementation and performance on change of location (Sally-Anne task) within the group with autism than the comparison group. However, they did not find a similar relationship for the unexpected-contents task (the Smarties task). The finding that a relationship between complement syntax competence in the context of communication verbs and false-belief tasks is only evident in the autism group is puzzling, given that participants were matched on verbal age. The authors explain this in terms of a possible developmental lag in the group with autism, and the idea that specific relationships between predictor variables and outcome variables might only hold at specific points in development. Indeed, this is a viable account consistent with similar findings in the domain of figurative-comprehension (Vulchanova et al. 2019) and the predictive role of deictic gestures for language outcomes (Ramos-Cabo et al. 2022). These findings are also compatible with neuro-constructivist accounts of developmental disorders (Westermann et al. 2010).

An interesting question concerning the above evidence, is whether it is overall language competence and complement syntax competence or only competence specifically in the domain of propositional attitude verbs which matters. Durrleman and Franck (2015) tested exactly this. For this purpose, they administered both a verbal and a nonverbal ToM task, and a number of different complement syntax clauses, such as after cognition verbs (*think, believe*) and verbs of communication (*say*), where the truth values are independent, and verbs of perception (*see, hear*), where the truth values are identical. In addition, they addressed the role of EF, following the findings in Pellicano (2007). The study sample included 17 autistic children and adolescents (age range 6–16 years) and a younger control group matched on non-verbal IQ (age range 4–9 years). The participants in both groups performed similarly on the complement syntax tasks, the non-verbal ToM task and the EF task. However, the autistic group displayed a significantly poorer performance on the verbal ToM task. Importantly, for both groups, correlations were established between understanding of the verbs of communication and cognition clauses and performance on the verbal ToM task, but not for verbs of perception. The EF indexes did not correlate with either task. These findings indicate that performance on ToM/false-belief tasks may depend on a specific type of language competence, namely complementation syntax expressing propositional attitudes. In addition, they support the findings by Lind and Bowler (2009) of a developmental lag in autism in that domain and this specific relationship applying only to a specific point in development, given that the control group in the Durrleman and Franck sample was much younger than the autistic group. Following up on the language lead, in a subsequent study, Durrleman et al. (2017) compared performance by children with autism to children with language impairment. The design of the complementation syntax was similar (with three types of verbs) and a picture sequencing task to assess ToM (based on Baron-Cohen et al. 1986). The results of this study revealed that both autistic participants, participants with language impairment, and typical controls who had similar scores on the complementation syntax also performed similarly on the ToM task. Importantly, the results confirm that only competence in the domain of the independent truth value clauses impacts on ToM ability.

The studies investigating the role of language in ability to solve ToM tasks indicate that language competence does indeed impact on the ability to mentalise. They also show that it is specific skills in the domain of complementation syntax which expresses propositional attitudes. In typical development, this type of syntax emerges relatively later, and has been shown to correlate strongly with ability to solve ToM tasks. An open question is what is the causal relationship between the two skills. Even though there is some evidence that it is complement syntax competence which predicts performance on ToM tasks (de Villiers and Pyers 2002), more research is needed to establish the direction of this relationship, especially in autistic children. It may be the case that both mature roughly around the same developmental window, and that both support each other by way of a scaffolding relationship. Furthermore, verbs of cognition and communication are only a small part of the group of verbs expressing propositional attitudes. The inventory of such verbs varies across languages and native speaker responses to such sentences have been attested to vary as shown in a recent EEG/ERP study from a Norwegian adult sample (Călinescu et al. 2020). Since the studies reviewed above have only tested autistic participants with English and French as their native languages, it is important to provide evidence from a wider selection of languages, both from autistic samples and samples involving matched controls, in order to establish the developmental trajectories of complementation syntax and its impact on ability to understand ToM contexts. There is also preliminary evidence that idiom comprehension may be related to ToM task performance in autistic children (Whyte et al. 2014). In this context and given that autistic samples are often matched to younger children for the purposes of matching either on intellectual or verbal ability, or both, more longitudinal evidence is needed from typical development, as well as more domains of language competence which develop relatively late (metacognition, metalinguistic awareness, pragmatic and discourse competence) and may exert an impact on ToM development.

6.7 'Pretending' Theory of Mind: Compensation and Camouflaging

Another reason why participants may be able to solve ToM tasks or everyday tasks is that they may learn certain strategies to do so. This does not necessarily mean that mentalizing occurs in the same way as for neurotypicals. Senju et al. (2009) explored the eye movements of individuals with Asperger syndrome while watching a task like the first-order false-belief test. In participants who were able to respond correctly to the classic version of these tests, they found atypical looking patterns that appeared to indicate that although their explicit responses were typical, their initial implicit ToM response was not. This would resonate with the idea that even autistic people who respond to ToM tasks (and everyday social situations) similarly to neurotypical participants do so using different strategies or approaches. In other words, they would compensate for difficulties in ToM, both in responding to experimental tasks and in real-life situations, by using alternative cognitive routes.

Compensation refers more to the use of cognitive routes to solve certain tasks or challenges, in a different way from typically developing individuals (Livingston et al. 2019). It is a concept related to the broader idea of *masking* and *adaptive morphing*, which involve "the employment of specific behavioural and cognitive strategies by autistic people to adapt to or cope within the predominantly non-autistic world" (Cook et al. 2021, p. 1). For example, not engaging in certain repetitive hand movements, forcing eye contact, and using learned rules about conversations or nonverbal behaviour might all come under masking strategies an autistic person might engage in.

The study of camouflaging in autism is a very active field at this moment. A recent review (Cook et al. 2021) found 29 studies in recent years. Although one could argue that everyone somehow adapts their social behaviour to the context in which they are, camouflaging appears to be more than this. Autistic people experience more frequently a mismatch between their behaviour and their understanding of social interactions, and point out that the way they behave does not represent their true character. It appears that adults with higher self-reported autistic traits report greater camouflaging, and so does having an autism diagnosis. Potentially, feelings of stigma and not being accepted seem to play a role. Another outcome is that reported camouflaging appears to be related to increased symptoms of mental health problems. This is a complex issue: More anxious people may experience the need to camouflage more, but also those who do camouflage may be under greater stress because of this effortful constant attempt to hide their true self. More importantly, the association between discrepancy measures and mental health difficulties is not consistently evident for adults or children, but it is for self-report measures. Therefore, the belief of people that they need to camouflage may be more related to these issues than their actual doing so or their ability to do so. In any case, it seems that camouflaging is not a cost-free and effortless endeavour.

Sex or gender differences have also been extensively explored in camouflaging (Cook et al. 2021). They have been proposed as a reason for reduced rates of detection or late diagnosis among women. It could be that autistic females have a greater capacity to camouflage or that they might experience greater pressure to do so. Both self-report and discrepancy measures indicate that women and girls camouflage more. But it is not exclusive to them: men also camouflage and the effect sizes between genders are often small to moderate.

There are various approaches to measure camouflaging, which can be broadly separated into discrepancy scores and observational measures (Hannon et al. 2022). In the first, researchers use a measure of autistic traits or theory-of-mind tasks, such as the eyes test, the animations, which are then compared to measures of observable autistic behaviours. Observational measures include self-report questionnaires that ask about strategies used to compensate for autistic behaviours or to fit in. Also in this category are scales for teachers or parents. The first group, the discrepancy measures, is of greater interest to us here because they directly relate to how ToM measured in the laboratory actually plays out in natural social interactions.

Livingston et al. (2019) carried out an interesting study with 136 autistic adolescents (aged 10–15). They measured ToM (with the animations test), anxiety and a test of autistic behaviours used in diagnostic procedures, the Autism Diagnostic Observation Schedule, ADOS (Lord et al. 2000). They categorized their participants into four groups: low compensation (poor ToM/poor ADOS), high compensation (poor ToM/good ADOS), deep compensation (good ToM/good ADOS), and unknown (good ToM/poor ADOS). Compared to low compensators, high compensators had a higher verbal IQ (but not nonverbal IQ), better EF and greater anxiety. These results indicate that EF could be an important help to compensate. But it might be the other way around and early compensation could support better social interaction and EF development.

The area of camouflaging still needs considerable research and there are several important limitations in the state-of-the-art evidence (Cook et al. 2021; Fombonne 2020; Lai et al. 2021). For a start, the concept and the validity of the construct and its relation to autism itself need clarification. The best way to measure it is also a matter of debate. The studies have looked at restricted categories of participants across ages. Most adults were white educated females with average IQ, many diagnosed in late life, while children and adolescent participants were mostly white males with average IQ. What happens outside these groups and in individuals with intellectual disabilities, or in nonautistic groups of other diagnostic categories but also atypical social interactions, is also unknown (Cook et al. 2021).

In any case, the idea that autistic children and adults actively respond to the social demands of a social context they do not fully share or understand is important from the perspective of understanding how atypical ToM develops and has an influence on the lives of autistic people. Also, it somehow incorporates their perspective and lived experiences into what happens during social interactions. But it still only looks at the individuals with autism, ignoring the social context in which they are immersed. More recent studies propose that this is a one-sided view which needs to be supplemented by a closer look at how mentalizing develops in neurotypical individuals with respect to their autistic peers.

6.8 When the Theory-of-Mind Deficit Happens to Everyone: The Double Empathy Perspective

All the research we have described up to now rests on the assumption that the place to look when thinking about how to best support the needs of autistic individuals is in the autistic individuals. The different studies have investigated how the autistic mind works and in what way it is different from that of neurotypicals when it comes to understanding others. The theory-of-mind account can be seen as central to this view: individuals with autism find it hard to understand the minds of others, and hence their atypicalities in development.

This perspective has recently been criticized on the grounds that it is too close to a medicalized and deficit-centred model of autism (López 2022; Mitchell et al. 2021). It comes from the view that if autistic persons find it hard to interact socially and are often isolated from others, it is because they are not capable of understanding them. But some have proposed that a more complete view of autism should not only focus on individuals, but also on their context. And this context is basically composed of and shaped by non-autistic people. In other words, it is all about adapting to the non-autistic rules and interpretations of how social interactions should work and be understood. A *transactional* view like this requires that both the individuals, their contexts, and the interaction between them should be the focus of attention.

Researchers in this area have proposed that the ToM deficit might be better understood as a form of cultural misunderstanding, rather than a developmental deficit. The rules of social interaction and the cues to mind reading might be different between autistics and neurotypicals. If this is the case, one would also expect to find a difficulty for neurotypicals to understand the minds of autistics. This is what autistic individuals indicate they perceive regularly (Mitchell et al. 2021). But there is also experimental evidence to support the idea. In a study by Sheppard et al. (2016), the experimenter greeted different autistic and non-autistic participants, and their reactions were recorded. The reactions were then presented to neurotypical perceivers (who did not know which group the reactors belonged to). Neurotypical participants were less accurate in determining the reaction seen in the videos of autistic persons than in non-autistic persons. This was the case even though autistic and non-autistic recordings were considered to be equally expressive. This would appear to indicate that not only do autistics find it hard to mindread (neurotypicals), but the reverse may also be true. This has been labelled as the *double-empathy* problem. It can occur whenever there is an issue with mutual understanding (and is therefore a difficulty for two people, not for one of the parties), and it is more likely when these persons are of different dispositions (Milton 2012; Milton et al. 2022).

The double-empathy hypothesis requires cross-group understanding to be harder, but also within-group social interaction to be easier. Shepperd's study shows the former. There is also some evidence that the interactions of autistic individuals are more successful with other autistics than between groups, although more research is needed (Mitchell et al. 2021). For example, Crompton et al. (2020) found that information was better communicated through a chain of all autistic people than a combination of autistic and non-autistic members. This would suggest that autistic interactions are more easily comprehensible to autistic people than to neurotypicals.

As with camouflaging, looking at ToM within the wider angle of double empathy broadens our view of social information processing in autism. Although this perspective is still needed of much research, it allows us to move beyond a purely deficit-centred approach which has governed the field and helps us consider the perspectives of autistics themselves.

6.9 Conclusions

Much has happened since Baron-Cohen et al.'s (1985) paper. The impact of the theory-of-mind account in the field of autism has been impressive, accumulating a phenomenal amount of research and debate, a small part of which we have reviewed here. Much of the past research may now appear to be out of touch with current thinking in the fields of social neuroscience or even of autism itself. For example, the search for a primary deficit does not fit in easily with current predominantly multicausal and interactive views of development (Fletcher-Watson and Happé 2019). Such a modular and unicausal model also appears to contrast with the current dimensional view of autism, where heterogeneity within the spectrum is the focus of much attention. Some argue that different profiles in autism, related to distinct cognitive and genetic underpinnings, might be more explicative of autism as we see it in the real world in all its diversity (Happé and Frith 2020).

The theory-of-mind account has also been seen often as a theory-of-mind *deficit* account of autism. This way of looking at autism is also currently challenged

(Silberman 2016) A neurodiversity perspective of autism attempts to view this condition as a different way of looking and understanding the world, rather than a delayed or inadequate approach, regardless of the support needs that the different individuals on the spectrum may have. From this point of view, developments such as the double empathy approach, although in its early stages and needed of much empirical work, are a way to incorporate the perspectives and concerns of autistic individuals who do not seem to 'suffer' from a disorder (see Fletcher-Watson and Happé 2019 for a very interesting integration of classic cognitive research with a neurodiversity perspective).

The role of ToM has also been questioned on its practical and applied implications. Although this model has helped to develop a great number of interventions and guides the work of many clinics and schools, the impact of programs based on teaching ToM on the quality of life and social interactions of people with autism appears to be limited (Fletcher-Watson et al. 2014). The necessary reflection on the focus of these interventions is equally important. If, as the camouflaging literature suggests, adapting to and dealing with the rules of the neurotypical world could come at a cost to mental health, interventions aimed in this direction should take note and incorporate the appropriate adjustments (Lai et al. 2021). It also calls for paying attention not only to teaching autistics to better understand the neurotypical world, but also to make this world more accessible and teach it to, in turn, understand autistics.

But even with all these considerations, the theory-of-mind account of autism is clearly central to how we see autism nowadays and has been an essential steppingstone to arrive at our current understanding of the needs of autistic individuals. Undoubtedly, it has pushed the field beyond an initial view in which they were seen as limited in their social interests, motivations, and abilities. It has also brought research forward in the study of typical development, notably in the cognitive and neuroimaging realm (Fletcher-Watson and Happé 2019). Above all, reflecting on ToM in autistic and neurotypical individuals has helped clinicians and educators all over the world to realize that perhaps a person with autism does not see the world the way they do. And this is the best starting point to understand how to make the neurotypical world more accessible to everyone.

Acknowledgments This work was supported by a Formación de Profesorado Universitario grant awarded to the first author by the Ministerio de Ciencia e Innovación (Spain); a Horizon 2020 Marie Slovowska-Curie (grant number grant 857897, eLADDA) to the second and third authors, and a grant awarded to the third author by the Ministerio de Ciencia e Innovación (Spain), via its Plan Estatal 2017-2020 – Proyectos Generación del Conocimiento (grant number: PGC2018-096094-B-I00), respectively.

References

Abell, F., F. Happé, and U. Frith. 2000. Do triangles play tricks? Attribution of mental states to animated shapes in normal and abnormal development. *Cognitive Development* 1–16. https://doi.org/10.1016/S0885-2014(00)00014-9.

American Psychiatric Association. 2013. *Diagnostic and statistical manual of mental disorders*. 5th ed. *DSM 5*. American Psychiatric Association.

Baron-Cohen, S. 1989. The autistic child's theory of mind: A case of specific developmental delay. *Journal of Child Psychology and Psychiatry, and Allied Disciplines* 30 (2): 285–297. https://doi.org/10.1111/j.1469-7610.1989.tb00241.x.

Baron-Cohen, S., and J. Hammer. 1997. Parents of children with Asperger syndrome: What is the cognitive phenotype? *Journal of Cognitive Neuroscience* 9 (4): 548–554. https://doi.org/10.1162/JOCN.1997.9.4.548.

Baron-Cohen, S., A.M. Leslie, and U. Frith. 1985. Does the autistic child have a 'theory of mind'? *Cognition* 21: 37–46.

———. 1986. Mechanical, behavioural and intentional understanding of picture stories in autistic children. *British Journal of Developmental Psychology* 4 (2): 113–125. https://doi.org/10.1111/j.2044-835X.1986.tb01003.x.

Baron-Cohen, S., T. Jolliffe, C. Mortimore, and M. Robertson. 1997. Another advanced test of theory of mind: Evidence from very high functioning adults with autism or Asperger syndrome. *Journal of Child Psychology and Psychiatry* 38 (7): 813–822. https://doi.org/10.1111/J.1469-7610.1997.TB01599.X.

Baron-Cohen, S., S. Wheelwright, J. Hill, Y. Raste, and I. Plumb. 2001a. The "Reading the mind in the eyes" test revised version: A study with normal adults, and adults with Asperger syndrome or high-functioning autism. *Journal of Child Psychology and Psychiatry* 42 (2): 241–251. https://doi.org/10.1111/1469-7610.00715.

Baron-Cohen, S., S. Wheelwright, R. Skinner, J. Martin, and E. Clubley. 2001b. The autism-spectrum quotient (AQ): Evidence from Asperger syndrome/high-functioning autism, males and females, scientists and mathematicians. *Journal of Autism and Developmental Disorders* 31: 5–17. https://doi.org/10.1023/a:1005653411471. 11439754.

Bowler, D.M. 1992. "Theory of mind" in Asperger's syndrome Dermot M. Bowler. *Journal of Child Psychology and Psychiatry* 33 (5): 877–893. https://doi.org/10.1111/j.1469-7610.1992.tb01962.x.

Călinescu, L., A. Giskes, M. Vulchanova, and G. Baggio. 2020. Embedding (im)plausible clauses in propositional attitude contexts: Modulatory effects on the N400 and late components. *Journal of Neurolinguistics* 53: 100877. https://doi.org/10.1016/j.jneuroling.2019.100877.

Castelli, F., F. Happé, U. Frith, and C. Frith. 2000. Movement and mind: A functional imaging study of perception and interpretation of complex intentional movement patterns. *NeuroImage* 12 (3): 314–325. https://doi.org/10.1006/nimg.2000.0612.

Castelli, F., C. Frith, F. Happé, and U. Frith. 2002. Autism, Asperger syndrome and brain mechanisms for the attribution of mental states to animated shapes. *Brain : A Journal of Neurology* 125 (Pt 8): 1839–1849.

Chahboun, S., V. Vulchanov, D. Saldaña, H. Eshuis, and M. Vulchanova. 2016. Can you play with fire and not hurt yourself? A comparative study in figurative language comprehension between individuals with and without autism Spectrum disorder. *PLoS One* 11 (12): e0168571. https://doi.org/10.1371/journal.pone.0168571.

Cook, J., L. Hull, L. Crane, and W. Mandy. 2021. Camouflaging in autism: A systematic review. *Clinical Psychology Review* 89: 102080. https://doi.org/10.1016/j.cpr.2021.102080.

Crompton, C.J., D. Ropar, C.V.M. Evans-Williams, E.G. Flynn, and S. Fletcher-Watson. 2020. Autistic peer-to-peer information transfer is highly effective. *Autism* 24 (7): 1704–1712. https://doi.org/10.1177/1362361320919286.

De Villiers, J.G. 1995. Questioning minds and answering machines. In *Proceedings of the 19th Boston University conference on language development*, ed. D. MacLaughlin and S. McEwen, 20–36. Somerville: Cascadilla.
de Villiers, J.G., and J.E. Pyers. 2002. Complements to cognition: A longitudinal study of the relationship between complex syntax and false-belief-understanding. *Cognitive Development* 17 (1): 1037–1060. https://doi.org/10.1016/S0885-2014(02)00073-4.
Dennett, D.C. 1978. Beliefs about beliefs. *Behavioral and Brain Sciences* 1 (4): 568–570. https://doi.org/10.1017/S0140525X00076664.
Durrleman, S., and J. Franck. 2015. Exploring links between language and cognition in autism spectrum disorders: Complement sentences, false belief, and executive functioning. *Journal of Communication Disorders* 54: 15–31. https://doi.org/10.1016/j.jcomdis.2014.12.001.
Durrleman, S., M. Burnel, and A. Reboul. 2017. Theory of mind in SLI revisited: Links with syntax, comparisons with ASD. *International Journal of Language & Communication Disorders* 52 (6): 816–830. https://doi.org/10.1111/1460-6984.12317.
Dwyer, P. 2022. Stigma, incommensurability, or both? Pathology-first, person-first, and identity-first language and the challenges of discourse in divided autism communities. *Journal of Developmental and Behavioral Pediatrics* 43 (2): 111–113. https://doi.org/10.1097/DBP.0000000000001054.
Fisher, N., F. Happé, and J. Dunn. 2005. The relationship between vocabulary, grammar, and false belief task performance in children with autistic spectrum disorders and children with moderate learning difficulties. *Journal of Child Psychology and Psychiatry* 46 (4): 409–419. https://doi.org/10.1111/J.1469-7610.2004.00371.X.
Flavell, J.H., P.J. Botkin, and C.L. Fry. 1968. *The development of role-taking and communication skills in children*. New York: Wiley.
Fletcher-Watson, S., and F. Happé. 2019. *Autism : A new introduction to psychological theory and current debates*. London: Routledge.
Fletcher-Watson, S., F. McConnell, E. Manola, and H. McConachie. 2014. Interventions based on the theory of mind cognitive model for autism spectrum disorder (ASD). *Cochrane Database of Systematic Reviews* 3. https://doi.org/10.1002/14651858.CD008785.pub2.
Fombonne, E. 2020. Camouflage and autism. *Journal of Child Psychology and Psychiatry* 61 (7): 735–738. https://doi.org/10.1111/jcpp.13296.
Frith, U. 2012. Why we need cognitive explanations of autism. *The Quarterly Journal of Experimental Psychology* 65 (11): 1–20. https://doi.org/10.1080/17470218.2012.697178.
Hannon, B., W. Mandy, and L. Hull. 2022. A comparison of methods for measuring camouflaging in autism. *Autism Research* 16 (1): 12–29. https://doi.org/10.1002/aur.2850.
Happé, F. 1994. An advanced test of theory of mind: Understanding of story characters' thoughts and feelings by able autistic, mentally handicapped, and normal children and adults. *Journal of Autism and Developmental Disorders* 24 (2): 129–154.
Happé, F., and U. Frith. 2020. Annual research review: Looking back to look forward – Changes in the concept of autism and implications for future research. *Journal of Child Psychology and Psychiatry* 61 (3): 218–232. https://doi.org/10.1111/JCPP.13176.
Hill, E.L. 2004. Executive dysfunction in autism. *Trends in Cognitive Sciences* 8 (1): 26–32.
Hogrefe, G.-J., H. Wimmer, and J. Perner. 1986. Ignorance versus false belief: A developmental lag in attribution of epistemic states. *Child Development* 57 (3): 567. https://doi.org/10.2307/1130337.
Joliffe, T., and S. Baron-Cohen. 1999. A test of central coherence theory: Linguistic processing in high-functioning adults with autism or Asperger syndrome: Is local coherence impaired? *Cognition* 71: 149–185.
Kliemann, D., and R. Adolphs. 2018. The social neuroscience of mentalizing: Challenges and recommendations. *Current Opinion in Psychology* 24: 1. https://doi.org/10.1016/J.COPSYC.2018.02.015.
Lai, M.C., L. Hull, W. Mandy, B. Chakrabarti, C.W. Nordahl, M.V. Lombardo, S.H. Ameis, P. Szatmari, S. Baron-Cohen, F. Happé, and L.A. Livingston. 2021. Commentary: 'Camouflaging' in autistic people – Reflection on Fombonne. *Journal of Child Psychology and Psychiatry and Allied Disciplines* 62 (8). https://doi.org/10.1111/jcpp.13344.

Leslie, A.M., and U. Frith. 1988. Autistic children's understanding of seeing, knowing and believing. *British Journal of Developmental Psychology* 6 (4): 315–324. https://doi.org/10.1111/j.2044-835X.1988.tb01104.x.

Lind, S.E., and D.M. Bowler. 2009. Language and theory of mind in autism Spectrum disorder: The relationship between complement syntax and false belief task performance. *Journal of Autism and Developmental Disorders* 39 (6): 929–937. https://doi.org/10.1007/s10803-009-0702-y.

Livingston, L.A., E. Colvert, P. Bolton, and F. Happé. 2019. Good social skills despite poor theory of mind: Exploring compensation in autism spectrum disorder. *Journal of Child Psychology and Psychiatry and Allied Disciplines* 60 (1): 102–110. https://doi.org/10.1111/jcpp.12886.

López, B. 2022. Commentary on autism and the double-empathy problem: Implications for development and mental health. *British Journal of Developmental Psychology* 40 (2): 368–370. https://doi.org/10.1111/bjdp.12410.

Lord, C., S. Risi, L. Lambrecht, E.H. Cook, B.L. Leventhal, P.C. Dilavore, A. Pickles, and M. Rutter. 2000. The autism diagnostic observation schedule-generic: a standard measure of social and communication deficits associated with the spectrum of autism. *Journal of Autism and Developmental Disorders* 30 (3): 205–223.

Micai, M., H. Joseph, M. Vulchanova, and D. Saldaña. 2017. Strategies of readers with autism when responding to inferential questions: An eye-movement study. *Autism Research* 10 (5): 888–900. https://doi.org/10.1002/aur.1731.

Milton, D. 2012. On the ontological status of autism: The 'double empathy problem'. *Disability & Society* 27 (6): 883–887. https://doi.org/10.1080/09687599.2012.710008.

Milton, D., E. Gurbuz, and B. López. 2022. The 'double empathy problem': Ten years on. *Autism* 26 (8): 1901–1903. https://doi.org/10.1177/13623613221129123.

Mitchell, P., E. Sheppard, and S. Cassidy. 2021. Autism and the double empathy problem: Implications for development and mental health. *British Journal of Developmental Psychology* 39 (1): 1–18. https://doi.org/10.1111/bjdp.12350.

Ozonoff, S., B.F. Pennington, and S.J. Rogers. 1991. Executive function deficits in high-functioning autistic individuals: Relationship to theory of mind. *Journal of Child Psychology and Psychiatry* 32 (7): 1081–1105. https://doi.org/10.1111/j.1469-7610.1991.tb00351.x.

Pellicano, E. 2007. Links between theory of mind and executive function in young children with autism: Clues to developmental primacy. *Developmental Psychology* 43 (4): 974–990. https://doi.org/10.1037/0012-1649.43.4.974.

———. 2010. Individual differences in executive function and central coherence predict developmental changes in theory of mind in autism. *Developmental Psychology* 46 (2): 530–544. https://doi.org/10.1037/a0018287.

Perner, J., and B. Lang. 1999. Development of theory of mind and executive control. *Trends in Cognitive Sciences* 3 (9): 337–344. https://doi.org/10.1016/S1364-6613(99)01362-5.

Perner, J., and H. Wimmer. 1985. "John thinks that Mary thinks that..." attribution of second-order beliefs by 5- to 10-year-old children. *Journal of Experimental Child Psychology* 39: 437–471.

Rajendran, G., and P. Mitchell. 2007. Cognitive theories of autism. *Developmental Review* 27 (2): 224–260. https://doi.org/10.1016/j.dr.2007.02.001.

Ramos-Cabo, S., J. Acha, V. Vulchanov, and M. Vulchanova. 2022. You may point, but do not touch: Impact of gesture-types and cognition on language in typical and atypical development. *International Journal of Language & Communication Disorders* 57 (2): 324–339. https://doi.org/10.1111/1460-6984.12697.

Russell, J., S. Hala, and E. Hill. 2003. The automated windows task: The performance of preschool children, children with autism, and children with moderate learning difficulties. *Cognitive Development* 18 (1): 111–137. https://doi.org/10.1016/S0885-2014(02)00163-6.

Schurz, M., J. Radua, M.G. Tholen, L. Maliske, D.S. Margulies, R.B. Mars, J. Sallet, and P. Kanske. 2021. Toward a hierarchical model of social cognition: A neuroimaging meta-analysis and integrative review of empathy and theory of mind. *Psychological Bulletin* 147 (3): 293–327. https://doi.org/10.1037/bul0000303.

Senju, A., V. Southgate, S. White, and U. Frith. 2009. Mindblind eyes: An absence of spontaneous theory of mind in Asperger syndrome. *Science* 325 (5942): 883–885. https://doi.org/10.1126/science.1176170.

Sheppard, E., D. Pillai, G.T.L. Wong, D. Ropar, and P. Mitchell. 2016. How easy is it to read the minds of people with autism Spectrum disorder? *Journal of Autism and Developmental Disorders* 46 (4): 1247–1254. https://doi.org/10.1007/S10803-015-2662-8/FIGURES/3.

Silberman, S. 2016. NeuroTribes : The legacy of autism and the future of neurodiversity. 548.

Tager-Flusberg, H. 2000. Understanding the language and communicative impairments in autism. In *International review of research in mental retardation*, vol. 23, 185–205. Academic. https://doi.org/10.1016/S0074-7750(00)80011-7.

Tager-Flusberg, H., and R.M. Joseph. 2005. How language facilitates the acquisition of false-belief understanding in children with autism. In *Why language matters for theory of mind*, ed. J.W. Astington and J.A. Baird, 298–318. New York: Oxford University Press. https://doi.org/10.1093/ACPROF:OSO/9780195159912.003.0014.

Umagami, K., A. Remington, B. Lloyd-Evans, J. Davies, and L. Crane. 2022. Loneliness in autistic adults: A systematic review. *Autism* 26 (8): 2117–2135. https://doi.org/10.1177/13623613221077721/ASSET/IMAGES/LARGE/10.1177_13623613221077721-FIG1.JPEG.

Vulchanova, M., D. Saldaña, S. Chahboun, and V. Vulchanov. 2015. Figurative language processing in atypical populations: The ASD perspective. *Frontiers in Human Neuroscience* 9 (FEB): 1–20. https://doi.org/10.3389/fnhum.2015.00024.

Vulchanova, M., S. Chahboun, B. Galindo-Prieto, and V. Vulchanov. 2019. Gaze and motor traces of language processing: Evidence from autism spectrum disorders in comparison to typical controls. *Cognitive Neuropsychology* 36 (7–8): 383–409. https://doi.org/10.1080/02643294.2019.1652155.

Wade, M., H. Prime, J.M. Jenkins, K.O. Yeates, T. Williams, and K. Lee. 2018. On the relation between theory of mind and executive functioning: A developmental cognitive neuroscience perspective. *Psychonomic Bulletin and Review* 25 (6): 2119–2140. https://doi.org/10.3758/s13423-018-1459-0.

Westermann, G., M.S.C. Thomas, and A. Karmiloff-Smith. 2010. Neuroconstructivism. In *The Wiley-Blackwell handbook of childhood cognitive development*, 723–748. Wiley. https://doi.org/10.1002/9781444325485.ch28.

White, S.J., D. Coniston, R. Rogers, and U. Frith. 2011. Developing the Frith-Happé animations: A quick and objective test of theory of mind for adults with autism. *Autism Research* 4 (2): 149–154. https://doi.org/10.1002/AUR.174.

Whyte, E.M., K.E. Nelson, and K.S. Scherf. 2014. Idiom, syntax, and advanced theory of mind abilities in children with autism spectrum disorders. *Journal of Speech, Language, and Hearing Research* 57 (1): 120–130. https://doi.org/10.1044/1092-4388(2013/12-0308).

Wilson, A.C. 2021. Do animated triangles reveal a marked difficulty among autistic people with reading minds? *Autism* 25 (5): 1175–1186. https://doi.org/10.1177/1362361321989152.

Wimmer, H., and J. Perner. 1983. Beliefs about beliefs: Representation and constraining function of wrong beliefs in young children's understanding of deception. *Cognition* 13 (1): 103–128. https://doi.org/10.1016/0010-0277(83)90004-5.

World Health Organization. 2018. ICD-11. Internacional Classification of Diseases. https://icd.who.int/browse11/l-m/en.

Zeidan, J., E. Fombonne, J. Scorah, A. Ibrahim, M.S. Durkin, S. Saxena, A. Yusuf, A. Shih, and M. Elsabbagh. 2022. Global prevalence of autism: A systematic review update. *Autism Research.* 15 (5): 778–790. https://doi.org/10.1002/AUR.2696.

Chapter 7
Theory of Mind After Acquired Brain Injury: Basic Aspects, Evaluation and Intervention

Inés Abalo-Rodríguez, Jesús Cabrera-Álvarez, Sandra Doval, Alberto Fernández Lucas, and Dolores Villalobos

Abstract Acquired brain injury (ABI) is a leading cause of death and disability worldwide. It presents with a wide range of symptoms affecting cognitive, behavioral, emotional, and social domains. The aim of this chapter is to analyze Theory of Mind (ToM) deficits in ABI and to offer different tools and considerations for clinicians due to their relevance. To do so, first a brief overview of the main considerations on ABI will be provided, addressing the demographics of this condition, as well as the main risk factors and the types into which ABI is classified. Subsequently, in the second section, the symptoms of ABI will be explored, looking at the different domains that are generally affected and placing special emphasis on its functional and social implications. In the third section, we will review the relationship between social cognition (SC) and ABI. We will define the most relevant constructs in SC: empathy, simulation, emotion recognition, and specially

I. Abalo-Rodríguez · J. Cabrera-Álvarez · S. Doval
Center for Cognitive & Computational Neuroscience, Complutense University of Madrid, Madrid, Spain

Department of Experimental Psychology, School of Psychology, Complutense University of Madrid, Madrid, Spain

A. F. Lucas
Center for Cognitive & Computational Neuroscience, Complutense University of Madrid, Madrid, Spain

Department of Legal Medicine, Psychiatry and Pathology (Faculty of Medicine), Complutense University of Madrid, Madrid, Spain

D. Villalobos (✉)
Center for Cognitive & Computational Neuroscience, Complutense University of Madrid, Madrid, Spain

Department of Experimental Psychology, School of Psychology, Complutense University of Madrid, Madrid, Spain

Institute of Knowledge Technology, Complutense University of Madrid, Madrid, Spain
e-mail: marvil17@ucm.es

T. Lopez-Soto et al. (eds.), *The Theory of Mind Under Scrutiny*, Logic, Argumentation & Reasoning 34, https://doi.org/10.1007/978-3-031-46742-4_7

ToM, among others. We will link those constructs to the underlying brain networks that have been proposed as supporting brain structures. Studies regarding ABI-derived deficits in those functions are reviewed and linked to brain networks. In the fourth part, we will explain the relationship between SC, especially ToM, and other cognitive processes being executive function (EF) the most important one. Self-awareness will be introduced as a primary ability in ABI patients that is related with EF and ToM abilities. Subsequently, communicative capacities will be also recognized as commonly impaired in ABI population with SC deficits, considering again their relationship with EF. In order to conceptualize and summarize the main tools and scales commonly used to assess ToM impairment in ABI population, a section will focus on introducing those that have been previously reported in literature and will provide a useful insight regarding the general state of ToM in ABI. This section is aimed at giving a general overview of those available scales that could be utilized by experts, including them in the clinical evaluation process of this population. The measured constructs and the type of injury will be also considered. Finally, in the last section of the chapter we are going to address specific intervention programs and techniques that focus on general SC abilities and especially ToM in ABI population. Although the rehabilitation of ToM has been more frequently implemented in other patients (autism and psychiatric disorders), recently designed specifics programs aimed to improve this important process in ABI patients are showing promised results.

Keywords Acquired brain injury · Theory of mind · Social cognition · Brain networks · Executive function · Self-awareness · ToM assessment · ToM intervention

7.1 Brief Introduction to Acquired Brain Injury (ABI)

7.1.1 Intro & Demographics

ABI can be defined as a brain damage that is caused by an event occurring after birth. It encompasses two different and broad groups depending on the nature of the injury. Thus, ABI includes both traumatic brain injury (TBI), where the damage is caused by an external traumatic event (e.g., a car accident, a fall, neurosurgery, etc.) and nontraumatic injury derived from either an external or internal source (e.g., stroke, brain tumors, infections, poisoning, substance abuse, etc.). ABI excludes congenital disorders, as they are present from birth, as well as brain damage resulting from neurodegenerative disorders, as ABI specifically referred to brain injury with sudden onset (Ciuffreda et al. 2012; Elbaum and Benson 2007).

This brain damage generally results in some alteration of cognitive, behavioral, emotional, and/or social functioning, that are either transient, long-lasting, or permanent depending on the severity of the injury (Roebuck-Spencer and Cernich 2014). ABI is one of the main causes of both dead and disability, and it affects persons of both sexes at all ages, ethnicities, and incomes (Rubiano et al. 2015). Moreover, several studies have shown that TBI patients usually face difficulties to return to previous levels of employment or to engage again in leisure activities (Draper et al. 2007; Engberg and Teasdale 2004; Griffen and Hanks 2014). This is especially true for individuals with more severe injuries (Andelic et al. 2008).

It is difficult to accurately quantify the number of people that are affected by ABI every year worldwide or to precisely measure its effect on people's well-being. Yet, its enormous impact in terms of incidence, disability, the financial cost of health care and the well-being of both patients and family members is undeniable.

Only for TBI, the existing data evince such a large impact that it has been named "the silent epidemic" (Coburn 1992). In the United States, for instance, TBI is the main cause of death for people under 44 years of age and it is estimated to cause disability in between 3.2 and 5.3 million of US citizens (1.1–1.7 % of the population) (Roebuck-Spencer and Cernich 2014). In Europe, a recent review estimated incidence rates ranging from 47.3 to 694 per 100.000 population per year for all ages and severities (Brazinova et al. 2021), and mortality rates that ranged from 9 to 28.10 per 100.000 population per year. These data reflected the results provided by studies performed at the country level; regional level studies reported incidence rates of 83.3–849 per 100.000 population per year and mortality rates of 3.3 to 24.4 per 100.000 population per year. Studies that estimate TBI impact by Years of Life Lost (YLL) report substantial harm at both the individual and population level in Europe as well, with an economic and societal burden that should not be ignored (Majdan et al. 2016).

In an attempt to estimate the global incidence of TBI, Dewan et al. (2019) conducted a study that also included the generally underrepresented low- and middle-income countries. According to their results, 69 million individuals suffer TBI globally every year (81% of which are mild in severity and 11% are moderate). The highest annual incidence per capita was reported in the region comprised of both the US and Canada (1299 cases per 100.000 people) and Europe (1012 cases per 100.00 people). However, when regional populations were considered, the largest number of cases appeared in the Southeast Asian Region and Western Pacific Region (18.3 and 17.3 million of TBI cases per year, respectively).

If we focus on nontraumatic brain injuries, strokes stand out as the most prevalent within this group. Estimating global incidence, prevalence and mortality data is also fraught with difficulties for strokes, as they vary greatly from region to region (Thrift et al. 2017). In the European Union, for example, this condition is the second most common cause of death and the leading cause of disability in adults (Wilkins et al. 2017). It is estimated to affect 1.1 million inhabitants each year (Béjot et al. 2016), causing 440000 deaths and 7.06 million disability-adjusted life years lost (Wafa et al. 2020). In 2017, the cost associated with stroke was estimated at €45 billion, including both direct and indirect costs of care provision and productivity losses (Wilkins et al. 2017). In the forthcoming years, stroke events and their long-term consequences are expected to increase dramatically, due to population growth and increased life expectancy (Bennett et al. 2014). Between 2017 and 2047, the number of people living with stroke is estimated to increase by 27%, mainly due to an ageing population and improved survival rates (Wafa et al. 2020).

Globally, stroke remains the second leading cause of death and disability, with approximately 13 million new cases each year (Lindsay et al. 2019). The estimated global cost of stroke is more than US $721 billion (Feigin et al. 2022) and the risk of stroke is nearly equal between women and men (Gorelick 2019). From 1990 to 2019, the absolute number of cases increased substantially (70.0% increase in incident stroke, 43.0% increase in stroke deaths, 102.0% increase in prevalent stroke and 143.0% increase in disability-adjusted life years lost) (Feigin et al. 2022).

However, overall stroke incidence rates decreased (estimated from 1990 to 2016) due to work done on prevention and better control of risk factors (Lindsay et al. 2019). Making interventions that decrease risk factors has proven to be effective, as it is estimated that ten risk factors were responsible for 90% of stroke events (Lanas and Seron 2021). Thanks to these interventions, mortality and age-standardized stroke incidence are declining significantly in high-income countries. However, it is important to note that 70% of strokes and 87% of both disability-adjusted life years and stroke-related deaths occur in low- and middle-income countries (Lanas and Seron 2021). It therefore becomes imperative to allocate resources to address the burden of stroke in these regions.

Taken together, these data reflect the importance and need of paying attention to this condition. Although the information provided here is not representative of all existing acquired brain injuries, it serves to evince how profound is ABI impact in our society. In many countries, and especially within the young populations, ABI kills more people than other diseases such as cancer with a relevant difference: that, unlike the others, the causes of ABI are well known and can be thus preventable (Reilly 2007). In order to develop policies capable to reduce their incidence and impact, understanding the causes and risk factors that lead to ABI becomes crucial.

7.1.2 Risk Factors

ABI presents a wide etiology, as there are different factors that can lead to damage in brain regions. Hence, while the most common cause of nonTBI is stroke, those more frequently linked to TBI are traffic accidents and falls (Brazinova et al. 2021). Importantly, the latter is mediated by age as a factor. Thus, falls are the main cause of TBI for individuals below 15 and above 40 years of age approximately, while motor vehicle accidents are the main cause for individuals ranging from 15 to 40 years (Roebuck-Spencer and Cernich 2014). In general, suffering from a TBI normally results in emergency department visits and/or hospitalizations; mortality as an outcome is especially found in severe TBI and elder age groups (see Roebuck-Spencer and Cernich 2014, for a more detailed description).

Even if everyone can suffer ABI, some groups are at higher risk than others. Importantly, it should be noticed that these factors are mediated by the context, as different regions have different needs, situations, and obstacles (Rubiano et al. 2015). These differences stand out, especially when comparing high- to low-income countries, since the latter present a higher TBI-related incidence and presence of risk factors while having at the same time a health system that is less prepared to deal with this condition (Hyder et al. 2007). It is thus crucial to contextualize ABI into the specificities of the region, especially when designing interventions.

Regarding strokes, the most common risk factors are cigarette smoking, alcohol consumption, physical inactivity, overweight, hypertension, diabetes, carotid stenosis, transient ischemic attack, and hypercholesterolemia (Elkind 1998; Romero et al. 2008). Besides, there are also other nonmodifiable risk factors, such as age, sex, ethnicity and family history (Romero et al. 2008). Due to their different origins, some authors classify them into two different categories: internal risk factors (e.g., genetic factors or hypertension) and external risk factors (e.g., smoking, air pollution or alcohol consumption) (Zhang et al. 2019). Some authors also highlight the importance of taking age into account, distinguishing risk factors among young and elderly population. In this vein, the former would include factors such as dyslipidemia, smoking and hypertension; while genetic, lifestyle, climatic or geographic diversity had no significant difference (Smajlovic 2015). On the other hand, elderly population risk factors would include obesity, vascular disease, and heart disease in the family (Costa et al. 2014).

For TBI, age, sex, economic status, and ethnic group have been proven to be the main risk factors (Cassidy et al. 2004; Krishnamoorthy et al. 2015; Roebuck-Spencer and Cernich 2014). In this vein, TBI incidence has been reported to be higher in children under 4, teenagers between 15 and 19, and adults over 65 (Roebuck-Spencer and Cernich 2014). Age also mediates the outcome from TBI, as elderly individuals often develop worse outcome and are more likely to suffer long-term disability and death due to TBI than younger ones (Flanagan et al. 2005; Geraldina et al. 2003; V. E. Johnson and Stewart 2015). Another variable that also represents a risk factor for TBI is sex, as men as a group have more probabilities of suffering a TBI than women (ratio: 1.6–2.8 approximately, Roebuck-Spencer and Cernich 2014). The differences between sexes regarding injury rates are more evident from 10 to 40 years of age, which is the age range in which TBI are more likely to be caused by traffic accidents. While death rates from TBI are also higher among men (ratio 1.17 for men's in-hospital mortality, Krishnamoorthy et al. 2015), the probability of long-term disability is significantly higher among women (49.5% compared to 39.9%, Roebuck-Spencer and Cernich 2014).

Besides, there are other risk factors regarding TBI incidence, such as socioeconomic status (lowest income levels have higher rates), racial/ethnic groups (minorities have higher rates) or alcohol consumption (positive associations between blood alcohol concentration and risk of injury) (Cassidy et al. 2004; Krishnamoorthy et al. 2015; Roebuck-Spencer and Cernich 2014). Recreational activities have also been associated with TBI, like bicycling or playground activities for younger children and football or basketball for older ones (10–19 years old children) (Clark and Guskiewicz 2016; Selassie et al. 2013).

7.1.3 Types of TBI According to Severity

There is great heterogeneity in the outcome of ABI patients, so that assessing the degree of severity of the brain injury becomes an essential task. For both TBI and non-traumatic brain injuries (e.g., stroke), the criteria used allow for classifying brain injury as "mild", "moderate" or "severe". Thus, different scales are used to assess the severity of brain injury, depending on the origin of the ABI. In the case of stroke, for example, the main scales used are the Scandinavian Stroke Scale (SSS), modified Rankin Scale (mRS), Barthel Index (BI) and modified National Institutes of Health Stroke Scale (mNIHSS). Although they employ different criteria, there is substantial to near-perfect agreement between the categorized scales (Govan et al. 2009). In the case of the TBIs, there are also several scales with similar criteria to each other. Due to the strong similarities that exist between these classification systems, in this section we will focus specifically on the criteria used to assess severity in TBI, reviewing them comprehensively.

To assess severity in TBI, three indicators are commonly used: (1) the individual's responsiveness at the time of the injury, (2) the duration of impaired consciousness, and (3) the duration of confusion (Roebuck-Spencer and Cernich 2014). Importantly, this categorization is supposed to classify the severity at the moment or shortly after the acquisition of the brain damage. This means that it should not be understood as the ultimate outcome at later timepoints, even if worse outcomes are expected with more severe injuries. The most common way to determine injury severity is the individual's responsiveness. To do so, Glasgow coma scale (GCS) is generally used. GCS scores range from 3 (no responsiveness) to 15 (full responsiveness), and measure responsiveness to stimuli in three different areas: verbalization, eye opening and motor response (Teasdale and Jennett 1974). Depending on the obtained score, the severity of the injury is classified into the aforementioned categories (see Table 7.1). Importantly, GCS may lead to overestimations of injury severity in individuals with alcohol and other substances intoxications. On the contrary, it can lead to underestimations in patients that present good responsiveness after injury but follow afterwards into neurological deterioration (as a consequence, for instance, of epidural hematoma).

Table 7.1 Classification of TBI severity and assessment

	Glasgow coma scale (GCS)	Time to follow commands	Russell posttraumatic amnesia (PTA)	Mississippi PTA classification scheme
Mild	13–15	≤ 20 min	<1 h	
Moderate	9–12	≤ 6 h	1–24 h	0–14 days
Moderate severe				15–28 days
Severe	6–8	> 6 h	1–7 days	29–70 days
Very severe	3–5		> 7 days	> 70 days

Adapted from Roebuck-Spencer and Cernich (2014)

Another indicator to assess injury severity is the duration of impaired consciousness, that is, the length of time that an individual takes to be conscious after the brain injury. This duration is usually measured through the time it takes to follow commands, the duration of unconsciousness or the length of coma, but there is no common agreement in the literature on the criteria that should be followed to classify injury severity depending on this parameter. Table 7.1 displayed the criteria suggested by Lezak and colleagues (Lezak 1995; Lezak et al. 2004).

Finally, the duration of confusion (also known as posttraumatic amnesia, PTA) can also be used to assess injury severity. This term refers to the recovery phase in which the patient is awake and responsive but confused, disoriented, and showing poor ability to retain new information (Roebuck-Spencer and Cernich 2014; Russell 1932). There have been different criteria to determine injury severity according to PTA. The Russell's criteria for PTA (Russell and Smith 1961) are the most cited ones. However, the great variability of outcomes for individuals classified with "severe TBI" according to these criteria led some authors to provide newer classification systems, like the Mississippi PTA classification scheme (Nakase-Richardson et al. 2009, 2011). The criteria recommended by both scales are displayed in Table 7.1. Despite the lack of agreement in a sole classification system for the duration of confusion, this indicator is considered to be a good predictor of short and long-term functional outcomes (Dikmen et al. 1995; Ellenberg 1996; Sherer et al. 2002).

As stated before, it is important to avoid the understanding of this classification as a categorization of the outcome at later time points in the recovery course. Even if the outcome in moderate and severe categories can be predicted by early injury indicators, this classification is less sensitive to predicting the outcome of milder damages, as they do not account for the existing variability among individuals. Furthermore, it is important to mention the existing large variability across subjects, as two individuals classified within the same category can end up reaching very different functional outcomes, independently of the indices used. Caution is thus needed when using this classification.

7.2 Symptoms and Implications of ABI

ABI results in a wide symptomatology that affects cognitive, emotional, behavioral, and social domains, and have critical implications for the person's well-being. The aim of this section is to revise the symptoms that are commonly derived from this condition attending to these domains. Even when the deficits associated with ABI will be revised, we consider that it is essential to address the concerns raised by some authors regarding a conceptualization of ABI solely based on the deficit-based model (Durham and Ramcharan 2018).

According to Durham & Ramcharan's view (Durham and Ramcharan 2018), ABI is most commonly conceptualized attending only to the impairments that it provokes, while leaving completely aside the 'insider's perspective' and without offering any kind of tool to overcome these deficits. As a consequence, a set of symptoms is the main thing found when looking for information on the topic. This is of relevance because searching for information regarding their new condition is precisely one of the first things that ABI patients or their families tend to do once they return home after hospitalization. Thus, the information that can be more easily found is mainly located in academic journals, written in a form that is not necessarily accessible to everyone, and that is generally limited to describing the deficit that they will experience or are experiencing as a consequence of their brain injury. Also, there is often little information about ways or tools to 'get better' or re-engage with the activities before the ABI onset.

This fact questionably helps them to feel better with their condition and it is detrimental for the recuperation for two main reasons. First of all, the first 6 months following the injury are crucial for recuperation, as during this period the greatest improvement is achieved, with slow progress thereafter (Durham and Ramcharan 2018; Kay and Lezak 1990). Thereby, it is of great interest to motivate and engage the person with their recuperation within this time window. As this is also the period when they tend to actively seek for information, it is crucial that the information they find helps them to commit to their recovery offering tools to this end in order to ensure the best possible outcome.

Secondly, several authors highlight the importance of obtaining self-awareness of the new condition as a crucial step in recuperation (Durham and Ramcharan 2018; Robertson and Schmitter-Edgecombe 2015). To achieve that, individuals need to accept the difficulties that they will have to face as a consequence of their new condition and find tools to cope with them. All of this requires the confidence and hope that they can in fact do something regarding these new difficulties. Unfortunately, this is questionably transmitted by the available information that treats ABI as a mere set of symptoms.

Taken together, these notions invite readers to reflect on the type of accessible information regarding ABI and the extent to which it is (or is not) of help in the recuperation process of individuals suffering from ABI and their families. During the last years, there has been a noticeable improvement in this regard, and many websites have engaged with this approach by making the information that they offer more accessible to the general population (e.g., Durham and Ramcharan 2018). Even though, there is still room for improvement, as this approach is not known by all practitioners and there is still a lack of information regarding coping mechanisms that can be adopted by ABI patients.

This chapter is directed to any reader interested in this population and/or specialists or professionals in the field, and this particular section specifically aims at addressing the symptomatology of this condition. Yet, we do consider that raising awareness of the problems derived from the deficit-based model is crucial, to take into consideration the well-being of the individuals affected by ABI, both directly and indirectly, and to ensure a better course of their recuperation. Different authors suggested that narratives written from the insider's perspective allow understanding how is it to have an ABI, as well as the emotional and psychological needs of these individuals. In this regard, some recommended readings are (Becker 2004; Bruner

1990; Calderwood 2003; Garrison 2007; Johnson 2010; Mason 2009; Osborne 1997; Strand 2004), among others.

7.2.1 *Symptoms Affecting Physical, Motor, Cognitive, Emotional, and Social Domains*

ABI often results in a set of symptoms with major repercussions on the person's well-being that sometimes derive in difficulties to recover the life they use to live before the onset of this condition. For the sake of clarity, we will revise them by attending to the different domains that can be affected. It should be noticed though, that the specific symptomatology displayed by an individual will depend on the brain regions affected as well as on the severity of the injury. First of all, physical problems derived from ABI have been constantly reported. They are broad and common and include symptoms such as headaches, nausea, blurred vision, decreased smell and/or taste and hearing loss. Also, there is generally some degree of motor affectation, as patients often present limb weakness, reduced strength, and coordination of their body or even paralysis. Some of them also present difficulty articulating words, which would in turn affect their communication with others having thus social consequences. Secondly, ABI also tends to result in cognitive problems. Patients often present with a slowing of cognitive abilities, such as processing information or thinking. They also have difficulties in understanding, planning or problem solving and often present rigid concrete thinking and short-memory affectation. Moreover, patients commonly report suffering general confusion and mental fatigue and to have problems maintaining concentration. Some studies even report losses in IQ (intellectual quotient) scales after suffering brain injury (Levine et al. 2005; Parker and Rosenblum 1996). Furthermore, acquired brain injury can also result in emotional problems and personality changes. Disinhibition, emotional fragility, restlessness, and difficulties in self-monitoring have been reported as symptoms after ABI onset. Also, some patients present both reduced self-control and social skills (Durham and Ramcharan 2018; Gasquoine 1997; Lundin et al. 2006). As a consequence of these symptoms, the patients' well-being and social life can get strongly impaired.

The aforementioned symptoms place the patient in a difficult context. As stated before, the improvement experienced during the first 6 months after the ABI is accompanied by a slowdown during the following years. Taken together, this leads patients to suffer a set of problems that, while not being a direct consequence of the brain injury, are indirectly provoked by it. For instance, it is very common that they experience frustration and anger about the problems derived from their ABI, which can be more intense and frequent than usual if the brain regions responsible for inhibition have been affected by the injury. Also, the vast majority of them experience grief and difficulties to cope with stress in everyday life (Durham and Ramcharan 2018; Gasquoine 1997; Lundin et al. 2006).

These problems cause difficulties when patients must engage in rehabilitation or try to successfully reintegrate into their communities, which in turn affect the success of rehabilitation, social relationships and the ability to live happily and independently (Durham and Ramcharan 2018). Taking all this into account, it is understandable why in the long term, ABI is indirectly related to increased rates of divorce or suicide (Humphreys et al. 2013; Ratcliff et al. 2005; Wadhawan et al. 2019). Importantly, it should be noted that all of these consequences affect not only the patient, but also the well-being of their family members (Durham and Ramcharan 2018; Humphreys et al. 2013).

In addition, ABI also has a strong impact on the professional life, being associated with high rates of unemployment. According to a meta-analysis including a total of 49 studies devoted to this matter, within 2 years post-injury only a 39.3% of individuals who have suffered non-traumatic ABI have returned to work; in the case of TBI, the data are 40.7% after one year and 40.8% after 2 years (van Velzen et al. 2009). In addition, they note, some of the individuals who returned to work were not able to sustain their job over time and point out that changes of occupation are common among people with ABI. In this vein, a systematic-review including a total of 27 articles reported that high education level and independence in activities of daily living are positively associated with return to work after suffering traumatic and non-traumatic ABI, respectively (Donker-Cools et al. 2016). On the other hand, they report, low education level, previous unemployment, and length of stay in rehabilitation were negatively correlated. Moreover, it is reported that the etiology of stroke is not significantly associated to unemployment.

As a consequence of this complicated context in which ABI patients are embedded, it is not difficult to understand why mental disorders have been reported to increase after the ABI onset. Thus, several studies report that people that have suffered an ABI have an 80% likelihood of being diagnosed with a mental illness, being especially common the diagnoses of severe depression, panic attacks, compulsive disorders and posttraumatic stress disorders (Durham and Ramcharan 2018). Some patients also present lower self-esteem, mood swings, anti-social behavior, delusional paranoia and anti-social behavior (Durham and Ramcharan 2018). Some studies also report that ABI patients have a greater risk of committing suicide (ratio of 1.4 of committed suicide and a ratio of 4.6 of suicide attempts (Simpson and Tate 2007)..Therefore, and considering their importance, the following section will focus on the ABI-derived deficits in social cognition abilities that could lead to all the aforementioned pathological situations.

7.3 Social Cognition in Acquired Brain Injury

Human beings are social animals with a natural inclination to cooperate with others to maximize their survival opportunities, and to compete within groups for status and power. To succeed in this context, humans develop a wide range of implicit and explicit cognitive abilities to anticipate others' thoughts, feelings, and actions (Tamir

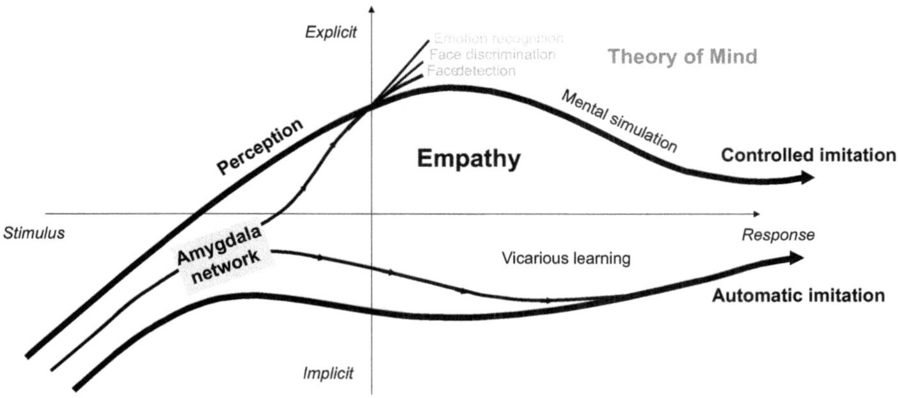

Fig. 7.1 Socio-cognitive constructs and networks that will be referred to during the chapter represented in the continuum stimulus–response. Colors representing the hot-cold axis (i.e., red-blue)

and Thornton 2018) and to modulate their own responses in social situations (Adolphs 2001). These abilities are labelled as SC defined as: *"the processes that subserve behavior in response to other individuals of the same species, and . . . the extremely diverse and flexible social behaviors that are seen in primates"* (Adolphs 1999). It is a wide concept that includes processes spanning from perception to information processing and behavior (Fig. 7.1).

There are two relevant dimensions that contribute to the conceptualization of SC: the first regards to *what* information is processed, while the second focuses on *how* that information is processed (Etchepare and Prouteau 2018). The former dimension is often described as a hot-cold axis which differentiates social abilities into two broad categories: hot abilities, related to *experiencing* others' feelings and states through mental simulation or emotion contagion; and cold abilities related to *understanding* others' emotions, beliefs, and thoughts in a more detached and objective manner. These two categories are closely linked to the concepts of empathy and ToM, respectively. We will see them in more detail below.

In addition to the hot-cold dimension, the depth of information processing is also a crucial aspect to understand SC. According to theories of dual cognition (Evans and Stanovich 2013) the first stage of external information processing is fast, adaptive, automatic, and largely unconscious. This idea has been extended to the psychosocial context to highlight the relevance of implicit mechanisms preceding the conscious experience during social interactions (C. D. Frith and Frith 2012; Greenwald and Lai 2020). For instance, when waiting at a crowded crosswalk, predicting the movements of others to avoid collisions tends to be an unconscious process, and yet a highly complex social interaction. Once information reaches consciousness, it has been extensively processed, and further steps are explicit, effortful, and slow. This second stage of processing allows for the necessary cognitive flexibility that enhances our social performance and adaptative behavior (Etchepare and Prouteau 2018).

Research has shown that SC may become altered in ABI patients. ABI can produce deficits in a variety of domains, such as understanding sarcasm and irony (Channon et al. 2005; Martin and McDonald 2005; Shamay-Tsoory et al. 2005), empathy (de Sousa et al. 2011; Shamay-Tsoory et al. 2004; Williams and Wood 2010; Wood and Williams 2008), social perspective-taking (Santoro and Spiers 1994), emotion recognition (Green et al. 2004; Ietswaart et al. 2008) and ToM (Martín and León-Carrión 2010). However, the severity and specific domains affected are highly dependent on the location, extent, and evolution of the lesions.

Within the context of ABI, stroke, tumor and TBI are the most studied etiopathologies. For instance, in a preoperative brain tumor sample, Goebel et al. (2018) showed that 83% of the patients exhibited deficits in at least one aspect of SC. Persistent deficits in SC can also be observed in stroke patients for several years following the onset of the lesion (Nijsse et al. 2019). TBI is a special case. In TBI, the typical etiopathological pattern is a rapid acceleration-deceleration movement that compresses the brain towards the skull and rotates it (Maas et al. 2008), leading to damage in the ventral surface of the prefrontal cortex and the lateral surface of the temporal cortex (Bigler 2007), areas that have been associated with the so called "social brain" (Kennedy and Adolphs 2012). Reports indicate that approximately, around 60–80% of relatives of TBI patients have reported changes in personality (McDonald and Genova 2021) linked to deficits in socio-cognitive abilities such as irritability, aggression, infantilism, and apathy (Maggio et al. 2022).

These changes in personality resulting from deficits in socio-cognitive abilities could lead to social isolation, which can in turn further exacerbate the patients' difficulties (C. E. Salas et al. 2018, 2021a), and may become an obstacle for patients' rehabilitation and their return to social life (Salas et al. 2021a, b; Yates 2003). Furthermore, these changes can also add an extra burden on relatives and caregivers (Gan et al. 2010; Kreitzer et al. 2018; Wells et al. 2005) as they may don't fully understand the observed changes in behavior.

7.3.1 Introduction to the Social Brain

The term "social brain" refers to a set of brain networks that are thought to be involved in social cognition. Several networks have been proposed as the underlying structures for the "social brain", including the social perception network, mirror network, emotional networks and mentalizing network (Kennedy and Adolphs 2012; Molapour et al. 2021). These networks are not isolated, they interact with one another to support social processing (Pessoa 2017). The social perception network is involved in processing and perceiving signals from the social context, while the mirror network is involved in observing and imitating others' actions. The emotional networks are involved in stimulus salience evaluation, threat detection, and empathy, and the mentalizing network is involved in supporting ToM processes (see Fig. 7.2) (Pessoa 2017). Each network comprises specific brain regions, which

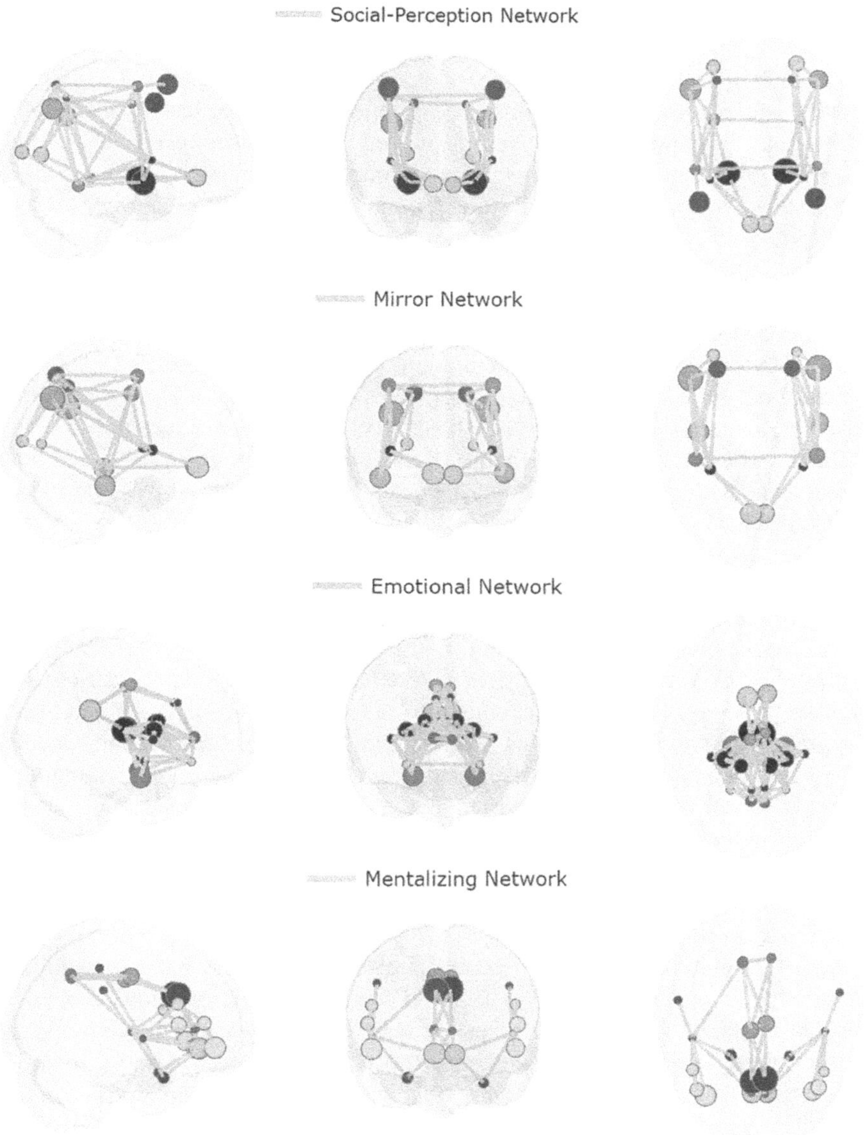

Fig. 7.2 Networks in the social brain. Schematic representation of the regions and networks involved in social cognition. Node size representing connectivity degree. Data provided by the Centre for Cognitive and Computational Neuroscience (Madrid). ROI coordinates derived from Desikan-Killiany atlas (Desikan et al. 2006)

will be discussed in the following subsections along with evidence regarding ABI-derived SC deficits associated to them. We will pay special attention to ToM and the mentalizing network.

7.3.1.1 Social-Perception Network and ABI-Related Deficits

To effectively navigate social contexts, an organism must possess the ability to capture and process signals from the environment, and particularly from other individuals. The human brain has developed a specialized perceptual system designed to process a wide range of multimodal social signals including facial expressions, prosody, body postures, language, and chemosensory signals (Molapour et al. 2021). Many of these social perceptual abilities are innate, and can be observed in newborns, such as their tendency to attend to faces, track others' gaze, or to recognize chemosensory signals from their mothers.

The social-perception network shows a main hub in the superior temporal sulcus (STS), that participates in processing multiple aspects of social stimuli. It connects to the fusiform gyrus, middle temporal visual area (MT/V5), premotor cortex, amygdala, intraparietal sulci, frontal eye field, insula, and medial prefrontal cortex (Lahnakoski et al. 2012). The network oversees the processing of facial and body expressions -with and without emotional content-, processing pain, motion and goal directed behaviors. Note that perception is a transversal process that interacts intensively with other social cognition domains (e.g., imitation) and therefore, the brain regions linked to social perception tend to partially overlap with those associated with other social cognition processes (Molapour et al. 2021).

One of the most widely studied processes mainly supported by this network, in collaboration with the emotional network (see below), is emotion recognition. This intricate process uses multimodal sources of information, mainly visual and auditory, to interpret the emotions that others might be feeling. Visual stimuli involve receiving information about face expressions and gestures while auditory stimuli are related to prosody (i.e., intonation, rhythm, etc.) and language. Interestingly, studies have shown that humans are better at recognizing emotions from voices than from faces (Kraus 2017). Both modalities require the successful completion of previous and simpler levels of processing such as detection (human faces and voices) and discrimination between stimuli categories.

In ABI, these processes may become affected. For example, patients with brain tumor often show impairments in facial differentiation and emotion recognition (Goebel et al. 2018), with more severe deficits observed when lesions are located in the anterior temporal cortex and amygdala (Campanella et al. 2014). In stroke patients, a review by Yuvaraj et al. (2013) showed that deficits in emotion perception were also present, and more frequently associated with lesions in the right hemisphere. This evidence is consistent with the right-hemispheric dominance hypothesis, which proposes that emotions are processed preferentially in the right hemisphere (further discussed in sect. 7.3.1.3 below).

Similarly, and again connecting with the emotional network, thalamic stroke patients have also exhibited deficits in emotion recognition (Cheung et al. 2006; Wilkos et al. 2015). The thalamus is a subcortical brain structure located in the center of the brain that acts as a relay node for sensory stimuli between the periphery and the cortex. Recent research has linked the thalamus to the emotional network of the social brain, underscoring the critical role this structure plays in mediating complex social processes (see below for a detailed discussion of the emotional network). Together, these studies emphasize the interconnectedness of the perception network with other neural networks involved in social cognition, and specifically with the emotion network for emotion recognition.

TBI can have a significant impact on the ability to recognize emotions from facial expressions. Studies have reported that among 12–39% of TBI patients experiencing difficulties in this area (Babbage et al. 2011). A recent meta-analysis conducted by Murphy et al. (2022) found that TBI patients exhibit impairments in recognizing emotions across all emotion categories and through different modalities. The study showed slightly worse performance with multimodal affective stimuli (i.e., face, voice, and body) than with only-face and only-voice modalities, but no significant difference was found between the latter two. Importantly, the study also showed a direct relationship between the severity of the lesion and the level of impairment.

In a review by McDonald and Genova (2021), they linked emotion recognition deficits in TBI patients to damage in the inferior longitudinal fasciculus and the inferior fronto-occipital fasciculus, tracts that specifically connect frontal and limbic areas to the occipital cortex. This suggests that damage to these tracts may hinder the interaction between the perception and emotional networks. Furthermore, D. Neumann et al. (2016) found reduced activation in the right fusiform gyrus during emotion recognition tasks in TBI patients, and this reduced activation was directly related to impaired performance. They interpreted these findings as evidence of impairments in more general face processing abilities, which may contribute to the emotion recognition deficits observed in TBI patients.

7.3.1.2 Mirror Network: ABI-Related Deficits in Action-Perception and Simulation

This network is active during the observation of others and allow us to mimic virtually or physically what they do and express. In psychological terms, this has been called *simulation* and it is a powerful cognitive tool to understand and experience others' thoughts, emotions, and behaviors (Fiske and Taylor 2017). When simulation is made explicit, we refer to the concepts of *imitation* and *mimicry*: the ability to reproduce the movements or expressions seen in others with one's own body.

Mirror mechanisms are partially innate (e.g., the tendency in newborns to imitate gestures) and gain complexity during development (Jones 2007). The role of mirror mechanisms in social binding and vicarious learning is well-established (Rizzolatti and Craighero 2004). Mimicry, whether conscious or unconscious, is a common way for people to communicate their likes and feelings of identification towards others, creating a sense of mutual understanding and social connection (van Schaik and Hunnius 2016). Similarly, vicarious learning involves learning from the actions of others by imitating them, in a process that can be automatic or controlled via top-down processes (C. D. Frith and Frith 2012).

Three brain regions have been classically linked to the mirror network: the inferior frontal gyrus, the ventral premotor cortex, and the inferior parietal lobule (Bonini et al. 2022; Caspers et al. 2010). Through meta-analysis, this set of regions was extended to at least nine, adding superior parietal lobule, dorsal premotor cortex, insula, and temporal gyri in its superior, middle, and inferior parts (Molenberghs et al. 2012). Some studies have also found significant activation in cerebellum and basal ganglia (Bonini et al. 2022).

Evidence regarding imitation deficits in ABI is scarce and very restricted to emotional mimicry. In a study by McDonald et al. (2011) they presented facial emotional expressions to participants while they measured their muscular activity with electromyography. Healthy participants exhibited spontaneous muscular activation 500ms after stimulus onset. In contrast, TBI patients did not show that activity pattern, particularly when presented with angry faces.

The same year de Sousa et al. (2011), published a study on the relationship between empathy and mimicry in TBI patients. They found that TBI patients who showed lower levels of empathy did not respond with automatic mimicry to angry faces, in contrast to control participants. Similarly, in a related study by Dethier et al. (2012), TBI patients exhibited lower intensity in their facial expressions, as judged by experts, when asked to pose emotions of sadness. However, no significant difference was found in the expression of emotions of anger or happiness.

Finally, and linked to the following subsection, Osborne-Crowley et al. (2019) found lower levels of skin conductance when eliciting specific emotional states in TBI patients. This suggests that difficulties in mirror behaviors may not only refer to the execution of movements but also to the physiological reactions elicited by them.

7.3.1.3 Emotional Network: ABI-Disrupted Responses to Affective Stimuli

Early theories on emotion processing proposed that the limbic system is a key brain network supporting these functions (Palomero-Gallagher and Amunts 2022). As mentioned above, two main hypotheses led the debate: the right-hemisphere dominance hypothesis, which suggests that all emotional information is processed by the right hemisphere; and the valence lateralization hypothesis, for which the left hemisphere is dominant for processing positive stimuli and the right hemisphere for negative stimuli. However, recent research has revealed that emotion processing is supported by an integrated system of distributed brain networks that contribute to specific aspects of emotion processing.

These networks are involved in stimuli salience evaluation, emotion processing, and empathy. They can be organized into two main subnetworks: one centered in the amygdala and the other centered in the hippocampus (Palomero-Gallagher and Amunts 2022). The amygdala subnetwork includes regions such as the insula, olfactory, orbitofrontal, anterior, and medial cingulate cortices, striatum, and nucleus accumbens (Adolphs 2010). It is related to triggering emotional responses, detecting threat and salient stimuli, perceiving emotional expressions, and contributing to reward learning. Some researchers have suggested that this network can be conceptualized as an emotional version of a mirror network because it partially works like an action-perception and simulation network but for emotional stimuli (Bonini et al. 2022).

The hippocampus subnetwork includes the entorhinal and retrosplenial cortices, areas of the anterior and posterior cingulate cortices, and the thalamus (Palomero-Gallagher and Amunts 2022). It mediates emotion processing by integrating information from other brain networks and cognitive domains through the entorhinal cortex. Both subnetworks overlap in the anterior cingulate cortex, where they exchange information creating a path for the amygdala subnetwork to modulate memory consolidation.

Both subnetworks are involved in empathy, that has been described as an isomorphic emotional response to the perceived or imagined affective states of another person, being aware that the source of the experience is not us (Cuff et al. 2016; de Vignemont and Singer 2006). This concept might be disentangled from sympathy, which refers to an emotional response characterized by feelings of sorrow and concern elicited by the apprehension of another person's emotional state. It is important to note that the emotional experience in sympathy is not isomorphic (i.e., same type of emotion) to the one experience by the eliciting persons.

In ABI the functions supported by these networks can become altered. For instance, in 2013, S. McDonald found that 60% of severe TBI patients reported little to no emotional empathy in self-reported questionnaires, compared to the 30% of matched healthy controls. Previously, Williams and Wood (2010) had showed similar rates (64.4% in patients vs 34.4% in controls) and interestingly, they had controlled for injury severity and the time since lesion onset, not finding any significant relationships with empathy. However, none of these studies reported lesion sites. In a study from de Sousa et al. (2012) 71% of TBI patients self-reported low empathy while only 16% of healthy controls did. They also measured less skin conductance in TBI patients when presented unpleasant videos. Neutral videos did not produce any significant difference between the groups.

Studies on emotional arousal have used the startle reflex, which is an automatic defensive response elicited by a sudden and threatening stimulus, resulting in a rapid contraction of certain muscles (e.g., those controlling the eye blink), as well as an increase in heart rate, blood pressure and skin conductance. This reflex has been used as a physiological measure of arousal in several research studies (Saunders et al. 2006; Williams and Wood 2012). In these studies, participants were presented with pleasant and unpleasant pictures followed by a sudden white noise to elicit the startle response and measure participant's eye blinks. Authors found lower startle responses in the TBI group after the presentation of unpleasant pictures compared to controls. In agreement, they rated those stimuli as less arousing than controls. No difference was found regarding the expected attenuation effect of pleasant stimuli.

This evidence points toward general deficits regarding emotion processing in ABI involving more than the explicit comprehension of emotional stimuli and also including deficits in the implicit mechanisms of physiological responses to those stimuli.

7.3.1.4 Mentalizing Network: ABI-Related Deficits in ToM

The mentalizing network comprises a set of structures that become active when individuals engage in thinking about the internal states of others. This network supports ToM, which is the ability to *understand* another person's perspective from their verbal and non-verbal behavior, including what they think, intend, feel, and believe. The process of mentalizing occurs at different levels of complexity. The simplest level occurs when individuals try to interpret the mental state of another person, which is referred as first order ToM. At higher levels of complexity, second order ToM refers to the attempt to interpret what one person is thinking about the state of a third person. In a third level, individuals wonder about another person's thoughts on the interpretation that a third person is doing of a fourth individual's mental state, which is known as third order ToM.

The mentalizing network is composed of regions in the medial frontal cortex, the temporo-parietal junction, inferior frontal gyri, temporal pole, cingulate cortex, and precuneus (Monticelli et al. 2021; Schurz et al. 2014). Some authors have proposed that anterior regions of the brain execute and monitor the mentalizing process while posterior regions are involved in actually representing others' mental states (Abu-Akel 2003; Saxe and Powell 2006).

TBI patients show ToM deficits across a wide range of tasks including verbal, visual and tasks related to other stimuli modalities (S. McDonald and Genova 2021). Goebel et al. (2018) showed deficits in ToM for the 77% of a sample of preoperative tumor patients. Honan et al. (2016) also found impairments in conversational skills that required ToM abilities in a sample of 160 people with ABI of various etiologies including stroke, tumor, TBI, encephalitis, and hypoxic injury.

Patients that underwent an anterior capsulotomy, a surgical procedure that consists in cutting the connections between orbitofrontal lobe and subcortical nuclei used in severe cases of obsessive-compulsive disorder, showed ToM impairments (F. Happé et al. 2001). Similarly, patients with highly selective frontal or temporo-parietal junction lesions exhibited worse performance in ToM tasks (Rowe et al. 2001; Samson et al. 2004).

In a meta-analysis conducted by Martín and León-Carrión (2010) on ABI studies, they observed moderate to large deficits in ToM abilities in ABI patients. The most complex ToM tasks, such as faux pass and indirect speech, showed higher effect sizes than the simpler first and second order ToM tasks (see section 10.5 for further details on the tasks). Notably, among these last two, the lowest effect size was associated to first order ToM. This finding is related to the hierarchical hypothesis of ToM tasks, which posits that ToM abilities develop constructively, such that performing a second order ToM task would require being able to perform a first order one. Additionally, they also found a correlation between the proportion of right hemisphere damaged patients and the size effect in faux pass and indirect speech tasks, suggesting a lateralization of ToM processes in the brain.

The changes in personality observed after an ABI is a major problem for patients and their social environment. According to McDonald and Genova (2021) these personality changes may be linked to the deficits in ToM. They argued that the emergence of childish behavior, insensitivity, disinterest, and a general reduction in social abilities could indicate a disregard or lack of awareness of others' perspectives. Interestingly, in their study, McDonald and Genova (2021) explicitly discussed and dismissed the possibility that TBI patients were highly linked to individuals with maladjusted behavior prior to their injury.

At this point, we have reviewed the evidence that directly relate SC deficits to ABI patients. However, it is important to note that this relationship is not always straightforward, as it could be mediated by difficulties in other cognitive processes. Therefore, in the following section we will examine the relationship between ToM and other higher-order cognitive domains.

7.4 Relationship Between ToM and Cognitive Processes

7.4.1 The Role of Executive Function (EF)

EF is a multifaceted neuropsychological construct consisting of a set of higher-order cognitive processes that allow higher organisms to make choices and to engage in purposeful, goal-directed, and future-oriented behavior. EF are related with regulatory behaviors, including planning, self-monitoring, problem solving, inhibiting responses, strategy development and implementation, and working memory, resulted in the comprehensive term EF clearly located in the frontal lobe and their connections.

EF can be considered as composed of different and connected components, but also some authors refer to them as a different group of interrelated processes. Ardila (2013) differentiates two components: 1. The "Metacognitive EF", which is the usual understanding of executive functions (generally measured in neuropsychology executive functions tests), that have been related with the dorsolateral area of the prefrontal cortex; and 2, The "Emotional/motivational EF" which is responsible for coordinating cognition and emotion. That means, the ability to fulfil basic impulses following socially acceptable strategies. Clinical evidence as well as experimental research suggest that the neural substrate for this function resides mainly in the medial and orbital portions of the prefrontal cortex.

Some of the most disturbing neurological disorders include those in which EF have been compromised, perhaps because persons affected lack characteristics that we associate with the "essence" of being human. These clinical disorders are termed the dysexecutive syndromes. They have been described and they can be roughly divided into "behavioral" and "cognitive" domains. Such a distinction is compatible with the two major functional/anatomical dissociations within the frontal lobes (Godefroy and Stuss 2007):

- Dorsolateral prefrontal regions evolving from the hippocampal-archicortical trend, primarily involved in spatial and conceptual reasoning processes.

– Ventral prefrontal cortex, part of the paleocortical trend emerging from the caudal orbitofrontal (olfactory) cortex, involved more in emotional and behavioral processing.

Considering the multiple connections that run between sites in the frontal cortex, basal ganglia, and thalamus, five major circuits can be described: (1) dorsolateral prefrontal circuit, (2) lateral orbitofrontal circuit, (3) anterior cingulate circuit, (4) motor circuit, and (5) oculomotor circuit. Lesions at any point within the first three frontal subcortical networks can lead to dysexecutive symptoms (Dafner and Searl 2008).

The behavioral dysexecutive syndrome is due to different underlying deficits being the most well documented: abulia, apathy, spontaneity, akinetic mutism, pseudodepressive state, lack of drive, poor motivation, inattention, indifference, euphoric state, distractibility, impulsivity, disinhibition, irritability, restlessness, moria, pseudopsychopathic state, anosognosia, indifference, confabulation, and perseveration.

The cognitive dysexecutive syndrome also presents impairments in different specific processes: response initiation and suppression, rule deduction, maintenance and shifting of set, planning and problem-solving, ability of information generation, divided attention and task coordination, sustained attention, working memory, strategic memory processes.

SC is sometimes considered an EF (difficult to include in one of the two groups previously presented) and other times it is considered as a more specific mental abilities but strongly connected with EF (Beaudoin and Beauchamp 2020). SC includes more basic abilities, such as face processing and joint attention, and more complex ones, such as ToM, moral reasoning, and social decision-making (Kilford et al. 2016). There is some suggestion in the literature that the ToM and humor may be related more to the polar frontal regions, where emotion and cognition might be integrated (Stuss and Alexander 2000). Some authors differentiate between cognitive and affective subcomponents. The affective component of ToM is defined as the ability to identify and understand the emotional states of others, whereas cognitive ToM requires knowledge about beliefs and intentions. Affective ToM is associated with a cognitive part of empathy ("*I understand what you feel*"), however empathy itself refers to "*I feel what you feel*". Finally, cognitive ToM is not related specifically with feelings, and it could be conceptualized as "*I understand what you think*" (Magno et al. 2022). The overlapping as well as distinct neural networks underlying these subcomponents have been described in a few studies (Schlaffke et al. 2015). In general, both components recruit some common structures but especially distinct regions in the prefrontal cortex. The orbitofrontal and ventromedial prefrontal cortices seem to be involved in affective ToM, whereas cognitive ToM abilities have been associated with the dorsolateral frontal regions (Sebastian et al. 2012; Shamay-Tsoory and Aharon-Peretz 2007). This disassociation, although must be considered cautiously, is strongly compatible with the different dysexecutive functions previously mentioned. Thus, the relation between ToM as a SC high-order ability and EF seems to be very close and have been deeply investigated.

Although the relationship between ToM and executive functions has been specially addressed in childhood pathologies (Wilson et al. 2021), particularly in autism spectrum disorder (Cardillo et al. 2021), their relevance in ABI patients is also considered in research. ABI patients allowed neurocognitive researchers to study the relationship between different cognitive processes and brain structures and connections. Dennis et al. (2009) studied in school-aged children with TBI the relations among ToM, EF -specially working memory and cognitive inhibition-, and frontal lesions. The relation between cognitive inhibition and ToM involved a single mediated path through working memory. Frontal injury had a direct impact on working memory, which then separately determined ToM performance. The expression of ToM in school-age children with TBI depends on the domain-general functions of working memory and cognitive inhibition.

The relation between ToM and EF can be appropriately investigated in older adults. Previous studies suggest that as most cognitive processes, ToM declines with age. But the main question could be related with the explanation of this decline, considering its relationship with EF. Cho and Cohen (2019) predicted that if the elderly has intact ToM competence but compromised EF, then older adults should perform similarly to younger adults when using ToM tasks with lower executive demands. Their results showed that, although older adults have impaired EF capacities (inhibition, shifting and updating), on tasks with EF reduced demands (spontaneous-response task using a ToM eye tracking task), older adults performed to the same extent as younger adults, despite their declining EF. However, other pieces of research are not consistent with the idea of all the EF processes are involved in ToM, and it is important to separate cognitive and affective ToM. Cognitive ToM might be more sensitive to aging than affective ToM. Inhibition seems to be the relevant EF process involved in affective ToM decline (Otsuka et al. 2021). Using the Reading the Mind in the Eyes Test (RMET) they found that only inhibition (assessed with the Stroop task) influenced the participants' performance on this task.

The relation between ToM and EF has also been addressed in adults with ABI providing further information on the relationship between the different components of SC and EF. For example, in a study with TBI patients, Henry et al. (2006) found that TBI participants' recognition of basic emotions, as well as their capacity for mental state attribution (ToM), was significantly reduced relative to controls. Performance on these measures was strongly correlated in the healthy control, (but not in the TBI sample). However, in the TBI (but not the control) sample, ToM was substantially correlated with performance on phonemic fluency, an EF measure considered to impose particular demands upon cognitive flexibility and self-regulation. These results are consistent with other evidence indicating that deficits in some aspects of EF may at least partially underlie deficits in SC following TBI.

Although TBI has probably been the most studied ABI etiology, several research exploring the relationship between ToM and EF has been performed in patients with stroke. Pluta et al. (2017) compared a group of patients with stroke with another group of healthy controls in different neuropsychological tests that assessed their pragmatic abilities, EF, attention, memory, psychomotor speed, and visuospatial abilities as well as a cognitive and affective ToM assessment. They found that patients with stroke demonstrated impaired performance in all ToM tasks. Pragmatic competence and, to a lesser degree, EF had the strongest contribution to ToM impairments. In contrast, attention and general cognitive functioning did not directly affect mentalizing abilities. They concluded that the different roles of cognitive functions in cognitive and affective components of ToM and that EF contributed specially to the cognitive components of ToM.

An important characteristic of this kind of patients is that stroke allows to correlate more accurately the damaged brain region (they are more localized than in TBI) and the cognitive impairments. Hamilton et al. (2017) found that right hemisphere stroke was associated with impaired ToM ability, but left hemisphere stroke was not. They also found a high correlation between performance on a specific ToM test (that measure social understanding, the RMET) and some measures of EF in participants with right hemisphere stroke only. Further analyses suggested that deficits in EF could not statistically explain all the difficulties shown by stroke participants on the mentioned ToM test.

In a review Aboulafia-Brakha et al. (2011) investigated the existence of particular relations between different EF domains and ToM tasks generally in adults with acquired neurological pathology. They considered EF components following Miyake´s approach (Miyake et al. 2000), that includes three often postulated EF: mental set shifting ("Shifting"), information updating and monitoring ("Updating"), and inhibition of prepotent responses ("Inhibition"). They also included a fourth process called "access" (process involved in verbal fluency tasks which is believed to mediate access to long-term memory representations) (Fisk and Sharp 2004). Authors studied the relations between different ToM tasks and different EF tasks, classified following the four mentioned processes. Out of 24 studies, 22 showed concomitant impairment in at least one ToM task and one executive domain. They concluded that EF and ToM appear tightly associated. However, the few dissociations observed suggest they cannot be reduced to a single function; and apparently no executive subprocess could be specifically associated with ToM performances.

Likewise, in a study performed by Bibby and McDonald (2005), participants with severe TBI were compared to a healthy group on verbal first-order and second-order ToM tasks, non-verbal ToM tasks, and on verbal and non-verbal tasks requiring them to make general/non-mental inferences. The clinical group performed worse than controls on all the mentioned tasks. This performance was not completely accounted for the working memory or implicit language demands of the tasks. Patients with TBI showed a general weakness in inference-making that, combined with linguistic and working memory limitations, impairs their performance on both non-verbal and second-order ToM tasks. However, a specific ToM impairment may underlie their poor performance on verbal first-order tasks. This finding probably supports the possibility of a separate cognitive module of ToM.

The psychosocial consequences of both EF and SC difficulties are now accepted as significant and enduring, with negative outcomes associated with such deficits including marital relationship quality, community integration and vocational roles. As previously described, return to work is one of the most important challenges for patients after suffering an ABI and it could be influenced by different personal and extra personal factors. Yeates et al. (2016), found significant associations between social skills and patient goal-orientated planning and implementation, mentalizing ability, recognition of positive and negative emotions, and recognition of simple sarcasm; their results highlight the relevance of patient EF and SC difficulties for the perceptions and appraisal of work colleagues and its relevance for patients' possibility to recover their vocational role.

7.4.2 *Self-Awareness (SA)*

Additionally, in the relationship among SC, ToM, and other cognitive processes in ABI patients, particular attention should be also paid to impaired SA. Indeed, in the conceptual biopsychosocial model of the processes involved in SC (Cassel et al. 2016) SA is presented in the center of the model bidirectionally connected with the different SC abilities.

SA is considered as the ability to be aware of one's own thoughts, feelings, and mental states (Morin 2006). SA is critical to empathy, enabling us to understand the difference between our own emotions and perspectives and those of others. SA is necessary to read our own internal and emotional states and to regulate these accordingly. It is also critical to monitoring our behavior in order to ensure that it meets our personal goals and falls in line with the expectations of others (McDonald and Genova 2021).

Indeed, SA is frequently impaired after ABI (Bach and David 2006). A closely-related concept is anosognosia, that is characterized by a partially or totally reduced ability to recognize deficits and disabilities caused by the brain lesion. Impaired SA or lack of insight into one's abilities and deficits, can be particularly problematic (Fleming et al. 1996). When recovering from ABI, SA is especially critical to understanding changes in abilities, limitations, and opportunities and to adjusting expectations and goals accordingly. People with poor SA of their abilities often exhibit poor judgment and require supervision during functional activities to ensure their safety. Additionally, impaired SA impacts on ABI rehabilitation process because patients are not motivated considering they do not need it as much as they need. Thus, implement a specific intervention for SA is crucial in patients with ABI and SA improvement also cause enhancement in functionality in their daily activities (Villalobos et al. 2019).

Two main models explaining possible patterns and mechanisms underlining impaired SA after ABI have been proposed in the scientific literature. The first model (Crosson et al. 1989) called "The pyramidal model of SA" proposes the existence of three interdependent and hierarchical levels of SA: intellectual, emergent, and anticipatory. At the basis of the pyramid there is the "intellectual SA," that is the patient's ability to understand that a particular function is impaired. "Emergent SA" is the further step, by which patients can realize that a function is impaired when a related problem occurs. Finally, at the top of this hierarchical model, the "anticipatory SA" allow patients anticipating likely problems due the related impaired functions. The model suggests that each level of SA is necessary to reach the following. On the other hand, a more recent model called "The dynamic comprehensive model of awareness" (Toglia and Kirk 2000) suggests that the different levels of SA are parallel and dynamic, rather than hierarchical. It presents the "offline SA" (comparable to the intellectual SA of the Crosson et al.'s model), consisting in the patients metacognitive SA; and an "online SA", which relates to both emergent (e.g., patients' ability of self-monitoring), and anticipatory (e.g., their ability to evaluate what a task requires while it is performed) aspects of SA (Bivona et al. 2022).

Decreased metacognitive SA is significantly correlated with increased problems in some components of EF (Bivona et al. 2008). Additionally, some cognitive predictors of SA (related with EF) could be considered in patients with ABI along with neuropsychological rehabilitation, especially verbal fluency was found to be the best predictor of SA, both at admission and discharge of the rehabilitation process. In addition, inhibition, and cognitive flexibility, as well as episodic memory, appeared as significant predictors of post-rehabilitation SA (Villalobos et al. 2020).

Patients with reduced SA can partially or totally omit the existence of deficits related to brain injury associated with their deficits in EF. In particular, they cannot accurately recognize functional consequences, including those related to emotional, personality, and social competency changes (Bivona et al. 2015) that are crucial for adjusted ToM. Relatedly, poorer SA of one's own emotional experiences (alexithymia) also influences one's ability to recognize the emotions of others after a TBI. Alexithymia is thought to affect nearly the 50% of patients with moderate to severe TBI, affecting more people with TBI than the general population, and being crucial in SC symptoms (Cassel et al. 2016).

In a specific study by Spikman et al. (2013), they explored the relationship between deficits in facial emotion recognition, as an important component of ToM, behavioral changes (common and disruptive symptoms after ABI) and SA in patients with TBI. They found that impaired emotion recognition in the patients, in particular sadness and anger, are significantly correlated with behavioral problems (as rated by proxies) and also with impaired SA. This study found these associations, strengthening the proposed recognition of social signals as a condition for adequate social functioning. Hence, deficits in emotion recognition can be conceived as markers for behavioral problems and closely related with lack of SA in TBI patients; thus, emotion recognition must be objectively measured early after injury.

7.4.3 Communicative Abilities

ABI patients usually present impaired communication skills (apart from patients that presents language impairments, i.e., aphasia). They show excessive talkativeness, poor topic maintenance, repetitiveness, difficulties in starting and maintaining a conversation, in the organization of narrative discourse, in the ability to understand sarcasm, irony, and indirect requests. Moreover, these patients often exhibit low levels of social appropriateness during their communicative interactions and an impaired ability to understand the prosodic aspects of speech. Recently, the term "social communication" has been addressed in this field to describe the specific impact of communication disorders on a person´s capacity to achieve personally relevant social goals across context (Byom et al. 2020). For the families of ABI patient`s apart from other deficits, communication difficulties impact upon family functioning and psychological wellbeing for several years post-injury. Changes to SC, insight and the "filter switch" of the patients are for them a key area of distress. Usually, familiars highlighted the need for information about communication changes to be provided at several time points post-injury. They also ask for peer support from therapist and other families with experience of communication difficulties (Grayson et al. 2021).

These difficulties seem to be related with other cognitive impairments, and EF and ToM could be specific processes related with them. So, the extent to which these communication deficits reflect ToM versus executive dysfunction is an interesting question to be investigated. In a study with TBI patients, S. McDonald et al. (2014) found that when communicating to meet another person's viewpoint, EF are required. In addition, when self-referential cognitions are activated, people with TBI have particular difficulty suppressing these in order to communicate another person's perspective. This suggests that inhibition has an important impact in this communicative context. The results support the view that ToM is an important ability that explains poor communicative output interrelated with other EF abilities.

Additionally, using everyday conversation task across four conditions (low cognitive load, high flexibility, high working memory and high inhibition), Honan et al. (2015) found that TBI individuals were impaired on high-ToM tasks only in the WM condition, so ToM impairments in everyday communication may arise due to EF abilities, especially working memory demands, in individuals with TBI.

Related with communication skills, pragmatics is a particularly interesting concept for this profile of patients. Pragmatics includes the study of meaning in relation to the use of language, as the relationship between signs and their users; the ability to use language and other means of expression, such as gestures and paralinguistic indicators, to convey communicative meaning; it is also the ability to manage conversations and discourse analysis. Quality of life and social integration are strongly influenced by this pragmatics abilities. As previous communication processes it seems to be related with ToM and these interactions has been recently investigated. Previously, it was common to mix pragmatic and ToM abilities and to conceive pragmatics as a sort of subcomponent of ToM, and thus to conflate or reduce the notion of pragmatics into the (wider) notion of ToM, but now it is known that it is not theoretically correct and a possible cause of methodological confusion in the relevant empirical research. Thus, it is necessary that the two faculties are investigated with separate theories as well as different experimental tasks (Bosco et al. 2018b).

Bosco et al. (2017) examined the relationship between EF (working memory, planning and flexibility) and ToM and communicative-pragmatic impairment in

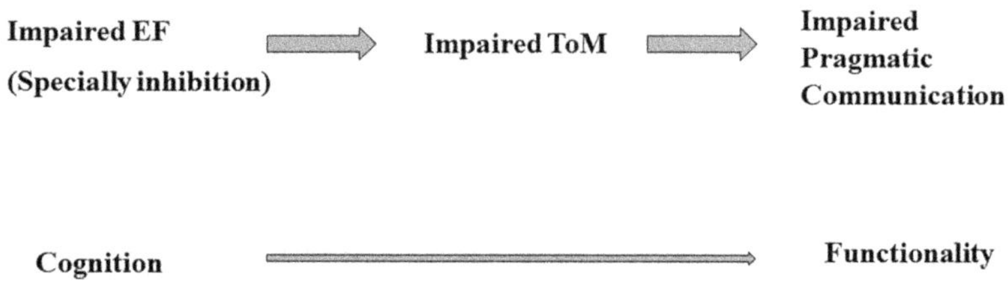

Fig. 7.3 Relationship between EF, ToM, and pragmatic communication skills, from cognitive to functional domains

patients with TBI. Their results showed that TBI participants performed poorly in comprehension and production tasks in the Assessment Battery of Communication (ABaCo), using both linguistic and extralinguistic means of expression, and that they were impaired in EF and ToM abilities. Cognitive difficulties were able to predict the pragmatic performance of TBI individuals, with both executive functions and ToM contributing to explaining patients' scores on the ABaCo. In a next study, the same authors again examined the relationship between the ability to understand and produce various kinds of communicative acts, (i.e., sincere, deceitful, and ironic) and cognitive (attention, memory, and EF) and ToM abilities following TBI. They found that ToM had a significant role in determining patients' performance in the extralinguistic production of sincere and deceitful communicative acts, linguistic and extralinguistic comprehension of deceit, and the linguistic production of irony. EF was able to explain the pattern of patients' performance in the linguistic and extralinguistic comprehension but not in production ability (Bosco et al. 2018a) (Fig. 7.3).

Also considering the relevance of SA in this complicated relationship, a recent study (Sherer et al. 2022) indicated that both the ability to interpret facial affect and self-awareness would be associated with communication ability. They also analyzed that facial affect recognition would influence self-awareness and that the effect of facial affect recognition on social communication would be partially mediated by self-awareness. They found an association between facial affect recognition and social communication with an effect of facial affect recognition on self-awareness. However, the effect of facial affect recognition on social communication was not mediated by self-awareness.

7.5 How to Assess ToM in ABI: Tools and Scales

As it has been previously introduced on this chapter, ABI is frequently followed by a deficiency in social functioning and cognition, including ToM. However, the evaluation of social cognition features in general, has not been traditionally consider as a part of the battery of cognitive assessment that ABI patients go through, highlighting both, non-social functions, and language-base skills, as the fundamental parts of the evaluation process. Specifically, in the study conducted by Kelly et al. 2017 that comprised a total of 443 clinicians, a 78% reported that they infrequently or never assessed SC domains using a formal assessment tool, and more particularly a 45% of them had never assessed ToM, arguing the lack of reliable test available as the greatest barrier to undertake them.

Nevertheless, numerous studies focused on measuring SC domains such as social communication, identity recognition, empathy, emotion perception and ToM in order to detail, understand, and describe changes in the functional outcomes linked to ABI. The lack of consensus regarding terminology, makes difficult to proper assess potential impairment in SC. For that reason, Wallis et al. 2021 conducted a scoping review aiming to determine the most frequently assessed domains of SC linked to ABI in literature (Fig. 7.4), as well as to map those measurement instruments (and its psychometrics characteristic) used to identify them. As it is shown in Fig. 7.4 Emotion Perception, ToM and Social Communication were the most reported ones across literature.

Considering that the previous quantitative review has unveiled the severity of ToM deficits in the ABI population (Martín and León-Carrión 2010), the literature discrepancies regarding the role of ToM impairment as one of the main underlying causes of these SC deficits in patients with ABI, might be the result of a non -appropriate use of the tools for its assessment. For that reason, it seems crucial to describe the tools and scales available to measure ToM impairment appropriately.

Remarkably, the proportion of the so-considered "subdomains" of the ToM construct reported in literature have been mainly focused on ToM itself as a general constructor and on social inference followed by sarcasm (Wallis et al. 2021).

In addition, ToM has been measured just once after brain injury, even when it is expectable that its impairment became severe over time due to its association with

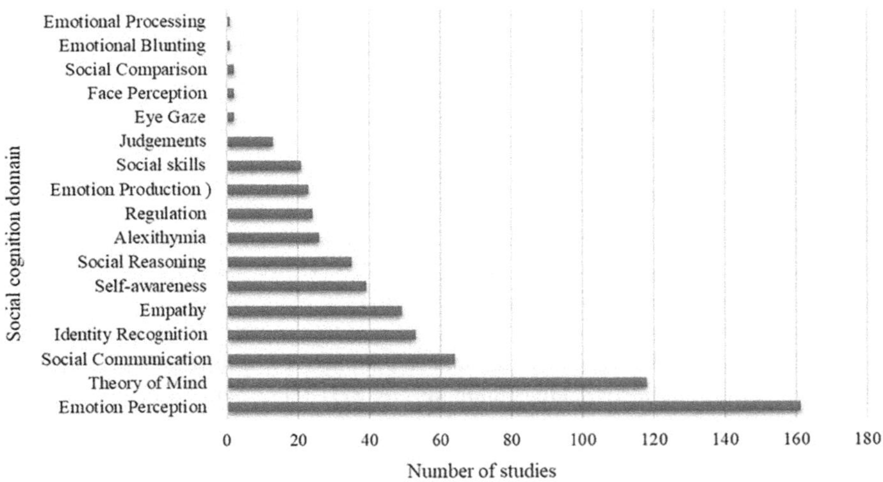

Fig. 7.4 The proportion of each SC domains reported in literature according to Wallis et al. (2021)

social communication. Also, it is known that there are changes in patients' social environment following the injury that result in a reduction of social communication as well as in the number of social interactions (Milders et al. 2006). Interestingly, Milders et al. (2006) by examining ToM impairments following TBI over time, found that ToM impairment remains stable after a 1-year follow-up. This fact highlights the necessity of measuring ToM after ABI, since an early intervention could provide a useful insight regarding the general state of the social impairment and contribute to reduce its impact in an individual's life throughout time.

7.5.1 Tools and Scales for ToM Assessment

According to Squire (2010) and Higginson and Carr (2001), clinicians should use those tools that meet the followings statements:

(a) Adequate to normative data, referring to what is typical in a specific population at a specific period.
(b) Useful to identify specifics deficit in ABI, to develop targeted interventions.
(c) Empirically validated.
(d) Documented reliability
(e) Simple and quick to be completed.
(f) Easy to score and evaluate.

The typical tools use to assess ToM in ABI population, that checks the mentioned statements will be introduced below.

7.5.1.1 Self-Report Questionnaires

The validity and reliability of self-report questionnaires has been criticized due to its sensitivity to language comprehension problems and insights, commonly related with ABI, and patient's difficulties in monitoring and characterized their condition (Bivona et al. 2018; McDonald 2008). Bivona et al. (2014) reported how most of the studies that administered self-reports to assess ToM competences did not consider/control self-awareness disorder. This could make difficult for ABI patients to recall the exposed situations in most self-report questionnaires in which they are required to report how they usually feel in those situations. However, they could provide relevant information if further analyses are conducted to measure the adaptability of these tools in ABI population.

An example of a self-report questionnaire that have been developed to measure ToM is "The Mentalization Questionnaire" (MZQ) (Hausberg et al. 2012), that measures:

(a) Ego-strength, measure the level of resilience and confidence to face conflict or stressful challenges.

(b) Relational attunement referring to the ability of recognize, understand, and engage with someone's emotional state, and deeply understand their experience.
(c) Reflexibility, in terms of being able to be introspective regarding your own feelings and motivations.
(d) Relational discomfort, referring to the perception of being misunderstood by others while dealing with interpersonal difficulties.
(e) Distrust, related to the degree of closed-mindedness and to the tendency to have a dichotomic point of view when facing a challenge.
(f) Emotional dyscontrol, measure the degree of difficulty to assess and manage own affective states and impulsivity.
(g) The first three measurements (a, b, and c) are positively related to highly functional components of mentalizing while the others are linked to failures and distortions (Gori et al. 2021).

In addition, the MZQ showed satisfactory psychometric properties, and sensitivity to distinguish between clinical samples and controls in the mentalizing ability. On the other hand, the Toronto Empathy Questionnaire (TEQ) (Spreng et al. 2009) that delves into a more specific aspect of ToM such as empathy, has also proved to be sensitive to empathy conceptualization as a primary emotional process.

7.5.1.2 Scales and Test

Perspective Taking (PT)

Cognitive empathy is a construct closely linked to ToM (Decety and Jackson 2004). The PT is one of the four subscales composing the Interpersonal Reactivity Index (IRI) that has been previously used to assess empathy deficit. This subscale is composed by 7-items (Davis and Association 1980) that measure the ability to adopt other's cognitive viewpoint in everyday life and present an acceptable reliability attending to Cronbach's alpha value ($\alpha = .77$) (Rosner and Cronbach 1960). In addition, it has been previously used to successfully measure ToM deficit in TBI (McDonald et al. 2014), and includes the following items that participants must rate from 0 (does not describe me well) to 4 (describes me very well):

(a) Before criticizing somebody, I try to imagine how I would feel if I were in their place.
(b) If I'm sure I'm right about something, I don't waste much time listening to other people's arguments. (Inverse item)
(c) I sometimes try to understand my friends better by imagining how things look from their perspective.
(d) I believe that there are two sides to every question and try to look at them both.
(e) I sometimes find it difficult to see things from the "other guy's" point of view. (Inverse item)
(f) I try to look at everybody's side of a disagreement before I make a decision.
(g) When I'm upset at someone, I usually try to "put myself in his/her shoes" for a while.

Social Inference – Minimal (SI-M) and Enriched (SI-E) (sub)test of TASIT and TASIT-S

The Awareness of Social Inference Test (TASIT) (McDonald et al. 2003) has been proven to be sensitive to brain injury (S. McDonald and Flanagan 2004). TASIT assess SC including emotion perception, ToM, and nonliteral inference in TBI. For the task, participants are required to watch a series of brief videos vignettes describing daily social situations and they must make judgements about a target character's feelings, intentions and thoughts based on the so-called paralinguistic cues, such as tone of voice, facial expressions and hand and body language. This test has three parts to be completed:

(a) Emotion Evaluation Test: basic emotion recognition in semantically ambiguous context (24 items).
(b) Social Inference – Minimal (SI-M): comprehension of sincere and sarcastic exchanges (15 items).
(c) Social Inference-Enriched (SI-E): differential detection and comprehension of sarcastic and deceptive communications (16 items).

Both, SI-M and SI-E measure ToM attributes.

Faux Pas Recognition Test

According to Baron-Cohen et al. (1999) a faux pas is defined as a "situation in which a speaker says something without considering if it is something that the listener might not want to hear or know, and which typically has negative consequences that the speaker never intended". Recognizing a faux pas is considered as an advanced ToM ability (Lee et al. 2010). The Faux Pas Test consist of 20 short stories, 10 of them contain no faux pas, whereas the other 10 describe a social faux pas. They are presented randomly, and participant answers questions with the printed story in front of them. Table 7.2 summarized the questions and the measured construct for each of them (Stone et al. 1998).

Table 7.2 Questions and measured construct for Faux Pas recognition test (Stone et al. 1998)

Measured construct	Questions
Faux Pas Detection	Did anyone say something they shouldn't have said or something awkward?
	Who said something they shouldn't have said or something awkward?
Understanding Inappropriateness	Why shouldn't he/she have said it or why was it awkward?
Intentions – the speaker's intentions/ motivations	Why do you think he/she said it?
Belief – either a false or true belief a story character holds	Did X know/realize that Y?
Empathy – does the respondent know how the people would feel?	How do you think X felt?
Story comprehension (Control questions)	One question about general story comprehension.
	Second question about general story comprehension

Generally, these stories firstly describe an event involving two individuals, that will meet afterwards. One of them forgot the prior event and says something awkward that could offend the other. The objective is to measure whether individuals can detect the faux pas, the person who made it, and understand both, the mental state of the speaker and the listener (Bivona et al. 2014).

The Faux Pas Recognition test has been previously used to assess ToM state in subjects with TBI. Muller et al. (2010) found that individuals with TBI displayed significantly lower scores on faux pas related questions on the faux pas stories in comparison to control groups, even though no significant differences were found regarding control questions in both, control stories and faux pas stories. This provides sensibility to ToM deficits in ABI.

False-Belief Stories

The false-belief stories have been traditionally used to measure ToM (Fine et al. 2001; Happé 1994), as reasoning require acknowledgement that a belief is false, and the participant cannot correctly answer from their perspective (Dennett 1978). This is to say, that the standard false-belief task involves the necessity to infer a false belief (Apperly et al. 2005). There are different orders of false belief-test; in first-order ones, individuals are asked about what a character believes or thoughts: "*Character A* thinks X is true". For second-order task there is an additional level of embedding: "*Character B* thinks that *character A* thinks that X is true". It is possible to add further levels of embeddings (e.g., third-order; *Character A* thinks that *character B* believes that *character A* does not know that *character B* is having an affair". However, it is important to consider that as the level of embedding increases, the working memory demand/load increases as well (Decety and Jackson 2004). In addition, it has been shown that second- or third-order beliefs provide a reliable measure of the ability to infer mental state (Stone 2007).

First-Order Stories

The first-order false belief task assesses the ability to differentiate between individual's own belief and other person's beliefs (Ho et al. 2014). The tasks consist of a false-belief story that describes an object being relocated from one place to another, without a character's knowledge, the so-called "location change tasks". In these stories a *character A* relocates one object and then leaves the room, then a *character B* moves that object into another place hidden from *character A*. Then *character A* comes back into the room. Some of the first-order false-belief stories commonly used are "Sally and Anne" (Baron-Cohen et al. 1985) and "Cigarettes" (Happé 1994) stories.

Then, the participant is required to answer to different questions related to the inference (Bivona et al. 2014; Tager-Flusberg et al. 1997):

(a) Mistaken belief: predict where the character will search for the object.
(b) Examination of reality: indicate where the object is.
(c) Memory accuracy: indicate where the object was at the beginning of the story (control question).

However, according to Siegal et al. (1996) these simplest false-belief tasks had been designed for children, so previous literature has developed a complex version for adults, including stories in which the main character has acted on a false-belief and the subject is required to explain the action of that character (Channon and Crawford 2000).

Second-Order Stories

Second-order false belief stories are more complex than first-order ones, participants are expected to infer what an individual is thinking on a third person's belief (Mazza et al. 2001). This ability also underlies the capability of distinguish between lies from jokes (Winner et al. 1998). In addition, it has been suggested that by measuring more advanced ToM tasks it is possible to obtain an index of the severity of the deficit; in fact, (Bibby and McDonald 2005; Hughes et al. 2000) reported that second-order false-belief tasks have good reliability. Some of the common second-order stories are the Ice Cream Van (Baron-Cohen 1989) and the Burglar story (U. Frith and Happé 1994), both include a series of cartoon drawings illustrating the action sequences, and then participants are required to answer questions regarding the basic mental state of one of the characters and about her/his false belief about the situation.

Muller et al. (2010) found that patients with TBI showed significantly lower scores in ToM questions for second-order belief stories in comparison to controls. Nevertheless, evidence regarding the efficiency of these tests to measure ToM in ABI are mixed, as some results might be due to the non-ToM factors that are involved in the performance of ToM tasks that could be specifically impaired for some ABI patients. For example, in those individuals with posterior lesions in temporal-parietal junction there are deficits in language control, working memory and inhibitory task demands that might influence their performance in false-belief stories (Samson et al. 2007).

Reading the Mind in the Eyes Test (RMET)

The RMET– Revised version (Baron-Cohen et al. 2001), was initially developed as a ToM task for Asperger's individuals, however, it proved to be also sensitive for TBI (Geraci et al. 2010). In this test, participants are presented with a series of 36 photographs of eyes, and they are asked to indicate which one of four given words better describes the mental state of the person who was photographed. As an example, for the male in Fig. 7.5 (left) the given words were serious (correct one), ashamed, alarmed, and bewildered; and for the female (right) were reflective (correct one), aghast, irritated, and impatient.

Guzmán et al. (2017) reported significant differences in the performance of the RMET in TBI individuals, unveiling and average lower performance on emotional recognition depending on the severity of the TBI symptoms.

Happé's Strange Stories Test

On this test developed by Happé (1994) individuals read a brief story in which a character lies with different intentions, the story remained in front of the subject to

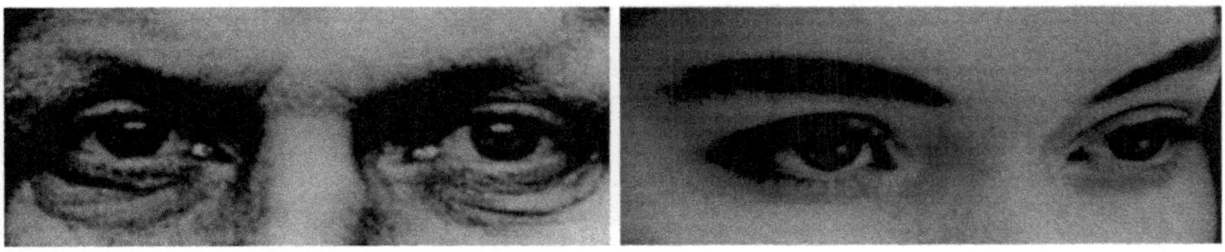

Fig. 7.5 Two examples of the RMET revised version (Baron-Cohen et al. 2001).

minimize memory load. For example, *Character A* says how good looks the new glasses of *character B* even though *character A* does not like them. Then participants are asked 2 questions, the first one is formulated to test comprehension (if incorrect, the story is again read until participant correctly answered), and the second one to infers character's intentions, feelings, or beliefs:

(a) Was it true, what X said?
(b) Why did X say that?

A new version of this approach was developed by Shamay-Tsoory and Aharon-Peretz (2007), considering that affective ToM and cognitive ToM are processed differently. This version considers:

(a) Cognitive ToM: Story character thought about another character's belief.
(b) Affective ToM: Story character though about other character´s feelings.

Results showed that individuals with ventromedial damage presented more impaired affective ToM, whereas cognitive ToM was mostly impairment by extensive prefrontal damage. This provided a meaningful insight about how measure ToM in the different ABI.

Social Problem

The *social problem resolution* task assessed the ability of participants to generate the best solution/appropriate course for a given social situation (Channon and Crawford 1999). The task comprises a total of 10 items, in order to reduce worming memory load, the stories remained in front of the participant.

> *"Example. Tony is always tired because he is kept awake by his new upstairs neighbours' noisy dogs. The neighbours are very pleasant but say that there is nothing they can do about the dogs. Question: What is the best thing for Tony to do in this situation?"*. Participants are asked to disclose the better solution that the character could choose for the given situation. Then responses are classified attending to both; (a) social sensitivity (b) practical effectiveness.

In addition, there is another evaluated construct, the social problems fluency, that assess the fluency in generating those ideas for solving the given social problems and the ability to identify the awkward elements of that situation (Channon and Crawford 1999). The task is composed of 10 scenarios that are comparable to the ones involved in social problem resolution. Then, participants must explain why the given situation might be awkward for the character and rate the awkwardness

(0–100). Channon and Crawford (2010) found that both, mentalizing task and problem solving were more sensitive to ABI patients, showing greater difficulties than controls in inferring character's mental state.

Moral Dilemmas
These dilemmas are composed of different hypothetical scenarios, in each of them participants must emit a dichotomous judgment about how they will face that dilemma by rejecting or accepting the proposed action (Greene et al. 2001). As an example, in the *Trolly dilemma* participants had to hypothetical face the following situation: "*There is a runaway trolley moving down railway tracks. In its path, there are five people tied up and unable to move and the trolley is heading straight for them. People are told that they are standing some distance off in the train yard, next to a lever. If they pull this lever, the trolley will switch to a different set of tracks—but will kill one person who is standing on the sidetrack. The people have the option to either do nothing and allow the trolley to kill the five people on the main track, or pull the lever, diverting the trolley onto the sidetrack where it will kill one person*" (Thomson 1985).

Martins et al. (2012) asked participants to judge 22 hypothetical dilemmas that were split into categories: non-moral, impersonal, and personal moral. Participants with TBI gave a higher rate of utilitarian responses for personal moral dilemmas in comparison to controls. There were a negative association between performance on social emotions recognition and affirmative responses rate, suggesting that the preference of utilitarian responses in these dilemmas might be the result of an impairment in social emotion recognition. In addition, Moretto et al. (2010) reported that ABI patients with injury in areas related with cognitive and affective information integration present utilitarian traits in their moral judgment.

Interestingly, Guzmán et al. (2017) conducted a study in which two personal dilemmas (Transplant and the Train) and two impersonal dilemmas (Norm against gas and Murderer Wagon) were applied in a TBI sample. The selection of the dilemmas was made considering those that have been used the most in other investigations. In the personal dilemma of the train, the people who responded with a non-utilitarian judgment were asked the question: what would you do if those on the track were your relatives? No significant differences were found into participants with mild, moderate, and severe TBI discussing that maybe is because participants were not impaired in areas commonly related to decision making, emotion and moral judgment such as the frontal lobe (Greene et al. 2004; Koenigs et al. 2007).

Cartoon Test
The Cartoon Test (Bibby and McDonald 2005; Gallagher et al. 2000) is a task with a non-verbal component. The test comprises 12 single-frame cartoons from magazines in which different humoristic situations are displayed. In six of them, the joke is based on a false belief or on a character's lack of knowledge regarding a physical situation (a brief caption is included), and participants are required to attribute mental states. Then participants must answer four questions that probe ToM ability

(Bivona et al. 2018). In the other 6 cartoons participants do not need to attribute mental state ones, as the joke is about a physical transgression or violation of a social norm.

Mental and physical cartoons are presented randomly, and participants are asked to explain why the introduced cartoon is funny. Once a response is obtained, they are separately summed into different subscores. For example, an explicit explanation received 3 points, 2 for an incomplete or implicit one, 1 when relevant parts are mentioned without further interpretation, and 0 for incorrect answers. F. Happé et al. (1999) found that patients with right hemisphere brain-damage were significantly worse at understanding mental state cartoons than controls. In addition, May et al. (2017) also reported that scores in cartoon test were significantly lower in the TBI as compared to the matched controls.

7.6 Specific Interventions in ToM in ABI Patients

Social-behavioral problems frequently occur, particularly in patients with ABI. They involve inappropriate, indifferent, or disinhibited interpersonal conducts and lead to unemployment, social isolation, and loneliness, thus limiting societal participation. Given these important consequences, evidence-based rehabilitation interventions focusing on SC with the explicit aim to improve everyday-life social behavior and participation are completely needed. In ABI population, and specially in TBI patients, the literature shows that treatments have focused on emotion perception skills; programs developed for other clinical populations (specially autism spectrum disorder and schizophrenia spectrum disorder) have had broader targets, focusing on ToM skills and/or modifying interpretational cognitive biases.

In the INCOG 2.0 Guidelines for cognitive rehabilitation following TBI part IV (Togher et al. 2023), authors present several recommendations for management of cognitive communication and SC disorders. In the recommendation regarding SC management, they conclude that clinician should consider evaluating aspect of SC ability, including emotion perception, ToM, and emotional empathy. Intervention which aims at improving emotion perception, perspective taking, ToM and social behavior are recommended. However, they suggest not to use computerized SC treatments given lack of evidence of generalization to real-life activities.

A recent systematic review by Cassel et al. (2016), explored the studies that focused on rehabilitation of SC in ABI patients. Authors found that some studies focused primarily on a particular skill associated with emotion perception, such as facial affect recognition, emotion prosody recognition and using context to infer emotional experience. In others, multiple aspects of emotion perception have been targeted, and addressed in a graded hierarchy. For instance, there has been work on "simple" cues to improve recognition, which are eventually incorporated into training on the perception of complex and dynamic emotions (i.e., judging dynamic audio-visual expressions of emotions through video, therapist modelling, and role

Table 7.3 Behavioral and cognitive rehabilitation techniques used to improve SC skills

Focus attention
Mimicry
Repetition/massed and distributed practice
Positive reinforcement
Errorless learning
Self-instructional training
Perspective-taking
Use of vanishing cues
Psychoeducation
Social skills training

play). Others have additionally related emotion perception to self-emotional processing through encouraging introspection and imitation.

In another recent systematic review Rodríguez-Rajo et al. (2018), authors explored rehabilitation of SC impairment after TBI with similar results. Most studies focus on emotion recognition and perception. Considering both reviews, which include coincident and also non-coincident studies, we can conclude that these programs have used specific behavioral and cognitive rehabilitation techniques to teach these skills (See Table 7.3).

In general, most studies have found some positive outcomes and treatment improved participants' SC, especially emotion perception abilities, on at least one outcome measure. In both systematics review only one study found not significant positive results (McDonald et al. 2009).

Probably, only two studies have developed and implemented an intervention program for ABI patients, specially focusing on ToM capacity. The first one, performed by Westerhof-Evers et al. (2017), evaluated the effects of a multifaceted Treatment for Social cognition and Emotion regulation (TScEmo) in patients with TBI. The program that was specifically created, contained three modules with the following goals: enhancing emotion perception (module 1), perspective taking and ToM (module 2), and basic and goal-directed social behavior (module 3). Thus, module 2 was specifically created to intervene ToM abilities, and consisted of psychoeducation, perspective taking, and self-monitoring strategies. In this module, patients learned that different viewpoints could coexist. They used a simplified thoughts-feelings-behavior triangle taken from cognitive behavioral therapy, in particular the strategies regarding the explicit communication about thoughts and feelings (Beck and Fernandez 1998). Patients were taught strategies to fill out the thoughts-feelings-behavior scheme (self and other), using hypothetical and real-life personal incidents. Additionally, patients were encouraged to ask significant others about their thoughts and feelings to improve insight and to prevent reaching inadequate conclusions about their motives (See Table 7.4).

Authors compared an experimental group following the TScEmo intervention, with a control group also formed by patients with TBI following an intervention with Cogniplus (computerized platform aimed at improving different cognitive process). They found that TScEmo was effective in improving specific aspects of SC (facial

Table 7.4 Treatment components of ToM module (module 2) of T-ScEmo

Perspective taking
Thoughts–Feelings–Behavior triangle (self, other)
Ask others about their thoughts and feelings
Attend to feelings of others

affect recognition and ToM), as well as (even more important) proxy-rated empathic behavior, quality of life, quality of the life partner relationship, and societal participation in patient with TBI. These treatment effects last for at least 5 months post-treatment (Westerhof-Evers et al. 2017).

As it was previously mentioned, there exist an important relation between EF, ToM and communication and pragmatic skills, which are usually impaired in ABI patients. Impaired EF and ToM, play a role in explaining communicative/pragmatic performance of patients with ABI. It has been suggested that a rehabilitation program should take these factors (EF and ToM) into consideration, to improve patients' communicative abilities. Thus, intervention programs with the aim of improving communication/pragmatic abilities have included specific components focusing on ToM. Gabbatore et al. (2015) designed and implemented the Cognitive Pragmatic Treatment (CPT) a program aimed at addressing all aspects of communicative-pragmatic competence, by also taking into consideration ToM and EF. Their results supported the efficacy of the CPT program in improving and enhancing communicative/pragmatic abilities in patients with TBI.

Two sessions of the CPT focused on ToM abilities. The tools and procedures included videotaped scenes and role playing. These components were not included in the ToM module developed by (Westerhof-Evers et al. 2017); however, they used role playing in the module focused on improving awareness and inhibition of undesired social behavior. These sessions seem to be an appropriate strategy to improve ToM abilities directly and through other components of SC.

Although specific interventions for ToM abilities in ABI population must be developed and probe their efficacy, it is also possible to consider some other strategies and tools used with this aim in other populations such as autisms spectrum disorder (Begeer et al. 2015; Lecheler et al. 2021) and schizophrenia (Bechi et al. 2013; Vass et al. 2021). For example, in the latest, authors utilized virtual reality, which is also a strategy used with ABI patients to intervene other cognitive aspects. In Vass et al. (2021) study, patients participated in simulated social interactions with an avatar in immersive virtual reality environments. The strength of the virtual conversations was based on prewritten and pre-recorded structured dialogue elements, which were designed to induce ToM impairments (by including double meaning sentences, overstatements, and irony), and to make them accessible for later cognitive interventions. Additionally, with the aim of improving affective ToM and self-reflectiveness, each simulation was followed by a task, where patients had to visualize the inferred emotions of the avatar. Considering the results of this study, in which the experimental group exhibited improvements in ToM and other important aspects such as negative symptoms and pragmatic language skills; these tools seem to be promising for ABI patients.

7.7 Conclusions

Patients with ABI are characterized by the presentation of a number of different symptoms that affect their daily life. Perhaps the most common deficits are those found in SC, such as impairments in ToM and empathy.

In general, SC seems to be affected in ABI, but the severity and evolution of the symptoms depends highly on the characteristics of the lesion. Using samples with heterogeneous lesions may obscure the specific deficits that may appear. Therefore, meta-analytic research is relevant in this field.

Notably, the reviewed evidence supports the idea that SC requires specific brain systems (McDonald and Genova 2021). In fact, social and non-social cognition can be dissociated in clinical conditions as autism spectrum disorder. And also, people with frontal damage have been reported to preserve or even enhance cognitive abilities while showing worse social functioning (Eslinger and Damasio 1985). Here, we have talked about psychological deficits due to brain lesions. But in terms of brain functioning, neuroimaging studies have shown that a lesion can increase brain activation in the altered networks compared to controls as a compensatory mechanism (Schmitz et al. 2006; Schroeter et al. 2010).

To correctly understand ToM deficits in ABI patients, it is interesting to explain its relationship with other cognitive processes. Particularly EF, as a complex group of higher-order cognitive and social abilities, located in the frontal cortex, has been associated with SC and specially with ToM. Different groups of patients with ABI demonstrated that impairment in EF components is related with ToM performance; in different studies, distinct subcomponents of EF have been proposed to account for this relation. Additionally, SA as a metacognitive process that allows patients to be aware of their deficits and disabilities caused by the brain lesion, has also been related with ToM probably through its well documented relationship with EF components. Finally, communicative abilities and pragmatics are also processes that are usually impaired in patients with deficits in ToM, and again this SC subcomponents seem to be an important ability that explains poor communicative performance interrelated with other EF abilities.

It is important to acknowledge that further research regarding ToM functioning after ABI should be conducted to improve our understanding of those processes that underlie this impairment and, accordingly, to choose the better tools for proper evaluation. Moreover, studies including neuroimaging techniques as well as behavioral information might contribute to better distinguish ABI subtypes, and to use the most suitable tools, attending to those clinical aspects that might affect results (Bivona et al. 2018):

- The severity of the injury.
- The neuropsychological and neuropsychiatric features of each individual.
- The level of self-awareness.

Finally, in ABI population, evidence-based intervention programs that focus on improving patients' disability, increasing their quality of life, are completely needed and they show significant results of improvement. Impairments in SC are one of the most important deficits referred by patients and also their caregivers. Thus, implementing specific rehabilitation focus on SC and ToM abilities seem crucial for these patients. The most common programs are related to emotion recognition and perception, and they have obtained significant positive results. Interventions focusing on specific ToM abilities in ABI patients are still not commonly applied but the results seem encouraging. Tools and techniques used in other populations to improve ToM abilities seem promising in ABI patients as well.

List of Abbreviatures:

Acquired brain injury	(ABI)
Assessment Battery of Communication	(ABaCo)
Barthel Index	(BI)
Executive function	(EF)
Glasgow coma scale	(GCS)
Interpersonal Reactivity Index	(IRI)
Mentalization Questionnaire	(MZQ)
Middle temporal visual area	(MT/V5)
Modified National Institutes of Health Stroke Scale	(mNIHSS).
Modified Rankin Scale	(mRS)
Perspective Taking	(PT)
Posttraumatic amnesia	(PTA)
Scandinavian Stroke Scale	(SSS),
Self-awareness	(SA)
Social Inference – Minimal	(SI-M)
Social Inference – Enriched	(SI-E)
Social cognition	(SC)
Superior temporal sulcus	(STS)
The Awareness of Social Inference Test	(TASIT)
The Reading the Mind in the Eyes test	(RMET)
Theory of Mind	(ToM)
Traumatic brain injury	(TBI)
Treatment for Social cognition and Emotion regulation	(TScEmo)
Toronto Empathy Questionnaire	(TEQ)
Years of Life Lost	(YLL)

References

Aboulafia-Brakha, T., B. Christe, M.-D. Martory, and J.-M. Annoni. 2011. Theory of mind tasks and executive functions: A systematic review of group studies in neurology. *Journal of Neuropsychology* 5 (1): 39–55. https://doi.org/10.1348/174866410X533660.

Abu-Akel, A. 2003. A neurobiological mapping of theory of mind. *Brain Research Reviews* 43 (1): 29–40. https://doi.org/10.1016/S0165-0173(03)00190-5.

Adolphs, R. 1999. Social cognition and the human brain. *Trends in Cognitive Sciences* 3 (12): 469–479. https://doi.org/10.1016/S1364-6613(99)01399-6.

Adolphs, R. 2001. The neurobiology of social cognition. *Current Opinion in Neurobiology* 11 (2): 231–239. https://doi.org/10.1016/S0959-4388(00)00202-6.

Adolphs, R. 2010. What does the amygdala contribute to social cognition? *Annals of the New York Academy of Sciences* 1191 (1): 42–61. https://doi.org/10.1111/j.1749-6632.2010.05445.x.

Andelic, N., N. Hammergren, E. Bautz-Holter, U. Sveen, C. Brunborg, and C. Roe. 2008. Functional outcome and health-related quality of life 10 years after Moderateto-Severe traumatic brain injury. *Acta Neurologica Scandinavica* 120 (1): 16–23.

Apperly, I.A., D. Samson, and G.W. Humphreys. 2005. Domain-specificity and theory of mind: Evaluating neuropsychological evidence. *Trends in Cognitive Sciences* 9 (12): 572–577. https://doi.org/10.1016/j.tics.2005.10.004.

Ardila, A. 2013. There are two different dysexecutive syndromes. *Journal of Neurological Disorders* 1 (1): 1–4. https://doi.org/10.4172/2329-6895.1000114.

Babbage, D.R., J. Yim, B. Zupan, D. Neumann, M.R. Tomita, and B. Willer. 2011. Meta-analysis of facial affect recognition difficulties after traumatic brain injury. *Neuropsychology* 25 (3): 277–285. https://doi.org/10.1037/a0021908.

Bach, L., and A. David. 2006. Self-awareness after acquired and traumatic brain injury. *Neuropsychological Rehabilitation* 16 (4): 397–414. https://doi.org/10.1080/09602010500412830.

Baron-Cohen, S. 1989. The autistic child's theory of mind: A case of specific developmental delay. *Journal of Child Psychology and Psychiatry* 30 (2): 285–297. https://doi.org/10.1111/j.1469-7610.1989.tb00241.x.

Baron-Cohen, S., A.M. Leslie, and U. Frith. 1985. Does the autistic child have a "theory of mind"? *Cognition* 21 (1): 37–46. https://doi.org/10.1016/0010-0277(85)90022-8.

Baron-Cohen, S., M. O'Riordan, V. Stone, R. Jones, and K. Plaisted. 1999. Recognition of faux pas by normally developing children and children with Asperger syndrome or high-functioning autism. *Journal of Autism and Developmental Disorders* 29 (5): 407–418. https://doi.org/10.1023/A:1023035012436.

Baron-Cohen, S., S. Wheelwright, J. Hill, Y. Raste, and I. Plumb. 2001. The "Reading the Mind in the Eyes" Test revised version: A study with normal adults, and adults with Asperger syndrome or high-functioning autism. *Journal of Child Psychology and Psychiatry and Allied Disciplines* 42 (2): 241–251. https://doi.org/10.1017/S0021963001006643.

Beaudoin, C., and M.H. Beauchamp. 2020. Social cognition. *Handbook of Clinical Neurology* 173: 255–264. https://doi.org/10.1016/B978-0-444-64150-2.00022-8.

Bechi, M., M. Spangaro, M. Bosia, A. Zanoletti, F. Fresi, M. Buonocore, F. Cocchi, C. Guglielmino, E. Smeraldi, and R. Cavallaro. 2013. Theory of Mind intervention for outpatients with schizophrenia. *Neuropsychological Rehabilitation* 23 (3): 383–400. https://doi.org/10.1080/09602011.2012.762751.

Beck, R., and E. Fernandez. 1998. Cognitive-behavioral therapy in the treatment of anger: A meta-analysis. *Cognitive Therapy and Research* 22 (1): 63–74. https://doi.org/10.1023/A:1018763902991.

Becker, S. 2004. *I had brain surgery, what's your excuse?* New York: Workman Publishing Co.

Begeer, S., P. Howlin, E. Hoddenbach, C. Clauser, R. Lindauer, P. Clifford, C. Gevers, F. Boer, and H.M. Koot. 2015. Effects and moderators of a short theory of mind intervention for children with autism spectrum disorder: A randomized controlled trial. *Autism Research: Official Journal of the International Society for Autism Research* 8 (6): 738–748. https://doi.org/10.1002/AUR.1489.

Béjot, Y., H. Bailly, J. Durier, and M. Giroud. 2016. Epidemiology of stroke in Europe and trends for the 21st century. *La Presse Médicale* 45 (12): e391–e398. https://doi.org/10.1016/j.lpm.2016.10.003.

Bennett, D.A., R.V. Krishnamurthi, S. Barker-Collo, M.H. Forouzanfar, M. Naghavi, M. Connor, C.M.M. Lawes, A.E. Moran, L.M. Anderson, G.A. Roth, G.A. Mensah, M. Ezzati, C.J.L. Murray, V.L. Feigin, and Study Stroke Expert Group, GBD. 2014. The global burden of ischemic stroke: Findings of the GBD 2010 Study. *Global Heart* 9 (1): 107. https://doi.org/10.1016/j.gheart.2014.01.001.

Bibby, H., and S. McDonald. 2005. Theory of mind after traumatic brain injury. *Neuropsychologia* 43 (1): 99–114. https://doi.org/10.1016/J.NEUROPSYCHOLOGIA.2004.04.027.

Bigler, E.D. 2007. Anterior and middle cranial fossa in traumatic brain injury: Relevant neuroanatomy and neuropathology in the study of neuropsychological outcome. *Neuropsychology* 21 (5): 515–531. https://doi.org/10.1037/0894-4105.21.5.515.

Bivona, U., P. Ciurli, C. Barba, G. Onder, E. Azicnuda, D. Silvestro, R. Mangano, J. Rigon, and R. Formisano. 2008. Executive function and metacognitive self-awareness after severe traumatic brain injury. *Journal of the International Neuropsychological Society* 14 (5): 862–868. https://doi.org/10.1017/S1355617708081125.

Bivona, U., A. Riccio, P. Ciurli, G.A. Carlesimo, V.D. Donne, E. Pizzonia, C. Caltagirone, R. Formisano, and A. Costa. 2014. Low self-awareness of individuals with severe traumatic brain injury can lead to reduced ability to take another person's perspective. *Journal of Head Trauma Rehabilitation* 29 (2): 157–171. https://doi.org/10.1097/HTR.0b013e3182864f0b.

Bivona, U., R. Formisano, S. De Laurentiis, N. Accetta, M. Rita Di Cosimo, R. Massicci, P. Ciurli, E. Azicnuda, D. Silvestro, U. Sabatini, C. Falletta Caravasso, G. Augusto Carlesimo, C. Caltagirone, and A. Costa. 2015. Theory of mind impairment after severe traumatic brain injury and its relationship with caregivers' quality of life. *Restorative Neurology and Neuroscience* 33 (3): 335–345. https://doi.org/10.3233/RNN-140484.

Bivona, U., R. Formisano, L. Mastrilli, S. Zabberoni, C. Caltagirone, and A. Costa. 2018. Theory of mind after severe acquired brain injury: Clues for interpretation. *In BioMed Research International* 2018: 5205642. https://doi.org/10.1155/2018/5205642.

Bivona, U., S. Lucatello, and A. Costa. 2022. Self-awareness after severe traumatic brain injury: From impairment of self-awareness to psychological adjustment. In *Cellular, molecular, physiological, and behavioral aspects of traumatic brain injury*, 539–552. Elsevier. https://doi.org/10.1016/B978-0-12-823036-7.00034-7.

Bonini, L., C. Rotunno, E. Arcuri, and V. Gallese. 2022. Mirror neurons 30 years later: Implications and applications. *Trends in Cognitive Sciences* 26 (9): 767–781. https://doi.org/10.1016/j.tics.2022.06.003.

Bosco, F.M., A. Parola, K. Sacco, M. Zettin, and R. Angeleri. 2017. Communicative-pragmatic disorders in traumatic brain injury: The role of theory of mind and executive functions. *Brain & Language* 168: 73–83.

Bosco, F.M., I. Gabbatore, R. Angeleri, M. Zettin, and A. Parola. 2018a. Do executive function and theory of mind predict pragmatic abilities following traumatic brain injury? An analysis of sincere, deceitful and ironic communicative acts. *Journal of Communication Disorders* 75: 102–117. https://doi.org/10.1016/j.jcomdis.2018.05.002.

Bosco, F.M., M. Tirassa, and I. Gabbatore. 2018b. Why pragmatics and theory of mind do not (Completely) overlap. *Frontiers in Psychology 9*: 1453. https://doi.org/10.3389/fpsyg.2018.01453.

Brazinova, A., V. Rehorcikova, M.S. Taylor, V. Buckova, M. Majdan, M. Psota, W. Peeters, V. Feigin, A. Theadom, L. Holkovic, and A. Synnot. 2021. Epidemiology of traumatic brain injury in Europe: A living systematic review. *Journal of Neurotrauma* 38 (10): 1411–1440. https://doi.org/10.1089/neu.2015.4126.

Bruner, J. 1990. *Acts of meaning*. Cambridge, MA: Harvard University Press.

Byom, L., T.M. O'Neil-Pirozzi, R. Lemoncello, S. MacDonald, P. Meulenbroek, B. Ness, and M.M. Sohlberg. 2020. Social communication following adult traumatic brain injury: A scoping review of theoretical models. *American Journal of Speech-Language Pathology* 29 (3): 1735–1748. https://doi.org/10.1044/2020_AJSLP-19-00020.

Calderwood, L. 2003. *Cracked: Recovering after traumatic brain injury*. Austin: Jessica Kingsley.

Campanella, F., T. Shallice, T. Ius, F. Fabbro, and M. Skrap. 2014. Impact of brain tumor location on emotion and personality: A voxel-based lesion-symptom mapping study on mentalization processes. *Brain* 137 (9): 2532–2545. https://doi.org/10.1093/brain/awu183.

Cardillo, R., I.C. Mammarella, E. Demurie, D. Giofrè, and H. Roeyers. 2021. Pragmatic language in children and adolescents with autism spectrum disorder: Do theory of mind and executive functions have a mediating role? *Autism Research: Official Journal of the International Society for Autism Research* 14 (5): 932–945. https://doi.org/10.1002/AUR.2423.

Caspers, S., K. Zilles, A.R. Laird, and S.B. Eickhoff. 2010. ALE meta-analysis of action observation and imitation in the human brain. *NeuroImage* 50 (3): 1148–1167. https://doi.org/10.1016/j.neuroimage.2009.12.112.

Cassel, A., S. McDonald, M. Kelly, and L. Togher. 2016. Learning from the minds of others: A review of social cognition treatments and their relevance to traumatic brain injury. *Neuropsychological Rehabilitation.*

Cassidy, J.D., L. Carroll, P. Peloso, J. Borg, H. von Holst, L. Holm, J. Kraus, and V. Coronado. 2004. Incidence, risk factors and prevention of mild traumatic brain injury: Results of the who collaborating center task force on mild traumatic brain injury. *Journal of Rehabilitation Medicine* 36 (43 Suppl): 28–60. https://doi.org/10.1080/16501960410023732.

Channon, S., and S. Crawford. 1999. Problem-solving in real-life-type situations: The effects of anterior and posterior lesions on performance. *Neuropsychologia* 37 (7): 757–770. https://doi.org/10.1016/S0028-3932(98)00138-9.

Channon, S., and Crawford, S. 2000. The effects of anterior lesions on performance on a story comprehension test: Left anterior impairment on a theory of mind-type task. *Neuropsychologia* 38 (7): 1006–1017. https://doi.org/10.1016/S0028-3932(99)00154-2.

Channon, S., and Crawford, S. 2010. Mentalising and social problem-solving after brain injury. *Neuropsychological Rehabilitation* 20 (5): 739–759. https://doi.org/10.1080/09602011003794583.

Channon, S., A. Pellijeff, and A. Rule. 2005. Social cognition after head injury: Sarcasm and theory of mind. *Brain and Language* 93 (2): 123–134. https://doi.org/10.1016/j.bandl.2004.09.002.

Cheung, C.C.Y., T.M.C. Lee, J.T.H. Yip, K.E. King, and L.S.W. Li. 2006. The differential effects of thalamus and basal ganglia on facial emotion recognition. *Brain and Cognition* 61 (3): 262–268. https://doi.org/10.1016/j.bandc.2006.01.008.

Cho, I., and A.S. Cohen. 2019. Explaining age-related decline in theory of mind: Evidence for intact competence but compromised executive function. *Plos One* 14 (9): e0222890. https://doi.org/10.1371/journal.pone.0222890.

Ciuffreda, K., N. Kapoor, M. Taub, M. Bartuccio, and D. Maino. 2012. Acquired brain injury. In *Visual diagnosis and care of the patient with special needs*, 95–100. New York: Wolters Kluwer Health/Lippincott Williams & Wilkins.

Clark, M., and K. Guskiewicz. 2016. Sport-related traumatic brain injury. In *Translational research in traumatic brain injury*, ed. D. Laskowitz and G. Grant. Boca Raton: CRC Press/Taylor and Francis Group.

Coburn, K. 1992. Traumatic brian injury: The silent epidemic. *AACN Advanced Critical Care* 3 (1): 9–18.

Costa, V. De S.P., P.S.R. Guimarães, K.B.P. Fernandes, V.S. Probst, A. De S. Marquez, and D.S. Fujisawa. 2014. Prevalence of risk factors for the occurrence of strokes in the elderly. *Fisioterapia Em Movimento* 27 (4): 555–563. https://doi.org/10.1590/0103-5150.027.004.AO07.

Crosson, B., P.P. Barco, C.A. Velozo, M.M. Bolesta, P.V. Cooper, D. Werts, and T.C. Brobeck. 1989. Awareness and compensation in postacute head injury rehabilitation. *Journal of Head Trauma Rehabilitation* 4 (3): 46–54. https://doi.org/10.1097/00001199-198909000-00008.

Cuff, B.M.P., S.J. Brown, L. Taylor, and D.J. Howat. 2016. Empathy: A review of the concept. *Emotion Review* 8 (2): 144–153. https://doi.org/10.1177/1754073914558466.

Dafner, K.R., and M.M. Searl. 2008. The dysexecutive syndromes. In *Handbook of clinical neurology*, ed. G. Goldenberg and B.L. Miller. Edinburgh: Elsevier.

Davis, M.H. 1980. A multidimensional approach to individual differences in empathy. *JSAS Catalog of Selected Documents in Psychology* 10: 85.

de Sousa, A., S. McDonald, J. Rushby, S. Li, A. Dimoska, and C. James. 2011. Understanding deficits in empathy after traumatic brain injury: The role of affective responsivity. *Cortex* 47 (5): 526–535. https://doi.org/10.1016/j.cortex.2010.02.004.

de Sousa, A., S. McDonald, and J. Rushby. 2012. Changes in emotional empathy, affective responsivity, and behavior following severe traumatic brain injury. *Journal of Clinical and Experimental Neuropsychology* 34 (6): 606–623. https://doi.org/10.1080/13803395.2012.667067.

de Vignemont, F., and T. Singer. 2006. The empathic brain: How, when and why? *Trends in Cognitive Sciences* 10 (10): 435–441. https://doi.org/10.1016/j.tics.2006.08.008.

Decety, J., and P.L. Jackson. 2004. The functional architecture of human empathy. *Behavioral and cognitive neuroscience reviews* 3 (2): 71–100. https://doi.org/10.1177/1534582304267187.

Dennett, D.C. 1978. Beliefs about beliefs [P&W, SR&B]. *Behavioral and Brain Sciences* 1 (4): 568–570. https://doi.org/10.1017/S0140525X00076664.

Dennis, M., A. Agostino, C. Roncadin, and H. Levin. 2009. Theory of mind depends on executive function in children with traumatic brain injury.pdf. *Journal of Clinical and Experimental Neuropsychology* 31 (7): 835–847.

Desikan, R.S., F. Ségonne, B. Fischl, B.T. Quinn, B.C. Dickerson, D. Blacker, R.L. Buckner, A.M. Dale, R.P. Maguire, B.T. Hyman, M.S. Albert, and R.J. Killiany. 2006. An automated labeling system for subdividing the human cerebral cortex on MRI scans into gyral based regions of interest. *NeuroImage* 31 (3): 968–980. https://doi.org/10.1016/j.neuroimage.2006.01.021.

Dethier, M., S. Blairy, H. Rosenberg, and S. Mcdonald. 2012. Spontaneous and posed emotional facial expressions following severe traumatic brain injury. *Journal of Clinical and Experimental Neuropsychology* 34 (9): 936–947. https://doi.org/10.1080/13803395.2012.702734.

Dewan, M.C., A. Rattani, S. Gupta, R.E. Baticulon, Y.-C. Hung, M. Punchak, A. Agrawal, A.O. Adeleye, M.G. Shrime, A.M. Rubiano, J.V. Rosenfeld, and K.B. Park. 2019. Estimating the global incidence of traumatic brain injury. *Journal of Neurosurgery* 130 (4): 1080–1097. https://doi.org/10.3171/2017.10.JNS17352.

Dikmen, S.S., B.L. Ross, J.E. Machamer, and N.R. Temkin. 1995. One year psychosocial outcome in head injury. *Journal of the International Neuropsychological Society* 1 (1): 67–77. https://doi.org/10.1017/S1355617700000126.

Donker-Cools, B.H.P.M., H. Wind, and M.H.W. Frings-Dresen. 2016. Prognostic factors of return to work after traumatic or non-traumatic acquired brain injury. *Disability and Rehabilitation* 38 (8): 733–741. https://doi.org/10.3109/09638288.2015.1061608.

Draper, K., J. Ponsford, and M. Schönberger. 2007. Psychosocial and emotional outcomes 10 years following traumatic brain injury. *Journal of Head Trauma Rehabilitation* 22 (5): 278–287. https://doi.org/10.1097/01.HTR.0000290972.63753.a7.

Durham, C., and P. Ramcharan. 2018. Acquired brain injury. In *Insight into acquired brain injury: Factors for feeling and faring better*, ed. C. Durham and P. Ramcharan, 1–29. Singapore: Springer Singapore. https://doi.org/10.1007/978-981-10-5666-6_1.

Elbaum, J., and D. Benson. 2007. *Acquired brain injury*. New York: Springer.

Elkind, M.S. 1998. Stroke risk factors and stroke prevention. *Seminars In Neurology* 18 (4): 12.

Ellenberg, J.H. 1996. Posttraumatic amnesia as a predictor of outcome after severe closed head injury. *Archives of Neurology* 53 (8): 782. https://doi.org/10.1001/archneur.1996.00550080104018.

Engberg, A.W., and T.W. Teasdale. 2004. Psychosocial outcome following traumatic brain injury in adults: A long-term population-based follow-up. *Brain Injury* 18 (6): 533–545. https://doi.org/10.1080/02699050310001645829.

Eslinger, P.J., and A.R. Damasio. 1985. Severe disturbance of higher cognition after bilateral frontal lobe ablation: Patient EVR. *Neurology* 35 (12): 1731–1741. https://doi.org/10.1212/wnl.35.12.1731.

Etchepare, A., and A. Prouteau. 2018. Toward a two-dimensional model of social cognition in clinical neuropsychology: A systematic review of factor structure studies. *Journal of the International Neuropsychological Society* 24 (4): 391–404. Cambridge University Press. https://doi.org/10.1017/S1355617717001163.

Evans, J.S.B.T., and K.E. Stanovich. 2013. Dual-process theories of higher cognition: Advancing the debate. *Perspectives on Psychological Science* 8 (3): 223–241. https://doi.org/10.1177/1745691612460685.

Feigin, V.L., M. Brainin, B. Norrving, S. Martins, R.L. Sacco, W. Hacke, M. Fisher, J. Pandian, and P. Lindsay. 2022. World Stroke Organization (WSO): Global Stroke Fact Sheet 2022. *International Journal of Stroke* 17 (1): 18–29. https://doi.org/10.1177/17474930211065917.

Fine, C., J. Lumsden, and R.J.R. Blair. 2001. Dissociation between "theory of mind" and executive functions in a patient with early left amygdala damage. *Brain* 124 (2): 287–298. https://doi.org/10.1093/brain/124.2.287.

Fisk, J.E., and C.A. Sharp. 2004. Age-related impairment in executive functioning: Updating, inhibition, shifting, and access. *Journal of Clinical and Experimental Neuropsychology* 26 (7): 874–890. https://doi.org/10.1080/13803390490510680.

Fiske, S.T., and S.E. Taylor. 2017. *Social cognition: From brains to culture*. 3rd ed. Sage.

Flanagan, S.R., M.R. Hibbard, and W.A. Gordon. 2005. The impact of age on traumatic brain injury. *Physical Medicine and Rehabilitation Clinics of North America* 16 (1): 163–177. https://doi.org/10.1016/j.pmr.2004.06.012.

Fleming, J., J. Strong, and R. Ashton. 1996. Self-awareness of deficits in adults with traumatic brain injury: How best to measure? *Brain Injury* 10 (1): 1–15. https://doi.org/10.1080/026990596124674.

Frith, C.D., and U. Frith. 2012. Mechanisms of social cognition. *Annual Review of Psychology* 63: 287–313. https://doi.org/10.1146/annurev-psych-120710-100449.

Frith, U., and F. Happé. 1994. Autism: Beyond "theory of mind.". *Cognition* 50 (1–3): 115–132. https://doi.org/10.1016/0010-0277(94)90024-8.

Gabbatore, I., K. Sacco, R. Angeleri, M. Zettin, B.G. Bara, and F.M. Bosco. 2015. Cognitive pragmatic treatment: A rehabilitative program for traumatic brain injury individuals. *Journal of Head Trauma Rehabilitation* 30 (5): E14–E28. https://doi.org/10.1097/htr.0000000000000087.

Gallagher, H.L., F. Happé, N. Brunswick, P.C. Fletcher, U. Frith, and C.D. Frith. 2000. Reading the mind in cartoons and stories: An fMRI study of "theory of mind" in verbal and nonverbal tasks. *Neuropsychologia* 38 (1): 11–21. https://doi.org/10.1016/S0028-3932(99)00053-6.

Gan, C., J. Gargaro, C. Brandys, G. Gerber, and K. Boschen. 2010. Family caregivers' support needs after brain injury: A synthesis of perspectives from caregivers, programs, and researchers. *NeuroRehabilitation* 27 (1): 5–18. https://doi.org/10.3233/NRE-2010-0577.

Garrison, J.F. 2007. *Don't leave me this way: Or when I get back on my feet, you'll be sorry*. London: Harper Paperbacks.

Gasquoine, P. 1997. Postconcussion symptoms. *Neuropsychology Review* 7 (2): 77–85.

Geraci, A., L. Surian, M. Ferraro, and A. Cantagallo. 2010. Theory of mind in patients with ventromedial or dorsolateral prefrontal lesions following traumatic brain injury. *Brain Injury* 24 (7–8): 978–987. https://doi.org/10.3109/02699052.2010.487477.

Geraldina, P., L. Mariarosaria, A. Annarita, G. Susanna, S. Michela, D. Alessandro, S. Sandra, and C. Enrico. 2003. Neuropsychiatric Sequelae in TBI: A comparison across different age groups. *Brain Injury* 17 (10): 835–846. https://doi.org/10.1080/0269905031000088612.

Godefroy, O., and D. Stuss. 2007. Dysexecutive syndromes. In *The behavioral and cognitive neurology of stroke*, ed. O. Godefroy and J. Bogousslavsky, 369–406. Cambridge University Press.

Goebel, S., H.M. Mehdorn, and C.D. Wiesner. 2018. Social cognition in patients with intracranial tumors: Do we forget something in the routine neuropsychological examination? *Journal of Neuro-Oncology* 140 (3): 687–696. https://doi.org/10.1007/s11060-018-3000-8.

Gorelick, P.B. 2019. The global burden of stroke: Persistent and disabling. *The Lancet Neurology* 18 (5): 417–418. https://doi.org/10.1016/S1474-4422(19)30030-4.

Gori, A., A. Arcioni, E. Topino, G. Craparo, and R. Lauro Grotto. 2021. Development of a new measure for assessing mentalizing: The Multidimensional Mentalizing Questionnaire (MMQ). *Journal of Personalized Medicine* 11 (4): 305. https://doi.org/10.3390/jpm11040305.

Govan, L., P. Langhorne, and C.J. Weir. 2009. Categorizing stroke prognosis using different stroke scales. *Stroke* 40 (10): 3396–3399. https://doi.org/10.1161/STROKEAHA.109.557645.

Grayson, L., M.C. Brady, L. Togher, and M. Ali. 2021. The impact of cognitive-communication difficulties following traumatic brain injury on the family; a qualitative, focus group study. *Brain Injury* 35 (1): 15–25. https://doi.org/10.1080/02699052.2020.1849800.

Green, R.E.A., G.R. Turner, and W.F. Thompson. 2004. Deficits in facial emotion perception in adults with recent traumatic brain injury. *Neuropsychologia* 42 (2): 133–141. https://doi.org/10.1016/j.neuropsychologia.2003.07.005.

Greene, J.D., R.B. Sommerville, L.E. Nystrom, J.M. Darley, and J.D. Cohen. 2001. An fMRI investigation of emotional engagement in moral judgment. *Science* 293 (5537): 2105–2108. https://doi.org/10.1126/science.1062872.

Greene, J.D., L.E. Nystrom, A.D. Engell, J.M. Darley, and J.D. Cohen. 2004. The neural bases of cognitive conflict and control in moral judgment. *Neuron* 44 (2): 389–400. https://doi.org/10.1016/j.neuron.2004.09.027.

Greenwald, A.G., and C.K. Lai. 2020. Implicit social cognition. *Annual Review of Psychology* 71 (1): 419–445. https://doi.org/10.1146/annurev-psych-010419-050837.

Griffen, J., and R. Hanks. 2014. Cognitive and behavioral outcomes from traumatic brain injury. In *Handbook on the neuropsychology of traumatic brain injury*, ed. M. Sherer and A.M. Sander, 25–45. New York: Springer.

Guzmán, K.P., J.O. González Ortiz, and M.R. Acosta Barreto. 2017. Comparación de los procesos de la cognición social entre adultos con trauma craneoencefálico leve, moderado y severo. *Psychologia* 11 (2): 57–68. https://doi.org/10.21500/19002386.2957.

Hamilton, J., B. Radlak, P.G. Morris, and L.H. Phillips. 2017. Theory of mind and executive functioning following stroke. *Archives of Clinical Neuropsychology* 32: 507–5018.

Happé, F.G.E. 1994. An advanced test of theory of mind: Understanding of story characters' thoughts and feelings by able autistic, mentally handicapped, and normal children and adults. *Journal of Autism and Developmental Disorders* 24 (2): 129–154. https://doi.org/10.1007/BF02172093.

Happé, F., H. Brownell, and E. Winner. 1999. Acquired "theory of mind" impairments following stroke. *Cognition* 70 (3): 211–240. https://doi.org/10.1016/S0010-0277(99)00005-0.

Happé, F., G.S. Malhi, and S. Checkley. 2001. Acquired mind-blindness following frontal lobe surgery? A single case study of impaired "theory of mind" in a patient treated with stereotactic anterior capsulotomy. *Neuropsychologia* 39: 83–90. www.elsevier.com/locate/neuropsychologia.

Hausberg, M.C., H. Schulz, T. Piegler, C.G. Happach, M. Klöpper, A.L. Brütt, I. Sammet, and S. Andreas. 2012. Is a self-rated instrument appropriate to assess mentalization in patients with mental disorders? Development and first validation of the Mentalization Questionnaire (MZQ). *Psychotherapy Research* 22 (6): 699–709. https://doi.org/10.1080/10503307.2012.709325.

Henry, J.D., L.H. Phillips, J.R. Crawford, M. Ietswaart, and F. Summers. 2006. Theory of mind following traumatic brain injury: The role of emotion recognition and executive dysfunction. *Neuropsychologia* 44 (10): 1623–1628. https://doi.org/10.1016/j.neuropsychologia.2006.03.020.

Higginson, I.J., and A.J. Carr. 2001. Measuring quality of life: Using quality of life measures in the clinical setting. *British Medical Journal* 322 (7297): 1297–1300. https://doi.org/10.1136/bmj.322.7297.1297.

Ho, K.K.Y., S.S.Y. Lui, K.S.Y. Hung, Y. Wang, Z. Li, E.F.C. Cheung, and R.C.K. Chanb. 2014. Theory of mind impairments in patients with first-episode schizophrenia and their unaffected siblings. *Schizophrenia Research* 166 (1–3): 1–8. https://doi.org/10.1016/j.schres.2015.05.033.

Honan, C.A., S. McDonald, A. Gowland, A. Fisher, and R.K. Randall. 2015. Deficits in comprehension of speech acts after TBI: The role of theory of mind and executive function. *Brain and Language* 150: 69–79. https://doi.org/10.1016/j.bandl.2015.08.007.

Honan, C.A., S. McDonald, C. Sufani, D.W. Hine, and F. Kumfor. 2016. The awareness of social inference test: Development of a shortened version for use in adults with acquired brain injury. *Clinical Neuropsychologist* 30 (2): 243–264. https://doi.org/10.1080/13854046.2015.1136691.

Hughes, C., A. Adlam, F. Happé, J. Jackson, A. Taylor, and A. Caspi. 2000. Good test-retest reliability for standard and advanced false-belief tasks across a wide range of abilities. *Journal of Child Psychology and Psychiatry and Allied Disciplines* 41 (4): 483–490. https://doi.org/10.1017/S0021963099005533.

Humphreys, I., R.L. Wood, C. Phillips, and S. Macey. 2013. The costs of traumatic brain injury: A literature review. *ClinicoEconomics and Outcomes Research* 5: 281–287. https://doi.org/10.2147/CEOR.S44625.

Hyder, A.A., C.A. Wunderlich, P. Puvanachandra, G. Gururaj, and O.C. Kobusingye. 2007. The impact of traumatic brain injuries: A global perspective. *NeuroRehabilitation* 22 (5): 341–353. https://doi.org/10.3233/NRE-2007-22502.

Ietswaart, M., M. Milders, J.R. Crawford, D. Currie, and C.L. Scott. 2008. Longitudinal aspects of emotion recognition in patients with traumatic brain injury. *Neuropsychologia* 46 (1): 148–159. https://doi.org/10.1016/j.neuropsychologia.2007.08.002.

Johnson, G. 2010. *Traumatic brain injury survival guide*. Traverse City.

Johnson, V.E., and W. Stewart. 2015. Age at injury influences dementia risk after TBI. *Nature Reviews Neurology* 11 (3): 128–130. https://doi.org/10.1038/nrneurol.2014.241.

Jones, S.S. 2007. Imitation in infancy the development of mimicry. *Psychological Science* 18 (7): 593–599. https://doi.org/10.1111/j.1467-9280.2007.01945.x.

Kay, T., and M. Lezak. 1990. The nature of head injury. In *Traumatic brain injury and vocational rehabilitation*, ed. D.W. Corthell, 22–65. Menomonie: University of Wisconsin-Stout, Research and Training Center.

Kelly, M., S. McDonald, and M.H.J. Frith. 2017. A survey of clinicians working in brain injury rehabilitation: Are social cognition impairments on the radar? *Journal of Head Trauma Rehabilitation* 32 (4): E55–E65. https://doi.org/10.1097/HTR.0000000000000269.

Kennedy, D.P., and R. Adolphs. 2012. The social brain in psychiatric and neurological disorders. *Trends in Cognitive Sciences* 16 (11): 559–572. https://doi.org/10.1016/j.tics.2012.09.006.

Kilford, E.J., E. Garrett, and S.J. Blakemore. 2016. The development of social cognition in adolescence: An integrated perspective. *Neuroscience & Biobehavioral Reviews* 70: 106–120. https://doi.org/10.1016/J.NEUBIOREV.2016.08.016.

Koenigs, M., L. Young, R. Adolphs, D. Tranel, F. Cushman, M. Hauser, and A. Damasio. 2007. Damage to the prefrontal cortex increases utilitarian moral judgements. *Nature* 446 (7138): 908–911. https://doi.org/10.1038/nature05631.

Kraus, M.W. 2017. Voice-only communication enhances empathic accuracy. *American Psychologist* 72 (7): 644–654. https://doi.org/10.1037/amp0000147.

Kreitzer, N., B.G. Kurowski, and T. Bakas. 2018. Systematic review of caregiver and dyad interventions after adult traumatic brain injury. *Archives of Physical Medicine and Rehabilitation* 99 (11): 2342–2354. W.B. Saunders. https://doi.org/10.1016/j.apmr.2018.04.016.

Krishnamoorthy, V., J.T. Distelhorst, M.S. Vavilala, and H. Thompson. 2015. Traumatic brain injury in the elderly. *Journal of Trauma Nursing* 22 (4): 204–208. https://doi.org/10.1097/JTN.0000000000000135.

Lahnakoski, J.M., E. Glerean, J. Salmi, I.P. Jääskeläinen, M. Sams, R. Hari, and L. Nummenmaa. 2012. Naturalistic fMRI mapping reveals superior temporal sulcus as the hub for the distributed brain network for social perception. *Frontiers in Human Neuroscience* 6: 233. https://doi.org/10.3389/fnhum.2012.00233.

Lanas, F., and P. Seron. 2021. Facing the stroke burden worldwide. *The Lancet Global Health* 9 (3): e235–e236. https://doi.org/10.1016/S2214-109X(20)30520-9.

Lecheler, M., J. Lasser, P.W. Vaughan, J. Leal, K. Ordetx, and M. Bischofberger. 2021. A matter of perspective: an exploratory study of a theory of mind autism intervention for adolescents. *Psychological Reports* 124 (1): 39–53. https://doi.org/10.1177/0033294119898120.

Lee, T.M.C., A.K.Y. Ip, K. Wang, C. Hua Xi, P. Pan Hu, H.K.F. Mak, S. Hui Han, and C.C.H. Chan. 2010. Faux pas deficits in people with medial frontal lesions as related to impaired understanding of a speaker's mental state. *Neuropsychologia* 48 (6): 1670–1676. https://doi.org/10.1016/j.neuropsychologia.2010.02.012.

Levine, S.C., R. Kraus, E. Alexander, L.W. Suriyakham, and P.R. Huttenlocher. 2005. IQ decline following early unilateral brain injury: A longitudinal study. *Brain and Cognition* 59 (2): 114–123. https://doi.org/10.1016/j.bandc.2005.05.008.

Lezak, M.D. 1995. *Neuropsychological assessment*. 3rd ed. New York: Oxford University Press.

Lezak, M.D., D.B. Howieson, D.W. Loring, H.J. Hannay, and J.S. Fischer. 2004. *Neuropsychological assessment*. 4th ed. New York: Oxford University Press.

Lindsay, M.P., B. Norrving, R.L. Sacco, M. Brainin, W. Hacke, S. Martins, J. Pandian, and V. Feigin. 2019. World Stroke Organization (WSO): Global Stroke Fact Sheet 2019. *International Journal of Stroke* 14 (8): 806–817. https://doi.org/10.1177/1747493019881353.

Lundin, A., C. de Boussard, G. Edman, and J. Borg. 2006. Symptoms and Disability until 3 Months after Mild TBI. *Brain Injury* 20 (8): 799–806. https://doi.org/10.1080/02699050600744327.

Maas, A.I., N. Stocchetti, and R. Bullock. 2008. Moderate and severe traumatic brain injury in adults. *The Lancet Neurology* 7 (8): 728–741. https://doi.org/10.1016/S1474-4422(08)70164-9.

Maggio, M.G., G. Maresca, M.C. Stagnitti, S. Anchesi, C. Casella, V. Pajno, R. de Luca, A. Manuli, and R.S. Calabrò. 2022. Social cognition in patients with acquired brain lesions: An overview on an under-reported problem. *Applied Neuropsychology: Adult* 29 (3): 419–431). Routledge. https://doi.org/10.1080/23279095.2020.1753058.

Magno, M.A., E. Canu, M. Filippi, and F. Agosta. 2022. Social cognition in the FTLD spectrum: Evidence from MRI. *Journal of Neurology* 269 (4): 2245–2258. https://doi.org/10.1007/s00415-021-10892-0.

Majdan, M., D. Plancikova, A. Brazinova, M. Rusnak, D. Nieboer, V. Feigin, and A. Maas. 2016. Epidemiology of traumatic brain injuries in Europe: A cross-sectional analysis. *The Lancet Public Health* 1 (2): e76–e83. https://doi.org/10.1016/S2468-2667(16)30017-2.

Martin, I., and S. McDonald. 2005. Evaluating the causes of impaired irony comprehension following traumatic brain injury. *Aphasiology* 19 (8): 712–730. https://doi.org/10.1080/02687030500172203.

Martín-Rodríguez, J.F., and J. León-Carrión. 2010. Theory of mind deficits in patients with acquired brain injury: A quantitative review. *Neuropsychologia* 48 (5): 1181–1191. https://doi.org/10.1016/j.neuropsychologia.2010.02.009.

Martins, A.T., L.M. Faísca, F. Esteves, A. Muresan, and A. Reis. 2012. Atypical moral judgment following traumatic brain injury. *Judgment and Decision Making* 7 (4): 478–487.

Mason, J. 2009. *Couldn't happen to me: A life changed by paralysis and traumatic brain injury*. Booksurge publishing.

May, M., M. Milders, B. Downey, M. Whyte, V. Higgins, Z. Wojcik, S. Amin, and S. O'Rourke. 2017. Social behavior and impairments in social cognition following traumatic brain injury. *Journal of the International Neuropsychological Society* 23 (5): 400–411. https://doi.org/10.1017/S1355617717000182.

Mazza, M., A. de Risio, L. Surian, R. Roncone, and M. Casacchia. 2001. Selective impairments of theory of mind in people with schizophrenia. *Schizophrenia Research* 47 (2–3): 299–308. https://doi.org/10.1016/S0920-9964(00)00157-2.

McDonald, J.D. 2008. Measuring personality constructs: The advantages and disadvantages of self-reports, informant reports and behavioral assessments. *Enquire* 1 (1): 75–94.

McDonald, S. 2013. Impairments in social cognition following severe traumatic brain injury. *Journal of the International Neuropsychological Society* 19 (3): 231–246). Cambridge University Press. https://doi.org/10.1017/S1355617712001506.

McDonald, S., and S. Flanagan. 2004. Social perception deficits after traumatic brain injury: Interaction between emotion recognition, mentalizing ability, and social communication. *Neuropsychology* 18 (3): 572–579. https://doi.org/10.1037/0894-4105.18.3.572.

McDonald, S., and H. Genova. 2021. The effect of severe traumatic brain injury on social cognition, emotion regulation, and mood. *Handbook of Clinical Neurology* 183: 235–260). Elsevier B.V. https://doi.org/10.1016/B978-0-12-822290-4.00011-6.

McDonald, S., S. Flanagan, J. Rollins, and J. Kinch. 2003. A new clinical tool for assessing social perception after traumatic brain injury. *Journal of Head Trauma Rehabilitation* 18 (3): 219–238. https://doi.org/10.1097/00001199-200305000-00001.

McDonald, S., C. Bornhofen, and C. Hunt. 2009. Addressing deficits in emotion recognition after severe traumatic brain injury: The role of focused attention and mimicry. *Neu- ropsychological Rehabilitation* 19 (3): 321–339. https://doi.org/10.1080/09602010802193989.

McDonald, S., S. Li, A. de Sousa, J. Rushby, A. Dimoska, C. James, and R.L. Tate. 2011. Impaired mimicry response to angry faces following severe traumatic brain injury. *Journal of Clinical and Experimental Neuropsychology* 33 (1): 17–29. https://doi.org/10.1080/13803391003761967.

McDonald, S., A. Gowland, R. Randall, A. Fisher, K. Osborne-Crowley, and C. Honan. 2014. Cognitive factors underpinning poor expressive communication skills after traumatic brain injury: Theory of mind or executive function? *Neuropsychology* 28 (5): 1–11. https://doi.org/10.1037/neu0000089.

Milders, M., M. Ietswaart, J.R. Crawford, and D. Currie. 2006. Impairments in theory of mind shortly after traumatic brain injury and at 1-year follow-up. *Neuropsychology* 20 (4): 400–408. https://doi.org/10.1037/0894-4105.20.4.400.

Miyake, A., N.P. Friedman, M.J. Emerson, A.H. Witzki, A. Howerter, and T.D. Wager. 2000. The unity and diversity of executive functions and their contributions to complex "Frontal Lobe" tasks: A latent variable analysis. *Cognitive Psychology* 41 (1): 49–100. https://doi.org/10.1006/cogp.1999.0734.

Molapour, T., C.C. Hagan, B. Silston, H. Wu, M. Ramstead, K. Friston, and D. Mobbs. 2021. Seven computations of the social brain. *Social Cognitive and Affective Neuroscience* 16 (8): 745–760. https://doi.org/10.1093/scan/nsab024.

Molenberghs, P., R. Cunnington, and J.B. Mattingley. 2012. Brain regions with mirror properties: A meta-analysis of 125 human fMRI studies. *Neuroscience and Biobehavioral Reviews* 36 (1): 341–349. https://doi.org/10.1016/j.neubiorev.2011.07.004.

Monticelli, M., P. Zeppa, M. Mammi, F. Penner, A. Melcarne, F. Zenga, and D. Garbossa. 2021. Where we mentalize: Main cortical areas involved in mentalization. *Frontiers in Neurology* 12: 712532. Frontiers Media S.A. https://doi.org/10.3389/fneur.2021.712532.

Moretto, G., E. Làdavas, F. Mattioli, and G. di Pellegrino. 2010. A psychophysiological investigation of moral judgment after ventromedial prefrontal damage. *Journal of Cognitive Neuroscience* 22 (8): 1888–1899. https://doi.org/10.1162/jocn.2009.21367.

Morin, A. 2006. Levels of consciousness and self-awareness: A comparison and integration of various neurocognitive views. *Consciousness and Cognition* 15 (2): 358–371. https://doi.org/10.1016/j.concog.2005.09.006.

Muller, F., A. Simion, E. Reviriego, C. Galera, J.M. Mazaux, M. Barat, and P.A. Joseph. 2010. Exploring theory of mind after severe traumatic brain injury. *Cortex* 46 (9): 1088–1099. https://doi.org/10.1016/j.cortex.2009.08.014.

Murphy, J.M., J.M. Bennett, X. de la Piedad Garcia, and M.L. Willis. 2022. Emotion recognition and traumatic brain injury: A systematic review and meta-analysis. *Neuropsychology Review* 32 (3): 520–536). Springer. https://doi.org/10.1007/s11065-021-09510-7.

Nakase-Richardson, R., A. Sepehri, M. Sherer, S.A. Yablon, C. Evans, and T. Mani. 2009. Classification schema of posttraumatic amnesia duration-based injury severity relative to 1-year outcome: Analysis of individuals with moderate and severe traumatic brain injury. *Archives of Physical Medicine and Rehabilitation* 90 (1): 17–19. https://doi.org/10.1016/j.apmr.2008.06.030.

Nakase-Richardson, R., M. Sherer, R.T. Seel, T. Hart, R. Hanks, J.C. Arango-Lasprilla, S.A. Yablon, A.M. Sander, S.D. Barnett, W.C. Walker, and F. Hammond. 2011. Utility of post-traumatic amnesia in predicting 1-year productivity following traumatic brain injury: Comparison of the Russell and Mississippi PTA classification intervals. *Journal of Neurology, Neurosurgery & Psychiatry* 82 (5): 494–499. https://doi.org/10.1136/jnnp.2010.222489.

Neumann, D., B.C. McDonald, J. West, M.A. Keiski, and Y. Wang. 2016. Neurobiological mechanisms associated with facial affect recognition deficits after traumatic brain injury. *Brain Imaging and Behavior* 10 (2): 569–580. https://doi.org/10.1007/s11682-015-9415-3.

Nijsse, B., J.M. Spikman, J.M.A. Visser-Meily, P.L.M. de Kort, and C.M. van Heugten. 2019. Social cognition impairments are associated with behavioral changes in the long term after stroke. *PLoS One* 14 (3): e0213725. https://doi.org/10.1371/journal.pone.0213725.

Osborne, C.L. 1997. *Over my head: A doctor's own story of head injury from the inside looking out*. Kansas City: Andrews McMeel.

Osborne-Crowley, K., E. Wilson, F. de Blasio, T. Wearne, J. Rushby, and S. McDonald. 2019. Subjective emotional experience and physiological responsivity to posed emotions in people with traumatic brain injury. *Neuropsychology* 33 (8): 1151–1162. https://doi.org/10.1037/neu0000580.

Otsuka, Y., M. Shizawa, A. Sato, and S. Itakura. 2021. The role of executive functions in older adults' affective theory of mind. *Archives of Gerontology and Geriatrics* 97: 104513. https://doi.org/10.1016/j.archger.2021.104513.

Palomero-Gallagher, N., and K. Amunts. 2022. A short review on emotion processing: A lateralized network of neuronal networks. *Brain Structure and Function* 227 (2): 673–684). Springer Science and Business Media Deutschland GmbH. https://doi.org/10.1007/s00429-021-02331-7.

Parker, R.S., and A. Rosenblum. 1996. IQ loss and emotional dysfunctions after mild head injury incurred in a motor vehicle accident. *Journal of Clinical Psychology* 52 (1): 32–43. https://doi.org/10.1002/(SICI)1097-4679(199601)52:1<32:AID-JCLP5>3.0.CO;2-Y.

Pessoa, L. 2017. A network model of the emotional brain. *Trends in Cognitive Sciences* 21 (5): 357–371). Elsevier Ltd. https://doi.org/10.1016/j.tics.2017.03.002.

Pluta, A., N. Gawron, M. Sobańska, A.D. Wójcik, and E. Lojek. 2017. The nature of the relationship between neurocognition and theory of mind impairments in stroke patients. *Neuropsychology* 31 (6): 666–681. https://doi.org/10.1037/neu0000379.

Ratcliff, G., A. Colantonio, M. Escobar, S. Chase, and L. Vernich. 2005. Long-term survival following traumatic brain injury. *Disability and Rehabilitation* 27 (6): 305–314. https://doi.org/10.1080/09638280400018338.

Reilly, P. 2007. The impact of neurotrauma on society: An international perspective. *Progress in Brain Research* 161: 3–9). Elsevier. https://doi.org/10.1016/S0079-6123(06)61001-7.

Rizzolatti, G., and L. Craighero. 2004. The mirror-neuron system. *Annual Review of Neuroscience* 27: 169–192. https://doi.org/10.1146/annurev.neuro.27.070203.144230.

Robertson, K., and M. Schmitter-Edgecombe. 2015. Self-awareness and traumatic brain injury outcome. *Brain Injury* 29 (7–8): 848–858. https://doi.org/10.3109/02699052.2015.1005135.

Rodríguez-Rajo, P., D. Leno Colorado, A. Enseñat-Cantallops, and A. García-Molina. 2018. Rehabilitation of social cognition impairment after traumatic brain injury: A systematic review. *Neurología.* 37 (9): 767–780. https://doi.org/10.1016/j.nrleng.2018.07.013.

Roebuck-Spencer, T., and A. Cernich. 2014. Epidemiology and societal impact of traumatic brain injury. In *Handbook on the neuropsychology of traumatic brain injury*, ed. M. Sherer and A. Sander, 3–23. Springer.

Romero, J.R., J. Morris, and A. Pikula. 2008. Review: Stroke prevention: Modifying risk factors. *Therapeutic Advances in Cardiovascular Disease* 2 (4): 287–303. https://doi.org/10.1177/1753944708093847.

Rosner, B., and L.J. Cronbach. 1960. Essentials of psychological testing. *The American Journal of Psychology* 73 (2): 323. https://doi.org/10.2307/1419921.

Rowe, A.D., P.R. Bullock, C.E. Polkey, and R.G. Morris. 2001. "Theory of mind" impairments and their relationship to executive functioning following frontal lobe excisions. *Brain* 124: 600–616.

Rubiano, A.M., N. Carney, R. Chesnut, and J.C. Puyana. 2015. Global neurotrauma research challenges and opportunities. *Nature* 527 (7578): S193–S197. https://doi.org/10.1038/nature16035.

Russell, W.R. 1932. Cerebral involvement in head injury. A study on the examination of two hundred cases. *Brain* 55: 549–603.

Russell, W.R., and A. Smith. 1961. Post-traumatic amnesia in closed head injury. *Archives of Neurology* 5: 4–17.

Salas, C.E., M. Casassus, L. Rowlands, S. Pimm, and D.A.J. Flanagan. 2018. "Relating through sameness": A qualitative study of friendship and social isolation in chronic traumatic brain injury. *Neuropsychological Rehabilitation* 28 (7): 1161–1178. https://doi.org/10.1080/09602011.2016.1247730.

Salas, C., M. Casassus, L. Rowlands, and S. Pimm. 2021a. Developing a model of long-term social rehabilitation after traumatic brain injury: The case of the head forward center. *Disability and Rehabilitation* 43 (23): 3405–3416. https://doi.org/10.1080/09638288.2020.1741697.

Salas, C.E., D. Rojas-Líbano, O. Castro, R. Cruces, J. Evans, D. Radovic, C. Arévalo-Romero, J. Torres, and Á. Aliaga. 2021b. Social isolation after acquired brain injury: Exploring the relationship between network size, functional support, loneliness and mental health. *Neuropsychological Rehabilitation* 32 (9): 2294–2318. https://doi.org/10.1080/09602011.2021.1939062.

Samson, D., I.A. Apperly, C. Chiavarino, and G.W. Humphreys. 2004. Left temporoparietal junction is necessary for representing someone else's belief. *Nature Neuroscience* 7 (5): 499–500. https://doi.org/10.1038/nn1223.

Samson, D., I.A. Apperly, and G.W. Humphreys. 2007. Error analyses reveal contrasting deficits in "theory of mind": Neuropsychological evidence from a 3-option false belief task. *Neuropsychologia* 45 (11): 2561–2569. https://doi.org/10.1016/j.neuropsychologia.2007.03.013.

Santoro, J., and M. Spiers. 1994. Social cognitive factors in brain injury-associated personality change. *Brain Injury* 8 (3): 265–276. https://doi.org/10.3109/02699059409150978.

Saunders, J.C., S. McDonald, and R. Richardson. 2006. Loss of emotional experience after traumatic brain injury: Findings with the startle probe procedure. *Neuropsychology* 20 (2): 224–231. https://doi.org/10.1037/0894-4105.20.2.224.

Saxe, R., and L.J. Powell. 2006. It's the thought that counts specific brain regions for one component of theory of mind. *Psychological Science* 17 (8): 692–699. http://www.fil.ion. https://doi.org/10.1111/j.1467-9280.2006.01768.x.

Schlaffke, L., S. Lissek, M. Lenz, G. Juckel, T. Schultz, M. Tegenthoff, T. Schmidt-Wilcke, and M. Brune. 2015. Shared and nonshared neural networks of cognitive and affective theory-of-mind: A neuroimaging study using cartoon picture stories. *Human Brain Mapping* 36 (1): 29–39. https://doi.org/10.1002/HBM.22610.

Schmitz, T.W., H.A. Rowley, T.N. Kawahara, and S.C. Johnson. 2006. Neural correlates of self-evaluative accuracy after traumatic brain injury. *Neuropsychologia* 44 (5): 762–773. https://doi.org/10.1016/j.neuropsychologia.2005.07.012.

Schroeter, M.L., B. Ettrich, M. Menz, and S. Zysset. 2010. Traumatic brain injury affects the frontomedian cortex-An event-related fMRI study on evaluative judgments. *Neuropsychologia* 48 (1): 185–193. https://doi.org/10.1016/j.neuropsychologia.2009.09.004.

Schurz, M., J. Radua, M. Aichhorn, F. Richlan, and J. Perner. 2014. Fractionating theory of mind: A meta-analysis of functional brain imaging studies. *Neuroscience and Biobehavioral Reviews* 42: 9–34). Elsevier Ltd. https://doi.org/10.1016/j.neubiorev.2014.01.009.

Sebastian, C.L., N.M.G. Fontaine, G. Bird, S.J. Blakemore, S.A. De Brito, E.J.P. Mccrory, and E. Viding. 2012. Neural processing associated with cognitive and affective Theory of Mind in adolescents and adults. *Social Cognitive and Affective Neuroscience* 7 (1): 53–63. https://doi.org/10.1093/SCAN/NSR023.

Selassie, A.W., D.A. Wilson, E.E. Pickelsimer, D.C. Voronca, N.R. Williams, and J.C. Edwards. 2013. Incidence of sport-related traumatic brain injury and risk factors of severity: A population-based epidemiologic study. *Annals of Epidemiology* 23 (12): 750–756. https://doi.org/10.1016/j.annepidem.2013.07.022.

Shamay-Tsoory, S.G., and J. Aharon-Peretz. 2007. Dissociable prefrontal networks for cognitive and affective theory of mind: A lesion study. *Neuropsychologia* 45 (13): 3054–3067. https://doi.org/10.1016/j.neuropsychologia.2007.05.021.

Shamay-Tsoory, S.G., R. Tomer, D. Goldsher, B.D. Berger, and J. Aharon-Peretz. 2004. Impairment in cognitive and affective empathy in patients with brain lesions: Anatomical and cognitive correlates. *Journal of Clinical and Experimental Neuropsychology* 26 (8): 1113–1127. https://doi.org/10.1080/13803390490515531.

Shamay-Tsoory, S.G., R. Tomer, B.D. Berger, D. Goldsher, and J. Aharon-Peretz. 2005. Impaired "Affective Theory of Mind" is associated with right ventromedial prefrontal damage. *Cognitive and Behavioral Neurology* 18 (1): 55–67. https://doi.org/10.1097/01.wnn.0000152228.90129.99.

Sherer, M., A.M. Sander, T.G. Nick, W.M. High, J.F. Malec, and M. Rosenthal. 2002. Early cognitive status and productivity outcome after traumatic brain injury: Findings from the TBI model systems. *Archives of Physical Medicine and Rehabilitation* 83 (2): 183–192. https://doi.org/10.1053/apmr.2002.28802.

Sherer, M., A.N. Clark, A.M. Sander, M.A. Struchen, J. Bogaards, L. Leon-Novelo, and E. Ngan. 2022. Relationships of self-awareness and facial affect recognition to social communication ability in persons with traumatic brain injury. *Neuropsychological Rehabilitation* 32 (8): 2013–2028. https://doi.org/10.1080/09602011.2022.2084118.

Siegal, M., J. Carrington, and M. Radel. 1996. Theory of mind and pragmatic understanding following right hemisphere damage. *Brain and Language* 53 (1): 40–50. https://doi.org/10.1006/brln.1996.0035.

Simpson, G., and R. Tate. 2007. Suicidality in people surviving a traumatic brain injury: Prevalence, risk factors and implications for clinical management. *Brain Injury* 21 (13–14): 1335–1351. https://doi.org/10.1080/02699050701785542.

Smajlovic, D. 2015. Strokes in young adults: Epidemiology and prevention. *Vascular Health and Risk Management* 11: 157–164. https://doi.org/10.2147/VHRM.S53203.

Spikman, J.M., M.V. Milders, A.C. Visser-Keizer, H.J. Westerhof-Evers, M. Herben-Dekker, and J. van der Naalt. 2013. Deficits in facial emotion recognition indicate behavioral changes and impaired self-awareness after moderate to severe traumatic brain injury. *PLoS ONE* 8 (6): e65581. https://doi.org/10.1371/journal.pone.0065581.

Spreng, R.N., M.C. McKinnon, R.A. Mar, and B. Levine. 2009. The Toronto empathy questionnaire: Scale development and initial validation of a factor-analytic solution to multiple empathy measures. *Journal of Personality Assessment* 91 (1): 62–71. https://doi.org/10.1080/00223890802484381.

Squire, L.R. 2010. Encyclopedia of neuroscience. *Encyclopedia of Neuroscience.* https://doi.org/10.1016/C2009-1-03742-3.

Stone, V.E. 2007. An evolutionary perspective on domain specificity in social intelligence. In *Social neuroscience: Integrating biological and psychological explanations of social behavior*, ed. E. Harmon-Jones and P. Winkielman, 316–349. The Guilford Press.

Stone, V.E., S. Baron-Cohen, and R.T. Knight. 1998. Frontal lobe contributions to theory of mind. *Journal of Cognitive Neuroscience* 10 (5): 640–656. https://doi.org/10.1162/089892998562942.

Strand, M. 2004. *Meditations on brain injury*. Zotolla Publishing.

Stuss, D.T., and M.P. Alexander. 2000. Executive functions and the frontal lobes: A conceptual view. *Psychological Research* 63 (3–4): 289–298. https://doi.org/10.1007/S004269900007.

Tager-Flusberg, H., K. Sullivan, and J. Boshart. 1997. Executive functions and performance on false belief tasks. *Developmental Neuropsychology* 13 (4): 487–493. https://doi.org/10.1080/87565649709540689.

Tamir, D.I., and M.A. Thornton. 2018. Modeling the predictive social mind. *Trends in Cognitive Sciences* 22 (3): 201–212). Elsevier Ltd. https://doi.org/10.1016/j.tics.2017.12.005.

Teasdale, G., and B. Jennett. 1974. Assessment of coma and impaired consciousness: A practical scale. *The Lancet* 304 (7872): 81–84.

Thomson, J.J. 1985. The trolley problem. *The Yale Law Journal* 94: 1395–1985.

Thrift, A.G., T. Thayabaranathan, G. Howard, V.J. Howard, P.M. Rothwell, V.L. Feigin, B. Norrving, G.A. Donnan, and D.A. Cadilhac. 2017. Global stroke statistics. *International Journal of Stroke* 12 (1): 13–32. https://doi.org/10.1177/1747493016676285.

Togher, L., J. Douglas, L.S. Turkstra, P. Welch-West, S. Janzen, A. Harnett, M. Kennedy, A. Kua, E. Patsakos, J. Ponsford, R. Teasell, M.T. Bayley, and C. Wiseman-Hakes. 2023. INCOG 2.0 Guidelines for cognitive rehabilitation following traumatic brain injury, Part IV: Cognitive-communication and social cognition disorders. *Journal of Head Trauma Rehabilitation* 38 (1): 65–82. https://doi.org/10.1097/HTR.0000000000000835.

Toglia, J., and U. Kirk. 2000. Understanding awareness deficits following brain injury. *NeuroRehabilitation* 15 (1): 57–70. https://doi.org/10.3233/NRE-2000-15104.

van Schaik, J.E., and S. Hunnius. 2016. Little chameleons: The development of social mimicry during early childhood. *Journal of Experimental Child Psychology* 147: 71–81. https://doi.org/10.1016/j.jecp.2016.03.003.

van Velzen, J.M., C.A.M. van Bennekom, M.J.A. Edelaar, J.K. Sluiter, and M.H.W. Frings-Dresen. 2009. How many people return to work after acquired brain injury? A systematic review. *Brain Injury* 23 (6): 473–488. https://doi.org/10.1080/02699050902970737.

Vass, E., V. Simon, Z. Fekete, L. Lencse, M. Ecseri, B. Kis, and L. Simon. 2021. A novel virtual reality-based theory of mind intervention for outpatients with schizophrenia: A proof-of-concept pilot study. *Clinical Psychology & Psychotherapy* 28 (3): 727–738. https://doi.org/10.1002/CPP.2519.

Villalobos, D., A. Bilbao, F. López-Muñoz, and J. Pacios. 2019. Improving self-awareness after acquired brain injury leads to enhancements in patients' daily living functionality. *Brain Impairment* 20: 268–275. https://doi.org/10.1017/BrImp.2019.10.

Villalobos, D., J.M. Caperos, Á. Bilbao, F. López-Muñoz, and J. Pacios. 2020. Cognitive predictors of self-awareness in patients with acquired brain injury along neuropsychological rehabilitation. *Neuropsychological Rehabilitation* 4 (23): 1–19. https://doi.org/10.1080/09602011.2020.1751663.

Wadhawan, A., J.W. Stiller, E. Potocki, O. Okusaga, A. Dagdag, C.A. Lowry, M.E. Benros, and T.T. Postolache. 2019. Traumatic brain injury and suicidal behavior: A review. *Journal of Alzheimer's Disease* 68 (4): 1339–1370. https://doi.org/10.3233/JAD-181055.

Wafa, H.A., C.D.A. Wolfe, E. Emmett, G.A. Roth, C.O. Johnson, and Y. Wang. 2020. Burden of stroke in Europe. *Stroke* 51 (8): 2418–2427. https://doi.org/10.1161/STROKEAHA.120.029606.

Wallis, K., M. Kelly, S.E. McRae, S. McDonald, and L.E. Campbell. 2021. Domains and measures of social cognition in acquired brain injury: A scoping review. *Neuropsychological Rehabilitation.* 32 (9): 2429–2463. https://doi.org/10.1080/09602011.2021.1933087.

Wells, R., J. Dywan, and J. Dumas. 2005. Life satisfaction and distress in family caregivers as related to specific behavioral changes after traumatic brain injury. *Brain Injury* 19 (13): 1105–1115. https://doi.org/10.1080/02699050500150062.

Westerhof-Evers, H.J., A.C. Visser-Keizer, L. Fasotti, M.C. Schönherr, M. Vink, J. Van Der Naalt, and J.M. Spikman. 2017. Effectiveness of a treatment for impairments in social cognition and emotion regulation (T-ScEmo) after traumatic brain injury: A randomized controlled trial. *Journal of Head Trauma Rehabilitation* 32 (5): 296–307. https://doi.org/10.1097/HTR.0000000000000332.

Wilkins, E., K. Wickramasinghe, P. Bhatnagar, J. Leal, R. Luengo-Fernandez, R. Burns, M. Rayner, and N. Townsend. 2017. *European cardiovascular disease statistics 2017*. Brussels: European Heart Network.

Wilkos, E., T.J.B. Brown, K. Slawinska, and K. Akucharska. 2015. Social cognitive and neurocognitive deficits in inpatients with unilateral thalamic lesions – Pilot study. *Neuropsychiatric Disease and Treatment* 11: 1031–1038. https://doi.org/10.2147/NDT.S78037.

Williams, C., and Wood, R.L. 2010. Alexithymia and emotional empathy following traumatic brain injury. *Journal of Clinical and Experimental Neuropsychology* 32 (3): 259–267. https://doi.org/10.1080/13803390902976940.

Williams, C., and Wood, R.L 2012. Affective modulation of the startle reflex following traumatic brain injury. *Journal of Clinical and Experimental Neuropsychology* 34 (9): 948–961. https://doi.org/10.1080/13803395.2012.703641.

Wilson, J., C. Hogan, S. Wang, G. Andrews, and D.H.K. Shum. 2021. Relations between executive functions, theory of mind, and functional outcomes in middle childhood. *Developmental Neuropsychology* 46 (7): 518–536. https://doi.org/10.1080/87565641.2021.1988086.

Winner, E., H. Brownell, F. Happé, A. Blum, and D. Pincus. 1998. Distinguishing lies from jokes: Theory of mind deficits and discourse interpretation in right hemisphere brain-damaged patients. *Brain and Language* 62 (1): 89–106. https://doi.org/10.1006/brln.1997.1889.

Wood, R.L.L., and C. Williams. 2008. Inability to empathize following traumatic brain injury. *Journal of the International Neuropsychological Society* 14 (2): 289–296. https://doi.org/10.1017/S1355617708080326.

Yates, P.J. 2003. Psychological adjustment, social enablement and community integration following acquired brain injury. *Neuropsychological Rehabilitation* 13 (1–2): 291–306. https://doi.org/10.1080/09602010244000408.

Yeates, G., M. Rowberry, S. Dunne, M. Goshawk, M. Mahadevan, R. Tyerman, M. Salter, M. Hillier, A. Berry, and A. Tyerman. 2016. Social cognition and executive functioning predictors of supervisors' appraisal of interpersonal behavior in the workplace following acquired brain injury. *NeuroRehabilitation* 38: 299–310.

Yuvaraj, R., M. Murugappan, M.I. Norlinah, K. Sundaraj, and M. Khairiyah. 2013. Review of emotion recognition in stroke patients. *Dementia and Geriatric Cognitive Disorders* 36 (3–4): 179–196. https://doi.org/10.1159/000353440.

Zhang, S., W. Zhang, and G. Zhou. 2019. Extended risk factors for stroke prevention. *Journal of the National Medical Association* 111 (4): 447–456. https://doi.org/10.1016/j.jnma.2019.02.004.

Chapter 8
Theory of Mind in Children Who Are Deaf: The Importance of Early Language and Conversational Access

Kimberly Peters and David B. Pisoni

Abstract The acquisition of a mature theory of mind (ToM) within a typical timeframe depends on the ability to converse early, easily, and proficiently about mental states with other skilled language users. Research shows that children who are deaf or hard of hearing who have parents with typical hearing are more likely to experience delays in spoken or sign language, and subsequent delays in ToM; while children whose hearing and communication status matches that of their parents are less likely to be language delayed and less likely to have deficits in ToM. Exposure to language per se seems insufficient to ensure timely ToM development. Rather, conversational access to, understanding of, and practice expressing mental, emotional, and cognitive terms as well as specific syntactic structures that describe the beliefs of others from an early age are important, regardless of communication mode. If children who are deaf or hard of hearing are identified late, receive hearing technology and other language interventions late, and do not develop strong early language and conversational skills, ToM is likely to be delayed in preschool and early elementary school. By contrast, children who are deaf and who are identified early, treated early, and acquire conversational competence within a typical time frame should demonstrate ToM development that more closely approximates ToM of their peers who have typical hearing.

Keywords Theory of mind · False belief · Social cognition · Language · Deaf and hard of hearing · Intersubjectivity · Cochlear implants

K. Peters (✉)
Western Washington University, Bellingham, WA, USA
e-mail: Petersk3@wwu.edu

D. B. Pisoni
Indiana University, Bloomington, IN, USA
e-mail: pisoni@indiana.edu

T. Lopez-Soto et al. (eds.), *The Theory of Mind Under Scrutiny*, Logic, Argumentation & Reasoning 34, https://doi.org/10.1007/978-3-031-46742-4_8

8.1 Introduction

Theory of mind (ToM) reflects a person's understanding of how mental and emotional states affect behavior (for reviews, see Wellman 2011, 2014). By 5 years of age, children with typical development understand that a person's thoughts and beliefs influence their actions (Apperly 2010; Bretherton and Beeghly 1982; Custer 1996; Gopnik et al. 1994; Perner 1991; Wellman 2002); this is referred to as false belief (FB) understanding and is one of the hallmarks of a mature ToM. False belief understanding is routinely mastered by the end of preschool in neurotypical children and has been studied extensively in both typically developing, and clinical populations via various experimental tasks. A developmental progression of ToM that extends beyond FB understanding has also been established in children who are typically developing and those who are neurodiverse (Wellman and Liu 2004) although historically most studies have interpreted understanding of FB as evidence of representational thinking, the signpost of a mature ToM.

Research involving children who are neurotypical suggests that language ability influences ToM acquisition (Astington and Baird 2005; Astington and Jenkins 1999; Milligan et al. 2007). Understanding of specific syntactic forms (de Villiers 1995; de Villiers and de Villiers 2000; de Villiers and Pyers 2002), use and understanding of mental state vocabulary (Grazzani and Ornaghi 2012; Peterson and Slaughter 2006; Ruffman et al. 2002); and understanding of intentional behavior in infancy (Wellman et al. 2008) have been found to be correlated with performance on ToM tasks.

In addition to language ability, early sociolinguistic environment and conversational access appear to play a role in the development of social cognition (see de Rosnay and Hughes 2006), particularly play and joint attention (Charman et al. 2000). Children's mental state language usage can be predicted from their mothers' tendency to use mental state language (Taumoepeau and Ruffman 2006, 2008), and care providers' talk about feelings, emotions, and thinking is correlated with performance on FB tasks (Adrian et al. 2005; Adrián et al. 2007; Meins et al. 2002; Labounty et al. 2008; Peterson and Slaughter 2003; Slaughter et al. 2007). Likewise, exposure to certain types of discourse and conversation predict theory of mind task performance (de Rosnay and Hughes 2006; Harris et al. 2005; Slaughter and Peterson 2012; Symons et al. 2005, 2006; Welch-Ross 1997; Ziv et al. 2013b). Children with more siblings have been found to acquire FB understanding earlier (Perner et al. 1994); and research shows a significant, positive correlation between ToM and number of months in a preschool setting (Altun 2019), and a positive relationship between social competence and peer play opportunities (Newton and Jenvey 2010).

Theory of mind is associated with a range of academic and social skills including reading comprehension (Atkinson et al. 2017; Dore et al. 2018; Figueroa et al. 2020; Holmer et al. 2016), responding appropriately to teacher criticism (Lecce et al. 2011) and using and understanding deceit and persuasion (Ding et al. 2015; Peterson et al. 2018; Slaughter et al. 2013). Children with better ToM demonstrate greater social acceptance and popularity (Peterson et al. 2016a; Slaughter et al. 2015), more pro-social behaviors (Eggum et al. 2011), better relationships with teachers and peers (Peterson et al. 2016b; Watson et al. 1999), more empathy (Peterson 2016), and less friendlessness and loneliness (Fink et al. 2014; Devine and Hughes 2013).

To understand both typical mechanisms of ToM development, as well as the specific contribution of language to ToM, researchers have been interested in the social cognitive development and FB understanding of children who are deaf and hard of hearing (DHH) since the 1990s. Children who are deaf with parents who are deaf (CDDP) are exposed to and acquire a natural sign language from birth; this provides a unique opportunity to study the effects of visual language on the developmental of mental representation (Courtin 2000; Courtin and Melot 2005) as well as the impact of auditory deprivation on ToM development. Children who are deaf with parents who have typical hearing (CDHP) (most children who are born deaf) provide an opportunity to study the effects of reduced auditory access and delayed language access on social cognitive development. Often CDHP are not exposed to a complete spoken or signed language until they are older, due to timing of identification of hearing loss, and lack of an easily accessible, shared language in infancy. For most of these children, but especially for those who are diagnosed later, language deficits can exist regardless of whether the child is learning Sign Language (Strong and Prinz 1997) or spoken language (P. de Villiers 2003; Geers et al. 1984). Late identification and delayed treatment of hearing loss has been found to result in long-term language learning delays (Mayberry et al. 2002; Mayberry and Lock 2003). Such delays create barriers to the rich bidirectional communication experiences that are believed to contribute to ToM acquisition. It is not uncommon for language delays to persist in later identified CDHP past the age at which many children with typical hearing have mastered early ToM skills (Peterson et al. 2005). Even in studies of early identified CDHP, language skills (when measured) are sometimes greater than 1.0 or even 1.5 standard deviations below average compared to same-age children with typical hearing. These children are often otherwise neurotypical, so cognitive and/or neurological differences can be ruled out as contributing to ToM gaps.

This paper reviews the literature on ToM acquisition in children who are deaf who have deaf parents (CDDP) and children who are DHH who have parents with typical hearing (CDHP). The preponderance of the evidence in the literature suggests that early access to high quality language models and conversation about the mind and mental states are important for language development and subsequent theory of mind acquisition in children who are DHH (Courtin 2000; Courtin and Melot 2005; Falkman et al. 2007; Peters et al. 2021; Remmel and Peters 2009; Pluta et al. 2021; Sundqvist et al. 2014; Yu et al. 2021; Ziv et al. 2013b).

8.2 Theory of Mind Assessment

Before describing the ToM research involving children who are DHH, it helps to have a basic understanding of the measures that have been used frequently in the assessment of ToM in this population. Often described as the 'litmus tests' of theory of mind" (Courtin 2000, p. 5) first order false belief is commonly measured using one or two experimental tasks—the "unexpected contents" FB test and the "change in location" FB test. In the former, the child is shown a container with obvious contents, for example, a band aid box, a crayon box, or a milk carton. They are asked what they *think* it is in the container and are then shown what is *really* in the container (some other object that does not make sense, a plastic pig, for example). After the container is closed, children are asked what a third party who has *not* looked inside the container will think is in it. "Passing" this task entails holding two mental representations simultaneously in mind —the child's own true belief (there is a pig in the band aid box) and the false belief, or mental *mis*representation, of the naïve third party (Mary will *think* there are band aids in the band aid box). The change in location FB task is based on the same premise, but the set-up is slightly different. The classic change in location task is often referred to in the literature as the "Sally Anne test," developed by Wimmer and Perner (1983). In this task, "Sally" takes a marble and hides it in her basket. She then leaves the room and goes for a walk. While she is away, Anne takes the marble out of Sally's basket and puts it in her own box. Sally comes back and the child is asked, "Where will Sally look for her marble?" To pass this task the child must answer that Sally will look for the marble in the basket, where she originally put it and believes it to be. Again, the child must be able to separate their own true belief from the false belief of the naive character—something that typically developing preschoolers can do by around 4–5 years of age (Miller 2009; Wellman et al. 2001).

Although false belief is often described as a landmark accomplishment in social cognitive development in children, it is well documented that theory of mind is multi-dimensional and develops over time, beginning in infancy and progressing through at least the teen years. Wellman and Liu (2004) developed a set of experimental tasks to measure early emerging ToM concepts and found that performance on these tasks follows a predictable sequence in typically developing preschoolers, and in children who are deaf (Peterson et al. 2005). Peterson and Wellman (2009) later modified the "Theory of Mind" Scale to include social pretense. In this task, an adult and the child pretend to paint a red toy car blue. After they are finished, the adult says, "Here comes X, he hasn't seen us pretending. What color will X say the car is?" Wellman and Peterson (2013) also developed a 6-item extended ToM scale that includes a measure of sarcasm understanding—a person can say something but mean something else—to enable research with older children, who tend to perform at ceiling on first order FB measures. The scales developed by Wellman and colleagues have been longitudinally validated and used extensively, albeit not exclusively, in ToM research involving children who are deaf, as well as children who have autism spectrum disorders, Down syndrome, and developmental language disability.

Table 8.1 describes the experimental ToM tasks commonly used in research with children who have normal hearing, and those who are DHH; as well as performance expectations for children who are typically developing. For a systematic review of ToM measures used with young children, see Beaudoin et al. (2020).

8.3 Children Who are Deaf with Deaf Parents (CDDP)

It is estimated that between 5% and 10% of children who are born DHH have parents who are also deaf and who use a natural sign language as their means of communicating (Mitchell and Karchmer 2004). One of the first studies of ToM in children who were deaf sign language users included *only* children whose parents had normal hearing and suggested that children who were deaf performed no differently than children with autism on FB tasks (Peterson and Siegal 1995); children in this study who were deaf passed a standard FB task at an average of 10-years-old, significantly later than children with typical hearing. Shortly thereafter, Courtin (2000) conducted the first study of ToM in children who were deaf with parents who were also deaf; these children had acquired French sign language naturally, in infancy, through interacting with their parents. Courtin assessed ToM understanding in four groups of 5- to 8-year-old children: (1) 37 children who were deaf with deaf parents, and who used French Sign Language from birth; (2) 54 children who were deaf with hearing parents, who learned French Sign Language later; (3) 45 children who were deaf with hearing parents who used spoken French; and (4) 39 children with typical hearing with typically hearing parents (Courtin 2000).

In Courtin's study, children were given two unexpected contents tasks and one change in location task. Children who were deaf with deaf parents outperformed all other groups in this study (including children with typical hearing) on the FB measures, suggesting that children who used a signed language fluently developed FB understanding earlier than typically developing children. In this study, children who were deaf and used spoken language performed the poorest on FB tasks; even at 8 years of age, many of these children did not perform as well as typically hearing 4-year-olds. Children who were deaf with hearing parents who did not have access to fluent sign language until they entered school performed similarly to younger children with typical hearing, suggesting that perhaps using sign language rather than spoken language could benefit social cognitive development in CDHP. "Sign language usage by deaf children promotes their understanding of the representational mind due to the need for visual (linguistic) perspective taking when communication occurs through sign language" (Courtin and Melot 2005, p.18).

Subsequent research conducted to replicate these findings did not produce the same advantage for sign language users, however. In a study of 88 5- to 7-year-old children who were deaf, CDDP and children with typical hearing performed similarly on FB tasks (Courtin and Melot 2005); there was no advantage for the sign language users. Children who were deaf with parents who had typical hearing demonstrated theory of mind delays but did not differ in their performance based on communication mode (spoke vs. sign language), suggesting that, for children

Table 8.1 Common ToM tasks with performance expectations by age, and references

Task Description	Performance Expectation	References
Diverse desires: Tests the child's understanding that different people may have different desires. The examiner asks the which of two foods (carrot or cookie) he/she would want for a snack. Then the examiner tells the child that a character (Mr. Jones) prefers the other food (e.g., carrot if the child prefers cookie). The examiner asks the child which food Mr. Jones will pick for his snack. The child is scored as correct if he/she chooses the food that Mr. Jones wants, rather than the food that the child wants.	2-year-olds can predict actions based on simple desires. 14-month-old children choose based on their desire (ego-centric) while 18-month-old children choose based on others' desires.	Wellman and Woolley (1990) and Repacholi and Gopnik (1997)
Diverse beliefs: Tests the child's understanding that different people can have different beliefs. The examiner tells the child that a character (Linda) wants to find her cat and asks the child to guess in which of two locations (bushes or garage) the cat is hiding. Then the examiner tells the child that Linda thinks the cat is in the other location (e.g., bushes if the child thinks garage). The examiner asks the child where Linda will look for the cat. The child is scored as correct if he/she chooses the location where Linda believes the cat is, rather than the location where the child believes the cat is (note: The true location of the cat is unknown).	3-year-old children understand that people can have different beliefs than their own, although they can't necessarily predict how someone will act based on a false belief.	Bartsch and Wellman (1989) and Wellman et al. (1996)
Knowledge access: Tests the child's understanding that perceptual access leads to knowledge. The examiner asks the child to guess what is inside an unmarked can. Then the examiner shows the child that the can contains a small toy dog. The examiner tells the child that a character (Polly) has never seen inside the can and asks if Polly knows what is inside. The child is scored as correct if he/she responds that Polly does not know, even though the child has seen inside and does know.	3- and 4- year-old children know that looking at something leads to knowledge. 3-year-olds understand that perception is a source of knowledge.	Pratt and Bryant (1990) and Pillow (1989)

(continued)

Table 8.1 (continued)

Task Description	Performance Expectation	References
Social pretense: Tests the child's ability to understand different mental states of people engaged or not engaged in pretend play. An examiner and the child pretend to paint a red toy car blue. After they are finished, they put away *all* evidence of painting materials. Then the examiner informs the child that another character (Peter) has not seen them pretending. The examiner asks the child what color Peter will say the car is. The child is scored as correct if he/she answers that Peter will say the car is red.	Between the ages of 3 and 4 years, children can assess others' thoughts about pretend happenings.	Hickling et al. (1997)
Appearance-reality: Tests the child's ability to hold two representations of the same object in mind while knowing which is pretend and which is real. The examiner shows the child a very realistic looking rock that is *actually* a sponge. Then the examiner asks the child what it is (the child usually says "a rock"). Then after the examiner shows the child that it is really a sponge, the examiner asks the child what he/she thought it was when he/she first saw it (memory question). The adult then asks the child what Peter will think it is (appearance question), and then "what is it really?" the child is scored as correct if he/she states that, "Peter will say it's a rock".	Many 3-yr-olds can make correct appearance–reality discriminations; this ability improves with age.	Flavell et al. (1983).
Unexpected contents first-order false belief: Tests the child's understanding that others can act on beliefs that are inaccurate. The examiner shows the child a container with obvious contents, for example, a band aid box. Then the examiner asks the child what he/she *thinks* it is in the band aid box (child says "band aids"). The examiner shows the child what is *really* in the band aid box (an object that does not make	False belief attribution is difficult for 3-year-olds but becomes accurate by 4- to 5-years-old.	Initially developed by Perner et al. (1987) and Wellman et al. (2001)

(continued)

Table 8.1 (continued)

Task Description	Performance Expectation	References
sense, a plastic pig, for example). Then the examiner asks what a third party (Sam) who has *not* looked inside the container will think is in it. The child is scored as correct if he/she recalls that Sam did not look in the container and that Sam will think there are band aids in the container.		
Change in location first-order false belief: Tests the child's understanding that others can act on beliefs that are inaccurate. The examiner tells the child a story about Sally and Anne, who are playing with a marble: Sally puts the marble in a basket when they are finished playing, and then leaves. While Sally is gone, Anne moves the marble from the basket to a box. The examiner asks the child where Sally will look for the marble when she returns. The child is scored as correct if he/she says that Sally will look in the basket for the marble.	False belief attribution is difficult for 3-year-olds but becomes accurate by 4- to 5-years-old.	Baron-Cohen et al. (1985)
Real-apparent emotion: Tests the child's understanding that people's facial expressions may not match how they feel inside. The examiner tells the child a story about a boy (Matt) who is being teased by some other children but does not want the other children to know that he is upset. The examiner shows the child drawings of a happy face, a sad face, and a neutral face and asks the child to indicate how Matt really feels and how Matt tries to look on his face. The child is scored as correct if he/she indicates that Matt feels more negative than he looks.	4-year-old children demonstrate emerging understanding and 6-year-olds demonstrate more consistent understanding of the difference between real and apparent emotion.	Harris et al. (1986)
Second-order false belief: Tests the child's awareness that people can have thoughts about others' thoughts and beliefs. The task is the same as the change in location	Children *begin* to show success at passing second order false belief tasks at age 6; by 7 to 9 years, most typically developing children are able to mentally represent	Miller (2009) and Perner and Wimmer (1985)

(continued)

Table 8.1 (continued)

Task Description	Performance Expectation	References
first order FB, except that Sally looks through a peep hole to see Anne move the marble. Then the examiner asks the child if Sally knows where the marble is *and* if Anne knows that Sally knows where the marble is. The child is scored as correct if he/she states that Sally knows where the marble is, and that Anne does not know that Sally knows. ("does Sally know where the marble is now? Does Anne know that Sally knows where the marble is?"). This involves comprehension of recursive language in addition to false belief.	second-order false belief. The grammar of expressing second-order false belief is complex, requiring recursion of complement phrases, which many 6- to 7-year-olds have not fully mastered.	
Sarcasm: tests the child's awareness that what a person says can be different than what they mean. The examiner tells the child a story about a girl and a boy going on a picnic: "It is the boy's idea. He says, 'it will be a lovely, sunny day'. But when they get the food out, big storm clouds come. It rains and all the food gets wet. The girl says, 'it's a lovely day for a picnic.'" The child is scored as correct if they mention sarcasm explicitly or allude in some other way (e.g., "joking," "doesn't mean it") to contrast between the literal meaning of the words "lovely day" and the speaker's intended meaning. Peterson et al. (2012), p. 474.	41% of typically developing children aged 7.5–11 years and 50% of typically developing children aged 6 years to 7.5 years passed the sarcasm task in the Peterson et al. (2012) study. 38% of typically developing children aged 6.6–9.7 years old passed the sarcasm story in the Happé (1994) study.	Happé (1994) and Peterson et al. (2012)

who are deaf who learn sign language later, the visible aspects of sign language do not appear to confer an advantage in understanding the mental representations, misrepresentations, or perspectives of others. Later studies that have included native sign language users have supported these findings—children who are deaf and who learn sign language from parents who are fluent sign language users do not demonstrate ToM delays (Edmondson 2006; Hao et al. 2010; Jackson 2001; Meristo and Hjelmquist 2009; Meristo and Strid 2020; Meristo et al. 2007; Schick et al. 2007;

Siegal and Peterson 2008; Peterson and Siegal 1999; Peterson et al. 2005; Woolfe et al. 2002, 2003). Children who are native users of *any* language (be it spoken or signed) outperform those who are not on first order theory of mind tasks—suggesting that access to fluent language input in infancy, and involvement in language-based dialogue, play an important role in the beginning stages of social cognitive learning.

While CDDP acquire early ToM skills in a timeframe comparable to that of their peers with typical hearing, CDDP do not appear to demonstrate the same trajectory for more advanced ToM skills, although very little research has been conducted with older CDDP. Second order FB tasks involve realizing it is possible to hold a false belief about someone *else's* false belief—a concept that is cognitively (and linguistically) more challenging. A study of Italian native sign language users showed that advanced ToM (second order ToM and irony) was delayed in CDDP compared to children with typical hearing (Giustolisi et al. 2018). In this study, the authors used the "birthday puppy" second order FB task (Sullivan et al. 1994), in which a mother deliberately misinforms her son about what she is going to give him for his birthday so she can surprise him later with a puppy. The child then discovers the true birthday present. Later, the mother is asked by the child's grandmother if the child knows what he is getting (this is referred to as second order ignorance) *and* what the child thinks he is getting (this is referred to as second order false belief). Whereas 80% of 9-year-old CDDP passed a first order FB task, only 20% of these children passed the second order FB task. This represented a significant lag in the acquisition of advanced ToM compared to children with typical hearing, most of whom can pass a second-order FB task by 6–7 years of age. Children who were deaf who were native sign language users also demonstrated delays in comprehension of ironic compliments compared to peers with typical hearing. These findings were consistent with research by O'Reilly, et al. (2014) who showed similar delays in second order ToM acquisition and ironic criticism by native sign language users. Giustolisi et al. (2018) found that a control group of deaf adults who were native sign language users passed the second order ToM and irony tasks, suggesting that these differences in advanced ToM resolve over time for individuals who learned sign language from their deaf parents. By contrast, O'Reilly, et al. (2014) found that, while adults who were not exposed to fluent sign until they entered school eventually acquired *first* order ToM (albeit significantly later than participants with typical hearing), only slightly greater than half of the late signers understood second order FB as adults; and fewer than a third understood sarcasm, even after many years of reportedly fluent sign language use. The authors argued that lack of access to fluent conversation with care providers early in life had persistent negative effects. "Clearly, the consequences of restricted early conversational experience as a function of upbringing in a hearing family are enduring, despite decades of evidently normal access to conversation in everyday adult life" (p. 1872). These findings were supported by later research showing that a large group of college students who were deaf did not understand sarcasm or advanced FB as well as their peers with typical hearing. This was true for college students who had learned sign language as children, as well as those who received cochlear implants in later childhood and used spoken language to communicate (Marschark et al. 2019).

In summary, CDDP who are exposed to a natural sign language from birth, do not demonstrate delays in early ToM skills or first order false belief reasoning compared to their peers with typical hearing. Some research suggests that more advanced ToM skills such as understanding of irony and sarcasm emerge later in CDDP than for children with typical hearing, however not enough research has been conducted with this population to say this conclusively.

8.4 Children Who are Deaf with Hearing Parents (CDHP)

Statistically, greater than 90% of children born deaf have parents who have typical hearing (Mitchell and Karchmer 2004). This means that early auditory access to the primary language used in the household is impoverished, even with use of appropriately fit technology (which cannot restore or replicate natural hearing); and delayed usually by several months (due to the steps involved in identifying hearing loss and appropriately fitting technology). Access to a complete visual language is also often delayed, as it takes time to enroll in sign language intervention services; and most parents and other family members of infants who are born deaf acquire sign language in tandem with their children (Mitchell and Karchmer 2005; Napier et al. 2007), and are therefore less able than deaf parents to provide a fluent language environment and adequate connected language models. In perhaps the earliest study of ToM in CDHP, Peterson and Siegal (1995) found that 65% of 8–13-year-old prelingually CDHP who used sign language failed a standard FB task. These authors were the first to point out that children who are deaf or hard of hearing whose parents are typically hearing seem to be at particularly high risk for ToM deficits, and they hypothesized that reduced conversational access in infancy, and resultant poor language and conversational fluency in early childhood, was at the heart of this ToM delay. While research has shown that CDHP could pass earlier acquired ToM tasks such as diverse desires and beliefs (Peterson and Wellman 2019b), these children are often in elementary school or even middle school before they can pass a standard FB task (Courtin 2000; Courtin and Melot 1998, 2005; de Villiers and de Villiers 2000, 2012; Figueras-Costa and Harris 2001; Gonzales et al. 2007; Jackson 2001; Jones et al. 2015; Lundy 2002; Peterson and Siegal 1997, 1998, 1999; Steeds et al. 1997; Woolfe et al. 2002) and exhibit significant delays in their comprehension of second order and advanced ToM concepts such as irony and sarcasm compared to their peers with typical hearing (Jones et al. 2015; Peterson and Wellman 2019b). In some cases, fewer than half of high school age DCHP could pass a standard, first order FB task (Russell et al. 1998). Frequently, the greatest ToM delays are observed in children classified as *late signers*—those children whose parents have typical hearing and who do not have access to fluent sign language until they enter school (Peterson and Siegal 1995).

Some researchers have suggested that the language demands of ToM tasks themselves might mask ToM competence in CDHP who have expressive or receptive language delays. The sophisticated linguistic structures necessary to follow directions ("What will Peter think is in this box?") and describe false belief reasoning ("She will think there are band-aids in the box, but really there's a pig") are advanced and challenging to parse and process. But even when the language demands of the ToM tasks are reduced, many CDHP show gaps in ToM. In a large study that included 176 4- to 8-year-old children who were deaf, Schick et al. (2007) included not only three standard false belief measures, but also a "low verbal" task called the *hidden sticker game,* the linguistic demands of which were reduced compared to a standard FB measure. In this task children play a game with an adult (described in detail below) in which they look for a sticker hidden in one of four white boxes:

> First the examiner pulled down the screen, hid a sticker in one of the four boxes, and then opened the screen. Then the examiner pointed to the box with the sticker and motioned to the child to take the box and keep the sticker. At least two more trials were completed until the child understood that it was advantageous always to pick the box that was pointed to. Then two adult confederates ("helpers") were introduced for the test phase of the game. One of the confederates (the knower) sat next to the examiner watching where the sticker was hidden; the other (the guesser) sat next to the child, screened from seeing the boxes like the child and wearing a blind fold. After the sticker was hidden, the screen was removed, and the confederates moved to the side of the table opposite the child. Each confederate pointed to a different box and hence the child received ambiguous advice about which box to choose. The knower pointed to the correct box, and the guesser pointed to an empty box according to a preordained rotation. The two adult confederates took turns in a random sequence at being the knower or the guesser, so the child could not solve the problem by just sticking with the same person's advice throughout the game. Before each trial the child was reminded about who was watching and who was wearing the blindfold. In addition, to reduce the load on the children's memory, the guesser lifted the blindfold, but kept it up on top of her forehead when she came round to the examiner's side of the table to point at a box. Ten test trials with the knower and guesser were run. In essence the child had to determine which helper knew where the sticker was hidden based on their visual access to the event (Schick et al. 2007, p. 383).

This low-verbal game has been shown to correlate with performance on verbal ToM tasks (P. de Villiers and Pyers 2001). Schick et al. found that, even for the *Hidden Sticker* task, CDHP (86 of whom used primarily spoken language and 41 of whom use primarily ASL) were delayed in their ToM compared to their peers with typical hearing and their peers who used sign language naturally from birth. "The nonverbal tasks were not passed easily by the DoH children, even though they can be argued to tap slightly earlier achievements such as seeing and knowing rather than false belief" (p. 390). Theory of mind performance was strongly correlated with children's memory for complement clauses ("He told the girl there was a bug in her hair, *but it was really a leaf*") (P. de Villiers and Pyers 2002) but not general language skill, advancing the notion that mastery of syntactic complements scaffolds mental representation of false beliefs, and that until children master the language to explain false belief, they cannot explicitly reason about false belief (de Villiers and de Villiers 2000).

Other researchers have attempted to control for language demands of ToM tasks with the same results. Levrez et al. (2012) evaluated a small group of 10-year-olds who were deaf and who used spoken French as their primary language. The authors used the *Non-verbal False Belief Task* (Forgeot d'Arc and Ramus 2011) in which children watched cartoons of daily activities and routines and determined the best ending out of two possible choices at the end of the cartoon. In each cartoon the child must recognize the false belief of the protagonist and choose how the protagonist will act based on their false belief. Like the CDHP in the Schick et al. study, the CDHP in this study were significantly delayed compared to the typically hearing 7-year-old controls on FB tasks, even though neither the task itself, nor the instructions for the task involved using or understanding language. de Villiers and de Villiers (2012) found that 4- to 8-year-old CDHP who used spoken English as their primary language were significantly delayed on both a verbal and low-verbal ("thought bubble") FB task compared to younger, typically hearing children, but not delayed on a true belief or deception measure. The authors hypothesized that false belief reasoning might be contingent upon language skill—specifically understanding of sentential complements—whereas deception might be more closely related to executive functioning (inhibitory control in particular). Children can lie successfully by following rules of behavior, whereas they can succeed at a false belief task *only* by accurately mentally representing another person's thoughts.

Cross-sectional research comparing ToM in younger and older DCHP indicates that ToM delays are ubiquitous in CDHP, at least for children who are late signers. Longitudinal research paints an unclear picture of how social cognitive skills develop over time as children become more linguistically proficient. Falkman et al. (2007), in a study of 10 late-signing DCHP between the ages of 7 and 10 years, found highly variable performance on FB tasks over a 2-year period, with some children experiencing *less* success at later test intervals. Performance on ToM tasks was correlated with child's sign language level at time of enrollment in the study. The authors found that "even high sign language proficiency was not enough for the deaf children to perform at the level of hearing children on the mentalizing tasks" (p. 193). By contrast, Peterson and Wellman (2019b) found that CDHP acquired ToM skills in the same *sequence* as their peers who acquire their first language naturally, but on a protracted timeline. They tested 27 CDHP children using their 6-item ToM Scale (Peterson et al. 2012) from age 3- to 13-years; all children in this study used Australian sign language as their preferred language. The authors found that the performance of the DCHP fit the developmental progression of the ToM scale at both test intervals (like the typically hearing children) but that DCHP were significantly delayed compared to children with typical hearing on all scale items, at all test intervals, performing comparably to participants with autism spectrum. Participants made steady, linear progress in their ToM development over time, however. This suggested that there is no critical or sensitive period for ToM acquisition and no developmental period at which rapid growth occurs for the skills measured by the 6-item scale, supporting a "delay" rather than "disorder" framework. Nonetheless, children who were profoundly deaf and who were late signers remained delayed compared to their peers with typical hearing until at least 10-years of age.

Not all children with hearing disorders are completely audiometrically deaf; children with hearing levels between 35 and 70 dB can often access the acoustic cues for speech using hearing aids. These children, who are classified as "hard of hearing", are more likely to develop expressive and receptive language skills that approximate those of their typically hearing peers, if they are amplified early and appropriately (Tomblin et al. 2015); these better language skills could support ToM development. Only two ToM studies to date have included children classified as hard of hearing. Those studies are worth mentioning here because they point to the role of language access in ToM acquisition and mastery. Netten, et al. (2017) tested 44 children with mild-to-severe hearing levels (thresholds between 35- and 70-dB HL in the better ear), the majority of whom used spoken language, on several ToM measures, including false belief. The children were between 3–1/2 and 7 years of age. Compared to a control group of 101 children with typical hearing, the children with mild-to-severe hearing loss were significantly delayed in both their language skills and their false belief understanding. As in previous research, performance on ToM tasks was correlated with language ability in the hard of hearing children. Similarly, Walker et al. (2017) conducted a longitudinal study of over 100 children whose hearing levels were in the mild-to-severe range and who used hearing aids. The goals of this study were to determine (a). if children with more residual hearing would demonstrate the same delays in ToM as their peers with profound hearing loss at ages 5 and 6 years; (b). if mental state input by caregivers at age 3 benefited children's ToM performance at age 5 and 6 years: (c). if language skills at age 3 years correlated with later ToM skills; and (d). if ToM delays would resolve by second grade in children who are hard of hearing, due to faster language growth trajectory compared to children who are severely to profoundly deaf. Like Netten, et al. (2017) the authors found that 5-year-old children who were hard of hearing and used hearing aids were significantly more likely to fail a standard false belief task than children with typical hearing; only 41% of the children who were hard of hearing passed the FB measure at age 5, compared to 81% of children with typical hearing. The likelihood of passing the false belief task at age 5 was predicted by maternal mental state language use at age 3 (supporting a conversational access theory), and language skills at age 3. By age 6, 79% of children who were hard of hearing passed the FB task, compared to 96% of children with typical hearing; and by second grade, children who were hard of hearing performed no differently than children with typical hearing on either false belief or language measures, showing that children who are hard of hearing will eventually catch up with their peers, and that having better incidental access to language and mental state terms might positively influence ToM acquisition in children who are DHH.

8.5 ToM in Children with Cochlear Implants

Optimizing language access and proficiency as early as possible is critically important for the development in ToM for children with severe to profound hearing loss whose parents have typical hearing. Until relatively recently, a limiting factor in achieving optimal auditory access to spoken language for CDHP has been technological feasibility—while hearing aids (acoustic amplifiers) can provide conversational access to the full range of acoustic cues for speech for children who are hard of hearing, they are of limited utility for children with hearing levels in the severe to profound range. Cochlear implants are an implantable technology that can provide auditory access to spoken language for children and adults who have moderate-severe to profound deafness through bypassing damaged receptor cells in the cochlea and stimulating neural structures in the auditory nerve directly. The first pediatric cochlear implant recipient was implanted in the United States at House Ear Institute in 1981 at the age of 3 years, although the United States Food and Drug Administration (FDA) did not approve cochlear implantation in children in the United States until 1990; in 2000, cochlear implants were FDA approved in children as young as 12-months-old. Many children who receive a cochlear implant early in development—by 12-months of age—and whose families enroll in appropriate early intervention and school supports develop intelligible spoken language that is on-par with that of their peers with typical hearing (Geers and Sedey 2011; Nicholas and Geers 2017; Percy-Smith et al. 2017; Spencer et al. 2012).

Based on spoken language outcomes observed in children with cochlear implants, it stands to reason that improved auditory access to language (and subsequent language gains) via cochlear implants might lead to improvements in social cognitive development. The earliest ToM studies of children using cochlear implants included children who were implanted well after the age of 2 years (late implantation, by current standards), whose language skills were delayed. Initial studies found mixed results on ToM measures in these "late language access" children, dependent on verbal abilities at the time of testing (Peterson 2004; Peterson 2009). In studies in which the majority of CDHP received cochlear implants after 2 years of age, ToM was delayed to varying degrees compared to children with typical hearing, regardless of communication mode or school environment (Peterson 2004). "The use of spoken modality does not seem to benefit ToM development.... Irrespective of whether they used cochlear implants or hearing aids, most of the oral deaf children were delayed in ToM development to the same extent as late signers" (Peterson 2009, p. 476).

Research that has included children who were first exposed to spoken language no later than 2–1/2 years of age via a cochlear implant indicate that earlier implanted CDHP are more likely to close social cognitive gaps than CDHP who are implanted later, and that performance on false belief tasks is correlated with language skills—but variability in performance remains the rule rather than the exception. Sundqvist et al. (2014) found that, while FB understanding was delayed in CDHP compared to children with typical hearing, there was no difference on an emotion understanding task (awareness that a character's emotions can be related to their belief or false belief) for children who received their cochlear implant(s) before 27 months of age; and FB understanding was significantly better than for children who were implanted after 27 months of age. Between group differences were observed despite both CI groups having similar cognitive and receptive vocabulary skills, pointing to duration of bidirectional conversational experiences as contributing factors in ToM development—the "early" implant group had more cumulative conversational exposure in conjunction with better language. In a large study of 72 preschoolers with cochlear implants (age at CI 6 months to 40 months) and 69 children with typical hearing, Ketelaar et al. (2012) found that CDHP performed similarly to their peers with typical hearing in their intention understanding (understanding the result a protagonist intended but failed to achieve), while they were delayed in their understanding of desires and beliefs compared to the control group, regardless of language modality. The authors reported that, even after correcting for language skills, CDHP were delayed in desire and belief understanding; none of the children with cochlear implants could pass a FB task, even the children with better (albeit delayed) language skills.

However, some children who receive cochlear implants early, and are enrolled in intensive intervention develop ToM skills that approximate that of their typically hearing peers. In perhaps the first study to examine ToM in children who used cochlear implants and listening and spoken language exclusively, Remmel and Peters (2009) found that 3- to 12-year-old CDHP who used cochlear implants were only slightly or not at all delayed on the Wellman and Liu (2004) scale; ToM performance was positively correlated with spoken language skill and duration of cochlear implant use (much like children in the Sundqvist et al. study). Unlike previous research with late signing children, in which some participants were unable to pass a standard FB task at 13-years-old, all children in the Remmel and Peters study over the age of 8 years performed at ceiling on the ToM Scale. The same cohort of children completed a video description task in which they explained the mistaken actions of characters and provided a rationale for their behavior; this was referred to as "false belief reasoning" and is a different task from the standard false belief task in that the children were not asked to predict what a character *would* do, but instead they were asked to describe and provide a justification for what the character had *already* done. Nearly all participants used FB reasoning when describing a character's anomalous actions ("she looked in the cabinet because *she thought the cake was there, but it wasn't*"). This pattern suggested intact FB *explanation*, despite comparatively poor FB *prediction* (Peters et al. 2009) consistent with findings of Peterson and Wellman (2019a) who found that *explaining* a person's anomalous actions based on a false belief was easier for young children than *predicting* a person's mistaken actions based on a false belief.

Recent cross-sectional and longitudinal studies support the hypothesis that early language access combined with duration of language experience leads to better social cognition in CDHP. In a large cross-sectional study of children who received their CIs between the ages of 6 and 27 months, Pluta et al. (2021) found that 4- to 5-year-old CDHP who used only spoken language at home were delayed in their FB reasoning compared to typically hearing peers, but that no significant differences were found between CDHP and a control group at 6- to 8-years old. Yu et al. (2021) found similar results in a longitudinal study of 84, 3- to 6-year-old DCHP who had received either hearing aids and/or a cochlear implant at 21 months of age on average. Children in this study were in preschool at both test intervals and used a variety of communication modes. A structural model that emphasized "hearing age" (duration of amplification use) best explained ToM skills for CDHP, while a model that emphasized language ability best accounted for ToM growth over time. Younger CDHP were more behind their typical peers than older CDHP, and children with better language skills had better ToM at the second test interval. These two studies suggest that ToM acquisition is associated with duration of accessible language exposure combined with language skill in CDHP who use spoken language—children who are implanted early and have good language are more likely to have better ToM, and even possibly catch up to their peers by about 6 to 8 years of age—much like children who are hard of hearing (Walker et al. 2017).

This story of "early, but not early enough" remains consistent across studies of CDHP: children who receive a cochlear implant or hearing aids by around 2 years of age demonstrate ToM delays at 4- to 5-years old compared to their typically hearing counterparts. By the age of 6- to 8-years, their ToM skills are comparable to peers with typical hearing, more in line with "language age", suggesting that duration of language access is important for ToM acquisition in these children. This seems to be the case *only* for children who make good gains in language. Children who are language delayed, remain delayed in ToM (Ziv et al. 2013a; Yu et al. 2021), although children with language delays who use spoken language demonstrate better ToM skills than children with language delays who are late signers (Peterson and Siegal 1999; Ziv et al. 2013a).

Research including only *very* early implanted children paints a much more optimistic picture. Peters, et al. (2021) found that a small group of 4- to 6-year-old CDHP all of whom had been implanted by 18-months of age and who used spoken language to communicate performed no differently than children with typical hearing on measures of expressive and receptive language and the 5-item ToM Scale (Wellman and Liu 2004). This result was observed even though the CDHP had fewer months of auditory language access than the hearing control group. Expressive and receptive language skills were strongly correlated with ToM in CDHP, even after controlling for age. A unique aspect of this study was that all CDHP attended mainstream, or co-enrolled/blended preschool programs that were co-taught by a teacher of the deaf and a general education teacher. This robust educational environment may have provided these CDHP with opportunities to interact with typical language and social peer models, and to observe and participate in typical and frequent conversational exchanges among other children.

While studies of children who received hearing technology in infancy suggest that access to spoken language by 2 years of age can ameliorate the negative effects of impoverished language input during the first months of life, ToM differences are evident even *before* the onset of verbal language in some CDHP—particularly for FB understanding. Novel testing paradigms that eliminate the need for a voluntary verbal or behavioral response allow researchers to measure awareness of false belief in babies who are deaf. Researchers track gaze, or "anticipatory looking" behaviors of infants to measure FB understanding. This procedure involves having child sit on their parent's lap while watching an animated clip of a cat ("Tom") chasing a mouse ("Jerry"). The child's gaze behaviors are recorded and measured using an infrared eye tracking device. During a familiarization phase, Tom chases Jerry through a Y-shaped tube with two exit points. When Jerry exits the tube and hides in one of two boxes, Tom enters the tube and finds Jerry in the appropriate box. In the "true belief" test trial, Jerry exits the tube and hides in one of the boxes. Tom momentarily leaves the screen but returns in time to see Jerry move into the other box. Tom then enters the Y-shaped tube. If the child looks towards the correct box, that is evidence that the child implicitly understands where Tom is going to look for Jerry. In the "false belief" test trial, Tom sees Jerry enter the first box, but then leaves the screen *while* Jerry moves to the second box (Tom does not know that Jerry has moved). Tom then returns and enters the tube. If the child looks to Jerry's *first* location, they have passed the FB trial (they realize that Tom is going to look in the box where he *thinks* Jerry is hiding, even though Jerry has moved) (Surian and Geraci 2012). In a study of anticipatory looking behaviors of ten 2-year-old CDHP who had received cochlear implants prior to 18 months of age, Meristo et al. (2012) found that CDHP demonstrated appropriate *true* belief attribution, but poor *false* belief attribution compared to their typically hearing counterparts. These finding were supported by later research (Meristo et al. 2016) showing that children with typical hearing demonstrated accurate "spontaneous" (implicit/non-verbal) FB attribution by the age of 2 years, whereas 5-year-old CDHP did not, regardless of technology used (hearing aids or cochlear implants). This research provides unique insights into implicit ToM understanding in typical children, and deficits in CDHP, and suggests that "a common first language with parents is conducive to very early mentalizing abilities among deaf infants" (Meristo et al. 2016, p. 143). However, later research that attempted to measure spontaneous ToM via anticipatory looking behaviors in two large groups of 2- to 3-year-old children with typical hearing resulted in very high fail rates during the familiarization phase of testing. The authors called into question the reliability of this method for measuring ToM in infants, and suggested a need to develop standardized, more controlled paradigms to study spontaneous/implicit ToM in young, pre-verbal children (Schuwerk et al. 2018).

8.6 Early Communication Interactions

To summarize, children who are born DHH, and whose parents are also deaf and use sign language have been found to demonstrate an early ToM trajectory that parallels that of children who are born with typical hearing and whose parents also have typical hearing—both groups of children have clear sensory access to a natural language from birth. Children who are born DHH and whose parents have typical hearing are much more likely to demonstrate ToM delays compared to their peers with typical hearing whether they use spoken language or sign language. "Thus, the severe ToM delays for deaf children with hearing parents are not a consequence of deafness per se, but rather of growing up deaf in the closed conversational world of a hearing family" (Wellman and Peterson 2013, p. 2). It seems, however, that earlier language access, better language proficiency, and opportunities to practice using language facilitates earlier ToM mastery in children who are deaf and whose parents have typical hearing.

Having a common, shared communication mode with care providers, family members, peers, and teachers from a very early age is instrumental in ToM development for all children, but especially for children who are DHH. "Even DoH children receiving their cochlear implants in toddlerhood are unlikely to have had as much access as hearing preschoolers to the informal causal-explanatory conversations about people that in hearing-only households are linked with early false belief success" (Peterson and Wellman 2019a, p. 656). Parental knowledge of mental state signs is correlated with performance on ToM tasks for children who are deaf (Moeller and Schick 2006), and parental use of a higher proportion of mental state terms in early childhood is associated with better ToM in children who are hard of hearing (Walker et al. 2017), but hearing parents of children who are profoundly deaf tend to use less mental state talk compared to parents of children with typical hearing (Morgan et al. 2014). Although very little research has been conducted directly examining the impact of conversational environments of CDHP on ToM, these findings reinforce the importance of parents of children who are DHH supporting not only language acquisition, but conversational fluency; and talking about non-observable mental state concepts such as "believing" "feeling" and "remembering", even in the very early stages of language development.

Beyond a shared language, the early reciprocal communication, and social experiences of CDHP may influence ToM acquisition. Early parental responsiveness and sensitivity, synchronous interactions, joint attention, and parental scaffolding support language development and executive skills such as self-regulation and inhibitory control in infants with typical hearing. Good "intersubjectivity" (intentional reciprocity between two individuals) is hypothesized to be a critical prerequisite not only in executive functioning and language acquisition for children who are deaf (Morgan et al. 2021) but for social cognitive development as well (Charman et al. 2000). These intersubjective interactions are frequently atypical or disrupted in CDHP (Morgan and Dye 2020). By contrast, the approximately 5% of children who are deaf with parents who are deaf and use sign language experience typical social exchanges in infancy and demonstrate normal intersubjectivity (Roos et al. 2016); language development and early ToM trajectories are likewise typical in these children. Supporting responsive, synchronous early parent-child interactions in CDHP dyads could positively influence executive functioning and language acquisition in children who are deaf (Szagun and Stumper 2012), which may positively impact ToM development over time.

Exposure to conversation among family members and peers also supports ToM acquisition. In one of the earliest studies of ToM in native sign language users, Woolfe et al. (2002) found that CDDP with older siblings who use British Sign Language (BSL) demonstrated ToM skills that were comparable to their peers with typical hearing. Similarly, Tomasuolo et al. (2012) found that children who were deaf attending a bilingual school setting with other children and teachers who were deaf demonstrated better FB understanding than students enrolled in a school setting in which only one other adult was able to communicate fluently in sign language. Other researchers, however, have found that later identified CDHP with older siblings have *poorer* FB reasoning, suggesting that competition for parental attention might have a negative impact on ToM development (Macaulay and Ford 2006).

8.7 Supporting ToM Development in CDHP

Identifying the mechanisms that underly ToM development and deficits in children who are deaf is important for understanding both typical and atypical social cognitive and linguistic processes. Identifying evidence-based intervention strategies to facilitate ToM acquisition in CDHP is also a priority. While there is a substantial literature on the benefits of ToM training for children with typical hearing (see Hofmann et al. 2016 for a review), much less is known about the impact of ToM training on children who are deaf. For CDHP with severe-to-profound hearing loss, early cochlear implantation may increase early auditory access to conversations about abstract concepts such as cognitive, emotional, and mental states. This is more likely to be the case for children who receive a cochlear implant at an early age (by 12- to 18-months of age). Similarly, if care providers and family members acquire conversational competence in sign language relatively quickly, including the vocabulary and syntax required to convey unobservable, more abstract concepts, this could be a reasonable avenue by which an infant who is deaf can be exposed to "mind-mindedness" in a comparable time frame to peers with typical hearing and CDDP.

Early intervention aimed specifically at improving intersubjectivity in deaf-hearing dyads has not been studied relative to its impact on later ToM; however, there is evidence that coaching parents towards using more supportive language strategies with their child who is deaf has positive effects on child language outcomes (Cruz et al. 2013). In the service of facilitating affective ToM (thinking about the emotions of others), parents can be coached in mirroring the expressions of their infants (or "sharing" their emotions) and responding in positive ways to infant emotions. Parents can also be trained to recognized stages of joint attention (an important precursor to ToM, per Charman et al. 2000; and Nelson et al. (2008), such as following the gaze of a care provider, using gaze to get something from a care provider, or understanding the emotions of a care provider. Helping parents learn how to follow the child's lead, join their child in play, and be flexible about how their child plays with toys rather than direct child play can also support positive dyadic interactions. Early intervention providers can use infant-parent observation tools, such as the *Parenting Interactions with Children: Checklist of Observations Linked to Outcomes* (PICCOLO) (Roggman et al. 2013) to document a variety of behaviors that support intersubjectivity, and to select targets for family coaching and education.

Pretend play is another important skill for children to develop on their path to a mature ToM. Research with typically developing children showed that preschoolers between 36 and 52 months old who engaged in more pretend play exhibited better performance on an appearance-reality measure (an object looks like a rock but is really a sponge) and in recognizing emotions in the eyes (Allen and Kinsey 2013). Children in this study participated for 4 weeks in a complex pretense interaction three times per week, for at least 15 min per session. During these play sessions, children engaged in role-plays, negotiated with other children. and visualized and acted out what different characters might do under a variety of circumstances. At the end of the study, children who participated in pretense groups performed significantly better on an appearance-reality ToM task and an emotion recognition task than children in the control group, although no changes were observed on the FB measure. In a study of typically hearing kindergarteners between 50 and 74 months, Qu et al. (2015) found that engaging in teacher supported sociodramatic play (pretend play that involved social role-taking with peers) for 2 months resulted in significant growth in children's ToM compared to a control group.

Parent coaching focused on increasing language that supports social cognitive skills might also benefit CDHP. A recent review of parent coaching strategies for CDHP suggested that "equipping caregivers to increase and improve language interactions with their children through coaching may facilitate spoken language development during early childhood" (Noll et al. 2021, p. 462), Care providers can be coached to increase their use of mental state and cognitive vocabulary, either in spoken language or sign language, a behavior that is associated with better ToM in DCHP (Moeller and Schick 2006; Stanzione and Schick 2014).) and children who have typical hearing (Ruffman et al. 2002; Taumoepeau and Ruffman 2006, 2008). Coaching parents to talk with their child about events that have happened in the past *and* what their child was thinking or how they and others felt, might further boost both affective and cognitive ToM. Supporting "autobiographical memory"—remembering oneself in the past—provides an opportunity for children to recall, visualize, and mentally represent past ideas and feelings that might facilitate ToM concepts such as understanding diverse thoughts (Welch-Ross 1997; Westby and Robinson 2014). Taking photographs and creating "experience books" is a simple way for parents to visually scaffold autobiographical memory.

Therapy aimed at improving expressive syntax has also proven effective in treating ToM deficits in school age children. Children aged 5 to 11 years who were deaf with ToM and expressive syntax delays in spoken French enrolled in an 8-week training program designed to teach them how to use and understand complement structures that contained verbs of communication such as "Jack says that witches are real". Children were *not* trained in ToM tasks specifically, and they were not trained on use of mental state complements, such as those that are required for FB prediction and explanation (e.g., "he *thinks* the ball is in the toybox"). At the end of the 8-week training, children demonstrated improvements both in use of complements (direct effect of training) and in understanding FB (transfer effect of training) (Durrleman et al. 2022). These findings align with earlier longitudinal research showing that use of complements is more strongly associated with comprehension of FB than general language skill (de Villiers and Pyers 2002) and that explicitly teaching children how to use complements (versus other syntactic structures such as relative clauses) improves ToM, but the reverse is not true (Hale and Tager-Flusberg 2003; Lohmann and Tomasello 2003).

Other studies have shown that simply talking about the anomalous actions of characters supports ToM development. Children demonstrate better FB explanation (explaining a character's anomalous actions by reference to false belief—"Mary looked for the cake in the cupboard because that's where she saw it last.") than FB prediction (predicting how a character will act based on a false belief— "Mary *will* look for the cake in the cupboard because that's where she thinks it is."). Repeated practice *explaining* the false beliefs of others resulted in improved and sustained FB understanding in CDHP (Peterson and Wellman 2019a). Similarly, deliberately exposing preschoolers with typical hearing to talk of mental states, deception, and FB through shared reading resulted in improvements in FB understanding (Guajardo and Watson 2002). The benefits of over-exposing children to mental state talk have been demonstrated as well in slighter older children with typical hearing (Lecce et al. 2014), and children who are deaf who use Signed Exact English (SEE) (Keddington 2011).

Reading books that provide examples of misrepresentations, misunderstanding, perspective-taking, and empathy enhances ToM awareness in children of all ages. A children's book like *Corduroy* in which a stuffed bear gets lost in a department store provides repeated opportunities for a preschooler to be exposed to and use advanced syntactic structures to talk about false belief ("Corduroy thinks it's a mountain, but it's really an escalator"). Graphic novels can be used to visually support a child's understanding that what an author writes is not necessarily intended to be interpreted literally. For example, it might be easier for a child to interpret sarcasm in the exclamation, "It looks like it's going to be a *great* day for a picnic!" if the written text it is accompanied by a picture of a frustrated character looking out the window at a looming storm. Graphic novels and "hybrid" graphic novels (ones that include complex illustrations and higher lexile text) are appropriate for children of all ages, including teenagers and young adults (Westby and Robinson 2014).

In an innovative training study that combined the use of syntactic structures supportive of ToM with visual representation of underlying linguistic structures via "thought bubbles", Peterson et al. (2012) observed significant improvements on ToM Scale performance among school age children who were deaf following 6 weeks of therapy. Children were first taught that when people look at things, they can get something like thought bubbles in their head and that when people look at things, they *think* about things. This was demonstrated using a two-dimensional doll (Sally) with moveable thought bubbles (for "Sally is thinking about a ball", the teacher moved a thought bubble with a ball near Sally's head). In the second stage of teaching, children were taught that a person can think about objects they no longer see. In stage 3, children learned that a person's thoughts about the world depend on what they see, and if something changes but the person did not see it change, their thoughts stay the same. In stage 4 children were taught that a person can use their thoughts to help them remember things (for example, where they put a toy). And in the final stage, children were taught that if a toy is moved, and the person does not see that it has been moved, they will look in the place where it was last—that is what the person will think, and what will be in their thought bubble. Later research using a slightly modified thought-bubble approach found similar results in preschool and kindergarten age CDHP after 15 to 18 weeks of treatment (Tucci et al. 2016), although progress was not uniform among participants, and children who were late signers did not achieve the same level of competence in ToM as children who used spoken language. The authors suggested this might be due to children who signed experiencing a different home language (non-fluent ASL) compared to their school language (fluent ASL). A simple version of thought bubble teaching can be used in preschool classrooms to provide both syntactic complement exposure and practice, and explicit visual representation of thinking. During prediction or guessing games, children can draw pictures of what they are thinking inside thought bubbles over the drawing of a head, and then talk about their own and each other's thoughts ("Kelli thinks there is a dog in the box, but I think there is a cow in the box"); after the true contents of the box is revealed, they can discuss the false beliefs of their friends ("Kelli thought it was a dog, but really it was a horse").

8.8 Future Research

Several gaps in our knowledge and understanding of theory of mind remain. Most ToM studies have not measured speech recognition, or connected expressive and receptive language directly, making it difficult to determine the specific underlying language access mechanisms associated with ToM growth (or lack thereof) for CDHP. Typically, "age at first access to language" is operationalized as "age at implantation/amplification"; however, in the absence of information about auditory speech recognition, quantity and quality of language access (and therefore conversational access) cannot be accurately ascribed to children who are deaf and who use spoken language. In studies in which language skill *was* measured, the majority of CDHP (either children who use spoken communication or children who acquired sign language late) were identified as deaf or hard of hearing later than 6 months of age.

Direct measures of bidirectional communication and social-emotional interactions between children and their family members would provide valuable measures of the early social and environmental experiences of CDHP for comparing to later ToM acquisition. The naturalness and effectiveness of these early reciprocal exchanges may provide additional insights into executive, language, and social cognitive development (Morgan et al. 2021), in addition to providing a roadmap for early treatment approaches.

Studies that include children who are identified as deaf or hard of hearing and amplified by 6 months of age and have access to spoken communication no later than 12-months of age are necessary. Children implanted by 9-months of age demonstrate significantly better long-term speech and language outcomes than children implanted later (Dettman et al. 2007; Dettman et al. 2021); better early language skills paired with earlier exposure to conversation might positively influence ToM acquisition. More studies that include preschoolers who are classified as hard of hearing might help clinicians and researchers understand the contribution of acoustic hearing (in particular, overhearing) to ToM development. There have been no studies of ToM in children who received technology very early, and whose families *also* use sign language (children who are bilingual-bimodal). Such research would be useful in determining if visual languages enhance access to social-linguistic cues beyond what hearing technology can provide.

Childhood deafness is a relatively low-incidence disability. As such, it can be a challenge to recruit large numbers of participants for research. Much of the ToM research that includes children who are deaf has been conducted at specialized, usually self-contained schools for the deaf. Children enrolled in self-contained school settings are more likely to be interacting with other children that have language and ToM delays (Boyle 1994), or concomitant disabilities affecting social and/or communicative competence (Shaver et al. 2013). This may limit opportunities to converse about the mind and may affect ToM acquisition (de Rosnay and Hughes 2006). Future research should aim to include children from a variety of home environments and educational settings.

Studies of teenagers and young adults who are deaf and received a cochlear implant at a very early age could provide valuable information about the longitudinal trajectory and developmental time course of ToM (second order ToM, irony and sarcasm, future thinking) in this population; as well as the impact of ToM on social skills in older children and young adults with cochlear implants. Language and learning gaps can show up later for children who are deaf, regardless of their verbal skills in elementary school (Marschark and Knoors 2019); and studies have shown that poorer ToM skills in early middle school are associated with poorer teacher-rated social competence in children with typical hearing (Devine et al. 2016).

The short and long-term ramifications of ToM delays in children who are born deaf remains relatively unexplored. Although the research on typically hearing children suggests that literacy, other academic skills, social relationships, mental health, and quality of life are negatively affected by poor understanding of others' intentions, perspectives and motives, these relationships have been less extensively examined in children who are deaf or hard of hearing—particularly those who have benefited from earlier access to language and communication.

While FB tasks have been described as the gold standard in ToM research, some researchers question the dependence on experimental FB measures as the only lens through which ToM mastery should be viewed. Westby (2014) described four ToM domains—cognitive, affective, interpersonal, and intrapersonal—and asserted that performance on standard first order false belief tasks reflect only the cognitive domain of ToM and may not be indicative of a child's affective social cognitive abilities. Affective ToM is important for developing and maintaining relationships, developing empathy for others, and understanding character traits when reading. There is limited research on affective ToM in children who are DHH; most research has focused on first- and second-order FB reasoning. Peterson (2016) compared teacher-rated empathy with performance on three standard FB tasks in a group of 4–13-year-old CDHP and a typically hearing control group and found that perceived

empathy correlated with FB, but only in the CDHP group. Similarly, Wiefferink et al. (2012) found that preschoolers who were deaf with cochlear implants were delayed compared to their peers with normal hearing on both verbal and non-verbal emotion recognition tasks. However, little has been done to further our knowledge or understanding of how children who are DHH interpret and infer the emotions of others, or how accurately children who are DHH ascribe others' behaviors to emotions (versus beliefs).

Written language may provide a unique window into ToM understanding in children who are DHH as well as a more integrated avenue for ToM teaching and learning since written language is woven throughout school curricula, beginning in elementary school. School age CDHP have been found to demonstrate evidence of first order, second order, and even advanced ToM understanding in their written narratives; writing may be a more ecologically valid method for assessing and observing advanced and nuanced ToM. Chilton et al. (2019) assessed 43 children who were DHH (hearing levels in the moderate-to-profound range) between 7- and 11-years-old who used spoken English and attended mainstream educational settings. Children were prompted to write a fictional narrative based on a wordless picture book that provided multiple opportunities to discuss thoughts and feelings of characters, as well as different perspectives. The authors found that 37 of the children used ToM in their writing, with 15 of the children demonstrating use of first order false belief, and 12 of the children demonstrating use of second or false belief. The authors made an important case for bridging the gap between cognitive science and educational practice stating, "Within education, there is little evidence that teachers overtly exploit ToM in teaching reading and writing with hearing or deaf children. Perhaps the most critical step in addressing this issue would be to make teachers more aware of ToM as a construct, how it is developmental, and why it is important" (p. 38).

Social cognitive reasoning is central to most academic activities such as reading comprehension, writing, and engaging in oral argument or negotiation—all tasks that require perspective-taking; awareness of others' knowledge and mental states; use and understanding of sarcasm and irony; and use and understanding figurative language such as similes, metaphors, idiomatic expressions, hyperbole, and personification. A review of the *Common Core State Standards*, a list of academic expectations by grade in English Language Arts and Math for children in the United States (Common Core State Standards Initiative 2010), reveals that ToM skills are implicitly embedded in many academic standards. For example, "Demonstrate understanding of figurative language", (Grade 5), "Analyze how an author develops and contrasts the points of view of different characters or narrators in a text" (Grade 7), "Analyze the interactions between individuals, events, and ideas in a text (e.g., how ideas influence individuals or events, or how individuals influence ideas or events." (Grade 7) and "Write for a range of tasks, purposes and audiences" (Grade 8). Assessing reading comprehension and evaluating written language for evidence of ToM would supply valuable information regarding the adverse educational impact of deafness beyond that which is provided via a standardized language assessment (the main tool that public schools in the U.S. use in determining eligibility for special education services). Further, gaps in ToM understanding should be addressed through specially designed instruction to ensure that children who are DHH have access to and can fully benefit from academic instruction.

Finally, it has been established in the research that children who are deaf and hard of hearing are at higher risk for verbal working memory and information processing speed differences compared to their typically hearing peers (see Pisoni et al. 2017) and that these differences affect language development (Romano et al. 2021). In addition, recent research has pointed to slower, more effortful linguistic processing as a compensatory strategy for reduced auditory access in children with cochlear implants (Pisoni and Kronenberger 2021). The efficiency and ease with which children who are deaf retain and process incoming linguistic information may influence ToM competence and/or performance, considering that language fluency appears to be essential for ToM functioning in this population. No research to date has examined the impact of verbal information processing speed or accuracy on social cognition in children who are deaf, although verbal processing speed and efficiency have been described as important foundational cognitive markers for language acquisition in typically hearing children (McMurray et al. 2022). Such a study might provide new insights into novel cognitive barriers children who are deaf and hard of hearing face and could also lead to supportive clinical interventions.

8.9 Conclusion

Theory of mind acquisition for children who are deaf is the result of multiple intersecting variables: age-appropriate expressive and receptive language skills, high-quality linguistic and social input by care providers, auditory and/or visual exposure to social interactions in a multitude of environments, opportunities to engage frequently and directly in conversation about the mind with adults and peers, and typical sensorimotor and neurocognitive abilities.

Research suggests that ToM acquisition can be supported through optimizing communication access and function from a very early age. Children who have access to fluent communication partners, and who are exposed to and engage in conversation about the mind from infancy appear to demonstrate fewer ToM gaps over time, whereas children who experience delays in linguistic and conversational access seem to experience subsequent and significant delays in ToM. The quality of early dyadic interactions (intersubjectivity) and its relationship to ToM development may hold important answers for families, teachers, and children.

What we have learned from ToM research as it pertains to children who are deaf and hard of hearing:

1. Children who have early access to a natural language (children who have typical hearing whose parents have typical hearing, or children who are deaf whose parents are deaf) do not exhibit developmental delays in early ToM skills.

2. Children who are deaf or hard of hearing whose parents have typical hearing are at high risk for ToM delays (specifically advanced ToM skills such as false belief and hidden emotion, but also earlier emerging ToM skills), regardless of hearing levels, communication modality, or preferred language.
3. Children who are deaf and native sign language users acquire second order and advanced ToM (irony and sarcasm) later, compared to children with typical hearing; these differences appear resolve over time.
4. Some children who are deaf who receive a cochlear implant very early (before 18 months of age) *and* have age-appropriate spoken language acquire first order ToM in a typical timeframe compared to peers with typical hearing. But having age-appropriate language appears to be insufficient for typical ToM acquisition.
5. Differences in ToM awareness appear early in children who are deaf with hearing parents, even on non-linguistic tasks.
6. The language learning environment of DHH children plays a role in ToM development, as children whose parents use less mental state talk and/or signs are at greater risk for ToM delays.
7. Many children who are deaf with typically hearing parents will eventually acquire first order and second order FB reasoning that is correlated with their language development, suggesting a delayed rather than discorded pattern in social cognitive development, and a positive relationship between language and ToM for these children.
8. Targeted speech and language therapy that includes an emphasis on complement syntax as well as explicit ToM practice can accelerate ToM acquisition in children who are deaf and hard of hearing. Therapies that teach specific linguistic forms alone, or in combination with explicit visual representations of thinking and other unobservable concepts have been shown to be effective in supporting ToM acquisition.

Longitudinal studies that include measures of language access, language use, and verbal processing of very early identified children who are deaf or hard of hearing would provide further insights into the developmental trajectory of ToM, the possible influence of language plateau on ToM development, and the social ramifications of ToM differences on other social and academic domains. The conversational environments of children who are deaf or hard of hearing should be studied systematically; direct observations of parent-child interactions would elucidate the role of early conversational and social experiences (intersubjectivity) in ToM development. The ToM trajectories of children who are deaf from diverse backgrounds, including children who are bilingual-bimodal remains unexplored. The effectiveness of therapy approaches to enhance ToM in children who are deaf should be critically reviewed and evaluated, as this remains a significant gap in the literature.

References

Adrian, J.E., R.A. Clemente, L. Villanueva, and C. Rieffe. 2005. Parent-child picture-book reading, mothers' mental state language and children's theory of mind. *Journal of Child Language* 32 (3): 673–686. https://doi.org/10.1017/s0305000905006963.

Adrián, J.E., R.A. Clemente, and L. Villanueva. 2007. Mothers' use of cognitive state verbs in picture-book reading and the development of children's understanding of mind: A longitudinal study. *Child Development* 78 (4): 1052–1067. https://doi.org/10.1111/j.1467-8624.2007.01052.x.

Allen, J.R., and K. Kinsey. 2013. Teaching theory of mind. *Early Education and Development* 24: 865–876.

Altun, D. 2019. Young children's theory of mind: Home literacy environment, technology usage, and preschool education. *Journal of Education and Training Studies* 7 (3): 86–98. https://doi.org/10.11114/jets.v7i3.4057.

Apperly, I.A. 2010. *Mindreaders: The cognitive basis of "theory of mind"*. Psychology Press. https://doi.org/10.4324/9780203833926.

Astington, J.W., and J.A. Baird, eds. 2005. *Why language matters for theory of mind*. Oxford University Press. https://doi.org/10.1093/acprof:oso/9780195159912.001.0001.

Astington, J.W., and J.M. Jenkins. 1999. A longitudinal study of the relation between language and theory-of-mind development. *Developmental Psychology* 35 (5): 1311–1320.

Atkinson, L., L. Slade, D. Powell, and J.P. Levy. 2017. Theory of mind in emerging reading comprehension: A longitudinal study of early indirect and direct effects. *Journal of Experimental Child Psychology* 164: 225–238. https://doi.org/10.1016/j.jecp.2017.04.007.

Baron-Cohen, S., A.M. Leslie, and U. Frith. 1985. Does the autistic child have a "theory of mind?". *Cognition* 21 (1): 37–46. https://doi.org/10.1016/0010-0277(85)90022-8.

Bartsch, K., and H. Wellman. 1989. Young children's attribution of action to beliefs and desires. *Child Development* 60 (4): 946–964. https://doi.org/10.2307/1131035.

Beaudoin, C., É. Leblanc, C. Gagner, and M.H. Beauchamp. 2020. Systematic review and inventory of theory of mind measures for young children. *Frontiers in Psychology* 10: 2905. https://doi.org/10.3389/fpsyg.2019.02905.

Boyle, S. 1994, November/December. Deaf and hearing students benefit by learning, communicating together. Special Edge, 8-13.

Bretherton, I., and M. Beeghly. 1982. Talking about internal states: The acquisition of an explicit theory of mind. *Developmental Psychology* 18 (6): 906–921. https://doi.org/10.1037/0012-1649.18.6.906.

Charman, T., S. Baron-Cohen, J. Swettenham, G. Baird, A. Cox, and A. Drew. 2000. Testing joint attention, imitation and play as infancy precursors to language and theory of mind. *Cognitive Development* 15: 481–498. https://doi.org/10.1016/S0885-2014(01)00037-5.

Chilton, H., C. Mayer, and W. McCracken. 2019. Evidence of theory of mind in the written language of deaf children. *Journal of Deaf Studies and Deaf Education* 24 (1): 32–40. https://doi.org/10.1093/deafed/eny027.

Common Core State Standards Initiative. 2010. *Common core state standards for English/language arts and literacy in history/social studies, science, and technical subjects*. Retrieved on 02/28/2023 from https://learning.ccsso.org/common-core-state-standards-initiative.

Courtin, C. 2000. The impact of sign language on the cognitive development of deaf children: The case of theories of mind. *Journal of Deaf Studies and Deaf Education* 5 (3): 266–276. https://doi.org/10.1093/deafed/5.3.266.

Courtin, C., and A.M. Melot. 1998. Development of theories of mind in deaf children. In *Psychological perspectives on deafness*, ed. M. Marschark and M. Clark, vol. 2, 79–102. Lawrence Erlbaum.

———. 2005. Metacognitive development of deaf children: Lessons from the appearance–reality and false belief tasks. *Developmental Science* 8 (1): 16–25. https://doi.org/10.1111/j.1467-7687.2005.00389.x.

Cruz, I., A.L. Quittner, C. Marker, and J.L. DesJardin. 2013. Identification of effective strategies to promote language in deaf children with cochlear implants. *Child Development* 84 (2): 543–559. https://doi.org/10.1111/j.1467-8624.2012.01863.x.

Custer, W.L. 1996. A comparison of young children's understanding of contradictory representations in pretense, memory, and belief. *Child Development* 67 (2): 678–688. https://doi.org/10.2307/1131840.

de Rosnay, M., and C. Hughes. 2006. Conversation and theory of mind: Do children talk their way to socio-cognitive understanding? *British Journal of Developmental Psychology* 24 (1): 7–37. https://doi.org/10.1348/026151005X82901.

de Villiers, J. (1995, April). Steps in the mastery of sentence complements. Biennial meeting of the society for research in child development.

de Villiers, P.A. 2003. Language of the deaf—Acquisition of English. In *The MIT encyclopedia of communication disorders*, ed. R. Kent, 336–338. Cambridge, MA: MIT Press.

de Villiers, J.G., and P.A. de Villiers. 2000. Linguistic determinism and the understanding of false beliefs. In *Children's reasoning and the mind*, ed. P. Mitchell and K. Riggs, 189–226. Psychology Press.

de Villiers, P.A., and J.G. de Villiers. 2012. Deception dissociates from false belief reasoning in deaf children: Implications for the implicit versus explicit theory of mind distinction. *British Journal of Developmental Psychology* 30: 188–209. https://doi.org/10.1111/j.2044-835X.2011.02072.x.

de Villiers, P., and J. Pyers. 2001. Complementation and false-belief representation. In *Research on child language acquisition: Proceedings of the 8th conference of the international association for the study of child language*, ed. M. Almgren, A. Barrena, M.J. Ezeizabarrena, I. Idiazabal, and B. MacWhinney, 984–1005. Somerville, MA: Cascadilla Press.

de Villiers, J.G., and J.E. Pyers. 2002. Complements to cognition: A longitudinal study of the relationship between complex syntax and false-belief-understanding. *Cognitive Development* 17 (1): 1037–1060. https://doi.org/10.1016/S0885-2014(02)00073-4.

Dettman, S.J., D. Pinder, R.J. Briggs, R.C. Dowell, and J.R. Leigh. 2007. Communication development in children who receive the cochlear implant younger than 12 months: Risks versus benefits. *Ear and Hearing* 28 (2 Suppl): 11S–18S.

Dettman, S., D. Choo, A. Au, A. Luu, and R. Dowell. 2021. Speech perception and language outcomes for infants receiving cochlear implants before or after 9 months of age: Use of category-based aggregation of data in an unselected pediatric cohort. *Journal of Speech Language Hearing Research* 64 (3): 1023–1039. https://doi.org/10.1044/2020_JSLHR-20-00228.

Devine, R.T., and C. Hughes. 2013. Silent films and strange stories: Theory of mind, gender, and social experiences in middle childhood. *Child Development* 84: 989–1003. https://doi.org/10.1111/cdev.12017.

Devine, R.T., N. White, R. Ensor, and C. Hughes. 2016. Theory of mind in middle childhood: Longitudinal associations with executive function and social competence. *Developmental Psychology* 52 (5): 758–771. https://doi.org/10.1037/dev0000105.

Ding, X.P., H.M. Wellman, Y. Wang, G. Fu, and K. Lee. 2015. Theory-of-mind training causes honest young children to lie. *Psychological Science* 26 (11): 1812–1821. https://doi.org/10.1177/0956797615604628.

Dore, R.A., S.J. Amendum, R.M. Golinkoff, and K. Hirsh-Pasek. 2018. Theory of mind: A hidden factor in reading comprehension? *Educational Psychology Review* 30: 1067–1089. https://doi.org/10.1007/s10648-018-9443-9.

Durrleman, S., A. Dumont, and H. Delage. 2022. Syntactic strategy training for theory of mind in deaf children. *The Journal of Deaf Studies and Deaf Education* 27 (1): 89–100. https://doi.org/10.1093/deafed/enab034.

Edmondson, P. 2006. Deaf children's understanding of other people's thought processes. *Educational Psychology in Practice* 22 (2): 159–169. https://doi.org/10.1080/02667360600720146.

Eggum, N.D., N. Eisenberg, K. Kao, T.L. Spinrad, R. Bolnick, C. Hofer, A.E. Kupfer, and W.V. Fabricius. 2011. Emotion understanding, theory of mind, and pro-social orientation: Relations over time in early childhood. *The Journal of Positive Psychology* 6 (1): 4–16. https://doi.org/10.1080/17439760.2010.536776.

Falkman, K.W., C. Roos, and E. Hjelmquist. 2007. Mentalizing skills of non-native, early signers: A longitudinal perspective. *European Journal of Developmental Psychology* 4 (2): 178–197. https://doi.org/10.1080/17405620600574867.

Figueras-Costa, B., and P. Harris. 2001. Theory of mind development in deaf children: A nonverbal test of false-belief understanding. *Journal of Deaf Studies and Deaf Education* 6 (2): 92–102. https://doi.org/10.1093/deafed/6.2.92.

Figueroa, M., S. Darbra, and N. Silvestre. 2020. Reading and theory of mind in adolescents with cochlear implant. *The Journal of Deaf Studies and Deaf Education* 25 (2): 212–223. https://doi.org/10.1093/deafed/enz046.

Fink, E., S. Begeer, C.C. Peterson, V. Slaughter, and M. de Rosnay. 2014. Friendlessness and theory of mind: A prospective longitudinal study. *British Journal of Developmental Psychology* 33 (1): 1–17. https://doi.org/10.1111/bjdp.12060.

Flavell, J.H., E.R. Flavell, and F.L. Green. 1983. Development of the appearance–reality distinction. *Cognitive Psychology* 15 (1): 95–120. https://doi.org/10.1016/0010-0285(83)90005-1.

Forgeot d'Arc, B., and F. Ramus. 2011. Belief attribution despite verbal interference. *Quarterly Journal of Experimental Psychology* 64 (5): 975–990. https://doi.org/10.1080/17470218.2010.524413.

Geers, A.E., and A.L. Sedey. 2011. Language and verbal reasoning skills in adolescents with 10 or more years of cochlear implant experience. *Ear and Hearing* 32 (Suppl): 39S–48S. https://doi.org/10.1097/AUD.0b013e3181fa41dc.

Geers, A.E., J. Moog, and B. Schick. 1984. Acquisition of spoken and signed English by profoundly deaf children. *Journal of Speech & Hearing Disorders* 49 (4): 378–388. https://doi.org/10.1044/jshd.4904.378.

Giustolisi, B., Mantovan, L., & Panzeri, F. 2018. Irony comprehension and theory of mind in young deaf igners. Conference proceedings of the 42nd annual Boston University conference on language development.

Gonzales, A., I. Quintana, C. Barajas, and M.J. Linero. 2007. The role of age and oral lexical competence in false belief understanding by children and adolescents with hearing loss. *Volta Review* 107 (2): 123–139.

Gopnik, A., V. Slaughter, and A.N. Meltzoff. 1994. Changing your views: How understanding visual perception can lead to a new theory of the mind. In *Children's early understanding of mind: Origins and development*, ed. C. Lewis and P. Mitchell, 157–181. Erlbaum.

Grazzani, I., and V. Ornaghi. 2012. How do use and comprehension of mental-state language relate to theory of mind in middle childhood? *Cognitive Development* 27 (2): 99–111. https://doi.org/10.1016/j.cogdev.2012.03.002.

Guajardo, N.R., and A.C. Watson. 2002. Narrative discourse and theory of mind development. *The Journal of Genetic Psychology* (163) 3: 305–325. https://doi.org/10.1080/00221320209598686.

Hale, C.M., and H. Tager-Flusberg. 2003. The influence of language on theory of mind: A training study. *Developmental Science* 6 (3): 346–359. https://doi.org/10.1111/1467-7687.00289.

Hao, J., Y. Su, and R.C. Chan. 2010. Do deaf adults with limited language have advanced theory of mind? *Research in Developmental Disabilities* 31 (6): 1491. https://doi.org/10.1016/j.ridd.2010.06.008.

Happé, F.G. 1994. An advanced test of theory of mind: Understanding of story characters' thoughts and feelings by able autistic, mentally handicapped, and normal children and adults. *Journal of Autism and Developmental Disorders* 24 (2): 129–154. https://doi.org/10.1007/BF02172093.

Harris, P.L., K. Donnelly, G.R. Guz, and R. Pitt-Watson. 1986. Children's understanding of the distinction between real and apparent emotion. *Child Development* 57 (4): 895–909. https://doi.org/10.1111/j.1467-8624.1986.tb00253.x.

Harris, P.L., M. de Rosnay, and F. Pons. 2005. Language and children's understanding of mental states. *Current Directions in Psychological Science* 14 (2): 69–73. https://doi.org/10.1111/j.0963-7214.2005.00337.x.

Hickling, A.K., H.M. Wellman, and G.M. Gottfried. 1997. Preschoolers' understanding of others' mental attitudes towards pretend happenings. *British Journal of Developmental Psychology* 15(3): 339–354.

Hofmann, S.G., S.N. Doan, M. Sprung, A. Wilson, C. Ebesutani, L.A. Andrews, J. Curtiss, and P.L. Harris. 2016. Training children's theory-of-mind: A meta-analysis of controlled studies. *Cognition* 150: 200–212. https://doi.org/10.1016/j.cognition.2016.01.006.

Holmer, E., M. Heimann, and M. Rudner. 2016. Theory of mind and reading comprehension in deaf and hard-of-hearing signing children. *Frontiers in Psychology* 7: 854. https://doi.org/10.3389/fpsyg.2016.00854.

Jackson, A.L. 2001. Language facility and theory of mind development in deaf children. *Journal of Deaf Studies and Deaf Education* 6 (3): 158–173. https://doi.org/10.1093/deafed/6.3.161.

Jones, A.C., R. Gutierrez, and A.K. Ludlow. 2015. Confronting the language barrier: Theory of mind in deaf children. *Journal of Communication Disorders* 56: 47–58. https://doi.org/10.1016/j.jcomdis.2015.06.005.

Keddington, Holly B. 2011. Effect of mental state discussion on theory of mind in students who are deaf. *Electronic Theses & Dissertations.*: 59. https://digitalcommons.tamuc.edu/etd/59.

Ketelaar, L., C. Rieffe, C.H. Wiefferink, and J.H. Frijns. 2012. Does hearing lead to understanding? Theory of mind in toddlers and preschoolers with cochlear implants. *Journal of Pediatric Psychology* 37 (9): 1041–1050. https://doi.org/10.1093/jpepsy/jss086.+.

Labounty, J., H.M. Wellman, S.L. Olson, K.H. Lagattuta, and D. Liu. 2008. Mothers' and Fathers' use of internal state talk with their young children. *Social Development* 17: 757–775. https://doi.org/10.1111/J.1467-9507.2007.00450.X.

Lecce, S., M. Caputi, and C. Hughes. 2011. Does sensitivity to criticism mediate the relationship between theory of mind and academic achievement? *Journal of Experimental Child Psychology* 110 (3): 313–331. https://doi.org/10.1016/j.jecp.2011.04.011.

Lecce, S., F. Bianco, R.T. Devine, C. Hughes, and R. Banerjee. 2014. Promoting theory of mind during middle childhood: A training program. *Journal of Experimental Child Psychology* 126: 52–67. https://doi.org/10.1016/j.jecp.2014.03.002.

Levrez, C., B. Bourdin, B. Le Driant, B. Forgeot d'Arc, and L. Vandromme. 2012. The impact of verbal capacity on theory of mind in deaf and hard of hearing children. *American Annals of the Deaf* 157 (1): 66–77. https://doi.org/10.1353/aad.2012.1610.

Lohmann, H., and M. Tomasello. 2003. The role of language in the development of false belief understanding: A training study. *Child Development* 74 (4): 1130–1144. https://doi.org/10.1111/1467-8624.00597.

Lundy, J. 2002. Age and language skills of deaf children in relation to theory of mind development. *Journal of Deaf Studies and Deaf Education* 7 (1): 41–56. https://doi.org/10.1093/deafed/7.1.41++.

Macaulay, C.E., and R.M. Ford. 2006. Language and theory-of-mind development in prelingually deafened children with cochlear implants: A preliminary investigation. *Cochlear Implants International* 7 (1): 1–14. https://doi.org/10.1179/cim.2006.7.1.1. PMID: 18792371.

Marschark, M., and H. Knoors. 2019. Sleuthing the 93% solution in deaf education. In *Evidence based practices in deaf education*, ed. Marschark and Knoors, 1–39. Oxford University Press.

Marschark, M., L. Edwards, C.C. Peterson, K. Crowe, and D. Walton. 2019. Understanding theory of mind in deaf and hearing college students. *Journal of Deaf Studies and Deaf Education* 24 (2): 104–118. https://doi.org/10.1093/deafed/eny039.

Mayberry, R.I., and E. Lock. 2003. Age constraints on first versus second language acquisition: Evidence for linguistic plasticity and epigenesis. *Brain and Language* 87: 369–384. https://doi.org/10.1016/S0093-934X(03)00137-8.

Mayberry, R.I., E. Lock, and H. Kazmi. 2002. Linguistic ability and early language exposure. *Nature* 417: 38. https://doi.org/10.1038/417038a.

McMurray, B., K.S. Apfelbaum, and J.B. Tomblin. 2022. The slow development of real-time processing: Spoken-word recognition as a crucible for new thinking about language acquisition and language disorders. *Current Directions in Psychological Science* 31 (4): 305–315. https://doi.org/10.1177/09637214221078325.

Meins, E., C. Fernyhough, R. Wainwright, M. Das Gupta, E. Fradley, and M. Tuckey. 2002. Maternal mind-mindedness and attachment security as predictors of theory of mind understanding. *Child Development* 73 (6): 1715–1726.

Meristo, M., and E. Hjelmquist. 2009. Executive functions and theory of mind deaf children: Different routes to understanding other minds? *Journal of Cognition and Development* 10 (1–2): 67–91. https://doi.org/10.1080/15248370902966552.

Meristo, M., and K. Strid. 2020. Language first: Deaf children from deaf families spontaneously anticipate false beliefs. *Journal of Cognition and Development* 21 (4): 622–630. https://doi.org/10.1080/15248372.2020.1749057.

Meristo, M., K.W. Falkman, E. Hjelmquist, M. Tedoldi, L. Surian, and M. Siegal. 2007. Language access and theory of mind reasoning: Evidence from deaf children in bilingual and oralist environments. *Developmental Psychology 43* (5): 1156. https://doi.org/10.1037/0012-1649.43.5.1156.

Meristo, M., G. Morgan, A. Geraci, L. Iozza, E. Hjelmquist, L. Surian, and M. Siegal. 2012. Belief attribution in deaf and hearing infants. *Developmental Science* 15 (5): 633–640. https://doi.org/10.1111/j.1467-7687.2012.01155.x.

Meristo, M., K. Strid, and E. Hjelmquist. 2016. Early conversational environment enables spontaneous belief attribution in deaf children. *Cognition* 157: 139–145.

Miller, S.A. 2009. Children's understanding of second-order mental states. *Psychological Bulletin* 135 (5): 749–773. https://doi.org/10.1037/a0016854.

Milligan, K., J.W. Astington, and L.A. Dack. 2007. Language and theory of mind: Meta-analysis of the relation between language ability and false-belief understanding. *Child Development* 78 (2): 622–646. https://doi.org/10.1111/j.1467-8624.2007.01018.x.

Mitchell, R.E., and M.A. Karchmer. 2004. Chasing the mythical ten percent: Parental hearing status of deaf and hard of hearing students in the United States. *Sign Language Studies* 4 (2): 138–163. https://doi.org/10.1353/sls.2004.0005.

———. 2005. Parental hearing status and signing among deaf and hard of hearing students. *Sign Language Studies* 5 (2): 231–244. http://www.jstor.org/stable/26190612.

Moeller, M.P., and B. Schick. 2006. Relations between maternal input and theory of mind understanding in deaf children. *Child Development* 77 (3): 751–766. https://doi.org/10.1111/j.1467-8624.2006.00901.x.

Morgan, G., and M.W.G. Dye. 2020. Executive functions and access to language: The importance of intersubjectivity. In *The Oxford handbook of deaf studies in learning and cognition*, ed. M. Marschark and H. Knoors, 268–284. Oxford University Press.

Morgan, G., M. Meristo, W. Mann, E. Hjelmquist, L. Surian, and M. Siegal. 2014. Mental state language and quality of conversational experience in deaf and hearing children. *Cognitive Development* 29: 41–49. https://doi.org/10.1016/j.cogdev.2013.10.002.

Morgan, G., M. Curtin, and N. Botting. 2021. The interplay between early social interaction, language and executive function development in deaf and hearing infants. *Infant Behavior & Development* 64: 101591. https://doi.org/10.1016/j.infbeh.2021.101591.

Napier, J., G. Leigh, and S. Nann. 2007. Teaching sign language to hearing parents of deaf children: An action research process. *Deafness & Education International* 9 (2): 83–100. https://doi.org/10.1179/146431507790560020.

Nelson, P.B., L.B. Adamson, and R. Bakeman. 2008. Toddlers' joint engagement experience facilitates preschoolers' acquisition of theory of mind. *Developmental Science* 11 (6): 847–852. https://doi.org/10.1111/j.1467-7687.2008.00733.x.

Netten, A.P., C. Rieffe, W. Soede, E. Dirks, A.M.H. Korver, S. Konings, J.J. Briaire, A.M. Oudesluys-Murphy, F.W. Dekker, J.H.M. Frinjns, and DECIBAL Collaborative study group. 2017. Can you hear what I think? Theory of mind in young children with moderate

hearing loss. *Ear and Hearing* 38 (5): 588–597. https://doi.org/10.1097/AUD.0000000000000427.
Newton, E., and V. Jenvey. 2010. Play and theory of mind: Associations with social competence in young children. *Early Child Development and Care* 18 (6): 761–773.
Nicholas, J.G., and A.E. Geers. 2017. Sensitivity of expressive linguistic domains to surgery age and audibility of speech in preschoolers with cochlear implants. *Cochlear Implants International* 19 (1): 26–37. https://doi.org/10.1080/14670100.2017.1380114.
Noll, D., D. DiFabio, S. Moodie, I.D. Graham, B. Potter, V. Grandpierre, and E.M. Fitzpatrick. 2021. Coaching caregivers of children who are deaf or hard of hearing: A scoping review. *Journal of Deaf Studies and Deaf Education* 26 (4): 453–468. https://doi.org/10.1093/deafed/enab018.
O'Reilly, K., C.C. Peterson, and H.M. Wellman 2014. Sarcasm and advanced theory of mind understanding in children and adults with prelingual deafness. *Developmental Psychology* 50 (7): 1862. https://doi.org/10.1037/a0036654.
Percy-Smith, L., T. Lindbjerg Tønning, J. Lignel Josvassen, J. Hølledig Mikkelsen, L. Nissen, E. Dieleman, M. Hallstrøm, and P. Cayé-Thomasen. 2017. Auditory verbal habilitation is associated with improved outcome for children with cochlear implant. *Cochlear Implants International* 19 (1): 38–45. https://doi.org/10.1080/14670100.2017.1389020.
Perner, J. 1991. *Understanding the representational mind*. Harvester.
Perner, J., and H. Wimmer. 1985. "John *thinks* that Mary *thinks* that...": Attribution of second-order beliefs by 5- to 10-year-old children. *Journal of Experimental Child Psychology* 39 (3): 437–471. https://doi.org/10.1016/0022-0965(85)90051-7.
Perner, J., S.R. Leekam, and H. Wimmer. 1987. Three-year-olds' difficulty with false belief: The case for a conceptual deficit. *British Journal of Developmental Psychology* 5: 125–137. https://doi.org/10.1111/j.2044-835X.1987.tb01048.x.
Perner, J., T. Ruffman, and S.R. Leekman. 1994. Theory of mind is contagious: You catch it from your sibs. *Child Development* 65 (4): 1228–1238. https://doi.org/10.1111/j.1467-8624.1994.tb00814.x.
Peters, K., E. Remmel, and D. Richards. 2009. Language, mental state vocabulary, and false belief understanding in children with cochlear implants. *Language, Speech, and Hearing Services in Schools* 40 (3): 245–255. https://doi.org/10.1044/0161-1461(2009/07-0079).
Peters, K., J. Beer, D. Pisoni, and Remmel. 2021. Theory of mind acquisition in children who are deaf: The importance of early identification and communication access. *Journal of Early Hearing Detection and Intervention* 6 (2): 101–113.
Peterson, C.C. 2004. Theory-of-mind development in oral deaf children with cochlear implants or conventional hearing aids. *Journal of Child Psychology and Psychiatry* 45 (6): 1096–1106. https://doi.org/10.1111/j.1469-7610.2004.t01-1-00302.x.
———. 2009. Development of social-cognitive and communication skills in children born deaf. *Scandinavian Journal of Psychology* 50 (5): 475–483. https://doi.org/10.1111/j.1467-9450.2009.00750.x.
———. 2016. Empathy and theory of mind in deaf and hearing children. *The Journal of Deaf Studies and Deaf Education* 21 (2): 141–147. https://doi.org/10.1093/deafed/env058.
Peterson, C.C., and M. Siegal. 1995. Deafness, conversation, and theory of mind. *Journal of Child Psychology & Psychiatry & Allied Disciplines* 36 (3): 459–474. https://doi.org/10.1111/j.1469-7610.1995.tb01303.x.
———. 1997. Psychological, biological, and physical thinking in normal, autistic, and deaf children. In *The emergence of core domains of thought*, ed. H.M. Wellman and K. Inagaki, 55–70. Jossey-Bass.
———. 1998. Changing focus on the representational mind: Deaf, autistic and normal children's concepts of false photos, false drawings and false beliefs. *British Journal of Developmental Psychology* 16 (3): 301–320. https://doi.org/10.1111/j.2044-835X.1998.tb00754.x.
———. 1999. Representing inner worlds: Theory of mind in autistic, deaf, and normal hearing children. *Psychological Science* 10 (2): 126–129. https://doi.org/10.1111/1467-9280.00119.

Peterson, C.C., and V.P. Slaughter. 2003. Opening windows into the mind: Mothers' preferences for mental state explanations and children's theory of mind. *Cognitive Development* 18 (3): 339–429. https://doi.org/10.1016/S0885-2014(03)00041-8.

———. 2006. Telling the story of theory of mind: Deaf and hearing children's narratives and mental state understanding. *British Journal of Developmental Psychology* 24 (1): 151–179. https://doi.org/10.1348/026151005X60022.

Peterson, C.C., and H.M. Wellman. 2009. From fancy to reason: Scaling deaf and hearing children's understanding of theory of mind and pretense. *British Journal of Developmental Psychology* 27 (2): 297–310. https://doi.org/10.1348/026151008X299728.

———. 2019a. Explaining the unpredictable: The development of causal theories of mind in deaf and hearing children. *Child Development* 90 (6): e654–e674. https://doi.org/10.1111/cdev.13092.

———. 2019b. Longitudinal theory of mind (ToM) development from preschool to adolescence with and without ToM delay. *Child Development* 90 (6): 1917–1934. https://doi.org/10.1111/cdev.13064.

Peterson, C.C., H.M. Wellman, and D. Liu. 2005. Steps in theory-of-mind development for children with deafness or autism. *Child Development* 76 (2): 502–517. https://doi.org/10.1111/j.1467-8624.2005.00859.x.

Peterson, C.C., H.M. Wellman, and V. Slaughter. 2012. The mind behind the message: Advancing theory-of-mind scales for typically developing children, and those with deafness, autism, or Asperger syndrome. *Child Development* 83 (2): 469–485. https://doi.org/10.1111/j.1467-8624.2011.01728.x.

Peterson, C.C., K. O'Reilly, and H.M. Wellman. 2016a. Deaf and hearing children's development of theory of mind, peer popularity, and leadership during middle childhood. *Journal of Experimental Child Psychology* 149: 146–158. https://doi.org/10.1016/j.jecp.2015.11.008.

Peterson, C.C., V.P. Slaughter, C. Moore, and H.M. Wellman. 2016b. Peer social skills and theory of mind in children with autism, deafness, or typical development. *Developmental Psychology* 51 (1): 46–67. https://doi.org/10.1037/a0039833.

Peterson, C.C., V. Slaughter, and H.M. Wellman. 2018. Nimble negotiators: How theory of mind (ToM) interconnects with persuasion skills in children with and without ToM delay. *Developmental Psychology* 54 (3): 494–509. https://doi.org/10.1037/dev0000451.

Pillow, B.H. 1989. Early understanding of perception as a source of knowledge. *Journal of Experimental Child Psychology* 47 (1): 116–129. https://doi.org/10.1016/0022-0965(89)90066-0.

Pisoni, D.B., and W.G. Kronenberger. 2021. Recognizing spoken words in semantically-anomalous sentences: Effects of executive control in early-implanted deaf children with cochlear implants. *Cochlear Implants International* 22 (4): 223–236. https://doi.org/10.1080/14670100.2021.1884433.

Pisoni, D.B., W.G. Kronenberger, M.S. Harris, and A.C. Moberly. 2017. Three challenges for future research on cochlear implants. *World Journal of Otorhinolaryngology-Head and Neck Surgery* 3: 240–254. https://doi.org/10.1016/j.wjorl.2017.12.010.

Pluta, A., M. Krysztofiak, M. Zgoda, J. Wysocka, K. Golec, J. Wojcik, E. Włodarczyk, and M. Haman. 2021. False belief understanding in deaf children with cochlear implants. *Journal of Deaf Studies and Deaf Education* 26 (4): 511–521. https://doi.org/10.1093/deafed/enab015.

Pratt, C., and P. Bryant. 1990. Young children understanding that looking leads to knowing (so long as they are looking into a single barrel). *Child Development* 61 (4): 973–982.

Qu, L., P. Shen, Y.Y. Chee, and L. Chen. 2015. Teachers' theory-of-mind coaching and children's executive function predict the training effect of sociodramatic play on children's theory of mind. *Social Development* 24 (4): 716–733.

Remmel, E., and K. Peters. 2009. Theory of mind and language in children with cochlear implants. *Journal of Deaf Studies and Deaf Education* 14 (2): 218–236. https://doi.org/10.1093/deafed/enn036.

Repacholi, B.M., and A. Gopnik. 1997. Early reasoning about desires: Evidence from 14- and 18-month-olds. *Developmental Psychology* 33 (1): 12–21. https://doi.org/10.1037//0012-1649.33.1.12.

Roggman, L.A., G.A. Cook, M.S. Innocenti, V.J. Norman, and K. Christiansen. 2013. Parenting interactions with children: Checklist of observations linked to outcomes (PICCOLO) in diverse ethnic groups. *Infant Mental Health Journal* 34 (4): 290–306. https://doi.org/10.1002/imhj.21389.

Romano, D.R., W.G. Kronenberger, S.C. Henning, C.J. Montgomery, A.M. Ditmars, C.A. Johnson, H.D. Bozell, A.D. Yates, and D.B. Pisoni. 2021. Verbal working memory error patterns and speech-language outcomes in youth with cochlear implants. *Journal of Speech, Language, and Hearing Research: JSLHR* 64 (12): 4949–4963. https://doi.org/10.1044/2021_JSLHR-21-00114.

Roos, C., E. Cramér-Wolrath, and K.W. Falkman. 2016. Intersubjective interaction between deaf parents/deaf infants during the infant's first 18 months. *Journal of Deaf Studies & Deaf Education* 21 (1): 11–22. https://doi.org/10.1093/deafed/env034.

Ruffman, T., L. Slade, and E. Crowe. 2002. The relation between children's and mothers' mental state language and theory-of-mind understanding. *Child Development* 73 (3): 734–751.

Russell, P.A., J.A. Hosie, C.D. Gray, C. Scott, and N. Hunter. 1998. The developmental of theory of mind in deaf children. *Journal of Child Psychology and Psychiatry* 39 (6): 903–910.

Schick, B., P. De Villiers, J. De Villiers, and R. Hoffmeister. 2007. Language and theory of mind: A study of deaf children. *Child Development* 78 (2): 376–396. https://doi.org/10.1111/j.1467-8624.2007.01004.x.

Schuwerk, T., B. Priewasser, B. Sodian, and J. Perner. 2018. The robustness and generalizability of findings on spontaneous false belief sensitivity: A replication attempt. *Royal Society Open Science* 5 (5): 1–16. https://doi.org/10.1098/rsos.172273.

Shaver, D.M., M. Marschark, L. Newman, and C. Marder. 2013. Who is where? Characteristics of deaf and hard of hearing students in regular and special schools. *Journal of Deaf Studies and Deaf Education* 19 (2): 203–219. https://doi.org/10.1093/deafed/ent056.

Siegal, M., and C.C. Peterson. 2008. Language and theory of mind in atypically developing children. In *Social cognition and developmental psychopathology*, ed. C. Sharp, P. Fonagy, and I. Goodyer, 81–112. Oxford University Press.

Slaughter, V., and C.C. Peterson. 2012. How conversational input shapes theory of mind development in infancy and early childhood. In *Access to language and cognition development*, ed. M. Siegal and L. Surian. Oxford University Press. https://doi.org/10.1093/acprof:oso/9780199592722.003.0001.

Slaughter, V., C.C. Peterson, and E. Mackintosh. 2007. Mind what mother says: Narrative input and theory of mind in typical children and those on the autism spectrum. *Child Development* 78 (3): 839–858. https://doi.org/10.1111/j.1467-8624.2007.01036.x.

Slaughter, V., C.C. Peterson, and C. Moore. 2013. I can talk you into it: Theory of mind and persuasion behavior in young children. *Developmental Psychology* 49 (2): 227–231. https://doi.org/10.1037/a0028280.

Slaughter, V., K. Imuta, C.C. Peterson, and J.D. Henry. 2015. Meta-analysis of theory of mind and peer popularity in preschool and early school years. *Child Development* 86 (4): 1159–1174. https://doi.org/10.1111/cdev.12372.

Spencer, L.J., B.J. Tomblin, and B.J. Gantz. 2012. Growing up with a cochlear implant: Education, vocation, and affiliation. *Journal of Deaf Studies and Deaf Education* 17 (4): 483–498. https://doi.org/10.1093/deafed/ens024.

Stanzione, C., and B. Schick. 2014. Environmental language factors in theory of mind development: Evidence from children who are deaf/hard-of-hearing or who have specific language impairment. *Topics in Language Disorders* 34 (4): 296–312. https://doi.org/10.1097/TLD.0000000000000038.

Steeds, L., K. Rowe, and A. Dowker. 1997. Deaf children's understanding of beliefs and desires. *Journal of Deaf Studies and Deaf Education* 2 (3): 185–195. https://doi.org/10.1093/oxfordjournals.deafed.a014324.

Strong, M., and P. Prinz. 1997. A study of the relationship between American sign language and English literacy. *Journal of Deaf Studies and Deaf Education* 2 (1): 37–46. https://doi.org/10.1093/oxfordjournals.deafed.a014308.

Sullivan, K., D. Zaitchik, and H. Tager-Flusberg. 1994. Preschoolers can attribute second-order beliefs. *Developmental Psychology* 30 (3): 395–402. https://doi.org/10.1037/0012-1649.30.3.395.

Sundqvist, A., B. Lyxell, R. Jönsson, and M. Heimann. 2014. Understanding minds: Early cochlear implantation and the development of theory of mind in children with profound hearing impairment. *International Journal of Pediatric Otorhinolaryngology* 78 (3): 538–544. https://doi.org/10.1016/j.ijporl.2013.12.039.

Surian, L., and A. Geraci. 2012. Where will the triangle look for it? Attributing false beliefs to a geometric shape at 17 months. *The British Journal of Developmental Psychology* 30 (Pt 1): 30–44. https://doi.org/10.1111/j.2044-835X.2011.02046.x.

Symons, D.K., C.C. Peterson, V. Slaughter, J. Roche, and E. Doyle. 2005. Theory of mind and mental state discourse during book reading and story-telling tasks. *British Journal of Developmental Psychology* 23: 81–102. https://doi.org/10.1348/026151004X21080.

Symons, D.K., K.-L.M. Fossum, and T.B.K. Collins. 2006. A longitudinal study of belief and desire state discourse during dother-child play and later false belief understanding. *Social Development* 15 (4): 676–691. https://doi.org/10.1111/j.1467-9507.2006.00364.x.

Szagun, G., and B. Stumper. 2012. Age or experience? The influence of age at implantation and social and linguistic environment on language development in children with cochlear implants. *Journal of Speech, Language, and Hearing Research* 55 (6): 1640–1654. https://doi.org/10.1044/1092-4388(2012/11-0119).

Taumoepeau, M., and T. Ruffman. 2006. Mother and infant talk about mental states relates to desire language and emotion understanding. *Child Development* 77 (2): 465–481. https://doi.org/10.1111/j.1467-8624.2006.00882.x.

———. 2008. Stepping stones to others' minds: Maternal talk relates to child mental state language and emotion understanding at 15, 24, and 33 months. *Child Development* 79 (2): 284–302. https://doi.org/10.1111/j.1467-8624.2007.01126.x.

Tomasuolo, E., G. Valeri, A. Di Renzo, P. Pasqualetti, and V. Volterra. 2012. Deaf children attending different school environments: Sign language abilities and theory of mind. *Journal of Deaf Studies and Deaf Education* 18 (1): 12–29. https://doi.org/10.1093/deafed/ens035.

Tomblin, J.B., M. Harrison, S.E. Ambrose, E.A. Walker, J.J. Oleson, and M.P. Moeller. 2015. Language outcomes in young children with mild to severe hearing loss. *Ear and hearing 36* (Suppl 1): 76S–91S. https://doi.org/10.1097/AUD.0000000000000219.

Tucci, S.L., S.R. Easterbrooks, and A.R. Lederberg. 2016. The effects of theory of mind training on the false belief understanding of deaf and hard-of-hearing students in prekindergarten and kindergarten. *Journal of Deaf Studies and Deaf Education* 21 (3): 310–325. https://doi.org/10.1093/deafed/enw031.

Walker, E.A., S.E. Ambrose, J. Oleson, and M.P. Moeller. 2017. False belief development in children who are hard of hearing compared with peers with normal hearing. *Journal of Speech Language and Hearing Research* 60: 3487–3506. https://doi.org/10.1044/2017_JSLHR-L-17-0121.

Watson, A.C., C.L. Nixon, A. Wilson, and L. Capage. 1999. Social interaction skills and theory of mind in young children. *Developmental Psychology* 35 (2): 386–391. https://doi.org/10.1037/0012-1649.35.2.386.

Welch-Ross, M.K. 1997. Mother-child participation in conversation about the past: Relationships to preschoolers' theory of mind. *Developmental Psychology* 33 (4): 618–629. https://doi.org/10.1037//0012-1649.33.4.618.

Wellman, H.M. 2002. Understanding the psychological world: Developing a theory of mind. In *Handbook of childhood cognitive development*, ed. U. Goswami, 167–187. Blackwell.

———. 2011. Developing a theory of mind. In *Blackwell handbook of childhood cognitive development—Second edition*, ed. U. Goswami, 258–284. http://onlinelibrary.wiley.com/doi/10.1002/9781444325485.ch/summary.

———. 2014. *Making minds: How theory of mind develops*. Oxford University Press.

Wellman, H.M., and D. Liu. 2004. Scaling of theory-of-mind tasks. *Child Development* 75 (2): 523–541. https://doi.org/10.1111/j.1467-8624.2004.00691.x.

Wellman, H.M., and C.C. Peterson. 2013. Deafness, thought bubbles, and theory-of-mind development. *Developmental Psychology* 49 (12): 2357–2367. https://doi.org/10.1037/a0032419.

Wellman, H.M., and J.D. Woolley. 1990. From simple desires to ordinary beliefs: The early development of everyday psychology. *Cognition* 35 (3): 245–275. https://doi.org/10.1016/0010-0277(90)90024-e.

Wellman, H.M., M. Hollander, and C.A. Schult. 1996. Young children's understanding of thought bubbles and of thoughts. *Child Development* 67: 768–788. https://doi.org/10.1111/j.1467-8624.1996.tb01763.x.

Wellman, H.M., D. Cross, and J. Watson. 2001. Meta-analysis of theory-of-mind development: The truth about false belief. *Child Development* 72 (3): 655–684. https://doi.org/10.1111/1467-8624.00304.

Wellman, H.M., S. Lopez-Durran, J. Labounty, and B. Hamilton. 2008. Infant attention to action predicts preschool theory of mind. *Developmental Psychology* 44 (2): 618–623. https://doi.org/10.1037/0012-1649.44.2.618.

Westby, C. 2014. Theory of mind deficits in students who are deaf or hard-of-hearing. *Word of Mouth* 26 (1): 9–12. https://doi.org/10.1177/1048395014541784b.

Westby, C., and L. Robinson. 2014. A developmental perspective for promoting theory of mind. *Topics in Language Disorders* 34 (4): 362–382. https://doi.org/10.1097/TLD.0000000000000035.

Wiefferink, C.H., C. Rieffe, L. Ketelaar, L. De Raeve, and J.H.M. Frijns. 2012. Emotion understanding in deaf children with a cochlear implant. *Journal of Deaf Studies and Deaf Education* 18: 175–186.

Wimmer, H., and J. Perner. 1983. Beliefs about beliefs: Representation and constraining function of wrong beliefs in young children's understanding of deception. *Cognition* 13 (1): 103–128. https://doi.org/10.1016/0010-0277(83)90004-5.

Woolfe, T., S. Want, and M. Siegal. 2002. Signposts to development: Theory of mind in deaf children. *Child Development* 73 (3): 768–778.

———. 2003. Siblings and theory of mind in native deaf signing children. *Journal of Deaf Studies and Deaf Education* 8 (3): 340–347. https://doi.org/10.1093/deafed/eng02.

Yu, C.-L., C.M. Stanzione, H.M. Wellman, and A.R. Lederberg. 2021. Theory-of-mind development in young deaf children with early hearing provisions. *Psychological Science* 32 (1): 109–119. https://doi.org/10.1177/0956797620960389.

Ziv, M., T. Most, and S. Cohen. 2013a. Understanding of emotions and false beliefs among hearing children versus deaf children. *Journal of Deaf Studies and Deaf Education* 18 (2): 161–174. https://doi.org/10.1093/deafed/ens073.

Ziv, M., M.L. Smadja, and D. Aram. 2013b. Mothers' mental-state discourse with preschoolers during storybook reading and wordless storybook telling. *Early Childhood Research Quarterly* 28 (1): 177–186. https://doi.org/10.1016/j.ecresq.2012.05.005.

Chapter 9
Analyzing the Dynamics Between Theory of Mind, Speech Disorders, and Brain Rewiring in Aphasia

Teresa Lopez-Soto

> *Any man could, if he were so inclined, be the sculptor of his own brain.*
>
> Santiago Ramón y Cajal.

Abstract Much has been discussed about the Theory of Mind (ToM) and language. Some authors suggest that activating one enlightens the other, while others view language merely as a tool for developing the most distinctly human ability known. When it comes to speech and communication disorders, tracing ToM can be challenging since patients lack the tools to demonstrate their cognitive performance. This chapter delves into the biological foundations of ToM (anatomical and neurochemical). It also explores the evolution of ToM and neuroplasticity along our lifespan so that the reader can understand that nothing is static, that our brain engages in an everlasting dynamic relationship to keep us healthy and safe. This understanding has significant implications in how we approach language pathologies and the cognitive decline they might hinder. Additionally, the chapter examines the pragmatic side of language, offering insights into the right hemisphere and its role in emotion and affective communication. It considers the production-perception loop in communication, acknowledging that language and cognition cannot always be solely measured from production or perception alone. Social, cultural, and pathological factors may intervene and disrupt this loop, prompting the need for careful consideration in ToM theories and methods, particularly in the context of language pathologies. All these considerations should not be considered as preliminary or complementary, but central in understanding the implications of language in ToM, what language must really entail, so that the best clinical assessment and intervention is possible for patients with language pathologies. Finally, the chapter blends all

T. Lopez-Soto (✉)
Laboratory of Speech and Phonetic Sciences, University of Seville, Seville, Spain

Department of Forensic Medicine, Psychiatry and Pathology, Complutense University of Madrid, Madrid, Spain
e-mail: teresals@us.es; teresl01@ucm.es

T. Lopez-Soto et al. (eds.), *The Theory of Mind Under Scrutiny*, Logic, Argumentation & Reasoning 34, https://doi.org/10.1007/978-3-031-46742-4_9

these analyses into offering a review of crucial aspects of ToM in patients with aphasia.

Keywords Theory of mind · Language · Production-perception integrated theory · Neuroplasticity · Right hemisphere · Language pathologies · Aphasia · Communication · Culture · Social context · Pragmatics

9.1 Linguistic Alignment and Theory of Mind

Theory of Mind (ToM) (or *mentalizing*) is the capacity to attribute and understand others' mental states. The term has also been used to relate to consciousness, the attribution of oneself. It should be considered a cognitive ability that let us process other people's beliefs, desires, intentions, and perspectives the same way we process ours. This "knowledge" helps us mingle with the group, it shapes our own personality and identity while playing a very important role in our own survival: ToM allows us to interpret and predict others' behaviors based on our "perception" of their mental states. ToM is also the capacity to attribute mental states to oneself and, for that reason, this area of research has attracted the interest of psychologists and philosophers alike. The development of this skill takes place early in life. Around the age of 2–3 years, children start to recognize beliefs and intentions in others different from their own. As we grow older, this ability becomes more efficient allowing us to successfully interact in complex social situations. Language (see Chaps. 17 and 18 for a review) and ToM are believed to go hand in hand in these early years, even though the development of ToM skills continues to readapt and readjust during our lifetime span while language skills remain intact (subtleties on this statement are still under continuous redefinition by researchers). All in all, without this capacity, we would not be able to enjoy a midsummer drive-in movie, appreciate the nuances of Lovecraft or Jane Austen's pieces of writing, or admire Leonardo da Vinci's masterpieces. ToM animates our humor, cynicism, suspicion, deceit, beliefs and intertwines our most complex and uncomplicated social relationships alike.

It was Premack and Woodruff (1978) who probably kicked off this amazing search into the complexities of human cognition laying the foundation for modern research on ToM. This research has potentially been increased in the past decades through studies that have tried to discover the linguistic and cognitive processes that underlie ToM reasoning as well as the neural systems involved. Many studies carried out in children, people with hearing loss, developmental and acquired language disorders, autism, and patients with brain lesions have guided investigation in the field.

In this sense, this volume presents a scrutiny of studies in all the aforementioned areas to better understand this fundamental aspect of human cognition.[1] In this

[1]ToM, in light of present research, is fundamentally a human skill. While numerous studies (for a general review and update see Kano et al. (2019) and Krupenye and Call (2019)) analyze the same

volume, Chaps. 12 and 13, authors discuss the idea of consciousness and "mind", one may even speak of soul, and the cognitive construct that let us build an image of ourselves. Is that image correct? Can we really predict other people's mental states? Many of us have experienced the contradiction that lies in not believing in ourselves, in the eyes of others. We have also attributed wrong states of mind to acquaintances, relatives, or our beloved ones. Are we attributing the same "defective" impressions to the consciousness that we believe defines us? When we dwell on the philosophical implications of ToM, we cannot but conclude that our cognitive capacity is very much limited. Jean-Paul Sartre made the statement that "we perceive what we need to", and that seems to suffice for our brain. Our perception of reality is not only biased by our own senses[2] but also by our own experience. The American philosopher and educator, John Dewey, was a strong believer of the inseparability of mind and body. He defended the idea that our brain is shaped according to our own life experience. Because of that, he was also a proponent of neuroplasticity and he argued that our cognitive skills, knowledge, behavior, and personality are not fixed traits but are rather shaped through our interactions with the environment.

All in all, what really matters when we approach ToM is that this skill helps us integrate ourselves and our fellow mates in our own construct of reality. This construct is ever-changing and although limited by our own physiology (read footnote 3 again), is what characterizes us as humans. This volume offers a very thorough review of pathological cases where ToM may be deficient: autism, brain injury, hearing loss among other psychopathologies. The common ground to those studies and reviews is that there exists a contextual processing impairment during the perception stage: individuals are not capable of analyzing others' mental states or their own mental state, for the matter. That impairment has serious consequences during the production stage whether it be behavioral or communicative. That is to say, our lack of ToM prevents us from surviving fully and healthily in our environment, it damages our own development as individuals and poses a certain risk to what we can provide for the group.

In this chapter, we will review important aspects related to ToM in patients where language has been impaired. I will not adopt a unique perspective: Is language and ToM indivisible systems or does ToM evolve into much more complex cognitive functions while the language system remains the same?

ToM is a complex cognitive system primarily located in the temporoparietal junction (TPJ) (shifts attention to unexpected stimuli, reorienting of attention) but that also revolves around a complex circuitry, including (a) the medial prefrontal cortex (executive functions), (b) the superior temporal sulcus (perception of faces

ability traced in primates, dolphins, birds, and dogs, primarily, there is still much to be discussed before completely disregarding ToM as a unique human trait.

[2]In our vision, for example, there is a specific area in both eyes, the "blind spot", where the optic nerve exits the retina, that is to say, there are no light-sensitive cells (photoreceptors) present. Actually, our eyes cannot detect every single aspect of light or form from our surroundings. How come we claim we can see, then? Well, our brain fills in the gap where actual visual perception did not take place by compensating with information based on the surrounding visual context.

and human gestures), (c) the posterior cingulate cortex (internal mentation, memory and episodic retrieval, spatial orientation and navigation, self-processing and social cognition, emotion and emotional regulation), and (d) the precuneus (spatial cognition, attention, and consciousness). ToM is related to executive functions, language and shows extensive connections with regions specifically involved in affective/emotional processing, such as the amygdala, temporal pole, and anterior insula (Shamay-Tsoory et al. 2006).

As a firm believer in **neuroplasticity** (Hebb 2005) and the neuropsychologist Donald Hebb's would popularly quote "neurons that fire together, wire together". ToM cannot be a brain circuitry that does not become stronger throughout our lifespan, just as much as language is not a monolithic that remains unaltered in its syntactic and lexical complexity from our early childhood. Speech therapy is rooted in this belief (Schlaug et al. 2009; Johnson and Pring 1990; Dollaghan 2007; Joyal et al. 2016; Clegg et al. 2007; Lorenzen and Murray 2008).

On the other hand, it is necessary to bear in mind that language is not restricted to the left hemisphere. The **right hemisphere** also plays a fundamental role in communication and is responsible of various functions related to nonverbal cues and spatial processing. Pragmatic information refers to the social and contextual aspects of language use and is necessary to understand the speaker's intentions based on the shared knowledge that speaker and listener share. Some of these functions are the following:

1. Nonverbal communication: Processing nonverbal cues (facial expressions, gestures, and body language).
2. Prosody: Intonation, stress patterns, and pitch variation.
3. Inference and contextual knowledge: Filling the gaps in conversation through shared knowledge and background understanding.

There may exist a pragmatic awareness that precludes the expression of ToM reasoning for which lateralization plays a definite role. Another important issue to be addressed here is the feedback that happens between language production and language perception. It is undeniably true that language is an important contributor to ToM, but language is a two-way process. Production and perception in language processing cannot be defined exclusively. We need to accept that the two systems are interwoven, and that this connection is used in predicting other people's behavior. The scientific basis behind is that we design our actions as a prediction of others' actions. **Predictive coding**, rooted in Bayesian inference, comes to say that our brain is constantly redefining our mental states by comparing the internal mental patterns with sensory information and readjusting them accordingly. In recent years, computational neuroscience has shifted to non-traditional models to understand how our brain works. New studies have been focusing on predictive coding and generative models. These new approaches aim to decipher how the brain interacts with our perception. In predictive coding, the brain is in a constant state of "guessing of what is going to happen next". The key to know whether predictions are solid and should be kept in store is feedback from new perception: the better the brain gets at predicting, the stronger these representations of the world wire in the brain. We can call this a "process of learning". Predictive coding is being used in Artificial Intelligence (AI) models of learning and is rooted in **Bayesian inference** and **Helmhortz's unconscious inference** (Barlow 1990). The hypothesis behind Bayesian inference, for example, is that our senses transduce noisy and ambiguous information, if not incomplete. In order to cope with this limitation, our nervous system works in terms of probability. The Bayes' theorem allows for a statistical model that can incorporate prior knowledge while updating it with new observed data so that a posterior distribution can be built. This new distribution represents an updated knowledge about a hypothesis/parameter and therefore allows for flexible and iterative updating of that same knowledge, handling uncertainty in an effective way (Colombo and Seriès 2012).

The Bayesian inference model has received acclaim and criticism in equal measure (Bowers and Davis 2012) while many authors defend a non-bayesian model for the human brain and still many more adopting a hybrid approach (Adler and Ma 2018; Tappin and Gadsby 2019; Bromiley et al. 2003).

In this chapter, I review the neural basis of ToM, emphasize the pragmatic hypothesis and the function of the right hemisphere in our understanding of others' mental states as well as I defend the hypothesis of an integrated theory of perception and production in language: I consider that actors construct forward models of their actions before actually executing them and that this also happens when we observe and imitate others' actions. This argumentation will precede an insight on ToM in communication disorders.

9.2 The Neural Basis of Theory of Mind (ToM)

9.2.1 The Anatomical Hypothesis of Theory of Mind

As with other cognitive skills and brain processes, the physiological basis of ToM is an ongoing topic of research. One thing is for sure: there does not exist a single region in charge of this neurological process. Evidence points in the direction of a "core" system that is distinctive as a neuronal circuitry. A failure in this core system will result in a failed ToM. Research carried out with patients with brain lesions show, for example, that ToM may become impaired, but not always. This fact makes us corroborate the existence of one core system which is, nevertheless, integrated in a more sophisticated circuitry for which neuroplasticity plays an essential role. For this reason, a failure in performing a ToM task may be due to 2 reasons:

A. An impairment at the level of that "core" system.
B. A failure to recruit a cognitive component essential to the ToM system but which has been "co-opted" to support performance in a different cognitive task, thus, promoting competition.

The study of the neural basis of ToM is a challenging matter. As already stated, imaging studies show contradictory outcomes in patients with brain damage. The main conclusion, so far, is that ToM is a central cognitive skill for humans and takes advantage of neuroplasticity to ensure its potential use. Patients with brain damage may show ToM skills in different areas (cortical lesions) indicating that ToM reasoning can be sustained no matter how severe or large these areas may be. The main conclusion is that there is no specific brain region that is dedicated solely to ToM. Several brain areas and neural networks play an important role and some central brain regions associated with this skill are the following:

1. **Temporoparietal Junction** (TPJ): The TPJ shifts attention to unexpected stimuli and is activated when attention is reoriented. The TPJ has been described as the most important brain region in ToM abilities.
2. **Medial prefrontal cortex** (mPFC): The mPFC and, more specifically, the ventromedial prefrontal cortex (vmPFC), is involved in the representation and processing of mental states, self-referential processing, and social cognition.
3. **Superior temporal sulcus** (STS): The STS is implied in processing social cues, including eye gaze, body gestures, and facial expressions. It is implied in perceiving and processing information on the beliefs, intentions, and mental states of others.
4. **Posterior cingulate cortex** (PCC) and precuneus: These regions are specialized in self-referential processing, introspection, and consciousness. They are involved in maintaining an image of self while the individual engages during social tasks that require ToM.
5. **Mirror neuron system**: The mirror neurons (MNs) (**premotor cortex, inferior parietal lobe**) relate to empathy and play an important role in imitation, and ToM processes.

Two distinctive areas are governed by this core system: (a) those that are specific to the understanding of the self and (b) those that are associated with social interaction and interpretation of others' mental states. This means that there must exist a competent outcome of both interpretation processes: one cannot understand others' mental states if introspection and self-referential processing is not complete. Another hypothesis could be that one cannot introspect and gain consciousness if interpretation of individuals around us is not satisfied completely.

I have mentioned that ToM may fail when an essential component to the ToM system has been "superseded" to support performance at an alternative cognitive task. It can fail if the "core" system is not tuned. The competition between the ToM task and other cognitive tasks happens because ToM is supported by 3 distinct neural mechanisms that are the following:

1. Language.
2. Executive Functions (EF).
3. Visuo-spatial processing in object and location tracking.

These are 3 essential neural systems in humans with independent functionality, away from the ToM task. For this reason, they may interfere with social processing and introspection. We use 2 criteria to distinguish 2 possible alternatives for the architecture of ToM in an individual: **dissociation** and **amelioration**.

A. **Dissociation** takes place when there is autonomy of the "core" ToM system from potential co-opted systems (whether it be language, EF, or visuo-spatial processing). This can be tested, for example, by examining patients with severe language disorders. Successful performance in ToM tasks will demonstrate this dissociation (Siegal and Varley 2002).
B. **Amelioration**, on the other hand, happens when successful performance of ToM tasks can only take place if visuo-spatial memory is facilitated using, for example, visual aids. If this is the case (as it happens with children younger than 4 years and adults with brain lesion). In this case, the "core" system of ToM is dependent on other cognitive neural systems in the brain.

At this point, the question is: is there a ToM "core" system? How do all these areas interact on ToM? Other areas would be the frontal lobes (for the EF) and amygdala circuits. The amount of research in delimiting the brain regions that take places at ToM is extensive. However, few studies account for the integration of these regions into a functionally interconnected circuit (Abu-Akel 2003a). Table 9.1 (from Abu-Akel and Shamay-Tsoory 2011, p. 2972) presents a good account of the various regions that researchers have found of relevance to identify ToM skills (collection extracted from the works of Abu-Akel (2003a), Brunet-Gouet and Decety (2006), Carrington and Bailey (2009), Frith and Frith (2006), Saxe (2006), Van Overwalle and Baetens (2009)).

In Fig. 9.1, one can appreciate a graphical view of the brain regions involved in ToM according to the review presented above. This figure is limited to cortical regions, but as can be seen, they are quite representative of language, EF, and

Table 9.1 Theory of mind brain regions (cf. Abu-Akel and Shamay-Tsoory 2011, p.2972)

Brain region	Brodmann's area
Posterior regions	
Temporo-parietal junction (including the inferior parietal lobe) (IPL/pSTS or TPJ)	39/40
Posterior cingulate/precuneus (PCC/PCun)	31/7
Superior temporal sulcus (STS)	21/22
Limbi-paralimbic regions	
Orbitofrontal (OFC)	11/12/47
Ventral medial prefrontal cortex (vMPFC)	10/32
Anterior cingulate/paracingulate cortex (ACC/PrCC)	24/32
Temporal pole (TP)	38
Amygdala	Subcortical
Striatum	Subcortical
Frontal regions	
Dorsal medial prefrontal cortex (dMPFC)	8/9
Dorsal lateral prefrontal cortex (DLPFC)	9/46
Inferior lateral frontal cortex (ILFC)	44/45/47

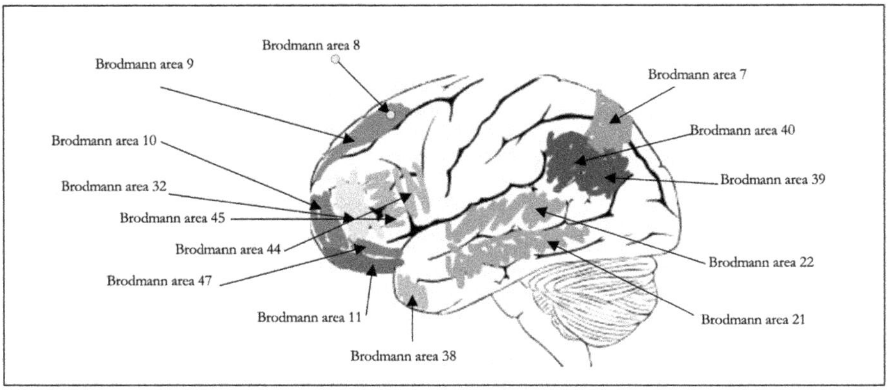

Fig. 9.1 Cortical areas involved in ToM according to Abu-Akel and Shamay-Tsoory's review (2011)

Table 9.2 Relation of Brodmann's areas involved in ToM and their functionality

Brodmann's area	Functionality
10	Risk and decision making, reward and conflict, pain, and working memory
32	It connects information between cognitive and emotional brain regions
45	Broca's area; semantic decision tasks (e.g., abstract vs. concrete interpretation of a word) and generation tasks (syntax)
44	Language processing (production and comprehension)
47	Empathy
11	Decision making and processing rewards, planning, encoding new information into long-term memory, and reasoning
38	Visual cognition, face recognition, and visual memory
9	Motor planning, organization, regulation, sustaining attention and working memory
8	Visual attention and control of eye movements
21	Secondary auditory cortex: auditory processing and language
22	Primary auditory cortex: auditory processing and language
39	Language and number processing, spatial cognition, memory retrieval, and attention
40	Reading (meaning and phonology)
7	Visuo-motor coordination

emotions. Below, in Table 9.2, a review of the main functionalities of the Brodmann areas involved.

All our cortical skills (albeit the motor area) are fundamental for ToM, but what is not so clear is how they are connected. The truth is that ToM works as a single unit which comprises both cognitive and affective processing. Authors such as Brothers and Ring (1992) relate to this twofold structure as the "cold-hot" dimension: the "cold" or cognitive dimension of ToM oversees the inference about knowledge and beliefs; the "hot" or affective dimension of ToM takes place with inferring emotions. Three groups can be distinguished in the anatomical differentiation displayed in Table 9.1:

A. Affective ToM processing can be found in:

  - Prefrontal Cortex (PFC)
  - Orbitofrontal (OFC)
  - Ventral medial prefrontal cortex (vMPFC)
  - Inferior lateral frontal cortex (ILFC)

  All these areas show a binding connection with the amygdala (Abu-Akel and Shamay-Tsoory 2011) and are believed to represent the affective parameter of the ToM system.

B. Cognitive ToM processing, on the other hand, can be found in:

  - Dorsal medial prefrontal cortex (dMPFC)
  - Dorsal lateral prefrontal cortex (DLPFC)

  Neither of these two show any connection to the limbic system, therefore exhibiting a clear cognitive function in the ToM system.

C. Some of the areas listed above, though, may have both a cognitive and affective parameter alike.

  - Anterior cingulate cortex (ACC). Evidence suggests that the ventral ACC (vACC) is involved in emotional aspects of self-reflection (Abu-Akel and Shamay-Tsoory 2011) while the dorsal ACC (dACC) would be involved in processing cognitive ToM, predominantly.

9.2.2 The Neurochemical Hypothesis of Theory of Mind

Investigations on impaired populations, such as autism and schizophrenia, have demonstrated that ToM abilities are subsidiary to the serotonergic and dopaminergic system. Abnormalities in these two systems can account for differences in ToM abilities among patients with different pathologies. The hypothesis behind lies in the fact that patients with autism and schizophrenia show deficits in the dopaminergic-serotonergic (DS) system (Abu-Akel 2003b).

Although autism and schizophrenia both exhibit dysregulated DS systems, schizophrenia primarily stems from dysfunctions in the dopaminergic system, whereas autism primarily stems from a dysfunction in the serotonergic system. Schizophrenic patients experience two different disruptions in the dopaminergic transmission system. The first disturbance involves an increasing activity in the mesolimbic subcortical component of the dopamine system. The second one involves a decreased activity in the mesocortical component of the dopaminergic transmission system.

In relation to autism, studies on neurochemicals reveal elevated average levels of 5-HT (5-hydroxytryptamine) in the bloodstream of children with autism compared to the general population. It is reported that 30–50% of children and adolescents with autism exhibit hyperserotonemia, which refers to increased serotonin levels (Muller et al. 2016).

Much can be said about the circumstantiality of these statements. However, it is only logical that the dopamine system is involved in the mechanism underlying mentalizing abilities since ToM implies that humans make predictions about the intentions or beliefs of others in order to be successful in the interaction, therefore, expecting some reward in return.

On the other hand, serotonin receptors affect different cognitive functions such as memory and EF (Buhot 1997; Roth et al. 2004; Bacqué-Cazenave et al. 2020; Cohen and Sherwood 2013).

The prefrontal cortex, temporo-parietal junction, and anterior cingulate cortex are brain regions that have been consistently implicated in mentalizing tasks based on various imaging and lesion studies (see Chap. 7 in this volume). Disruptions in both the dopaminergic and serotonergic systems can lead to impairments in cognitive abilities, including language use, which relay on ToM abilities or cognitive abilities that influence ToM, such as EF. Additionally, the dopaminergic system is believed to play a role in predicting the consequences of future events, making it a potentially important factor in the development of ToM abilities (Abu-Akel 2003b).

However, a limitation of this hypothesis is that individuals with schizophrenia and autism exhibit similar performance levels in ToM abilities. This finding raises questions about the neurochemical basis of the theory, as no correlation has been found between specific cognitive or affective processing and the underlying neurochemical origins of this pathological conditions.[3,4]

Deficits in the DS system may not be alone as part of the neurochemical hypothesis of Theory of Mind. Other neurotransmitters and their associated pathways in the brain may play a role in the functioning and development of ToM abilities. Domes et al. (2007) carried out a double-blind, placebo-controlled, within-subject design in which 30 healthy males were tested using the Reading the Mind in the Eyes Test (RMET) after intranasal administration of 24 IU oxytocin. The results

[3]The disruption of the DS systems in autism and schizophrenia have given rise to the development of different psychopharmacological interventions using atypical neuroleptics. The results have proven effective in the absence of early disruption to the DS system and extensive duration of illness. Still, atypical neuroleptics have been shown to improve cognitive functioning in patients with schizophrenia, such as attention, memory, and EF (Meltzer and McGurk 1999; Berman et al. 1999), where restoration of ToM abilities were difficult to retrieve as well as to restore language abilities (Baron-Cohen et al. 2013).

[4]Present-day neurobiological studies have improved our knowledge in the neurochemical etiology of schizophrenia. The DS system may not be the only parameter to take into account in order to understand the lack of social abilities in this pathology. Studies on the N-methyl-D-aspartate receptor (NMDA) (a glutamate receptor and ion channel found in neurons) have been shown to actively participate in synaptic processes that have a direct link with ToM abilities in schizophrenia (Coyle 2006).

demonstrate that the neuropeptide oxytocin plays a central role in social cognition since the administration of this neurotransmitter improved performance on the RMET compared with control group. More on the relationship of the hypothalamus-pituitary-adrenal axes with stress, mood, personality, and neurocognitive functioning can be found in Chap. 11 in this volume.

9.2.3 *The Motor Theory of Social Cognition*

ToM abilities are all about understanding other people's behavior. To be so, one would also have to be capable of interpreting one's own behavior. In other words, for ToM, we would need to activate the same system that we trigger when we are about to behave: the motor system. The Motor Theory of Social Cognition has much to do with the mirror neurons. In this section, I explain, briefly, what the theory is about and how much it can substantiate the ToM field of research.

For many researchers, our ToM abilities would have a specific sensorimotor foundation. We need to use the same cognitive and neural mechanisms to understand other people's behavior as the ones we use to devise our own plan of behavior. The caveat to the theory is that no anatomical study has proven evidence of an overlapping of neural systems between ToM systems and the motor system. As we have already seen in previous sections, ToM interacts with three other neural circuits: EF, language, and visuo-spatial abilities. Anatomically, neuroimaging studies have not been able to find the overlapping activation during action perception and action execution of either the ventral premotor cortex, inferior frontal gyrus and right inferior parietal cortex,[5] which are responsible for action planning, and ToM abilities.

Electrophysiology demonstrates that there exist neurons in our brain that respond to both motor and sensory events. These co-called **mirror neurons** fire in response to the observation of specific actions. The very interesting detail about this type of neurons is that they fire both when the individual is executing an action or when the individual sees some action being executed: yawning when someone else is doing it is a good example of the activation of our MNs. It comes without saying that researchers have had to give a special attention to the role of MNs and the ToM system. Are they somehow interconnected? Do we lack evidence but should continue pursuing in this direction? MNs were first identified in the macaque monkeys

---

[5]There are primary and subsidiary brain areas related to the motor system, namely:

- Primary Motor Cortex (M1): Precentral Gyrus
- Premotor Cortex: Superior Frontal Gyrus, Middle Frontal Gyrus
- Supplementary Motor Area (SMA)
- Basal Ganglia: Caudate Nucleus, Putamen, Pallidum, Thalamus
- Cerebellum
- Brainstem: Medulla, Pons, Midbrain

(premotor cortex) (Rizzolatti et al. 1996) and have inspired researchers to study where they would be located in the human brain.

Among the first electrophysiological studies on MNs in the human brain we can find the EEG study by Cochin et al. (1998, 1999) who report that observation and execution of finger movements result in reduced power of alpha wave activity (7.5–10.5 Hz). They also identified the localization of this activity in the motor and frontal cortices. Magnetoencephalography (MEG) studies carried out by Salmelin and Hari (1994) had already shown that the stimulation of the median nerves when manipulating an object result in suppression of a post-stimulus rebound effect when recording precentral motor cortex. However, the suppression is smaller when the post-stimulus rebound event is recorded in response to action observation alone. That is to say, there seems to be strong neural feedback when one is observing the performance of an action vs. when one is performing the action on their own. So far, the anatomical localization of the MNs appears to happen in the motor system.

Studies such as that by Rizzolatti and Craighero (2004) suggest that the human mirror system may be in the inferior parietal lobe, the inferior frontal gyrus (including Broca's area) and superior temporal sulcus. They employed functional Magnetic Resonance Imaging (fMRI).[6] One interesting aspect of MNs is that the system they encompass is sensitive to all the movements implied in the action being imitated, not just the action as a whole. So, for example, if someone yawns in front of us, we may as well just yawn and not just rub our eyes or verbalize that we are tired.[7]

So, going back to the connection of the Motor Theory of Social Cognition and its implications in ToM, some researchers were fast to establish a connection between MNs and our understanding of others (Gallese and Goldman 1998). These authors support the idea that "humans' mind-reading abilities rely on the capacity to adopt a simulation routine (p. 493).

> (. . .) ventral premotor cortex (referred to also as inferior area 6) is composed of two distinct areas, designated as F4 and F5. Area F5 occupies the most rostral part of inferior area 6, extending rostrally within the posterior bank of the inferior limb of the arcuate sulcus. Area F5 is reciprocally connected with the hand field of the primary motor cortex3–5 and has direct, although limited, projections to the upper cervical segments of the spinal cord. Microstimulation in F5 evokes hand and mouth movements at thresholds generally higher than in the primary motor cortex. (p. 493)

The anatomical distribution of these visuomotor neurons have been established by authors such as Matelli and Luppino (1997). They established 2 types of neurons in the F5 area after experimentation to response to visual stimuli: (a) **canonical neurons**, which are activated when observing "graspable objects", and (b) **mirror neurons**, which fire when the individual observes an action being performed (either by themselves or by other individuals). Gallese and Goldman (1998) are clear about

[6] fMRI is an imaging technique that detects oxygen levels and blood flow changes in the brain. The neural basis of the BOLD (Blood Oxygen Level Dependent signal) response has received some critical interpretation (Logothetis 2008).

[7] Given the world we live in is full of social media interaction, repetition, and action reinforcement, we should consider the power of mirror neurons for learning as individuals but also as a society.

that when they say: "every time we are looking at someone performing an action, the same motor circuits that are recruited when we ourselves perform that action are concurrently activated (p. 495)". A dazzling aspect of MNs is that **they do not fire when mimicking/using tools to substitute the original grasping "device"** (for example, when using knife and fork instead of using the fingers to grab and eat food).

We can find 2 different approaches to understand the main role of MNs in the human brain: (a) for some researchers, everything has to do with **learning by imitation** (Ramachandran 2000; Rizzolatti and Craighero 2004; Borenstein and Ruppin 2005; Arbib et al. 2000; Gallese and Stamenov 2002); (b) for others, MNs are the basis of our social cognition and, therefore, **connect with ToM**.

Rizzolatti and Craighero (2004) state that we learn by imitation in 2 different ways: (a) we change a motor pattern after observation or a more adequate pattern; and (b) we create a new motor pattern after observation to fulfill an action in a more effective way. They also established a phylogenetic paradigm of the MN system regarding communication and language. According to these authors, MNs are specialized neurons in the brain that fire when an individual performs an action and when they observe someone else performing the same action. Thus, the mirroring effect may suggest a direct link between the sender (the person that performs the action in the first place) and the receiver (the person that observes the action). The action now belongs to both of them, and a message is created.[8] In other words, by imitation and observation, the receiver is capable of understanding the sender's message. These authors support the idea that language evolved from this mirroring mechanism (see Rizzolatti and Arbib 1998). By observing others' behavior, early humans started to understand and imitate the gestures and actions of others, forming a basis for communication and, therefore, the development of language. According to this theory, then, language evolved from gestural communication, first, to vocalizations later. Other authors (Armstrong et al. 1995; Corballis 2003) have also supported the significant role of vocalizations gradually emerging as a secondary form of communication, after gestures imitation. When a gesture/vocalization is imitated, a **signifier** is created (De Saussure 1916) for a **signified**. The term "mind-reading" would have a clear explanation here, since, in this context, MNs are seen as a crucial component in bridging the gap between the actions of individuals and the understanding of those actions by others. By observing and mirroring the actions of others, humans can effectively communicate and convey information without the need for explicit verbalization.

[8]To learn more about the origins of language in the human mammal see: Seyfarth et al. (2005); Dor et al. (2014); Gasser (2004); Corballis (2002); Bickerton (1990); Hauser et al. (2002); Deacon (1997); Tomasello (2010); Pinker and Bloom (1990); Hurford (1990); Fitch (2017); Berwick and Chomsky (2016).

9.3 Neuroplasticity

Neuroplasticity refers to the capacity of the human brain to modify, substitute and anatomically and functionally create new neural pathways according to experience, stimulation, observation, and imitation. Gu and Kanai (2014) give a thorough review of the variability that exists in neuroplasticity from individual to individual. In a section above, I specified not to agree with proposals that make single-cut statements between language use and ToM, and that is very much related to the principle of neuroplasticity: if old brain pathways are pruned off and new neural pathways are being constantly created, any cognitive skill can be enhanced along our lifespan. Therefore, language skills must evolve along our lifetime and so does ToM. Or better said, any cognitive skill may continue to develop through a combination of behavioral skills and changes in neural connections. The real question lies in the fact that "having ToM" abilities may not be equivalent to "using ToM" abilities, in the same way as "having language" abilities may not be equivalent to "using language" abilities, at least, to our best benefit and their optimal potential. What are the parameters and mechanisms underlying ToM that best contribute to our environment adaptation and well-being? In other words, what are the best circumstances (such as the nature of our emotional relationships) that lay the ground for an optimal implementation of ToM? Also, what brain processes and neural interconnections should occur for ToM to fulfill its goals?

When it comes to ToM and neuroplasticity, we can understand that age plays a central role in "using ToM" or, rather, succeeding in ToM, as much as we succeed (or fail) in memory, attention, inhibition, or any other cognitive skill for the matter as we get older. Many studies have been focused on the effect of ageing in ToM. In this section, you will find a review of some important studies in the field.

The studies in the area of aging and cognitive decline do not let us infer that ToM must always be affected by the effect of the passing of life. Still, most ToM studies have been carried out in children, and many of them in pathological population (autism, schizophrenia, etc.). However, it is common now to find more studies that try to discover how ToM evolves with age because it is important to detect what factors may contribute to its involution/deterioration, with important therapeutic connotations, especially in neurodegenerative diseases and, definitely, in the well-being of older adults.

The loss of cognitive skills with age has received extensive attention in science[9] (3285 publications in 2022 in the topic "ageing and cognitive decline" according to PubMed, last retrieved July 16th, 2023, in comparison with 1088 publications in

[9]For more extensive reading on aging and cognitive decline see: Bishop et al. (2010); Deary et al. (2009); Salthouse (2009, 2011, 2012); Salthouse et al. (2003); Hänninen et al. (1996); Lenehan et al. (2015); O'Sullivan et al. (2001); Glisky (2007); Park and Reuter-Lorenz (2009); Nyberg et al. (2012); Fjell and Walhovd (2010); Raz et al. (2005).

2012).[10] Raz et al. (2005), carried out a longitudinal study where they analyzed regional brain volumes over five years in healthy adults. Shrinkage of the brain was observed in the caudate, cerebellum, hippocampus, and association cortices, while minimal change occurred in the entorhinal cortex and none in the primary visual cortex. Individual differences in volume change were present in all regions except the inferior parietal lobule. Sex, however, was not found to be a differentiating factor and education did not prove any kind of neuroprotective influence.

Let's go back to footnote 11 to review some of the probable risk factors associated with cognitive decline. We may take them into consideration and hypothesize they play a role in the deterioration of ToM in old adults. The discoveries in this area and population may be difficult to interpret, as we enter a realm of comorbidities that may not be easy to ponder. Happé et al. (1998) found out that older adults outperformed the younger adults using ToM tasks that included *double bluffs*, *mistakes*, *persuasion*, and *white lies*. Still, the study was replicated by Maylor et al. (2002) showing the opposite results. This has been a trend in the study of ToM in older adults: mixed results and an extensive variability that requires redefinition. One hypothesis could be education: older adults who have led a very intellectual life, have attended college, or read profusely may keep ToM abilities with a higher probability when compared with the population who had to spend more time doing physical work. Raz et al. (2005) did not find education to be a significant factor in deterioration of cognitive skills. Other studies, such as the one made by Li et al. (2013) profusely analyzed how the educational level could derive in cognitive processing that may affect ToM abilities, too.

In their study, Li et al. (2013) found significant differences in *false belief* and *faux-pas* tasks depending on age. Their results show low performances of the old adults in these tasks when compared with young adults. Among the old population, participants with a low educational level performed worse than old participants with a high educational level. However, they did not separate the young subjects into high and low educational level and so their results may be insufficient as to determine how education influences ToM abilities. A good outcome of their study is that if we compare the results between the old population with high education and the young population, age does not seem to be a detrimental factor in ToM abilities. This result corroborates the importance of education in preserving the cognitive reserve.

Slessor et al. (2007) or Bailey and Henry (2008) discovered that when young and old adults in a ToM study are recruited with similar education level, results favor the

---

[10] Studies in the area report different causes for cognitive decline with age. Some are the following:

- Sleep duration (Ma et al. 2020; Suh et al. 2018; Pace-Schott and Spencer 2015).
- Obesity (Dye et al. 2017).
- Hypertension (Ma et al. 2021).
- Hearing loss and depression (Rutherford et al. 2018).
- Physical activity and nutrition (Dominguez et al. 2021; Erickson et al. 2022; Lu et al. 2021).
- Cancer and cancer treatment (Ahles and Root 2018).
- Blood pressure (Li et al. 2021).
- Oxidative stress (Ionescu-Tucker and Cotman 2021).
- Cardiovascular disease (Bauer et al. 2023).
- Dyssomnia (Zhou et al. 2022).
- Alcohol consumption (Listabarth et al. 2022).

young population. A conclusion here would be that the younger the better, but a high education probably promotes healthier cognitive skills when aging.

Regarding executive functions tasks, Li et al. (2013) found out that education may be considered an ameliorating factor but not definitive regarding ToM abilities: in their study, young adults performed better than old adults, corroborating similar studies.

In a comprehensive meta-analysis, Henry et al. (2013) conducted a study comprising 1462 subjects (790 young adults and 672 old adults). They analyzed 6 distinct categories of ToM tasks: *stories*, *eyes*, *videos*, *false-belief*, and *paux pas*. Each task was further classified based on its domain (affective, cognitive, or mixed) and modality (verbal, visual-static, visual-dynamic, verbal and visual-static, or verbal and visual-dynamic). Results demonstrate that, when considering all task types, old adults outperformed younger counterparts, with a moderate level of difficulty ($r = .41$).

At this point, one may believe that ToM abilities and our capacity to socially integrate with others and succeed in those interactions is irremediably going to deteriorate with age. The results presented in the previous paragraphs seem to go in that direction. However, we may not be asking the right questions here. Other researchers have, interestingly, dug into more productive paths. For instance, Zhang et al. (2018) attempted to explain ToM abilities evolution with age from the perspective of motivation. These authors start with a review of several studies[11] taken from the perspective of Cattell's (Cattell 1941) theory of crystalized and fluid intelligence.[12]

Motivation, types of intelligence, are among the key factors in understanding how ToM evolves with age. Static studies which compare young and old adults may not suffice in understanding what ToM abilities imply, why and why not we use them and what factors may determine an optimized use of this highly complex skill. When it comes to motivation, selection is the key. The Selection, Optimization, and Compensation (SOC) model was proposed by Baltes et al. (1999) in order to explain the conceptions of Lifespan Psychology (LP) (Mayer 2003; Peterson 2013) state that cognitive development is never completed but rather "extends across the entire life course and that from conception onward lifelong adaptive processes of acquisition, maintenance, transformation, and attrition in psychological structures and functions are involved" (Baltes et al. 1999, p. 472).

---

[11] Bottiroli et al. 2016; Charlton et al. 2009; Kemp et al. 2012; Li et al. 2013; Rakoczy et al. 2012; Slessor et al. 2007; von Hippel and Dunlop 2005.

[12] The theory of *fluid* vs. *crystallized intelligence* was developed by the psychologist Cattell to explain the integration of our biological-genetic inheritance in relation with life experience and intelligence and how much we can develop cognitive traits that may remain permanent over time. For Cattell's theory, *fluid intelligence* has to do with a somehow innate capacity to process information spontaneously in our everyday life, the little tracks that make up the sum of our experience. On the other hand, *crystallized intelligence* refers to our ability to create an intelligence reserve that can be enhanced throughout life and can really make a difference when one consider the development of our mental skills, redefining our personality and reshaping our level of intelligence.

LP speaks of culture and biology (in a way, the *genoma* concept) in a way that there is an overall interplay between them that lays the foundation of the paths of cognitive development as well as the adaptive challenges we need to face as we age. There would 3 primary goals of ontogenetic development:

Growth: It refers to behaviors and efforts aimed at achieving higher levels of functioning, thus demonstrating adaptation to new environments.

Maintenance: It encompasses behaviors geared towards the preservation of already existing levels of functioning despite new challenges or after experiencing a life setback (*resilience*).

Regulation of loss. Readjustment of older levels of functioning by replacement in face of the depletion of external or internal resources (lack of stimulation, new stimulation, facing new environments, processing new factors).

A key idea here is that age does only affect the first goal (growth) while, as the need of adaptation never ceases, resilience and regulation of loss tend to increase (Staudinger et al. 1995; Baltes et al. 1998). The authors defend the idea of *culture* and its implications for the continuation of *maintenance* and *regulation of loss.* There is an interaction, then, between our biological makeup and the influence of culture in the way we develop resources and face challenges in our lives. The decrease in growth is very much connected with other biological factors, such as reproduction and consolidation of achievements. We could say that once the family is created, the human mammal concentrates on the defense of that territory to the detriment of the search of new horizons.

SOC introduces the idea that age-related declines in cognitive abilities lead to a selective allocation of the cognitive resources we use for growth towards a concentration of other tasks that are useful for maintaining overall functioning: from a more broad-based resource allocation in young adulthood towards focus and specialization. Freund (2006) demonstrated that older adults are more likely to deploy greater effort in tasks related to maintenance rather than performance optimization or growth. Adulthood comes with a physical cost derived from the cognitive effort: (a) older adults tend to show stronger cortisol responses during cognitive tests when compared to younger adults and (b) older adults take longer to recover from stress-related responses (Crosswell et al. 2021; Schlotz et al. 2011). This physical and cognitive expense make older adults prioritize tasks that are oriented towards their individual survival, a trait that is emphasized the older we age. Hess and colleagues (Germain and Hess 2007; Hess et al. 2001, 2005) found out that the cognitive and memory performance of older adults is disproportionately enhanced when the task materials are more personally relevant or when there is social accountability involved.

All these studies are, after all, seeking important factors that contribute to neuroplasticity even in adverse situations. Looking at the question from that perspective, we should refer to motivation which can be defined as the process that drives us to initiate, guide, and sustain goal-oriented behaviors. In short, the process of selection, as a motivational strategy, plays a significant role in how older adults allocate their cognitive resources, thus allowing for neuroplasticity to take place when growth is no longer available. Hess and colleagues realized that older participants in their study performed poorly in comparison with younger adults. What stroke them was that they had reported not to be interested in the tests they had to take in the lab. That triggered the question: were the results an honest outcome of these older adults' cognitive measure or were they not truly motivated by the task, and this did not allow for them to show their full potential? (Hess 2014). The Selective Engagement Hypothesis state that, as we age, there is a general decline in our cognitive abilities (processing speed and working memory) that has an impact on the efficiency with which we process information at an old age. In order to cope with that situation, healthy older adults maximize their mental energy, by prioritizing, concentrating on personally relevant matters that are meaningful to them. This selection, according to some theories, promotes motivation. Carstensen (2006) proposed the Socioemotional Selectivity Theory suggesting that, as people age, their focus shifts from future-oriented goals to present-oriented emotional goals: young adults prioritize future (career, family, partnership, education) as they have a perception of a future time that is perceived as much shorter by older adults. So, as we become increasingly aware of our limited time left, we start to prioritize emotionally meaningful objectives: positive social relationships, stable emotional fulfillment and any other activity that can bring joy and a sense of purpose so that they contribute to our overall well-being. This would mean an **allocation of cognitive resources**.

It is important to study the effects of aging on the functional and structural characteristics of the brain, more in particular in relation to language networks, because most patients with aphasia are older adults. Current knowledge about brain maps comes from data from young healthy adults, more should be investigated on the neural differences in older adults (Cahana-Amitay and Albert 2015) and, specifically, longitudinal studies across the lifespan (Grady 2012). Even though it seems that most language processes seem to remain robust to brain aging (Shafto and Tyler 2014), some aspects of language production may be impaired. Hoffman and Morcom (2018) found out that younger adults outperform elderly adults in semantic tasks, and the latter show different patterns of brain activation, with less activity in certain semantic areas of the left hemisphere while increasing right frontal and parietal regions. Agarwal et al. (2016) compared young and older adults who displayed similar cognitive performance during a word retrieval task. They discovered that, when cognitive reserve is intact, the brain can adapt and reorganize the way it processes language, particularly in the dominant hemisphere, which is typically responsible for language functions. This preservation of cognitive reserve and the subsequent reorganization of dominant hemisphere networks play a crucial role in enabling older adults to maintain their language skills. For some authors, language engagement, such as bilingualism[13] (see Chap. 18 in this volume), seems

[13]For a review of aging and bilingualism, see Gallo et al. (2022); Bialystok (2021); Borsa et al. (2018); Craik and Bialystok (2005); Mungas et al. (2018); Sundaray et al. (2018); Rossi and Diaz (2016); Heredia et al. (2020); Birdsong (2018); Chan et al. (2020).

to have a beneficial effect on executive control systems, preventing loss of cognitive reserve and suggesting mutual interactions between the brain networks that are involved in cognitive, affective, praxic, and linguistic functions (cf. Nasios et al. 2019).

In summary, neuroplasticity is present in the human mammal throughout our lifespan. It involves growth at early stages so that strong personal and social networks can be created, and it remains as we age to guarantee that those networks are strong enough for us to feel fulfilled in life. Allocation of cognitive resources must be a common trait in the human brain that can enhance all complex cognitive skills. What is of utmost importance is understanding the factors that hinder the neuroplasticity and healthy development of our cognitive capacity.

9.4 Pragmatic Awareness that Precludes the Expression of ToM Reasoning: The Right Hemisphere

Language processing in the human mammal is a complex system that entails different layers. From a psycholinguistic perspective, language happens when a series of individual units (words) are inserted in higher structures (phrases, sentences) in a bidirectional context (conversation). Words carry the target load of linguistic exchange, and they display a literal meaning as retrieved from the mental lexicon which is enriched with context or specified: one word can have different meanings depending on its position in the sentence and the syntactic relationship it conveys. That is to say, words may be central to express meaning (such as verbs) or peripheral (such as complements to the verbs). Compare the 2 sentences below:

(e) *The lion ate the deer.*
(f) *The deer ate the lion.*

Words carry out semantic information which is expressed orally, with phonetic cues coming to add or modify that semantic information. Phonetic cues include silence (pauses), intonation and stress. Compare the 2 sentences below:

(g) *I love that man, who is always there for me.*
(h) *I love that man who is always there for me.*

Language control in the brain is localized in the frontal, temporal and parietal lobe.[14] Below, a summary of the cytoarchitectonic structure of the left hemisphere using Brodmann areas.

Studies in the language control area of the human brain (Friederici 2011) have found that different aspects of language functions are distributed between the left and right hemispheres. For example, networks in the temporal cortex and the inferior

[14]The occipital lobe, especially Brodmann areas 19, 18 and 17, also play a role in language control when analyzing body gestures.

frontal cortex in the left hemisphere have a clear dominance in understanding sentence structure (syntactic processes). On the other hand, the processing of the meaning of language (semantic processes) is not so lateralized and involve both the left and right temporofrontal areas. Electrophysiological measurements indicate that, within these language networks, the brain first focuses on local structure building for sentences' grammar and semantic relationships. Additionally, prosodic information (intonation of speech) in the right temporofrontal network processes extra information. The integration of both hemispheres in language function is clearly seen in studies on patients with lesions in the corpus callosum (the structure that connects the two brain hemispheres). These studies reveal that the posterior part of this structure plays a significant role in integrating the syntactic and prosodic information, becoming an essential structure in the language control system in the brain.

One important region in the language control system is the auditory Cortex (A1, A2) located in the temporal lobe, Brodmann areas 41, 42 (Fig. 9.2).

Both the left and right hemispheres are involved in processing speech and tonal pitch (intonation), but there seems to be some preferences. The left A1-A2 are more specialized in responding to characteristics of speech sounds that are produced with the phonatory system (phonemes), while the right A1-A2 are more sensitive to the overall intonational pattern present in the phonemic sequence (F0, or absolute

Fig. 9.2 Brodmann areas

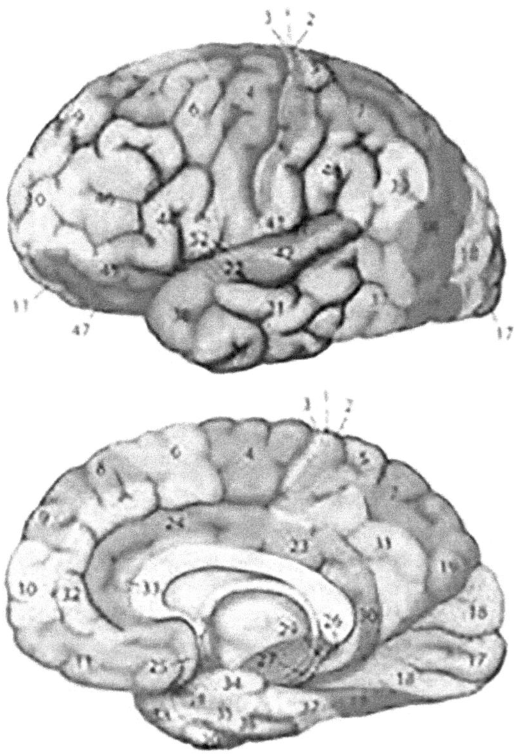

frequency, that derives from the vocal cords. Intonation, as a melody, requires to be built over a longer period of time before it creates a unit that can be interpreted by the A1-A2 as so. This may imply that the A1-A2 in the right hemisphere is more sensitive to longer stretches of time when compared with the A1-A2 in the left hemisphere, more sensitive to small cycles. The specialization of the two auditory cortices is not related to the type of stimulus, but rather to different temporal and spectral characteristics. While the left hemisphere is specialized in processing rapidly changing information with limited frequency resolution, the opposite seems to be the case of the right hemisphere, more sensitive to long cycles of time with high frequency resolution. Also, the left hemisphere's system requires a precise time resolution of around 20–50 ms. while the right hemisphere's system necessitates a time resolution of around 150–300 ms. Hickok and Poeppel (2007) proposed that this differentiation in frequency operation between the 2 hemispheres leads to a relative lateralization of functions: the left hemisphere works predominantly at gamma frequencies, while the right hemisphere works in the theta range (Giraud et al. 2007).

This neural basis of prosodic information processing in the brain has been traditionally studied through observation of patients with brain lesions in the right/left hemisphere. Some studies (Brådvik et al. 1991; Weintraub et al. 1981) have observed a predominant right hemisphere activity in prosody processing, while others have found deficits in processing prosody when the lesion was in the left and right hemisphere alike. Bryan (1989) filtered out segmental information from the stimulus in his study and discovered that patients with right hemisphere lesions exhibited notably poorer performance in prosody information when compared with patients with lesions in the left hemisphere. Overall, the right hemisphere cannot be denied significant importance in the processing of prosody over the left hemisphere.

Prosody is related to the field of pragmatics in the sense that it complements linguistic information with cues that comes from the context of the speaker-listener setting, background, and common vital experience. According to Huang (2012) "[...] pragmatics may be defined as the systematic study of meaning by virtue of, or dependent on, the use of language. The central topics of inquiry include implicature, presupposition, speech acts, deixis, and reference (p.1)". Prosody and pragmatics complement each other to enable effective communication by adding emotional expression, speech acts and social meaning.

A. **Emotional Expression**: The interpretation of enthusiasm, anger, irony, sarcasm, politeness, etc., are determined by certain prosodic patterns. Pragmatics is in charge of its interpretation so that there is an understanding of the speaker's emotional state or attitude.
B. **Speech Acts**: Examples of speech acts are requests, commands, questions, apologies, promises, etc. The use of different intonational patterns may establish the difference between a statement and a question, clarification, etc. (compare *You are tired,* and *You are tired?* and think of irony being used between 2 partners after dinner and a loaded dishwasher).

C. **Emphasis and Focus**: Prosody can be used to highlight certain words/phrases through stress or pitch changes, focusing or shifting the attention to some semantic information in a sentence (compare the emphasis in the sentence *You are **not** tired. **I** am tired* in the same discourse context described above).
D. **Turn-Taking and Discourse Structure**: When a speaker is done and it is the listener's turn to respond, we find all kind of prosody markers, such as pauses, rising intonation (as in yes-no questions) that help elicit a response.
E. **Politeness and Social Meaning**: In many linguistic communities, women speak with a higher pitch level than men to show social differences, for example.

In summary, prosody is central in determining extra meaning to the sentence: the interpretation of a sentence is not just the adding of lexical terms.

Because of the importance of pragmatics in language, many scholars have defended a component view of pragmatics in the brain. The component view of pragmatics would imply that there is a modular area in the brain that is specialized in pragmatic processing ("modular pragmatics") (Fodor 1983a). Kasher and Sagi (2010) considered that modular pragmatics consists of a pragmatic central system of pragmatic knowledge that governs both the production and comprehension of speech acts and, especially, indirect speech acts.[15] This cognitive system is described by Kasher as the **Modular Speech Act Theory**. Wilson and Sperber (2004) go beyond in their theory and defend that language comprehension implies a modular ability for mind-reading or Theory of Mind, which involves the ability to inferentially attribute mental states or intentions to others based on their linguistic cues, as measured by intonation, body language and other extralinguistic units.

In summary, prosody and pragmatics complement each other and make communication more effective. Recent approaches to the area of pragmatics adopt a ToM perspective since pragmatics refers to our capacity to interpret the linguistic cues in language that help us understand the speaker's state of mind and attitude. From the studies cited in this chapter, prosody may lie in the right hemisphere. Is pragmatics also in the right hemisphere of the human brain? Numerous studies have been carried out on what we can call **apragmatism**, a communication disorder associated with lesions in the right hemisphere. Minga et al. (2023) "define apragmatism as a disorder in conveying and/or comprehending meaning or intent through linguistic, paralinguistic and/or extralinguistic modes of context-dependent communication. The context includes (among other things) the conversational partner(s), environment, cultural considerations and goal of the interaction" (p. 656). The term must be differentiated from **pragmatic aphasia** which should be restricted to language disorder, while apragmatism would specifically cover cognitive-communication disorders in which the ability to interpret language in context would be impaired.

Ross (1981) stated that "the affective components of language, encompassing prosody and emotional gesturing, are a dominant function of the right hemisphere, and that their functional-anatomic organization in the right hemisphere mirrors that

[15] As seen above, an example of direct act would be *Pass me the salt* while expressions such as *Could you pass me the salt?* would be considered an indirect act.

of propositional language in the left hemisphere" (p. 561). The term coined to define this disorder would be **aprosodia**. Aprosodia, in Ross's perspective, would be symmetrical to Broca's or Wernicke's aphasia, but in the right hemisphere. Aprosodia can be classified into 'receptive aprosodia' and 'expressive aprosodia'. In **expressive aprosodia**, patients are uncapable of using prosodic cues in language and so that they lack the ability to transmit emotions (Rosenbek et al. 2004; Cohen et al. 1994; Patel 2010). On the other hand, **receptive aprosodia** has clear consequences in ToM for patients, as they show an impairment at the level of acoustic analysis but also a domain-general emotion recognition deficit, as explained in Sheppard et al. (2021).

In this section, we have paid attention to the functionality of the right hemisphere in language and, hence, in ToM from the phonetic-semantic module of pragmatics and prosody processing. Studies carried out on patients with lesions in the right hemisphere (see more in Chap. 7 in this same volume) clearly demonstrate the importance of that hemisphere in language and communication disorders. ToM is definitely a complex circuit that is not privative of one hemisphere, the implications of *corpus callosum* in ToM would require a section in this volume (see the importance of corpus callosum in ToM in Melogno et al. 2021; Symington et al. 2010; Booth et al. 2011; Paul et al. 2014; Bridgman et al. 2014) but when it comes to communication and/or speech disorders, language must be carefully considered and the role hemisphere in language and, hence, ToM must be born in mind when we are dealing with social impairment associated to language pathologies. In the next section, I briefly discuss the hypothesis of an integrated theory of perception and production in language.

9.5 Hypothesis: An Integrated Theory of Perception and Production in Language

A garden-path sentence is a sentence that is grammatically correct but that breaks the interpretation predictions made by the reader/listener. An example could be:

(i) *The old man boys the boat.*

The beginning of the sentence makes all sense and creates an expectation of a verb after the phrase *The old man*. Because the 'path' is broken, the reader/listener must revise her expectations. The technique is used in Linguistics to determine the function of speech acts in verbs, for example, and comes to demonstrate that our brain has already 'walked' certain paths prior to actually receiving the input. Examples like the garden-path sentences have made researchers believe that perception and production are interwoven, and that language processing is predictive (Kuperberg and Jaeger 2016).

Traditionally, language perception and language production has been seen as 2 independent systems. The Broca-Wernicke connectionism model

(Broca-Wernicke-Lichtheim-Geschwind classical model) has offered substantial knowledge about the brain and language and aphasiology during the twentieth century was based on the Wernicke-Geschwind model. However, language is not restricted to the left hemisphere only, as we read in the section above. Besides, other cognitive functions such as attention, emotion, memory, and the executive functions are also involved. In the past decades, the use of neuroimaging tools has given rise to the so-called **dual stream model** (Hickok and Poeppel 2004, 2007), consisting of 2 integrating networks or 'streams (Nasios et al. 2019). The ventral stream is specialized in language comprehension while the dorsal stream, left hemisphere dominant, is focused on production. This new model of neuromodulation is a new attempt to understand why, for example, patients with lesions in similar areas have very different outcomes when it comes to language. Functional organization of the brain is a formidable challenge when it is necessary to understand both language and cognition measures in the area of ToM in patients with speech and/or communication disorders.

Pickering and Garrod (2013) further proposed an alternative perspective to the traditional theory that treats production and comprehension as 2 distinct entities. For them, producing and understanding language are intimately bound, allowing individuals to anticipate their own actions in view of those of the others. In this regard, their hypothesis of an integrated theory of production and perception in language strongly resembles the functionality of ToM. They recognize production and perception as types of action (production) and action perception. They explore the evidence that supports the connection of action, action perception, and **joint action**. The interpretation is done through prediction. For them, actors in a conversation create predictive models of their own actions (production) even before they execute them. At the same time, they unconsciously observe the action perception of others and simultaneously develop predictive models of those observed behaviors. This model implies that the language we produce is determined by the language we unconsciously perceive, simultaneously favoring the creation of a communication that interweaves the speaker and the listener into a single discursive unit.

Grimshaw (1998) indicates that studies in neuroimaging show that both hemispheres of the brain are active in almost all tasks, although not necessarily at the same competence level. There seems to be a division in the task they are assigned to. For example, the right hemisphere deals with information from the left visual field, while the left hemisphere deals with information from the right visual field but with specialized attributions. The left hemisphere extracts local features, while the right hemisphere focuses on the form globally. A similar division of task distribution is seen in speech processing, where the left hemisphere specializes in linguistic units (syntax and lexicon) while the right hemisphere specializes in prosody (intonation and rhythm). Also, it has been proposed that it is the *corpus callosum* the structure that aids parallel processing by separating each hemisphere from the other until integration needs to take place. They claim that **lateralization** may seem as a benefit in individuals who are capable of processing each dimension in opposite hemispheres while lack of lateralization should be considered an impediment for those individuals who process both dimensions in the same hemisphere.

The two hemispheres play a distinct and specialized role in any cognitive task, even though the success of the task relies on their interconnection. This fact is clearly expressed in the so-called **disconnection syndrome**. The disconnection syndrome (Geschwind 1965) refers to the effects of lesions in either hemisphere by the alteration of the association pathways connecting the different parts of the brain. For example, the **Collosal Syndrome**, which occurs when the *corpus callosum* (a large bundle of nerve fibers that connects the two cerebral hemispheres) is partially severed, either surgically or because a stroke. This has effects in patients who show difficulties in following complex verbal commands and may respond with short, one-word answers. Some consequences of this split-brain state are the following:

1. **Hemialexia**. Patients can't read or verbally describe individual words that are presented to the left half-field.
2. **Auditory Suppression**. Right-handed individuals identify single words only when presented to one ear at a time, favoring the right ear and with poor report from the left ear.
3. **Left-sided apraxia**. Poor comprehension by the right hemisphere and poor control by the left hemisphere.
4. **Right-hemisphere verbal comprehension**. The patient can comprehend some words auditorily but with limitations in their syntactic ability and with a significant reduction of lexical capability when compared with the intact hemisphere.[16]

Neurological intervention is possible, thanks to neural plasticity, and patients may show remarkable adaptive capacities to compensate this disconnection over time. This, however, does not eliminate the hypothesis of distinct functions in the two cerebral hemispheres and the necessity to establish an integrative bridge between them.

All these aspects (neuroplasticity, the role of the right hemisphere, the role of the left hemisphere, the integration between them, lateralization, prosody, and pragmatics) presented in the sections are above should necessarily be born in mind when one approaches ToM in patients with speech or communication disorders. I depart from the hypothesis that all cognitive skills, including language and social abilities, evolve during our lifespan and that could be in an ascending or in a descending direction. However, when language is involved, as is the case with language pathologies, and its competence is affected and thus other cognitive skills such as ToM, it is the same use of language that will facilitate the intervention and eventual rehabilitation, which is just a hopeful hypothesis to keep in mind.

In the last section of this chapter, I review the effect of ToM in language-related pathologies (aphasia).

[16]Studies in the area usually deal with right-handed patients.

9.6 The Role of Language in Theory of Mind: Insights from Aphasia and Adult Development

Some researchers have proposed that natural language is the central scaffolding system for thinking (Clark 1998).[17] ToM reasoning has been claimed to mediate in the ability to create sentences (Carruthers 1998; Nedergaard et al. 2023) (see more about the relationship between language and reasoning in Chap. 17 in this volume). Astington and Jenkins (1999), in a longitudinal study, revealed a significant correlation between language and ToM in children: as their language skills advanced, so did their ToM abilities, thus establishing the groundwork for a hypothesis where language and ToM are irremediably interwoven.

Language development plays a central role in developing ToM in children (but it is also true that most of the research in the area has been done with children). We use language to rationalize and mentalize. However, it remains unclear whether language keeps its crucial role once beliefs and desires are established, once ToM is established in the adult population. Patients with ToM impairment will probably show impairment at the language level, at the cognitive level (EF, for example) and at the affective and social levels. However, ToM is not centralized on the left hemisphere (temporal lobe). Patients with lesions in that area may keep their ToM abilities (see Chap. 7 in this volume). Also, studies such as Varley and Siegal (2000) explored the consequences of late acquired aphasia, specifically focusing on the loss of grammatical skills and how it may affect ToM. Unexpectedly, their findings indicate that brain regions within the ToM network do include the linguistic temporal lobe, but ToM can still be substantiated despite damage in that area. These results propose an intriguing possibility that language may be necessary for the initial development of ToM, but they state the strength of the ToM system even in the absence of this important human ability. Evidence from deaf children, however, shows that grammatical knowledge is not enough to account for reasoning (see Chap. 8 in this volume). A possible interpretation is that children's reasoning is guided by inferences about other's communicative actions, a plausible explanation to explain why language and reasoning evolve similarly at early stages in life.

There are some aspects of ToM that seem to be language-related because grammatical capacity is necessary to solve the particular problems that ToM tasks present (i.e., without language, we cannot measure ToM abilities) (Perner et al. 1987). However, other features of Tom that involve eye gaze or emotion interpretation can be seen as 'social-perceptual' which is a language-independent component with its own development (Tager-Flusberg and Sullivan 2000).

Linguistic alignment refers to the ability to integrate linguistic comprehension and production so that ToM abilities can be refined and tuned. Individuals with a weak linguistic alignment will show a weak ToM skill with consequences in false

[17] Andy Clark goes beyond traditional Cognition as seen in Philosophy and Psychology and analyzes how everything technological is an extension of our cognitive resources. Interesting reading to understand how AI and Neuroscience are interwoven.

attributions of beliefs, desires, and perspectives in others as well as a biased impression of their own consciousness and self. The correlation here implies that an impairment in sensitivity to contextual cues from others' mental states will disrupt spontaneous speech in dialog. Slocombe et al. (2013) define **linguistic alignment** as "the process by which interlocutors converge with their conversational partner on a number of different linguistic levels" (p. 1424). This means that lexical, syntactic, and prosodic features, to mention a few, must be paralleled to that of the listener so that the conversation has a unified and integrative tone. Failure in linguistic alignment breaks interaction and influences social, cognitive, and emotional levels for both speakers.

When it comes to speech/communication disorders in ToM, it is important to understand the following premises:

1. What brain regions are involved in language.
2. How language and ToM are interconnected (Fodor 1983b; Hermer-Vazquez et al. 1999; Pinker 2003).
3. Language reorganization in aging.
4. The connection between perception and production in language.

9.6.1 Language Is a Bilateral System in the Human Brain

Broca's and Wernicke's areas, connected by the *arcuate fasciculus* have traditionally been identified as crucial regions for language function in the brain. However, recent research has shown that language processing is far more complex and may be distributed among different regions (for a review, see Table 9.3) including other non-language-specific areas that are involved with various cognitive tasks, including EF, memory, or emotion. There is also uncertainty as to the unique function of the Wernicke's area in language comprehension, with other areas, including the Broca's area, being shown competent in understanding both at the syntactic and semantic areas, together with the dorsal premotor cortex (Siegal et al. 2001). If ToM measures rely on language abilities for the completion of the tests but the individual has lesion on the language regions, it may be difficult to test their ToM abilities: language is a means to measure ToM. However, the traditional modular model (Broca-Wernicke-Geschwind) may not account for the right categorization of aphasic syndromes according to specific brain regions. Functional neuroimaging studies show that there is more than the Broca-Wernicke-Geschwind areas including both temporal lobes during comprehension and a wide range of frontal and parietal regions with subcortical regions and connections also found to be involved. If that is the present trend, then a patient with any kind of aphasia would probably have an impaired ToM, too. The challenging issue here is also the assessment method: is language before reasoning so that aphasia will definitely affect ToM or is reasoning independent from language and, therefore, ToM will be unaffected despite language impairment?

Table 9.3 Relation of Brodmann's areas involved in language control and their functions

Frontal lobe	Brodmann's area	Functions
	47 (pars orbitalis)	Semantic processing in context
	45 (pars Triangularis)	**Broca's area**
		(i) Speech production
		(ii) Grammatical processing
	45 (**right hemisphere**)	Prosody and emotion in speech
	44 (pars Opercularis)	Processing of sounds (phonemes) into intelligible speech
	44 (**right hemisphere**)	Prosody and non-literal language (metaphors)
Temporal Lobe	**Brodmann's area**	**Functions**
	41 (primary auditory cortex)	A1
		Frequency, intensity, duration
		Tonotopic organization
		Contralateral auditory processing
		Auditory localization
	42 (auditory association cortex)	A2
		Memory & classification of sounds
	22 (superior temporal gyrus) STG	**Wernicke's area**
		Auditory short-term memory
		Language comprehension
	21	Auditory processing
	37	Object naming
		Recognition memory
Parietal Lobe	**Brodmann's area**	**Functions**
	40	Reading (meaning & phonology)
	39 (angular gyrus)	Language and number processing
		Spatial cognition
		Memory retrieval
		Attention

Both language and ToM circuits have brain regions in common. They actually display an interconnected function in ToM (pragmatics in language, prosody, body language, and the common knowledge speakers of a linguistic community have in common). However, impairment at the level of language does not necessarily imply an equal reaction in cognitive and social abilities. In this section, I review the specificities of this disparity: when and how, in language pathologies, especially aphasia, ToM has been affected.

The first assumption to bear in mind is that **grammar is a distinct function from other cognitive systems, including pragmatic competence** (Bloom 1999; Chomsky 1980). Let us remember the importance of the right hemisphere (pragmatics, prosody) in ToM: if grammar is located, mostly, in the Broca's area and neighboring regions, then, so long the right hemisphere is preserved, the likelihood of a reserve of cognitive and social abilities is guaranteed. Further, with an unimpaired right hemisphere, the visuo-spatial representation and memory of the location and features of objects is available, and so is ToM (Jonides et al. 1993).

Varley, Siegal and colleagues, from the University of Sheffield, have extensive research in the field of ToM and aphasia. Regarding grammatical aphasia, they state that certain level of proficiency in grammar may be necessary to succeed in ToM tasks although in a study with late-signing deaf children from hearing families (Woolfe et al. 2002) they found out that deaf children who achieve the same syntax and morphology competence as native-signing deaf children from deaf families, still display difficulties on ToM tasks. That entails a strong correspondence between the foundation of language (if language is acquired at late stages in development) and the foundation of ToM (more on ToM and deaf children can be found in Chap. 8 in this volume). This correlation, however, only demonstrate that at critical stages of cognitive development, an impairment in language can have strong impact on the development of cognitive skills. This must be accepted because many studies have demonstrated that patients with left-hemisphere damage (LHD) still perform optimally in ToM (Stuss et al. 2001; Surian and Siegal 2001). If research on patients with LHD perform comparatively well in ToM even though they obviously present damaged language regions, the main conclusion is that **language does not lie on the left hemisphere only** damage on the right hemisphere has proven to cause more serious cognitive and social impairment.

De Villiers and Pyers (2002) associated the ability to process rules of embedding complement clauses under verbs of speech or cognition as being the key element in ToM reasoning. Have a look at the following sentences:

(e) *Adam said that he would buy his toy later in the evening.*
(f) *She thought that some boundaries should not be trespassed.*

The first sentence (e) includes a verb of speech ("say", "tell", "demand", "explain", etc.) followed by a subordinate clause. The subordinate clause functions as the complement of the verb of speech "say". In the second sentence (f) we can find a verb of cognition ("think", "believe", "imagine", "idealize", etc.) followed by another subordinate clause. That subordinate clause is again the complement that the verb of cognition requires. This type of construction is very frequent to be used when testing ToM abilities. According to numerous studies on language as one of several facilitators of social cognition (De Villiers and De Villiers (2014); Sabbagh and Shafman 2009; Astington and Jenkins 1999; Hale and Tager-Flusberg 2003) embedded clauses is a key linguistic construction that helps us identify false belief. In studies like these, ToM reasoning appears to depend on the developing of complex syntactic abilities that permit false propositions (the subordinate clause) within true statements (Siegal and Varley 2006).

On the other hand, studies with patients with right-hemisphere damage (RHD) show independence of ToM from language. Siegal et al. (1996) carried out an experiment of contrast between LHD and RHD patients and discovered that only patients with RHD presented difficulties in simple ToM tasks such as "Sam thinks his puppy is in the garage, but his puppy instead is really in the kitchen" (Happé et al. 1999; Stuss et al. 2001; Shamay-Tsoory et al. 2005; Cutica et al. 2006; Martin and McDonald 2003; Stemmer et al. 1994). Caramazza and Mahon (2006) associated the right-sided network to the computation of mental states and Surian and Siegal (2001) stated that patients with RHD require greater explicitness through language and visual aids to understand other people's mental states. The main conclusion here is that **language is not restricted to the left hemisphere** and that **the language component in the right hemisphere is dedicated to cognitive and socio-affective features**. So, maybe, the controversy should not be laid on whether language or reasoning (ToM) is first but understanding that language is a necessarily bilateral function in the human being.

9.6.2 The Frontal and Prefrontal Cortex (Executive Functions) Are Decisive for ToM[18]

Fletcher et al. (1995), using fMRI (functional Magnetic Resonance Imaging), early observed evident activation in the medial prefrontal cortex (mPFC) and the right temporoparietal junction (RTPJ) in tasks where participants were engaged in ToM processes and had to understand the characters' thoughts and intentions. Schurz et al. (2021) in a study on the regions that are involved in understanding others' mental and emotional states discovered that both empathy and ToM tasks are engage in overlapping brain regions including the mPFC, the temporoparietal junction (TPJ), the superior temporal sulcus (STS), and the precuneus. Brunet et al. (2000) used PET scans to understand the brain regions involved in the attribution of intentions to the actor's actions and they found out increased activity in the inferior frontal gyrus (IFG) in the frontal lobe when using nonverbal tasks. The separation of the verbal activity from the equation in ToM research makes it easier to understand the brain regions that are involved in ToM. The preserved ToM ability in LHD might suggest multiple functionalities of the same brain regions in language, cognitive and socio-affective tasks. A good conclusion for this is the need to analyze human cognition not only from a mapping perspective, but mostly from the intersection between regions.

Further evidence on the role of EF in ToM was also clear in the 90s. In studies with autistic population (Ozonoff et al. 1991; Frye et al. 1995, 1996; Perner and Lang 1999) report parallel impairment on tests of executive function when ToM was impaired. Still, some studies followed which disagreed with this hypothesis of a

[18]The amygdala is another important region with implications in ToM and other cognitive and social pathologies. See more in Hiser and Koenigs (2018); Koubiyr et al. 2022; Del Sette et al. 2023; Adolphs et al. 2002; Baron-Cohen et al. 2001; Kennedy and Adolphs 2012; Mendez et al. 2005.

connection between EF and ToM (Baron-Cohen et al. 1999; Fine et al. 2001; Lough et al. 2001).

EF and social cognition have been connected in numerous studies (Hiser and Koenigs 2018): some regarding empathy (Häusser 2012), epilepsy (Wang et al. 2022; Stewart et al. 2016; Thompson 2014; Giovagnoli et al. 2011); autism (Zemestani et al. 2022); schizophrenia (Hirao et al. 2008); and frontotemporal dementia (Younes et al. 2022). The coincidence of multiple cognitive-associated pathologies and EF cannot be casual so, together with the temporoparietal junction, this hypothesis of the decisiveness of the frontal and prefrontal cortexes in ToM should be not underrated.

9.6.3 Insights from Aphasia

Aphasia is an interesting area of research for ToM because of different reasons, according to Hallowell (2017, p. 4):

1. It is acquired.
2. It has a neurological cause.
3. It affects production and perception of language (oral and written modalities alike).
4. It is not a sensory, psychiatric, or intellectual disorder.

There are different types of aphasia which result from damage to specific areas of the brain, usually in the left hemisphere, but also in the right hemisphere (**aprosodia**). A general classification would be as follows:

A. **Broca's Aphasia (Expressive Aphasia).** The patient has difficulties with language production. Comprehension might be preserved but they find it difficult to express themselves.
B. **Wernicke's Aphasia (Receptive Aphasia).** The patient has fluent speech but does not understand and usually give answers that have no correlation with what is going on in the conversation (nonsensical words or inappropriate lexical terms for the context).
C. **Global Aphasia.** It affects all aspects of language.
D. **Anomic Aphasia.** Difficulties in finding a lexical term or retrieve words, especially nouns and verbs.
E. **Conduction Aphasia.** The patient cannot repeat words.
F. **Transcortical Motor Aphasia.** It is one type of Broca's Aphasia in which the patient can only produce correct language when they repeat the words that are given.
G. **Transcortical Sensory Aphasia.** One type of Wernicke's Aphasia in which the patient can produce correct answers when prompted with them (similar to F).
H. **Mixed Transcortical Aphasia.** (Mixture between F and G).
I. **Subcortical Aphasia.** Damage to the basal ganglia or thalamus.

Table 9.4 Classification of primary progressive Aphasia (Fittipaldi et al. 2019)

Type of PPA	Acronym	Symptoms
Non-fluent variant PPA	nfvPPA	Effortful and/or agrammatic language production
Semantic variant PPA	svPPA	Anomia & single-word comprehension deficit
"Semantic dementia"		
"Temporal variant frontotemporal dementia"		
Logopenic variant PPA	lvPPA	Word retrieval and sentence repetition deficit

In this section, I will review 2 important subtypes of aphasia, namely, Primary Progressive Aphasia (PPA) and Agrammatic Aphasia or **agrammatism:** Broca's syndrome (Tesak and Code 2008); "motor aphasia" (Goldstein 1948); "syntactic aphasia" (Wepman and Jones 1964); "efferent motor aphasia" (Luria 1970); "non-fluent aphasia" (Goodglass et al. 1964).

PPA is associated with linguistic deficits at symptom onset and in early stages (Mesulam 2003). In Table 9.4, a summary of the different types of PPAs:

The brain regions damaged in aphasia are not restricted to language areas (Hillis 2007; Hickok and Poeppel 2007; Turkeltaub et al. 2012; Fridriksson et al. 2011; Saur et al. 2008). There is evidence of deterioration in other domains, such as memory (Mesulam et al. 2012), EF (Macoir et al. 2017), and praxis (Johnen et al. 2018) (cf. Fittipaldi et al. 2019). PPA variants have typically been subsumed under the family of frontotemporal dementia (Younes and Miller 2020; Olney et al. 2017; Younes et al. 2022). As already mentioned in Sect. 5.2, damage at the frontal and prefrontal areas has a direct impact on ToM impairment. For that reason, the study of the connection between language and ToM in aphasia is better accomplished for when we analyze studies on "agrammatic aphasia". The studies of Varley and Siegal, at the University of Sheffield, is a good departure to offer novel insights into the relationship between grammar and ToM.

Agrammatic aphasia is an impairment of language due to a lesion of the perisylvian areas of the language-dominant hemisphere in the brain. In a study with an adult (53-year-old right-handed male) with a severe motor speech disorder (apraxia) and aphasia, Varley and Siegal (2000) measured false-belief and true-belief tasks in 2 experiments (Experiment 2 took place 20 months after Experiment 1). Their findings can be summarized as follow:

1. **Unimpaired Executive Functions** (91st percentile of a normative sample matched for age and education on the Wisconsin card sorting test (Heaton and Staff 1993).
2. **Impaired language** taken from the PALPA battery (chance level on sentence comprehension in both speech and reading) (Kay et al. 1996).

3. **Unimpaired in false-belief and reality control questions** (The WAIS picture-arrangement test resulted in 16/20, 84th percentile of a normative sample matched for age) (Wechsler 1981).

Their results are decisive to understand that language impairment does not necessarily imply cognitive dysfunction. It is also interesting to see that EF were not impaired, which coincide with conclusions already stated in Sect. 5.2 of this chapter: **ToM implies malfunctioning of frontal and prefrontal functions of the brain**. A conclusion by the authors is that "grammar is distinct and separate from other cognitive systems" (Varley and Siegal 2000, p. 725). For the authors, language is a "performance factor" in ToM, but not a core component (Siegal and Varley 2002).

Other authors (Baldo et al. 2005) have suggested that if aphasic patients fail on the Wisconsin Card Sorting Test (Heaton and Staff 1993), they may not be able to infer the mental states of others because any problem solving that implies language transcends to other problem-solving tasks for which language is equally necessary. Works by Baldo have inspired many other pieces of research that have attempted to describe **inner speech and aphasia** (Fama and Turkeltaub 2020; Stark et al. 2017; Langland-Hassan et al. 2015; Fama et al. 2017, 2019a, b; Hayward et al. 2016). Inner speech (inner voice or self-talk) is defined as the process of silently talking to oneself without vocalizing the words aloud. It is said to be used in the regulation of thoughts, feelings, and actions. It also serves to solve problems and plays, therefore, a central role in cognitive processes (memory, attention, and self-regulation). Inner speech's father is probably Vygotsky (1934) (you can read more in Chap. 12 in this volume and in Fernyhough 2016; Morin 2005; Alderson-Day et al. 2016; Moseley et al. 2013).

Finally, for authors such as Bánréti et al. (2016), aphasic patients, especially Broca's aphasics, may be able to make correct ToM inferences despite their inability to access recursive sentence embedding.

When looking at ToM and aphasia, the analysis of the role of language is important. In this chapter, I have reviewed the hypothesis of language perception and production and the fact that language cannot be restricted to the left hemisphere. The processing of pragmatic and prosodic information in the right hemisphere is crucial for ToM abilities, so language is important. A possible summary of the role of language and ToM in aphasia could be as follows:

A. **Role of frontal and temporal lobes**. EF play a decisive role in making judgments in grammatical constructions. Also, the ventromedial prefrontal cortex has been linked to impaired ToM abilities as it is involved in social and affective processing.
B. **Unclear role of language related regions**. Research is not conclusive as to what extent damage in the left hemisphere can affect ToM.
C. **The role of the superior temporal sulcus (STS) and the temporoparietal junction (TPJ).** Damage in this area is involved in social cognition, emotion processing, and ToM.
D. **The role of the right hemisphere.** The left hemisphere, far beyond the Broca's and the Wernicke's area, interact with the right hemisphere in understanding and

attributing mental states to others. A comprehensive ToM ability should incorporate the action of both hemispheres of the brain.

It would be interesting to analyze intervention approaches and neuroplasticity to further understand the development of ToM in patients with aphasia. Language and ToM have common brain regions in their functionalities. They interact, too, but more has to be investigated away from brain mapping so that we can understand what mechanisms or paths allow for an optimal use of ToM abilities. And the same should be done to excel in the use of our language abilities because both skills make us different from other mammals.

Acknowledgments This work has been partially funded by FEDER Funds (EU) US-1381050 with the project title "Abduction and Medical Diagnosis".

References

Abu-Akel, A. 2003a. A neurobiological mapping of theory of mind. *Brain Research Reviews* 43 (1): 29–40.

———. 2003b. The neurochemical hypothesis of 'theory of mind'. *Medical Hypotheses* 60 (3): 382–386.

Abu-Akel, A., and S. Shamay-Tsoory. 2011. Neuroanatomical and neurochemical bases of theory of mind. *Neuropsychologia* 49 (11): 2971–2984.

Adler, W.T., and W.J. Ma. 2018. Comparing Bayesian and non-Bayesian accounts of human confidence reports. *PLoS Computational Biology* 14 (11): e1006572.

Adolphs, R., S. Baron-Cohen, and D. Tranel. 2002. Impaired recognition of social emotions following amygdala damage. *Journal of Cognitive Neuroscience* 14 (8): 1264–1274.

Agarwal, S., E.A. Stamatakis, S. Geva, and E.A. Warburton. 2016. Dominant hemisphere functional networks compensate for structural connectivity loss to preserve phonological retrieval with aging. *Brain and Behavior* 6 (9): e00495.

Ahles, T.A., and J.C. Root. 2018. Cognitive effects of cancer and cancer treatments. *Annual Review of Clinical Psychology* 14: 425–451. https://doi.org/10.1146/annurev-clinpsy-050817-084903.

Alderson-Day, B., S. Weis, S. McCarthy-Jones, P. Moseley, D. Smailes, and C. Fernyhough. 2016. The brain's conversation with itself: Neural substrates of dialogic inner speech. *Social Cognitive and Affective Neuroscience* 11 (1): 110–120.

Arbib, M.A., A. Billard, M. Iacoboni, and E. Oztop. 2000. Synthetic brain imaging: Grasping, mirror neurons and imitation. *Neural Networks* 13 (8–9): 975–997.

Armstrong, D.F., W.C. Stokoe, and S.E. Wilcox. 1995. *Gesture and the nature of language*. Cambridge: Cambridge University Press.

Astington, J.W., and J.M. Jenkins. 1999. A longitudinal study of the relation between language and theory-of-mind development. *Developmental Psychology* 35 (5): 1311.

Bacqué-Cazenave, J., R. Bharatiya, G. Barrière, J.P. Delbecque, N. Bouguiyoud, G. Di Giovanni, et al. 2020. Serotonin in animal cognition and behavior. *International Journal of Molecular Sciences* 21 (5): 1649.

Bailey, P.E., and J.D. Henry. 2008. Growing less empathic with age: Disinhibition of the self-perspective. *The Journals of Gerontology Series B: Psychological Sciences and Social Sciences* 63 (4): P219–P226.

Baldo, J.V., N.F. Dronkers, D. Wilkins, C. Ludy, P. Raskin, and J. Kim. 2005. Is problem solving dependent on language? *Brain and Language* 92 (3): 240–250.

Baltes, P.B., U. Lindenberger, and U.M. Staudinger. 1998. Life-span theory in developmental psychology. In *Handbook of child psychology*, 1029–1143. New York, USA: Wiley.
Baltes, P.B., U.M. Staudinger, and U. Lindenberger. 1999. Lifespan psychology: Theory and application to intellectual functioning. *Annual Review of Psychology* 50 (1): 471–507.
Bánréti, Z., I. Hoffmann, and V. Vincze. 2016. Recursive subsystems in aphasia and Alzheimer's disease: Case studies in syntax and theory of mind. *Frontiers in Psychology* 7: 405. https://doi.org/10.3389/fpsyg.2016.00405.
Barlow, H. 1990. Conditions for versatile learning, Helmholtz's unconscious inference, and the task of perception. *Vision Research* 30 (11): 1561–1571.
Baron-Cohen, S., S. Wheelwright, V. Stone, and M. Rutherford. 1999. A mathematician, a physicist and a computer scientist with Asperger syndrome: Performance on folk psychology and folk physics tests. *Neurocase* 5 (6): 475–483.
Baron-Cohen, S., S. Wheelwright, J. Hill, Y. Raste, and I. Plumb. 2001. The "reading the mind in the eyes" test revised version: A study with normal adults, and adults with Asperger syndrome or high-functioning autism. *The Journal of Child Psychology and Psychiatry and Allied Disciplines* 42 (2): 241–251.
Baron-Cohen, S., H. Tager-Flusberg, and M. Lombardo, eds. 2013. *Understanding other minds: Perspectives from developmental social neuroscience*. Oxford: OUP Oxford.
Bauer, P., L. Kraushaar, and M. Jünemann. 2023. Vascular ageing and the complex relationship between cognitive decline and cardiovascular disease. *European Journal of Preventive Cardiology* 30 (7): 533–534. https://doi.org/10.1093/eurjpc/zwad027.
Berman, I., D. Klegon, H. Fiedosewicz, and H. Chang. 1999. The effect of novel antipsychotics on cognitive function. *Psychiatric Annals* 29 (11): 643–646.
Berwick, R.C., and N. Chomsky. 2016. *Why only us: Language and evolution*. MIT press.
Bialystok, E. 2021. Bilingualism: Pathway to cognitive reserve. *Trends in Cognitive Sciences* 25 (5): 355–364. https://doi.org/10.1016/j.tics.2021.02.003.
Bickerton, D. 1990. *Language and species*. Chicago: University of Chicago Press.
Birdsong, D. 2018. Plasticity, variability and age in second language acquisition and bilingualism. *Frontiers in Psychology* 9: 81. https://doi.org/10.3389/fpsyg.2018.00081.
Bishop, N.A., T. Lu, and B.A. Yankner. 2010. Neural mechanisms of ageing and cognitive decline. *Nature* 464 (7288): 529–535.
Bloom, P. 1999. Language capacities: Is grammar special? *Current Biology* 9 (4): R127–R128.
Booth, R., G.L. Wallace, and F. Happé. 2011. Connectivity and the corpus callosum in autism spectrum conditions: Insights from comparison of autism and callosal agenesis. *Progress in Brain Research* 189: 303–317.
Borenstein, E., and E. Ruppin. 2005. The evolution of imitation and mirror neurons in adaptive agents. *Cognitive Systems Research* 6 (3): 229–242.
Borsa, V.M., D. Perani, P.A. Della Rosa, G. Videsott, L. Guidi, B.S. Weekes, R. Franceschini, and J. Abutalebi. 2018. Bilingualism and healthy aging: Aging effects and neural maintenance. *Neuropsychologia* 111: 51–61. https://doi.org/10.1016/j.neuropsychologia.2018.01.012.
Bottiroli, S., E. Cavallini, I. Ceccato, T. Vecchi, and S. Lecce. 2016. Theory of mind in aging: Comparing cognitive and affective components in the faux pas test. *Archives of Gerontology and Geriatrics* 62: 152–162.
Bowers, J.S., and C.J. Davis. 2012. Bayesian just-so stories in psychology and neuroscience. *Psychological Bulletin* 138 (3): 389.
Brådvik, B., C. Dravins, S. Holtås, I. Rosén, E. Ryding, and D.H. Ingvar. 1991. Disturbances of speech prosody following right hemisphere infarcts. *Acta Neurologica Scandinavica* 84 (2): 114–126.
Bridgman, M.W., W.S. Brown, M.L. Spezio, M.K. Leonard, R. Adolphs, and L.K. Paul. 2014. Facial emotion recognition in agenesis of the corpus callosum. *Journal of Neurodevelopmental Disorders* 6: 1–14.

Bromiley, P.A., N.A. Thacker, M.L. Scott, M. Pokrić, A.J. Lacey, and T.F. Cootes. 2003. Bayesian and non-Bayesian probabilistic models for medical image analysis. *Image and Vision Computing* 21 (10): 851–864.

Brothers, L., and B. Ring. 1992. A neuroethological framework for the representation of minds. *Journal of Cognitive Neuroscience* 4 (2): 107–118.

Brunet, E., Y. Sarfati, M.C. Hardy-Baylé, and J. Decety. 2000. A PET investigation of the attribution of intentions with a nonverbal task. *NeuroImage* 11 (2): 157–166.

Brunet-Gouet, E., and J. Decety. 2006. Social brain dysfunctions in schizophrenia: A review of neuroimaging studies. *Psychiatry Research: Neuroimaging* 148 (2–3): 75–92.

Bryan, K.L. 1989. Language prosody and the right hemisphere. *Aphasiology* 3 (4): 285–299.

Buhot, M.C. 1997. Serotonin receptors in cognitive behaviors. *Current Opinion in Neurobiology* 7 (2): 243–254.

Cahana-Amitay, D., and M. Albert. 2015. *Redefining recovery from aphasia*. New York: Oxford University Press.

Caramazza, A., and B.Z. Mahon. 2006. The organisation of conceptual knowledge in the brain: The future's past and some future directions. *Cognitive Neuropsychology* 23 (1): 13–38.

Carrington, S.J., and A.J. Bailey. 2009. Are there theory of mind regions in the brain? A review of the neuroimaging literature. *Human Brain Mapping* 30 (8): 2313–2335.

Carruthers, P. 1998. *Language, thought and consciousness: An essay in philosophical psychology*. Cambridge: Cambridge University Press.

Carstensen, L.L. 2006. The influence of a sense of time on human development. *Science* 312 (5782): 1913–1915.

Cattell, R.B. 1941. Some theoretical issues in adult intelligence testing. *Psychological Bulletin* 38 (7): 592.

Chan, C.G.H., W.Q. Yow, and A. Oei. 2020. Active bilingualism in aging: Balanced bilingualism usage and less frequent language switching relate to better conflict monitoring and goal maintenance ability. *The Journals of Gerontology. Series B, Psychological Sciences and Social Sciences* 75 (9): e231–e241. https://doi.org/10.1093/geronb/gbaa058.

Charlton, R.A., T.R. Barrick, H.S. Markus, and R.G. Morris. 2009. Theory of mind associations with other cognitive functions and brain imaging in normal aging. *Psychology and Aging* 24 (2): 338.

Chomsky, N. 1980. Rules and representations. *Behavioral and Brain Sciences* 3 (1): 1–15.

Clark, A. 1998. *Being there: Putting brain, body, and world together again*. Cambridge, MA: MIT Press.

Clegg, J., S. Brumfitt, R.W. Parks, and P.W. Woodruff. 2007. Speech and language therapy intervention in schizophrenia: A case study. *International Journal of Language & Communication Disorders* 42 (S1): 81–101.

Cochin, S., C. Barthelemy, B. Lejeune, S. Roux, and J. Martineau. 1998. Perception of motion and qEEG activity in human adults. *Electroencephalography and Clinical Neurophysiology* 107 (4): 287–295.

Cochin, S., C. Barthelemy, S. Roux, and J. Martineau. 1999. Observation and execution of movement: Similarities demonstrated by quantified electroencephalography. *European Journal of Neuroscience* 11 (5): 1839–1842.

Cohen, M.J., C.A. Riccio, and A.M. Flannery. 1994. Expressive aprosodia following stroke to the right basal ganglia: A case report. *Neuropsychology* 8 (2): 242.

Colombo, M., and P. Seriès. 2012. Bayes in the brain—On Bayesian modelling in neuroscience. *The British Journal for the Philosophy of Science* 63: 697–723.

Corballis, M.C. 2002. *From hand to mouth: The origins of language*. Princeton: Princeton University Press.

———. 2003. From hand to mouth: The gestural origins of language. *Studies in the Evolution of Language* 3: 201–218.

Coyle, J.T. 2006. Glutamate and schizophrenia: Beyond the dopamine hypothesis. *Cellular and Molecular Neurobiology* 26: 363–382.

Craik, F., and E. Bialystok. 2005. Intelligence and executive control: Evidence from aging and bilingualism. *Cortex: A Journal Devoted to the Study of the Nervous System and Behavior* 41 (2): 222–224. https://doi.org/10.1016/s0010-9452(08)70899-2.

Crosswell, A.D., L. Whitehurst, and W.B. Mendes. 2021. Effects of acute stress on cognition in older versus younger adults. *Psychology and Aging* 36 (2): 241.

Cutica, I., M. Bucciarelli, and B.G. Bara. 2006. Neuropragmatics: Extralinguistic pragmatic ability is better preserved in left-hemisphere-damaged patients than in right-hemisphere-damaged patients. *Brain and Language* 98 (1): 12–25.

De Saussure, F. 1916. Nature of the linguistic sign. *Course in General Linguistics* 1: 65–70.

De Villiers, J.G., and J.E. Pyers. 2002. Complements to cognition: A longitudinal study of the relationship between complex syntax and false-belief-understanding. *Cognitive development* 17 (1): 1037–1060.

De Villiers, J.G., and P.A. De Villiers. 2014. Linguistic determination and the understanding of false beliefs. In *Children's reasoning and the mind*, 205–242. Hove, East Sussex, UK: Psychology Press.

Deacon, T.W. 1997. *The co-evolution of language and the brain*. New York, USA: WW Norlon, Nueva Ymk.

Deary, I.J., J. Corley, A.J. Gow, S.E. Harris, L.M. Houlihan, R.E. Marioni, et al. 2009. Age-associated cognitive decline. *British Medical Bulletin* 92 (1): 135–152.

Del Sette, P., M. Veneruso, R. Cordani, S. Lecce, G. Varallo, C. Franceschini, C. Venturino, F. Pizza, G. Plazzi, and L. Nobili. 2023. Narcolepsy and emotions: Is there a place for a theory of mind approach? *Sleep Medicine* 102: 84–89. https://doi.org/10.1016/j.sleep.2022.12.013.

Dollaghan, C.A. 2007. *The handbook for evidence-based practice in communication disorders*. Baltimore: Paul H Brookes Publishing.

Domes, G., M. Heinrichs, A. Michel, C. Berger, and S.C. Herpertz. 2007. Oxytocin improves "mind-reading" in humans. *Biological Psychiatry* 61 (6): 731–733.

Dominguez, L.J., N. Veronese, L. Vernuccio, G. Catanese, F. Inzerillo, G. Salemi, and M. Barbagallo. 2021. Nutrition, physical activity, and other lifestyle factors in the prevention of cognitive decline and dementia. *Nutrients* 13 (11): 4080. https://doi.org/10.3390/nu13114080.

Dor, D., C. Knight, and J. Lewis, eds. 2014. *The social origins of language* (Vol. 19). Oxford, UK: Oxford University Press.

Dye, L., N.B. Boyle, C. Champ, and C. Lawton. 2017. The relationship between obesity and cognitive health and decline. *The Proceedings of the Nutrition Society* 76 (4): 443–454. https://doi.org/10.1017/S0029665117002014.

Erickson, K.I., S.D. Donofry, K.R. Sewell, B.M. Brown, and C.M. Stillman. 2022. Cognitive aging and the promise of physical activity. *Annual Review of Clinical Psychology* 18: 417–442. https://doi.org/10.1146/annurev-clinpsy-072720-014213.

Fama, M.E., and P.E. Turkeltaub. 2020. Inner speech in aphasia: Current evidence, clinical implications, and future directions. *American Journal of Speech-Language Pathology* 29 (1S): 560–573. https://doi.org/10.1044/2019_AJSLP-CAC48-18-0212.

Fama, M.E., W. Hayward, S.F. Snider, R.B. Friedman, and P.E. Turkeltaub. 2017. Subjective experience of inner speech in aphasia: Preliminary behavioral relationships and neural correlates. *Brain and Language* 164: 32–42. https://doi.org/10.1016/j.bandl.2016.09.009.

Fama, M.E., M.P. Henderson, S.F. Snider, W. Hayward, R.B. Friedman, and P.E. Turkeltaub. 2019a. Self-reported inner speech relates to phonological retrieval ability in people with aphasia. *Consciousness and Cognition* 71: 18–29. https://doi.org/10.1016/j.concog.2019.03.005.

Fama, M.E., S.F. Snider, M.P. Henderson, W. Hayward, R.B. Friedman, and P.E. Turkeltaub. 2019b. The subjective experience of inner speech in aphasia is a meaningful reflection of lexical retrieval. *Journal of Speech, Language, and Hearing Research: JSLHR* 62 (1): 106–122. https://doi.org/10.1044/2018_JSLHR-L-18-0222.

Fernyhough, C. 2016. *The voices within: The history and science of how we talk to ourselves*. New York: Basic Books.

Fine, C., J. Lumsden, and R.J.R. Blair. 2001. Dissociation between 'theory of mind' and executive functions in a patient with early left amygdala damage. *Brain* 124 (2): 287–298.
Fitch, W.T. 2017. Empirical approaches to the study of language evolution. *Psychonomic Bulletin & Review* 24: 3–33.
Fittipaldi, S., A. Ibanez, S. Baez, F. Manes, L. Sedeno, and A.M. García. 2019. More than words: Social cognition across variants of primary progressive aphasia. *Neuroscience & Biobehavioral Reviews* 100: 263–284.
Fjell, A.M., and K.B. Walhovd. 2010. Structural brain changes in aging: Courses, causes and cognitive consequences. *Reviews in the Neurosciences* 21 (3): 187–222.
Fletcher, P.C., F. Happe, U. Frith, S.C. Baker, R.J. Dolan, R.S. Frackowiak, and C.D. Frith. 1995. Other minds in the brain: A functional imaging study of "theory of mind" in story comprehension. *Cognition* 57 (2): 109–128.
Fodor, J.A. 1983a. *The modularity of mind*. Cambridge, MA: MIT Press.
———. 1983b. *Representations: Philosophical essays on the foundations of cognitive science*. Cambridge, MA: MIT Press.
Freund, A.M. 2006. Age-differential motivational consequences of optimization versus compensation focus in younger and older adults. *Psychology and aging* 21 (2): 240.
Fridriksson, J., J.D. Richardson, J.M. Baker, and C. Rorden. 2011. Transcranial direct current stimulation improves naming reaction time in fluent aphasia: A double-blind, sham-controlled study. *Stroke* 42 (3): 819–821.
Friederici, A.D. 2011. The brain basis of language processing: From structure to function. *Physiological Reviews* 91 (4): 1357–1392.
Frith, C.D., and U. Frith. 2006. The neural basis of mentalizing. *Neuron* 50 (4): 531–534.
Frye, D., P.D. Zelazo, and T. Palfai. 1995. Theory of mind and rule-based reasoning. *Cognitive Development* 10 (4): 483–527.
Frye, D., P.D. Zelazo, P.J. Brooks, and M.C. Samuels. 1996. Inference and action in early causal reasoning. *Developmental Psychology* 32 (1): 120.
Gallese, V., and A. Goldman. 1998. Mirror neurons and the simulation theory of mind-reading. *Trends in Cognitive Sciences* 2 (12): 493–501.
Gallese, V., and M.I. Stamenov. 2002. Mirror neurons and the evolution of brain and language. *Mirror Neurons and the Evolution of Brain and Language*: 1–400.
Gallese, V., L. Fadiga, L. Fogassi, and G. Rizzolatti. 1996. Action recognition in the premotor cortex. *Brain* 119 (2): 593–609.
Gallo, F., V. DeLuca, Y. Prystauka, T. Voits, J. Rothman, and J. Abutalebi. 2022. Bilingualism and aging: Implications for (delaying) neurocognitive decline. *Frontiers in Human Neuroscience* 16: 819105. https://doi.org/10.3389/fnhum.2022.819105.
Gasser, M. 2004. The origins of arbitrariness in language. In *Proceedings of the annual meeting of the cognitive science society* (Vol. 26, No. 26).
Germain, C.M., and T.M. Hess. 2007. Motivational influences on controlled processing: Moderating distractibility in older adults. *Aging, Neuropsychology, and Cognition* 14 (5): 462–486.
Geschwind, N. 1965. Disconnexion syndromes in animals and man. *Brain* 88 (3): 585–585.
Giovagnoli, A.R., S. Franceschetti, F. Reati, A. Parente, C. Maccagnano, F. Villani, and R. Spreafico. 2011. Theory of mind in frontal and temporal lobe epilepsy: Cognitive and neural aspects. *Epilepsia* 52 (11): 1995–2002. https://doi.org/10.1111/j.1528-1167.2011.03215.x.
Giraud, A.L., A. Kleinschmidt, D. Poeppel, T.E. Lund, R.S. Frackowiak, and H. Laufs. 2007. Endogenous cortical rhythms determine cerebral specialization for speech perception and production. *Neuron* 56 (6): 1127–1134.
Glisky, Elizabeth L. 2007. Changes in cognitive function in human aging. *Brain Aging: Models, Methods, And Mechanisms* 1: 3–20.
Goldstein, K. 1948. *Language and language disturbances; Aphasic symptom complexes and their significance for medicine and theory of language*.
Goodglass, H., F.A. Quadfasel, and W.H. Timberlake. 1964. Phrase length and the type and severity of aphasia. *Cortex* 1 (2): 133–153.

Grady, C. 2012. The cognitive neuroscience of ageing. *Nature Reviews Neuroscience* 13 (7): 491–505.

Grimshaw, G.M. 1998. Integration and interference in the cerebral hemispheres: Relations with hemispheric specialization. *Brain and Cognition* 36 (2): 108–127.

Gu, J., and R. Kanai. 2014. What contributes to individual differences in brain structure? *Frontiers in Human Neuroscience* 8: 262.

Hale, C.M., and H. Tager-Flusberg. 2003. The influence of language on theory of mind: A training study. *Developmental Science* 6 (3): 346–359.

Hallowell, B. 2017. *Aphasia and other acquired neurogenic language disorders*. San Diego: Plural Publishing.

Hänninen, T., K. Koivisto, K.J. Reinikainen, E.L. Helkala, H. Soininen, L. Mykkänen, et al. 1996. Prevalence of ageing-associated cognitive decline in an elderly population. *Age and Ageing* 25 (3): 201–205.

Happé, F.G., E. Winner, and H. Brownell. 1998. The getting of wisdom: Theory of mind in old age. *Developmental Psychology* 34 (2): 358.

Happé, F., H. Brownell, and E. Winner. 1999. Acquired 'theory of mind' impairments following stroke. *Cognition* 70 (3): 211–240.

Hauser, M.D., N. Chomsky, and W.T. Fitch. 2002. The faculty of language: What is it, who has it, and how did it evolve? *Science* 298 (5598): 1569–1579.

Häusser, L.F. 2012. Empathie und Spiegelneurone. Ein Blick auf die gegenwärtige neuropsychologische Empathieforschung [Empathy and mirror neurons. A view on contemporary neuropsychological empathy research]. *Praxis der Kinderpsychologie und Kinderpsychiatrie* 61 (5): 322–335. https://doi.org/10.13109/prkk.2012.61.5.322.

Hayward, W., S.F. Snider, G. Luta, R.B. Friedman, and P.E. Turkeltaub. 2016. Objective support for subjective reports of successful inner speech in two people with aphasia. *Cognitive Neuropsychology* 33 (5–6): 299–314. https://doi.org/10.1080/02643294.2016.1192998.

Heaton, R.K., and P.A.R. Staff. 1993. Wisconsin card sorting test: Computer version 2. *Odessa: Psychological Assessment Resources* 4: 1–4.

Hebb, D.O. 2005. *The organization of behavior: A neuropsychological theory*. Hove, East Sussex, UK: Psychology Press. (From unpublished papers dated 1932–1945).

Henry, J.D., L.H. Phillips, T. Ruffman, and P.E. Bailey. 2013. A meta-analytic review of age differences in theory of mind. *Psychology and Aging* 28 (3): 826.

Heredia, R.R., A.M. Blackburn, and L.A. Vega. 2020. Moderation-mediation effects in bilingualism and cognitive reserve. *Frontiers in Psychology* 11: 572555. https://doi.org/10.3389/fpsyg.2020.572555.

Hermer-Vazquez, L., E.S. Spelke, and A.S. Katsnelson. 1999. Sources of flexibility in human cognition: Dual-task studies of space and language. *Cognitive Psychology* 39 (1): 3–36.

Hess, T.M. 2014. Selective engagement of cognitive resources: Motivational influences on older adults' cognitive functioning. *Perspectives on Psychological Science* 9 (4): 388–407.

Hess, T.M., D.C. Rosenberg, and S.J. Waters. 2001. Motivation and representational processes in adulthood: The effects of social accountability and information relevance. *Psychology and Aging* 16 (4): 629.

Hess, T.M., C.M. Germain, D.C. Rosenberg, C.M. Leclerc, and E.A. Hodges. 2005. Aging-related selectivity and susceptibility to irrelevant affective information in the construction of attitudes. *Aging, Neuropsychology, and Cognition* 12 (2): 149–174.

Hickok, G., and D. Poeppel. 2004. Dorsal and ventral streams: A framework for understanding aspects of the functional anatomy of language. *Cognition* 92 (1–2): 67–99.

———. 2007. The cortical organization of speech processing. *Nature Reviews Neuroscience* 8 (5): 393–402.

Hillis, A.E. 2007. Aphasia: Progress in the last quarter of a century. *Neurology* 69 (2): 200–213.

Hirao, K., J. Miyata, H. Fujiwara, M. Yamada, C. Namiki, M. Shimizu, N. Sawamoto, H. Fukuyama, T. Hayashi, and T. Murai. 2008. Theory of mind and frontal lobe pathology in schizophrenia: A voxel-based morphometry study. *Schizophrenia Research* 105 (1–3): 165–174. https://doi.org/10.1016/j.schres.2008.07.021.

Hiser, J., and M. Koenigs. 2018. The multifaceted role of the ventromedial prefrontal cortex in emotion, decision making, social cognition, and psychopathology. *Biological Psychiatry* 83 (8): 638–647. https://doi.org/10.1016/j.biopsych.2017.10.030.
Hoffman, P., and A.M. Morcom. 2018. Age-related changes in the neural networks supporting semantic cognition: A meta-analysis of 47 functional neuroimaging studies. *Neuroscience & Biobehavioral Reviews* 84: 134–150.
Huang, Y. 2012. *The Oxford dictionary of pragmatics*. Oxford: Oxford University Press.
Hurford, J.R. 1990. Nativist and functional explanations in language acquisition. *Logical Issues in Language Acquisition* 85: 136.
Ionescu-Tucker, A., and C.W. Cotman. 2021. Emerging roles of oxidative stress in brain aging and Alzheimer's disease. *Neurobiology of Aging* 107: 86–95. https://doi.org/10.1016/j.neurobiolaging.2021.07.014.
Johnen, A., S. Reul, H. Wiendl, S.G. Meuth, and T. Duning. 2018. Apraxia profiles—A single cognitive marker to discriminate all variants of frontotemporal lobar degeneration and Alzheimer's disease. *Alzheimer's & Dementia: Diagnosis, Assessment & Disease Monitoring* 10: 363–371.
Johnson, J.A., and T.R. Pring. 1990. Speech therapy and Parkinson's disease: A review and further data. *British Journal of Disorders of Communication* 25 (2): 183–194.
Jonides, J., E.E. Smith, R.A. Koeppe, E. Awh, S. Minoshima, and M.A. Mintun. 1993. Spatial working memory in humans as revealed by PET. *Nature* 363 (6430): 623–625.
Joyal, M., A. Bonneau, and S. Fecteau. 2016. Speech and language therapies to improve pragmatics and discourse skills in patients with schizophrenia. *Psychiatry Research* 240: 88–95.
Kano, F., C. Krupenye, S. Hirata, M. Tomonaga, and J. Call. 2019. Great apes use self-experience to anticipate an agent's action in a false-belief test. *Proceedings of the National Academy of Sciences* 116 (42): 20904–20909.
Kasher, A., and I. Sagi. 2010. Speech act theory and pragmatics. In *In handbook of pragmatics*, vol. 5, 29–54. Amsterdam/Philadelphia: John Benjamins Publishing Company.
Kay, J., R. Lesser, and M. Coltheart. 1996. Psycholinguistic assessments of language processing in aphasia (PALPA): An introduction. *Aphasiology* 10 (2): 159–180.
Kemp, J., O. Després, F. Sellal, and A. Dufour. 2012. Theory of mind in normal ageing and neurodegenerative pathologies. *Ageing Research Reviews* 11 (2): 199–219.
Kennedy, D.P., and R. Adolphs. 2012. The social brain in psychiatric and neurological disorders. *Trends in Cognitive Sciences* 16 (11): 559–572.
Koubiyr, I., C. Dulau-Metras, M. Deloire, J. Charré-Morin, A. Saubusse, B. Brochet, and A. Ruet. 2022. Amygdala network reorganization mediates the theory of mind performances in multiple sclerosis. *Journal of Neuroscience Research* 100 (2): 537–550. https://doi.org/10.1002/jnr.24986.
Krupenye, C., and J. Call. 2019. Theory of mind in animals: Current and future directions. *Wiley Interdisciplinary Reviews: Cognitive Science* 10 (6): e1503.
Kuperberg, G.R., and T.F. Jaeger. 2016. What do we mean by prediction in language comprehension? Language. *Cognition and Neuroscience* 31 (1): 32–59.
Langland-Hassan, P., F.R. Faries, M.J. Richardson, and A. Dietz. 2015. Inner speech deficits in people with aphasia. *Frontiers in Psychology* 6: 528. https://doi.org/10.3389/fpsyg.2015.00528.
Lenehan, M.E., M.J. Summers, N.L. Saunders, J.J. Summers, and J.C. Vickers. 2015. Relationship between education and age-related cognitive decline: A review of recent research. *Psychogeriatrics* 15 (2): 154–162.
Li, X., K. Wang, F. Wang, Q. Tao, Y. Xie, and Q. Cheng. 2013. Aging of theory of mind: The influence of educational level and cognitive processing. *International Journal of Psychology* 48 (4): 715–727.
Li, C., Y. Ma, R. Hua, Z. Yang, B. Zhong, H. Wang, and W. Xie. 2021. Dose-response relationship between long-term blood pressure variability and cognitive decline. *Stroke* 52 (10): 3249–3257. https://doi.org/10.1161/STROKEAHA.120.033697.

Listabarth, S., M. Groemer, T. Waldhoer, B. Vyssoki, N. Pruckner, S. Vyssoki, A. Glahn, D.M. König-Castillo, and D. König. 2022. Cognitive decline and alcohol consumption in the aging population-a longitudinal analysis of the survey of health, ageing and retirement in Europe. *European Psychiatry: The Journal of the Association of European Psychiatrists* 65 (1): e83. https://doi.org/10.1192/j.eurpsy.2022.2344.

Logothetis, N.K. 2008. What we can do and what we cannot do with fMRI. *Nature* 453 (7197): 869–878.

Lorenzen, B., and L.L. Murray. 2008. Bilingual aphasia: A theoretical and clinical review. *American Journal of Speech-Language Pathology* 17 (3): 299–317.

Lough, S., C. Gregory, and J.R. Hodges. 2001. Dissociation of social cognition and executive function in frontal variant frontotemporal dementia. *Neurocase* 7: 123–130.

Lu, Y., X. Gwee, D.Q. Chua, T.S. Lee, W.S. Lim, M.S. Chong, P. Yap, K.B. Yap, I. Rawtaer, T.M. Liew, F. Pan, and T.P. Ng. 2021. Nutritional status and risks of cognitive decline and incident neurocognitive disorders: Singapore longitudinal ageing studies. *The Journal of Nutrition, Health & Aging* 25 (5): 660–667. https://doi.org/10.1007/s12603-021-1603-9.

Luria, A.R. 1970. The functional organization of the brain. *Scientific American* 222 (3): 66–79.

Ma, Y., L. Liang, F. Zheng, L. Shi, B. Zhong, and W. Xie. 2020. Association between sleep duration and cognitive decline. *JAMA Network Open* 3 (9): e2013573. https://doi.org/10.1001/jamanetworkopen.2020.13573.

Ma, Y., R. Hua, Z. Yang, B. Zhong, L. Yan, and W. Xie. 2021. Different hypertension thresholds and cognitive decline: A pooled analysis of three ageing cohorts. *BMC Medicine* 19 (1): 287. https://doi.org/10.1186/s12916-021-02165-4.

Macoir, J., M. Lavoie, R. Laforce Jr., S.M. Brambati, and M.A. Wilson. 2017. Dysexecutive symptoms in primary progressive aphasia: Beyond diagnostic criteria. *Journal of Geriatric Psychiatry and Neurology* 30 (3): 151–161.

Martin, I., and S. McDonald. 2003. Weak coherence, no theory of mind, or executive dysfunction? Solving the puzzle of pragmatic language disorders. *Brain and Language* 85 (3): 451–466.

Matelli, M., and G. Luppino. 1997. Functional anatomy of human motor cortical areas. *Handbook of Neuropsychology* 11: 9–26.

Mayer, K.U. 2003. The sociology of the life course and lifespan psychology: Diverging or converging pathways? In *Understanding human development: Dialogues with lifespan psychology*, 463–481. Boston: Springer US.

Maylor, E.A., J.M. Moulson, A.M. Muncer, and L.A. Taylor. 2002. Does performance on theory of mind tasks decline in old age? *British Journal of Psychology* 93 (4): 465–485.

Melogno, S., M.A. Pinto, T.G. Scalisi, F. Badolato, and P. Parisi. 2021. Case report: Theory of mind and figurative language in a child with agenesis of the corpus callosum. *Frontiers in Psychology* 11: 596804.

Meltzer, H.Y., and S.R. McGurk. 1999. The effects of clozapine, risperidone, and olanzapine on cognitive function in schizophrenia. *Schizophrenia Bulletin* 25 (2): 233–256.

Mendez, M.F., E. Anderson, and J.S. Shapira. 2005. An investigation of moral judgement in frontotemporal dementia. *Cognitive and Behavioral Neurology* 18 (4): 193–197.

Mesulam, M.M. 2003. Primary progressive aphasia—A language-based dementia. *New England Journal of Medicine* 349 (16): 1535–1542.

Mesulam, M.M., C. Wieneke, C. Thompson, E. Rogalski, and S. Weintraub. 2012. Quantitative classification of primary progressive aphasia at early and mild impairment stages. *Brain* 135 (5): 1537–1553.

Minga, J., S.M. Sheppard, M. Johnson, R. Hewetson, P. Cornwell, and M.L. Blake. 2023. Apragmatism: The renewal of a label for communication disorders associated with right hemisphere brain damage. *International Journal of Language & Communication Disorders* 58 (2): 651–666.

Morin, A. 2005. Possible links between self-awareness and inner speech theoretical background, underlying mechanisms, and empirical evidence. *Journal of Consciousness Studies* 12 (4–5): 115–134.

Moseley, P., C. Fernyhough, and A. Ellison. 2013. Auditory verbal hallucinations as atypical inner speech monitoring, and the potential of neurostimulation as a treatment option. *Neuroscience & Biobehavioral Reviews* 37 (10): 2794–2805.
Muller, C.L., A.M. Anacker, and J. Veenstra-VanderWeele. 2016. The serotonin system in autism spectrum disorder: From biomarker to animal models. *Neuroscience* 321: 24–41.
Mungas, D., D.R. Early, M.M. Glymour, A. Zeki Al Hazzouri, and M.N. Haan. 2018. Education, bilingualism, and cognitive trajectories: Sacramento area Latino aging study (SALSA). *Neuropsychology* 32 (1): 77–88. https://doi.org/10.1037/neu0000356.
Nasios, G., E. Dardiotis, and L. Messinis. 2019. From Broca and Wernicke to the neuromodulation era: Insights of brain language networks for neurorehabilitation. *Behavioural Neurology* 2019.
Nedergaard, J.S., M. Wallentin, and G. Lupyan. 2023. Verbal interference paradigms: A systematic review investigating the role of language in cognition. *Psychonomic Bulletin & Review* 30 (2): 464–488.
Nyberg, L., M. Lövdén, K. Riklund, U. Lindenberger, and L. Bäckman. 2012. Memory aging and brain maintenance. *Trends in Cognitive Sciences* 16 (5): 292–305.
O'Sullivan, M.R.C.P., D.K. Jones, P.E. Summers, R.G. Morris, S.C.R. Williams, and H.S. Markus. 2001. Evidence for cortical "disconnection" as a mechanism of age-related cognitive decline. *Neurology* 57 (4): 632–638.
Olney, N.T., S. Spina, and B.L. Miller. 2017. Frontotemporal Dementia. *Neurologic Clinics* 35 (2): 339–374. https://doi.org/10.1016/j.ncl.2017.01.008.
Ozonoff, S., B.F. Pennington, and S.J. Rogers. 1991. Executive function deficits in high-functioning autistic individuals: Relationship to theory of mind. *Journal of Child Psychology and Psychiatry* 32 (7): 1081–1105.
Pace-Schott, E.F., and R.M. Spencer. 2015. Sleep-dependent memory consolidation in healthy aging and mild cognitive impairment. *Current Topics in Behavioral Neurosciences* 25: 307–330. https://doi.org/10.1007/7854_2014_300.
Park, D.C., and P. Reuter-Lorenz. 2009. The adaptive brain: Aging and neurocognitive scaffolding. *Annual Review of Psychology* 60: 173–196.
Patel, R. 2010. Treating expressive aprosodia: A case study. *Journal of Medical Speech-Language Pathology* 18 (4): 115–119.
Paul, L.K., C. Corsello, D.P. Kennedy, and R. Adolphs. 2014. Agenesis of the corpus callosum and autism: A comprehensive comparison. *Brain* 137 (6): 1813–1829.
Perner, J., and B. Lang. 1999. Development of theory of mind and executive control. *Trends in Cognitive Sciences* 3 (9): 337–344.
Perner, J., S.R. Leekam, and H. Wimmer. 1987. Three-year-olds' difficulty with false belief: The case for a conceptual deficit. *British Journal of Developmental Psychology* 5 (2): 125–137.
Peterson, C.C. 2013. *Looking forward through the lifespan: Developmental psychology*. Australia: Pearson Higher Education AU.
Pickering, M.J., and S. Garrod. 2013. An integrated theory of language production and comprehension. *Behavioral and Brain Sciences* 36 (4): 329–347.
Pinker, S. 2003. *The language instinct: How the mind creates language*. London: Penguin UK.
Pinker, S., and P. Bloom. 1990. Natural language and natural selection. *Behavioral and Brain Sciences* 13 (4): 707–727.
Premack, D., and G. Woodruff. 1978. Does the chimpanzee have a theory of mind? *Behavioral and Brain Sciences* 1 (4): 515–526.
Rakoczy, H., A. Harder-Kasten, and L. Sturm. 2012. The decline of theory of mind in old age is (partly) mediated by developmental changes in domain-general abilities. *British Journal of Psychology* 103 (1): 58–72.
Ramachandran, V.S. 2000. Mirror neurons and imitation learning as the driving force behind "the great leap forward" in human evolution.

Raz, N., U. Lindenberger, K.M. Rodrigue, K.M. Kennedy, D. Head, A. Williamson, et al. 2005. Regional brain changes in aging healthy adults: General trends, individual differences and modifiers. *Cerebral Cortex* 15 (11): 1676–1689.

Rizzolatti, G., and M.A. Arbib. 1998. Language within our grasp. *Trends in Neurosciences* 21 (5): 188–194.

Rizzolatti, G., and L. Craighero. 2004. The mirror-neuron system. *Annual Review of Neuroscience* 27: 169–192.

Rizzolatti, G., L. Fadiga, V. Gallese, and L. Fogassi. 1996. Premotor cortex and the recognition of motor actions. *Cognitive Brain Research* 3 (2): 131–141.

Rosenbek, J.C., G.P. Crucian, S.A. Leon, B. Hieber, A.D. Rodriguez, B. Holiway, et al. 2004. Novel treatments for expressive aprosodia: A phase I investigation of cognitive linguistic and imitative interventions. *Journal of the International Neuropsychological Society* 10 (5): 786–793.

Ross, E.D. 1981. The aprosodias: Functional-anatomic organization of the affective components of language in the right hemisphere. *Archives of Neurology* 38 (9): 561–569.

Rossi, E., and M.T. Diaz. 2016. How aging and bilingualism influence language processing: Theoretical and neural models. *Linguistic Approaches to Bilingualism* 6 (1–2): 9–42. https://doi.org/10.1075/lab.14029.ros.

Roth, B.L., S.M. Hanizavareh, and A.E. Blum. 2004. Serotonin receptors represent highly favorable molecular targets for cognitive enhancement in schizophrenia and other disorders. *Psychopharmacology* 174: 17–24.

Rutherford, B.R., K. Brewster, J.S. Golub, A.H. Kim, and S.P. Roose. 2018. Sensation and psychiatry: Linking age-related hearing loss to late-life depression and cognitive decline. *The American Journal of Psychiatry* 175 (3): 215–224. https://doi.org/10.1176/appi.ajp.2017.17040423.

Sabbagh, M.A., and D. Shafman. 2009. How children block learning from ignorant speakers. *Cognition* 112 (3): 415–422.

Salmelin, R., and R. Hari. 1994. Characterization of spontaneous MEG rhythms in healthy adults. *Electroencephalography and clinical neurophysiology* 91 (4): 237–248.

Salthouse, T.A. 2009. When does age-related cognitive decline begin? *Neurobiology of Aging* 30 (4): 507–514.

———. 2011. Neuroanatomical substrates of age-related cognitive decline. *Psychological Bulletin* 137 (5): 753.

Salthouse, T. 2012. Consequences of age-related cognitive declines. *Annual Review of Psychology* 63: 201–226.

Salthouse, T.A., T.M. Atkinson, and D.E. Berish. 2003. Executive functioning as a potential mediator of age-related cognitive decline in normal adults. *Journal of Experimental Psychology: General* 132 (4): 566.

Saur, D., B.W. Kreher, S. Schnell, D. Kümmerer, P. Kellmeyer, M.S. Vry, et al. 2008. Ventral and dorsal pathways for language. *Proceedings of the National Academy of Sciences* 105 (46): 18035–18040.

Saxe, R. 2006. Uniquely human social cognition. *Current Opinion in Neurobiology* 16 (2): 235–239.

Schlaug, G., S. Marchina, and A. Norton. 2009. Evidence for plasticity in white-matter tracts of patients with chronic Broca's aphasia undergoing intense intonation-based speech therapy. *Annals of the New York Academy of Sciences* 1169 (1): 385–394.

Schlotz, W., K. Hammerfald, U. Ehlert, and J. Gaab. 2011. Individual differences in the cortisol response to stress in young healthy men: Testing the roles of perceived stress reactivity and threat appraisal using multiphase latent growth curve modeling. *Biological Psychology* 87 (2): 257–264.

Schurz, M., J. Radua, M.G. Tholen, L. Maliske, D.S. Margulies, R.B. Mars, et al. 2021. Toward a hierarchical model of social cognition: A neuroimaging meta-analysis and integrative review of empathy and theory of mind. *Psychological Bulletin* 147 (3): 293.

Seyfarth, R.M., D.L. Cheney, and T.J. Bergman. 2005. Primate social cognition and the origins of language. *Trends in Cognitive Sciences* 9 (6): 264–266.

Shafto, M.A., and L.K. Tyler. 2014. Language in the aging brain: The network dynamics of cognitive decline and preservation. *Science* 346 (6209): 583–587.

Shamay-Tsoory, S.G., R. Tomer, B.D. Berger, D. Goldsher, and J. Aharon-Peretz. 2005. Impaired "affective theory of mind" is associated with right ventromedial prefrontal damage. *Cognitive and Behavioral Neurology* 18 (1): 55–67.

Shamay-Tsoory, S.G., Y. Tibi-Elhanany, and J. Aharon-Peretz. 2006. The ventromedial prefrontal cortex is involved in understanding affective but not cognitive theory of mind stories. *Social Neuroscience* 1 (3–4): 149–166.

Sheppard, S.M., E.L. Meier, A.Z. Durfee, A. Walker, J. Shea, and A.E. Hillis. 2021. Characterizing subtypes and neural correlates of receptive aprosodia in acute right hemisphere stroke. *Cortex* 141: 36–54.

Siegal, M., and R. Varley. 2002. Neural systems involved in 'theory of mind'. *Nature Reviews Neuroscience* 3 (6): 463–471.

———. 2006. Aphasia, language, and theory of mind. *Social Neuroscience* 1 (3–4): 167–174.

Siegal, M., J. Carrington, and M. Radel. 1996. Theory of mind and pragmatic understanding following right hemisphere damage. *Brain and Language* 53 (1): 40–50.

Siegal, M., R. Varley, and S.C. Want. 2001. Mind over grammar: Reasoning in aphasia and development. *Trends in Cognitive Sciences* 5 (7): 296–301.

Slessor, G., L.H. Phillips, and R. Bull. 2007. Exploring the specificity of age-related differences in theory of mind tasks. *Psychology and Aging* 22 (3): 639.

Slocombe, K.E., I. Alvarez, H.P. Branigan, T. Jellema, H.G. Burnett, A. Fischer, et al. 2013. Linguistic alignment in adults with and without Asperger's syndrome. *Journal of Autism and Developmental Disorders* 43: 1423–1436.

Stark, B.C., S. Geva, and E.A. Warburton. 2017. Inner speech's relationship with overt speech in Poststroke aphasia. *Journal of Speech, Language, and Hearing Research: JSLHR* 60 (9): 2406–2415. https://doi.org/10.1044/2017_JSLHR-S-16-0270.

Staudinger, U.M., M. Marsiske, and P.B. Baltes. 1995. Resilience and reserve capacity in later adulthood: Potentials and limits of development across the life span. *Development and Psychopathology* 2: 801–847.

Stemmer, B., F. Giroux, and Y. Joanette. 1994. Production and evaluation of requests by right hemisphere brain-damaged individuals. *Brain and Language* 47 (1): 1–31.

Stewart, E., C. Catroppa, and S. Lah. 2016. Theory of mind in patients with epilepsy: A systematic review and meta-analysis. *Neuropsychology Review* 26 (1): 3–24. https://doi.org/10.1007/s11065-015-9313-x.

Stuss, D.T., G.G. Gallup Jr., and M.P. Alexander. 2001. The frontal lobes are necessary for "theory of mind". *Brain* 124 (2): 279–286.

Suh, S.W., J.W. Han, J.R. Lee, S. Byun, S.J. Kwon, S.H. Oh, K.H. Lee, G. Han, J.W. Hong, K.P. Kwak, B.J. Kim, S.G. Kim, J.L. Kim, T.H. Kim, S.H. Ryu, S.W. Moon, J.H. Park, J. Seo, J.C. Youn, D.Y. Lee, et al. 2018. Sleep and cognitive decline: A prospective nondemented elderly cohort study. *Annals of Neurology* 83 (3): 472–482. https://doi.org/10.1002/ana.25166.

Sundaray, S., T. Marinis, and A. Bose. 2018. Comprehending non-literal language: Effects of aging and bilingualism. *Frontiers in Psychology* 9: 2230. https://doi.org/10.3389/fpsyg.2018.02230.

Surian, L., and M. Siegal. 2001. Sources of performance on theory of mind tasks in right hemisphere-damaged patients. *Brain and Language* 78 (2): 224–232.

Symington, S.H., L.K. Paul, M.F. Symington, M. Ono, and W.S. Brown. 2010. Social cognition in individuals with agenesis of the corpus callosum. *Social Neuroscience* 5 (3): 296–308.

Tager-Flusberg, H., and K. Sullivan. 2000. A componential view of theory of mind: Evidence from Williams syndrome. *Cognition* 76 (1): 59–90.

Tappin, B.M., and S. Gadsby. 2019. Biased belief in the Bayesian brain: A deeper look at the evidence. *Consciousness and Cognition* 68: 107–114.

Tesak, J., and C. Code. 2008. *Milestones in the history of aphasia: Theories and protagonists*. Hove, East Sussek, UK: Psychology Press.
Thompson, P. 2014. The importance of theory of mind in epilepsy. *Epilepsy & Behavior: E&B* 39: 143–144. https://doi.org/10.1016/j.yebeh.2014.07.006.
Tomasello, M. 2010. *Origins of human communication*. Cambridge, MA: MIT Press.
Turkeltaub, P.E., H.B. Coslett, A.L. Thomas, O. Faseyitan, J. Benson, C. Norise, and R.H. Hamilton. 2012. The right hemisphere is not unitary in its role in aphasia recovery. *Cortex* 48 (9): 1179–1186.
Van Overwalle, F., and K. Baetens. 2009. Understanding others' actions and goals by mirror and mentalizing systems: A meta-analysis. *NeuroImage* 48 (3): 564–584.
Varley, R., and M. Siegal. 2000. Evidence for cognition without grammar from causal reasoning and 'theory of mind' in an agrammatic aphasic patient. *Current Biology* 10 (12): 723–726.
von Hippel, W., and S.M. Dunlop. 2005. Aging, inhibition, and social inappropriateness. *Psychology and Aging* 20 (3): 519.
Vygotsky, L. 1934. 1962. Thought and language. *Trans. E.*
Wang, H., P. Zhao, J. Zhao, J. Zhong, P. Pan, G. Wang, and Z. Yi. 2022. Theory of mind and empathy in adults with epilepsy: A meta-analysis. *Frontiers in Psychiatry* 13: 877957. https://doi.org/10.3389/fpsyt.2022.877957.
Wechsler, D. 1981. The psychometric tradition: Developing the Wechsler adult intelligence scale. *Contemporary Educational Psychology* 6: 82–85.
Weintraub, S., M.M. Mesulam, and L. Kramer. 1981. Disturbances in prosody: A right-hemisphere contribution to language. *Archives of Neurology* 38 (12): 742–744.
Wepman, J.M., and L.V. Jones. 1964. Five aphasias. *Disorders of communication*.
Wilson, D., and D. Sperber. 2004. La teoría de la relevancia. *Revista de investigación lingüística* 7 (1): 233–283.
Woolfe, T., S.C. Want, and M. Siegal. 2002. Signposts to development: Theory of mind in deaf children. *Child Development* 73 (3): 768–778.
Younes, K., and B.L. Miller. 2020. Frontotemporal dementia: Neuropathology, genetics, neuroimaging, and treatments. *The Psychiatric Clinics of North America* 43 (2): 331–344. https://doi.org/10.1016/j.psc.2020.02.006.
Younes, K., V. Borghesani, M. Montembeault, S. Spina, M.L. Mandelli, A.E. Welch, E. Weis, P. Callahan, F.M. Elahi, A.Y. Hua, D.C. Perry, A. Karydas, D. Geschwind, E. Huang, L.T. Grinberg, J.H. Kramer, A.L. Boxer, G.D. Rabinovici, H.J. Rosen, W.W. Seeley, et al. 2022. Right temporal degeneration and socioemotional semantics: Semantic behavioural variant frontotemporal dementia. *Brain: A Journal of Neurology* 145 (11): 4080–4096. https://doi.org/10.1093/brain/awac217.
Zemestani, M., O. Hoseinpanahi, M.A. Salehinejad, and M.A. Nitsche. 2022. The impact of prefrontal transcranial direct current stimulation (tDCS) on theory of mind, emotion regulation and emotional-behavioral functions in children with autism disorder: A randomized, sham-controlled, and parallel-group study. *Autism Research: Official Journal of the International Society for Autism Research* 15 (10): 1985–2003. https://doi.org/10.1002/aur.2803.
Zhang, X., S. Lecce, I. Ceccato, E. Cavallini, L. Zhang, and T. Chen. 2018. Plasticity in older adults' theory of mind performance: The impact of motivation. *Aging & Mental Health* 22 (12): 1592–1599.
Zhou, Y., W. Ran, Z. Luo, J. Wang, M. Fang, K. Wei, J. Sun, and M. Lou. 2022. Impaired peri-olfactory cerebrospinal fluid clearance is associated with ageing, cognitive decline and dyssomnia. *eBioMedicine* 86: 104381. https://doi.org/10.1016/j.ebiom.2022.104381.

Chapter 10
Sleep and Its Disorders: When ToM Is Not Awake

Enrique Montes-Latorre and Paolo Porcacchia

'O sleep, O gentle sleep, Nature's soft nurse, how have I frightened thee. That thou no more will weigh my eyelids down. And steep my senses in forgetfulness?'

William Shakespeare, Henry IV, Part 2

'For all life is a dream, and dreams themselves are only dreams'.

Pedro Calderón de la Barca, Life is a Dream, Act 2-Scene 19

Abstract Sleep is a physiological behaviour, cyclical and reversible, that occupies one third of our lives, even if some consider this time as regrettably wasted. But its ultimate explanation is still unknown, although multiple theories have tried to clarify it throughout the centuries. The issue of sleep and its abnormalities has attracted man's interest from the beginning of time and many philosophers, poets and artists have addressed the topic. But the actual science of sleep, as we know it today, is relatively young. If we give Hans Berger, the father of electroencephalography, the merit of introducing a new technique to study the brain of the sleeper in 1929, then the history of somnology is less than 100 years old. Another significant discovery was the dichotomy between REM sleep, NREM sleep and their relation to wakefulness, encompassing the three stages of human consciousness. Sleep disorders comprise manifold phenomena affecting not only sleep itself but daytime functioning and some of them can implicate injuries or serious health outcomes. Some of these issues are reviewed in this chapter.

Keywords Sleep · Polysomnography · Sleep Disorders · Human Physiology · Theory of Mind

E. Montes-Latorre (✉)
Department of Clinical Neurophysiology, University Hospital Virgen del Rocío, Seville, Spain

P. Porcacchia
Department of Clinical Neurophysiology, Sleep Pathology Unit, Hospital Virgen del Rocío, Seville, Spain

T. Lopez-Soto et al. (eds.), *The Theory of Mind Under Scrutiny*, Logic, Argumentation & Reasoning 34, https://doi.org/10.1007/978-3-031-46742-4_10

10.1 Introduction

If Theory of Mind is about our own consciousness, it seems necessary to introduce the term from the point of view of the Science of Sleep. Many authors have started to analyse ToM and its relationship with characters in our dreams (McNamara et al. 2007). Even today, in the third decade of the twenty-first century, sleep remains a big mystery. We can describe sleep as a physiological behaviour, characterised by its periodicity and reversibility, but its ultimate sense remains unknown. We all know of its importance: a 60-year-old person has spent some 20 years under the influence of Hypnos, the Greek god of Sleep; it seems to be essential for life: its complete deprivation can lead to death; it is also a universal behaviour in warm-blooded animals, including birds and mammals (Zimmer 1993). In this chapter we will outline some of the main features of sleep in humans, the methods of study and the different conditions, physiological and pathological, that may interfere with it. With this approach, we would like to consider the importance of a new type of consciousness: when ToM is not awake.

10.2 The Science of Sleep

There are some significant milestones in the history of the Science of Sleep. Our current understanding of sleep is intimately related to electroencephalography (EEG). Hans Berger, a German psychiatrist, recorded electrical activity from the human brain and published his results in 1929, although he dated the discovery of human electroencephalography on July sixth, 1924. He also demonstrated that this electrical activity was clearly different considering whether the subject was awake or asleep. (İnce et al. 2021; Millett 2001) Some facts were remarkable: the presence of electrical activity confirmed that sleep is an active process, in opposition to the extended theory that it was a state of profound rest and lack of reactiveness. The other main issue is that sleep activity could be studied without disturbing the sleeper.

Alfred Loomis -attorney, banker, inventor, and physicist- described a classification system of EEG patterns during sleep in the 1930s. In 1953, Eugene Aserinsky and Nathaniel Kleitmann published that during sleep there are several phases in which ocular movements appeared. This phase was baptised REM (Rapid Eye Movements), and this marked a difference between two different states in normal sleep: one was called REM (Rapid Eye Movements) and the other phase were called NREM (Non-Rapid Eye Movements). They assumed that the "Rapid Eye Movements" period, which was associated with irregular breathing and changes in the heart rate, should correspond to dreaming. Dement (2005) and Kleitman reported that there was a cyclical succession of EEG patterns occurring at intervals of some 90 min from one rapid eye movement period to the next one. Thus, it became clear that sleep did not lie in a homogeneous EEG state, but in two different patterns (REM and NREM) which were also different to wakefulness. REM sleep was characterized by rapid eye movements, atonia (active motor inhibition) and intensive brain activity (Dement 2005; Stone and Hughes 2013).

In the following years some other important discoveries were published. Endogenous depression is associated with a shortened REM latency, but the distinctive pattern of narcolepsy, in which REM periods appear shortly after sleep onset, was reported at the beginning of the 1960s. And although most of the sleep research had been carried out in the United States, the merits of the description of the Sleep Apnea Syndrome returned to Europe with the Lugaresi and Tassinari group (Dement 2005). An Australian pneumologist, Dr. Colin Sullivan, deserves the credit for the application of Continuous Positive Nasal Airway Pressure (CPAP) for the effective treatment of Sleep Apnea (Berry et al. 2002; Dement 2005; Guileminault et al. 1986).

In 1968 the results of a committee of sleep researchers were published as a manual for scoring sleep (Rechtschaffen and Kales); this system has become the widely accepted standard for scoring sleep stages (Anderer et al. 2007; Himanen and Hasan 2000). Most of the sleep centres now use the scoring rules of the American Academy of Sleep Medicine. Sleep is analysed by epochs, each one lasting 30 s, and each one is classified according to the predominant pattern (Berry et al. 2017; Silber et al. 2007; Danker-Hopfe et al. 2009).

10.3 Polysomnographic Scoring

The term polysomnography (PSG), that encloses recording of the EEG and several other physiological parameters as eye movements (electro-oculogram), respiratory, cardiac and electromyography during sleep, was coined by Dr. Holland, a member of the Stanford University Sleep Group in 1974. As recorded in PSG, normal sleep shows a well-defined architecture (Bloch 1997; Orr 1985; Patil 2010). A detailed description of EEG during the vigil state and sleep is beyond the scope of this chapter, but we will try to outline some features (Berry et al. 2017; Kirsch 2022).

Sleep does not start abruptly. The waking state (Stage W) is characterized by low-voltage mixed frequencies and a posterior dominant alpha rhythm when the eyes are closed. Eye blinks, rapid eye movements and normal tonic EMG activity are registered. Gradually, the eye movements become slow, the tonic EMG activity diminishes, and the EEG background activity desynchronizes, with loss of the posterior alpha rhythm. Stage 1 is scored when more than fifty percent of the activity is of low voltage and mixed frequency, and there are not rapid eye movements; it lasts a few minutes, usually less than 10. This stage is easily interrupted (i.e., a low arousal threshold) and should the subject be awakened, they will probably claim not to be really asleep. Stage 2 NREM follows this short Stage 1 transition and corresponds to light sleep. EEG background is composed of a mixture of frequencies, delta waves are present throughout this activity but not as prominent as in Stage 3. Against this background, two distinctive attributes appear: a) 12 to 14 Hz sleep spindles and b) K complexes. In Stage 2 more intense stimuli are needed to produce an arousal.

In the first cycle, Stage 2 goes on for 10–25 min and the gradual emergence of high-voltage waves in the delta range (below 4 Hz) gives way to Stage 3 NREM, corresponding to "slow wave" or "deep" sleep. Previously NREM sleep was subdivided into four stages, namely Stages 3 and 4 representing slow wave sleep. These Stages are now merged into a single Stage 3. Arousal threshold in this stage is high.

The transition from NREM to REM sleep is not abrupt. Usually, in the first sleep cycle, the subject enters in REM sleep after he passed through NREM stages. REM sleep EEG is characterized by low-middle amplitude 3–7 Hz activity (sawtooth waves are common), more similar to that of wakefulness or Stage 1 than to the other NREM Stages. Ocular movements are also quite similar to those of a vigil subject. But in REM sleep there is atonia of all striated muscles, with the exception of the diaphragm, just briefly interrupted by episodic bursts of extraocular muscles (producing the rapid eye movements) and, occasionally, of striated muscle activity. One can correctly describe REM sleep as richness of brain electrical activity within a paralysed body, hence the term "paradoxical sleep". Blood pressure, heart rate and respiratory rate become irregular. Vivid dreams occur in this phase, which usually lasts for a few minutes, ending with a body movement and getting into NREM sleep again. This combination of NREM and one REM segment is called a Sleep Cycle and in a young normal adult, some four or five cycles with an average duration of 90 min are repeated through the night. However, the cycles also vary quantitatively: deep NREM sleep is more prominent during the first two cycles, the largest amount of Stage 3 in the first third, and REM sleep periods become progressively longer in the last two thirds of the night.

As mentioned before, this accounts for the standard sleep architecture of a normal adult but, as we all know, sleep patterns differ with age. Babies sleep longer but also in a different way to children, adults, and the elderly. From the twelfth week of gestation two distinctive patterns can be distinguished: one is "active" sleep, with irregular respiration, ocular movements and muscle activity, precursor of REM sleep; the other is "quiet" sleep, an embrionic form of delta sleep. At birth, these patterns alternate in a one-to-one ratio, although in periods of 50–60 min instead of the 90-min average of adults, and for a daily time of sleep of some 16 h. First day/night cycles appear around 3 months of age. At 14 months, sleep is plainly focused on night-time, but the infant still needs a nap. Total sleep time decreases slowly reaching the standard adult parameters after adolescence: sleep cycles lengthen to the average of 90-min duration and the REM proportion diminishes. In the elderly, night sleep is fragmented and holds a smaller slow-wave ratio. While there is no difference in sleep scoring in adult and elderly people, specific criteria may apply when scoring sleep in children. In addition, some cut-off value may differ in this specific population. For example, the cut-off value to confirm a pathologic apnea/hypopnea index or to confirm a pathologic periodic limb movement index is different in children with respect to the adult population (Grigg-Damberger et al. 2007).

On the whole, Stage 1 accounts for 5 to 10% of total sleep time in young adults; Stage 2 comprises the largest percentage: about 45 to 55. Stage 3 accounts for 10 to 20% and REM sleep is usually 20–25% (AASM 2014).

10.4 Sleep Requirements: How Much Sleep Is Required?

The American Association of Sleep recommends that adults aged 18–60 years old should sleep at least 7 h per night (Watson et al. 2015). But this is not a universal rule for everybody: short sleepers (less than 6 h per night) may not show or report daytime signs of fatigue or sleepiness. On the other hand, some individuals spend more time in bed than necessary for sleep and may refer long periods of wakefulness during the night, but do not show daytime impairment. Long sleepers are people who sleep more than the average in their age group, and this is related to their natural biological cycles. All these ranges of sleep can be considered of good quality, but the complaint turns up when patients do not sleep the hours they need according to their own biology. Insufficient sleep is usually caused by the (voluntary or involuntary) reduction of sleep hours and result in daytime sleepiness, associating fatigue, lack of energy, decreased attention and an increased risk of driving and work accidents.

10.5 Sleep Disorders

What do Parkinson's disease, involuntary but periodic limb movements, jet lag and apnea have in common? Well, all these conditions share a link with sleep, in some of them sole or distinctive, in others as part of a broader clinical spectrum. Sleep disorders are common and vary in frequency and severity. Some of them are life-threatening conditions, some are not clinically evident by the subject but referred by a partner or a roommate. Some have a strong genetic base and others are acquired.

A first approach to sleep disorders is based upon the main complaint: difficulties in initiating or maintaining sleep, excessive somnolence during the day, disorders of sleep-wake schedule or behavioural disorders during sleep. In this section we review some of the most important sleep disorders in the clinic.

Insomnia The subjective complaint of insufficient or inadequate sleep, insomnia is the most common sleep disorder among the population. More a symptom than a disease, it may result from a broad variety of causes and conditions. Subjects refer difficulty in falling asleep, frequent awakenings or early-morning awakening. These symptoms occur despite adequate circumstances for sleep. Traditionally, insomnia has been categorized as primary or secondary, but in the most recent classification (AASM 2014) three subtypes are considered:

1. Chronic insomnia disorder, occurring at least 3 times per week and present for at least 3 months.
2. Short-Term insomnia disorder, when symptoms have been present for less than 3 months, and
3. Other insomnia disorder. It must also be stressed that the complaint of lack of sufficient sleep is subjective and relies on the subject's perception and also on the grade of impairment of daily activity. In some cases, there is a mismatch between

the subjective perception of sleep quantity and quality and the amount of sleep registered in the PSG laboratory. This variant is called paradoxical insomnia or sleep state misperception.

Another frequent symptom is **sleepiness**. If not relieved by increasing the sleep duration it usually reflects a sleep disorder, especially when it happens in an inappropriate place or under undesirable circumstances (social events, driving or at work). Additional factors may help to determine the cause: snoring and referred apnoeic episodes in obstructive sleep apnea; cataplexy, sleep attacks and hallucinations in narcolepsy.

10.6 Sleep and the COVID Pandemic

Recently, the COVID-19 pandemic had an impact on the prevalence of insomnia and sleep problems, not only in patients with COVID-19 infection, but also in health care workers and up to 30–40% of the general population. The most affected group were patients infected with COVID-19, with an estimated prevalence of sleep problems that reached 74.8% in some series (Jahrami et al. 2021). Also, healthcare workers showed a slightly higher prevalence of sleep problems (36.0%) (Salari et al. 2020). The increase burden of these problems was frequently associated to a parallel increase of anxiety and depression symptoms, often in a bi-directional manner (Pappa et al. 2020; Deng et al. 2021).

During COVID-19 outbreak, many behavioural, demographic, social and psychological factors could modify sleep habits and so alter sleep quality. In an interesting work distinct profiles of changes in sleep behaviours were identified (Robillard et al. 2021). It was found that some factors were associated to worse sleep quality, as female sex, chronic illnesses, being employed, family responsibilities, earlier wake-up times, or higher stress levels. In addition, maladaptive strategies as heavier alcohol use or more television exposure were related with sleep difficulties. In another study (Dai et al. 2021) it was observed how maintaining regular sleep habits could be a protective factor against sleep problems.

10.7 Sleep-Related Breathing Disorders

Sleep-related breathing disorders are characterized by dysfunction of respiration during sleep, although abnormalities may also occur during wakefulness. There is increasing evidence of the relation between sleep-disordered breathing and cardiovascular conditions, encompassing hypertension, chronic heart failure and arrhythmia. One of the main disorders -*sleep apnea*- can be subdivided into obstructive or central, although many patients may have a combination of both. Obstructive sleep apnea features repetitive episodes of complete (apnea) or partial (hypopnea) upper airway blockage during sleep. This recurrent obstruction results in hypoxia and sleep-fragmentation. By definition, apnea and hypopnea episodes must persist a minimum of 10 s, but it frequently may last for more than 30 s. Snoring, apnoeic episodes and sleepiness are the classical markers. Most patients awaken feeling unrefreshed and tired, even if the time in bed has been appropriate, and excessive sleepiness during the day is often reported. Nowadays, CPAP is an effective treatment maintaining upper airway permeability during sleep, therefore the importance of the diagnosis.

Central sleep apnea during sleep is a less frequent condition, with a different physiopathologic mechanism. Although in obstructive apnea there is an obstruction/collapse of upper airway during sleep, in central apnea the major aspect is a decrease of central respiratory drive, resulting in reduced or absent ventilatory effort. Clinically, two groups are recognised: central sleep apnea with or without hypercapnia (a condition of abnormally elevated carbon dioxide levels in the blood) (Dauvilliers 2019). The group of central sleep apnea without hypercapnia comprises central sleep apnea associated with chronic heart failure, primary central sleep apnea, central sleep apnea due to high altitude and treatment-emergent central sleep apnea. The forms with hypercapnia include congenital central hypoventilation syndrome, and conditions related to brain stem dysfunction (due to cerebrovascular accident, Arnold-Chiari malformation, or tumour) neuromuscular disease (Steinert disease), obesity hypoventilation syndrome and central sleep apnea due to a medication or substance (morphine overdose).

10.8 Narcolepsy

Narcolepsy, first described by the French physician Gélineau in 1880, is characterized by a set of clinical symptoms. The classic tetrad of sleep attacks, excessive daytime somnolence, cataplexy, and hypnagogic/hypnopompic hallucinations is completed with a fifth typical feature: night sleep disruption (AASM 2014). Sleep attacks or unwanted episodes of sleep recur several times a day and may last from over a few minutes to an hour. These refreshing naps diminish the somnolence and patients wake up reinvigorated, until the next episode occurs. Automatic behaviour is another expression: there is a partial disconnection of the environment, but patients are able to carry on tasks that cannot remember afterwards. Cataplexy –a cardinal symptom that characterizes type 1 narcolepsy- is a sudden and reversible loss of muscle tone elicited by emotions: laughter, anger, or unexpected strain. Just certain muscles or the entire striated musculature may be involved. The presence of cataplexy is nearly pathognomonic of narcolepsy, nearly because it has been reported in some other rare neurological diseases (Martínez and Santamaría 2007). Patients with type 2 narcolepsy do not have cataplexy.

Two other features are common during the transition sleep-wakefulness: sleep paralysis and hallucinations. In the first, the patient is conscious but unable to move or to speak; hallucinations may occur during sleep onset (hypnagogic) or while awakening (hypnopompic) in the morning. Both manifestations, involving the abnormal intromission of REM sleep, are usually fearful experiences, but also short-lived and patients learn to recognize them. Disturbance of night sleep is also a common feature: not only the wakefulness is divided by the sleep attacks, but the night rest is fragmented. Frequently, narcolepsy leads to personal and social complications: driving or work-related accidents, job dismissal, impoverishment of social relationships and depression. (Scammel 2022; Martínez and Santamaría 2007).

In type 1 narcolepsy, there is a deficiency of the neuropeptides hypocretin-1 and hypocretin-2 (also known as orexins). In contrast, type 2 narcolepsy (cataplexy is absent) is a heterogeneous disorder and only one third or one fourth can have low hypocretin concentration. At the genetic level, narcolepsy type 1 is related with an area of chromosome 6 known as the human leukocyte antigen (HLA) complex. The DQB1*06.02 haplotype has been found in more than 90% of patients with cataplexy. This genetic variation can increase the risk of an auto-immune response against hypocretin-releasing neurons in the hypothalamus, a response that might be triggered by an external agent: there is a link between type 1 narcolepsy and a nucleoprotein of the H1N1 Influenza virus. An increased risk of narcolepsy after natural H1N1 infection was reported from China, but this nucleoprotein was present in Pandemrix, an H1N1 influenza vaccine and it was found that the number of cases of narcolepsy increased in vaccinated patients with the DQB*06.02 haplotype (Sarkanen et al. 2018).

Narcolepsy treatment is symptomatic and directed toward three principal targets: excessive day somnolence, cataplexy, and sleep fragmentation. Stimulant medication (amphetamine- and non-amphetamine-like), sodium oxybate, antidepressants, histamine 3 (H_3) receptor antagonist/inverse agonist are some of the medications used in these cases.

10.9 Sleep Wake Rhythm Disorders

Biological rhythms are periodic variations that exist in all living organisms, from plants to human beings. Many of these rhythms are related to the environment, like tides influence the aquatic specimen. Others are more properly endogenous and similar for all the members of one species. In humans, the internal clock synchronizes to the 24-h light-dark cycle. Disruption of this system or misalignment with the external environment results in the so-called circadian rhythm sleep-wake disorders (*circa* = near; *dies* = day).

In delayed sleep-wake phase disorder, the postponement of sleep in relation to the required or conventional period and consequent difficulty awakening at a desired time, leads to excessive sleepiness and may impair occupational or social performance. In contrast, advanced sleep-wake disorder is featured by an early timing of sleep with disturbance staying awake until conventional bedtime and maintaining sleep until the desired time for awakening. Irregular sleep-wake rhythm disorder is often seen in patients with neurodegenerative conditions such as Alzheimer, Parkinson, or Huntington diseases and also in children with developmental disorders.

Jet lag disorder, associated with transmeridian travel across at least two time zones, reflects a temporary imbalance between the endogenous circadian clock and the appropriate sleep-wake pattern in the new environment. Adaptation from eastward travel is frequently more difficult than westward travel. The severity and duration of symptoms are related to the number of time zones travelled, the exposure to light or other circadian time cues in the new destination, but also to factors inherent to the journey such as decreased mobility, consumption of alcohol or caffeine, and the overall tiredness.

Shift work sleep disorder is also a well-known factor of mismatch, related to work schedules that overlap the usual time of sleep. Not only sleep quality is impaired, but reduced alertness may result in safety consequences at work and increased risk of motor accidents.

10.10 Restless Legs Syndrome

Another common disorder of sleep is Restless Legs Syndrome (RLS). Sir Thomas Willis, in 1672, made the broadly regarded first account, but the actual description of the clinical entity has been draught by the Swedish physician Ekbom in 1945 (Allen et al. 2014). It is a complex condition described as an abnormal sensation in the limbs, particularly the legs that provoke uneasiness and an irresistible urge to move them. Symptoms appear at rest, usually at bedtime, and interfere with sleep onset. Usually, patients are obliged to get out of bed several times at night; the disagreeable sensation is fleetingly relieved by walking, rubbing, or moving the legs vigorously. These symptoms have a clear circadian pattern: usually they appear at dusk or at night-time, just at rest or when the patient goes to bed. In severe cases, the symptoms may also be present during the day. As a consequence, insomnia is common, not only affecting sleep onset but also fragmenting sleep. Activities that take place in the last hours of the day or at night-time such as sedentary work, travelling or leisure are also impaired. Most of the RLS cases are idiopathic, but there are also secondary causes (Caminero and García-Borreguero 2007). Central nervous system iron depletion is a frequent cause and RLS severity correlates with low serum ferritin levels, often without criteria of anaemia. Other conditions are pregnancy, Parkinson disease, peripheral neuropathy, and uraemia. Some medications such as antihistamine drugs (often used as over-the-counter sleep inducers), antipsychotics and antidepressants are known to worsen the symptoms. RLS is a treatable condition: iron replacement is recommended in patients with low ferritin levels. Dopamine agents and alfa-2-delta calcium channel ligands are the preferred lines of treatment, but long-term dopaminergic therapy is associated with augmentation, a worsening of RLS symptoms, increasing in intensity and appearing earlier (Garcia-Borreguero et al. 2016).

Many RLS patients (up to 80%) also have periodic limb movements during sleep. These are involuntary repetitive stereotyped limb movements, typically of the lower extremities, with a duration of 0.5–10 s on intervals of 5–90 s as registered in PSG. They are considered relevant, that is establishing a disorder, if occurring more than 5 per hour in children or more than 15 per hour in adults and cause sleep disturbance or produce daytime dysfunction (Caminero and García-Borreguero 2007; AASM 2014).

Nocturnal leg cramps are caused by sudden involuntary contractions usually on the calf or foot and interfere with sleep. Most are idiopathic but can also be related to structural metabolic abnormalities or as a side-effect of medications. They must be considered in the differential diagnosis of RLS and periodic limb movements (AASM 2014).

10.11 Parasomnias

Parasomnias are complex abnormal phenomena or behaviours occurring during sleep or during sleep-wake transitions, and comprise behavioural, motor, perceptive or sensory and autonomic features (Vaughn 2022). Its clinical relevance affects not only the patient but also the bed partner or roommate, causing sleep disruption, psychosocial distress or even injuries. (Pareja 2007). Currently, parasomnias are subdivided into three major groups: NREM sleep, REM sleep and other parasomnias (AASM 2014). A good clinical history is essential for the diagnosis and bedpartners or witnesses offer invaluable clues as the subject is either asleep or cannot recall the event. The timing of the episodes is also important: NREM events, such as a disorder of arousal, tend to occur in the first 2 h of sleep, whereas REM sleep behaviour disorder predominates in the second half. Overnight video PSG also provides significant information, but it is only required as an essential diagnostic criterion in REM sleep behaviour disorder (AASM 2014). NREM sleep parasomnias include disorders of arousal such as confused arousal, sleepwalking (somnambulism) and sleep terrors (Foldvary-Schaefer 2022). They are considered as a mixed state between wakefulness and NREM sleep, and are more frequent in children, resolving by puberty, although in some cases may persist through adolescence and adulthood. Confusional arousal, also called Elpenor syndrome, occurs during the transition between sleep and wakefulness and consists of mental confusion upon arousal from slow-wave sleep. In the Odyssey, Elpenor became drunk and climbed onto the roof to sleep. The next morning, he fell and broke his neck, dying in the act (Homer 1919). Sleepwalking is characterized by a complex automatic behaviour, leaving the bed, with eyes open and a surprised expression on the face. The patient is difficult to awaken, remains confused when aroused and does not remember the episode. Inappropriate behaviour can occur, and the patient may suffer or provoke injuries during the episode. Sleep terrors have an abrupt onset, commonly with a cry, and dramatic autonomic output (tachycardia, tachypnoea, diaphoresis, and mydriasis). Often the subject sits up in bed, with a fearful expression; if awakened is confused; children usually do not remember any contents of the episode, but adults can recall dreams of imminent danger.

REM sleep parasomnias include REM sleep Behaviour Disorder, recurrent isolated sleep paralysis and nightmare disorder. REM sleep Behaviour Disorder is characterized by complex motor comportment and vocalization appearing during REM sleep. The subject acts out the dream (dream-enactment) due to a pathological loss of the muscle atonia which accompanies REM sleep. Eyes are usually closed. At the end of the episode, the patient recovers full wakefulness quickly and can recall the contents of the dream. This entity is associated with Parkinson disease, dementia with Lewy bodies, and multiple system atrophy.

Nightmares are vivid dreams with a threatening content producing anxiety and distress, resulting in rapid and complete awakening. They tend to occur during the second half of the night and can be recalled in detail, with difficulty returning to sleep. These features can help to distinguish nightmares from sleep terrors (AASM 2014). Other parasomnias include the exploding head syndrome, nocturnal enuresis, and sleep-related hallucinations.

10.12 Conclusion

In this chapter we have reviewed the architecture of normal sleep and we also have leaned out to the rich but complex world of sleep disorders. Acknowledging that sleep is essential for life, we have not discussed the proposals on the functions of sleep nor the entangled theories of dreams and dreaming. The answers to these questions must be sought somewhere else. Sleep is one of the essential keystones of a healthy lifestyle. Unfortunately, it is often menaced by several disorders, some of them very common, as insomnia; other are rare, but can have significant health consequences. Shakespeare's accurate metaphor, "Nature's soft nurse", is still absolutely relevant today.

References

Allen, R.P., D.L. Picchietti, D. Garcia-Borreguero, W.G. Ondo, A.S. Walters, J.W. Winkelman, M. Zucconi, R. Ferri, C. Trenkwalder, H.B. Lee, and International Restless Legs Syndrome Study Group. 2014. Restless legs syndrome/Willis-Ekbom disease diagnostic criteria: Updated international restless legs syndrome study group (IRLSSG) consensus criteria--history, rationale, description, and significance. *Sleep Medicine* 15 (8): 860–873. https://doi.org/10.1016/j.sleep.2014.03.025.

American Academy of Sleep Medicine (AASM). 2014. International Classification of Sleep Disorders, 3rd ed.

Anderer, P., Gruber, G., Parapatics, S., and G. Dorffner. 2007. Automatic sleep classification according to Rechtschaffen and kales. Annual international conference of the IEEE engineering in medicine and biology society. IEEE engineering in medicine and biology society. Annual international conference, 2007, 3994–3997. https://doi.org/10.1109/IEMBS.2007.4353209.

Berry, R.B., J.M. Parish, and K.M. Hartse. 2002. The use of auto-titrating continuous positive airway pressure for treatment of adult obstructive sleep apnea. An American Academy of sleep medicine review. *Sleep* 25 (2): 148–173.

Berry, R.B., R. Brooks, C. Gamaldo, S.M. Harding, R.M. Lloyd, S.F. Quan, M.T. Troester, and B.V. Vaughn. 2017. AASM scoring manual updates for 2017 (version 2.4). *Journal of Clinical Sleep Medicine: JCSM: official publication of the American Academy of Sleep Medicine* 13 (5): 665–666. https://doi.org/10.5664/jcsm.6576.

Bloch, K.E. 1997. Polysomnography: A systematic review. *Technology and Health Care: official journal of the European Society for Engineering and Medicine* 5 (4): 285–305.

Caminero, A.B., and D. García-Borreguero. 2007. Síndrome de piernas inquietas y movimientos periódicos de las extremidades durante el sueño. In *Continua Neurologica. Trastornos del sueño*, 30–42. Madrid. isbn:978-84-9751-237-4.

Dai, W., J. Zhou, G. Li, B. Zhang, and N. Ma. 2021. Maintaining normal sleep patterns, lifestyles and emotion during the COVID-19 pandemic: The stabilizing effect of daytime napping. *Journal of Sleep Research* 30 (4): e13259. https://doi.org/10.1111/jsr.13259.

Danker-Hopfe, H., P. Anderer, J. Zeitlhofer, M. Boeck, H. Dorn, G. Gruber, E. Heller, E. Loretz, D. Moser, S. Parapatics, B. Saletu, A. Schmidt, and G. Dorffner. 2009. Interrater reliability for sleep scoring according to the Rechtschaffen & Kales and the new AASM standard. *Journal of Sleep Research* 18 (1): 74–84. https://doi.org/10.1111/j.1365-2869.2008.00700.x.

Dauvilliers. 2019. Les troubles du sommeil. 3[e] ed. Elsevier Masson SAS. ISBN: 978-2-294-74892-9.

Dement, W.C. 2005. History of sleep physiology and medicine. In *Principles and practice of sleep medicine*, ed. M.H. Kryger, T. Roth, and W.C. Dement, 4th ed. Saunders: Elsevier. https://doi.org/10.1016/B0-7216-0797-7/X5001-0.

Deng, J., F. Zhou, W. Hou, Z. Silver, C.Y. Wong, O. Chang, E. Huang, and Q.K. Zuo. 2021. The prevalence of depression, anxiety, and sleep disturbances in COVID-19 patients: A meta-analysis. *Annals of the New York Academy of Sciences* 1486 (1): 90–111. https://doi.org/10.1111/nyas.14506.

Foldvary-Schaefer, N. 2022. Disorders of arousal from non-rapid eye movement sleep in adults. ed. A.Y. Avidan, and A.F., Eichler. Uptodate. Retrieved 30 December 2022. https://www-uptodate-com.bvsspa.idm.oclc.org/contents/disorders-of-arousal-from-non-rapid-eye-movement-sleep-in-adults?

Garcia-Borreguero, D., M.H. Silber, J.W. Winkelman, B. Högl, J. Bainbridge, M. Buchfuhrer, G. Hadjigeorgiou, Y. Inoue, M. Manconi, W. Oertel, W. Ondo, J. Winkelmann, and R.P. Allen. 2016. Guidelines for the first-line treatment of restless legs syndrome/Willis-Ekbom disease, prevention and treatment of dopaminergic augmentation: A combined task force of the IRLSSG, EURLSSG, and the RLS-foundation. *Sleep Medicine* 21: 1–11. https://doi.org/10.1016/j.sleep.2016.01.017.

Grigg-Damberger, M., D. Gozal, C.L. Marcus, S.F. Quan, C.L. Rosen, R.D. Chervin, M. Wise, D.L. Picchietti, S.H. Sheldon, and C. Iber. 2007. The visual scoring of sleep and arousal in infants and children. *Journal of Clinical Sleep Medicine: JCSM official publication of the American Academy of Sleep Medicine* 3 (2): 201–240.

Guilleminault, C., G. Nino-Murcia, G. Heldt, R. Baldwin, and D. Hutchinson. 1986. Alternative treatment to tracheostomy in obstructive sleep apnea syndrome: Nasal continuous positive airway pressure in young children. *Pediatrics* 78 (5): 797–802.

Himanen, S.L., and J. Hasan. 2000. Limitations of Rechtschaffen and Kales. *Sleep Medicine Reviews* 4 (2): 149–167. https://doi.org/10.1053/smrv.1999.0086.

Hirshkowitz, M. 2016. Polysomnography challenges. *Sleep Medicine Clinics* 11 (4): 403–411. https://doi.org/10.1016/j.jsmc.2016.07.002.

Homer. 1919. *Odyssey*. Book 10.

İnce, R., S.S. Adanır, and F. Sevmez. 2021. The inventor of electroencephalography (EEG): Hans Berger (1873–1941). *Child's Nervous System: ChNS: official journal of the International Society for Pediatric Neurosurgery* 37 (9): 2723–2724. https://doi.org/10.1007/s00381-020-04564-z.

Jahrami, H., A.S. BaHammam, N.L. Bragazzi, Z. Saif, M. Faris, and M.V. Vitiello. 2021. Sleep problems during the COVID-19 pandemic by population: A systematic review and meta-analysis. *Journal of Clinical Sleep Medicine: JCSM official publication of the American Academy of Sleep Medicine* 17 (2): 299–313. https://doi.org/10.5664/jcsm.8930.

Kirsch, D. 2022. Stages and architecture of normal sleep. In ed. S.M. Harding and A.F. Eichler. Uptodate. Retrieved 30 December 2022. https://www-uptodate-com.bvsspa.idm.oclc.org/contents/stages-and-architecture-of-normal-sleep

Martínez, J.E., and J. Santamaría. 2007. Narcolepsia e hipersomnia idiopática. In *Continua Neurologica. Trastornos del sueño. Ars Medica*, 43–48. Madrid. isbn:978-84-9751-237-4.

McNamara, P., D. McLaren, S. Kowalczyk, and E.F. Pace-Schott. 2007. "Theory of Mind" in REM and NREM dreams. In *The new science of dreaming: vol. 1. Biological aspects*, ed. D. Barrett and P. McNamara, 201–220. Westport/London: Praeger Publishers/Greenwood Publishing Group.

Millett, D. 2001. Hans Berger: From psychic energy to the EEG. *Perspectives in Biology and Medicine* 44 (4): 522–542. https://doi.org/10.1353/pbm.2001.0070.

Orr, W.C. 1985. Utilization of polysomnography in the assessment of sleep disorders. *The Medical Clinics of North America* 69 (6): 1153–1167. https://doi.org/10.1016/s0025-7125(16)30980-4.

Pappa, S., V. Ntella, T. Giannakas, V.G. Giannakoulis, E. Papoutsi, and P. Katsaounou. 2020. Prevalence of depression, anxiety, and insomnia among healthcare workers during the COVID-19 pandemic: A systematic review and meta-analysis. *Brain, Behavior, and Immunity* 88: 901–907. https://doi.org/10.1016/j.bbi.2020.05.026.

Pareja, J.A. 2007. Parasomnias. In *Continua Neurologica. Trastornos del sueño. Ars medica*, 11–21. Madrid. isbn:978-84-9751-237-4.

Patil, S.P. 2010. What every clinician should know about polysomnography. *Respiratory Care* 55 (9): 1179–1195.

Robillard, R., K. Dion, M.H. Pennestri, E. Solomonova, E. Lee, M. Saad, A. Murkar, R. Godbout, J.D. Edwards, L. Quilty, A.R. Daros, R. Bhatla, and T. Kendzerska. 2021. Profiles of sleep changes during the COVID-19 pandemic: Demographic, behavioural and psychological factors. *Journal of Sleep Research* 30 (1): e13231. https://doi.org/10.1111/jsr.13231.

Salari, N., H. Khazaie, A. Hosseinian-Far, H. Ghasemi, M. Mohammadi, S. Shohaimi, A. Daneshkhah, B. Khaledi-Paveh, and M. Hosseinian-Far. 2020. The prevalence of sleep disturbances among physicians and nurses facing the COVID-19 patients: A systematic review and meta-analysis. *Globalization and Health* 16 (1): 92. https://doi.org/10.1186/s12992-020-00620-0.

Sarkanen, T., A. Alakuijala, I. Julkunen, and M. Partinen. 2018. Narcolepsy associated with Pandemrix vaccine. *Current Neurology and Neuroscience Reports* 18 (7): 43. https://doi.org/10.1007/s11910-018-0851-5.

Scammel, T.E. 2022. Clinical features and diagnosis of narcolepsy in adults. In ed. F. Benca and A Eichler. Uptodate. Retrieved 30 December 2022. https://www-uptodate-com.bvsspa.idm.oclc.org/contents/clinical-features-and-diagnosis-of-narcolepsy-in-adults

Silber, M.H., S. Ancoli-Israel, M.H. Bonnet, S. Chokroverty, M.M. Grigg-Damberger, M. Hirshkowitz, S. Kapen, S.A. Keenan, M.H. Kryger, T. Penzel, M.R. Pressman, and C. Iber. 2007. The visual scoring of sleep in adults. *Journal of Clinical Sleep Medicine: JCSM: official publication of the American Academy of Sleep Medicine* 3 (2): 121–131.

Stone, J.L., and J.R. Hughes. 2013. Early history of electroencephalography and establishment of the American clinical neurophysiology society. *Journal of Clinical Neurophysiology: official publication of the American Electroencephalographic Society* 30 (1): 28–44. https://doi.org/10.1097/WNP.0b013e31827edb2d.

Vaughn, B. D. Approach to abnormal movements and behaviours during sleep. In ed. A.Y. Avidan and F. Eichler. Uptodate. Retrieved 30 December 2022. https://www-uptodate-com.bvsspa.idm.oclc.org/contents/approach-to-abnormal-movements-and-behaviors-during-sleep

Watson, N.F., M.S. Badr, G. Belenky, D.L. Bliwise, O.M. Buxton, D. Buysse, D.F. Dinges, J. Gangwisch, M.A. Grandner, C. Kushida, R.K. Malhotra, J.L. Martin, S.R. Patel, S.F. Quan, and E. Tasali. 2015. Recommended amount of sleep for a healthy adult: A joint consensus statement of the American Academy of sleep medicine and Sleep Research Society. *Sleep* 38 (6): 843–844. https://doi.org/10.5665/sleep.4716.

Zimmer, D.E. 1993. Dormir y soñar. La mitad nocturna de nuestras vidas. Salvat. ISBN 84-345-8890-0.

Chapter 11
Hypotalamus-Pituitary-Adrenal (HPA) Axes and Their Relationship with Stress, Mood, Personality, and Neurocognitive Functioning

Ana María Martínez Robayo

Abstract Contemporary brain research has expanded its field of study from that of pathological processes in relation to nervous and endocrine functions involved, up to the study of the normal relations of these two systems and the behavior. From the psychological perspective, it is important to address the study of the hypothalamic-pituitary-adrenal (HPA) axis. The HPA axis is connected to the central nervous system (CNS) and the endocrine system (ES). The HPA axis is a major neuroendocrine system that controls reactions to stress and regulates many body processes, including digestion, the immune system, mood and emotions, sexuality, and energy storage and expenditure. The evidence also informs of the role of HPA axes in neurocognitive changes. The parts of the HPA axis include the hypothalamus, limbic system, the pituitary gland, and the adrenal glands. The two systems work together to adjust the balance of the hormones and the stress response. It is well documented that the disruption of the HPA axes and changes in the neurocognitive and psychological state of individuals produce high levels of stress. This chapter presents an overview of brain structures and neurohormonal agents involved in HPA axes: Limbic System, hypothalamus, regulation of pituitary secretion and relationship with, stress, mode, personality and neurocognition.

Keywords Hypothalamic-pituitary-adrenal (HPA) axis · Neuroendocrine system · Stress response · Limbic system · Neurocognitive changes

---

A. M. M. Robayo (✉)
Department of Forensics, Psychiatry and Pathology, School of Medicine, Complutense University of Madrid, Madrid, Spain
e-mail: anamam25@ucm.es; amartinezrobayo@us.es

T. Lopez-Soto et al. (eds.), *The Theory of Mind Under Scrutiny*, Logic, Argumentation & Reasoning 34, https://doi.org/10.1007/978-3-031-46742-4_11

11.1 The Limbic System

The limbic system is made up of some structures under the cerebral cortex. Surrounding the subcortical limbic areas is the limbic cortex, composed of a ring of cerebral cortex beginning in the orbitofrontal area on the ventral surface of the frontal lobes. It extends upward into the subcallosal gyrus below the border anterior to the corpus callosum. It continues the highest part of it, in the most internal cerebral hemisphere, in the gyrus of the corpus callosum (cingulum). It goes backwards the corpus callosum and downwards, on the ventromedial surface of the temporal lobe, in direction to the parahippocampal gyrus and the uncus (See Fig. 11.1).

The general areas included in the limbic system are the cingulate gyrus (CG), subcallosal (SG) and parahippocampal (PG). The cingulate gyrus contains a prelimbic area, anterior cingulate and retrosplenial cortex (Fig. 11.1). In the limbic system Brodmann (BA) areas 24 and 32 and 25, 28, 27 and 28, 35, 36, 48 and 49 are included. The olfactory bulb (olfactory tubercle/olfactory tubercle) is also part of the limbic system), amygdala or tonsillar body (central and basolateral nuclei), uncus (gyrus intralaminar, uncinate and part of the dentate gyrus), part of the hypothalamus (mainly the mammillary bodies) and thalamus (anterior and dorsal nuclei).

The formation of hippocampus referring to the hippocampus, the dentate gyrus, subincular complex and entorhinal cortex, anterior parahippocampal gyrus and the uncus. Also included are the alveus and the habenular nucleus. The mammillary bodies send information to the anterior nuclei of the thalamus (beam mammillothalamus) and the hypothalamus sends most of its projections to the mammillary bodies, also receiving afferents from the medial prefrontal cortex.

In the hippocampus, small multipolar neurons and some neurons are identified. They are spindle-shaped and together constitute the layer of polymorphic cells. This layer lies underlying the layer of pyramidal cells, which are large, with an apical

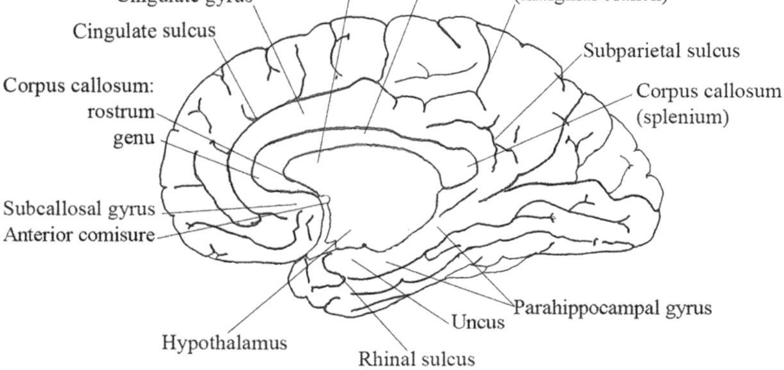

Fig. 11.1 Medial section of the brain: main turns, fissures of the brain

dendritic arborization and basal dendrites. The cylindrical axis of these neurons is directed to the depth, to pass through the layer of polymorphic cells and enter the white substance. These fibers are arranged in increasingly compact groups to form a band of white matter called alveus.

The fibers course above the hippocampus, immediately below the splenium of the corpus callosum. In the most superficial part of the hippocampus is the layer molecular formed by small neurons and neuropil, where they make synaptic contact through numerous afferent fibers to the hippocampus, mainly from the gyrus of the cingulate The dentate gyrus is a formation of the hippocampus that lies dorsal to the parahippocampal gyrus and subinculum, consisting mainly of cells of the molecular layer, some pyramidal and medium cells, and numerous granular cells. The latter have a characteristic of dendritic arborization towards the molecular layer and its axis cylinders come out towards the white matter to also form part of the elements of the fornix. The fibers of the trigone cerebrum are directed forward, and in the anterior end of the thalamus, descend to form the pillars of the fornix, enter the hypothalamus and synapse with neurons in the mammillary nuclei. The cylinders of these neurons form the anteroventral nucleus of the thalamus (mammillothalamic tract of Vicq d'Azyr).

Some fibers from the fornix terminate in the reticular formation of the midbrain and others in the medial and preoptic regions of the hypothalamus. Others in the pericommissural fornix, others in the septal area and some more in the cingulate gyrus. The cylinders of neurons in the anteroventral nucleus of the thalamus synapse on neurons cortical cells of the cingulate gyrus, whose axons pass backward to terminate in the neural elements of the hippocampus. These kinds of connections are called the Papez circuit, which is directly related to the regulation of the emotional expression. Other types of limbic system connections include hippocampal-septal fibers, the connections of the parahippocampal gyrus and the neocortex, with the cerebral amygdala and from it to the dorsomedial nucleus of the thalamus, as well as the connections established by the fibers from the olfactory bulb with the various components of the limbic system, mainly the cerebral amygdala, the anterior pre-frontal substance, the uncus, the preoptic region and the septal area (See Fig. 11.1).

The cerebral amygdala is a complex of nuclei located immediately under the cortex of the anteromedial pole of each temporal lobe. It has many bidirectional connections with the hypothalamus. In lower animals, the amygdala is associated with the association of olfactory stimuli. The amygdala receives neural signals from all portions of the cortex: limbic and neocortex of the temporal, parietal and occipital lobes and especially from the auditory and visual association areas. The amygdala transmits signals again towards the same cortical areas: to the hippocampus, the septum, the thalamus, and especially to the hypothalamus. In general, stimulation of the amygdala can practically cause the same effects as those triggered by the hypothalamus plus additional ones. Effects measured by the hypothalamus include increased or decreased blood pressure blood pressure, heart rate, motility and gastrointestinal secretion, defecation and urination, pupillary dilation or rarely constriction, piloerection, and the secretion of different anterior pituitary hormones, especially those gonadotropins and adrenocorticotropic hormone (ACTH).

In addition to these effects measured by the hypothalamus, stimulation of the amygdala can also produce different types of involuntary movements that include tonic movements, such as straightening the head or tilting the body, circular movements, rhythmic movements and, different movements associated with smelling and eating, such as licking, chewing, and swallowing. In addition, stimulation of certain tonsillar nuclei can rarely cause a pattern of anger, escape, punishment, fear like the pattern of anger triggered by hypothalamus. Stimulation of other nuclei can trigger reward reactions and pleasure. Also, excitation of other portions of the amygdala may produce sexual activities that include erection, copulatory movements, ejaculation, ovulation, uterine activity, and premature labor.

11.2 The Hypothalamus

The hypothalamus is a structure of the brain involved in the main circuits that form the components of the limbic system of the forebrain and the limbic system of the mid brain. The hypothalamus comprises the ventral walls of the third ventricle, below the hypothalamic sulcus and the structures of the ventricular floor including the tuber cinereum and infundibulum and the mammillary bodies. It continues without demarcation accurate with basal olfactory area (diagonal twist of anterior perforated space), and backwards it continues with the central gray matter and the mesencephalic tegmentum (See Fig. 11.2).

The hypothalamus is in close anatomical and functional relationship with the preoptic area, which is part of the forebrain. Some authors distinguish two large zones, based on a sagittal line passing through the anterior crus of the fornix or trigone: the medial hypothalamic area and the lateral hypothalamic area (Truex and

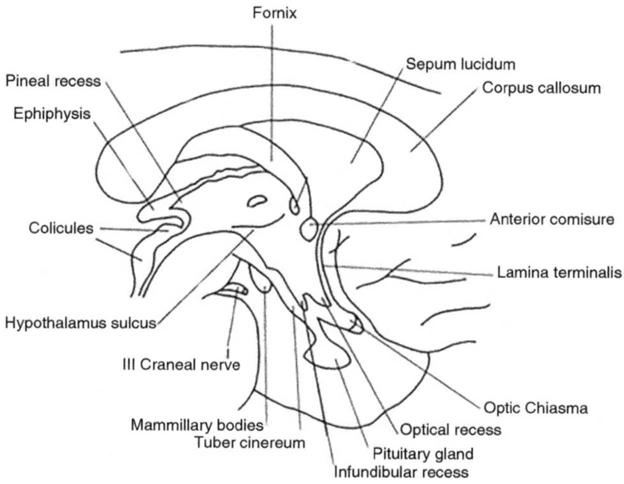

Fig. 11.2 General view of the hypothalamus in a sagittal section

Carpenter 1969; López 1980; Feria Velasco and Tapia Arizmendi 1986; Beatty 1995; Kandel et al. 2013), whose neurons have been the subject of numerous studies.

The hypothalamus is divided into nuclei and into 4 main areas: rostral or preoptic, the supraoptic, (on the optic chiasm), the tuberal and the mammillary or caudal area. It is further divided into 3 nuclei-related regions: supraoptic, tuberal, and mamillary. The supraoptic region includes the suprachiasmatic, supraoptic, paraventricular and anterior: The supraoptic nucleus consists largely of large neurons with highly-branched dendrites, while in the paraventricular nucleus there are large, often bipolar neurons and numerous small multipolar cells between an intricate network of blood capillaries (Sawchenko and Swanson 1983). The suprachiasmatic nucleus is made up of a reduced group of small neurons, which receive fibers from the retina (Moore and Lenn 1972; Nolte 2013) and whose function is related to the establishment of biological rhythms and fluid intake (Drucker-Colín et al. 1984).

The axons of the large neurons of both nuclei form bundles, which together constitute the hypothalamic-pituitary fasciculus, which ends in the neurohypophysis where they discharge their neurosecretion. Between both nuclei we can find nerve cells singly or in small groups that appear to form a bridge of incomplete gray matter between the two and that some authors refer to as *intermediate nucleus*. The axons of these neurons also contribute to the formation of the fasciculus in the hypothalamic-pituitary. The anterior hypothalamic nucleus, poorly differentiated in the human mammal and that imperceptibly continues with the preoptic area.

The tuberal region is formed by the ventromedial, dorsomedial, and infundibular (arcuate), in which there are mainly small ovoid neurons, multipolar, with short dendritic extensions. The mammillary region includes the nuclei posterior and the medial, intermediate, and lateral mammillary (mamillary tubercle). Most of the mammillary tubercles are made up of small neurons, with prolongations moderately branched short dendritic veins, while the lateral mammillary nucleus is formed by large multipolar neurons, with a soma of more than about 15 to 20 mm in length diameter. The posterior hypothalamic nucleus is also composed of large neurons, with a long cylinder axis and numerous small ovoid neurons with a short cylinder axis.

The axons of neurons of the preoptic, hypothalamic, and limbic structures descend in an orderly fashion. The two fundamental principles on the organization of the defending axons through the hypothalamus and medial forebrain bundle refer, on the one hand, to a tendency of a simple medial-lateral organization and dorso-ventral, and the other with a trend towards an anteroposterior organization. In this way, the axons of the neurons closest to the midline descend medially while the axons of ventrally placed neurons descend more ventrally, and the axons of neurons located in more cephalic areas descend laterally to the axons of caudally placed neurons (Pfaff 1980).

The hypothalamus communicates with cortical and subcortical areas via afferent and efferent connections (see Table 11.1). The medial bundle of the forebrain contains ascending (Fig. 11.3) and descending (Fig. 11.4) fibers and conducts impulses from areas via sept hypothalamic fibers, olfactory bulb and tubercle, cortex pear-shaped and from the hippocampal formation, by olfactory-hypothalamic fibers.

Table 11.1 Main afferent and efferent connections of the hypothalamus (Palckovits and Zaborszky 1979; López 1980)

Afferent connections	Efferent connections
Medial trac of forebrain Fornix or cerebral trigone, Stria terminalis, Peduncle-mammillary fascicle Reticular loop Pallid- hypothalamic fasciculus Subthalamo-hypothalamic fasciculus Retino-hypothalamic fibers.	Mammillothalamic trac-Vicq d' Azyr Mamillo-tegmental bundle Dorsal longitudinal fasciculus of Schatz Diffuse projections hypothalamus-thalamus-cortical, hypothalamic-tegmental hypothalamus-reticular

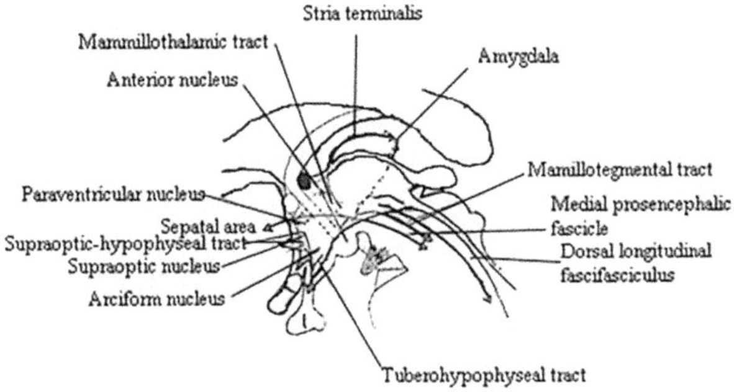

Fig. 11.3 Afferent connections to the hypothalamus through the mammillary peduncle, the fornix, and the stria terminalis

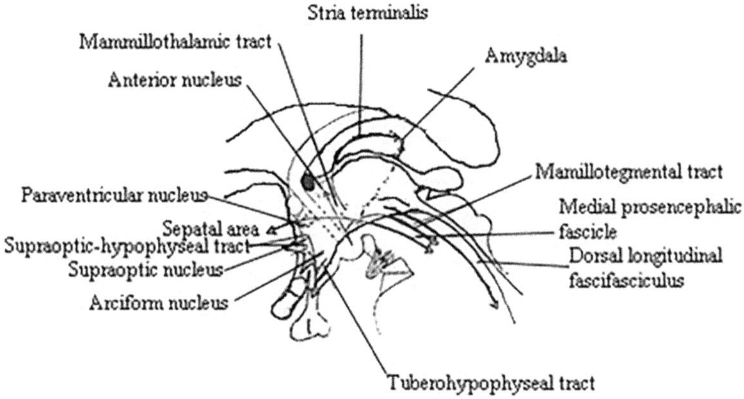

Fig. 11.4 Main efferent pathways from the hypothalamus

Probably others that come from the brain stem with visceral afferent impulses, reach the hypothalamus incorporated into this bundle. It also contains fibers that establish connections intrahypothalamic and other fibers in the bundle continue to the roof of the midbrain, where the signals are relayed to the visceral control nuclei of the brain stem.

The fornix is another fiber system that conducts nerve impulses from the hippocampus and probably from the cingulate gyrus, hypothalamus, and other structures. It is constituted by axons of the pyramidal cells of the hippocampus, which first course through the fimbria and then they are incorporated into the fornix. Its initial part is formed by posterior columns that start from the fimbriae, surround the thalamus, and lie under the corpus callosum, where they constitute the body of the fornix and descend to the hypothalamus. Upon reaching the anterior commissure, each anterior column is divided into a pericommissural component that passes in front of it, to be distributed throughout the preoptic region and septal areas; and a post commissural component, which runs down behind the commissure anterior and carries fibers destined for the anterior and lateral hypothalamic areas, to the anterior nucleus of thalamus and the mammillary region of the hypothalamus. There is evidence that part of the fornix fibers pass beyond the mamillary bodies, intersect at the supramamillary commissure, and reach the midbrain tegmentum (Nauta 1958) and it is also possible that the fornix conducts fibers that reach the habenula in the epithalamus through of the stria medullaris.

The stria terminalis originates in the tonsillar nucleus, courses through the temporal horn and the body of the lateral ventricle, contiguous with the caudate nucleus, until it reaches the anterior commissure, where some fibers separate to distribute to the preoptic area, the anterior hypothalamic area, and the nucleus ventromedially of the hypothalamus (See Fig. 11.4).

The remainder pass through the anterior commissure follows a reversed path to the opposite side until reaching the corresponding amygdaline nucleus. The latter seems to be connected to the hypothalamus by fibers that come through the preoptic space.

The hypothalamus also receives impulses through the mammillary peduncle, which contains fibers originating in the neuron of the brain stem, close to the nucleus of the solitary bundle (See Fig. 11.4). Fibers from the globe also reach the hypothalamus through the lenticular loop and the pale-hypothalamic fasciculus. It has been assumed that the fibers running from the brain to the thalamus give off collateral branches to the hypothalamus. Likewise, it has been proposed that there of retino-hypothalamic connections, then after bilateral ocular enucleation.

The mammillothalamic track originates for the most part in the medial mammillary nucleus and ascends to terminate in the anterior nucleus of the thalamus, especially in the anteroventral portion. These fibers seem to be organized in a somatotopic ally (Powell and Cowan 1954) and perhaps constitutes the most important connection between the hypothalamus and the dorsal thalamus. The mammillotegmental track has a common origin with the mammillothalamic bundle, to which it is connected. After its origin in the mammillary nuclei, it descends to the tegmentum of the midbrain and terminates primarily in the dorsal tegmental nucleus, where fibers originate, they are incorporated into the dorsal longitudinal fasciculus and descend to some nuclei of the brain stem.

The dorsal longitudinal fasciculus of Schutz constitutes one of the most important systems that relate the hypothalamus to the reticular formation and nuclei of the brain stem. They contain fibers from the periventricular system and, in turn, discharge impulses from almost all hypothalamic regions to the brain stem. It is in the periaqueductal gray matter, at the level of the Edingen-Westphal nucleus. It is divided into a lateral and a dorsal component. In its course it gives fibers to the midbrain tectum, especially to the superior colliculus, to the Edinger-Westhal, salivary, lower dorsal to the motor nuclei of the V and VII cranial nerves, to the nucleus of the hypoglossal nerve and nucleus ambiguus.

In addition, it carries impulses to the nucleus of the reticular formation, such as the dorsal raphe nucleus, and to the reticular formation of the dorsal part of the pons, and the medulla oblongata. It reaches the caudal limit of the medulla oblongata. It Contains fibers ascending which come from neurons located at the level of the solitary bundle and the correspondent nucleus. In addition to the fascicles, there are diffuse connections that different parts of the hypothalamus descend to the tegmentum of the midbrain and perhaps to regions beyond the caudal brain stem. Some of them are incorporated into the medial bundle of the forebrain (hypothalamic-tegmental bundles). Probably, these fibers establish multi-synaptic relays in the reticular formation through which impulses are carried to the spinal cord itself through the reticulospinal tracts.

The tuberoinfundibular system has been characterized by the tuberoinfundibular, whose neural cell bodies are in the medial hypothalamic region, in the arcuate and ventromedial muscles. They receive afferent fibers from the amygdala, the preoptic area, the septum, from the anterior hypothalamic region and from short-axis cylindrical neurons, located in the same ventromedial nucleus. Some of these afferent fibers are excitatory and others inhibitory (Renaud 1979).

This system is of fundamental importance in the production of hormones and releasing factors, and the axons of their neurons, in addition to making synapses with the neurons of the intrahypothalamic or extrahypothalamic nuclei, establish special channels with the neural processes and complex forms in the blood capillaries of the median eminence. The collateral branches of the axons of these tuberoinfundibular neurons synapse with cells of the anterior hypothalamic region, paraventricular nucleus, preoptic areas of the brain anterior, dorsal medial nuclei of the thalamus and cerebral amygdala.

The hypothalamus, which is one of the afferent and efferent pathways of the limbic system, has as important functions cardiovascular regulation: regulation of body temperature, regulation of body fluids, regulation of reproductive functions, gastrointestinal and alimentary regulation, and, finally, the hypothalamic control of the anterior pituitary (Ganong 1999; Guyton 1991). The hypothalamus also has monoaminergic (noradrenergic and serotonergic) connections. [Sensory (nipples and genital organs, olfactory cortex, and retina)].

The functional state of the hypothalamus is related to the autonomic, endocrine, motor, and behavioral systems, largely reflecting activation or inhibition of ventral mechanisms represented in limbic and reticular circuits, and hypothalamus pathways in the forebrain (Nauta 1958; Janing 1978; Guyton 1991). The hypothalamus exerts influence on the pituitary. This influence is through the neural pathways, and the biochemical release of thyrotropin (TRH), into the distal neurohypophysis from the anterior pituitary (Guyton 1991). The highest concentration of TRH is found in the interstitial space and concentrations progressively smaller are found in the interstitial space of the ventrodorsally hypothalamus. It is also found in lower concentrations in extrahypothalamic brain areas (Berne and Levy 1992; Carpenter 1991).

11.3 Regulation of Pituitary Secretion

The pituitary is located at the base of the skull, in a cavity of the sphenoid called the Sella turcica, separated from the cranial content by the sellar diaphragm, which is made up of an extension of the dura. The pituitary gland is attached to the diencephalon via the pituitary stalk, which passes through a hole in the diaphragm. The pituitary gland is ovoid in shape and its size it varies with age, sex, and some special states such as pregnancy. The pituitary gland consists of two parts: the anterior (adenohypophysis) and the posterior (neurohypophysis) (Morán and Zárate 1993), each one with different embryological and morphological origin.

The adenohypophysis is made up of at least five types of secretory cells. Which produce six structurally and functionally recognized hormones. These cells are called: corticotropes, which secrete corticotropin (ACTH) and peptides originating from proopiomelanocortin (POMC), somatotrophs that secrete the growth hormone (GH), gonadotrophs, which secrete luteinizing hormone (LH), and follicle stimulating hormone (FSH), thyrotropes, thyrotropin (TSH) producing, and lactotrophs, that synthesize prolactin (PRL) (Douglas et al. 1984; Moran et al. 1986; Guillemin 1978).

The neurohypophysis shows a network of capillaries, pituitary cells, and unmyelinated nerve fibers. Which contain many electron-dense granules. The neurohypophyses are made up of the part distal nerve bundle originating from the supraoptic and paraventricular nuclei of the hypothalamus oxytocin and vasopressin are synthesized in these nuclei, they bind to certain proteins carriers called macrophysics and travel to the neurohypophysis, where they are stored in granules before going into circulation (Moran et al. 1986; Guillemin 1978; Janson 2019).

Corticotropin (ACTH) is synthesized in corticotropes and is a single chain polypeptide containing 39 amino acids, releases cortisol, with molecular weight (MW) of 4500 and effector organ of the adrenal cortex. ACTH is derived from the precursor molecule proopiomelanocortin (POMC) which is detected in the pituitary but not in the serum and gives rise to other polypeptides such as b Lipotropin (b -LPH), g -LPH, b -endorphin and a-melanocyte-stimulating hormone (a – 2 MSH). The pituitary gland synthesizes b -LPH and this gives rise to a -LPH, and b-endorphin, but a - MSH is only detected in the fetal pituitary (Krieger 1983; Douglas et al. 1984; Braak and Braak 1987.

The main function of ACTH is the stimulation of cortisol, the adrenal cortex, the secretion of glucorticoids, mineral corticoids and androgens in the adrenal cortex (stimulation of the cortisol by the adrenal cortex), where ACTH binds to the receptor found on the membrane and stimulates steroidogenesis with the measurement of cyclic adenosine monophosphate (cAMP), (Braak and Braak 1987). ACTH acts by binding to a specific cell membrane receptor and by activation of adenyl cyclase, which raises intracellular cAMP levels. This action initiates a series of biochemical events that culminate in the synthesis and release of cortisol (Hersman et al. 1972).

The hypothalamus regulates, through corticotropin-releasing hormone (CRH), the secretion of ACTH pulsatile. Glucocorticoids exert inhibitory feedback on the secretion of the ACTH, both at the hypothalamic and pituitary level. In addition, ACTH itself can exert negative feedback on CRH secretion. Hypothalamic hormone-hormone corticotropin-releasing hormone (CRH), 41 amino acids, effector cells: corticotrope and hormone pituitary ACTH.

It is observed in the adrenal insufficiency changes in the Nervous System (NS) that are relevant only to glucocorticoids. These changes include the appearance of EEG waves slower than the normal alpha rhythm, changes in personality or increased sensitivity to olfactory and taste stimuli. The slight but definite personality changes include irritability, apprehension, and inability to consent (Guillemin 1978; Ganong 1999; Janson 2019).

11.4 Physiological Effects of Glucocorticoids

When a person or an animal is exposed to any noxious or potentially noxious stimulus, there is an increase in ACTH secretion and, consequently, an elevation in the concentration of circulating glucocorticoids. These stress stimuli that elevate ACTH secretion also activate the sympathetic suprarenal medulla system, and part of the function of glucocorticoids circulating may be the maintenance of vascular reactivity to catecholamines. In addition, the glucocorticoids are necessary for catecholamines to exert their full mobilizing action on free fatty acids, and these are an important source of energy in emergencies. Other 3 theory holds that glucocorticoids prevent other stress-induced changes from causing increase in plasma glucocorticoids at high drug concentrations than in the short term are lifesaving, but in the long term they are damaging and unbalancing (Ganong 1999).

11.5 Circadian Rhythm of ACTH

The secretion of ACTH and therefore of cortisol is episodic and these episodes follow a circadian rhythm. Adrenocorticotropin hormone is secreted in irregular discharges throughout the day and plasma cortisol concentration tends to rise and to descend in response to these discharges. In men, discharges are more frequent in the early morning and less frequent in the evening (Ganong 1999). But if the sleep-wake pattern is reversed, the secretion rhythm is also changed and there happens an adaptation in terms of two or three weeks (Hersman et al. 1972; Ganong 1999). This diurnal or circadian rhythm of the ACTH secretion is found in patients with adrenal insufficiency receiving doses glucocorticoid. It is not due to the stress of getting up in the morning, traumatic as it may be, because the increase in ACTH secretion occurs before awakening. if the day lengthens experimentally more than 24 hours, the adrenal cycle is also lengthened, but the increase in the ACTH secretion also occurs during the sleep period. The biological clock in charge of ACTH diurnal rhythm is in the suprachiasmatic nuclei of the hypothalamus (Ganong 1999). When there is hypersecretion of ACTH, the levels of these hormones and of cortisol (if the adrenal glands are intact) is often elevated at baseline; and elevation persists despite large doses of exogenous glucocorticoids (Hersman and Poe 1981).

11.6 Stress Response

Morning ACTH concentration plasma in a healthy person in rest is around 25 pg./ml (5.5 mol/lt.). During intense stress, the amount of ACTH secreted exceeds the amount needed to produce maximum excretion of glucocorticoids (Ganong 1999). Increases in ACTH secretion to meet emergency situations are measured almost exclusively by the hypothalamus through the release of CRH. The median eminence is the one that secretes this polypeptide that is transported by the portal vessels pituitary glands to the anterior pituitary, the site where ACTH secretion is stimulated.

Afferent neural pathways from many parts of the brain converge toward the median eminence. Fibers from the amygdaloid nuclei mediate the response to emotional stress, and fear, anxiety, and apprehension cause marked increases in ACTH secretion. The impulses from the suprachiasmatic nuclei also provide the stimulus for the rhythm circadian. The impulses that ascend to the hypothalamus to the macroceptive pathways and the reticular formation trigger ACTH secretion in response to injury (Ganong 1999). Stimulation of the baroreceptors produce inhibitory impulses that act through the nucleus of the solitary fasciculus. Circulating adrenaline and norepinephrine do not increase the ACTH secretion in man and corticoid suprarenal and medulla suprarenal secretions are independently regulated from the hypothalamus (Ganong 1999). Injection of CRH into the cerebral ventricles produces hyperglycemia, increased cardiac output, decreased reproductive function, decreased gastrointestinal function, and other changes that are also observed in the presence of stress. But impossible to ensure that the Endogenously secreted CRH may be responsible for these changes during stress (de Kloet et al. 2015).

It was previously believed that stress stimulated ACTH secretion by initially decreasing the plasma corticosteroid concentration, but this has been shown to be incorrect. Untreated adrenalectomized animal, whose circulating concentration of these hormones is zero, reacts to stress as a much greater than normal increase in plasma ACTH. Thus, the magnitude of ACTH secretion depends on two opposing factors: the sum of the neural and possibly other stimuli that converge across the median eminence to increase ACTH secretion, the magnitude of the braking action of glucocorticoids on ACTH secretion, which is proportional to its concentration in the circulating blood (Ganong 1999). Glucocorticoid excess accelerates basic EEG rhythms and produces mental disorders, ranging from increased appetite, insomnia, euphoria, to outright toxic psychosis. Glucocorticoid deficiency is also accompanied by mental symptoms, but produced by excess of these compounds are more serious. The central hypothesis in the 1970s was that if the chemical substrate of the mechanisms of hypothalamic-pituitary regulation was like that altered in limbic structures when some affective disorders appeared, it was possible to identify neuroendocrine effects in these patients (De la Fuente and Wells 1981; Guillemin 1978; Janson 2019).

11.7 Cortisol and Its Relationship on Stress, Mood, Personality, and Cognitive Functioning

Cortisol is a key biological response to stress hormone; it is a glucocorticoid produced primarily through the hypothalamic-pituitary-adrenal (HPA) axes. It seems to be one of the most studied biomarkers in psychobiological research (Adam et al. 2018) and plays a very important role in mood, personality, and neurocognitive functioning. Experimental work on stimulation and lesions in the hypothalamic area evidenced in the 50 s that some pituitary gland substance was released here that stimulated the secretion of pituitary adrenocorticotropic hormone (ACTH) (Yasuda et al. 1982). The hypothalamic physiological factor of ACTH is corticotropin-releasing factor (CRF), CRF has been evidenced in the external area of the median eminence, the greatest richness in cell bodies with CFR have been in the paraventricular nucleus (PVN), which coincides with the remarkable physiological significance that had been attributed to this nucleus in the regulation of ACTH (Vale et al. 1977, 1983; Yasuda et al. 1982).

Exposure to stress activates the HPA axes, which triggers physiological changes in the brain and throughout the body through the secretion of glucocorticoids (cortisol and corticosterone). To help an organism overcome stress, multiple physiological systems need to be activated. Proper activation of these systems is important for the response to threat enough for an organism to return to biological equilibrium once the stressor ends. Thus, many pathological conditions are characterized by an inappropriate response to stress (McEwen 1998; De Kloet et al. 2006; de Kloet et al. 2015; de Kloet et al. 2018). In the stress response mediated by the HPA axes, the activation of corticotropin-releasing hormone (CRF), neurons of the paraventricular nucleus of the hypothalamus (PVN), stimulates the secretion of the adrenocorticotropic hormone (ACTH) in the anterior pituitary, which in turn promotes the synthesis and release of glucocorticoids, mainly cortisol by the adrenal cortex (Lupien 2009; Ulrich-Lai and Herman 2009; Gunn et al. 2015; Herman et al. 2016). The CRF gives as a result the release of ACTH and this immediately activates the cortex adrenal gland that leads to the synthesis and secretion of glucocorticoids, i.e., cortisol. When there is an increase in cortisol and other glucocorticoids, this increase activates mineralocorticoids in the pituitary and hypothalamus, decreasing release of CRF, which is given a negative feedback system, reducing release of ACTH from the anterior pituitary, and thus decreased secretion of glucocorticoids.

The PVN is a region of the brain that is central in the initiation of the response neuroendocrine and autonomic to a stressful challenge (Herman and Cullinan 1997). The activity of the PVN is subject to regulation by GABA, the dominant inhibitory neurotransmitter in the hypothalamus (Decavel and Van Den Pol 1990; Gunn et al. 2015). Serotonin (5-HT) which acts on several types of receptors is a robust/potent activator of the HPA axis and the direct synaptic interaction of serotonergic axons with CRH neurons in the hypothalamic PVN (Liposits et al. 1987; Chaouloff 1993; Carrasco and Van de Kar 2003). Substantial evidence from a series of immunohistochemical studies and combined with retrograde screening methods indicates that the serotonergic neurons that innervate the PVN are preferentially located within the dorsal midbrain nuclei (DRN) and median raphe nuclei (MRN) (Sawchenko and Swanson 1983; Lowry 2002). In turn, CRH input to serotonergic neurons located in the caudal part of the DRN may be critical for activation of the serotonergic mesocorticolimbic system (Lowry 2002; Hammack et al. 2002).

The hair cortisol concentrations (HCC) is a method that allows to measure the la long-term cortisol secretion in both humans (Kirschbaum et al. 2009; Meyer and Novak 2012; Stalder et al. 2012; Stalder and Kirschbaum 2012; Pulopulos et al. 2014), as well as in animals (Meyer and Novak 2012).

The HCC, providing a retrospective index of cortisol secretion during periods of several months (Gow et al. 2010; Stalder et al. 2012; Dettenborn et al. 2010). He HCC is a recent development and therefore most systematic studies have not included HCC; however, reference will also be made to the secretion of cortisol in blood, saliva, or urine. For several authors the cortisol awakening response (CAR) is a good indicator of the performance of the HPA axes (Pruessner et al. 1997; Van Zuiden et al. 2011; Laceulle et al. 2015).

The HPA axes play an important role in responses to stress both short and long term. In the long term, therefore, it is not surprising that dysfunction of this system has negative health implications, including links to Post-Traumatic Stress Disorder (PTSD), anxiety and depression disorders, personality, and changes in neurocognitive functioning.

There is substantial evidence (Lupien et al. 2009; Hill-Soderlund et al. 2015; Harrewijn et al. 2020). Environmental stress activates a cascade in the limbic-axis hypothalamic-pituitary-adrenal leading to an increase in the secretion of cortisol (Bremner et al. 1999). PTSD has been associated with impaired functioning of the HPA axes (for review: De Kloet et al. 2006; Heim and Nemeroff 2009; Van Zuiden et al. 2011). Different studies have shown elevated cortisol levels in saliva at bedtime in youth with maltreatment-related PTSD (Carrion et al. 2010; De Bellis et al. 1999) and without abuse (Carrion et al. 2007). The PTDE in adolescents and adults is also related to low indicators in the CAR (Pruessner et al. 1997; Van Zuiden et al. 2011; Laceulle et al. 2015).

A synthesis of the existing literature suggests that exposure to trauma leads to increased cortisol secretion, resulting in increased negative feedback sensitivity of the HPA axes; and on the other hand, long term to a blunted HPA axes with lower basal cortisol levels in patients diagnosed with PTDE (Miller et al. 2007; Steudte-Schmiedgen et al. 2016). Likewise, it is argued that the context plays an important role, with a greater delayed cortisol production in patients with PTDE, especially in settings of stress or trauma (Miller et al. 2007; Steudte-Schmiedgen et al. 2016). The CAR levels are lower in adult patients diagnosed with PTSD (Rohleder et al. 2004; Neylan et al. 2005; Wessa et al. 2006; De Kloet et al. 2007). The studies with HCC, report a small significant positive relationship between exposure to trauma and HCC (Khoury et al. 2019) or chronic neuroendocrine dysregulation mediated by the stress response and more associated with the clinical phenotype, rather than with exposure to trauma (Van den Heuvel et al. 2020). HCC levels altered are also associated with stress-related mental illness (Steudte et al. 2011a, b).

The literature is also extensive, where it is reflected that in major depression, desensitization of glucocorticoid receptors occurs. That is, the negative feedback regulation fails and hypersecretion of the cortisol (Bhagwagar et al. 2005; Burke et al. 2005; Halbreich et al. 1985; Pfohl et al. 1985; Stetler and Miller 2005; Ströhle and Holsboer 2003). If the system negative feedback does not work properly, exposure to a stressful factor could cause an excessive response, since the response would not be seen buffered by negative feedback loops. Depression is generally associated with sustained hyperactivity of the stress axis caused by the neurotransmitter corticotropin-releasing factor (CRF) (Leonard and Myint 2009). Major depression is associated with dysregulation in the functioning of the axis HPA and altered cortisol negative feedback (Leonard and Myint 2009). The literature suggests that the distinctive characteristics of major depression are dysfunction of the HPA axes, an overactive hormone-releasing system corticotropin (CRH) and blunted adrenocorticotropic (ACTH) response to CRF exogenous (Ströhle and Holsboer 2003; Plotsky et al. 1998; Krempel et al. 2022).

Depressed people may be hyper-responsive to challenges for a wide range of reasons. Another alternative explanation of exaggerated reactivity of neurohormonal and inflammatory markers, is that people with depression may have prolonged involuntary processing of emotional information, including stress-induced negative emotions (Lyubomirsky et al. 1998), peripheral physiological signals such as pupil dilation or central activation in areas of the brain, such as the prefrontal cortex or the amygdala, this latter responsible for encoding emotional characteristics (Siegle et al. 2002).

In addition, the participation of brain structures such as the locus coeruleus and the central nucleus of the amygdala may be the same in depression and stress acute, structures innervated by CRF-containing nerve endings (see Pittenger and Duman 2008; de Kloet et al. 2018). Similarities have also been documented between the response to acute stress and depression, including elevated levels of neurohormones, high blood pressure blood pressure and heart rate, as well as greater activation and mobilization of the energy reserves. We know that the CRF initiates the modifications in the HPA axis, which it also stimulates the sympathetic-adrenal (SAM) system, therefore depressed people in addition to show hyperresponsiveness of the HPA axis, they may also show hyperresponsiveness of the SAM, as a result of negative feedback regulation failure (Ströhle and Holsboer 2003; Leonard 2001; Weinstein et al. 2010).

On the other hand, cortisol alterations in the HPA axis have been identified in different disorders such as suicide attempters (Plotsky et al. 1998; Lindqvist et al. 2008; Chatzittofis et al. 2016; Keilp et al. 2016; O'Connor et al. 2016; Courtet and Oli'e 2020), impulsivity, depression and schizophrenia (Lewitzka et al. 2018). Thus, the dysregulation of the stress response, particularly at the level of the HPA axis, could represent a vulnerability factor for suicide (Alacreu-Crespo et al. 2020; O'Connor et al. 2020), as in other stress-related disorders such as depression, anxiety, TSSD (Zorn et al. 2017) or borderline personality disorder (Drews et al. 2019).

Suicide attempters are more vulnerable to social stress and studies show an altered response to cortisol. The emotional consequences of stress response to stress are mediated by deregulation of the HPA axes. According to the stress-diathesis model (Mann and Rizk 2020) HPA deregulation is part of a diathesis that makes people more vulnerable to suicide. In the brain-centered model, it is assumed that genetic and epigenetic factors such as childhood or adult stressors and DNA methylation are the cause of suicidal diathesis. Courtet and Oli'e (2020) state that when faced with stressors, such as rejection, suicidal patients show difficulties interpreting and adapting to the situation, partly due to the deregulation of the HPA axes. Some researchers suggest that this dysregulation could be related to traits of impulsiveness. Impulsivity traits play an important role in the suicidal vulnerability under stress conditions. Greater impulsiveness can increase sensitivity to emotional distress that results in responses to an inadequate physiology. The physiological response is observed in the deregulation of cortisol in saliva in response to emotional response to social stress (Alacreu-Crespo et al. 2020, 2022).

Studies of association between the facets of neuroticism, extraversion and consciousness and the activity of the HPA axis, refer that the basal cortisol levels are related to facets of personality traits, while CAR and cortisol induced by stress were not related to personality (Laceulle et al. 2015; Kaess et al. 2017; Funder and Ozer 2019). Also, significant positive relationships are evident between extraversion, neurotism, and the personality traits of aggressiveness and cortisol in saliva (LeBlanc and Ducharme 2005; Ouanes et al. 2017; Parent-Lamarche and Marchand 2015) and significant negative relationships between neurotism and cortisol in saliva (LeBlanc and Ducharme 2005), and small effect (Funder and Ozer 2019).

In clinical application, high sensation seeking has also been linked to hypothalamic-pituitary axis dysfunction. High sensation seeking is a function of strong focus and weak excitation and inhibition systems (Shabani et al. 2011). That said, some researchers have shown strong relationships between cortisol secretion and sensation seeking and disinhibition behaviors (Shabani et al. 2011; Frenkel et al. 2018). Significant negative relationships between sensation seeking and HCC secretions (Funder and Ozer 2019; Ucheagwu et al. 2021) or a medium effect size (Funder and Ozer 2019) are also reported.

In Borderline Personality Disorder (BPD) it has also been observed greater vulnerability to stress, which would indicate that in BPD it would be related to bad HPA functioning. Overall, the findings suggest that the pathology of BPD is related to HPA axis abnormalities (i.e., increased CAR). It is assumed that the changes observed in CAR would reflect neuroendocrine maladaptive processes. Most of the studies on BPD are with the CAR, where an increased CAR in adult patients (Lieb et al. 2004; Rausch et al. 2015). In adolescents, CAR is higher in people with a history of child abuse and pronounced symptoms of BPD (Kaess et al. 2017) and in those who committed nonsuicidal self-injury (NSSI), that these behaviors are important precursors of development of the BPD (Reichl et al. 2016). In people of different ages with BDP gives the increase in CAR (Rausch et al. 2021; Carvalho et al. 2012) and elevated salivary cortisol levels over the day (Lieb et al. 2004). It seems that whereas basal cortisol may be more trait-like in nature and may be more informative than stress-induced cortisol when trait-like characteristics are studied, such as facets of personality (Laceulle et al. 2015; Kaess et al. 2017), and conscientiousness, extraversion, and neuroticism (Laceulle et al. 2015; Funder and Ozer 2019).

Emotional intelligence (EI) has also been studied as a predictor of the response to stress and with contradictory data between laboratory results or in applied environments (school tasks). Research suggests that EI can promote better health through its action of moderating the relationship between stress and health (Oginska-Bulik 2005; Salovey et al. 2002; Mikolajczak et al. 2006, 2007), either through its influence on the behavior or physiology. Laboratory studies on EI and reactivity physiological stress reveal that trait EI is associated with less deterioration of mood and is a significant moderator of the relationship between exposure to stress and cortisol reactivity (Landa et al. 2008; Laceulle et al. 2015; Kaess et al. 2017). Evidence supporting that notion includes findings from a negative relationship between EI and self-reported feelings of stress (Wilbraham et al. 2018; Liu et al. 2019) and feelings of inability to control life events (Gohm et al. 2005). Meanwhile in applied settings, the relationship between IE and salivary cortisol in educational settings, after stressful school tasks there is no significant relationship (Resnik and Dewaele, 2020; Wilbraham et al. 2018; Liu et al. 2019).

HPA axes have also been linked to related cognitive changes with age. It is important to mention that aging is associated with the decrease of different cognitive functions, however, the magnitude of this cognitive decline depends on different factors. Cortisol can affect the cognitive performance acutely through the activation of receptors located in the prefrontal cortex, hippocampus, and amygdala (Lupien et al. 2007; de Kloet et al. 1999; McEwen et al. 2016; van den Heuvel et al. 2022) carried out a review where they stated that the low occupation of cortisol is detrimental to the proper functioning of these brain structures. For decades, the activity of the HPA axes has been associated with cognitive decline in old adults, given the marked increase in basal cortisol levels with age and reduction in hippocampal volume (Lupien et al. 1998; Li et al. 2006). The low performance in working memory, learning and verbal memory in the short and long term, it is related to low levels in HCC (Shields et al. 2015, 2017; Pulopulos et al. 2014; Akan et al. 2023), on the other hand, and high cortisol levels in daytime saliva are related to a worse performance on tasks that assess attention and short-term verbal memory (Pulopulos et al. 2014; Law and Clow 2020).

Several investigators have reported elevated levels of basal cortisol in patients with Alzheimer's disease and mild cognitive impairment (Arsenault-Lapierre et al. 2010; Venero et al. 2013) and healthy older people (Hodgson et al. 2004; Karlamangla et al. 2005; MacLullich et al. 2005; Li et al. 2006; Kuningas et al. 2007; Lee et al. 2008; Beluche et al. 2010; Comijs et al. 2010; Franz et al. 2011; Gerritsen et al. 2011; Johansson et al. 2011; Ouanes and Popp 2019; Liu et al. 2023). Other researchers have not found a significant relationship between cortisol and cognition. (Peavy et al. 2009; Köhler et al. 2010; Schrijvers et al. 2011; Liu et al. 2019; Feeney et al. 2020; van den Heuvel et al. 2022; Liu et al. 2023).

Animal studies have shown that low long-term exposure to cortisol (HCC) can have a negative effect on cognition, an effect related to the low occupancy of mineralocorticoid receptors in the central nervous system. These receptors are located primarily in the hippocampus and prefrontal cortex (Sloviter et al. 1993; Stienstra et al. 1998; Berger et al. 2006).

Although all the results are not revealing for us to affirm the disruption of the HPA axes in all disorders of mood, personality, and changes in neurocognitive functioning. However, we cannot step aside the differences in the hypothesis of this relationship, raised in the investigations, as well as the use of different techniques to measure cortisol, tests to assess personality, mood, or cognitive aspects. In most of these studies researchers used saliva, blood, or urine samples to measure the activity of HPA axes. In addition, cortisol levels measured in blood, saliva, or urine can be highly variable, as they are likely to be affected by different factors that may occur prior to sampling (Stalder and Kirschbaum 2012). Biological samples reflect specific measurements in plasma or saliva, or in levels of integral cortisol for a few hours in urine (Pulopulos et al. 2014).

However, the health of the HPA axes plays a crucial role in mood, personality, and cognition. This information is highly relevant in clinical neuropsychology, where the state of mind, personality, and cognitive processes are considered. It is important to acknowledge that the explanation of changes in these behaviors goes beyond the production of neurotransmitters in the brain or other external factors. Unfortunately, in psychological clinical practice, psychoneuroendocrinology is often overlooked and receives less attention in understanding brain functioning and behavior.

References

Adam, K.M., M.E. Quinn, R. Tavernier, M.T. McQuillan, K.A. Dahlke, and K.E. Gilbert. 2018. Diurnal cortisol slopes and mental and physical health outcomes: A systematic review and meta-analysis. *Psychoneuroendocrinology* 83: 25–41.

Akan, O., A. Bierbrauer, L. Kunz, P.D. Gajewski, S. Getzmann, J.G. Hengstler, and O.T. Wolf. 2023. Chronic stress is associated with specific path integration deficits. *Behavioural Brain Research* 442: 114305.

Alacreu-Crespo, A., E. Oli'e, S. Guillaume, C. Girod, A. Cazals, I. Chaudieu, and P. Courtet. 2020. Dexamethasone suppression test may predict more severe/violent suicidal behavior. *Frontiers in Psychiatry* 11: 97.

Alacreu-Crespo, A., V. Hidalgo, G. Girod, E. Oli'e, and Ph. Courtet. 2022. The impulsiveness level influences the salivary cortisol response and social stress sensitivity in suicidal patients. *Journal of Psychiatric Research* 156: 159–167.

Arsenault-Lapierre, G., H. Chertkow, and S. Lupien. 2010. Seasonal effects on cortisol secretion in normal aging, mild cognitive impairment and Alzheimer's disease. *Neurobiology of Aging* 31 (6): 1051–1054.

Beatty, J. 1995. *Principles of behavioral neuroscience*. Brown & Benchmark Publishers.

Beluche, I., I. Carrière, K. Ritchie, and M.L. Ancelin. 2010. A prospective study of diurnal cortisol and cognitive function in community-dwelling elderly people. *Psychological Medicine* 40 (6): 1039–1049.

Berger, S., D.P. Wolfer, O. Selbach, H. Alter, G. Erdmann, H.M. Reichardt, and G. Schütz. 2006. Loss of the limbic mineralocorticoid receptor impairs behavioral plasticity. *Proceedings of the National Academy of Sciences* 103 (1): 195–200.

Berne, D., and H. Levy. 1992. *Fisiología médica*. London: Mosby Yearbook.

Bhagwagar, Z., S. Hafizi, and P.J. Cowen. 2005. Increased salivary cortisol after waking in depression. *Psychopharmacology* 182 (1): 54–57.

Braak, H., and E. Braak. 1987. Argyrophilic grains: Characteristic pathology of cerebral cortex in cases of adult onset dementia without Alzheimer changes. *Neuroscience Letters* 76 (1): 124–127.

Bremner, J.D., S.M. Southwick, and D.S. Charney. 1999. The neurobiology of posttraumatic stress disorder: An integration of animal and human research. In *Posttraumatic stress disorder: A comprehensive text*, ed. J.D. Bremner and P.A. Saigh, 103–143. Allyn & Bacon.

Burke, H.M., M.C. Davis, C. Otte, and D.C. Mohr. 2005. Depression and cortisol responses to psychological stress: A meta-analysis. *Psychoneuroendocrinology* 30 (9): 846–856.

Carpenter, M.B. 1991. Gross anatomy and internal structure. In *Core Text of Neuroanatomy*. Williams & Wilkins.

Carrasco, G.A., and L.D. Van de Kar. 2003. Neuroendocrine pharmacology of stress. *European Journal of Pharmacology* 463 (1–3): 235–272.

Carrion, V.G., C.F. Weems, and A.L. Reiss. 2007. Stress predicts brain changes in children: A pilot longitudinal study on youth stress, PTSD, and the hippocampus. *Pediatrics* 119: 509–516.

Carrion, V.G., C.F. Weems, K. Richert, B.C. Hoffman, and A.L. Reiss. 2010. Decreased prefrontal cortical volume associated with increased bedtime cortisol in traumatized youth. *Biological Psychiatry* 68 (5): 491–493.

Carvalho, F.S., T. Beblo, N. Schlosser, K. Terfehr, C. Otte, B. Löwe, O.T. Wolf, C. Spitzer, M. Driessen, and K. Wingenfeld. 2012. Associations of childhood trauma with hypothalamic-pituitary-adrenal function in borderline personality disorder and major depression. *Psychoneuroendocrinology* 37 (10): 1659–1668.

Chaouloff, F. 1993. Physiopharmacological interactions between stress hormones and central serotonergic systems. *Brain Research Reviews 18* (1): 1–32.

Chatzittofis, A., S. Arver, K. Öberg, J. Hallberg, P. Nordström, and J. Jokinen. 2016. HPA axis dysregulation in men with hypersexual disorder. *Psychoneuroendocrinology* 63: 247–253.

Comijs, H.C., L. Gerritsen, B.W. Penninx, M.A. Bremmer, D.J. Deeg, and M.I. Geerlings. 2010. The association between serum cortisol and cognitive decline in older persons. *The American Journal of Geriatric Psychiatry* 18 (1): 42–50.

Courtet, P., and E. Oli'e. 2020. Neurobiology of suicidal behavior and social stress. *Journal of Neuroscience Research* 98: 579–580.

De Bellis, M.D., M.S. Keshavan, D.B. Clark, B.J. Casey, J.N. Giedd, A.M. Boring, and N.D. Ryan. 1999. Developmental traumatology part II: Brain development. *Biological Psychiatry* 45 (10): 1271–1284.

De Kloet, E.R., M.S. Oitzl, and M. Joëls. 1999. Stress and cognition: Are corticosteroids good or bad guys? *Trends in Neurosciences* 22 (10): 422–426.

De Kloet, C.S., E. Vermetten, E. Geuze, A. Kavelaars, C.J. Heijnen, and H.G.M. Westenberg. 2006. Assessment of HPA-axis function in posttraumatic stress disorder: Pharmacological and non-pharmacological challenge tests, a review. *Journal of Psychiatric Research* 40 (6): 550–567.

De Kloet, E.R., R.H. DeRijk, and O.C. Meijer. 2007. Therapy insight: Is there an imbalanced response of mineralocorticoid and glucocorticoid receptors in depression? *Nature Clinical Practice Endocrinology & Metabolism* 3 (2): 168–179.

de Kloet, S.F., H.D. Mansvelder, and T.J. De Vries. 2015. Cholinergic modulation of dopamine pathways through nicotinic acetylcholine receptors. *Biochemical Pharmacology* 97 (4): 425–438.

de Kloet, E.R., O.C. Meijer, A.F. de Nicola, R.H. de Rijk, and M. Joëls. 2018. Importance of the brain corticosteroid receptor balance in metaplasticity, cognitive performance and neuro-inflammation. *Frontiers in Neuroendocrinology* 49: 124–145.

De la Fuente, J.R., and L.A. Wells. 1981. Human growth hormone in psychiatry disorders. *The Journal of Clinical Psychiatry* 42: 270.

Decavel, C., and A.N. Van Den Pol. 1990. GABA: A dominant neurotransmitter in the hypothalamus. *Journal of Comparative Neurology* 302 (4): 1019–1037.

Dettenborn, L., A. Tietze, F. Bruckner, and C. Kirschbaum. 2010. Higher cortisol content in hair among long-term unemployed individuals compared to controls. *Psychoneuroendocrinology* 35 (9): 1404–1409.

Douglas, J., E. Civelli, and F. Herbert. 1984. Polyprotein gene expression: Generation of diversity of neuroendocrine peptides. *Annual Review of Biochemistry* 53: 665.

Drews, E., E.A. Fertuck, J. Koenig, M. Kaess, and A. Arntz. 2019. Hypothalamic-pituitary-adrenal axis functioning in borderline personality disorder: A meta-analysis. *Neuroscience and Biobehavioral Reviews* 96: 316–334.

Drucker-Colín, R., J. Bernal-Pedraza, F. Fernández-Cancino, and A. Oksenberg. 1984. Is vasoactive intestinal polypeptide (VIP) a sleep factor? *Peptides* 5 (4): 837–840.

Feeney, J.C., A.M. O'Halloran, and R.A. Kenny. 2020. The association between hair cortisol, hair cortisone, and cognitive function in a population-based cohort of older adults: Results from the Irish longitudinal study on ageing. *The Journals of Gerontology: Series A* 75 (2): 257–265.

Feria Velasco, A., and G. Tapia Arizmendi. 1986. Estructura de los elementos tisulares centrales que intervienen en el fenómeno consultivo. En A. Feria-Velasco, Martínez de Muñoz y F. Rubio-Donnadieu (comps.), *Epilepsia. Un enfoque multidisciplinar*. Mexico. Ed. Trillas, cap. 2, p. 28.

Franz, C.E., R.C. O'Brien, R.L. Hauger, S.P. Mendoza, M.S. Panizzon, E. Prom-Wormley, and W.S. Kremen. 2011. Cross-sectional and 35-year longitudinal assessment of salivary cortisol and cognitive functioning: The Vietnam era twin study of aging. *Psychoneuroendocrinology* 36 (7): 1040–1052.

Frenkel, M.O., R.B. Heck, and H. Plessner. 2018. Cortisol and behavioral reaction of low and high sensation seekers differ in responding to a sport-specific stressor. *Anxiety, Stress, & Coping* 31 (5): 580–593.

Funder, D.C., and D.J. Ozer. 2019. Evaluating effect size in psychological research: Sense and nonsense. Advances in methods and practices in psychological. *Science* 2 (2): 156–168.

Ganong, W.F. 1999. *Review of medical physiology*. Appleton & Lange Stamford (CT).

Gerritsen, L., H.C. Comijs, D.J. Deeg, B.W. Penninx, and M.I. Geerlings. 2011. Salivary cortisol, APOE-ε4 allele and cognitive decline in a prospective study of older persons. *Neurobiology of Aging* 32 (9): 1615–1625.

Gohm, C.L., G.C. Corser, and D.J. Dalsky. 2005. Emotional intelligence under stress: Useful, unnecessary, or irrelevant? *Personality and Individual Differences* 39 (6): 1017–1028.
Gow, R., S. Thomson, M. Rieder, S. Van Uum, and G. Koren. 2010. An assessment of cortisol analysis in hair and its clinical applications. *Forensic Science International* 196 (1–3): 32–37.
Guillemin, R. 1978. Peptides in the Brain? The new endocrinology of the neuron. *Science* 202 (4366): 390–402.
Gunn, B.G., L. Cunningham, S.G. Mitchell, J.D. Swinny, J.J. Lambert, and D. Belelli. 2015. GABAA receptor-acting neurosteroids: A role in the development and regulation of the stress response. *Frontiers in Neuroendocrinology* 36: 28–48.
Guyton, A. 1991. *Basic neuroscience: Anatomy and physiology*. Philadelphia.
Halbreich, U., G.M. Asnis, R. Shindledecker, B. Zumoff, and R.S. Nathan. 1985. Cortisol secretion in endogenous depression. I. Basal plasma levels. *Archives of General Psychiatry* 42 (9): 904–908.
Hammack, S.E., K.J. Richey, M.J. Schmid, M.L. LoPresti, L.R. Watkins, and S.F. Maier. 2002. The role of corticotropin-releasing hormone in the dorsal raphe nucleus in mediating the behavioral consequences of uncontrollable stress. *Journal of Neuroscience* 22 (3): 1020–1026.
Harrewijn, A., P. Vidal-Ribas, K. Clore-Gronenborn, S.M. Jackson, S. Pisano, D.S. Pine, and A. Stringaris. 2020. Associations between brain activity and endogenous and exogenous cortisol–A systematic review. *Psychoneuroendocrinology* 120: 104775.
Heim, C., and C.B. Nemeroff. 2009. Neurobiology of posttraumatic stress disorder. *CNS Spectrums* 14 (1 Suppl 1): 13–24.
Herman, J.P., and W.E. Cullinan. 1997. Neurocircuitry of stress: Central control of the hypothalamo–pituitary–adrenocortical axis. *Trends in Neurosciences* 20 (2): 78–84.
Herman, J.P., J.M. McKlveen, S. Ghosal, B. Kopp, A. Wulsin, R. Makinson, J. Scheimann, and B. Myers. 2016. Regulation of the hypothalamic-pituitary-adrenocortical stress response. *Comprehensive Physiology* 6: 603–621.
Hersman, M.S., and G.A. Poe. 1981. Sensitivity of the total power radiometer with periodic absolute calibration. *IEEE Transactions on Microwave Theory and Techniques* 29 (1): 32–40.
Hersman, R., L.J. Kleine, and C.E. Gilmore. 1972. A clinical evaluation of Propyliodone Bronchography 1. *Veterinary Radiology* 13 (1): 27–35.
Hill-Soderlund, A.L., S.J. Holochwost, M.T. Willoughby, D.A. Granger, J.L. Gariépy, W.R. Mills-Koonce, and M.J. Cox. 2015. The developmental course of salivary alpha-amylase and cortisol from 12 to 36 months: Relations with early poverty and later behavior problems. *Psychoneuroendocrinology* 52: 311–323.
Hodgson, N., V.A. Freedman, D.A. Granger, and A. Erno. 2004. Biobehavioral correlates of relocation in the frail elderly: Salivary cortisol, affect, and cognitive function. *Journal of the American Geriatrics Society* 52 (11): 1856–1862.
Janing, W. 1978. The anatomic nervous system. En Schmidt, R.F. (Comp), *Fundamentals of neurophysiology*. Cap. 8, 220.
Janson, J. 2019. *Variation in hypothalamus-pituitary-adrenal Axis responses to acute psychosocial stress as a function of stress protocol variations, coping, and emotion regulation*. Germany: Friedrich-Alexander-Universität Erlangen-Nürnberg.
Johansson, P., J.O. Johansson, F. Labrie, N. Mattsson, O. Hansson, K. Blennow, and J. Svensson. 2011. Mild dementia is associated with increased adrenal secretion of cortisol and precursor sex steroids in women. *Clinical Endocrinology* 75 (3): 301–308.
Kaess, M., S. Whittle, J.G. Simmons, M. Jovev, N.B. Allen, and A.M. Chanen. 2017. The interaction of childhood maltreatment, sex, and borderline personality features in the prediction of the cortisol awakening response in adolescents. *Psychopathology* 50 (3): 188–194.
Kandel, E.R., H. Markram, P.M. Matthews, R. Yuste, and C. Koch. 2013. Neuroscience thinks big (and collaboratively). *Nature Reviews Neuroscience* 14 (9): 659–664.
Karlamangla, A.S., B.H. Singer, J. Chodosh, B.S. McEwen, and T.E. Seeman. 2005. Urinary cortisol excretion as a predictor of incident cognitive impairment. *Neurobiology of Aging* 26 (1): 80–84.

Keilp, J.G., B.H. Stanley, S.R. Beers, N.M. Melhem, A.K. Burke, T.B. Cooper, et al. 2016. Further evidence of low baseline cortisol levels in suicide attempters. *Journal of Affective Disorders* 190: 187–192.

Khoury, J.E., M.B. Enlow, A. Plamondon, and K. Lyons-Ruth. 2019. The association between adversity and hair cortisol levels in humans: A meta-analysis. *Psychoneuroendocrinology* 103: 104–117.

Kirschbaum, C., A. Tietze, N. Skoluda, and L. Dettenborn. 2009. Hair as a retrospective calendar of cortisol production—Increased cortisol incorporation into hair in the third trimester of pregnancy. *Psychoneuroendocrinology* 34 (1): 32–37.

Köhler, S., A.J. Thomas, A. Lloyd, R. Barber, O.P. Almeida, and J.T. O'Brien. 2010. White matter hyperintensities, cortisol levels, brain atrophy and continuing cognitive deficits in late-life depression. *The British Journal of Psychiatry* 196 (2): 143–149.

Krempel, R., D. Schleicher, I. Jarvers, A. Ecker, R. Brunner, and S. Kandsperger. 2022. Sleep quality and neurohormonal and psychophysiological accompanying factors in adolescents with depressive disorders: Study protocol. *Bjpsych Open* 8 (2). https://doi.org/10.1192/bjo.2022.29.

Krieger, D.T. 1983. Brain peptides: What, where, and why? *Science*. 222: 975.

Kuningas, M., R.H. de Rijk, R.G. Westendorp, J. Jolles, P. Eline Slagboom, and D. van Heemst. 2007. Mental performance in old age dependent on cortisol and genetic variance in the mineralocorticoid and glucocorticoid receptors. *Neuropsychopharmacology* 32 (6): 1295–1301.

Laceulle, O.M., E. Nederhof, M.A.G. Aken, and J. Ormel. 2015. Adolescent personality: Associations with basal, awakening, and stress-induced cortisol responses. *Journal of Personality* 83 (3): 262–273.

Landa, J.M.A., E. López-Zafra, M.P.B. Martos, and M. del Carmen Aguilar-Luzon. 2008. The relationship between emotional intelligence, occupational stress and health in nurses: A questionnaire survey. *International Journal of Nursing Studies* 45 (6): 888–901.

Law, R., and A. Clow. 2020. Stress, the cortisol awakening response and cognitive function. *International Review of Neurobiology* 150: 187–217.

LeBlanc, J., and M.B. Ducharme. 2005. Influence of personality traits on plasma levels of cortisol and cholesterol. *Physiology & Behavior* 84 (5): 677–680.

Lee, B.K., T.A. Glass, G.S. Wand, M.J. McAtee, K. Bandeen-Roche, K.I. Bolla, and B.S. Schwartz. 2008. Apolipoprotein e genotype, cortisol, and cognitive function in community-dwelling older adults. *American Journal of Psychiatry* 165 (11): 1456–1464.

Leonard, B.E. 2001. Stress, norepinephrine and depression. *Journal of Psychiatry and Neuroscience* 26 (Suppl): S11–S16.

Leonard, B.E., and A. Myint. 2009. The psychoneuroimmunology of depression. *Human Psychopharmacology* 24 (3): 165–175.

Lewitzka, U., M. Bauer, B. Ripke, T. Bronisch, and L. Günther. 2018. Impulsivity and saliva cortisol in patients with suicide attempt and controls. *Neuropsychobiology* 75 (4): 162–168.

Li, G., M.M. Cherrier, D.W. Tsuang, E.C. Petrie, E.A. Colasurdo, S. Craft, and C.W. Wilkinson. 2006. Salivary cortisol and memory function in human aging. *Neurobiology of Aging* 27 (11): 1705–1714.

Lieb, K., M.C. Zanarini, C. Schmahl, M.M. Linehan, and M. Bohus. 2004. Borderline personality disorder. *The Lancet* 364 (9432): 453–461.

Lindqvist, D., L. Träskman-Bendz, and F. Vang. 2008. Suicidal intent and the HPA-axis characteristics of suicide attempters with major depressive disorder and adjustment disorders. *Archives of Suicide Research* 12 (3): 197–207.

Liposits, Z.S., C. Phelix, and W.K. Paull. 1987. Synaptic interaction of serotonergic axons and corticotropin releasing factor (CRF) synthesizing neurons in the hypothalamic paraventricular nucleus of the rat: A light and electron microscopic immunocytochemical study. *Histochemistry* 86 (6): 541–549.

Liu, T., D. Li, F. Shangguan, and J. Shi. 2019. The relationships among testosterone, cortisol, and cognitive control of emotion as underlying mechanisms of emotional intelligence of 10-to 11-year-old children. *Frontiers in Behavioral Neuroscience* 13: 273.

Liu, M.S., Z.Y. Tian, Z. Zhang, F. Yang, Y. Lou, Y.J. Wang, et al. 2023. Impaired cognitive function in patients with autonomous cortisol secretion in adrenal incidentalomas. *The Journal of Clinical Endocrinology & Metabolism* 108 (3): 633–641.

López, A.L. 1980. Anatomia Funcional del Sistema Nervioso. Méjico. Ed. Limusa, pp. 474 y596.

Lowry, C.A. 2002. Functional subsets of serotonergic neurons: Implications for control of the hypothalamic-pituitary-adrenal axis. *Journal of Neuroendocrinology* 14 (11): 911–923.

Lupien, S.J. 2009. Brains under stress. *The Canadian Journal of Psychiatry* 54 (1): 4–5.

Lupien, S.J., M. De Leon, S. De Santi, A. Convit, C. Tarshish, N.P.V. Nair, and M.J. Meaney. 1998. Cortisol levels during human aging predict hippocampal atrophy and memory deficits. *Nature Neuroscience* 1 (1): 69–73.

Lupien, S.J., F. Maheu, M. Tu, A. Fiocco, and T.E. Schramek. 2007. The effects of stress and stress hormones on human cognition: Implications for the field of brain and cognition. *Brain and Cognition* 65 (3): 209–237.

Lupien, S.J., B.S. McEwen, M.R. Gunnar, and C. Heim. 2009. Effects of stress throughout the lifespan on the brain, behaviour and cognition. *Nature Reviews Neuroscience* 10 (6): 434–445.

Lyubomirsky, S., N.D. Caldwell, and S. Nolen-Hoeksema. 1998. Effects of ruminative and distracting responses to depressed mood on retrieval of autobiographical memories. *Journal of Personality and Social Psychology* 75 (1): 166.

MacLullich, A.M., I.J. Deary, J.M. Starr, K.J. Ferguson, J.M. Wardlaw, and J.R. Seckl 2005. Plasma cortisol levels, brain volumes and cognition in healthy elderly men. *Psychoneuroendocrinology* 30 (5): 505–515.

Mann, J.J., and M.M. Rizk. 2020. A brain-centric model of suicidal behavior. *American Journal of Psychiatry* 177: 902–916.

McEwen, B.S. 1998. Stress, adaptation, and disease. Allostasis and allostatic load. *Annals of the New York Academy of Sciences* 840: 33–44.

McEwen, B.S., C. Nasca, and J.D. Gray. 2016. Stress effects on neuronal structure: Hippocampus, amygdala, and prefrontal cortex. *Neuropsychopharmacology* 41 (1): 3–23.

Meyer, J.S., and M.A. Novak. 2012. Minireview: Hair cortisol: A novel biomarker of hypothalamic-pituitary-adrenocortical activity. *Endocrinology* 153 (9): 4120–4127.

Mikolajczak, M., O. Luminet, and C. Menil. 2006. Predicting resistance to stress: Incremental validity of trait emotional intelligence over alexithymia and optimism. *Psicothema* 18: 79–88.

Mikolajczak, M., E. Roy, O. Luminet, C. Fillée, and P. De Timary. 2007. The moderating impact of emotional intelligence on free cortisol responses to stress. *Psychoneuroendocrinology* 32 (8–10): 1000–1012.

Miller, G.E., E. Chen, and E.S. Zhou. 2007. If it goes up, must it come down? Chronic stress and the hypothalamic-pituitary-adrenocortical axis in humans. *Psychological Bulletin* 133 (1): 25–45.

Moore, R.Y., and N.J. Lenn. 1972. A retinohypothalamic projection in the rat. *Journal of Comparative Neurology* 146 (1): 1–14.

Morán, C., and A. Zárate. 1993. Regulación de la hipófisis. En *Fundamentos de Neuroendocrinología*, ed. A. Zárate, C. Morán, A. Feria, and C. Kubli, 109–122. Cap. VII. México: Secretaría de Salud, Fondo de Cultura Económica.

Moran, C., E. Delcid, C. Cano, M.E. Fonseca, M. Loyo, and A. Zárate. 1986. Estudio Hormonal en 100 pacientes con adenohiposfisiario tratado por medios quirúrgicos. *Archivos de investigación médica (Mexico)* 17: 55.

Nauta, W. 1958. Hippocampal projection and related neural pathways to the midbrain in the cat. *Brain* 81: 319.

Neylan, T.C., A. Brunet, N. Pole, S.R. Best, T.J. Metzler, R. Yehuda, and C.R. Marmar. 2005. PTSD symptoms predict waking salivary cortisol levels in police officers. *Psychoneuroendocrinology* 30 (4): 373–381.

Nolte, J. 2013. *The human brain in photographs and diagrams E-book*. Elsevier Health Sciences.

O'Connor, D.B., E. Ferguson, J.A. Green, R.E. O'Carroll, and R.C. O'Connor. 2016. Cortisol levels and suicidal behavior: A meta-analysis. *Psychoneuroendocrinology* 63: 370–379.

O'Connor, D.B., N. Gartland, and R.C. O'Connor. 2020. Stress, cortisol and suicide risk. In *International review of neurobiology* 152: 101–130.

Oginska-Bulik, N. 2005. Emotional intelligence in the workplace: Exploring its effects on occupational stress and health outcomes in human service workers. *International Journal of Occupational Medicine and Environmental Health* 18 (2): 167–175.
Ouanes, S., and J. Popp. 2019. High cortisol and the risk of dementia and Alzheimer's disease: A review of the literature. *Frontiers in Aging Neuroscience* 11: 43.
Ouanes, S., E. Castelao, S. Gebreab, A. von Gunten, M. Preisig, and J. Popp. 2017. Life events, salivary cortisol, and cognitive performance in nondemented subjects: A population-based study. *Neurobiology of Aging* 51: 1–8.
Palckovits, M., and L. Zaborszky. 1979. Neural connections of the hypothalamus. In *Handbook of the hypothalamus*, ed. P.J. Morgane and Panksepp, 379. New York: Marcel Dekker, Inc., cap. 6.
Parent-Lamarche, A., and A. Marchand. 2015. The moderating role of personality traits in the relationship between work and salivary cortisol: A cross-sectional study of 401 employees in 34 Canadian companies. *BMC Psychology* 3 (1): 1–13.
Peavy, G.M., D.P. Salmon, M.W. Jacobson, A. Hervey, A.C. Gamst, T. Wolfson, and D. Galasko. 2009. Effects of chronic stress on memory decline in cognitively normal and mildly impaired older adults. *American Journal of Psychiatry* 166 (12): 1384–1391.
Pfaff, D. 1980. *Estrogens and Brain function neural analysis of a hormone controlled mammalian reproductive behavior*. New York: Springer.
Pfohl, B., B. Sherman, J. Schlechte, and G. Winokur. 1985. Differences in plasma ACTH and cortisol between depressed patients and normal controls. *Biological Psychiatry* 20 (10): 1055–1072.
Pittenger, C., and R.S. Duman. 2008. Stress, depression, and neuroplasticity: A convergence of mechanisms. *Neuropsychopharmacology* 33 (1): 88–109.
Plotsky, P.M., M.J. Owens, and C.B. Nemeroff. 1998. Psychoneuroendocrinology of depression: Hypothalamic-pituitary-adrenal axis. *Psychiatric Clinics of North America* 21 (2): 293–307.
Powell, T.P.S., and W.M. Cowan. 1954. The connexions of the midline and intralaminar nuclei of the thalamus of the rat. *Journal of Anatomy* 88 (Pt 3): 307.
Pruessner, J.C., J. Gaab, D.H. Hellhammer, D. Lintz, N. Schommer, and C. Kirschbaum. 1997a. Increasing correlations between personality traits and cortisol stress responses obtained by data aggregation. *Psychoneuroendocrinology* 22 (8): 615–625.
Pulopulos, M.M., V. Hidalgo, M. Almela, S. Puig-Perez, C. Carolina Villada, and A. Alicia Salvador. 2014. Hair cortisol and cognitive performance in healthy older people. *Psychoneuroendocrinology* 44: 100–111.
Rausch, J., A. Gäbel, K. Nagy, N. Kleindienst, S.C. Herpertz, and K. Bertsch. 2015. Increased testosterone levels and cortisol awakening responses in patients with borderline personality disorder: Gender and trait aggressiveness matter. *Psychoneuroendocrinology* 55 (Supplement C): 116–127.
Rausch, J., E. Flach, A. Panizza, R. Brunner, S.C. Herpertz, M. Kaess, and K. Bertsch. 2021. Associations between age and cortisol awakening response in patients with borderline personality disorder. *Journal of Neural Transmission: Translational Neuroscience, Neurology and Preclinical Neurological Studies, Psychiatry and Preclinical Psychiatric Studies* 128 (9): 1425–1432.
Reichl, C., A. Heyer, R. Brunner, P. Parzer, J.M. Völker, F. Resch, and M. Kaess. 2016. Hypothalamic-pituitary-adrenal axis, childhood adversity and adolescent nonsuicidal self-injury. *Psychoneuroendocrinology* 74: 203–211.
Renaud, L.P. 1979. Neurophysiology and neuropharmacology of medial hypothalamic neurons and their extra hypothalamic connections. In *Handbook of the hypothalamus*, ed. P.J. Morgane and J. Panksepp, 593. Nueva York: Marcel Dekker, Inc., cap. 9.
Resnik, P., and J.M. Dewaele. 2020. Trait emotional intelligence, positive and negative emotions in first and foreign language classes: A mixed-methods approach. *System* 94: 102324.
Rohleder, N., U.M. Nater, J.M. Wolf, U. Ehlert, and C. Kirschbaum. 2004. Psychosocial stress-induced activation of salivary alpha-amylase: An indicator of sympathetic activity? *Annals of the New York Academy of Sciences* 1032 (1): 258–263.

Salovey, P., L.R. Stroud, A. Woolery, and E.S. Epel. 2002. Perceived emotional intelligence, stress reactivity, and symptom reports: Further explorations using the trait meta-mood scale. *Psychology and Health* 17 (5): 611–627.

Sawchenko, P.E., and L.W. Swanson. 1983. The organization and biochemical specificity of afferent projections to the paraventricular and supraoptic nuclei. *Progress in Brain Research* 60: 19–29.

Schrijvers, E., N. Direk, P.J. Koudstaal, C. Kirschbaum, A. Hofman, H. Tiemeier, and M. Breteler. 2011. Associations of serum cortisol with cognitive function and dementia: The Rotterdam study. *Journal of Alzheimer's Disease* 25 (4): 671–677.

Shabani, S., M. Dehghani, M. Hedayati, and O. Rezaei. 2011. Relationship of serum serotonin and salivary cortisol with sensation seeking. *International Journal of Psychophysiology* 81 (3): 225–229.

Shields, G.S., J.C. Bonner, and W.G. Moons. 2015. Does cortisol influence core executive functions? A meta-analysis of acute cortisol administration effects on working memory, inhibition, and set-shifting. *Psychoneuroendocrinology* 58: 91–103.

Shields, G.S., M.A. Sazma, A.M. McCullough, and A.P. Yonelinas. 2017. The effects of acute stress on episodic memory: A metaanalysis and integrative review. *Psychological Bulletin* 143: 636–675.

Siegle, G.J., S.R. Steinhauer, M.E. Thase, V.A. Stenger, and C.S. Carter. 2002. Can't shake that feeling: Event-related fMRI assessment of sustained amygdala activity in response to emotional information in depressed individuals. *Biological Psychiatry* 51 (9): 693–707.

Sloviter, R.S., E. Dean, and S. Neubort. 1993. Electron microscopic analysis of adrenalectomy-induced hippocampal granule cell degeneration in the rat: Apoptosis in the adult central nervous system. *Journal of Comparative Neurology* 330 (3): 337–351.

Stalder, T., and C. Kirschbaum. 2012. Analysis of cortisol in hair–state of the art and future directions. *Brain, Behavior, and Immunity* 26 (7): 1019–1029.

Stalder, T., S. Steudte, N. Alexander, R. Miller, W. Gao, L. Dettenborn, and C. Kirschbaum. 2012. Cortisol in hair, body mass index and stress-related measures. *Biological Psychology* 90 (3): 218–223.

Stetler, C., and G.E. Miller. 2005. Blunted cortisol response to awakening in mild to moderate depression: Regulatory influences of sleep patterns and social contacts. *Journal of Abnormal Psychology* 114 (4): 697–705.

Steudte, S., I.T. Kolassa, T. Stalder, A. Pfeiffer, C. Kirschbaum, and T. Elbert. 2011a. Increased cortisol concentrations in hair of severely traumatized Ugandan individuals with PTSD. *Psychoneuroendocrinology* 36 (8): 1193–1200.

Steudte, S., T. Stalder, L. Dettenborn, E. Klumbies, P. Foley, K. Beesdo-Baum, and C. Kirschbaum. 2011b. Decreased hair cortisol concentrations in generalised anxiety disorder. *Psychiatry Research* 186 (2–3): 310–314.

Steudte-Schmiedgen, S., C. Kirschbaum, N. Alexander, and T. Stalder. 2016. An integrative model linking traumatization, cortisol dysregulation and posttraumatic stress disorder: Insight from recent hair cortisol findings. *Neuroscience & Biobehavioral Reviews* 69: 124–135.

Stienstra, C.M., F. Van Der Graaf, A. Bosma, Y.J.G. Karten, W. Hesen, and M. Joëls. 1998. Synaptic transmission in the rat dentate gyrus after adrenalectomy. *Neuroscience* 85 (4): 1061–1071.

Ströhle, A., and F. Holsboer. 2003. Stress responsive neurohormones in depression and anxiety. *Pharmacopsychiatry* 36 (Suppl. 3): S207–S214.

Truex, R.C., and M.B. Carpenter. 1969. *Human neuroanatomy*. 6th ed, 488–518. Baltimore: Williams & Wilkins.

Ucheagwu, V.A., J.O. Ugokwe, and R.N. Ugokwe-Joseph. 2021. Relationship between the alternative five factor model of personality and hair cortisol concentrations among university students from Nigeria. *IFE Psychologia: An International Journal* 29 (1): 32–42.

Ulrich-Lai, Y.M., and J.P. Herman. 2009. Neural regulation of endocrine and autonomic stress responses. *Nature Reviews. Neuroscience* 10: 397–409.

Vale, W., C. Rivier, and M. Brown. 1977. Regulatory peptides of the hypothalamus. *Annual Review of Physiology* 39 (1): 473–527.

Vale, W., C. Rivier, M.R. Brown, J. Spies, G. Koob, L. Swanson, and J. Rivier. 1983, January. Chemical and biological characterization of corticotropin releasing factor. In *Proceedings of the 1982. Laurentian hormone conference*, 245–270. Academic Press.

Van den Heuvel, L.L., T. Stalder, S. du Plessis, S. Suliman, C. Kirschbaum, and S. Seedat. 2020. Hair cortisol levels in posttraumatic stress disorder and metabolic syndrome. *Stress* 23 (5): 577–589.

van den Heuvel, L.L., S. Suliman, E. Bröcker, S. Kilian, T. Stalder, C. Kirschbaum, and S. Seedat. 2022. The association between hair *cortisol levels*, inflammation and cognitive functioning in females. *Psychoneuroendocrinology* 136: 105619.

Van Zuiden, M., A. Kavelaars, A.R. Rademaker, E. Vermetten, C.J. Heijnen, and E. Geuze. 2011a. A prospective study on personality and the cortisol awakening response to predict posttraumatic stress symptoms in response to military deployment. *Journal of Psychiatric Research* 45 (6): 713–719.

Van Zuiden, M., E. Geuze, H.L. Willemen, E. Vermetten, M. Maas, C.J. Heijnen, and A. Kavelaars. 2011b. Pre-existing high glucocorticoid receptor number predicting development of posttraumatic stress symptoms after military deployment. *American Journal of Psychiatry* 168 (1): 89–96.

Venero, C., C. Díaz-Mardomingo, I. Pereda-Pérez, S. García-Herranz, L. Utrera, A. Valencia, and H. Peraita. 2013. Increased morning salivary cortisol levels in older adults with nonamnestic and multidomain mild cognitive impairment. *Psychoneuroendocrinology* 38 (4): 488–498.

Weinstein, A.A., P.A. Deuster, J.L. Francis, R.W. Bonsall, R.P. Tracy, and W.J. Kop. 2010. Neurohormonal and inflammatory hyper-responsiveness to acute mental stress in depression. *Biological Psychology* 84 (2): 228–234.

Wessa, M., N. Rohleder, C. Kirschbaum, and H. Flor. 2006. Altered cortisol awakening response in posttraumatic stress disorder. *Psychoneuroendocrinology* 31 (2): 209–215.

Wilbraham, S.J., P. Qualter, and M.P. Roy. 2018. Emotional intelligence and cortisol responses: Can laboratory findings be replicated in classrooms and using other EI measures? *Personality and Individual Differences* 120: 58–64.

Yasuda, N., M.A. Greer, and T. Aizawa. 1982. Corticotropin-releasing factor. *Endocrine Reviews* 3 (2): 123–140.

Zorn, J.V., R.R. Schür, M.P. Boks, R.S. Kahn, M. Joels, and C.H. Vinkers. 2017. Cortisol stress reactivity across psychiatric disorders: A systematic review and meta-analysis. *Psychoneuroendocrinology*.

Part III
ToM from the Perspective of Philosophy of Mind

Chapter 12
What We Are for Us, What We Are for Others: Consciousness and Identity

Pilar López Segura and Tomás Ortiz Alonso

Heed these words, you who wish to probe the depths of nature:
If you do not find within yourself that which you seek, neither will you find it outside.
In you is hidden the treasure of treasures.
Know Thyself and you will know the Universe and the Gods.

Temple of Apollo, Delphi, Greece.

Abstract This chapter deals with consciousness and its relation with identity, or with what we perceive as identity. To explore this topic from a broad perspective, we will first take a tour of the different explanations with which philosophy, throughout history, has tried to explain the phenomenon of consciousness. Secondly, we will go into the neurofunctional and neuroanatomical perspective, to set the precedents on which we will expose the current theories with which neuroscience tries to explain consciousness. Finally, we will recapitulate on what has been seen so far and try to reconstruct, with it, a possible explanation of the relationship between consciousness and identity, and the relationship that it has not only with ourselves but with our environment – we will try to explore whether consciousness and identity are individual phenomena or also dependent on others; If our identity is something that relies on us, or that also relies on what others think about us.

Keywords Identity · Self-awareness · Self-identity · Deception · Self-deception · Lying and identity

P. L. Segura (✉) · T. O. Alonso
Department of Forensic Medicine, Psychiatry and Pathology, Universidad Complutense de Madrid, Madrid, Spain
e-mail: tortiz@ucm.es

T. Lopez-Soto et al. (eds.), *The Theory of Mind Under Scrutiny*, Logic, Argumentation & Reasoning 34, https://doi.org/10.1007/978-3-031-46742-4_12

Preface

If someone were to ask us about the meaning of our life and what we want to make of ourselves as a person and our life as a project, we would perfectly know what we are being asked. Perhaps we would not know what to answer; perhaps we may not have thought about it for a long time; perhaps we would answer very different things depending on the period of our life. All this is possible. But we would always know what this hypothetical person wants to know. We might not know the answer, but we would not be puzzled by the question itself.

This question appeals to our whole being and all its stages (past, present, and possible futures). It appeals to all our beliefs, learning, capacities, and desires. It involves everything we know we are, everything we have been and done, everything we would like to be and do; and it also involves everything that makes us up as people but that we are unaware of (everything that has shaped us without our knowledge). That is why there are almost infinite answers to this kind of question, so it is not surprising not to know the answer, or not to know it completely. Quite often, it is surprising that we know the answer.

However, it is not only the answer that deserves analysis. The question itself is also worth analysing, so is our ability to understand it. To understand this question, one must identify oneself as a "character", not just as an individual with agent capacity. It is a necessary premise that the individual agent answering this question identifies himself as a personality, with an internal narrative, with a determinate and determinable personality.

The human being knows that she is a person, what being a person implies, and that persons are "selfs", with identity and character; she knows she has interests and goals, character traits, dreams about what to become; knows that she has a life that belongs to her. But how is this possible? How, starting from a certain specialization or adaptation of cells, even from a certain macroscopic organism with the capacity to receive the demands of its environment and to adapt to them, does something as abstract as a character and a meta-identification of it, something like a vital sense, emerge?

This approach is tricky, given that one does not transition from an amoeba to a child dreaming of being a football player without some intermediate stages. But it is also evident that the child's dream does not have the same characteristics as her arms, her hair, or her eyes. This dual nature of humans (which is not unique to them and has undergone a certain evolution that comparative psychology deals with) has raised the most fundamental questions of our species, our science, our literature, our art, and even our religion. What is this dual nature, physical-mental? How and why do we have the capacity to ask ourselves about it? Are they separate from each other, or do they respond to the same principles? How do they relate to and generate each other? This enigma about the dual nature that we can perceive within ourselves is one of the foundational mysteries of human philosophy and science, and one that remains unresolved to this day.

From now on, we will try to build a chapter based on these questions. We will revise some possible answers to the question of consciousness as they are sought today. Though this is a question that cannot be answered from a single discipline –

dealing with at least two ontologically different realities makes a multidisciplinary approach advisable. It is a question that has accompanied humans throughout their history, taking and continuing to take infinite and infinitely different forms. We will have a neuroscientific and philosophical focus; but since it is one of the core questions of our species, it accepts almost infinite approaches, a characteristic of the subject that this book, which brings together a wide range of disciplines, brings up.

The problem we face can be summarised, at least to begin with, as the problem of consciousness. We will try to make a modest approximation to those which, in our view, are most unique to humans: consciousness as awareness, consciousness as self-awareness, and consciousness of self about community.

The organization of our chapter will be as follows. First, we will briefly show how this has been a question that has accompanied us throughout our history, and we will summarise some of the pre-scientific answers it has received. Secondly, we will look at the current state of consciousness research: what are the most widespread scientific hypotheses that currently account for consciousness, its reason for existence, and its characteristics, and what answers do they offer to the most important questions about consciousness. Thirdly, we will discuss what is known about the neurofunctional basis of consciousness, research which has been growing impressively over the last decades and which is grouped under the name "Neural Correlates of Consciousness" (NCC from now on). Fourth, we will turn to a particular "network" mode of brain functioning that has been generating a lot of interest lately, as it is the one that can seemingly answer some of these puzzles: the idea of the self as an individual, the narrative that permeates the idea of the self, and how it relates to our abstract thinking (Default Mode Network, or DMN from now on). Fifth, we will look at the relationship, offered by this same brain network or mode, that this idea of ourselves has to our community; what we learn from others and what they learn from us, and how this influences the generation of an internal narrative. Sixthly, and as a very relevant capacity for this, we will investigate lying and try to elucidate how it might influence our idea of ourselves because of the influence it has on what other individuals think of us. Finally, we will summarise all the above and offer our point of view on the subject.

12.1 Introduction

12.1.1 Consciousness: A Formal Definition

Turning our gaze to the etymology of the word 'consciousness' (Harper n.d.) we find an overlapping of meanings. In its original form, this word, born around 1630, originally meant 'inner knowledge'. Although it seems to refer to the most popular meanings of the word (being aware of oneself, of what is going on inside, of one's sense of self and identity), this is not the case if we look back even further. It is postulated that the term 'consciousness' is derived from the older *conscious*, which meant ‘knowing, privy to’ in the poetic sense; it comes in turn from the word *conscire*, ‘be (mutually) aware’, where we can already find reference to shared or communal knowledge. It is also believed that the Latin terms are a loan from ancient Greek, specifically of the word *syneidos*, which meant being aware of one’s thoughts but also to the consciousness of right or wrong; that is, to the *moral* consciousness (Marietta 1970). Going even deeper we find the word *súnoida*, the root of *syneidos*, which in Ancient Greek meant ‘to be conscious or cognizant of’; but also ‘to share in the knowledge of’ (Liddell et al. 1996). It is therefore a term that, from its origin (or what has been traced as such) is ambiguous and elusive. It includes references to interiority and subjectivity, but also the relationship of the latter with the community; and, above all, with knowledge. Although the Greek moral note is the least common in the term used today, it retains much of its etymological overtones: always referring to a particular kind of inner knowledge with some relation to the external world.

Using a more recent source, according to the Oxford Dictionary of English (in its 2011 edition; Stevenson and Waite 2011), the term ‘consciousness’ has plenty of meanings:

1. Internal knowledge or conviction; the state or fact of being mentally conscious or aware of something.
2. In *Philosophy* and *Psychology*:
    (a) The faculty or capacity from which awareness of thought, feeling, and volition and the external world arises; the exercise of this. In *Psychology* also: *spec.* The aspect of the mind made up of operations that are known to the subject.
    (b) As a count noun. A state or form of consciousness.
3. Shared or mutual knowledge. *Obsolete. Rare.*
4. Others:
    (a) The totality of the impressions, thoughts, and feelings, which make up a person’s sense of self or define a person’s identity.
    (b) Attributed as a collective faculty to an aggregate of people, a period, etc., a set of shared defining ideas and beliefs.
    (c) With adjective specifying an area of operation, such as moral consciousness, religious consciousness, etc.
5. The state of being aware of and responsive to one’s surroundings, regarded as the normal condition of waking life.
6. As the second element of compounds with the sense ‘consciousness of ——, awareness of ——’.

The dictionary entry seems relevant to us because it faithfully portrays the elusive nature of the term which occupies this chapter. Thus, consciousness is identified with the subjective sensation of *knowledge* about something – generally, about an object or state, be it mental or external (*I am aware that the sun has risen or that it is night*); the capacity according to which our thoughts, desires or wills present themselves within us, according to which we can know that they exist and that they belong to us;

the set of operations occurring in our 'mind' that we are able to perceive; shared or mutual knowledge (although an odd usage today, it reflects an interesting change in the point of view, as well as in the content of what is known); corpus of shared ideas or beliefs which define a group of people, a certain period of time, etc.; the totality of 'notes' that make up the subjective feeling of individuality, or that define the identity of a particular person; a state, referred to a specific person, who is awake, who knows the place and time in which she is and is able to respond to her environment, interpreting it appropriately (being located, so to speak, in waking life). There are more definitions and more specifications of them, but the ones reflected here are enough to get the point across.

Consciousness is identified with a kind of interpretative and subjective knowledge; it generally refers to a more or less concrete and determined state or conformation (be it the knowledge of oneself, of the environment, be it a certain group of people who share this knowledge, or of the processes occurring in one's mind), and which seems to be attributed – insofar as it relates to some kind of knowledge – exclusively to subjective interiority (consciousness is not an attribute of the environment, nor knowledge: it is always, so to speak, how a concrete subject, an aggregate of persons, etc., thinks about, feels or knows something).

These are some of the characteristics that, a priori, seem to be shared by all definitions of consciousness. But it is the differences that are striking. Consciousness is identified with a state of alertness and normality within waking life, a state in which the person can interpret and respond to their environment and the circumstances that compose it. In this case, the term 'consciousness' is therefore associated with a momentary picture of the environment as how the person perceives it; in sum, how their interiority perception relates to the exterior. In this sense, 'being conscious' and 'being aware' would be comparable expressions. In another definition, referring to the person being aware of the existence of something concrete that he or she has become aware of, these two expressions would no longer be comparable: the person is aware that it is raining, but that is something very different from the person being self-aware. One cannot be aware that it is raining without being aware of herself; but we realize that both phrases, using the adjective 'aware' in them, similarly, do not mean the same thing at all.

The complications grow if we continue with our analysis since another meaning of consciousness is that which identifies the narrative that unites the defining notes of a person and turns them into meaning, into an identity. That is: if, playing a guessing game about our group of friends, someone was to give us the clues 'she is very hospitable', 'she is very cheerful' and 'she does not like to have her plans changed', we could (easily or hardly depending on how concrete those defining notes are) identify our friend Juana within the group. Therefore, this meaning of consciousness is identifiable with the concept of identity (as long as it refers to us). It could be understood when applied to oneself as being aware of what our essential and defining characteristics are, those that make up our personality and our identity, and how these characteristics are woven together to form a whole that is what we understand as our 'self'.

Earlier we noted that consciousness, being conscious or aware at a given moment, seemed to refer to a kind of momentary picture, as well as to the capacity to respond to it. However, this meaning has the opposite temporal sense: in this case, it refers to the sensation of meaning, of identity, which is produced by the fact that certain notes of a person are maintained through time, context, and its changes. In this meaning, it refers precisely to the fact that the temporal stability of certain characteristics is what makes them define a person, and that when these characteristics are united and threaded together, they form a narrative with meaning: an identity. And finally, one of the most interesting aspects of bringing up the definition of the term is the change in point of view: two of the definitions refer to a group of people, not to a person observed as an individual. Consciousness can thus be predicated on social groupings, and again identified with essential, narrative, or defining overtones (the group of ideas and beliefs, for example, around which a certain grouping of people is built).

Therefore, we could summarise consciousness as a kind of theoretical, narrative, and common thread that the subject(s) know and can identify as fundamental, essential, or characteristic of him/herself/themselves. From then on, terminological differences and diversifications are multiplying.

But whether outwards or inwards, momentarily, or continuously, individually, or communally, if we put all definitions of 'consciousness' together it seems to be always present if we are. Just as 'the worst of evils, death, means nothing', according to Epicurus, 'for if we are, death is not, and if death is, we are not' (Epicurus, 341–270 BC, *Letter to Meneceus*); thus consciousness, in these terms, would have to mean for us the whole: for whenever there is us, there is consciousness.

12.1.2 Consciousness: A Mythical Origin

As we have pointed out, to speak of consciousness is, in general, to speak of people and their states as such, of their identity and their environment. Therefore, to speak of consciousness implies going back to the early moments of humans in the sense in which we know them today: people with the capacity for language, abstract thought, feelings, the capacity to communicate them and with each others, etc. Language is a particularly interesting case, as it is the link between a person's interiority and his or her community. It is through communication with others (verbal and non-verbal, but it must be kept in mind that non-verbal communication has in any case a significant correlate that is secondarily linguistic) that we presuppose consciousness for them and assume an identity from them; it is also through our "inner voice" that we perceive it in ourselves. We will not be able to explore in depth the fertile and interesting relationship between language and consciousness, given the demands of the present work; but we will briefly return to it later. For now, it serves as one of the identifying features of the type of human we will be analysing.

It is on this basis that we think primarily about the most ancient civilizations; here Ancient Greece takes the lead. It could be considered as the origin – or one of the most important origins – of what we mean by 'consciousness'. Our term is, presumably, different from what the ancient Greeks could understand by consciousness; but it is on its basis that our conception of it is built.

Even though some have argued that consciousness, as we know it today, is a relatively recent historical development that arose sometime after the Homeric era (Jaynes 1974), this term or conception came about because the groundwork was laid for it. According to Jaynes's view, early historical humans did not experience themselves as unified internal subjects of their thoughts and actions, at least not in the ways we do today. Others have claimed that even during the classical period, no word of ancient Greek corresponds to 'consciousness' (Wilkes 1984, 1988, 1995), but in the sense that we currently give to the term (which, as we have seen here, brings together very different points of view). This does not mean, however, that the ancient Greeks lacked the notion of identity or character, which they not only knew, but which was one of the main notions in their conception of the world, society, philosophy, and myths.

In Ancient Greece, however, we do not find words referring to the 'being aware' or 'being aware of X' senses discussed above; this is a field of study much more suited to modern neuroscience than to mythical literature. Instead, we do find attention to our idea of consciousness as self-awareness, relating to the self and its character, in sum to the concrete, narrative identity of a person. This sense of the word 'consciousness', which is indeed more similar to what the term 'soul' (without its connotation of immortality) captures, is what was known in Ancient Greece by the word *psyché* (Bremmer 2016). The discussion of the soul as a vital principle – even if it is sometimes considered to be material, as in the atomist school (Berryman 2022) – goes back to the beginnings of philosophy, being already to be found in the pre-Socratic doctrines (Lorenz 2009). For some of the pre-Socratic philosophers, such as Thales, anything that had a principle of change or motion had a soul, so even metals were considered animate (O'Grady n.d.).

Homer will echo what his time considered in this respect; we find an immortal considered soul, which plays the role of a vital or animating principle, but also gathers and reflects the true essence of the person. Ulysses, descending into Hades, finds the souls of his friends; the original term for these 'souls' found in the Greek realm is *psyché* (Homer and Lawrence 1981). The personification of the soul in Ancient Greece – whose story, although recorded in art from about the fourth century BC, is narrated at length in Apuleius' *Metamorphoses* (late second century AD) – was precisely the goddess Psyché. In her we find particularly clear the more rational notes, which establish her relation to what we now know as 'consciousness'; and it is her union with Eros, son of Aphrodite, and personification of romantic love and desire, which symbolizes the immortal, non-corporeal nature of humans. Since then, the concept of *psyché* has not ceased to evolve (Katona 2002), giving a mythical answer to questions that today form part of the basis of neuroscience, psychology, and philosophy (Crivellato and Ribatti 2007); as the very words 'psychology' and 'psychiatry' indicate, both coming from *psyché* (Antonakou and Triarhou 2017).

Originally, the Greek term *psyché* meant 'breath, wind, or vital breath' (Autenrieth et al. 2007), because people exhaled one last breath when they died;

this was the breath in which their vital principle departed. Thus, *psyché* was translated by the Latins as *anima*: this word encompasses the connotation of wind, breath, and vital breath (Lewis et al. 1969b, 'anima'). This word is derived from *animus*, the Latin form of *ἄνεμος*, which meant 'wind' (Lewis et al. 1969a, 'animus'). Thus, everything that was alive possessed a soul. Animals and plants (Aristotle and Lawson-Tancred 1986) although substantially different from humans, also had a vital animating principle or soul. And so natural beings were divided into two categories according to whether they possessed this vital principle or not: animate beings (possessing anima or soul) and inanimate beings (lacking it). The fundamental difference, then, between the human soul and the animal or vegetable soul was what brings it to the present day as fertilizer for philosophy, psychology, and neuroscience: its rational, ethical, and abstract capacity.

12.1.3 Consciousness: Philosophical Roots

We have shown that, although our conception of 'consciousness' is by no means identical to the Greek one, it does find its roots in our ancient predecessor. The question of our capacity to think, to feel, to find in ourselves motivations and desires, and abstract attitudes (appreciation of beauty, morality, etc.), is a direct legacy of Ancient Greece. How it is that we are, in short, able to identify ourselves collectively as persons, distinct from any other element in creation; how to analyse what surrounds us, to analyse ourselves, and to find ourselves distinct; and how to identify ourselves individually as a concrete person, as a being, as an individual and as a story, is indeed what once gave rise to the questions about the soul that comes to us today in such different forms as the philosophy of mind, of language, of beauty, psychology, sociology, anthropology, linguistics, and neuroscience. But this path has yielded countless intermediate fruits, which we certainly do not have the opportunity to show in full in this chapter. We will nevertheless try to make a modest and grossly simplistic summary of the evolution of the term that can give an account, today, of what we are looking for in it.

Plato inherited the mythical approach to consciousness. Is well-known his defence of the expulsion of the poets from his ideal city (Plato et al. 2007); he did not condemn poetry itself, but because it was used as a source of knowledge and wisdom (Espíndola 2016; Villar 2017). It had to remain as a form of recreation, of artistic amusement, but never to be considered as truth or dogma (Leszl 2016).

He developed the first truly philosophical and rational study of the soul and reason, from which he found to be from which he drew his well-known theory of the soul (Plato and Waterfield 2010). He thus defended a dualistic theory (Olshewsky 1976; Zucca and Medda 2019): the body was the physical and perishable 'shell' of a soul of an ethereal, rational, and immortal nature, which was part of the body as a 'stage' on its way to knowledge. The body represented on the one hand a prison role – the soul was imprisoned in it until its release by death – but on the other hand, it was the only tool for accessing knowledge. The soul, instead, comprised everything else of a person: her/his essence, her/his wisdom, her/his, in sum, liveliness, and identity. Plato thought this to be immortal, whereas the body was perishable and mortal; he shared this vision with Orphic traditions (Edmonds III 2015).

Aristotle came up with a quite different theory of the soul (Aristotle and Lawson-Tancred 1986). According to his view, the soul and the body could not be considered equal; but they were answerable to each other. The soul was the essence of the body, but in a considerably different sense from Plato: the soul gave form and existence to the body, whilst the body gave substance to the soul; it was its embodiment, rather than a shell on which the soul incarnates (Olshewsky 1976). Matter (which was the body) took a particular form as dictated by its theoretical essence (which was the soul): this theory, held afterward by Scholastics, was called hylomorphism (see Aristotle's *Metaphysics* for broader information). Thus, in no way could the soul exist without its corporeal counterpart, just as, so to speak, Michelangelo's Pietà, not being only marble, could not exist if marble did not also exist. This example is directly taken from the one used by Aristotle himself in his to explain his theory of the four causes (Todd 1976), with which the soul is directly related (for more on the subject, see Aristotle's *Physics* II.3 and *Metaphysics* I.3–7). The soul is found as one of the four causes (specifically, as a formal cause) in all living beings. This explains Aristotle's view – discussed above – of the soul as something proper to every living thing: as a form, or essence, of being, it must necessarily be found in everything alive. It is the soul, for Aristotle, and its different natures that give rise to the diversity of living things found in the universe (Vieira 2022). It is the soul that contains the essential information of the living being and the form to which matter obeys; therefore, plants are plants precisely because they respond to a soul proper to plants, animals to a soul proper to animals, and humans to a soul proper to humans. The human soul is fundamentally distinguished from the other two by a component that has brought its study to the present day: humans, unlike any other being in creation, possess a rational soul (Aristotle and Lawson-Tancred 1986). Humans can think: to think about the world, about themselves, and about their thinking – which today is called metacognitive ability. This was, for Aristotle, its essential marker, and what made it different from the others. And it would be, from then on, the main inquiry about the mind – or the soul, or consciousness.

After the medieval pause, in which no novel theory is developed – since questions about the soul are answered by the Christian filter – we reach the first truly rationalist stage of philosophy. It takes root directly, then, with the notes that most drew Aristotle's attention to our particular soul: it is, in fact, a philosophy about reason – which occupies the place the soul previously occupied. Modern and considerably more familiar terminology is found in Descartes' philosophy; we do not find references to the soul, but to reason. In this early modern era (seventeenth century) consciousness – human reason – had become the very central issue of study whether talking about what used to be called the 'soul', about humans, about philosophy, or the universe. Human reason is the fundamental question to be solved; and consciousness, as how our reason presents itself to us, was the best way to approach the study. Thus, Descartes does not hesitate only to identify but to *define* reason in terms of self-consciousness or self-awareness (Descartes 1970; Van Gulick 2022). Cartesian philosophy identifies two ultimate and irreducible principles, opposed in nature, that make up the human being: the *res cogitans*, or thinking matter, and the *res extensa*, or corporeal matter (for more on the subject, see *Meditations on First Philosophy*, R. Descartes, included on the compilation referred above). The *res extensa* is the necessary basis for reason's presence in the world; but the touchstone of his philosophy is the reason, which allows us to find the truth (Descartes 1641/1993; Descartes 1644/1911; Descartes 1970).

Descartes bases his whole philosophy on the analysis of reason as it presents itself to us – self-awareness –, and how we can recognize the truth of the world and knowledge thanks to it. That is why he constructs his methodical doubt method (which is explained as a project on his *Discourse on Method* and developed on his *Meditations on First Philosophy*, both included in the compilation cited), with which he dismantles all of reality, the universe, and everything within his reach through the tools and rules that reason places at his disposal. Thus, according to him, we can be certain that we exist, as the world-famous Cartesian quotation dictates, only because we think; since we cannot be certain about our body, as it comes to us through the senses that so often confuse us. We can be certain, moreover, of everything that is presented to reason as 'clear and distinct' truth; not at all of what is presented to us by the senses, which is generally 'obscure and confused' information (for more on the subject, see (Descartes 1970, 2019)).

Thus, we see that for rationalist philosophy, initiated mainly by Descartes but continued by other great philosophers such as John Locke, not only was reason essential to humans, but it was the essential question of the universe: it accounted for truth, for our existence and, indirectly, also for God. It is also worth pointing out that although Cartesian reason maintains the characteristic of immortality – with Platonic roots and more characteristic of what was traditionally understood as the soul –, it has lost its mythical notes. It is no longer the form, the essence, or the source of life. From then on, which remains central in the question about the soul, mind or consciousness is its rational facet. In this way, what Descartes calls 'reason' gradually comes to resemble what will later be grouped under the term 'consciousness' or 'self-awareness'.

A little more than 30 years after Descartes' works, another great work on the subject was published: Spinoza's *Ethics* (1677; published posthumously). He is the first author to propose a vision that deeply resembles what we currently understand by consciousness.

In Spinoza's view and against Descartes one, soul (or mind) and body are by no means distinct substances. On the contrary, they are the same; they are just approached from different points of view. What appears to us as the mind, or soul, is only the image of the body from its capacity of understanding; it is the rational representation, the mental model, of the body. The same is true of nature, which may be thought of from the material or mental point of view; but, even if we change the attribute from which we conceive it, the same rules, the same connections and the same order of causes will always be found, since we are thinking of the same thing – though we think of it each time according to one of its attributes (Spinoza 1996).

Thus, mind and body are the same thing only approached in terms of their rational and physical attributes, respectively. This division can also be seen in knowledge: we can have 'first-order knowledge', or perceptual one; and 'second-order knowledge', or rational one. It comes from the observation of nature in terms of its rational laws, which leads us to ascertain the internal causes and relations of things, for which their essence is responsible.

Hence in Spinoza, we find a definition of the mind or soul as a representation, or model, of the body; moreover, of the same very nature. The soul is just an approach to the unity a person conforms to from the mental point of view. This consideration of humans as a unity, being soul only its mental representation, is an approach that, as we shall see, will appear in the hands of current authors – bridging terminological gaps.

Almost contemporary to Spinoza, another author was proposing a considerably modern approach: Gottfried Leibniz. Leibniz's theory of mind – which is also his theory of soul, as he identifies them –establishes that the most profound distinction between the human soul and other beings is its capacity for *perception*. For Leibniz, the fact that the human soul, which he will defend as an immaterial, eternal, and immortal entity, is also perceptive – that is: it relates to the world, has the possibility for the world to enter, and has the possibility of understanding it – is what makes it different in nature from any other element of creation. Just as he equates mind with soul, he equates thinking with perceiving: the most proper activity of this soul-mind is perception, both external world and of itself (which, in a way, is exactly what is proposed by current theories that equate consciousness with a particular mode of attention, point to which we will return later).

As mentioned, he holds that the soul is an immaterial, indivisible substance that is the ultimate cause of all human thoughts and actions; being to that extent the ultimate cause of all actions – insofar as they are effects of the main cause. Considering the soul as the 'first principle' or 'first cause' of humans, he offers an explanation about it aligned with the metaphysical principle described by Aristotle as the 'first unmoved mover' (Kenny and Amadio 2022), in which an effect-producing cause must necessarily be distinct from effects. The soul is both the source of the mind and the source of its own life: it can exist independently of the body. In this line, the soul – once created by God – is immortal and eternal. It can exist in its own right, with no dependence on anything, containing thus the principle of free will. A soul existentially different from the body, with diverse natures and ontologies, is what Leibniz presents to us – which is one of the foundations of his *Monadology* (Leibniz 2014). With the mechanical explanation, perceptions – mind, or soul – cannot be reached nor explained. He shows it with his well-known analogy of the mill; he asks us to imagine a machine 'whose structure makes it think, feel, have perceptions'. Now, if this – being it the brain – were of the size of a mill, and we could enter it, we would find inside it interconnected parts that do not explain the phenomenon, the action, the simple substance – and as such non-mechanical – of perception. Perception is a simple substance, ontologically different from the complex – thus mechanical – substance that sustains it.

This soul-mind is a metaphysical, simple, and irreducible substance, to which all human capacities ultimately refer perceptions; the understanding of perceptions; and separate from but also proper to this soul-mind, the capacity of consciousness. Leibniz is in this sense extremely modern, for in addition to recognizing consciousness as an activity separate from other perceptions or mere understanding, he also contemplates what occurs outside this consciousness. He is thus the first author to consider 'unconscious' or 'subconscious' thoughts, which he calls '*petite perceptions*' (Leibniz 1953). On his view, consciousness is a continuum that allows infinite grades; also contains those in which there is mental activity without awareness of it (unconscious thoughts) (Morejón 2022).

The fact that he considers consciousness as a particular kind of perceptive ability of the soul leads him to another modern conclusion: to distinguish, albeit in a general way, between consciousness and self-consciousness. That is: between awareness of the world and being conscious about the proper consciousness, or self-awareness. In this sense is again aligned with recent theories that defend that consciousness is a particular mode of attention: that which is directed towards us, rather than towards external objects.

Another interesting contribution to the field of consciousness that came from philosophy was made by Kant. Immanuel Kant's theory of the soul – framed in his general theory, known as transcendental idealism – is a metaphysical theory, but one that anchors the human consciousness and soul to its experiential facet.

In Kant's view, the soul is the metaphysical entity from which all human experience and knowledge sources. It is divided into two parts or aspects, which, although operating in different spheres and according to very different rules, are combined in the soul (following the same combinatorial principle between the purely rational and the purely experiential, which it takes from the example of the mind, its entire theory is constructed).

The empirical part of the soul is the one attached to the surrounding world; is responsible for the individual's perceptions and experiences. Enables us to endow our perceptions with space and time. This is indispensable for the development of consciousness and thought, given that perception is considered by Kant as the first necessary step for it; the rational part comes later and is constructed on the perceptual ground. Without a concrete Spatiotemporal location, no perceptions are possible; hence, neither is consciousness.

Furthermore, the rational part of the soul is responsible for understanding and making sense of the world; creating, by combining perceptions and giving them meaning, what he called knowledge. Kant ascribes to this part the capacity of forming opinions and making judgments, being the rational basis of knowledge. This, in a way, is also very much in line with theories of attention, which postulate (we will expand on this point later) that we construct the idea of our consciousness by modelling what we infer from the outside.

In any case, the most important thing about the theory of the soul and the mind – which Kant, in line with Leibniz, equates – is that it is composed of an externally oriented part and an inwardly oriented part. The two are indivisibly linked, for it is only together that a unified experience is created, both of the universe and the self.

And this two conform to the soul, which has shown herein is largely equated with the mind; this latter one, with consciousness. This equality between mind and consciousness was to be maintained throughout the nineteenth century, in the context of the rise of modern scientific psychology.

The last event that is worth commenting on in the field of modern philosophy of consciousness is the decline of the concept of consciousness in the field of scientific psychology during the twentieth century – the same one that had defended and sustained it during the nineteenth century. A chapter could be devoted exclusively to this transformation, to the content and progress of scientific psychology and contemporary theories of the philosophy of mind. But since the present volume is devoted to the theory of mind and entire chapters will expand on these points, we do not consider it pertinent to expand on this in the framework of an introduction to our subject.

In any case, the ancient concept of consciousness, although it had suffered from a certain cracking – as has been exposed along these lines – witnessed its disappearance in the field of philosophy, as we said, in the twentieth century. This decline was due to the emergence of behaviourism in modern psychology, with a particular impact in the United States (Watson 1924; Skinner 1953; Braat et al. 2020); only certain movements as Gestalt psychology maintained the ancient idea of consciousness as a matter of scientific and psychological interest, keeping it in the focus of psychological research. During the twentieth century, however, behaviourism also had to witness its decline (Braat et al. 2020), giving way to cognitive psychology (Barone et al. 1997) – which approaches the mind from a computational point of view, emphasizing the emphasis on information processing capacity and possible models of internal mental processes (Braisby and Gellatly 2012).

As long as consciousness is lived from inside as a kind of mental process or internal state, that accompanies all our perceptions and thoughts, the emergence of cognitive psychology provoked a resurgence of research and scientific, psychological, and philosophical interest in consciousness. In the field of philosophy of mind, some of the most important books of the last decades have been written on these topics (Chalmers 1997; Dennett and Weiner 2007; Parfit 1987); and in the field of neuroscience and neuropsychology, thousands of articles are published every year on the subject, being now – for the last 4 decades – one of the main themes of neuroscientific interest.

That is why in the following chapter, we will focus on outlining some of the threads and discoveries about consciousness that neuroscience has contributed, as well as the most important theories currently on the subject.

12.2 Science and Consciousness: An Overview

As previously commented, consciousness has suffered several lurches throughout its history. It has represented – perhaps under other names – one of the most important concepts of certain epochs; it has also suffered a radical lack of interest from others.

In our time we are experiencing a boom in research and interest in consciousness, as well as a shift in the dominant discipline that deals with its study. Philosophy has traditionally been compromised with this. Now, although still an important topic in the field, it certainly generates more research in the field of neuroscience.

Consciousness has been the subject of an active debate in the neuroscientific world for some 30 years. Thanks to the effort and interest devoted to its study, an overwhelming number of theories have been developed that attempt to answer not only what consciousness is, but also what its neurological correlates are, what its evolutionary function is and what its scheme of functioning is. There are several theories, from this corpus of possible answers, which are considered to be the most relevant (del Pin et al. 2021; Northoff and Lamme 2020; Wahbeh et al. 2022; Yurchenko 2022).

Some of the most current and discussed neuroscientific theories about consciousness are the following: Integrated Information Theory (IIT) (Tononi 2008; Tononi et al. 2016); Global Neuronal Workspace (Dehaene et al. 2006; Dehaene and Changeux 2005); Recurrent Processing Theory (RPT) (Lamme 2006, 2010; Lamme and Roelfsema 2000); Higher-Order Theories of Consciousness (HOT) (Rosenthal 2005; Brown et al. 2019); Attention Schema Theory (AST) (Graziano 2013, 2019; Graziano and Kastner 2011a, b); Synchrony Theory (Engel and Singer 2001); Predictive Coding Theory (PCT) (Hohwy 2013; Jehee et al. 2006; Spratling 2010); Entropy Theory of Consciousness (ETC) (Carhart-Harris 2018; Carhart-Harris et al. 2014; Erra et al. 2016); Embodied Theory (Park and Blanke 2019; Park and Tallon-Baudry 2014; Thompson and Varela 2001; Shapiro and Spaulding 2021); Temporo-spatial Theory of Consciousness (TTC) (Northoff 2013, 2016; Northoff et al. 2020; Northoff and Huang 2017; Northoff and Zilio 2022); and Operational Space-time theory (OST) (Fingelkurts et al. 2010, 2012; Northoff and Lamme 2020).

The diversity is such that, in addition to numerous and varied secondary and alternative theories to those mentioned here, there are already several versions of each of them, as well as combinations of one or the other. Each one proposes one cause as the key process that gives rise to consciousness. On its basis, each one argues for a neurological correlate, an evolutionary function, and a scheme of working for consciousness. But as Northoff and Lamme (2020) propose, they may not be mutually exclusive. As long as they opt for diverse features as crucial for the arising of consciousness, but also focus on different processes, they may not be contradicting each other. All these theories could be explaining different stages of the global process, thus opening the door to a possible global and inclusive explanation of this great unknown.

We are aligned with this synthetic approach. If Asterix and Obelix were teleported to an airport and suddenly saw a plane take off and fly, perhaps their first intention would be to explain this amazing phenomenon. They would reach the conclusion that the flight of an airplane cannot be explained by the presence of screws – which, however, are indispensable for flight. Nor can it be explained by the presence of engines alone – which again are indispensable for flight, but not sufficient.

They might ask: what is the necessary but also sufficient element for the fly? In our view, it is the combination of all the necessary factors that makes up the sufficient circumstance. The seats are not necessary, the plane can fly without them. Presumably neither are toilets, nor luggage racks. But some parts and factors are necessary for flight – the existence of the plane; the engines; the fuel; a certain capacity for acceleration; the pilot; etc. It is not one of them that makes the difference between flying and not flying: it is the combination of the necessary factors that produce the necessary, but also sufficient, combination for flight. In short, we consider that consciousness is indeed a property, or capacity – as it would be the ability to fly for aircrafts – of certain physical combinations.

12.2.1 Neural Correlates of Consciousness (NCC)

As defined by their proponents (Crick and Koch 1998, 2003), the NCC are 'the minimum neural mechanisms jointly sufficient for any one specific conscious experience' (Koch et al. 2016). There has been and is active debate about the location of the NCC – as will be noted through the analysis of each specific theory, given that each has its proposal about the structures on which consciousness relies. But broadly the discussion is essentially polarised between those who argue in favour of a frontoparietal communication network; and those who do so in favour of a smaller, posterior area of the cortex (a posterior cortical 'hot zone').

There is general agreement on that, for consciousness to exist, certain but large parts of the brain – especially located in the midline – need to function properly. But a correlative structure is not from any point of view explicative. As Wu (2018) notes, what is being looked for in neuroscience is not the circumstances that are correlated with consciousness, but the features that causally explain it. And correlation does not imply causality.

Defenders of the NCC try to find the neural mechanisms that *explain* behavioural consciousness, irrespectively from the content of the conscious experience (Koch et al. 2016). They defend that through the finding of a neural conformation that appears exclusively when the person is acting like she was conscious, consciousness could be explained.

One of the main problems of this approach is that it excludes disorders of consciousness – coma or vegetative states. In these states it is not possible to know if the person is or is not conscious, but because what is absent is the capacity to *report* consciousness or to act like they were conscious – the reason why these conditions are known as unresponsive wakefulness syndromes (Gosseries et al. 2014; Overgaard 2009). The absence of consciousness is generally determined by the absence of certain neuroimaging signals thought of as 'consciousness markers'. This is problematic, as no neuroimaging-identifiable marker has been identified that systematically relates even to behavioural consciousness (Boly 2011; Storm et al. 2017).

Yet is problematic in another sense: finding the neurological correlates of consciousness, although very important from a clinical point of view, does not explain the process according to which a certain neurological combination or conformation correlates with the process of consciousness (Schurger and Graziano 2022; Wu 2018). Chalmers (1997) defends that the NCC are a mapping possibility, to say so: under a certain state of a neural system there is a certain state of behavioural consciousness; under another state of this neural system, there would be another state of behavioural consciousness. Therefore, for the NCC defenders, the main goal would be to find which one is this minimal neural system responsible, simply through the adoption of different states of activity, of all possible states of consciousness. Those would be a direct consequence (what they call a 'correlate') of these states of activity.

It is not the same to claim that certain neural combinations are the basis from which consciousness can occur, that the process of consciousness and the neural structure that produces it can be correlated. The following neuroscientific theories instead defend that consciousness has some neural basis, which enable the processes and events that lead to conscious experiences. But in addition to the certain basis, they propose an explanation of how consciousness occurs.

12.2.2 Integrated Information Theory (IIT)

According to IIT, consciousness is a phenomenon that occurs in any system that carries or transmits effective and integrated information – it does not occur if the information transmitted has no sense or effective meaning. Consciousness is a kind of consequence that occurs from a certain level of integration of the information (Tononi 2008; Tononi et al. 2016; Tononi and Koch 2015). This level of integration, or amount of integrated information, is represented by a measure called 'phi'. Phi represents the level of complexity, parallelisable with the level of consciousness and derived from how integrated the transported information is, that the system possesses.

But what this notion of 'integration' represents? 'Phi' is calculated by computing the information that at any given moment is being carried by the system. If the resulting value of effective information or content is greater than the sum of its informational parts, the system is transporting integrated information; and this means it is conscious. Integration is a property that is predicable only about certain combinations of information units, and which derives from the connection, or relationship, between them. Integration is thus defined as a feature of certain sets of information; a sort of 'emergent property' of the complex, whose informational value is greater than what would be the sum of its parts (Tononi 2008; Tononi and Koch 2015).

The essential feature of consciousness – or rather of conscious information – is integration. It is this that distinguishes systems with consciousness from those without. Thus, for example, Tononi et al. (Northoff and Lamme 2020; Tononi

2012) give the example of the cerebellum compared to the cortex: the former carries a high density of information but given the lower sophistication of its structures it cannot integrate it. Thus, the information carried by the system of the cerebellum, spinal cord, etc., is not conscious. The information carried by the more sophisticated cortical structures, capable of integrating it, is integrated and conforms a set of information. Therefore, is conscious.

In this context, integration is a concept derived from the uniqueness of experiences (Tononi and Koch 2015). That is: what the subject consciously experiences present itself as a whole. In this sense, it is a theory that is rooted in both Kant and Leibniz. Kant argued that for there to be experience, an integration, an indissoluble link between the rational and experiential aspects of reason was necessary. That only by taking the data of experience, anchored in each space-time, and passing them through the filter of reason and judgment, could we obtain an experience. And this, in turn, was unitary and indivisible.

In the structure proposed by Leibniz, the monads are irreducible units of meaning - traditionally equated with the soul (Furth 1967). They are not atoms, since in Leibniz's terminology they have appetites, are changeable, etc. Although in Leibniz monads are indivisible, what Tononi postulates is somewhat comparable: experiences are irreducible units of meaning. They are conscious because of this inherent unity, their integration – which would disappear if we were to divide experience into parts. This is why the informational content is indeed greater in the case of an integrated complex of information: because the information itself can be divided into parts, but integration is a *property* of the combination of *those parts of information.* It therefore cannot be divided by definition. Integration is a mode of relationship between informational units, which is therefore not contained in them but derives from their combination.

An example that may be intuitive is the following. Imagine a set of fridge magnets with words written on them. We see ‘bus’, ‘couple’, ‘time’, ‘station’, ‘a lot’, ‘to go’, ‘to come’, and ‘after’. We place the words in the following way: ‘to go’, ‘station’, ‘bus’, ‘couple’, ‘to come’, ‘after’, ‘a lot’, and ‘time’. In the former situation, we had the same informational units, but they were not integrated. In the latter our consciousness, with its characteristic integrative capacity, has led us to a concrete situation, to a unitary and integrated experience: that time we went to look for our partner, when she/he came to see us at X place, after a long time without seeing each other. What we can feel or perceive – in short: what we *experience* – in this second situation is not contained in the sum of the informational units; it does not exist in them. The information is the content; the integration – and thus consciousness – is *our experience* of a certain combination of that content. This is also aligned with what Gestalt psychology proposes (Bedau and Humphreys 2008; Gibb et al. 2019; Yurchenko 2022), based on what we experience from an introspective point of view.

It is correct to assume that their proposers pass through a kind of panpsychism, but it is not to ascribe them to a usual one. They establish several conditions, for a system to be conscious, that have to be met in addition to the integrated information – which separates them from claiming that any organism or object that is capable of integrating information is conscious (Tononi and Koch 2015). This, however, has been argued extensively (Cerullo 2015; Mindt 2017). In any case, this is their proposal about how consciousness emerges; they argue for some neural basis – or correlates, given that they consider that any unified experience is related to a certain neural conformation – which will be revised later. In sum, Tononi proposes that the key factor of consciousness is its ability to integrate information and make the parts into a whole.

12.2.3 Global Neuronal Workspace Theory (GNWT)

First proposed by Baars (Baars 1988, 2005) and developed by Dehaene and colleagues (Dehaene et al. 1998, 2011; Dehaene and Changeux 2005; Mashour et al. 2020), this theory has considerably evolved since its origins. In its first version, it proposed a computational, cognitive model for explaining how consciousness could occur (Baars 1988). This was transformed by Dehaene et al. (Dehaene et al. 1998, 2003; Dehaene and Changeux 2005; Dehaene and Naccache 2001) into an explanation, adapted to brain mechanisms and with a concrete neuroanatomical correlate: they called it a 'neuronal model of a global workspace'. Firstly, they related it with 'conscious effortful tasks', those which require focused attention and conscious control; after it was simply related to conscious experiences, but always maintaining its pragmatical facet.

For Dehaene and colleagues, consciousness is quite a simple process, useful and explainable based on tasks ongoing. This model or theory establishes the presence of a global neural workspace – composed of a certain type of cortical neurons (Dehaene and Christen 2011) that are particularly highly connected– which is responsible for conscious experiences.

This neural workspace coordinates with another, less integrated and sophisticated brain system, composed of sets of neurons specialized in certain functions. These function unconsciously, as they deal with specific functions for which they contain encapsulated information. The sets of neurons belonging to the global neuronal workspace, however, are essentially characterized – as well as anatomically and topographically – by their ability to send and receive information to and from each other within their workspace. Thus, at a certain moment, certain information is 'selected', because of its importance or its relevance to the task at hand – in this sense, it is a pragmatic theory, as it remains aligned with the task the person is carrying out – and enters the neuronal workspace, where it is amplified and becomes conscious.

Thus, there is no constant stream of consciousness that exists *per se* – as IIT advocates – but rather the possibility of 'conscious access' to certain information. The information is 'accessed' once has entered the global neuronal workspace and has been amplified on it. After being selected and amplified by the global neuronal workspace it becomes 'accessible content' to other cognitive functions (Dehaene et al. 1998, 2003, 2011). From the selection moment, this information is conscious; consciousness here explained is not 'generic' nor 'phenomenal' consciousness

(Wu 2018), that is, our 'being conscious'. Rather, what Dehaene and colleagues explain is *access* consciousness: they propose a pathway through which certain information would become conscious. And actually, as Schurger and Graziano (Schurger and Graziano 2022) point out, they do not explain how this happens: they simply describe that through the 'selection' by a massively interconnected and recurrently active network (so that information travels constantly, in loops, in bottom-up and top-down directions), certain information – one that relates relevantly to an issue in which the subject is involved – becomes conscious and therefore accessible to our cognitive machinery.

On its more recent version, GNWT proposes that this recurrent and connected activity produces a 'non-linear network ignition' (Mashour et al. 2020) which amplifies and sustains a neural representation of the corresponding information – that was initially encoded by the lower-level processors commented above –, allowing it to be accessed by other cognitive systems. Though it needs to be amplified and sustained by the neuronal workspace – that does so thanks to its recurrent activation and loops, providing an electrical infrastructure with which to multiply and maintain the signal – the information was already on the processing network. By being selected by the neuronal workspace the representation of this signal is amplified extraordinarily thanks to the co-ordinated, non-linear activation of the neurons that make up the workspace – the process called 'ignition' –, making it a piece of information, by its very presence as a stimulus, accessible to the cognitive systems. The access of the representations is made competitively: there is a lot of information that at any given time can access the workspace; but depending on the circumstances, it is only one representation does so and becomes conscious.

Therefore, in GNWT's view consciousness is a certain mode to access information; it does not relate to an experience, a unified complex of information, or the relationship in virtue of which some parts are transformed into a whole.

This theory has had a lot of impact in the neuroscientific world, as well as many applications in artificial intelligence (Goyal et al. 2021; Signa et al. 2021; Van Rullen and Kanai 2021), which in fact was seconded by the theory proponents who defended the possibility of a kind of machine-consciousness (Dehaene et al. 2017). But we must bear in mind that it was *created* from computational models and not vice versa. It was a theory whose starting point was the functioning of computational networks, and it was from these that consciousness was explained.

Because of this radically computational and physicalist approach, this theory has also been criticised (Seth and Bayne 2022) or, merely, considered not as an explanation of consciousness but perhaps as a model for certain processes of its functioning (Wahbeh et al. 2022) but which leaves aside many others, more psychological but equally characteristic of it, such as the privacy of consciousness (Yurchenko 2022), the unity of conscious experiences or its self-referentiality.

12.2.4 Recurrent Processing Theory (RPT)

Proposed by Victor Lamme (Lamme 2015, 2006, 2010, 2020; Lamme and Roelfsema 2000), this theory also relies on the concept of recurrent processing as essential for the transition from unconscious to conscious stream, but in a considerably different manner to that proposed by the GNWT.

In Lamme and colleagues' view, what is necessary and sufficient for conscious experience is the recurrent processing that takes place, and this is the crucial difference, in the *sensory areas*; particularly in the primary visual cortex (V1) and its connections with the secondary visual cortex (V2) and high-level processing areas, on which he relies to explain and construct all his argumentation (Lamme 2006, 2020).

Authors divide the process of conscious identification of a visual stimulus into four stages – from the presentation of the stimulus and its being visually registered without the subject's awareness of having seen it until the subject reaches full awareness of having seen something and of what she has seen. The transition from unconscious states to conscious experiences has been studied and approached, in this theory, by the transition from unconscious to conscious vision (Lamme 2010, 2020; Lamme and Roelfsema 2000; Northoff and Lamme 2020; Seth and Bayne 2022). For which they argue that only the recurrent processing of a certain signal, relative to a visual stimulus, in sensitive areas is necessary; no transmission of the sensitive information to any other network or area is required for it to become conscious (Northoff and Lamme 2020; Wu 2018).

The involvement of higher cognitive areas such as the prefrontal cortex has no influence on the conscious experience of the stimulus; it influences its further processing or modulation (Lamme 2015, 2020; Seth and Bayne 2022; Yurchenko 2022). But the essential transition from the unconscious to the conscious occurs through the integration and organisation of perceptual information, bringing together all the factors that make up the stimulus. In this way, Lamme, in line with Tononi, also considers that the essential characteristic of consciousness is its uniqueness or 'wholeness' (Northoff and Lamme 2020), but differs from the latter in that he considers that this is already achieved in the recurrent processing in the sensory cortex. Thanks to this recurrent processing, through which information travels both 'upstream' (V1 to V2) and 'downstream' (V2 to V1), perceptual information is organized, its different qualities are linked together and the experiential 'whole' is formed. This happens without the need for any other processing area. 'Wholeness' and thus consciousness of what is being perceived is achieved, according to Lamme and colleagues, at the moment when the sensory cortex unifies all the perceived factors (hair+ movement + paws + face + . . . is consciously perceived as 'a dog'). Hence the sensory cortex, as responsible for the unifying process, is the source of conscious experiences.

In this theory, we find no reference to our self-awareness or the psychological facets of our introspective experience; is a theory focused on the transition whereby sensory stimulus transition from being unconsciously perceived as consciously experienced.

12.2.5 Synchrony Theory (ST)

The synchrony theory (Crick and Koch 1990, 2003; Engel et al. 2001; Engel and Singer 2001) proposes that the key mechanism underlying consciousness is the neural synchronization.

This theory focuses on the resolution of a specific problem, the same one that Lamme was trying to solve with his RPT: the wholeness of perceptual consciousness. Specifically, the theory of synchrony was proposed to solve the so-called 'binding problem' (Bayne and Chalmers 2003; Hardcastle 2008; Revonsuo and Newman 1999; Smythies 1994; Treisman 1996), referring to the unknown process by which one has a unified experience, from the introspective point of view, of any stimulus.

In this theory the linking of the different characteristics or qualities of an experienced object is again identified as the essential mechanism that explains consciousness, and again the example that serves as the basis for the construction of the whole theory is visual perceptual experience (Engel et al. 1992, 1999; Fries et al. 1997; Womelsdorf et al. 2007). It is known that the process of viewing an object is a divided and hierarchical process, based on the specialized activity of neuronal sets, each specialized in the perception of a single quality of the perceived stimulus (Crick and Koch 2003; Larock 2006).

Therefore, the perception of visual stimuli is organized according to a principle of functional specialization (Bartels and Zeki 2006; Livingstone and Hubel 1988; Zeki et al. 1991): each visual attribute is 'perceived' and processed by neuronal systems, anatomically independent from each other – one set of neurons processes colour, another group of neurons processes shape... This functional division and our lack of awareness of it raise the 'binding problem'. The binding problem encompasses the gap between the process of perception and what we consciously perceive. Consciously, we perceive a unified experience of seeing a table. But the fact of perceiving it is a kind of 'assembly line' between several neuronal sets; a process of which we are not consciously aware.

Why are we not aware of the process of constructing the experience? And how are the different perceptions of the concrete qualities linked – or bonded – to give rise to the unity of the visual experience of the table, if each area perceives one quality and is segregated from the others?

Advocates of the synchrony theory propose an answer to this question; the problem of the uniqueness, the wholeness, of the sensory experience; or the binding problem. In line with Lamme, they consider it to be the crucial mechanism responsible for the transition from unconscious to conscious vision. And also aligned with Lamme their proposal for visual awareness is afterward postulated as a global explanation that accounts for general consciousness.

The answer they defend is that neural synchrony – that is, neuronal paired firings when perceiving a certain feature – is the solution to the binding problem, and also the mechanism on which visual awareness relies (Engel et al. 1992, 1999, 2001; Engel and Singer 2001; Fries et al. 1997). This synchronization would be a temporal one, in the millisecond range (Engel and Singer 2001), and would be the communicatory mechanism through which different sets of neurons, perceiving differentiated aspects of reality, could be paired with each other. Thus, being a process in which, through the synchronization of form – the electrical information of each neuronal set – the synchronization of content – the specific perceptual information perceived by each anatomical group – would also be assumed.

An example might be a room with a group of radios. They would all be playing, each with different and repetitive choppy sounds. According to ST, if we programmed them all to the same frequency, we would get the message through the composite sound of all the radios: in one, only the nouns would sound; in another, the articles; in another, the adjectives; in another, the silences... And so, all being on the same frequency, if we were to close the door – because we are not aware of the assembly line of the message, as we have said – and listen from outside, we would hear only one speaker, perfectly normal and unitary, transmitting a specific message.

This theory has been broadly criticized (Bartels and Zeki 2006; Dong et al. 2008; Larock 2006; Merker 2013; Palanca and DeAngelis 2005) but the empirical results remain incomplete and inconclusive (Boly et al. 2013; Brosch et al. 1997; Castelo-Branco et al. 2000; Freiwald et al. 1995; Ward et al. 2006).

However, it is a theory of consciousness that is built *ad hoc* for resolving one of its concrete problems, so it does not refer to self-awareness, intentionality or so many other aspects that are evident and striking from the experience of consciousness.

12.2.6 Higher-Order Theories (HOT)

Generally speaking, HO theories claim that a conscious mental state is definable by the presence about it of a 'reflexive meta-mental self-awareness' (Van Gulick 2022). That is to say: for there to be consciousness about one's state, two simultaneous mental states are required. One being the state as such: nervousness about a performance, for example. The second being the awareness about the state itself: 'I know I am nervous'. The first one of them is the mental state X in which the subject finds herself – which also can occur unconsciously, that is, without the subject being aware of being in it. The second one of them is what advocates of HO theories call a 'non-inferential, higher-order state' (Van Gulick 2022).

There are different versions of HO theories we will summarize below; but all HO theorists agree on the premise that the second, higher-order mental state must be about the first, lower-order mental state (Carruthers and Gennaro 2020; Seth and Bayne 2022). The former of the states may be of any type – deductive, perceptual, volitional, intentional, etc. But the latter must concern, us for to be exposing a HO theory of consciousness, the former. It is the latter, reflexive and meta-cognitive state that is considered the 'consciousness mark': if the latter mental state appears, we are conscious about the former – for the second state of mind, which is reflective and meta-mental, consists precisely in being aware of one's own state of mind. Latter and former, first and second, are terms here used only referring to the order in which the mental states have been presented, not to anything that occurs in our perception. They are higher and lower mental states: the higher one being about the lower one.

HO proponents defend those states to be simultaneous: consciousness is thus defined as a reflexive mental state, whose content is another mental state. If we lack the reflexive, meta-mental awareness state about some mental state – that is, the higher-order mental state, whose content is another mental state in which we find ourselves, or the lower-order mental state– then we will not be aware of that mental state.

As showed herein this theory accounts for what is called 'phenomenal consciousness' (Brown et al. 2019; Carruthers and Gennaro 2020; Chalmers 1997; Nagel 1974) which is defined as the introspective characteristics of being aware of something concrete: what are the subjective qualities of the experience of drinking a hot chocolate (Northoff and Huang 2017; Northoff and Zilio 2022; Wahbeh et al. 2022). It opposes to considering 'being conscious' as the 'condition of merely being awake and alert and behaviourally responsive to external stimuli' (Brown et al. 2019).

HOT theories are characterized by two assumptions: firstly, the necessity of a second-order (or higher-order) representational mechanism for there to be a conscious experience. Since first-order mental states are not sufficient for there to be conscious experience, although they are sufficient for there to be an organismic response to them (Brown et al. 2019; Rosenthal 2005). In this sense, HOT are different from theories like RPT, and ST, which defend that some kinds of first-order states are sufficient for there to be conscious experience about them.

The second assumption is that if an organism has no inner awareness of being in a certain first-order state cannot be phenomenally conscious about its being in that first-order state. That is to say that conscious experiences rely on a high-order representation, as explained herein, of a low-level state. This assumption distinguishes HOT from cognitive theories as GNWT: there is no need of a cognitive mechanism nor network for consciousness to arise. The only requirement for having a conscious experience is that there is some kind of inner awareness of one's mental state.

This theory has many versions (Brown et al. 2019; Lau and Rosenthal 2011; Rosenthal and Weisberg 2008), but the main division could be made between the perceptual and thoughtful higher-order theories. Higher-order Perception Theory claims that the higher-order mental state is perceptual. Phenomenal consciousness relies on a second-order inner sense which perceives our first-order sensations or perceptions (Carruthers and Gennaro 2020). Those inner, second-order senses are responsible, hence, for our conscious experience of first-order perceptions; we are aware of a first-order mental state if we *reflexively perceive it*. Higher-order Thought Theory (HOTT), instead, claims for the higher-order mental state to be a *thought:* we are conscious about a first-order mental state if it is the subject of a metacognitive thought about it; if it produces a *higher-order thought* (Carruthers and Gennaro 2020; Lau and Rosenthal 2011; Rosenthal 2005; Rosenthal and Weisberg 2008).

In any case, all kinds of higher-order theory have their roots in a more subjective and philosophical conception of consciousness – roots of this theory can be found in Locke as well as in Descartes and even Kant: consciousness about us is equated with our capacity for meta-reflection, or meta-perception, or both, of our own mental state or mental existence; consciousness is the 'brain's [. . .] theory about itself' (Cleeremans et al. 2020; Wahbeh et al. 2022).

In this sense and as Brown et al. (2019) argue, HOT is a theory that not only proposes a theoretical structure about the transition from non-consciousness to consciousness but is adapted to how we live everyday experiences (such as memories, emotions, etc.), given that it is based upon introspective meta-awareness of something being experienced. It avoids explaining, however, the 'unity' or 'wholeness' of the conscious experience (Lamme 2020), as it is assumed on the phenomenal consciousness – we are aware of a state of mind because it is represented by a state of mind of a higher order, so it is already represented as a unity. But if they were simply different stages of the conscious experience? We will find later this proposal.

12.2.7 Predictive Coding Theory (PCT)

Considering the brain as a predictive processing machine (Hohwy 2013; Whyte 2019; Whyte and Smith 2021) is one of the most candent and debated theories about brain function. This approach presents the brain as a machine that generates perceptual hypotheses – based on internal models, generated by our social learning (Otten et al. 2017) – which are tested with the data obtained through the senses (Marvan and Havlík 2021).

Here appears the concept of the 'Bayesian brain' (Hohwy 2014; Marvan and Havlík 2021; Seth 2013; Whyte 2019) given that the conscious and perceptual experiences and their content are seen as a result of a statistical operation based on internal probabilistic models, sustained by the brain and that develop with similar methods that Bayesian probability does (Doya et al. 2011; Knill and Pouget 2004).

Thus, according to this theory, the key process that explains the functioning of the brain and our conscious experiences is its functioning as a probabilistic machine. It is explained by its capacity to generate, inferentially, predictive perceptual models about reality; its capacity to receive inputs about reality and insert them into the predictive model previously generated; its capacity to analyse the differences between what is predicted and what is received, known as 'prediction error'; and, finally, its capacity to propagate this prediction error through the model, accordingly, adjusting it. Though cortical processing consists, mainly, of contrasting sensory inputs with the theoretical expectations posed by the neural model; and then adjusting the model in order to minimize predictive error (Northoff and Lamme 2020; Otten et al. 2017; Seth and Bayne 2022; Spratling 2010; Wahbeh et al. 2022; Whyte and Smith 2021).

Consciousness is seen as an inferential property (Hohwy 2012, 2016); or the inferential adjustment of the predictive model (Hobson and Friston 2014, 2016; Marvan and Havlík 2021). This trial-error view is applied also to interoception in order to explain our capacity to perceive also our own internal states. In this view, emotional content responds to an inferential model based on the interoceptive signals – those would be the 'inputs', and the emotional content would be the 'theoretical model'. Thus, what we experience as self-awareness would only respond to an interoceptive predictive model able to evaluate the physical and physiological conditions and changes in the body; this 'interoceptive inference' would give rise to emotional content and self-awareness(Feldman Barrett 2017; Seth 2013).

This theory, besides being a computational one that aims to explain the brain's capacities as if it were only a computational machine, leaves aside so many aspects of conscious experience; in fact, it leaves aside, as (Marvan and Havlík 2021), the most primary of all, which is how consciousness itself appears; how the transition from the unconscious to the conscious experience takes place, why the world is experienced as it is from our subjective experience, and why is subjectively experienced at all (Clark 2019). It has been also criticized for its vagueness, its incapacity to provide a unifying theory, and for its inaccuracy in seeking the emergence of conscious experience – entirely subjective – in prominently external stimuli (Schlicht et al. 2021).

12.2.8 Embodied Theory (ET) or Embodied Cognition (EC)

From ET's point of view, what is analysed is phenomenal consciousness; what it is like to be conscious. And for the ET/EC proponents, the main issue to explain about what it is like to be conscious is the ability to answer that question: the possibility of identifying *someone* inside us that *experiences* things.

Hence the main question that ET aims to resolve is the reflexive capacity that our consciousness implies about what we experience. That is: the proposition 'I have seen the sunset' implies a self-consciousness, a reflexive point of view that turns on what has been experienced and identifies, in a metacognitive way, that something has been subjectively experienced by someone. Consciousness has, as it has been widely shown, an aspect that consists of being aware of the sunset, and another that consists of being self-aware of oneself being aware of the sunset. So, for the ET advocates, the main issue to resolve is: from where this 'I' emerges? On what this subjective dimension of our experiences, this introspective capacity, relies? (Park and Blanke 2019; Park and Tallon-Baudry 2014).

Neuroscientific approaches have, generally, a computational bias, considering the brain as the main and almost only source of consciousness; some by their information integration functions; others, by certain sets of neurons that synchronize their electrical signals; others, by the generation of a trial-and-error model; others, by the possession of an area of amplification of the representation that makes it conscious; and so on, all the theories that we have discussed. EC theory was born with the purpose of dissociating itself from this computational approach to mental processes and a simplistic approach to the origin of consciousness (Shapiro and Spaulding 2021). Is the necessary wholeness of the study of persons around which defence EC theory is built it argues that human being is a whole composed of two different realities ('corporeality', so to speak, and 'selfness', or to be a 'self'). Trying to understand persons from only one of these approaches is limiting, simplistic and impossible. According to EC theory, humans must be understood as a unity composed from different parts and studied from all their dimensions at the same time. If we have corporeality and consciousness is because both respond of each other; its combination is what explains our uniqueness (López-Ibor et al. 2011).

The main interest of the ET proponents (Clark 2001; Damasio 2005) was to consider consciousness for what it is: as a capacity of a particular mind, with a particular body, with a particular relationship to its environment (Clark 2001). Thus, ET proposes that we are human, that we have a body and brain, that we have a mind and consciousness; and that consciousness is an incomprehensible phenomenon if one tries to disentangle one of these parts from the others or to study it as a property of a kind of brain-computer. To understand consciousness, ET proponents argue, we must study ourselves as a whole: as human animals, with language, with a social community in which we are shaped, with an environment to which we relate; with a brain, yes, but also with a body; with a mind, yes, but also with ethics, with empathy, with sensitivity (Clark 2001). This account for consciousness challenges not only the neuroscientific rationalist or computational theories; it also takes a stand against rationalist philosophy, first of all, as Damasio himself points out (Damasio 2005), against its main follower, Descartes. In this sense, it is a much more Spinozian approach: what we are is explained by the whole that we form, not by one of its parts separated from the others.

Thus, the basic idea of ET or EC is that what essentially identifies our consciousness is our first-person experience (Van Elk & Blanke 2011), which derives from our first-person experience of *our bodies* (Damasio 2005). Computers have no self-awareness nor first-person experience; and we, who have self-awareness, do not have only brains. ET proponents believe that these questions, ignored in other theories that account for consciousness, are the ones that contain the answer: we must look for the answers in the structure we have – in our body, our language, our environment, which is what gives rise – phenomenologically and factually – to our unique kind of self-awareness (Shapiro and Spaulding 2021). Is *from* our body that rise our feelings; and is *from* the combination of this our body and its relationships with the environment, without feelings and emotions, that emerge cognition. As Damasio (2005) puts it: being, feeling, and knowing are three distinct and consecutive stages.

According to ET, our subjective self-awareness is created through the basis given of a referential 'neural subjective frame': a 'constantly updated neural map of the internal state of the body' (Park and Tallon-Baudry 2014). This neural map is composed of the information of our entire body; this is why we speak of the 'embodied' mind or cognition since the first-person view is constructed through this internal representation of our own body and our own anatomy (Chemero 2013; Park and Tallon-Baudry 2014).

It is through the possibility of interoception and our body map that we know positively and experientially that we are distinct from the outside; it is through the senses that information from the external world affects us, and yet it is from our interoceptive capacity that we are aware of our own body and our 'being in it'. Proponents of ET argue that this low-level interoceptive mapping, involving various brain areas and functions, is not sufficient for first-person perception, but is necessary (Park and Tallon-Baudry 2014). It is on this basis that cognition and other mechanisms rely on, and with whose combination our conscious experience and our social ability are constructed (Chemero 2016). We have a body we know as ours; though it we experience the world, we feel things and have emotions; and on all that cognition is constructed (Damasio 2005) and consciousness is achieved.

We will therefore speak of 'embodied cognition', as the first step is knowing our own body and our interoception. A map of that is configured, from which we relate to the outside: so, we apprehend ourselves as distinct of the outside. It is from this distinction from the environment that self-awareness emerges (Baggs and Chemero 2021; Chemero 2013; Thompson and Varela 2001).

Is through this map that we distinguish ourselves from the outside. Thus is this process that opens the possibility of considering an 'inside'. Though it would be the source of our self-consciousness: our body is felt as something distinct from the environment, something that is *sensed*, while the environment is observed or perceived. And so, self-awareness would have arisen from our ability to feel our own body, while the rest of the reality cannot be felt but observed (or listened, or perceived). Our body is the source of our self-awareness; and therefore, our cognition must be comprehended as an *embodied* cognition.

The subjective neural map that our mind constructs from our body experiences must also be responsible for our homeostatic changes related to alertness and vigilance. As it is born from our differentiation from the environment, and those changes are necessary in order to make us able to respond to the environmental variations. Thus, the authors suggest, this model of subjective neural representation could 'link state and content consciousness' (i.e., between consciousness as a generalised state of response and alertness, and consciousness as a state of alertness) (Park and Tallon-Baudry 2014). ET defenders suggest that it is this perceptual map from which the first person is configured that is the mechanism underlying self-consciousness and the emotions we experience. And it is this first material, not the other way around, that underlies cognition: emotions and bodily experiences are the basis of our decisions, our self-awareness, and our cognition (Damasio 2005).

It is a conception of consciousness and cognition that considers not only the body but also the surrounding world. Since it is our distinction from it that allows us to identify a 'self' within ourselves, it is the external world that shapes self-awareness; and our body, as the only way to experience the environment, is therefore the basis of the idea of self-awareness, knowledge, and emotions (Barsalou 1999, 2008; Chemero 2013, 2016).

This theory is particularly interesting because of its consideration of self-awareness, or the meta-perception of our own presence within ourselves, as the key to understanding consciousness (Chemero 2016); and it is also particularly interesting because a necessary, not contingent, step in developing that self-awareness is the external world and its stimuli.

Consciousness is not a consequence in this theory – be it of a network, of the integration of information, or the binding of features of perceptual experience. Consciousness is the essential characteristic of a being that develops socially with its environment, and therefore needs to be clearly distinguished from it.

12.2.9 Attention Schema Theory (AST)

The AST was developed by Graziano and colleagues (Graziano 2019, 2020; Graziano and Kastner 2011a, b; Webb and Graziano 2015; Wilterson et al. 2020). In this theory we find bases shared with embodied cognition; also, in Graziano's opinion, it all starts with the meta-representation of our body and our anatomy (Graziano and Botvinick 2002). This is the first thing we have to learn, as children, to relate to the world. Our body teaches us that there is a difference between the external world and us; that we are able to feel our interiority, while we are not able to feel what is going on inside the 'mind' of a plant, a fish or our parents, siblings, or friends. This is why our body schema is the first experience of self-awareness we have. Graziano, therefore, shares with the advocates of embodied cognition the need to look first and foremost at our bodily structure and the mental schema we have of it.

There is considerable evidence for the brain existence of this 'body schema'(Berlucchi and Aglioti 1997, 2010; Cardinali et al. n.d.; Graziano and Botvinick 2002; Hoffmann et al. 2010; Maravita and Iriki 2004), which involves very different brain areas and systems: it integrates information from all the systems that deal with the relationship of the embodied agent with its environment (i.e. information from visual, auditory, tactile, proprioceptive, vestibular systems and everything that has to do with the body relating to its environment) (Berlucchi and Aglioti 2010; Hoffmann et al. 2010; Maravita and Iriki 2004). From the integrated information of all these systems, what is called the 'body schema' is constructed, consisting of a mental model, constantly updated, that represents the state of our body at any given moment, which allows us to evaluate the situation in which we find ourselves-both interoceptive and externally; whether we are clear or whether we are too close to someone, whom we will hit if we move too fast. But, in addition to the present situation, it allows us to plan future situations: from an evaluation of our interior and the present state we can, for example, plan a future movement – a mental act of well-known brain representation, so much so that we have an area, the premotor area, named after it.

On this basis, Graziano makes his proposal. The brain must construct a mental model, what we call a body schema, to know its current position and state, as well as those of its environment. This is crucial for the development of any activity, however tiny it may be, in relation to an external environment: we must know where our arms, feet, and head are. If we were running through the countryside and an embankment appeared in front of us, our body schema, and the ability of our brain to adjust it would serve us to calculate at what approximate distance it is, whether we have time to brake or whether it is better to turn to the left because our arm is long enough to reach the branch of that tree. In sum, the body schema is essential for a being that lives in a certain environment to which she relates.

Just as our body is essential for relating to our environment, so is our ability to pay attention or not to pay attention to something. Attention is the ability to direct the mind towards a particular stimulus among others is a process 'by which signals compete for the brain's limited computing resources' (Webb and Graziano 2015). The process of attention is, hence, selective: in order to pay attention to some stimulus there must exist a suppression of all the others we may be perceiving (Beck and Kastner 2009; Desimone and Duncan 1995). This suppression of non-relevant stimuli and the subsequent selection of relevant stimuli is something our brain is constantly doing. It is a process that occurs in both directions in which we relate to the environment: both in the bottom-up direction (stronger stimuli tend, at least intuitively, to attract our attention more); and in the top-down direction (our mind chooses one for its relevance and suppresses, or tries to suppress, the others) (Beck and Kastner 2009; Webb and Graziano 2015).

Both the body schema and attention are essential for us to relate effectively to our environment. We have a body schema for reasons of simplicity and processing efficiency (Webb and Graziano 2015). The body schema allows us to integrate all the information that our sensory and proprioceptive systems provide us with about the environment and our inner self because it is much simpler to use a model in which all this is integrated than to waste conscious resources on knowing all the processes that precede the integration of all this information. As an example, Webb and Graziano (2015) ask us to consider the visualization of a colour. The act of seeing a colour involves a number of processes (reflection of light, activation of retinal neurons, travel along the optic nerve, decussation of images, and a long etcetera) of which we are not aware. Only the 'red' experience reaches our consciousness, as this is the only thing, we need to know consciously in order to act accordingly.

Graziano proposes that the same thing happens with attention: we have an internal model of how our attention works, the 'attention schema', analogous in structure and functioning to the 'body schema'. This attention schema is what we subjectively perceive as self-awareness.

Thus, according to Graziano and colleagues, what is actually happening is not any awareness of ourselves *per se*, but rather our brain building a model of its own top-down regulation, that is, of its ability to select one stimulus as relevant and suppress others. Thus, the brain does not invest processing resources in integrating all the processes involved in attention (reception of various stimuli, selection of one, suppression of others, maintaining secondary attention to the environment, etc.). It has a model of its own attention that is simpler and more efficient since the results of the attention process are already integrated into it: self-awareness. With it, we can more than sufficiently deal with the world, without needing to resort to all the processes that underlie its construction – in the same way that to deal with colours we do not need to be aware of all the processes that have led to our being able to see something as red.

However, we know somehow how it is possible for the brain to construct a corporeal scheme, and that is that we have a body. We have a material body endowed with nerves and senses; that it ends at certain points where the external world begins, and if we are healthy, we are able to feel and identify it; that the nerves that run through our body send electrical signals that actually reach our brain, which is able to integrate and represent them; and so forth. However, this is not the case with attention. We neither feel it, nor does it relate to the exterior, nor does it begin or end... So, how is it that we can construct a model of our attention? We construct it with the very same machinery with which we do it every day: with our social perception machinery (Graziano 2013, 2019; Graziano and Kastner 2011a, b; Webb and Graziano 2015; Wilterson et al. 2020).

The attention schema is primarily used to interact with other human beings (Graziano and Kastner 2011b; Webb and Graziano 2015; Wilterson et al. 2020). The way we interact with other humans is based on analyse their behaviour – and their linguistic messages, but those would be included in their behaviour –, to construct an attentional model based on it and to predict their behaviour basing our conclusions on the model we just constructed. To all this information, integrated into the attentional model of person X, is to which we adapt our behaviour in social contexts (Graziano 2013, 2019; Guterstam et al. 2019; Kelly et al. 2014; Pesquita et al. 2016; Wilterson et al. 2020).

From this point of view, the attention schema (seen from inside the 'self-awareness') would serve not one but two important evolutionary functions: firstly, to reduce the brain's resources involved in attention and attention control thanks to a schema in which the important information is integrated and included, and which serves effectively to our everyday lives; secondly, to construct available attentional models of others' minds with which predict their behaviour and adapt ours.

This theory, although again computationally focused and in that sense limited – which has been widely criticised; see (Graziano 2020) for a review of the criticisms with a response from AST advocates – gives an essential role in the development of self-awareness to both attention and the social environment. These two factors have been shown to be closely related to self-awareness, but not to fairness. It is possible to dissociate attention and awareness – it is possible to pay attention to a stimulus without conscious experience of it (Hsieh et al. 2011; Jiang et al. 2006; Lin and Murray 2015; Travers et al. 2018; Webb et al. 2016a, b; Wilterson et al. 2020). And also, has been shown an overlapping between the brain regions involved with self-awareness and theory of mind and those involved in social cognition and evaluating others' behaviour or attention (Guterstam et al. 2019; Igelström et al. 2016; Igelström and Graziano 2017; Kelly et al. 2014; Mitchell 2008; Webb et al. 2016a, b).

A posterior overview of brain regions will be made, but we would like to highlight the approximation to more integrative theories, with attention to the influence the environment and the social community may have on our shaping as beings with self-awareness.

12.2.10 Temporo-Spatial Theory of Consciousness (TTC)

The TTC has been recently developed by Northoff and colleagues (Northoff 2014, 2018; Northoff 2013, 2014, 2018, 2021; Northoff and Huang 2017; Northoff and Lamme 2020; Northoff and Zilio 2022). It aims to be a sort of integrative theory concerning other theories shown herein (Frewen et al. 2020; Northoff et al. 2020; Northoff and Lamme 2020).

According to TTC, there are four different consciousness dimensions, each one corresponding to a different neural activity and a subjective experience (Northoff and Huang 2017):

- State consciousness: consciousness of us generally considered. Corresponds to the spontaneous activity of our brains, thus the task-free or resting state paradigms.
- Content consciousness: consciousness related to its certain contents at a certain moment. Corresponds to the pre-stimulus state.
- Phenomenal or experienced consciousness: related to the subjective experience of a certain situation. Corresponds to the post-stimulus non-report paradigms.
- Cognitive processing consciousness: related to the subjective experience of that certain situation when reported to others. Corresponds to the post-stimulus report paradigms.

A distinction must be made between the report and no-report paradigms, because as recently discovered, the neural markers or signals we trace with neuroimaging are different when the subject has been asked to report her subjective experience and when has been not (Block 2019; Whyte et al. 2022).

The essential proposal of TTC is that the key underlying consciousness is the temporospatial dynamics of the brain, in a Kantian sense: time and space, being the primary conditions of any experience, are the ones on which consciousness is sustained and by which the different conscious states named above are bonded. Those four states are related to consciousness: the key question is how they bind with each other to configure subjective (or phenomenal) consciousness; and how is related this phenomenal consciousness to the brain's activity.

The stimulus-related states (both pre- and post-stimulus, report, and no-report) relate to consciousness to the extent that we can assess whether we have been aware of the stimulus. The pre-stimulus consciousness state shapes what we perceive when exposed to some stimulus (Northoff and Lamme 2020; Northoff and Zilio 2022). For example, I was waiting for my mother to come by car, but she finally came by bicycle; so, I did not notice she had arrived until she spoke to me. In this sense what is expected to happen shapes what we perceive and how we do so. In post-stimulus consciousness states, it is concerning the stimulus that consciousness is configured –

the experience relates to that stimulus. Finally, the resting, no-task, state consciousness relates to the conscious activity the brain spontaneously generates, when is not involved with any stimulus (Northoff and Huang 2017; Northoff and Lamme 2020).

The question TTC aims to answer is how these states are unified and integrated for the conscious experience, or phenomenal consciousness, to arise. The empirical data are varied and non-conclusive (Northoff et al. 2020; Northoff and Lamme 2020). It has been found through functional magnetic resonance imaging (fMRI) and magnetoencephalography (MEG) exams contradictory results. On one hand, it has been found that the spontaneous brain's dynamics and activity were correlated with the 'degree of non-additivity during pre-post-stimulus interaction' (Northoff and Lamme 2020; Aru et al. 2019; Huang et al. 2017; Wainio-Theberge et al. 2020; Wolff et al. 2019). But on the other hand, it has also been found that this spontaneous brain's activity and its posterior shaping of the conscious reception of the stimulus is related mainly to self-consciousness and self-reflection (Huang et al. 2016a, b; Kolvoort et al. 2020; Scalabrini et al. 2017; Wolff et al. 2019) but also with activities as diverse as social tactile interaction (Scalabrini et al. 2019) or complex cognitive processes (Bongers et al. 2020).

Facing these diverse neuroimaging conclusions, Northoff and Lamme (2020) propose the possibility of this division being unreal from the functional point of view – artificially imposed by the needs of the empirical study of consciousness. That division may be only theoretical, thus not truly reflecting the brain's activity (Northoff 2014, 2018). So TTC, in an attempt to circumvent this potentially unrealistic and functionally false division, proposes to link these four states and try to generate a comprehensive and integrated view of the phenomenon of consciousness both when it is related to a stimulus and when it is not.

This is the core proposal: these four consciousness states are not four but one, and this one phenomenon occurs both at the neural basis (the brain) and the phenomenal experience (subjective consciousness). They link each other through what TTC defends as the key concepts for understanding consciousness: the spatiotemporal brain's frame, i.e., not concerning external time and space but the spatiotemporal coordinates the brain generates itself and in which frames its activity (Northoff 2013, 2021; Northoff and Huang 2017; Northoff and Lamme 2020; Northoff and Zilio 2022).

As said herein, this is a prominently Kantian approach: time and space are the necessary categories for perceptual and conscious experience to arise; we talk of the spatial topography and the temporal dynamics of the brain. It is from there that our subjective experience of consciousness – being it level, phenomenal, access, or form consciousness – is constructed. In the frame by them constructed is that both spontaneous and stimulus-related activities are held.

These basic categories are the structure on which consciousness relies, and that explains the possibility of having conscious experience of both the stimulus-related states and the spontaneous, resting states' activity of the brain. But the key feature of these as being the crucial explanation of consciousness is that they are shared not only by the stimulus and resting states; they are also shared by the neural and phenomenal aspects of consciousness, that is, by the brain and subjective consciousness (Kolvoort et al. 2020; Northoff et al. 2020; Northoff and Huang 2017; Northoff and Lamme 2020; Northoff and Zilio 2022). As long as they are shared, they would bridge the abyss from the neural to the phenomenal experience: the brain has its own temporospatial frames in which both the neural activity and the phenomenological experience are held and developed (Northoff and Zilio 2022).

It is according to these shared times and spaces that both aspects of consciousness – neural basis and phenomenological experience – are held and bonded.

Advocates of TTC have always tried to offer a theory in which all the others could be integrated. But as some have argued (Schurger and Graziano 2022) they offer mostly a *law* about consciousness' working or arising, not an explanation of how this occurs.

12.2.11 Conclusion

As shown above, there is a plethora of theories about consciousness, or aiming to explain consciousness, or trying to find its necessary and sufficient factors, and trying to correlate all this with some neural regions – and we have only overviewed some of them; for reviews with alternative approaches or wider revisions see (Gamboa 2022; Lake 2022; Northoff and Lamme 2020; Rahimian 2022; Seth and Bayne 2022).

If there is no accordance yet, some think is because there are some false assumptions generally accepted within the field or because of a generalized inaccuracy (Rahimian 2022); others because there is no equilibrate research between different approaches – namely, physicalist and non-physicalist accounts for consciousness (Seth and Bayne 2022); because of a too narrow point of view and strict materialistic paradigms (Lake 2022). In sum, almost as much disagreement is found among the answers on why there is no agreement as on the theories among which there is no agreement.

We find ourselves aligned with a more integrative, ecological, and social approach to consciousness.

We will now turn very briefly to a review of the brain regions that each theory proposes – on which we find some further agreement. In this sense we will also discuss below the Default Mode Network, a brain network that seems to be related to processes of self-awareness, as well as general or state consciousness (the consciousness of the outside, we could call it); and even social awareness, related to others and our social relations. In this sense it is a network that comprises what we think about how consciousness could work.

12.3 Brain and Consciousness: An Overview

The first thing to put under view is what brain regions are considered crucial for consciousness arising, in the line stablished by the research of the NCC (Koch et al. 2016). Once postulating a key function for consciousness and settling on its basis a whole theory, the next required step in neuroscience is to test it experimentally. Theories that account for consciousness have been not an exception. In this sense, we can find three main types of theories:

- Cognitive theories: these claim the prefrontal cortex – in sum, the frontal regions of the brain – to be the crucial anatomic area for consciousness' rising.
- Integrative theories: these claim for a 'posterior hot zone', in which is included the somatosensory cortex, is sufficient for consciousness. This hot zone will include also the parietal and temporal cortices as crucial for integration.
- Processing theories: these will be unspecific in defending a particular area as necessary and sufficient for consciousness since their basic postulate is that it arises from some processing or recurrence of processing, a process that takes place between several brain regions.

Finally, we will comment on the Default Mode Network (DMN from now on), its relationship with conscious processing and self-reflection, and, strikingly, also with social cognition and relationship. It has been lately generating considerable interest, with which we are aligned. The DMN is the brain basis that we find most interesting to the consciousness extent, and which can provide the most clues about our conscious processing, its reasons, and way of functioning (although it is like trying to infer the nut only through knowing the spanner).

12.3.1 Cognitive and Integrative Theories: Frontal vs Posterior Regions

Depending on what is considered the crucial feature or function of consciousness, a certain brain region will be considered prominent and responsible for consciousness. The first and rough division must be made, therefore, between the defenders of frontal vs posterior areas (Boly et al. 2017; Northoff and Lamme 2020; Odegaard et al. 2017; Overgaard 2018; Storm et al. 2017).

Traditionally there was agreement that the prefrontal cortex was essential for consciousness, which is now being replaced by the posterior hot zone (Andersen et al. 2016; Boly et al. 2017; Frässle et al. 2014; Overgaard 2018; Railo et al. 2011). This shift is supported, among others, by studies of patients who, after bilateral frontal lobectomy (Boly et al. 2017), maintain consciousness – in both awakeness and self-awareness senses.

12.3.1.1 Frontal Regions: Cognitive Theories of Consciousness

As the theories defending a prefrontal source of consciousness, we find GNWT and HOT (Northoff and Lamme 2020); for another perspective, see (Wu 2018).

GNWT postulates, as previously shown, an ignition process through which a certain signal is amplified, so that the information contained in it becomes accessible to higher cognitive centres. It is therefore not exactly a cognitive theory of consciousness, since the only thing that involves the higher cognitive centres is access to the information once it has already been adapted for access to consciousness by the global neural workspace.

The GNW itself involves large brain regions (inferior parietal cortex, mid-temporal cortex, precuneus and anterior cingulate, as well as their thalamic connections) (Aru et al. 2019, 2020; Northoff and Lamme 2020; Sanchez-Vives et al. 2021) but the area with the most central role in the ignition process discussed above, and therefore essential for the conversion of certain information from non-conscious to conscious, is the prefrontal cortex (DLPFC) (Bellet et al. 2022; Dehaene et al. 2003, 2014, 2017; Dehaene and Christen 2011; Mashour et al. 2020; Northoff and Lamme 2020; Sergent and Dehaene 2004).

HOT offers a similar view, whether based on different arguments. Its main theoretical defence, as shown herein, is that a sensory representation, whether integrated or not, is not enough to generate the emergence of a conscious experience of it. In order for a stimulus to be consciously perceived, on the one hand, the perception, or the stimulus, and simultaneously a higher-order thought whose object was precisely the perception, or the lower-level state, was required.

In this sense, higher-order thoughts are similar – not identical – to metacognitive thoughts about states that are occurring. And therefore most versions of the HOT argue that the prefrontal cortex is the necessary and sufficient basis for consciousness, because even if there is full and complete perception in the posterior areas of the brain, there will be no awareness of it without the prefrontal cortex, and no awareness of it without the prefrontal cortex (Afrasiabi et al. 2021; Brown et al. 2019; Lau and Rosenthal 2011; Northoff and Lamme 2020; Rosenthal and Weisberg 2008).

They also considered the thalamo-cortical connections and subcortical regions are also considered essential, at least as prerequisites, for consciousness to arise (Aru et al. 2019, 2020; Li et al. 2021; Sanchez-Vives et al. 2021).

12.3.1.2 Posterior Regions: Integrative Accounts for Consciousness

The theories that clearly claim a posterior source of consciousness are IIT, RPT, and ST.

Integration Information Theory

From the IIT's point of view, the regions that are claimed to be essential are covered by the complex known as 'posterior hot zone' (Koch et al. 2016; Northoff and Lamme 2020; Tononi 2012; Tononi et al. 2016; Tononi and Koch 2015) including parietal, temporal and occipital areas. This is so because IIT argues that consciousness occurs from the moment that certain information changes from being a set of juxtaposed units of information, unrelated to each other, to an integrated whole in which the parts are related, bonded, and integrated with each other, achieving a greater amount of information than the sum of its parts. For in the integrated complex we must include, to the sum of information given by the separated parts, the information of their relationship and the sense composed by this relationship. Furthermore, the IIT bases its entire argument on the conjugation and union of units of visual information. The whole theory is based on the transition between the data collected by different areas of the vision process towards the perceived totality of the image, a step that occurs when all the units of information are integrated, and which implies that the image becomes conscious.

Therefore, from the IIT point of view, the necessary and sufficient region for consciousness will be the one in which all this information is integrated, which they argue is in the so-called 'posterior hot zone' – which includes the primary and secondary visual areas in the occipital area, where visual information is indeed received and then integrated. This posterior hot zone involves other sensory areas, but particularly the parietal cortices, in charge of integrating other kinds of information. This would give them not the highest value of Phi – the amount of integrated information, thus the 'consciousness level' – but the sufficiently high one to support consciousness. This account of consciousness avoids the involvement of frontal regions, given that they are not defending the area with the highest phi – case in which, as Northoff and Lamme (2020) point out, higher phi values would indeed be reached – but the one with *sufficient* phi to give rise to consciousness (Northoff and Lamme 2020; Tononi et al. 2016; Tononi and Koch 2015).

Recurrent Processing Theory and Synchrony Theory

A similar defence comes by the hands of the proponents of RPT theory (Lamme 2006, 2015, 2020; Lamme and Roelfsema 2000). The RPT has been developed in a way that is even more dependent on the conscious vision process (Lamme 2006, 2015; Lamme and Roelfsema 2000), making it even more limited in its defence of NCC. Given that RPT theorises almost exclusively on the transition from unconscious to conscious vision, the areas they consider sufficient by themselves to give rise to conscious experience are the visual integrative areas. Those are located on the occipital lobe, within the 'posterior hot zone' previously named. Thus for the RPT proponents, the visual cortices are sufficient for consciousness to arise, as long as there is recurrent activity on them – that is, whenever signals are sent in the different visual areas in both directions, both from V1 to V2, V4, etc., and from V4 to V2 and V1 (Fahrenfort et al. 2007, 2012; Hurme et al. 2017; Lamme et al. 1998, 2001; Northoff and Lamme 2020; Wokke et al. 2012). The prefrontal areas are neither needed for subjective conscious, having at best a modulatory role.

Lastly, ST is also in this group, more aligned with RPT. From the ST's point of view consciousness also arises from the synchrony achieved between signals from different neurons – as discussed above, they argue that it is synchrony in the form of the signal that leads in turn to synchrony in the content. They also argue that the temporal synchrony and organisation between neural signals necessary for conscious experiences is achieved in the posterior areas of the brain (Engel et al. 2001; Engel and Singer 2001).

12.3.2 The Whole Brain Consciousness: Processing Accounts for Consciousness

In this section, we find the least of the theories, but it is a pathway that slowly gains momentum. Are included the AST, the TTC, and the empirical findings that defend the DMN area as the anatomical basis for consciousness.

From TTC's view, the neural sufficient basis for consciousness should include not only the neural structures that appear to be directly connected with conscious experience but also the neural prerequisites for it. This would involve all structures indirectly related to the emergence of consciousness (Northoff 2021; Northoff et al. 2020; Northoff and Huang 2017; Northoff and Zilio 2022). That would be all the structures that need to be intact for the ones directly related with consciousness (Northoff and Lamme 2020), which has been also contrasted empirically (Afrasiabi et al. 2021).

In the AST's view, for consciousness to appear it is necessary for the brain to construct a functional model of its own process of attention. In such are included the attentional processes – therefore the attentional selection of the relevant stimulus, both on the bottom-up and top-down directions – and the model generation. This involves the attentional networks and, presumably, the DMN (Ding et al. 2019; Huang et al. 2020; Kleckner et al. 2017; Yang et al. 2022). As will be shown below, the DMN is in fact the large system that interconnects both the perceptions of social and self-consciousness. It is related to social thoughts, judgments, and analysis of other people; to self-conception, internal narrative, motives, and meanings; and, finally, the relationship with others – that is, not only their analysis but how we connect with them. So, the DMN system is consistent with the core AST's proposal: that our conception of ourselves directly relies on our social connections, conceptions, and perceptions.

12.4 The Land of the Self, the Land of the You: The Default Mode Network, the Land of Us

12.4.1 What Is the Default Mode Network

The Default Mode Network is a brain system that has recently attracted enormous interest (Buckner et al. 2008; DeSerisy et al. 2021; Mancuso et al. 2022; Raichle 2015; Simony et al. 2016; Smallwood et al. 2021; Xu et al. 2016). This was because it was the system that was found to be activated when the subjects were not involved in the performance of any task. That is: it was, and so it was so named, the system that the subject's brain, when left alone and undisturbed, was set in motion by default (Buckner et al. 2008; Buckner and Carroll 2007; Mazoyer et al. 2001; Raichle et al. 2001; Raichle and Snyder 2007; Shine and Breakspear 2018). It was thus seen to be effectively detached from task performance and thus from attention directed outward. But gradually, however, it began to be seen not only as the system engaged with the 'freethinking' or default brain's functioning, but the system related to self-awareness. This is the reason for the interest since it was brought to light (for a historical review, see (Buckner et al. 2008)): that the brain has a system, when it is not occupied with external tasks, for thinking about ourselves from many different perspectives; remembering our past or planning our future being just some of the possibilities of this system, namely, the DMN (Allen et al. 2014; Buckner et al. 2008; Buckner and Carroll 2007; Raichle and Snyder 2007).

12.4.2 Anatomy of the DMN

Anatomically, the DMN is a complex system involving different regions which generate diverse functional subsystems, also having a number of interacting hubs. Broadly, the DMN involves particularly the medial prefrontal cortex (mPFC) – concretely, the ventromedial and dorsomedial prefrontal cortices (vmPFC and dmPFC, respectively) –, the posterior cingulate cortex (PCC) and precuneus (Prec), the inferior parietal lobule (IPL), and the bilateral temporoparietal junction (TPJ). Besides those, also the hippocampal formation seems to be involved in the network with less robust data, either directly or indirectly – through its functional connections to the regions known to be part of it (Buckner et al. 2008). Thus, also the medial temporal lobe would be included in the DMN. The same happens with the lateral temporal cortex, which connections are also less robust than the first ones but consistent on the different analysis (Buckner et al. 2008; Konu et al. 2020; Li et al. 2021; Medea et al. 2018; Raichle 2015; Simony et al. 2016; Smallwood et al. 2016, 2021; Smith et al. 2021; Sormaz et al. 2018; Stawarczyk et al. 2021; Van Kesteren et al. 2013; Xu et al. 2016; Yeshurun et al. 2021).

From the early days of the DMN's fame, its role was strongly investigated. We will base on the wondrous review of the DMN's functions made by Yeshurun et al. (2021) for our explanation.

12.4.3 The DMN and the Conscious Self

The DMN has now been extensively proven to be related to self-awareness, self-conception and self-directed thoughts (Buckner and Carroll 2007; Buckner and DiNicola 2019; Li et al. 2021; Qin and Northoff 2011; Raichle 2015; Raichle and Snyder 2007), integration of emotional, self-referential, temporal and spatial information (Lanzoni et al. 2020; Smith et al. 2018, 2021) – necessary to compose an accurate scene of the proper situation –, autobiographical memory retrieval (Addis et al. 2007; Diana et al. 2007; Schacter et al. 2007; Smith et al. 2021; Vilberg and Rugg 2012), contextual knowledge and context or memory-guided cognition (Aminoff et al. 2007; Bar 2007, 2009; Murphy et al. 2018; Ranganath and Ritchey 2012; Sormaz et al. 2018; Vatansever et al. 2017; Yeshurun et al. 2021), future-oriented thought (Schacter et al. 2007; Smallwood et al. 2021; Xu et al. 2016) and development of concrete personal goals (Medea et al. 2018), spontaneous thought (Smallwood et al. 2016) and mind-wandering (Christoff et al. 2009), scene and context representation (Baldassano et al. 2017, 2018; Hasson et al. 2015; Vodrahalli et al. 2018).

Thus, the DMN is related to every mental activity that can be included under the umbrella of 'self-awareness'. In this sense is a system worth studying for understanding its real implications for consciousness generation, especially self-consciousness. In fact, it has been also related to consciousness in its generic sense. It is known that only the DMN does not suffice to produce conscious experience (Achard et al. 2012; Bodien et al. 2019; Demertzi et al. 2015; Li et al. 2021; Norton et al. 2012). But it has been also proven that its dynamic functional connections and operational synchrony – including subcortical connections – directly relate to consciousness, posing thus the DMN as one of the neural correlates of consciousness and self-consciousness (Bodien et al. 2017; Edlow et al. 2021; Hutchison et al. 2013; Li et al. 2021).

But all those are not the unique functions of this astonishing discovery of neuroimaging. This information alone is impressive, given that there is indeed a brain system dedicated to self-awareness; indeed, it has long been regarded as an 'intrinsic system' (Yeshurun et al. 2021), as it seemed to relate only to our self-consciousness and self-projection. However, there are yet two whole facets of it that have not been discussed: its facet as a sense-making system, and its social projections. Crucially, in the same ways that this system can make sense of us as a person or self, is so to make sense of reality.

12.4.4 The DMN as a Sense-Making Network

It is the brain's system that makes us able to find a meaning in a story, an event, or, widely, of our life and our world; to find it in others, in our community, or in social relationships. Strikingly – or not striking at all according to AST – the system in charge of our self-awareness, our goals, our memories, our context-guided cognition, and our mind-wandering. In sum, the system in charge of being 'I', is also in charge of social cognition and social relationships; is also in charge of 'us'. We will review these functions shortly; but we must bear in mind that the DMN is sort of the 'sense system', which makes us able to find it inside of us, inside of others, and in our surrounding world.

We first comment on the DMN as the system in charge of making sense of the world (Yeshurun et al. 2021) or the stimulus receipt. It has been shown that DMN responds to the *senses* of what it receives from reality; it is able to identify the content, despite the form under which it appears. The results that give rise to this hypothesis come from inter-subject studies: in these studies, the functional parallelisms are calculated across all the participants' brains, instead of within each participant's brain (Hu and Yang 2021; Simony et al. 2016; Yeshurun et al. 2021).

By conducting narrative experiments – subjecting the participants to the identification of a narrative meaning, whether in a story told or read, a film, etc. – it was found that the DMN had certain connectivity patterns dependent on the sense of the story receipt. That is: those patterns appeared, crucially, only when an 'intact, coherent and temporally extended spoken story' was presented (Yeshurun et al. 2021). They did not appear, instead, neither when the participants were not subjected to the interpretation of a story – that is, they were in a state of rest –, nor when they were presented with disjointed fragments, without order or concert, haphazard or incoherent pieces of stories; or only random words (Simony et al. 2016; Yeshurun et al. 2021).

Another type of experiment supported these results. Participants were presented with a number of narratives, always the same ones, in a variety of formats. Strikingly, was discovered that, although there were substantial differences in the low-level perceptual areas, these did not change the meaning perceived by the participants; and entailed similar inter-subject temporal responses within the DMN when the same narrative was presented (Simony et al. 2016; Yeshurun et al. 2021). Some of the formats were: written text and spoken narrated stories (Regev et al. 2013; Wilson et al. 2018; Yeshurun et al. 2021); different levels of representation, such as verbal description or script lectures versus viewing the movies (Baldassano et al. 2017; Tikka et al. 2018; Vodrahalli et al. 2018; Yeshurun et al. 2021; Zadbood et al. 2017); telling the same story but using different synonyms (Yeshurun et al. 2017a, 2021); tell the same story in different languages to native speakers of each one (Dehghani et al. 2017; Honey et al. 2012; Yeshurun et al. 2021); listening the story versus reading it (Deniz et al. 2019); and different levels of abstraction (Nguyen et al. 2019; Yeshurun et al. 2021).

Yet, the inter-subject activation patterns of the DMN remained correlatable and with similar temporal patterns of response, showing thus a meaning-dependent signal. There were, however, substantial and expected differences in the low-level brain systems involved in perception, which confirmed that the DMN did not show differences because the meaning was always the same despite the format was not.

There has been also proven an integrating role of the DMN, as commented previously, not only with respect to our own reality – for example, making predictions based on our memorized experiences – but regarding the external world. The DMN is engaged, as said above, with contextual knowledge, for which is necessary the ability to integrate the information received through the senses (Margulies et al. 2016). But it can be done also the other way around – not integrating sensorial information and then composing the contextual scene but having the sense and looking for its physical correlates. Some hubs of the DMN have been proven able to represent word concepts through the integration of sensory and motor data, giving therefore weight to the embodied cognition argument (Fernandino et al. 2016; Lanzoni et al. 2020).

Another experiment along this line consisted of telling all participants the same story, with a priori the same possible interpretation. However, for half of the participants, a word crucial to the meaning of the story was replaced by its opposite: 'for example, "laughing" by "sobbing"' (Fernandino et al. 2016; Yeshurun et al. 2021). This change resulted in two very different stories, which led to large different neural responses in accordance with the vast difference in the meaning (Fernandino et al. 2016), whereas it entailed small changes in the early auditory areas – as expected since the change of a single word produces a very similar auditory impression between the original and the modified story. These results are consistent with the DMN as the system in charge of sense comprehension since the change in the heard was very small, while the change in the understood was vast (Yeshurun et al. 2021).

The DMN's capacity of making sense, or finding meaning in narratives, stories, or events in the real world or in us, denotes its constant creation or identification of schemas or narratives with order and meaning. This ability is in fact necessary for our life to unfold over time, for our personal development, and, above all, to *integrate* within a concrete reality and a certain community. It is also consistent with the other roles that have been proven of the DMN to have – contextual knowledge, imaging the future and retrieving memories, making predictions, memory guided-cognition, etc. Indeed, it has been suggested by neuroimaging studies that the creation and adjusting of those mental models or schemas rely on the functional connections between regions either that compose the DMN or are closely related to it (precuneus, bilateral temporoparietal junction and the hippocampal region (Baldassano et al. 2018; Gilboa and Marlatte 2017; Yeshurun et al. 2021).

In one study, two meaningful situations or schemas were presented to participants. It resulted in two patterns of DMN's activation: one homogeneously distributed among participants who had listened to one of the stories, and the other distributed among the other group of participants (Baldassano et al. 2018), leading then to the conclusion that each neural response was effectively associated to each one of the meanings. But this was not the only prove of DMN's implication with the meaning and the mental schemas: it was found to be involved when the unfolding of events in reality contradicted the schema that the person had constructed, or the expectations they had about what was going to happen (e.g. an abrupt and unexpected change in the plot of a film) (Brandman et al. 2021; Dohmatob et al. 2020; Yeshurun et al. 2021).

Therefore the DMN seems to be in charge of identifying the sense of current situation, of traducing it to a mental schema which includes predictions about the unfolding of that situation, and of adjusting it in accordance with reality; thus, it seems to be the brain's system that makes us able to compose mental schemas while understand and process narratives, finding on them a kind of common thread, significance and meaning (Lee et al. 2020). And in the same way the DMN does with reality does with us and with others: it finds meanings, composes mental schemas with a certain narrative and updates it as life unfolds.

Consequently, it was also found that prior beliefs about the development of a situation modulated the DMN's responses to it, given that the previous beliefs about an event substantially change our interpretation of it. Two groups of participants in one study were told the same story, which was that late at night, a husband calls a friend asking if he knows where his wife is. However, each group had been conditioned with different contextual information before hearing the story. Thus, one group – with a similar pattern of neural responses within the group, but different from the other group – reported believing that the husband was paranoid and jealous. While the other group – with brain responses similar to each other within the group, but different from the other – reported believing that the wife was being unfaithful. Thus, the activation pattern of the DMN was strongly aligned within each group, in which the participants shared the belief; while it differed substantially from the activation pattern of the other group, with which they differed on the interpretation (Yeshurun et al. 2017b, 2021).

These findings suggest the DMN's capacity of apprehending the sense through integrating it within a corpus of previous beliefs. This process is conscious but temporally large: it requires many minutes to unfold, and thus to maintain the mental schema active, constantly present on the conscious stream and in a constant process of readjusting. This has been supported by other findings, observed in similar situations (Yeshurun et al. 2021), such as: the change and manipulation of the perspective from which history is received (Bacha-Trams et al. 2017; Lahnakoski et al. 2014) – and, crucially, the synchronous brain activity was correspondent with similar interpretations and psychologies (Lahnakoski et al. 2014) –; the manipulation of the focus of the story (Cooper et al. 2011); the ability of interpreting a situation, through the comprehension of the context, as an ironic one (Uchiyama et al. 2012); or the change in the interpretation of the situation according to the intentions attributed to it (Bašnáková et al. 2014; Koster-Hale and Saxe 2013).

12.4.5 The DMN and the Shared Life, the Social Being

Given all this, the last aspect of the DMN, also surprising, which we have not yet commented on, has become evident: its relationship with the community and with the social context. Not only the DMN's responses are conditioned and shaped by our autobiographical memories and prior beliefs – which makes them idiosyncratic within the individual self and different across different individuals. It has been also proven that the areas responsible for the internal representation of emotions, intentions or thoughts of others are also part of this network, which is an essential component in the generation of personal relationships (Arioli et al. 2021; Iacoboni 2011). But the crucial finding is that DMN's responses are consistent and parallelisable between individuals with the same beliefs or with the same responses to situations, whereas they are different between individuals who interpret situations differently. Thus, the DMN is also a kind of community-forming system: alignment occurs between neural response patterns – a correlation of the form - between individuals in whom there is a parallelism in psychological interpretation or meaning-making – that is, a correlation of the content (Yeshurun et al. 2021).

It was found also a similar pattern of neural response through individuals that interpreted on the same way an ambiguous animation in which a delicate and complex social narrative was represented through abstract forms (Nguyen et al. 2019; Yeshurun et al. 2021). While the pattern of neural response differed when comparing individuals with a different interpretation of the given situation – thus the similarity of the neural pattern underlined the similarity of the psychological interpretation.

This crucial finding has been supported by others, strengthening the idea that DMN responses are more similar the more aligned the psychological interpretations or sense-making of the situation are (Yeshurun et al. 2021). This alignment has been found in similar interpretations and emotional responses to spoken storytelling (Saalasti et al. 2019; Smirnov et al. 2019), in similar senses of humour (Jääskeläinen et al. 2016), in similar sexual desires when seeing an erotic movie (Chen et al. 2020), similar decision-making when confronting a moral dilemma (Bacha-Trams et al. 2017; Tei et al. 2019; Van Baar et al. 2019) and, lastly, with personality, demographic or cultural alignments – which also means that different interpretations and diverse activation patterns have been linked with differences on these factors (Finn et al. 2020; Nummenmaa et al. 2018; Yeshurun et al. 2021).

This great body of results – and there are thousands of them that we are not able to revise on this chapter – suggests that the DMN is indeed a high-level processing system that is concerned with the establishment of a narrative: it is an 'active sense-making network' (Yeshurun et al. 2021). It is not only concerned with the unfolding of real-life events but also turned inwards, thus presumably allowing us the aspects of consciousness that are of most interest – our self-identification with a meaning, with a character, with desires, with a past, with an idea of the future; in short, self-consciousness. As we have seen, however, it is also capable of turning towards others. People with the same interpretations of things have even similar activation patterns of the DMN, resulting not only in a network of sense-making but also a community-creation, around common senses, system. But how is this supposed to happen? How is it possible to have similar neural patterns relying upon coupled psychological interpretations?

In their incredible work, Yeshurun et al. (2021) propose an interpretation with which we are aligned. They propose the question is answered by walking the pathway the other way around; is the dynamic interaction with our community, therefore our adaption to the senses we are presented to, which shapes our DMN patterns of response – not excluding that more physical factors, such as genetics, could be also involved. But their proposal is that the essential shaping of the neural responses within our DMN is made by dynamic social interaction. It has been indeed empirically shown. It can occur in two directions: the sender's brain's activations shape the receiver's, or backward (Friston et al. 2020; Hasson et al. 2012; Hasson and Frith 2016).

This correlates with the process named 'brain-brain coupling'. In a crucial study, in which this property was christened, it was found that the neural activation pattern of a person telling a story was coupled by the brains of the people listening to the story (Stephens et al. 2010). What was found was that early auditory areas were synchronically coupled between the speaker and the listeners, showing only the acoustic properties of the stimulus. But the DMN and the language areas of the listeners coupled the activation pattern of the speaker with a certain delay with respect to the speaker. This finding suggested a causal relationship between the pattern neural activation of the speaker and the subsequent but correlatable pattern neural activation of the listeners. This mechanism, called brain-brain coupling and supported by DMN, is what underlies, according to the authors, successful communication processes (Stephens et al. 2010; Yeshurun et al. 2021).

This has been followed by other crucial discoveries along the same line. For example, in order to check whether the brain-brain coupling was indeed due to the sense of the narrative or whether it was only to its acoustic properties, it was tried to insert into the communication process, from time to time, single words without any meaning or relation to what was being narrated. The result confirmed expectations, as the synchronisation of primary auditory areas remained almost intact, while the pairing in DMN activation patterns was disrupted (Silbert et al. 2014; Stephens et al. 2010).

The same occurred when the story was told in a language the listener did not know (Stephens et al. 2010) or when the stimuli were scrambled (Nguyen et al. 2019). Furthermore, the more coupled the DMN patterns of activation, the more comprehension achieved between the speaker and the listener and the more successful the communication process was considered (AbdulSabur et al. 2014; Dikker et al. 2014; Heidlmayr et al. 2020; Kuhlen et al. 2012; Liu et al. 2017; Nguyen et al. 2019; Silbert et al. 2014; Stephens et al. 2010; Yeshurun et al. 2021).

Finally, the brain-brain coupling has been also observed during non-verbal communication, if comprehension between the sender and the receiver is successfully achieved (Anders et al. 2011; Schippers et al. 2009, 2010). It has been also shown that it is possible to change the interpretation some group gives to an ambiguous movie, entailing changes also in the DMN's activation patterns, through conversation. A movie was projected to a group of participants; similar DMN's responses were found across participants coming from similar social networks. After the first visualization of this ambiguous movie, participants with different responses were told to gather in conversational groups, with the aim of reaching an agreement and consensus on the interpretation of the film. After the consensus conversation, they watched the film a second time. Impressively, the constructive and consensual conversation changed the participants' interpretation, and with it, their neural responses (Sievers et al. 2020; Yeshurun et al. 2021).

Lastly, it has been also found neuroscientific prove that the neural structures and alignments within the DMN – therefore our interpretation of our context and our lives – is shaped by the people we consider close to us, the people we choose to keep in our lives and with whom we find ourselves connected in our daily lives. It has been observed that the neural responses of a group of students were more parallelisable within people with closer relationships, while less correlatable among people who had a less close relationship (Hyon et al. 2020; Parkinson et al. 2018).

There are several striking results regarding this DMN's coupling, sense-dependent, among different subjects' brains, which unfortunately we cannot review in this chapter, given the length of the topic. But we believe we have made our point: there is a narrative brain system, capable of finding senses, of understanding them, of filing them in an even larger network of senses obtained and stored through memory, previous experiences, and expectations about the concrete situation, of making predictions on the basis of this general network or scheme of senses. Which, in addition, oversees finding meanings in relation to our inner self, to our character, and which therefore and presumably may be responsible for that inner narrative we call 'identity'.

12.5 Conclusion

It is this same system that is responsible for making sense of the reality that surrounds us. Which, even if a priori it has none, is an essential activity for our own meaning since it allows us to integrate the aforementioned reality into our particular path. It is also the ability underlying our capacity to integrate or not into an environment, or, presumably, to choose whether an environment suits us in relation to what we think about the world and what we want to make of our lives and ourselves.

And it turns out, excitingly, that this same network, perhaps responsible for making us meaningful beings, couples with the corresponding network of other people's brains, people whose senses are similar or twinned with ours. That is: this network couples with the networks of those who are psychologically compatible with us, with whom we share psychological interpretations of the world.

Regarding finally with the theories herein reviewed, the theory that comes closest to these neurological data is the attention schema theory, which in fact proposes that our idea of identity is a brain model of our attention derived from our capacity to model other's intentions and desires basing ourselves on their attention direction.

However, it is also known that attention does not explain, but *modulates*, the integration within the DMN (Yeshurun et al. 2021). It has been shown that more shocking or surprising stimuli, which attract attention more powerfully, provoke greater shared responses across participants, always within the DMN – in regions such as the medial prefrontal cortex (Schmälzle et al. 2015). Additionally, when comparing neurological response patterns among listeners with different levels of attention, those who were more alert and attentive to the stimulus showed greater shared activation patterns than those who were less attentive (Cohen and Parra 2016; Ki et al. 2016; Yeshurun et al. 2021). In one study a comparison between two comparable stimuli was made to contrast if, effectively, the DMN was modulated by attention (Regev et al. 2013, 2019; Yeshurun et al. 2021). Two unrelated narratives were presented to participants at the same time, each in a different modality – one narrated, the other written. In each trial, participants' attention was biased toward one of the two formats. Notably, the processing of the stimulus toward which the attention has been biased was found to occur within the DMN and dorsal attention networks, also entailing a conscious perception of it by the subject. Instead, the non-attended stimulus processing was found to be constrained to early, primary sensory cortices, entailing also a subject unaware of the narrative she has not attended to.

These results suggest, indeed, a modulation of the DMN response by attention. However, studies have been carried out with the aim of quantifying the degree of coupling between the brains of people with similar psychological positions when watching a film. The result is that it is a complex, large coupling, and involves multidimensional spatial aligned patterns, for which only the dorsal attention networks cannot account (Chen et al. 2017).

All these data confirm some assumptions of some other theories, such as the process of ignition and amplification postulated by GNWT (since, as mentioned above, it is the process of attention that seems to enable the transition of stimulus processing from the primary sensory areas to the DMN, where the stimulus accesses consciousness and is understood in terms of its meaning or sense) or the integration process proposed by IIT (since when the stimulus accesses the DMN, it is integrated into our schema of the world in relation also to our memory and to our future predictions). But on our view, the theory that comes closest to explaining the processes taking place in the DMN is indeed the AST. Although it does not seem that subjectivity and self-consciousness are a model of attention at all. Attention, in relation to DMN, selects the stimulus according to how striking it is; but it is DMN that then, independently of attention, underlies its processing it in terms of sense and integrates it into a model of meaning that is not only about our attention, but about the surrounding world. Furthermore, as shown, attention cannot explain only by itself the coupling process that is proven to occur across people with the same psychological approach to reality, the same interpretation about narratives or events, the greater shared brain responses when emotional responses are parallel… In sum, it offers a view of our self-consciousness detached from emotions, from the past, from the future, from the way we connect with others or not, from the way we integrate into the world or not.

AST proposes a solution to the process by which we build a model of ourselves. But it is not a model of us that DMN shows; DMN shows that there is a brain system of a narrative and interpretative nature, a network of *sense-making*, not models, of the world. A model is a representation; a sense is an *intentional* representation, a model *directed* towards something and *understood* on the basis of something. And it is this something that escapes the AST, and yet seems essential and intrinsic to the DMN. Essentially, it is this meaning around which the model is constructed that separates some psychological interpretations from others. Both the people who believed that the wife in the previous example had been unfaithful and those who believed that the husband was paranoid had constructed a model of reality; that is precisely where they were *similar*, not distinct. What differentiated the two groups in a completely essential way is the *meaning*, the *interpretation*, around which that model had been constructed; and that, it has been shown, is precisely what gives rise to the differences or similarities in terms of DMN activity. This exceeds the account the AST proposes for self-consciousness.

Attention can therefore be a process of selection and model building. But the DMN is a system that deals with the construction of models based on meanings; and it happens to be the one that also deals with our self-consciousness, our history, and our personality. This is obviously a powerful line for future investigations. It accounts for many of the processes that we normally identify with our self-consciousness and those that generate the most mystery. It is the system that underlies our self-awareness as identities, as people with a history and a future; that is activated when we think freely, when we design future goals or when we interpret the world around us and integrate ourselves into it. It is not already clear what are the reasons underlying these capacities, but the DMN seems to be the responsible system, and that offers cues about its nature – the coupling with others and with the context, its sense-making nature, its relationship with meaning and narratives (linguistic and non-linguistic).

In sum, we can search for an evolutionary role for self-consciousness we are not able to confirm; but we can also look around us and see what is that identity brings out to us. This sense-making ability makes us narrative beings, therefore *creative* beings: only the ability to know who we are gives us the possibility of being something else. The same applies to reality: the capacity to identify narratives within reality, to understand the reasons in virtue of which events occur, gives us the possibility to comprehend, predict and change our surrounding world.

Lastly, humans are social beings; to have a community is an essential need of human beings, and it has been shown that it is also the fact that seems to shape our way of confronting the world and understanding ourselves. It is the social tissue what shapes us, and which conforms all these abilities; therefore, what shapes us as narrative and interpretative beings.

The ability, therefore, to identify meanings and meanings in our world; to plan future goals in the light of past events; to construct a narrative of our own and the world; and, above all, the ability to share this with others – to the extent that even the brain patterns of those with whom we understand each other align with our own – opens infinite possibilities. For it opens the possibility not only of changing our world according to its meanings and senses, or our own; it opens the possibility of building communities around common meanings. Which – as we said in the beginning of this long journey –, it may be worth remembering, was contained in the Greek word from which 'conscious' seems to derive, *súnoida*: 'to share in the knowledge of'.

References

AbdulSabur, N.Y., Y. Xu, S. Liu, H.M. Chow, M. Baxter, J. Carson, and A.R. Braun. 2014. Neural correlates and network connectivity underlying narrative production and comprehension: A combined fMRI and PET study. *Cortex; a Journal Devoted to the Study of the Nervous System and Behavior* 57: 107–127. https://doi.org/10.1016/J.CORTEX.2014.01.017.

Achard, S., C. Delon-Martin, P.E. Vértes, F. Renard, M. Schenck, F. Schneider, C. Heinrich, S. Kremer, and E.T. Bullmore. 2012. Hubs of brain functional networks are radically reorganized in comatose patients. *Proceedings of the National Academy of Sciences of the United States of America* 109 (50): 20608–20613. https://doi.org/10.1073/pnas.1208933109.

Addis, D.R., A.T. Wong, and D.L. Schacter. 2007. Remembering the past and imagining the future: Common and distinct neural substrates during event construction and elaboration. *Neuropsychologia* 45 (7): 1363–1377. https://doi.org/10.1016/J.NEUROPSYCHOLOGIA.2006.10.016.

Afrasiabi, M., M.J. Redinbaugh, J.M. Phillips, N.A. Kambi, S. Mohanta, A. Raz, A.M. Haun, and Y.B. Saalmann. 2021. Consciousness depends on integration between parietal cortex, striatum, and thalamus. *Cell Systems 12* (4): 363–373.e11. https://doi.org/10.1016/J.CELS.2021.02.003.

Allen, E.A., E. Damaraju, S.M. Plis, E.B. Erhardt, T. Eichele, and V.D. Calhoun. 2014. Tracking whole-brain connectivity dynamics in the resting state. *Cerebral Cortex (New York, N.Y.: 1991) 24* (3): 663–676. https://doi.org/10.1093/CERCOR/BHS352.

Aminoff, E., N. Gronau, and M. Bar. 2007. The parahippocampal cortex mediates spatial and nonspatial associations. *Cerebral Cortex* 17 (7): 1493–1503. https://doi.org/10.1093/CERCOR/BHL078.

Anders, S., J. Heinzle, N. Weiskopf, T. Ethofer, and J.D. Haynes. 2011. Flow of affective information between communicating brains. *NeuroImage* 54 (1): 439–446. https://doi.org/10.1016/J.NEUROIMAGE.2010.07.004.

Andersen, L.M., M.N. Pedersen, K. Sandberg, and M. Overgaard. 2016. Occipital MEG activity in the early time range. *Cerebral Cortex (New York, N.Y.: 1991) 26* (6): 2677–2688. https://doi.org/10.1093/CERCOR/BHV108.

Antonakou, E.I., and L.C. Triarhou. 2017. Soul, butterfly, Mythological Nymph: Psyche in Philosophy and neuroscience. *Arquivos de Neuro-Psiquiatria* 75 (3): 176–179.

Arioli, M., Z. Cattaneo, E. Ricciardi, and N. Canessa. 2021. Overlapping and specific neural correlates for empathizing, affective mentalizing, and cognitive mentalizing: A coordinate-based meta-analytic study. *Human Brain Mapping* 42 (14): 4777–4804. https://doi.org/10.1002/hbm.25570.

Aristotle. 1986. *De Anima (On The soul)*. (H. Lawson-Tancred, Trans.). Penguin Classics.

Aru, J., M. Suzuki, R. Rutiku, M.E. Larkum, and T. Bachmann. 2019. Coupling the state and contents of consciousness. *Frontiers in Systems Neuroscience* 13: 43. https://doi.org/10.3389/FNSYS.2019.00043/BIBTEX.

Aru, J., M. Suzuki, and M.E. Larkum. 2020. Cellular mechanisms of conscious processing. *Trends in Cognitive Sciences* 24 (10): 814–825. https://doi.org/10.1016/j.tics.2020.07.006.

Autenrieth, G., R.P. Keep, and I. Flagg. 2007. *A Homeric dictionary for schools and colleges. Based upon the German of Dr. Georg Autenrieth*. Kessinger Publishing.

Baars, B.J. 1988. *A cognitive theory of consciousness*. https://philpapers.org/rec/BAAACT.

———. 2005. Global workspace theory of consciousness: Toward a cognitive neuroscience of human experience. *Progress in Brain Research* 150: 45–53. https://doi.org/10.1016/S0079-6123(05)50004-9.

Bacha-Trams, M., E. Glerean, R. Dunbar, J.M. Lahnakoski, E. Ryyppö, M. Sams, and I.P. Jääskeläinen. 2017. Differential inter-subject correlation of brain activity when kinship is a variable in moral dilemma. *Scientific Reports* 7 (1): 1–16. https://doi.org/10.1038/s41598-017-14323-x.

Baggs, E., and A. Chemero. 2021. Radical embodiment in two directions. *Synthese* 198 (s9): 2175–2190. https://doi.org/10.1007/s11229-018-02020-9.

Baldassano, C., J. Chen, A. Zadbood, J.W. Pillow, U. Hasson, and K.A. Norman. 2017. Discovering event structure in continuous narrative perception and memory. *Neuron 95* (3): 709–721. e5. https://doi.org/10.1016/j.neuron.2017.06.041.

Baldassano, C., U. Hasson, and K.A. Norman. 2018. Representation of real-world event schemas during narrative perception. *Journal of Neuroscience* 38 (45): 9689–9699. https://doi.org/10.1523/JNEUROSCI.0251-18.2018.

Bar, M. 2007. The proactive brain: Using analogies and associations to generate predictions. *Trends in Cognitive Sciences* 11 (7): 280–289. https://doi.org/10.1016/j.tics.2007.05.005.

———. 2009. The proactive brain: Memory for predictions. *Philosophical Transactions of the Royal Society B: Biological Sciences* 364 (1521): 1235. https://doi.org/10.1098/RSTB.2008.0310.

Barone, D.F., J.E. Maddux, and C.R. Snyder. 1997. *Social cognitive psychology: History and current domains*. New York: Springer Science+Business Media, LLC.

Barrett, L.F. 2017. The theory of constructed emotion: An active inference account of interoception and categorization. *Social Cognitive and Affective Neuroscience* 12 (1): 1–23. https://doi.org/10.1093/scan/nsw154.

Barsalou, L.W. 1999. Perceptual symbol systems. *Behavioral and Brain Sciences* 22 (4): 577–609. https://doi.org/10.1017/S0140525X99002149.

———. 2008. Grounded cognition. *Annual Review of Psychology* 59: 617–645. https://doi.org/10.1146/annurev.psych.59.103006.093639.

Bartels, A., and S. Zeki. 2006. The temporal order of binding visual attributes. *Vision Research* 46 (14): 2280–2286. https://doi.org/10.1016/j.visres.2005.11.017.

Bašňáková, J., K. Weber, K.M. Petersson, J. van Berkum, and P. Hagoort. 2014. Beyond the language given: The neural correlates of inferring speaker meaning. *Cerebral Cortex (New York, N.Y.: 1991)* 24 (10): 2572–2578. https://doi.org/10.1093/CERCOR/BHT112.

Bayne, T., and D.J. Chalmers. 2003. What is the unity of consciousness? *The Unity of Consciousness: Binding, Integration, and Dissociation*. https://doi.org/10.1093/ACPROF:OSO/9780198508571.003.0002.

Beck, D.M., and S. Kastner. 2009. Top-down and bottom-up mechanisms in biasing competition in the human brain. *Vision Research* 49 (10): 1154–1165. https://doi.org/10.1016/J.VISRES.2008.07.012.

Bedau, M., & Humphreys, P. (2008). *Emergence: Contemporary readings in philosophy and science*. https://doi.org/10.7551/MITPRESS/9780262026215.001.0001.

Bellet, J., M. Gay, A. Dwarakanath, B. Jarraya, T. van Kerkoerle, S. Dehaene, and T.I. Panagiotaropoulos. 2022. Decoding rapidly presented visual stimuli from prefrontal ensembles without report nor post-perceptual processing. *Neuroscience of Consciousness* 2022 (1): 1–12. https://doi.org/10.1093/nc/niac005.

Berlucchi, G., and S. Aglioti. 1997. The body in the brain: Neural bases of corporeal awareness. *Trends in Neurosciences* 20 (12): 560–564. https://doi.org/10.1016/S0166-2236(97)01136-3.

Berlucchi, G., and S.M. Aglioti. 2010. The body in the brain revisited. *Experimental Brain Research* 200 (1): 25–35. https://doi.org/10.1007/s00221-009-1970-7.

Berryman, S. 2022. *Ancient atomism, Stanford Encyclopedia of Philosophy*. Stanford University. Retrieved from: https://plato.stanford.edu/entries/atomism-ancient/.

Block, N. 2019. What is wrong with the no-report paradigm and how to fix it. *Trends in Cognitive Sciences* 23 (12): 1003–1013. https://doi.org/10.1016/j.tics.2019.10.001.

Bodien, Y.G., C. Chatelle, and B.L. Edlow. 2017. Functional networks in disorders of Consciousness. *Seminars in Neurology* 37 (5): 485–502. https://doi.org/10.1055/s-0037-1607310.

Bodien, Y.G., Z.D. Threlkeld, and B.L. Edlow. 2019. Default mode network dynamics in covert consciousness. *Cortex* 119: 571–574. https://doi.org/10.1016/j.cortex.2019.01.014.

Boly, M. 2011. Measuring the fading consciousness in the human brain. *Current Opinion in Neurology* 24 (4): 394–400. https://doi.org/10.1097/WCO.0B013E328347DA94.

Boly, M., A.K. Seth, M. Wilke, P. Ingmundson, B. Baars, S. Laureys, D.B. Edelman, and N. Tsuchiya. 2013. Consciousness in humans and non-human animals: Recent advances and future directions. *Frontiers in Psychology 4*: 1–20. https://doi.org/10.3389/fpsyg.2013.00625.

Boly, M., M. Massimini, N. Tsuchiya, B.R. Postle, C. Koch, and G. Tononi. 2017. Are the neural correlates of consciousness in the front or in the back of the cerebral cortex? Clinical and neuroimaging evidence. *Journal of Neuroscience* 37 (40): 9603–9613. https://doi.org/10.1523/JNEUROSCI.3218-16.2017.

Bongers, A., A.B. Flynn, and G. Northoff. 2020. Is learning scale-free? Chemistry learning increases EEG fractal power and changes the power law exponent. *Neuroscience Research* 156: 165–177. https://doi.org/10.1016/j.neures.2019.10.011.

Braat, M., et al. 2020. The rise and fall of behaviourism: The narrative and the numbers. *History of Psychology* 23 (3): 252–280. Retrieved from: https://doi.org/10.1037/hop0000146.

Braisby, N., and A. Gellatly. 2012. *Cognitive psychology*. Oxford: Oxford University Press.

Brandman, T., R. Malach, and E. Simony. 2021. The surprising role of the default mode network in naturalistic perception. *Communications Biology* 4 (1) https://doi.org/10.1038/s42003-020-01602-z.

Bremmer, J.N. 2016, March 7. *Psychē*. Oxford Classical Dictionary. Retrieved from: https://oxfordre.com/classics/display/10.1093/acrefore/9780199381135.001.0001/acrefore-9780199381135-e-5407;jsessionid=DDFBF36A39D5078F6FEF916B83EB3EBE.

Brosch, M., R. Bauer, and R. Eckhorn. 1997. Stimulus-dependent modulations of correlated high-frequency oscillations in cat visual cortex. *Cerebral Cortex* 7 (1): 70–76. https://doi.org/10.1093/cercor/7.1.70.

Brown, R., H. Lau, and J.E. LeDoux. 2019. Understanding the higher-order approach to Consciousness. *Trends in Cognitive Sciences* 23 (9): 754–768. https://doi.org/10.1016/J.TICS.2019.06.009.

Buckner, R.L., and D.C. Carroll. 2007. Self-projection and the brain. *Trends in Cognitive Sciences* 11 (2): 49–57. https://doi.org/10.1016/J.TICS.2006.11.004.

Buckner, R.L., and L.M. DiNicola. 2019. The brain's default network: Updated anatomy, physiology and evolving insights. *Nature Reviews Neuroscience* 20 (10): 593–608. https://doi.org/10.1038/s41583-019-0212-7.

Buckner, R.L., J.R. Andrews-Hanna, and D.L. Schacter. 2008. The brain's default network: Anatomy, function, and relevance to disease. *Annals of the New York Academy of Sciences* 1124 (1): 1–38. https://doi.org/10.1196/annals.1440.011.

Cardinali, L., A.E. Claudio, B. Ae, and A. Farnè. n.d. *Peripersonal space and body schema: Two labels for the same concept?* https://doi.org/10.1007/s10548-009-0092-7.

Carhart-Harris, R.L. 2018. The entropic brain – Revisited. *Neuropharmacology* 142: 167–178. https://doi.org/10.1016/J.NEUROPHARM.2018.03.010.

Carhart-Harris, R.L., R. Leech, P.J. Hellyer, M. Shanahan, A. Feilding, E. Tagliazucchi, D.R. Chialvo, and D. Nutt. 2014. The entropic brain: A theory of conscious states informed by neuroimaging research with psychedelic drugs. *Frontiers in Human Neuroscience 8*: 20. https://doi.org/10.3389/FNHUM.2014.00020/BIBTEX.

Carruthers, P., and R. Gennaro. 2020. *Higher-order theories of consciousness*. Stanford Encyclopedia of Philosophy. Retrieved from https://plato.stanford.edu/entries/consciousness-higher/.

Castelo-Branco, M., R. Goebel, S. Neuenschwander, and W. Singer. 2000. Neural synchrony correlates with surface segregation rules. *Nature* 405 (6787): 685–689. https://doi.org/10.1038/35015079.

Cerullo, M.A. 2015. *The problem with Phi: A critique of integrated information theory*. https://doi.org/10.1371/journal.pcbi.1004286.

Chalmers, D.J. 1997. *The conscious mind*. New York: Oxford University Press.

Chemero, A. 2013. Radical embodied cognitive science. *Review of General Psychology* 17 (2): 145–150. https://doi.org/10.1037/a0032923.

———. 2016. Sensorimotor empathy. *Journal of Consciousness Studies* 23 (5–6): 138–152.

Chen, J., Y.C. Leong, C.J. Honey, C.H. Yong, K.A. Norman, U. Hasson, and N.N. Author. 2017. Shared memories reveal shared structure in neural activity across individuals HHS public access author manuscript. *Nature Neuroscience* 20 (1): 115–125. https://doi.org/10.1038/nn.4450.

Chen, P.H.A., E. Jolly, J.H. Cheong, and L.J. Chang. 2020. Intersubject representational similarity analysis reveals individual variations in affective experience when watching erotic movies. *NeuroImage* 216. https://doi.org/10.1016/J.NEUROIMAGE.2020.116851.

Christoff, K., A.M. Gordon, J. Smallwood, R. Smith, and J.W. Schooler. 2009. Experience sampling during fMRI reveals default network and executive system contributions to mind wandering. *Proceedings of the National Academy of Sciences of the United States of America* 106 (21): 8719–8724. https://doi.org/10.1073/PNAS.0900234106.

Clark, A. 2001. *Being there: Putting brain, body, and world together again*. Cambridge, Massachusetts, US: The MIT Press.

———. 2019. Consciousness as generative entanglement. *Journal of Philosophy* 116 (12): 645–662. https://doi.org/10.5840/jphil20191161241.

Cleeremans, A., D. Achoui, A. Beauny, L. Keuninckx, J.R. Martin, S. Muñoz-Moldes, L. Vuillaume, and A. de Heering. 2020. Learning to be conscious. *Trends in Cognitive Sciences* 24 (2): 112–123. https://doi.org/10.1016/J.TICS.2019.11.011.

Cohen, S.S., and L.C. Parra. 2016. Memorable Audiovisual narratives synchronize sensory and supramodal neural responses. *ENeuro* 3 (6) https://doi.org/10.1523/ENEURO.0203-16.2016.

Cooper, E.A., U. Hasson, and S.L. Small. 2011. Interpretation-mediated changes in neural activity during language comprehension. *NeuroImage* 55 (3): 1314–1323. https://doi.org/10.1016/J.NEUROIMAGE.2011.01.003.

Crick, F., and C. Koch. 1990. *Towards a neurobiological theory of consciousness*. https://resolver.caltech.edu/CaltechAUTHORS:20130816-103136937.

———. 1998. Consciousness and neuroscience. *Cerebral Cortex (New York, N.Y.: 1991)* 8 (2): 97–107. https://doi.org/10.1093/CERCOR/8.2.97.

———. 2003. A framework for consciousness. *Nature Neuroscience* 6 (2): 119–126. https://doi.org/10.1038/nn0203-119.

Crivellato, E., and D. Ribatti. 2007. Soul, mind, brain: Greek philosophy and the birth of neuroscience. *Brain Research Bulletin* 71 (4): 327–336.

Damasio, A.R. 2005. *Descartes' error: Emotion, reason, and the human brain*. London, UK: Penguin Publishing Group.

Dehaene, S., and J.P. Changeux. 2005. Ongoing spontaneous activity controls access to Consciousness: A neuronal model for inattentional blindness. *PLoS Biology* 3 (5): e141. https://doi.org/10.1371/JOURNAL.PBIO.0030141.

Dehaene, S., J.P. Changeux, L. Naccache, Jérôme, S., and S. Claire. 2006. Conscious, preconscious, and subliminal processing: A testable taxonomy. *Trends in Cognitive Sciences* 10 (5): 204–211. https://doi.org/10.1016/j.tics.2006.03.007.

Dehaene, S., and Y. Christen. 2011. Characterizing consciousness: From cognition to the clinic? *Research and Perspectives in Neurosciences* 18: 55–84. https://doi.org/10.1007/978-3-642-18015-6.
Dehaene, S., and L. Naccache. 2001. Towards a cognitive neuroscience of consciousness: Basic evidence and a workspace framework. *Cognition* 79 (1–2): 1–37. https://doi.org/10.1016/S0010-0277(00)00123-2.
Dehaene, S., M. Kerszberg, and J.P. Changeux. 1998. A neuronal model of a global workspace in effortful cognitive tasks. *Proceedings of the National Academy of Sciences of the United States of America* 95 (24): 14529–14534. https://doi.org/10.1073/pnas.95.24.1452.
Dehaene, S., C. Sergent, and J.P. Changeux. 2003. A neuronal network model linking subjective reports and objective physiological data during conscious perception. *Proceedings of the National Academy of Sciences of the United States of America* 100 (14): 8520–8525. https://doi.org/10.1073/pnas.1332574100.
Dehaene, S., J.P. Changeux, and L. Naccache. 2011. The global neuronal workspace model of conscious access: From neuronal architectures to clinical applications. *Research and Perspectives in Neurosciences* 18: 55–84. https://doi.org/10.1007/978-3-642-18015-6_4/COVER.
Dehaene, S., L. Charles, J.R. King, and S. Marti. 2014. Toward a computational theory of conscious processing. *Current Opinion in Neurobiology* 25: 76–84. https://doi.org/10.1016/J.CONB.2013.12.005.
Dehaene, S., H. Lau, and S. Kouider. 2017. What is consciousness, and could machines have it? *Science* 358 (6362): 486–492. https://doi.org/10.1126/SCIENCE.AAN8871.
Dehghani, M., R. Boghrati, K. Man, J. Hoover, S.I. Gimbel, A. Vaswani, J.D. Zevin, M.H. Immordino-Yang, A.S. Gordon, A. Damasio, and J.T. Kaplan. 2017. Decoding the neural representation of story meanings across languages. *Human Brain Mapping* 38 (12): 6096–6106. https://doi.org/10.1002/HBM.23814.
del Pin, S.H., Z. Skóra, K. Sandberg, M. Overgaard, and M. Wierzchoń. 2021. Comparing theories of consciousness: Why it matters and how to do it. *Neuroscience of Consciousness* 2021 (2): 1–8. https://doi.org/10.1093/nc/niab019.
Demertzi, A., G. Antonopoulos, L. Heine, H.U. Voss, J.S. Crone, C. de Los Angeles, M.A. Bahri, C. di Perri, A. Vanhaudenhuyse, V. Charland-Verville, M. Kronbichler, E. Trinka, C. Phillips, F. Gomez, L. Tshibanda, A. Soddu, N.D. Schiff, S. Whitfield-Gabrieli, and S. Laureys. 2015. Intrinsic functional connectivity differentiates minimally conscious from unresponsive patients. *Brain* 138 (9): 2619–2631. https://doi.org/10.1093/BRAIN/AWV169.
Deniz, F., A.O. Nunez-Elizalde, A.G. Huth, and J.L. Gallant. 2019. The representation of semantic information across human cerebral cortex during listening versus reading is invariant to stimulus modality. *Journal of Neuroscience* 39 (39): 7722–7736. https://doi.org/10.1523/JNEUROSCI.0675-19.2019.
Dennett, D.C., and P. Weiner. 2007. *Consciousness explained.* Little, Brown and Company: New York etc.
Descartes, R. 1644/1911. *The principles of philosophy* (trans: Haldane, E., and Ross, G.) Cambridge: Cambridge University Press.
Descartes, R. 1970. *The philosophical works of Descartes.* Edited by G.R.T. Ross (trans: E.S. Haldane). Cambridge: Cambridge University Press.
———. 2019. *Principles of philosophy* (trans: Veitch, J.) Lightning Source UK Ltd.
Descartes, R. 1641/1993. In *Meditations on first philosophy*, ed. Donald A. Cress. Indianapolis, Indiana: Hackett Publishing Co, Inc.
DeSerisy, M., B. Ramphal, D. Pagliaccio, E. Raffanello, G. Tau, R. Marsh, J. Posner, and A.E. Margolis. 2021. Frontoparietal and default mode network connectivity varies with age and intelligence. *Developmental Cognitive Neuroscience* 48: 100928. https://doi.org/10.1016/j.dcn.2021.100928.
Desimone, R., and J. Duncan. 1995. Neural mechanisms of selective visual attention. *Annual Review of Neuroscience* 18: 193–222. https://doi.org/10.1146/ANNUREV.NE.18.030195.001205.

Diana, R.A., A.P. Yonelinas, and C. Ranganath. 2007. Imaging recollection and familiarity in the medial temporal lobe: A three-component model. *Trends in Cognitive Sciences* 11 (9): 379–386. https://doi.org/10.1016/J.TICS.2007.08.001.

Dikker, S., L.J. Silbert, U. Hasson, and J.D. Zevin. 2014. On the same wavelength: Predictable language enhances speaker-listener brain-to-brain synchrony in posterior superior temporal gyrus. *The Journal of Neuroscience: The Official Journal of the Society for Neuroscience* 34 (18): 6267–6272. https://doi.org/10.1523/JNEUROSCI.3796-13.2014.

Ding, Y., Manaenko, A., Ho, A., Zhu, J., Zhang, J. H., Zhao, T., Zhu, Y., Tang, H., & Xie, R. (2019). Consciousness: New concepts and neural networks. https://doi.org/10.3389/fncel.2019.00302.

Dohmatob, E., G. Dumas, and D. Bzdok. 2020. Dark control: The default mode network as a reinforcement learning agent. *Human Brain Mapping* 41 (12): 3318. https://doi.org/10.1002/HBM.25019.

Dong, Y., S. Mihalas, F. Qiu, R. von der Heydt, and E. Niebur. 2008. Synchrony and the binding problem in macaque visual cortex. *Journal of Vision* 8 (7). https://doi.org/10.1167/8.7.30.

Doya, K., S. Ishii, A. Pouget, and R.P.N. Rao, eds. 2011. *Bayesian brain: Probabilistic approaches to neural coding*. Cambridge, Massachusetts, US: MIT Press.

Edlow, B.L., J. Claassen, N.D. Schiff, and D.M. Greer. 2021. Recovery from disorders of consciousness: Mechanisms, prognosis and emerging therapies. *Nature Reviews Neurology* 17 (3): 135–156. https://doi.org/10.1038/s41582-020-00428-x.

Edmonds III, R.G. 2015. A lively afterlife and beyond: The soul in Plato, Homer, and the Orphica, *Études platoniciennes*. Société d'Études platoniciennes. Available at: https://journals.openedition.org/etudesplatoniciennes/517.

Engel, A.K., and W. Singer. 2001. Temporal binding and the neural correlates of sensory awareness. *Trends in Cognitive Sciences* 5 (1): 16–25. https://doi.org/10.1016/S1364-6613(00)01568-0.

Engel, A.K., P. König, A.K. Kreiter, T.B. Schillen, and W. Singer. 1992. Temporal coding in the visual cortex: New vistas on integration in the nervous system. *Trends in Neurosciences* 15 (6): 218–226. https://doi.org/10.1016/0166-2236(92)90039-B.

Engel, A.K., P. Fries, P. König, M. Brecht, and W. Singer. 1999. Temporal binding, binocular rivalry, and consciousness. *Consciousness and Cognition* 8 (2): 128–151. https://doi.org/10.1006/ccog.1999.0389.

Engel, A.K., P. Fries, and W. Singer. 2001. Dynamic predictions: Oscillations and synchrony in top–down processing. *Nature Reviews Neuroscience* 2 (10): 704–716. https://doi.org/10.1038/35094565.

Erra, R.G., D.M. Mateos, R. Wennberg, and J.L.P. Velazquez. 2016. Towards a statistical mechanics of consciousness: Maximization of number of connections is associated with conscious awareness. *Physical Review E* 94 (5). https://doi.org/10.1103/PhysRevE.94.052402.

Espíndola, L. 2016. Plato on the political role of poetry. The expulsion of the traditional poets and the reform of poetry. *Praxis Filosófica Nueva Serie* 43: 37–56.

Fahrenfort, J.J., H.S. Scholte, and V.A.F. Lamme. 2007. Masking disrupts reentrant processing in human visual cortex. *Journal of Cognitive Neuroscience* 19 (9): 1488–1497. https://doi.org/10.1162/JOCN.2007.19.9.1488.

Fahrenfort, J.J., T.M. Snijders, K. Heinen, S. van Gaal, H.S. Scholte, and V.A.F. Lamme. 2012. Neuronal integration in visual cortex elevates face category tuning to conscious face perception. *Proceedings of the National Academy of Sciences of the United States of America* 109 (52): 21504–21509. https://doi.org/10.1073/PNAS.1207414110/SUPPL_FILE/PNAS.201207414SI.

Fernandino, L., J.R. Binder, R.H. Desai, S.L. Pendl, C.J. Humphries, W.L. Gross, L.L. Conant, and M.S. Seidenberg. 2016. Concept representation reflects multimodal abstraction: A framework for embodied semantics. *Cerebral Cortex (New York, N.Y.: 1991)* 26 (5): 2018–2034. https://doi.org/10.1093/CERCOR/BHV020.

Fingelkurts, A.A., A.A. Fingelkurts, and C.F.H. Neves. 2010. Natural world physical, brain operational, and mind phenomenal space–time. *Physics of Life Reviews* 7 (2): 195–249. https://doi.org/10.1016/J.PLREV.2010.04.001.

Fingelkurts, A.A., A.A. Fingelkurts, S. Bagnato, C. Boccagni, and G. Galardi. 2012. DMN operational synchrony relates to self-consciousness: Evidence from patients in vegetative and minimally conscious states. *The Open Neuroimaging Journal* 6: 55–68. https://doi.org/10.2174/1874440001206010055.

Finn, E.S., E. Glerean, A.Y. Khojandi, D. Nielson, P.J. Molfese, D.A. Handwerker, and P.A. Bandettini. 2020. Idiosynchrony: From shared responses to individual differences during naturalistic neuroimaging. *NeuroImage* 215. https://doi.org/10.1016/J.NEUROIMAGE.2020.116828.

Frässle, S., J. Sommer, A. Jansen, M. Naber, and W. Einhäuser. 2014. Binocular rivalry: Frontal activity relates to introspection and action but not to perception. *Journal of Neuroscience* 34 (5): 1738–1747. https://doi.org/10.1523/JNEUROSCI.4403-13.2014.

Freiwald, W.A., A.K. Kreiter, and W. Singer. 1995. Stimulus dependent intercolumnar synchronization of single unit responses in cat area 17. *Neuroreport* 6 (17): 2348–2352. https://doi.org/10.1097/00001756-199511270-00018.

Frewen, P., M.L. Schroeter, G. Riva, P. Cipresso, B. Fairfield, C. Padulo, A.H. Kemp, L. Palaniyappan, M. Owolabi, K. Kusi-Mensah, M. Polyakova, N. Fehertoi, W. D'Andrea, L. Lowe, and G. Northoff. 2020. Neuroimaging the consciousness of self: Review, and conceptual-methodological framework. *Neuroscience and Biobehavioral Reviews* 112: 164–212. https://doi.org/10.1016/j.neubiorev.2020.01.023.

Fries, P., P.R. Roelfsema, A.K. Engel, P. König, and W. Singer. 1997. Synchronization of oscillatory responses in visual cortex correlates with perception in interocular rivalry. *Proceedings of the National Academy of Sciences* 94 (23): 12699–12704. https://doi.org/10.1073/PNAS.94.23.12699.

Friston, K.J., T. Parr, Y. Yufik, N. Sajid, C.J. Price, and E. Holmes. 2020. Generative models, linguistic communication and active inference. *Neuroscience & Biobehavioral Reviews* 118: 42–64. https://doi.org/10.1016/J.NEUBIOREV.2020.07.005.

Furth, M. 1967. Monadology. *The Philosophical Review* 76 (2): 169. Available at: https://doi.org/10.2307/2183641.

Gamboa, L.D. 2022. The role of the will in Chatton's and Ockham's theories of consciousness. *Vivarium* 60 (4): 273–295. https://doi.org/10.1163/15685349-06004002.

Gibb, S., R. Findlay Hendry, and T. Lancaster. 2019. The Routledge handbook of emergence. *The Routledge Handbook of Emergence*: 1–420. https://doi.org/10.4324/9781315675213/ROUTLEDGE-HANDBOOK-EMERGENCE-ROBIN-FINDLAY-HENDRY-TOM-LANCASTER-SOPHIE-GIBB.

Gilboa, A., and H. Marlatte. 2017. Neurobiology of schemas and schema-mediated memory. *Trends in Cognitive Sciences* 21 (8): 618–631. https://doi.org/10.1016/J.TICS.2017.04.013.

Gosseries, O., H. Di, S. Laureys, and M. Boly. 2014. Measuring consciousness in severely damaged brains. *Annual Review of Neuroscience* 37: 457–478. https://doi.org/10.1146/ANNUREV-NEURO-062012-170339.

Goyal, A., Didolkar, A., Lamb, A., Badola, K., Ke, N. R., Rahaman, N., Binas, J., Blundell, C., Mozer, M., & Bengio, Y. (2021). *Coordination among neural modules through a shared global workspace*. http://arxiv.org/abs/2103.01197.

Graziano, M.S.A. 2013. *Consciousness and the social brain*. Oxford: Oxford University Press.

———. 2019. Attributing awareness to others: The attention schema theory and its relationship to behavioural prediction. *Journal of Consciousness Studies* 26 (3–4): 17–37.

———. 2020. Consciousness and the attention schema: Why it has to be right. *Cognitive Neuropsychology* 37 (3–4): 224–233. https://doi.org/10.1080/02643294.2020.1761782.

Graziano, M.S.A., and M.M. Botvinick. 2002. How the brain represents the body: Insights from neurophysiology and psychology. In *Common mechanisms in perception and action (attention and performance XIX)*, 136–157. https://www.princeton.edu/~graziano/Papers/Attn_Perf19.pdf.

Graziano, M.S.A., and S. Kastner. 2011a. Awareness as a perceptual model of attention. *Cognitive Neuroscience* 2 (2): 125–127. https://doi.org/10.1080/17588928.2011.585237.
———. 2011b. Human consciousness and its relationship to social neuroscience: A novel hypothesis. *Cognitive Neuroscience* 2 (2): 98–113. https://doi.org/10.1080/17588928.2011.565121.
Guterstam, A., H.H. Kean, T.W. Webb, F.S. Kean, and M.S.A. Graziano. 2019. Implicit model of other people's visual attention as an invisible, force-carrying beam projecting from the eyes. *Proceedings of the National Academy of Sciences* 116 (1): 328–333. https://doi.org/10.1073/PNAS.1816581115.
Hardcastle, V.G. 2008. The binding problem. *A Companion to Cognitive Science*: 553–565. https://doi.org/10.1002/9781405164535.CH43.
Harper, D. n.d.. *Etymology of consciousness*. Online Etymology Dictionary. Retrieved from: https://www.etymonline.com/word/consciousness.
Hasson, U., and C.D. Frith. 2016. Mirroring and beyond: Coupled dynamics as a generalized framework for modelling social interactions. *Philosophical Transactions of the Royal Society of London. Series B, Biological Sciences 371* (1693). https://doi.org/10.1098/RSTB.2015.0366.
Hasson, U., A.A. Ghazanfar, B. Galantucci, S. Garrod, and C. Keysers. 2012. Brain-to-brain coupling: A mechanism for creating and sharing a social world. *Trends in Cognitive Sciences* 16 (2): 114. https://doi.org/10.1016/J.TICS.2011.12.007.
Hasson, U., J. Chen, and C.J. Honey. 2015. Hierarchical process memory: Memory as an integral component of information processing. *Trends in Cognitive Sciences* 19 (6): 304–313. https://doi.org/10.1016/j.tics.2015.04.006.
Heidlmayr, K., K. Weber, A. Takashima, and P. Hagoort. 2020. No title, no theme: The joined neural space between speakers and listeners during production and comprehension of multi-sentence discourse. *Cortex* 130: 111–126. https://doi.org/10.1016/J.CORTEX.2020.04.035.
Hobson, J.A., and K.J. Friston. 2014. Consciousness, dreams, and inference: The Cartesian theatre revisited. *Journal of Consciousness Studies* 21 (1–2): 6–32.
———. 2016. *A response to our theatre critics*. https://philpapers.org/rec/HOBART.
Hoffmann, M., H. Marques, A. Arieta, H. Sumioka, M. Lungarella, and R. Pfeifer. 2010. Body schema in robotics: A review. *IEEE Transactions on Autonomous Mental Development* 2 (4): 304–324. https://doi.org/10.1109/TAMD.2010.2086454.
Hohwy, J. 2012. Attention and conscious perception in the hypothesis testing brain. *Frontiers in Psychology 3*: 96. https://doi.org/10.3389/fpsyg.2012.00096.
———. 2013. *The predictive mind*. The predictive mind. https://doi.org/10.1093/ACPROF:OSO/9780199682737.001.0001.
———. 2014. *The predictive mind*. The predictive mind. https://doi.org/10.1093/acprof:oso/9780199682737.001.0001.
———. 2016. Prediction error minimization, mental and developmental disorder, and statistical theories of consciousness. *Disturbed Consciousness* 293–324. https://doi.org/10.7551/mitpress/9780262029346.003.0012.
Homer. 1981. *The odyssey (English and Ancient Greek Edition)* (trans: Lawrence, T.E.). New York: Oxford University Press.
Honey, C.J., C.R. Thompson, Y. Lerner, and U. Hasson. 2012. Not lost in translation: Neural responses shared across languages. *The Journal of Neuroscience* 32 (44): 15277. https://doi.org/10.1523/JNEUROSCI.1800-12.2012.
Hsieh, P.J., J.T. Colas, and N. Kanwisher. 2011. Pop-out without awareness: Unseen feature singletons capture attention only when top-down attention is available. *Psychological Science* 22 (9): 1220–1226. https://doi.org/10.1177/0956797611419302.
Hu, Y., and Z. Yang. 2021. Impact of inter-individual variability on the estimation of default mode network in temporal concatenation group ICA. *NeuroImage* 237: 118114. https://doi.org/10.1016/J.NEUROIMAGE.2021.118114.
Huang, Z., N. Obara, H.H. Davis, J. Pokorny, and G. Northoff. 2016a. The temporal structure of resting-state brain activity in the medial prefrontal cortex predicts self-consciousness. *Neuropsychologia* 82: 161–170. https://doi.org/10.1016/J.NEUROPSYCHOLOGIA.2016.01.025.

Huang, Z., J. Zhang, J. Wu, P. Qin, X. Wu, Z. Wang, R. Dai, Y. Li, W. Liang, Y. Mao, Z. Yang, J. Zhang, A. Wolff, and G. Northoff. 2016b. Decoupled temporal variability and signal synchronization of spontaneous brain activity in loss of consciousness: An fMRI study in anesthesia. *NeuroImage* 124: 693–703. https://doi.org/10.1016/J.NEUROIMAGE.2015.08.062.

Huang, Z., J. Zhang, A. Longtin, G. Dumont, N.W. Duncan, J. Pokorny, P. Qin, R. Dai, F. Ferri, X. Weng, and G. Northoff. 2017. Is there a nonadditive interaction between spontaneous and evoked activity? Phasedependence and its relation to the temporal structure of scale-free brain activity. *Cerebral Cortex* 27: 1037–1059. https://doi.org/10.1093/cercor/bhv288.

Huang, Z., J. Zhang, J. Wu, G.A. Mashour, and A.G. Hudetz. 2020. Temporal circuit of macroscale dynamic brain activity supports human consciousness. *Science Advances* 6 (11): 87–98. https://doi.org/10.1126/SCIADV.AAZ0087/SUPPL_FILE/AAZ0087_SM.PDF.

Hurme, M., M. Koivisto, A. Revonsuo, and H. Railo. 2017. Early processing in primary visual cortex is necessary for conscious and unconscious vision while late processing is necessary only for conscious vision in neurologically healthy humans. *NeuroImage* 150: 230–238. https://doi.org/10.1016/J.NEUROIMAGE.2017.02.060.

Hutchison, R.M., T. Womelsdorf, J.S. Gati, S. Everling, and R.S. Menon. 2013. Resting-state networks show dynamic functional connectivity in awake humans and anesthetized macaques. *Human Brain Mapping* 34 (9): 2154–2177. https://doi.org/10.1002/HBM.22058.

Hyon, R., A.M. Kleinbaum, and C. Parkinson. 2020. Social network proximity predicts similar trajectories of psychological states: Evidence from multi-voxel spatiotemporal dynamics. *NeuroImage* 216: 116492. https://doi.org/10.1016/J.NEUROIMAGE.2019.116492.

Iacoboni, M. 2011. Hybrid social cognitive models, meta-consciousness, and representations. *Cognitive Neuroscience* 2 (2): 118–119. https://doi.org/10.1080/17588928.2011.585236.

Igelström, K.M., and M.S.A. Graziano. 2017. The inferior parietal lobule and temporoparietal junction: A network perspective. *Neuropsychologia* 105: 70–83. https://doi.org/10.1016/J.NEUROPSYCHOLOGIA.2017.01.001.

Igelström, K.M., T.W. Webb, Y.T. Kelly, and M.S.A. Graziano. 2016. Topographical organization of attentional, social, and memory processes in the human temporoparietal cortex. *ENeuro* 3 (2): 1363–1377. https://doi.org/10.1523/ENEURO.0060-16.2016.

Jääskeläinen, I.P., J. Pajula, J. Tohka, H.J. Lee, W.J. Kuo, and F.H. Lin. 2016. Brain hemodynamic activity during viewing and re-viewing of comedy movies explained by experienced humor. *Scientific Reports* 6 (1): 1–14. https://doi.org/10.1038/srep27741.

Jaynes, J. 1974. *The origins of consciousness in the breakdown of the bicameral mind*. Boston: Houghton Mifflin.

Jehee, J.F.M., C. Rothkopf, J.M. Beck, and D.H. Ballard. 2006. Learning receptive fields using predictive feedback. *Journal of Physiology, Paris 100* (1–3): 125. https://doi.org/10.1016/J.JPHYSPARIS.2006.09.011.

Jiang, Y., P. Costello, F. Fang, M. Huang, and S. He. 2006. A gender- and sexual orientation-dependent spatial attentional effect of invisible images. *Proceedings of the National Academy of Sciences of the United States of America* 103 (45): 17048–17052. https://doi.org/10.1073/pnas.0605678103.

Katona, G. 2002. The evolution of the concept of Psyche from homer to Aristotle. *Journal of Theoretical and Philosophical Psychology* 22 (1): 28–44. https://doi.org/10.1037/h0091193.

Kelly, Y.T., T.W. Webb, J.D. Meier, M.J. Arcaro, and M.S.A. Graziano. 2014. Attributing awareness to oneself and to others. *Proceedings of the National Academy of Sciences of the United States of America* 111 (13): 5012–5017. https://doi.org/10.1073/pnas.1401201111.

Kenny, A.J.P., and A.H. Amadio. Last review: 2022. *The unmoved mover*. Aristotle. Encyclopædia Britannica. Encyclopædia Britannica, Inc. Available at: https://www.britannica.com/biography/Aristotle/The-unmoved-mover.

Ki, J.J., S.P. Kelly, and L.C. Parra. 2016. Attention strongly modulates reliability of neural responses to naturalistic narrative stimuli. *The Journal of Neuroscience: The Official Journal of the Society for Neuroscience* 36 (10): 3092–3101. https://doi.org/10.1523/JNEUROSCI.2942-15.2016.

Kleckner, I.R., J. Zhang, A. Touroutoglou, L. Chanes, C. Xia, W.K. Simmons, K.S. Quigley, B.C. Dickerson, and L. Feldman Barrett. 2017. Evidence for a large-scale brain system supporting allostasis and interoception in humans. *Nature Human Behaviour* 1 (5). https://doi.org/10.1038/s41562-017-0069.

Knill, D.C., and A. Pouget. 2004. The Bayesian brain: The role of uncertainty in neural coding and computation. *Trends in Neurosciences* 27 (12): 712–719. https://doi.org/10.1016/J.TINS.2004.10.007.

Koch, C., M. Massimini, M. Boly, and G. Tononi. 2016. Neural correlates of consciousness: Progress and problems. In *Nature reviews neuroscience*, Vol. 17, Issue 5, 307–321. Nature Publishing Group. https://doi.org/10.1038/nrn.2016.22.

Kolvoort, I.R., S. Wainio-Theberge, A. Wolff, and G. Northoff. 2020. Temporal integration as "common currency" of brain and self-scale-free activity in resting-state EEG correlates with temporal delay effects on self-relatedness. *Human Brain Mapping* 41: 15. https://doi.org/10.1002/hbm.25129.

Konu, D., A. Turnbull, T. Karapanagiotidis, H.T. Wang, L.R. Brown, E. Jefferies, and J. Smallwood. 2020. A role for the ventromedial prefrontal cortex in self-generated episodic social cognition. *NeuroImage* 218 (January): 116977. https://doi.org/10.1016/j.neuroimage.2020.116977.

Koster-Hale, J., and R. Saxe. 2013. Theory of mind: A neural prediction problem. *Neuron* 79 (5): 836. https://doi.org/10.1016/J.NEURON.2013.08.020.

Kuhlen, A.K., C. Allefeld, and J.D. Haynes. 2012. Content-specific coordination of listeners' to speakers' EEG during communication. *Frontiers in Human Neuroscience* 6 (September): 266. https://doi.org/10.3389/fnhum.2012.00266.

Lahnakoski, J.M., E. Glerean, I.P. Jääskeläinen, J. Hyönä, R. Hari, M. Sams, and L. Nummenmaa. 2014. Synchronous brain activity across individuals underlies shared psychological perspectives. *NeuroImage* 100 (100): 316–324. https://doi.org/10.1016/J.NEUROIMAGE.2014.06.022.

Lake, J. 2022. Implications of postmaterialist theories of consciousness for psychiatry: Towards an integral paradigm. *International Journal for Transformative Research* 9 (1): 49–61. https://doi.org/10.2478/IJTR-2022-0006.

Lamme, V.A.F. 2006. Towards a true neural stance on consciousness. *Trends in Cognitive Sciences* 10 (11): 494–501. https://doi.org/10.1016/j.tics.2006.09.001.

Lamme, V.A.F., 2010. How neuroscience will change our view on consciousness. *Cognitive Neuroscience 1* (3): 204–220. https://doi.org/10.1080/17588921003731586.

Lamme, V. 2015. The crack of dawn – Perceptual functions and neural mechanisms that mark the transition from unconscious processing to conscious vision. *Open Mind* 22: 1–34. https://doi.org/10.15502/9783958570092.

Lamme, V.A.F. 2020. Visual functions generating conscious seeing. *Frontiers in Psychology* 11: 83. https://doi.org/10.3389/fpsyg.2020.00083.

Lamme, V.A.F., and P.R. Roelfsema. 2000. The distinct modes of vision offered by feedforward and recurrent processing. *Trends in Neurosciences* 23 (11): 571–579. https://doi.org/10.1016/S0166-2236(00)01657-X.

Lamme, V.A.F., H. Supèr, and H. Spekreijse. 1998. Feedforward, horizontal, and feedback processing in the visual cortex. *Current Opinion in Neurobiology* 8 (4): 529–535. https://doi.org/10.1016/S0959-4388(98)80042-1.

Lamme, V.A.F., K. Zipser, and H. Spekreijse. 2001. Masking interrupts figure-ground signals in V1. *Journal of Vision* 1 (3). https://doi.org/10.1167/1.3.32.

Lanzoni, L., D. Ravasio, H. Thompson, D. Vatansever, D. Margulies, J. Smallwood, and E. Jefferies. 2020. The role of default mode network in semantic cue integration. *NeuroImage* 219. https://doi.org/10.1016/J.NEUROIMAGE.2020.117019.

Larock, E. 2006. Why neural synchrony fails to explain the unity of visual consciousness. *Behavior and Philosophy* 34: 39–58.

Lau, H., and D. Rosenthal. 2011. Empirical support for higher-order theories of conscious awareness. *Trends in Cognitive Sciences* 15 (8): 365–373. https://doi.org/10.1016/J.TICS.2011.05.009.

Lee, H., B. Bellana, and J. Chen. 2020. What can narratives tell us about the neural bases of human memory? *Current Opinion in Behavioral Sciences* 32: 111–119. https://doi.org/10.1016/j.cobeha.2020.02.007.

Leibniz, G.W. 1953. *Discourse on metaphysics* (trans: Lucas, P.G., L. Grint, and G.W. Leibniz). Manchester: Manchester University Press.

———. 2014. *Leibniz's Monadology: A new translation and guide* (trans: L. Strickland). Edinburgh: Edinburgh University Press.

Leszl, W.G. 2016. Plato's attitude to poetry and the fine arts, and the origins of aesthetics, *Études platoniciennes*. Société d'Études platoniciennes. Retrieved from: https://journals.openedition.org/etudesplatoniciennes/1121.

Lewis, C.T., W. Freund, E.A. Andrews, and C. Short. 1969a. *Animus. A Latin dictionary, founded on Andrews' edition of Freund' Latin dictionary*. Oxford: Clarendon Press.

———. 1969b. *Anima. A Latin dictionary, founded on Andrews' edition of Freund' Latin dictionary*. Oxford: Clarendon Press.

Li, J., W.H. Curley, B. Guerin, D.D. Dougherty, A.v. Dalca, B. Fischl, A. Horn, and B.L. Edlow. 2021. Mapping the subcortical connectivity of the human default mode network. *NeuroImage* 245: 118758. https://doi.org/10.1016/j.neuroimage.2021.118758.

Liddell, H.G., R. Scott, and H.S. Jones. 1996. σύνοιδα (súnoida). *A Greek-English lexicon. with a revised supplement*. Oxford, UK: Clarendon Press.

Lin, Z., and S.O. Murray. 2015. More power to the unconscious: Conscious, but not unconscious, exogenous attention requires location variation. *Psychological Science* 26 (2): 221–230. https://doi.org/10.1177/0956797614560770.

Liu, Y., E.A. Piazza, E. Simony, P.A. Shewokis, B. Onaral, U. Hasson, and H. Ayaz. 2017. Measuring speaker–listener neural coupling with functional near infrared spectroscopy. *Scientific Reports* 7 (1): 1–13. https://doi.org/10.1038/srep43293.

Livingstone, M., and D. Hubel. 1988. Segregation of form, color, movement, and depth: Anatomy, physiology, and perception. *Science* 240 (4853): 740–749. https://doi.org/10.1126/SCIENCE.3283936.

López-Ibor, J.J., Tomás, O., and María I. L.-I., 2011. Perception, experience and body identity. *Actas Españolas de Psiquiatría* 39 (supl. 3): 3–118.

Lorenz, H. 2009. *Ancient theories of soul*, ed. E.N. Zalta. Retrieved from: https://plato.stanford.edu/entries/ancient-soul/.

Mancuso, L., S. Cavuoti-Cabanillas, D. Liloia, J. Manuello, G. Buzi, F. Cauda, and T. Costa. 2022. Tasks activating the default mode network map multiple functional systems. *Brain Structure and Function* 227 (5): 1711–1734. https://doi.org/10.1007/s00429-022-02467-0.

Maravita, A., and A. Iriki. 2004. Tools for the body (schema). *Trends in Cognitive Sciences* 8 (2): 79–86. https://doi.org/10.1016/J.TICS.2003.12.008.

Margulies, D.S., S.S. Ghosh, A. Goulas, M. Falkiewicz, J.M. Huntenburg, G. Langs, G. Bezgin, S.B. Eickhoff, F.X. Castellanos, M. Petrides, E. Jefferies, and J. Smallwood. 2016. Situating the default-mode network along a principal gradient of macroscale cortical organization. *Proceedings of the National Academy of Sciences of the United States of America* 113 (44): 12574–12579. https://doi.org/10.1073/PNAS.1608282113/SUPPL_FILE/PNAS.201608282SI.

Marietta, D.E. 1970. Conscience in Greek stoicism. *Numen* 17 (3): 176–187. https://doi.org/10.2307/3269701.

Marvan, T., and M. Havlík. 2021. Is predictive processing a theory of perceptual consciousness? *New Ideas in Psychology* 61. https://doi.org/10.1016/j.newideapsych.2020.100837.

Mashour, G.A., P. Roelfsema, J.P. Changeux, and S. Dehaene. 2020. Conscious processing and the global neuronal workspace hypothesis. *Neuron* 105 (5): 776–798. https://doi.org/10.1016/j.neuron.2020.01.026.

Mazoyer, B., L. Zago, E. Mellet, S. Bricogne, O. Etard, O. Houdé, F. Crivello, M. Joliot, L. Petit, and N. Tzourio-Mazoyer. 2001. Cortical networks for working memory and executive functions sustain the conscious resting state in man. *Brain Research Bulletin* 54 (3): 287–298. https://doi.org/10.1016/S0361-9230(00)00437-8.

Medea, B., T. Karapanagiotidis, M. Konishi, C. Ottaviani, D. Margulies, A. Bernasconi, N. Bernasconi, B.C. Bernhardt, E. Jefferies, and J. Smallwood. 2018. How do we decide what to do? Resting-state connectivity patterns and components of self-generated thought linked to the development of more concrete personal goals. *Experimental Brain Research* 236 (9): 2469–2481. https://doi.org/10.1007/s00221-016-4729-y.

Merker, B. 2013. Cortical gamma oscillations: The functional key is activation, not cognition. *Neuroscience & Biobehavioral Reviews* 37 (3): 401–417. https://doi.org/10.1016/J.NEUBIOREV.2013.01.013.

Michel, M. 2017. A role for the anterior insular cortex in the global neuronal workspace model of consciousness. *Consciousness and Cognition* 49: 333–346. https://doi.org/10.1016/j.concog.2017.02.004.

Mindt, G. 2017. The problem with the 'Information' in integrated information theory. *Journal of Consciousness Studies* 24: 130–154.

Mitchell, J.P. 2008. Activity in right temporo-parietal junction is not selective for theory-of-mind. *Cerebral Cortex* 18 (2): 262–271. https://doi.org/10.1093/cercor/bhm051.

Morejón, Gil. 2022. *The unconscious of thought in Leibniz, Spinoza, and Hume*. Edinburgh: Edinburgh University Press.

Murphy, C., E. Jefferies, S.A. Rueschemeyer, M. Sormaz, H.T. Wang, D.S. Margulies, and J. Smallwood. 2018. Distant from input: Evidence of regions within the default mode network supporting perceptually-decoupled and conceptually-guided cognition. *NeuroImage* 171: 393. https://doi.org/10.1016/J.NEUROIMAGE.2018.01.017.

Nagel, T. 1974. What is it like to be a bat? *The Philosophical Review* 83 (4): 435. https://doi.org/10.2307/2183914.

Nguyen, M., T. Vanderwal, and U. Hasson. 2019. Shared understanding of narratives is correlated with shared neural responses. *NeuroImage* 184: 161–170. https://doi.org/10.1016/J.NEUROIMAGE.2018.09.010.

Northoff, G. 2013. What the brain's intrinsic activity can tell us about consciousness? A tri-dimensional view. *Neuroscience and Biobehavioral Reviews* 37 (4): 726–738. https://doi.org/10.1016/j.neubiorev.2012.12.004.

———. 2014. *Unlocking the brain*. Oxford, UK: Oxford University Press.

———. 2016. Slow cortical potentials and "inner time consciousness" — A neuro-phenomenal hypothesis about the "width of present". *International Journal of Psychophysiology* 103: 174–184. https://doi.org/10.1016/j.ijpsycho.2015.02.012.

———. 2018. *The spontaneous brain: From the mind-body to the world-brain problem*. The MIT Press.

———. 2021. Why is there sentience?: A temporo-spatial approach to consciousness. *Journal of Consciousness Studies* 28 (7–8): 67–82.

Northoff, G., and Z. Huang. 2017. How do the brain's time and space mediate consciousness and its different dimensions? Temporo-spatial theory of consciousness (TTC). *Neuroscience and Biobehavioral Reviews* 80: 630–645. https://doi.org/10.1016/j.neubiorev.2017.07.013.

Northoff, G., and V. Lamme. 2020. Neural signs and mechanisms of consciousness: Is there a potential convergence of theories of consciousness in sight? *Neuroscience and Biobehavioral Reviews* 118 (August): 568–587. https://doi.org/10.1016/j.neubiorev.2020.07.019.

Northoff, G., and F. Zilio. 2022. Temporo-spatial theory of consciousness (TTC) – Bridging the gap of neuronal activity and phenomenal states. *Behavioural Brain Research* 424. https://doi.org/10.1016/j.bbr.2022.113788.

Northoff, G., S. Wainio-Theberge, and K. Evers. 2020. Is temporo-spatial dynamics the "common currency" of brain and mind? In quest of "spatiotemporal neuroscience". *Physics of Life Reviews* 33: 34–54. https://doi.org/10.1016/j.plrev.2019.05.002.

Norton, L., R.M. Hutchison, G.B. Young, D.H. Lee, M.D. Sharpe, and S.M. Mirsattari. 2012. Disruptions of functional connectivity in the default mode network of comatose patients. *Neurology* 78 (3): 175–181. https://doi.org/10.1212/WNL.0b013e31823fcd61.

Nummenmaa, L., J.M. Lahnakoski, and E. Glerean. 2018. Sharing the social world via intersubject neural synchronisation. *Current Opinion in Psychology* 24: 7–14. https://doi.org/10.1016/J.COPSYC.2018.02.021.

O'Grady, P. n.d. *Thales of Miletus*. Internet encyclopedia of philosophy. Retrieved from: https://iep.utm.edu/thales/#H7.

Odegaard, B., R.T. Knight, and H. Lau. 2017. Should a few null findings falsify prefrontal theories of conscious perception? *The Journal of Neuroscience* 37 (40): 9593–9602. https://doi.org/10.1523/jneurosci.3217-16.2017.

Olshewsky, T.M. 1976. On the relations of soul to body in Plato and Aristotle. *Journal of the History of Philosophy* 14 (4): 391–404. Retrieved from: https://doi.org/10.1353/hph.2008.0163.

Otten, M., A.K. Seth, and Y. Pinto. 2017. A social Bayesian brain: How social knowledge can shape visual perception. *Brain and Cognition* 112: 69–77. https://doi.org/10.1016/J.BANDC.2016.05.002.

Overgaard, M. 2009. How can we know if patients in coma, vegetative state or minimally conscious state are conscious? *Progress in Brain Research* 177 (C): 11–19. https://doi.org/10.1016/S0079-6123(09)17702-6.

———. 2018. Phenomenal consciousness and cognitive access. *Philosophical Transactions of the Royal Society B: Biological Sciences* 373 (1755). https://doi.org/10.1098/rstb.2017.0353.

Palanca, B.J.A., and G.C. DeAngelis. 2005. Does neuronal synchrony underlie visual feature grouping? *Neuron* 46 (2): 333–346. https://doi.org/10.1016/J.NEURON.2005.03.002.

Parfit, D. 1987. *Reasons and persons*. Oxford: Clarendon Press.

Park, H.D., and O. Blanke. 2019. Coupling inner and outer body for self-consciousness. *Trends in Cognitive Sciences* 23 (5): 377–388. https://doi.org/10.1016/J.TICS.2019.02.002.

Park, H.D., and C. Tallon-Baudry. 2014. The neural subjective frame: From bodily signals to perceptual consciousness. *Philosophical Transactions of the Royal Society B: Biological Sciences* 369 (1641). https://doi.org/10.1098/RSTB.2013.0208.

Parkinson, C., A.M. Kleinbaum, and T. Wheatley. 2018. Similar neural responses predict friendship. *Nature Communications* 9 (1) https://doi.org/10.1038/S41467-017-02722-7.

Pesquita, A., C.S. Chapman, and J.T. Enns. 2016. Humans are sensitive to attention control when predicting others' actions. *Proceedings of the National Academy of Sciences of the United States of America* 113 (31): 8669–8674. https://doi.org/10.1073/pnas.1601872113.

Plato. 2007. *The republic* (M.S. Lane, Intr.) (trans: Pritchard, L.H.D.). London: Penguin.

———. 2010. *Phaedrus* (trans: Waterfield, R.). Oxford University Press.

Qin, P., and G. Northoff. 2011. How is our self related to midline regions and the default-mode network? *NeuroImage* 57 (3): 1221–1233. https://doi.org/10.1016/J.NEUROIMAGE.2011.05.028.

Rahimian, S. 2022. The myth of when and where: How false assumptions still haunt theories of consciousness. *Consciousness and Cognition* 97: 103246. https://doi.org/10.1016/j.concog.2021.103246.

Raichle, M.E. 2015. The Brain's default mode network. *Annual Review of Neuroscience* 38: 433–447. https://doi.org/10.1146/annurev-neuro-071013-014030.

Raichle, M.E., and A.Z. Snyder. 2007. A default mode of brain function: A brief history of an evolving idea. *NeuroImage* 37 (4): 1083–1090. https://doi.org/10.1016/J.NEUROIMAGE.2007.02.041.

Raichle, M.E., A.M. MacLeod, A.Z. Snyder, W.J. Powers, D.A. Gusnard, and G.L. Shulman. 2001. A default mode of brain function. *Proceedings of the National Academy of Sciences of the United States of America* 98 (2): 676–682. https://doi.org/10.1073/PNAS.98.2.676.

Railo, H., M. Koivisto, and A. Revonsuo. 2011. Tracking the processes behind conscious perception: A review of event-related potential correlates of visual consciousness. *Consciousness and Cognition* 20 (3): 972–983. https://doi.org/10.1016/J.CONCOG.2011.03.019.

Ranganath, C., and M. Ritchey. 2012. Two cortical systems for memory-guided behaviour. *Nature Reviews Neuroscience* 13 (10): 713–726. https://doi.org/10.1038/NRN3338.

Regev, M., C.J. Honey, E. Simony, and U. Hasson. 2013. Selective and invariant neural responses to spoken and written narratives. *Journal of Neuroscience* 33 (40): 15978–15988. https://doi.org/10.1523/JNEUROSCI.1580-13.2013.

Regev, M., E. Simony, K. Lee, K.M. Tan, J. Chen, and U. Hasson. 2019. Propagation of information along the cortical hierarchy as a function of attention while reading and listening to stories. *Cerebral Cortex (New York, N.Y.: 1991)* 29 (10): 4017–4034. https://doi.org/10.1093/CERCOR/BHY282.

Revonsuo, A., and J. Newman. 1999. Binding and Consciousness. *Consciousness and Cognition* 8 (2): 123–127. https://doi.org/10.1006/CCOG.1999.0393.

Rosenthal, D.M. 2005. *Consciousness and Mind.* 378. https://philpapers.org/rec/ROSCAM-8.

Rosenthal, D., and J. Weisberg. 2008. Higher-order theories of consciousness. *Scholarpedia* 3 (5): 4407. https://doi.org/10.4249/SCHOLARPEDIA.4407.

Saalasti, S., J. Alho, M. Bar, E. Glerean, T. Honkela, M. Kauppila, M. Sams, and I.P. Jääskeläinen. 2019. Inferior parietal lobule and early visual areas support elicitation of individualized meanings during narrative listening. *Brain and Behavior* 9 (5). https://doi.org/10.1002/BRB3.1288.

Sanchez-Vives, M.V., A. Barbero-Castillo, M. Perez-Zabalza, and R. Reig. 2021. GABAB receptors: Modulation of thalamocortical dynamics and synaptic plasticity. *Neuroscience* 456: 131–142. https://doi.org/10.1016/J.NEUROSCIENCE.2020.03.011.

Scalabrini, A., Z. Huang, C. Mucci, M.G. Perrucci, A. Ferretti, A. Fossati, G.L. Romani, G. Northoff, and S.J.H. Ebisch. 2017. How spontaneous brain activity and narcissistic features shape social interaction. *Scientific Reports* 7 (1). https://doi.org/10.1038/s41598-017-10389-9.

Scalabrini, A., S.J.H. Ebisch, Z. Huang, S. di Plinio, M.G. Perrucci, G.L. Romani, C. Mucci, and G. Northoff. 2019. Spontaneous brain activity predicts task-evoked activity during animate versus inanimate touch. *Cerebral Cortex* 29 (11): 4628–4645. https://doi.org/10.1093/cercor/bhy340.

Schacter, D.L., D.R. Addis, and R.L. Buckner. 2007. Remembering the past to imagine the future: The prospective brain. *Nature Reviews Neuroscience* 8 (9): 657–661. https://doi.org/10.1038/NRN2213.

Schippers, M.B., V. Gazzola, R. Goebel, and C. Keysers. 2009. Playing charades in the fMRI: Are mirror and/or mentalizing areas involved in gestural communication? *PLoS One* 4 (8): e6801. https://doi.org/10.1371/JOURNAL.PONE.0006801.

Schippers, M.B., A. Roebroeck, R. Renken, L. Nanetti, and C. Keysers. 2010. Mapping the information flow from one brain to another during gestural communication. *Proceedings of the National Academy of Sciences of the United States of America* 107 (20): 9388–9393. https://doi.org/10.1073/pnas.1001791107.

Schlicht, T., K. Dołęga, S.B. Fink, and Y.-T. Lin. 2021. You can't always get what you want. *Philosophy and the Mind Sciences* 2. https://doi.org/10.33735/PHIMISCI.2021.80.

Schmälzle, R., F.E.K. Häcker, C.J. Honey, and U. Hasson. 2015. Engaged listeners: Shared neural processing of powerful political speeches. *Social Cognitive and Affective Neuroscience* 10 (8): 1137. https://doi.org/10.1093/SCAN/NSU168.

Schurger, A., and M. Graziano. 2022. *Special issue: Consciousness science and its theories consciousness explained or described?* (December 2021), 1–9.

Sergent, C., and S. Dehaene. 2004. Neural processes underlying conscious perception: Experimental findings and a global neuronal workspace framework. *Journal of Physiology Paris 98* (4–6 SPEC ISS): 374–384. https://doi.org/10.1016/J.JPHYSPARIS.2005.09.006.

Seth, A.K. 2013. Interoceptive inference, emotion, and the embodied self. *Trends in Cognitive Sciences* 17 (11): 565–573. https://doi.org/10.1016/j.tics.2013.09.007.

Seth, A.K., and T. Bayne. 2022. Theories of consciousness. *Nature Reviews Neuroscience* 23 (7): 439–452. https://doi.org/10.1038/S41583-022-00587-4.

Shapiro, L., and S. Spaulding. 2021. In *Embodied cognition, Stanford Encyclopedia of Philosophy*, ed. E.N. Zalta. Stanford University. Retrieved from: https://plato.stanford.edu/entries/embodied-cognition/.

Shine, J.M., and M. Breakspear. 2018. Understanding the brain, by default. *Trends in Neurosciences* 41 (5): 244–247. https://doi.org/10.1016/j.tins.2018.03.004.

Sievers, B., C. Welker, U. Hasson, A. Kleinbaum, and T. Wheatley. 2020. How consensus-building conversation changes our minds and aligns our brains. *PsyArxiv*. https://doi.org/10.31234/OSF.IO/562Z7.

Signa, A., A. Chella, and M. Gentile. 2021. Cognitive robots and the conscious mind: A review of the global workspace theory. *Current Robotics Reports* 2 (2): 125–131. https://doi.org/10.1007/s43154-021-00044-7.

Silbert, L.J., C.J. Honey, E. Simony, D. Poeppel, and U. Hasson. 2014. Coupled neural systems underlie the production and comprehension of naturalistic narrative speech. *Proceedings of the National Academy of Sciences of the United States of America* 111 (43): E4687–E4696. https://doi.org/10.1073/PNAS.1323812111/SUPPL_FILE/PNAS.201323812SI.

Simony, E., C.J. Honey, J. Chen, O. Lositsky, Y. Yeshurun, A. Wiesel, and U. Hasson. 2016. Dynamic reconfiguration of the default mode network during narrative comprehension. *Nature Communications* 7. https://doi.org/10.1038/ncomms12141.

Skinner, B.F. 1953. *Science and human behavior*. New York: Macmillan.

Smallwood, J., T. Karapanagiotidis, F. Ruby, B. Medea, I. de Caso, M. Konishi, H.T. Wang, G. Hallam, D.S. Margulies, and E. Jefferies. 2016. Representing representation: Integration between the temporal lobe and the posterior cingulate influences the content and form of spontaneous thought. *PLoS One* 11 (4). https://doi.org/10.1371/journal.pone.0152272.

Smallwood, J., B.C. Bernhardt, R. Leech, D. Bzdok, E. Jefferies, and D.S. Margulies. 2021. The default mode network in cognition: A topographical perspective. *Nature Reviews Neuroscience* 22 (8). https://doi.org/10.1038/s41583-021-00474-4.

Smirnov, D., H. Saarimäki, E. Glerean, R. Hari, M. Sams, and L. Nummenmaa. 2019. Emotions amplify speaker-listener neural alignment. *Human Brain Mapping* 40 (16): 4777–4788. https://doi.org/10.1002/HBM.24736.

Smith, V., D.J. Mitchell, and J. Duncan. 2018. Role of the default mode network in cognitive transitions. *Cerebral Cortex* 28: 3685–3696. https://doi.org/10.1093/cercor/bhy167.

Smith, V., J. Duncan, and D.J. Mitchell. 2021. Roles of the default mode and multiple-demand networks in naturalistic versus symbolic decisions. *Journal of Neuroscience* 41 (10): 2214–2228. https://doi.org/10.1523/JNEUROSCI.1888-20.2020.

Smythies, J. (1994). The walls of Plato's Cave: The Science and Philosophy of (Brain, consciousness and perception). Aldershot, UK: Avebury.

Sormaz, M., C. Murphy, H.T. Wang, M. Hymers, T. Karapanagiotidis, G. Poerio, D.S. Margulies, E. Jefferies, and J. Smallwood. 2018. Default mode network can support the level of detail in experience during active task states. *Proceedings of the National Academy of Sciences of the United States of America* 115 (37): 9318–9323. https://doi.org/10.1073/pnas.1721259115.

Spinoza, B. 1996. II. In *Ethics/Benedict de Spinoza, edited and translated by Edwin Curley with an introduction by Stuart Hampshire*. London: Penguin Books.

Spratling, M.W. 2010. Predictive coding as a model of response properties in cortical area V1. *The Journal of Neuroscience* 30 (9): 3531. https://doi.org/10.1523/JNEUROSCI.4911-09.2010.

Stawarczyk, D., M.A. Bezdek, and J.M. Zacks. 2021. Event representations and predictive processing: The role of the midline default network Core. *Topics in Cognitive Science* 13 (1): 164–186. https://doi.org/10.1111/TOPS.12450.

Stephens, G.J., L.J. Silbert, and U. Hasson. 2010. Speaker-listener neural coupling underlies successful communication. *Proceedings of the National Academy of Sciences of the United States of America* 107 (32): 14425–14430. https://doi.org/10.1073/PNAS.1008662107/-/DCSUPPLEMENTAL/PNAS.201008662SI.

Stevenson, A., and M. Waite. 2011. Consciousness. In *Concise Oxford English dictionary*. Oxford, UK: Oxford University Press.

Storm, J.F., X.M. Boly, A.G. Casali, M. Massimini, U. Olcese, X. Cyriel, M.A. Pennartz, and M. Wilke. 2017. Symposium consciousness regained: Disentangling mechanisms, brain systems, and behavioral responses. *Journal of Neuroscience* 37 (45): 10882–10893. https://doi.org/10.1523/JNEUROSCI.1838-17.2017.

Tei, S., J.P. Kauppi, J. Fujino, K.F. Jankowski, R. Kawada, T. Murai, and H. Takahashi. 2019. Inter-subject correlation of temporoparietal junction activity is associated with conflict patterns during flexible decision-making. *Neuroscience Research* 144: 67–70. https://doi.org/10.1016/J.NEURES.2018.07.006.

Thompson, E., and F.J. Varela. 2001. Radical embodiment: Neural dynamics and consciousness. *Trends in Cognitive Sciences* 5 (10): 418–425. https://doi.org/10.1016/S1364-6613(00)01750-2.

Tikka, P., J. Kauttonen, and Y. Hlushchuk. 2018. Narrative comprehension beyond language: Common brain networks activated by a movie and its script. *PLoS One* 13 (7): e0200134. https://doi.org/10.1371/JOURNAL.PONE.0200134.

Todd, R.B. 1976. The four causes: Aristotle's exposition and the ancients. *Journal of the History of Ideas* 37 (2): 319. Retrieved from: https://doi.org/10.2307/2708827.

Tononi, G. 2008. Consciousness as integrated information: A provisional manifesto. *Biological Bulletin* 215 (3): 216–242. https://doi.org/10.2307/25470707.

———. 2012. The integrated information theory of Consciousness: An updated account. *Archives Italiennes de Biologie* 150 (2/3): 56–90. https://doi.org/10.4449/AIB.V149I5.1388.

Tononi, G., and C. Koch. 2015. Consciousness: Here, there and everywhere? *Philosophical Transactions of the Royal Society B: Biological Sciences* 370 (1668) Royal Society of London. https://doi.org/10.1098/rstb.2014.0167.

Tononi, G., M. Boly, M. Massimini, and C. Koch. 2016. Integrated information theory: From consciousness to its physical substrate. *Nature Reviews Neuroscience* 17 (7): 450–461. https://doi.org/10.1038/nrn.2016.44.

Travers, E., C.D. Frith, and N. Shea. 2018. Learning rapidly about the relevance of visual cues requires conscious awareness. *Quarterly Journal of Experimental Psychology* 71 (8): 1698–1713. https://doi.org/10.1080/17470218.2017.1373834.

Treisman, A. 1996. The binding problem. *Current Opinion in Neurobiology* 6 (2): 171–178. https://doi.org/10.1016/S0959-4388(96)80070-5.

Uchiyama, H.T., D.N. Saito, H.C. Tanabe, T. Harada, A. Seki, K. Ohno, T. Koeda, and N. Sadato. 2012. Distinction between the literal and intended meanings of sentences: A functional magnetic resonance imaging study of metaphor and sarcasm. *Cortex: A Journal Devoted to the Study of the Nervous System and Behavior* 48 (5): 563–583. https://doi.org/10.1016/J.CORTEX.2011.01.004.

Van Baar, J.M., L.J. Chang, and A.G. Sanfey. 2019. The computational and neural substrates of moral strategies in social decision-making. *Nature Communications* 10 (1): 1–14. https://doi.org/10.1038/s41467-019-09161-6.

Van Elk, M., and O. Blanke. 2011. Bodily self-consciousness, and the primacy of self related signals such as the 1stperson perspective and self-location. *Cognitive Neuroscience* 2 (2): 123–124. https://doi.org/10.1080/17588928.2011.588488.

Van Gulick, R. 2022. *Consciousness*. Stanford Encyclopedia of Philosophy. Retrieved from https://plato.stanford.edu/entries/consciousness/.

Van Kesteren, M.T.R., S.F. Beul, A. Takashima, R.N. Henson, D.J. Ruiter, and G. Fernández. 2013. Differential roles for medial prefrontal and medial temporal cortices in schema-dependent encoding: From congruent to incongruent. *Neuropsychologia* 51 (12): 2352–2359. https://doi.org/10.1016/j.neuropsychologia.2013.05.027.

Van Rullen, R., and R. Kanai. 2021. Deep learning and the global workspace theory. *Trends in Neurosciences* 44 (9): 692–704. https://doi.org/10.1016/J.TINS.2021.04.005.

Vatansever, D., D.K. Menon, and E.A. Stamatakis. 2017. Default mode contributions to automated information processing. *Proceedings of the National Academy of Sciences of the United States of America* 114 (48): 12821–12826. https://doi.org/10.1073/PNAS.1710521114/SUPPL_FILE/PNAS.1710521114.SAPP.PDF.

Vieira, P.S. 2022. *Aristotle on the soul: Vegetative, sensitive, and intellectual, Filosofia do Início*. Retrieved from: https://filosofiadoinicio.com/en/2021/07/aristotle-on-the-soul.html.

Vilberg, K.L., and M.D. Rugg. 2012. The neural correlates of recollection: Transient versus sustained fMRI effects. *The Journal of Neuroscience* 32 (45): 15679. https://doi.org/10.1523/JNEUROSCI.3065-12.2012.

Villar, B.J. 2017. *La poesía en 'La República' de Platón. Un exilio interior, Ensayos de Filosofia. Ensayos de Filosofía – Spain*. Retrieved from: https://www.ensayos-filosofia.es/archivos/articulo/la-poesia-en-la-republica-de-platon-un-exilio-interior.

Vodrahalli, K., P.H. Chen, Y. Liang, C. Baldassano, J. Chen, E. Yong, C. Honey, U. Hasson, P. Ramadge, K.A. Norman, and S. Arora. 2018. Mapping between fMRI responses to movies and their natural language annotations. *NeuroImage* 180: 223–231. https://doi.org/10.1016/j.neuroimage.2017.06.042.

Wahbeh, H., D. Radin, C. Cannard, and A. Delorme. 2022. What if consciousness is not an emergent property of the brain? Observational and empirical challenges to materialistic models. *Frontiers in Psychology* 13. https://doi.org/10.3389/FPSYG.2022.955594.

Wainio-Theberge, S., A. Wolff, and G. Northoff. 2020. Bridging the gap–spontaneous fluctuations shape stimulus-evoked spectral power. *bioRxiv*. 2020-06.

Ward, L.M., S.M. Doesburg, K. Kitajo, S.E. MacLean, and A.B. Roggeveen. 2006. Neural synchrony in stochastic resonance, attention, and consciousness. *Canadian Journal of Experimental Psychology* 60 (4): 319–326. https://doi.org/10.1037/CJEP2006029.

Watson, J. 1924. *Behaviorism*. New York: W. W. Norton.

Webb, T.W., and M.S.A. Graziano. 2015. The attention schema theory: A mechanistic account of subjective awareness. *Frontiers in Psychology 6*. https://doi.org/10.3389/fpsyg.2015.00500.

Webb, T.W., K.M. Igelström, A. Schurger, and M.S.A. Graziano. 2016a. Cortical networks involved in visual awareness independent of visual attention. *Proceedings of the National Academy of Sciences of the United States of America* 113 (48): 13923–13928. https://doi.org/10.1073/pnas.1611505113.

Webb, T.W., H.H. Kean, and M.S.A. Graziano. 2016b. Effects of awareness on the control of attention. *Journal of Cognitive Neuroscience* 28 (6): 842–851. https://doi.org/10.1162/jocn_a_00931.

Whyte, C.J. 2019. Integrating the global neuronal workspace into the framework of predictive processing: Towards a working hypothesis. *Consciousness and Cognition* 73: 102763. https://doi.org/10.1016/j.concog.2019.102763.

Whyte, C.J., and R. Smith. 2021. The predictive global neuronal workspace: A formal active inference model of visual consciousness. *Progress in Neurobiology* 199: 101918. https://doi.org/10.1016/j.pneurobio.2020.101918.

Whyte, C.J., J. Hohwy, and R. Smith. 2022. An active inference model of conscious access: How cognitive action selection reconciles the results of report and no-report paradigms. *Current Research in Neurobiology* 3: 100036. https://doi.org/10.1016/j.crneur.2022.100036.

Wilkes, K.V. 1984. Is consciousness important? *British Journal for the Philosophy of Science* 35: 223–243.

———. 1988. Yishi, duo, us and consciousness. In *Consciousness in contemporary science*, ed. A. Marcel and E. Bisiach. Oxford: Oxford University Press.

———. 1995. Losing consciousness. In *Conscious Experience*, ed. T. Metzinger. Ferdinand Schöningh: Paderborn.

Wilson, S.M., A. Bautista, and A. McCarron. 2018. Convergence of spoken and written language processing in the superior temporal sulcus. *NeuroImage* 171: 62–74. https://doi.org/10.1016/J.NEUROIMAGE.2017.12.068.

Wilterson, A.I., C.M. Kemper, N. Kim, T.W. Webb, A.M.W. Reblando, and M.S.A. Graziano. 2020. Attention control and the attention schema theory of consciousness. *Progress in Neurobiology* 195 (June): 101844. https://doi.org/10.1016/j.pneurobio.2020.101844.

Wokke, M.E., A.R.E. Vandenbroucke, H.S. Scholte, and V.A.F. Lamme. 2012. Confuse your illusion. *Psychological Science 24* (1): 63–71. https://doi.org/10.1177/0956797612449175.

Wolff, A., L. Yao, J. Gomez-Pilar, M. Shoaran, N. Jiang, and G. Northoff. 2019. Neural variability quenching during decision-making: Neural individuality and its prestimulus complexity. *NeuroImage* 192: 1–14. https://doi.org/10.1016/j.neuroimage.2019.02.070.

Womelsdorf, T., J.M. Schoffelen, R. Oostenveld, W. Singer, R. Desimone, A.K. Engel, and P. Fries. 2007. Modulation of neuronal interactions through neuronal synchronization. *Science* 316 (5831): 1609–1612. https://doi.org/10.1126/SCIENCE.1139597.

Wu, W. 2018. In *The neuroscience of consciousness, Stanford Encyclopedia of Philosophy*, ed. E.N. Zalta. Stanford University. Available at: https://plato.stanford.edu/entries/consciousness-neuroscience/#RecuProcTheo.

Xu, X., H. Yuan, and X. Lei. 2016. Activation and connectivity within the default mode network contribute independently to future-oriented thought. *Scientific Reports* 6. https://doi.org/10.1038/srep21001.

Yang, H.-X., H.-X. Hu, Y.-J. Zhang, Y. Wang, S.S.Y. Lui, and R.C.K. Chan. 2022. A network analysis of interoception, self-awareness, empathy, alexithymia, and autistic traits. *European Archives of Psychiatry and Clinical Neuroscience 272*: 199–209. https://doi.org/10.1007/s00406-021-01274-8.

Yeshurun, Y., M. Nguyen, and U. Hasson. 2017a. Amplification of local changes along the timescale processing hierarchy. *Proceedings of the National Academy of Sciences of the United States of America* 114 (35): 9475–9480. https://doi.org/10.1073/pnas.1701652114.

Yeshurun, Y., S. Swanson, E. Simony, J. Chen, C. Lazaridi, C.J. Honey, and U. Hasson. 2017b. Same story, different story: The neural representation of interpretive frameworks. *Psychological Science* 28 (3): 307. https://doi.org/10.1177/0956797616682029.

Yeshurun, Y., M. Nguyen, and U. Hasson. 2021. The default mode network: Where the idiosyncratic self meets the shared social world. *Nature Reviews Neuroscience* 22 (3): 181–192. https://doi.org/10.1038/S41583-020-00420-W.

Yurchenko, S.B. 2022. From the origins to the stream of consciousness and its neural correlates. *Frontiers in Integrative Neuroscience* 16. https://doi.org/10.3389/FNINT.2022.928978.

Zadbood, A., J. Chen, Y.C. Leong, K.A. Norman, and U. Hasson. 2017. How we transmit memories to other brains: Constructing shared neural representations via communication. *Cerebral Cortex* 27 (10): 4988–5000. https://doi.org/10.1093/cercor/bhx202.

Zeki, S., J.D.G. Watson, C.J. Lueck, K.J. Friston, C. Kennard, and R.S.J. Frackowiak. 1991. A direct demonstration of functional specialization in human visual cortex. *Journal of Neuroscience* 11 (3): 641–649. https://doi.org/10.1523/jneurosci.11-03-00641.1991.

Zucca, D., and R. Medda, eds. 2019. On soul-body 'dualism' in Plato. In *The soul/body problem in Plato and Aristotle*, 75–94. Retrieved from: https://doi.org/10.5771/9783896657510-75.

Chapter 13
The Mind-Body Problem: An Overview of Proposed Solutions

Javier Alejandro Galadí

Abstract The Philosophy of Mind consists of problems concerning aspects and properties of the human mind. The most important of these problems is that of the relation between mind and body, or, more generally, between mental and physical phenomena. Usually referred to as the *mind-body problem*, this has been one of the fundamental problems in Philosophy since René Descartes (1596–1650) and his critics introduced it four centuries ago. The mental seems, at first glance, completely different from the physical. Physical properties are public, i.e., equally observable by everyone, but mental properties are not. It can be deduced that someone feels pain by his behaviour, but only that person can feel it directly. Conscious mental events are private in the sense that the subject has privileged access to them that no one has for the physical. Conscious experiences, such as the smell of jasmine, are completely different from the configurations and movements, however complex, of particles, atoms and molecules, or the physical changes of cells and tissues. Despite this, conscious phenomena do not seem to arise out of nothing, but from physical-biological processes in the body, especially from neural processes in the brain. But how can physical-biological systems have states such as thoughts, fears and hopes?

Keywords Mind-body problem · Dualism · Interactionism · Idealism · Epiphenomenalism · Theory of mind · Parallelism · Russellian monism · Neutral monism

13.1 Introduction

Naturalism, an increasingly widespread school in Philosophy, claims that everything that exists can be explained in physical and natural terms. At the same time, since René Descartes (1637), many philosophers perceive human consciousness as the most self-evident reality:

J. A. Galadí (✉)
Centre for Brain and Cognition, Universitat Pompeu Fabra, Barcelona, Spain
e-mail: javier.galadi@upf.edu

T. Lopez-Soto et al. (eds.), *The Theory of Mind Under Scrutiny*, Logic, Argumentation & Reasoning 34, https://doi.org/10.1007/978-3-031-46742-4_13

> And so, because our senses sometimes deceive us, I decided to suppose that nothing was such as they lead us to imagine it to be. And because there are men who make mistakes in reasoning, even about the simplest elements of geometry, and commit logical fallacies, I judged that I was as prone to error as anyone else, and I rejected as false all the reasoning I had hitherto accepted as valid proof. Finally, considering that all the same thoughts which we have while awake can come to us while asleep without any one of them then being true, I resolved to pretend that everything that had ever entered my head was no truer than the illusions of my dreams. But immediately afterwards I noted that, while I was trying to think of all things being false in this way, it was necessarily the case that I, who was thinking them, had to be something; and observing this truth: I am thinking therefore I exist, was so secure and certain that it could not be shaken by any of the most extravagant suppositions of the sceptics, I judged that I could accept it without scruple, as the first principle of the philosophy I was seeking.

Nevertheless, many authors have expressed that maintaining both claims simultaneously —naturalism and the existence of consciousness—is not easy, as Sellars (1922) expresses it:

> No problem is more crucial for a naturalistic view of the world than the mind-body problem.

We are puzzled that the mental has such different properties from the physical because these differences defy the naturalistic stance. Thus, for example, McGinn (1989) writes:

> Somehow, we feel, the water of the physical brain is turned into the wine of consciousness, but we draw a total blank on the nature of this conversion.

We therefore assume that there is an unresolved tension, conflict, or problem. In this chapter, we will analyse how this problem continues to challenge humanity four centuries after its birth.

13.1.1 Mind-Body Problem and Theory of Mind

There are multiple connections between the general theme of this volume—the theory of mind (ToM)—and the mind-body problem. If the mind-body problem analyses the relation between the mental and the physical, a prior question would be the existence of each of the two spheres. The existence of the physical world is usually assumed to be less problematic than the existence of minds. Facing the problem of the existence of minds is a task that can be subdivided into two: the problem of the existence of one's own mind and the problem of other minds. The latter philosophical problem is the one that seems closer to the ToM.

While the mind-body problem is the main problem in philosophy of mind, *the problem of other minds* is another classic problem in the discipline. There is a problem with other minds because everyone can doubt the existence of the minds of others. And this is so because although there are various arguments in favour of their existence, absolute certainty does not seem possible. Since one can come to think that others do not think or feel, this could lead to *solipsism*: the school of thought according to which I am alone in the universe. One argument in favour of the existence of other minds is that of analogy. According to this argument, since others are similar to me in the appearance they present—speaking, walking, etc.—they must possess thoughts and sensations, i.e., they must possess minds. So argues John Stuart Mill (1873):

> ... by what considerations am I led to believe, that there exist other sentient creatures; that the walking and speaking figures which I see and hear, have sensations and thoughts, or, in other words, possess Minds? The most strenuous Intuitionist does not include this among the things that I know by direct intuition. I conclude it from certain things, which my experience of my own states of feeling proves to me to be marks of it. These marks are of two kinds, antecedent and subsequent: the previous conditions requisite for feeling, and the effects or consequences of it. I conclude that other human beings have feelings like me, because, first, they have bodies like me, which I know, in my own case, to be the antecedent condition of feelings; and because, secondly, they exhibit the acts, and other outward signs, which in my own case I know by experience to be caused by feelings. I am conscious in myself of a series of facts connected by a uniform sequence, of which the beginning is modifications of my body, the middle is feelings, the end is outward demeanor.

However, it can be retorted that there is no way to prove it and that it is an induction from a single case —the subject's own— (Locke 1968). Another argument in favour is that of best explanation. This is a common type of reasoning in science: it is that the most rational thing to do is to believe the hypothesis that provides the best available explanation of a particular phenomenon at a given time. This argument, applied to accepting the existence of other minds as the best explanation for the observed evidence, can avoid the objections raised by the analogy argument (Pargetter 1984).

In any case, from an early age human beings develop the ability to attribute thoughts and sensations to other human beings. This is what is known as ToM (Premack and Woodruff 1978). Therefore, theory of mind is prior, in the phylogenetic development of the individual, to the philosophical problem of other minds. In fact, the vast majority of human beings develop the ToM and do not raise the philosophical problem of other minds in the course of their lives.

It should be noted that while the problem of other minds is a philosophical problem, theory of mind is, in principle, a human capacity that increases the probability of achieving success in everyday social interactions. It would therefore be a categorical error to place them on the same ontological plane. But it cannot be overlooked that there is also no universal agreement on the nature of ToM. The school of thought that characterizes ToM as a theory proper is called Theory-Theory. Theory-Theory asserts that from the time we are children we naturally try to construct theories to explain our observations. Thus, ToM would be a consequence of trying to find explanations that help us understand our environment. In particular, for eliminativists such as Paul Churchland, ToM is a folk theory, but an empirical one, analogous to the old Aristotelian physics (see Sect. 13.4.2.6). An alternative school to Theory-Theory is Simulation Theory, which holds that ToM is the result of projecting our own mental states onto others. Its origin is in the discovery of mirror neurons, and it proposes that ToM is based on the empathic capacity that these neurons provide us with for understanding the actions and emotions of others.

It may seem that having a ToM is a stronger claim than having a mind, since having a ToM would imply not only the requirement of having a mind but also the concept of mind. Having this concept could be an exclusive capacity of the human being—although there is a whole philosophical and scientific debate about it (Premack and Woodruff 1978; Penn and Povinelli 2007). However, from some schools of thought such as, for example, eliminativism, the opposite is true, since our capacity to attribute minds to others is admitted by means of a ToM, but, at the same time, the real existence of such minds is questioned (see Sect. 13.4.2.6).

Later in this chapter, we will return to ToM in the context of folk psychology, precisely in the Section devoted to eliminativism (see Sect. 13.4.2.6). This school of thought doubts the existence of minds and, therefore, is the one most focused on analysing the nature of ToM.

13.2 Origin of the Mind-Body Problem

It is possible that humanity has reflected on the mind-body problem since its origins. In fact, beliefs of a spiritual character may be associated with a reflection on the nature of consciousness. This would be reflected in the sacred texts of antiquity and even in earlier evidence:

> Neolithic burial practices appear to express spiritual beliefs and provide early evidence for at least minimally reflective thought about the nature of human consciousness . . . Preliterate cultures have similarly been found invariably to embrace some form of spiritual or at least animist view that indicates a degree of reflection about the nature of conscious awareness (Van Gulick 2021).

But it was in ancient Greece that signs of the formulation of the problem in terms like those of the modern period appear. In Homer and Hesiod, the notion of mind or soul appears opposed to that of the body, since the latter needs the former in order to pass from simple inert matter to a living organism. Thus, when the former abandons it, that is, with the separation of the body and the soul or mind, death ensues. In the pre-Socratics, the soul or mind has a material nature, although more subtle—for example, the air in Anaximenes—than that of the bodies. In Orphism and Pythagoreanism, the soul has a divine origin, and is eternal and immaterial (Russell 1945). This could be considered the origin of dualism and will have a continuity in Plato. Thus, in the *Phaedo* we find reflected the Orphic-Pythagorean tradition of the divine and immortal soul that connects the human being with the world of ideas while the body is corruptible. In the *Republic*, the *Phaedrus* and the *Timaeus*, Plato presents a soul with three parts: concupiscent, irascible, and rational. Only the latter will exist after the death of the body. Let us remember that in Platonic metaphysics the truly real has no material nature. The material is ephemeral, changeable, and deceptive.

In Aristotle, the dualism sensible-intelligible is replaced by the dualism matter-form. The soul becomes the form of the body, but its properties continue to be that of a life-giving principle and regulator of vital functions. Aristotle distinguishes three types of soul: the vegetative, the sensitive and the rational. The latter is present only in the human being and gives us the capacity for reasoning. Matter is for Aristotle something that can be anything, indeterminate. But all indeterminacy must be determined by a form. Thus, matter and form are inseparable, although a certain dualism is not avoided since the soul lacks a bodily organ.

In medieval philosophy, the influence of the Christian religion produces that the soul is considered a particular divine creation for everyone. The notion of soul has for the Christian, then, a personal and independent character of the biological or intellective processes to which Aristotle associated it, appearing in a third plane called spiritual.

In modern philosophy, Descartes will be the first who, beyond the religious or theological sphere, seeks relations between the intellectual mind and a body governed by physical mechanisms. And he will do so by detaching the body from a mind that is no longer the principle of life and movement, but rather inextensive thought (Russell 1945). Descartes in the seventeenth century was able to reduce all that exists (except God himself) to two types of substance: the *res extensa* and the *res cogitans*. At that time, this was a formidable philosophical achievement because for the first time the diversity of the known world could be simplified to this duality (second meditation of Descartes 1641). The *res extensa* is composed of the bodies, animate or not, that occupy extension in space, while the *res cogitans* can be identified with the soul, mind, or consciousness of thinking subjects:

> But what shall I now say that I am, when I am supposing that there is some supremely powerful and, if it is permissible to say so, malicious deceiver, who is deliberately trying to trick me in every way he can? Can I now assert that I possess even the most insignificant of all the attributes which I have just said belong to the nature of a body? I scrutinize them, think about them, go over them again, but nothing suggests itself; it is tiresome and pointless to go through the list once more. But what about the attributes I assigned to the soul? Nutrition or movement? Since now I do not have a body, these are mere fabrications. Sense-perception? This surely does not occur without a body, and besides, when asleep I have appeared to perceive through the senses many things which I afterwards realized I did not perceive through the senses at all. Thinking? At last, I have discovered it—thought; this alone is inseparable from me. I am, I exist—that is certain. But for how long? For as long as I am thinking. For it could be that were I totally to cease from thinking, I should totally cease to exist. At present I am not admitting anything except what is necessarily true. I am, then, in the strict sense only a thing that thinks; that is, I am a mind, or intelligence, or intellect, or reason—words whose meaning I have been ignorant of until now. But for all that I am a thing which is real, and which truly exists. But what kind of a thing? As I have just said—a thinking thing (Descartes 1641).

To arrive at this simplification, Descartes relied on the reduction to mechanistic explanations that at that time were beginning to describe much of the physiology and function of living beings. However, everything related to the human soul appeared to Descartes as irreducible to such explanations and constitutive of an independent and differentiated type of substance. One way to summarise the history of the mind-body problem is to say that in these four centuries, we have been unable to complete the reductive program. Simplifying the two substances to a single substance is, from a philosophical point of view, undeniably attractive.

Soon after the publication of Descartes' ideas, criticisms appeared concerning the way in which mind and body interact. This issue has been called *the problem of causal interaction*. Gassendi (1592–1655) was the first to point this out in 1641 and can be said to be the true creator of the mind-body problem, since Descartes had proposed a substantial dualism which he had not, in principle, seen as problematic. The main objection to the Cartesian proposal is that if the two substances are distinct and independent, it is difficult to explain the interaction between them. However, the body and the mind seem to interact in such a way that, for example, blows received by the body are perceived with a subjective sensation of pain and, conversely, our will is translated into movements of the body.

13.3 Definition of the Problem

Although I have defined the mind-body problem as the issue of the relation between the mental and the physical, the definition in the literature is multiple and not without some complexity. Sometimes the term 'problem of consciousness' is used as equivalent to the mind-body problem, but depending on the author, both problems may not be exactly the same. In fact, some authors speak of problems, in the plural, of consciousness. If we add to this Chalmers' hard problem (Chalmers 1995) as a reformulation that might not coincide with any of the previous ones, we can see that the definition of the problem is not simple.

In recent decades, some have renamed the problem the 'mind-brain problem', as the capacity to produce thought is more specifically attributed to the human brain. When the focus is on the 'problem of consciousness', the brain and the body are not seen as problematic physical realities, whereas the emergence of consciousness is. In 1995 David Chalmers renamed the problem the 'hard problem of consciousness'. The hard problem is the problem of experience, subjective experience, and the difficulty of explaining it in terms of physical events.

The ambiguity of the term 'consciousness' is often exploited by both philosophers and scientists writing on the subject. It is common to see a paper on consciousness begin with an invocation of the mystery of consciousness, noting the strange intangibility and ineffability of subjectivity, and worrying that so far, we have no theory of the phenomenon. Here, the topic is clearly the hard problem—the problem of experience. In the second half of the paper, the tone becomes more optimistic, and the author's own theory of consciousness is outlined. Upon examination, this theory turns out to be a theory of one of the more straightforward phenomena—of reportability, of introspective access, or whatever. At the close, the author declares that consciousness has turned out to be tractable after all, but the reader is left feeling like the victim of a bait-and-switch. The hard problem remains untouched (Chalmers 1995).

In any case, there are authors who consider the hard problem as a reformulation of the classical mind-body problem. The new 'hard problem' would be no more than an improved version of an old problem that appeared with Descartes and his critics in

1641, and the hard problem is hard simply because it is the mind-body problem (Westphal 2016).

When considering problems, in plural, of consciousness, these problems can be classified into three families (Van Gulick 2021): i) the *descriptive* questions as "What is consciousness?", "What are its principal features?", and "By what means can they be best discovered, described and modelled?"; ii) the *explanatory* questions as "How does consciousness of the relevant sort come to exist?", "Is it a primitive aspect of reality, and if not, how does (or could) consciousness in the relevant respect arise from or be caused by nonconscious entities or processes?"; and iii) the *functional* questions as "Why does consciousness of the relevant sort exist?", "Does it have a function, and if so, what it is it?", "Does it act causally, and if so with sorts of effects?", "Does it make a difference to the operation of systems in which it is present?, and if so, why and how?"

Finally, when we face the philosophical problem, mind or consciousness can be characterised either by focusing on its representational aspects, as Rosenthal (1997) or Dennett (1993) do, or on its experiential or phenomenal aspects, as Chalmers (1995) or Nagel (1974) do. The first characterisation is usually considered relatively more manageable in terms of cognitive explanations. Here I will therefore refer specifically to the second characterisation to directly address the perplexity it entails.

13.4 Proposed Solutions

The following is a sample of the main solutions proposed and the main objections they raise. As an open problem, new solutions continue to be proposed today, and it is impossible to list them all. Nor is there space here to consider the replies and counter-replies to the objections raised to each solution. Moreover, I will focus on the proposed solutions to the philosophical problem, leaving aside the scientific theories of consciousness that have proliferated in recent years.

Solutions to the mind-body problem can be classified according to whether they maintain the existence of the mental and the physical separately, or whether they try to reduce all that exists to a single type of entity. The former is called *dualisms* and the latter *monisms*. Dualists assert that both the mental and the physical are real, and that neither can be reduced to the other. Therefore, mental phenomena would be, at least in some respect, non-physical. Monism, on the other hand, does not accept fundamental divisions. Some authors have focused on the nature of the problem itself, instead of looking for solutions. Thus, the so-called New mysterians (Nagel 1974; McGinn 1989) attack physicalist positions and adopt an epistemic approach, arguing that the mind-body problem is currently unsolvable, and will perhaps always remain unsolvable for human beings. There are also proposed solutions that are not easily classifiable under one of the monism-dualism labels. Such is the case with panpsychism, emergentism, and functionalism.

13.4.1 Dualist Solutions

Dualism is roughly the thesis that not everything is fundamentally physical, and things that are not fundamentally physical are fundamentally mental (Chalmers 2013). Dualist proposals place at least some aspects of consciousness outside the realm of the physical, but the specific forms of dualism differ in what those aspects are. They can be divided into substance dualisms and property dualisms.

13.4.1.1 Substance Dualism

Substance dualism involves the existence of non-physical minds or selves as entities. One of the problems posed by substance dualism is that of causal interaction, i.e., how the two different substances posited by the theory, the mental and the physical, can have an impact on each other. Thus, substance dualism is subdivided into three forms according to the directions in which the causal interaction between the mental and the physical takes place: interactionism, epiphenomenalism, and parallelism.

Interactionism

According to interactionism, the mental and the physical, although distinct substances, can interact in some way. For example, Descartes (1664) speculated on the idea that this interaction could take place through the pineal gland. More recently, Popper and Eccles (2012) have considered the mind as an independent entity:

> . . . the hypothesis is that the self-conscious mind is an independent entity . . . that is actively engaged in reading out from the multitude of active centres in the modules of . . . areas of the dominant cerebral hemisphere. The self-conscious mind selects from these centres in accord with its attention and its interests and integrates its selection to give the unity of conscious experience from moment to moment. It also acts back on the neural centres . . . Thus, it is proposed that the self-conscious mind exercises a superior interpretive and controlling role upon the neural events by virtue of a two-way interaction . . . It is proposed that the unity of conscious experience comes not from an ultimate synthesis in the neural machinery but in the integrating action of the self-conscious mind on what it reads out from the immense diversity of neural activities in the . . . brain . . .

The main objection to interactionism is that physical science has shown us that the physical world seems to be self-sufficient in explaining natural phenomena. It is the so-called *principle of physical causal closure* which also seems to be applicable to a physical system such as the brain. Moreover, it is difficult to imagine how two different and ontologically independent substances could ever interact with each other (Stoljar 2010).

Epiphenomenalism

According to epiphenomenalism, physical events are causal with respect to mental events. But the reverse is not true: the mental has no causal power over the physical. Epiphenomenalism tries to respect the causal closure of the physical world and proposes that the mental is an epiphenomenon, i.e., a non-reducible secondary phenomenon that accompanies the physical without influencing it:

> Behavior is caused by muscles that contract upon receiving neural impulses, and neural impulses are generated by input from other neurons or from sense organs. On the epiphenomenalist view, mental events play no causal role in this process. (Robinson 2019)

There are two main objections to epiphenomenalism. The first is that epiphenomenalism seems to be incompatible with being aware that we are conscious, since for us to know that we have consciousness would have to produce some change in our brain. The second problem is that of the emergence of consciousness in biological evolution. If consciousness is epiphenomenal, then it has no effect on an organism's adaptive capacity and should not have been selected (Bailey 2006).

Parallelism

Parallelism is the school of thought according to which the mental and physical realms function synchronously without the need for to interact causally with the other. If epiphenomenalism respects the causal closure of the physical, parallelism preserves both the closure of the physical and the closure of the mental (Robinson 2002). An example of a parallelist proposal is Malebranche's *occasionalism*, according to which the soul and the body do not act directly on each other, but it is God who produces a sensation in the soul when the body experiences it, and who gives the body a movement when the soul desires it (Nadler 2010). Another example of parallelism is the *pre-established harmony* proposed by Gottfried W. Leibniz:

> ... one can say that no created substance exerts a metaphysical action or influx on any other thing. For, not to mention the fact that one cannot explain how something can pass from one thing into the substance of another ... (Leibniz 1989).

According to pre-established harmony, God arranged things from the beginning of creation so that both substances behave as if they were interacting, without the need for God's intervention for events (Craig 1998).

The main objection to parallelism is that the theory otherwise requires belief in a deity who intervenes in physical and mental events or programmes them in advance. In fact, Leibniz himself accuses Malebranche of *deus ex machina*, not realising that the same could be said of his solution (Clatterbaugh 1999).

13.4.1.2 Property Dualism

According to property dualism, there is only material substance, but it can instantiate two essentially different kinds of properties: physical properties and mental properties. One of the advantages of this type of dualism is that by not positing an immaterial substance it avoids religious connotations. Another is that it seems to avoid the problem of mental causation; there is no interaction between two different kinds of things. Moreover, mental properties are accepted as real and distinct from physical properties (Vintiadis 2019). In this school of thought, it is generally accepted that mental characteristics are supervenient on physical characteristics:

> supervenience might be taken to mean that there cannot be two events alike in all physical respects but differing in some mental respects, or that an object cannot alter in some mental respects without altering in some physical respects (Davidson 2001).

The dualism of properties that accepts the supervenience of the mental in the physical is often called *non-reductive physicalism*, since phenomenal properties supervene on physical properties but cannot be reduced to them (Kim 2007; Vintiadis 2019). In general, the dependence of the mental on the physical expressed by supervenience can be of many different types. As an extreme case, the mental could be identical to the physical (Francescotti 2011).

The main objection to property dualism is that if we respect the closure of the physical domain, and causal exclusion, i.e., that no event can have more than one sufficient cause, mental properties have no causal efficacy. Therefore, the conclusion is that phenomenal properties that are irreducibly mental are also merely epiphenomenal, i.e., they have no causal effect on physical events (Kim 2007).

Up to this point we have dealt with dualistic theories that refuse to reduce the duality of the mental and the physical to a single ontological category or to a single type of properties. In the next Section, we will approach the theories that have tried to unify the two realities, although, as we will see, each one does so in a different way.

13.4.2 Monisms

Dualisms become monisms when we reduce one of the two substances to the other. Thus, physicalism and idealism are the main forms of monism, although there are other options such as dual-aspect theories, neutral monism, and anomalous monism. An extreme case of physicalism is eliminativism.

13.4.2.1 Dual-Aspect Theories

The first monist response to Cartesian dualism is due to Baruch Spinoza in 1677, for whom thinking substance and extended substance are one and the same substance, which is now understood under this attribute, now under that one (Spinoza 1996). Spinoza identified this one substance with God or nature (*Deus sive natura*). It is one and the same thing that thinks and is extensive: the mental and the material would be from this perspective a single thing seen from different angles. In this way, the two very different spheres are coordinated: the correspondence between the two worlds would be due, in the end, to the fact that in fact there is only one.

Thus, Spinoza inaugurated the dual-aspect monism: the single substance of the world has a mental (experiential, intentional) aspect, just as it has a physical aspect. Dual-aspect monism respects both the physical and mental dimensions of existence equally (Skrbina 2014). Atmaspacher (2014) has analysed several variants of twentieth-century dual-aspect monism reaching the following conclusion:

> The one feature that ... variants ... have in common is that they regard the mental and the physical as two aspects of one underlying reality that itself is neutral with respect to the mind-matter split. This is the key point of dual-aspect approaches. They combine an (epistemic) dualism with an (ontic) monism and, in this way, suggest an alternative to the conventional physicalist program of naturalizing the mind. In fact, dual-aspect approaches consider both mind and matter to be naturalized by their underlying reality.

In the twentieth century, Julian Huxley (1942) defended monism against dualism based on the progress of science and the theory of evolution. But the substance of which the world is made reveals material or mental properties depending on the point of view: when the world is seen from the outside, we have matter and when it is seen from the inside, we have mind.

But if there is an underlying reality that we can understand as either mental or physical, depending on the point of view from which we observe it, neutral monism and dual-aspect theory share a central claim: there is an underlying reality that is neither mental nor physical. If the dual-aspect theory insists that the two aspects are fundamental and irreducible to each other, we would fall into panpsychism (see Sect. 13.4.3.3). If not, it would be closely associated with emergentism (see Sect. 13.4.3.1). In each case, the most challenging criticisms would be those of neutral monism, panpsychism or emergentism, respectively (Atmaspacher 2014; Skrbina 2014).

13.4.2.2 Idealism

Idealists say that the physical can be reduced to the mental, since the supposed physical world is empirical and therefore a social construct created from shared subjective experiences. This school of thought has its classical example in George Berkeley (1685–1753) for whom the objects of human knowledge are ideas which would be equivalent, in today's language, to the contents of consciousness or mental objects in the broad sense:

> Anyone who surveys the objects of human knowledge will easily see that they are all ideas that are either actually imprinted on the senses or perceived by attending to one's own emotions and mental activities or formed out of ideas of the first two types, with the help of memory and imagination, by compounding or dividing or simply reproducing ideas of those other two kinds. By sight I have the ideas of light and colours with their different degrees and variations. By touch I perceive hard and soft, heat and cold, motion and resistance, and so on; and each of these also admits of differences of quantity or degree. Smelling supplies me with odours; the palate with tastes; and hearing conveys sounds to the mind in all their variety of tone and composition. And when a number of these are observed to accompany each other, they come to be marked by one name and thus to be thought of as one thing. Thus, for example, a certain colour, taste, smell, shape, and consistency having been observed to go together, they are taken to be one distinct thing, called an 'apple'. Other collections of ideas constitute a stone, a tree, a book, and similar perceptible things; and these can arouse the emotions of love, hate, joy, grief, and so on, depending on whether they please or displease us. (Berkeley 2013)

For Berkeley, to exist is to be perceived, and although it is possible to conceive of something existing other than in a mind that perceives it—for example, we can imagine trees in a park and nobody that perceives them—it would just be framing ideas in your mind that you call trees and framing the idea that no one perceives them. This just goes to show that you have the power to form ideas in your mind. When we strive to conceive of the existence of external bodies, we are merely contemplating our own ideas. But the mind, which has no regard for itself, deludes itself into thinking that it can conceive, and does conceive, that bodies exist without being thought or without the mind (Berkeley 2013).

Even Berkeley himself was aware of several objections to his idealism: for example, it makes real things no different from imaginary things. It also seems absurd to suppress natural causes and attribute everything to the immediate operation of the mind. We could no longer say that fire heats, or that water cools, but that the mind heats or cools. On the other hand, we have the persistence of objects: do things continue to exist when no one perceives them? Another objection is the distinction between error and truth: since we judge the reality of things by our senses, how does one distinguish error from truth in situations such as when one thinks that an oar is crooked because one end is underwater? Finally, it also seems difficult from idealism to explain the similarity of specific objects of perception: why do certain things seem the same to all of us? (Berkeley 2013).

13.4.2.3 Neutral Monism

According to neutral monism ultimate reality is intrinsically neither mental nor physical but neutral. For neutral monists the difference between the physical and the psychological lies not in the object but in the direction of investigation. Ernst Mach (1838–1916), the father of modern neutral monism, calls neutral entities *events/sensations*. For him, reality consists of a viscous mass of events, which in some places (as in the ego) is more firmly coherent than in others (Mach 1959) and the differences between the mental and the physical come from the direction of investigation:

> Thus, the great gulf between physical and psychological research persists only when we acquiesce in our habitual stereotyped conceptions. A color is a physical object as soon as we consider its dependence, for instance, upon its luminous source, upon other colors, upon temperatures, upon spaces, and so forth. When we consider, however, its dependence upon the retina . . . it is a psychological object, a sensation. Not the subject matter, but the direction of investigation, is different in the two domains.
>
> Both in reasoning from the observation of the bodies of other men or animals, to the sensations which they possess, as well as in investigating the influence of our own body upon our own sensations, we have to complete observed facts by analogy. This is accomplished with much greater ease and certainty, when it relates, say, only to nervous processes, which cannot be fully observed in our own bodies—that is, when it is carried out in the more familiar physical domain—than when it is extended to the psychical domain, to the sensations and thoughts of other people. Otherwise, there is no essential difference. (Mach 1959)

Another paradigm of neutral monism is represented by William James (1842–1910), for whom consciousness is a non-entity, a mere echo of the archaic concept of the soul. The only thing that exists is *pure experience* in each present instant (James 1904). In the case of perceptual knowledge, the perceived object and its perception are only two names for an indivisible fact: experience. The object is in the mind, and the mind is around the object. Experience is part of a wider world, and its connections can be traced in different directions, which are known as the physical and the mental (James 1895). In the case of conceptual knowledge, two experiences are interrelated in the same subject, where the second piece is representative of the first in the practical sense of substituting it in various operations, sometimes physical and sometimes mental, that lead to its associates and results (James 1904).

The last representative of classical neutral monism is Bertrand Russell (1872–1970) for whom the *sensation* we have when we see an object is simply that object. The object and our sensation when we perceive it are the same thing (Russell 2022). For Russell, all that physics gives us are certain equations which give abstract quantitative properties of their changes. The qualitative aspect of mental objects stems from the fact that they are but sensations which reveal their intrinsic character, and which offer the most indubitable knowledge of the world. On the contrary, our knowledge of the physical world is purely abstract, since we know only certain logical features of its structure, but nothing of its intrinsic character (Russell 1995). As to what changes, what it changes from and what it changes to, physics is silent on this point (Russell 1959).

One objection to neutral monism is that we have no indication of what these neutral entities are. In some versions of neutral monism, neutral entities seem to have a mixture of physical characteristics—such as being in space—and phenomenological characteristics—qualitative character. And this makes them appear to be entities that are both physical and mental, rather than neither physical nor mental. As neutral entities little can be said about them, and to the extent that their qualities are described they appear to be either physical or mental. Moreover, their supposedly neutral elements can be interpreted as mental because the way in which physical objects are constructed from neutrals is reminiscent of Berkeley's subjective idealism (Popper and Eccles 2012). The fact that there are intrinsic properties that explain the phenomenal and extrinsic relationships that construct the physical can be seen as a metaphysical speculation that is difficult to prove (Chalmers 1996). Ordinary material objects must be constructed from sensations. However, neutral monism was never able to show the method of construction and produced no more than sketches of how it should proceed, but never a set of working plans (Tully 2003). Chalmers has objected that even if fundamental neutral entities had constitutively phenomenal qualities there need not necessarily be conscious experience of those qualities. He relies on the quality/awareness gap, analogous to the physical/awareness gap when attacking physicalism. No instantiation of qualities requires awareness of them. It is conceivable that all those qualities and properties are instantiated without any awareness of them. And this leads us to doubt that there is room for consciousness in a neutral universe (Chalmers 2013).

13.4.2.4 Mind/Brain Identity Theory

A special type of physicalism is that proposed by identity theories. These theories argue that the states and processes that we call mental are states and processes of the nervous system. Instead of speaking of correlation or correspondence, they assert directly that the former are identified with the latter. In this sense, they oppose Cartesian dualism, since the mental and the physical are not ontologically exclusive categories, but mental phenomena occur as states and processes in certain physical systems, specifically, in our nervous system. Thus, the mind-body problem disappears as the mental is identified with a region of the physical world and there is no need for interaction between two different things.

For Ullin T. Place, pioneer of identity theory, behaviourism did not explain conscious experience:

> What I do want to assert, however, is that the statement 'consciousness is a process in the brain', although not necessarily true, is not necessarily false. 'Consciousness is a process in the brain', on my view is neither self-contradictory nor self-evident; it is a reasonable scientific hypothesis, in the way that the statement 'lightning is a motion of electric charges' is a reasonable scientific hypothesis. (Place 1956)

Thus, by presenting his thesis as a scientific hypothesis, it could not be refuted *a priori* because of purely logical arguments. The logical objections that could be made to the identity contained in the first statement are equivalent to those that could be made to that expressed in the second. Mental concepts such as 'sensation' and physical concepts such as 'neurophysiological activity' are different in the same way that 'lightning' and 'movement of electric charges' are different, and we identify different concepts thanks to scientific research. Let us note that competent speakers who use the expression 'lightning' meaningfully do not need to translate it into scientific language. Thus, 'sensation' and 'neurophysiological processes' can be identified, even though they are associated with different epistemic pathways: ordinary processes of observation, on the one hand, and scientific procedures, on the other. Other identity advocates as Feigl (1958) and Smart (1959) reach similar conclusions by making use of the distinction between reference and sense by arguing that 'sensation' and 'neurophysiological activity' may have different senses but could have the same referent.

In what can be seen as a last theoretical proposal within identity, David Lewis (1966) stresses that the defining characteristic of the mental is its causal role and distinguishes between *type-type* identity theories (which would be the identity theories of the authors analysed so far) and his *token-token* identity theory. According to type-type identity theories, when two people share the same mental state, they also share the same nervous system state. According to token-token or case identity theories, two people who share the same mental state may not share the same type of state of their nervous systems. This distinction is the starting point for later schools such as functionalism and anomalous monism.

The most common objection to identity theory is the *multiple realizability argument*. This claims that if mental states can be realized in other kinds of systems than brains, identity theory is false (Putnam 1967). Another frequent objection is that identity theories do not account for *qualia*, the qualitative components of conscious experiences, such as those we experience when we see the colour red, for example (Chalmers 1996).

13.4.2.5 Anomalous Monism

Mental events resist being explained by physical theory. How can this fact be reconciled with the causal role of mental events in the physical world? On the assumption that both the causal role and the anomaly of mental events are undeniable facts, Donald Davidson's aim as the creator of anomalous monism was to explain how they can be compatible with the physical world. Davidson (2001) formulates this apparent contradiction by considering three principles:

> The first principle asserts that at least some mental events interact causally with physical events. (We could call this the Principle of Causal Interaction) …
>
> The second principle is that where there is causality, there must be a law: events related as cause and effect fall under strict deterministic laws. (We may term this the Principle of the Nomological Character of Causality) …
>
> The third principle is that there are no strict deterministic laws on the basis of which mental events can be predicted and explained (the Anomalism of the Mental). (Davidson 2001)

How can the three principles be reconciled? Causally interacting mental events (first principle) must instantiate some property of strict law (second principle), but mental properties are not suitable for inclusion in strict laws (third principle). Therefore, mental events must instantiate some other property that is suitable for such inclusion, and this other property must be physical. Consequently, causally interacting mental events must be identical to physical events. The conclusion Davidson (2001) reached is that a distinction had to be made between *type identity* and *token identity*: although the class or type of mental events cannot be reduced to the class of neural events, each individual mental event—each case or token—is nevertheless identical to a physical event.

It is often objected to anomalous monism that the identity of two individual events is not compatible with the types or classes by which they are characterised being irreducibly different (Leder 1985). It has also been blamed on anomalous monism, which implies an absence of causal power of mental properties. If we assume that a given event, by virtue of its mental property, causes a physical event, the causal closure of the physical domain says that this physical event must also have a physical cause. We can consider the possibility that each of them is only a partial cause, and that the two together constitute a complete or sufficient cause. But this violates the principle of causal closure of the physical since a complete causal story of how this physical event comes about is at least partially outside the physical realm. Could it be that the mental cause and the physical cause are sufficient? In that case, the physical effect is overdetermined. Moreover, the idea of overdetermination also seems to violate the principle of causal closure of the physical (Kim 1989).

13.4.2.6 Eliminativism

The most radical physicalist variant is eliminativism which denies the existence of consciousness. In addition, it advocates eliminating the mentalistic vocabulary and replacing it with the neuroscientific one. The two ideas from which eliminativism starts are that our notions about our mental life may derive from our cultural heritage and that the reference of the words we use to denote mental states are physiological states.

In philosophy of mind, it is often claimed that one of the distinguishing characteristics of the mental from the physical is that the mental is *intentional*. The misnomer *intentionality* referring to mental states means that these states are always *about something*. In many cases, that 'something' is a proposition, i.e., the meaning of a declarative sentence such as 'my arm is broken'. In these cases, the fundamental units of thought are called *propositional attitudes*. Thus, the content of a propositional attitude is a proposition that can be true or false from the perspective of the subject. And the subject in turn can have different attitudes towards that proposition such as belief, desire, or fear. An example of a propositional attitude is that the subject fears that her/his arm is broken.

This view of other minds (and our own) composed of propositional attitudes as units is *theory of mind*, also called *folk psychology* (FP) by eliminativists as Paul Churchland (1981):

> Each of us understands others, as well as we do, because we share a tacit command of an integrated body of lore concerning the law-like relations holding among external circumstances, internal states, and overt behavior. Given its nature and functions, this body of lore may quite aptly be called 'folk psychology'.

FP is embedded in our common sense and constitutes the shared body of wisdom that allows us to explain and predict other people's behaviour, desires, beliefs, fears, intentions, perceptions, etc. However, *eliminativism* argues that FP is fundamentally false, i.e., that common sense misleads us about psychological phenomena and that we will need future neuroscience to truly understand them (Churchland 1981). Eliminativism is an extreme physicalism that asserts that the neuroscience of the future will eventually be integrated into physical science by eliminating references to mind or consciousness. For eliminativists, intentionality and propositional attitudes are at the core of FP and what makes the mental seem so different from physical phenomena. Therefore, propositional attitudes are the target of their criticism.

The eliminativists argue that FP is an empirical science analogous to the old Aristotelian physics, which expressed our common sense of the physical. For eliminativists, intentionality would not be a mystery but a structural feature of FP. Thus, FP and mathematical physics are sciences whose only difference is the abstract entities they handle numbers in the case of physics and propositions in the case of FP (Churchland 1981).

But, from the eliminativist point of view, as an empirical theory FP is false since, for example, conceiving learning as the manipulation and storage of propositional attitudes we would be unable to explain pre-linguistic learning. Nor does it have explanatory power for phenomena such as the nature and dynamics of mental illness, creative imagination, differences in intelligence between individuals, the nature and functions of sleep, the construction of three-dimensional visual images from two-dimensional stimuli, perceptual illusions, or memory. In addition, the attribution of propositional attitudes has lost strength over the course of human history as we have moved from a generalised *animistic* approach to nature to one restricted to higher animals. FP as a research programme would be stagnant. Finally, materialistic neuroscience fits better than FP into the framework of natural history and the physical sciences. FP cannot be part of this framework because its intentional categories are not reducible to it (Churchland 1981).

A major objection to eliminative materialism is that it is self-refuting. If there really are no propositional attitudes such as beliefs, then the eliminativists' belief that there are supposedly no beliefs would not exist (Reppert 1992). One can also refute eliminative materialism by arguing that FP is highly successful in predicting human behaviour. Its success could be compared to that of the natural sciences and improves on that of most recent psychological and neurobiological theories. Moreover, FP not only predicts but also justifies, evaluates, praises, and rationalises (Lahav 1992). Finally, all eliminativist reasoning is based on FP being an empirical theory subject to refutation, but there is an alternative view that FP is more a simulation our mind makes of what the other would do with the beliefs and desires we think they have, i.e., a putting ourselves in the other's situation rather than a complete theory of mind (Goldman 1992).

Finally, it is worth emphasizing the differences between eliminativism and identity theories, which can sometimes be unclear. For the former, conscious sensations simply do not exist; for the latter, they exist but are no more than neurophysiological processes. According to eliminativism the mentalistic notions of FP are irreferential, they do not designate anything real. On the contrary, the theories of identity are a reductionist materialism in which mentalistic notions do indeed designate something, but something different from what FP designated, and to correct their reference, we simply must reduce them to the vocabulary of neuroscience.

13.4.2.7 General Physicalism

We have seen peculiar forms of Physicalism such as Eliminativism, Anomalous monism, or Identity theories. But there is a more generally accepted physicalism that does not deny consciousness, nor does it use either token-token or type-type identities between consciousness phenomena and neurological phenomena. This is the physicalism that expects that sooner or later we will find a scientific explanation of consciousness. This Physicalism (or materialism) is, broadly speaking, the thesis that everything is fundamentally physical (Chalmers 2013). Physicalists claim that, despite appearances, mental states are only physical states. Physicalism offers a simple and unified view of the world but seems to have difficulties in offering a satisfactory explanation of consciousness.

These difficulties are expressed in the following well-known arguments against physicalism:

Objections to Physicalism: New mysterianism

The notion of new mysterians groups authors who defend the idea that consciousness cannot be explained in physical terms (Flanagan 1991). The expression 'Mysterians' is based on the distinction that Chomsky (1976) drew between problems—questions that can be understood and solved—and mysteries—incomprehensible and unsolvable questions. The old Mysterians would be the dualists for whom consciousness is not a natural phenomenon (Flanagan 1991). For the new Mysterians (or, for brevity, simply Mysterians), consciousness is a natural phenomenon, but scientifically unapproachable. One of the main representatives of Mysterianism, Thomas Nagel (1987), puts it this way:

> Dualism is the view that you consist of a body plus a soul, and that your mental life goes on in your soul. Physicalism is the view that your mental life consists of physical processes in your brain. But another possibility is that your mental life goes on in your brain, yet that all those experiences, feelings, thoughts, and desires are not physical processes in your brain. This would mean that the grey mass of billions of nerve cells in your skull is not just a physical object. It has lots of physical properties—great quantities of chemical and electrical activity go on in it—but it has mental processes going on in it as well.

The following are the main Mysterian arguments:

A) Nagel: External Objective Third-person and Internal First-person Subjective Points of View

For Nagel (1974), an organism has conscious mental states if and only if there is something that is like to be that organism. The physicalist view would require at least an idea of consciousness as the subjective character of experience. But the subjective character is not captured by reductive analyses of the mental:

> I do not deny that conscious mental states and events cause behavior, nor that they may be given functional characterizations. I deny only that this kind of thing exhausts their analysis. Any reductionist program must be based on an analysis of what is to be reduced. If the analysis leaves something out, the problem will be falsely posed. It is useless to base the defence of materialism on any analysis of mental phenomena that fails to deal explicitly with their subjective character. For there is no reason to suppose that a reduction which seems plausible when no attempt is made to account for consciousness can be extended to include consciousness. Without some idea, therefore, of what the subjective character of experience is, we cannot know what is required of a physicalist theory.
>
> While an account of the physical basis of mind must explain many things, this appears to be the most difficult. It is impossible to exclude the phenomenological features of experience from a reduction in the same way that one excludes the phenomenal features of an ordinary substance from a physical or chemical reduction of it—namely, by explaining them as effects on the minds of human observers. If physicalism is to be defended, the phenomenological features must themselves be given a physical account. But when we examine their subjective character, it seems that such a result is impossible. The reason is that every subjective phenomenon is essentially connected with a single point of view, and it seems inevitable that an objective, physical theory will abandon that point of view. (Nagel 1974)

To illustrate the connection between subjectivity and point of view, and the divergence between subjective and objective conceptions, Nagel proposes the example of the bat. Its experiences have a specific subjective character that is beyond our ability to conceive. Even the subjective character of the experience of a person born deaf and blind and mine are mutually inaccessible. And that affects the mind-body problem because it does not seem possible that the character of subjective experiences can be revealed from the physical functioning of the organism as an objective fact that can be observed and understood externally. In physical science, the aim is to know the thing by eliminating the subjectivities of the scientist's particular point of view. But what would be left of what it was like to be a bat if the bat's point of view were eliminated? If the subjective character of experience is comprehensible only from one point of view, then any shift towards greater objectivity removes us from the real nature of the phenomenon (Nagel 1974).

Daniel Dennett (1993) has objected to Nagel that the subjective experience of the bat may be accessible to the extent that its characteristics can be studied scientifically. We know, for example, that echolocation has a limited range in terms of distance from detected objects.

B) Jackson: Epistemologic Argument

The limits of the objective point of view are also emphasised by Frank Jackson (1982) who illustrates his epistemological argument or *knowledge argument* through the fictional character Mary. She is a scientist who investigates the world from a black and white room through a black and white television monitor. She specialises in the neurophysiology of vision and has all the physical information that can be obtained about what happens when we see colours. Jackson concludes that physicalism is false by the following reasoning:

> What will happen when Mary is released from her black and white room or is given a colour television monitor? Will she learn anything or not? It seems just obvious that she will learn something about the world and our visual experience of it. But then it is inescapable that her previous knowledge was incomplete. But she had all the physical information. Ergo there is more to have than that, and Physicalism is false. (Jackson 1982)

For Jackson (1982), the strength of this argument lies in the fact that you can have all the physical information without having all the information you need to have.

Daniel Dennett (1993) has objected that if Mary already really knew all about colour, that knowledge would necessarily include a deep understanding of why and how human neurology causes us to perceive colour *qualia*. Although she has not been in the physical state of seeing the colour red, she knows what it would be like to be in that state since she also knows the laws that relate that state to other states of seeing different colours (Maloney 1985). It has also been argued that Mary does not acquire a new factual knowledge, but a new ability (Lycan and Prinz 2008). It can even be argued that she acquires knowledge by acquaintance of a qualia but does not acquire any new propositional knowledge by doing so (Conee 1994).

C) Levine: The Explanatory Gap

Joseph Levine (1983) provides an epistemological argument to prove that any attempt to find psychophysical laws leaves an explanatory gap. This is because there is no way to determine exactly which statements about such laws are true. Consider, for example, the case of heat and the motion of molecules. Everything that needs to be explained about heat is explained as the motion of molecules. So, it is logical to conclude that heat and the motion of molecules are the same thing. On the contrary, there is nothing that we can determine about the physical substrate that explains why a conscious experience has the qualitative character that it has. Or, to put it another way, understanding its physical or functional properties does not explain or make intelligible what that particular experience is. It is therefore conceivable that there is a physical substrate without the usually associated experience, and vice versa:

> The idea is this. If there is nothing we can determine about C-fiber firing that explains why having one's C-fibers fire has the qualitative character that it does—or, to put it another way, if what it's particularly like to have one's C-fibers fire is not explained, or made intelligible, by understanding the physical or functional properties of C-fiber firings—it immediately becomes imaginable that there be C-fiber firings without the feeling of pain, and vice versa. We don't have the corresponding intuition in the case of heat and the motion of molecules—once we get clear about the right way to characterize what we imagine—because whatever there is to explain about heat is explained by its being the motion of molecules. So, how could it be anything else? (Levine 1983)

Against the explanatory gap, Papineau has argued that it is our inability to free ourselves from dualistic thinking that makes us think that there is something inexplicable in mind-brain cases. One reason for this is the cultural inheritance we received since Descartes (Papineau 2011). Another reason for such inability would be that we have two different cognitive systems for thinking about mental and material processes. On the one hand, we have *folk psychology* for attributing mental states and, on the other hand, we have *folk physics* for reasoning about the material world (Papineau 2011; Bloom 2005).

D) McGinn: Cognitive Closure

We have seen how Levine's explanatory gap asserts that there is a practical limit to our current explanatory capacities. Colin McGinn (1989) goes further by claiming that, given our human cognitive limits, we will never be able to bridge the gap. For this, he introduces the idea of cognitive closure: A type of mind M is cognitively closed with respect to a property P (or theory T) if and only if the concept formation procedures available to M cannot be extended to an understanding of P (or an understanding of T). Human beings would be cognitively closed to a natural explanation of consciousness, since we would always be puzzled as to how any property, we discover instantiated in the brain could give rise to consciousness. For McGinn (1989) our inability to solve the scientific problem solves the philosophical problem:

> My position is both pessimistic and optimistic at the same time. It is pessimistic about the prospects for arriving at a constructive solution to the mind-body problem, but it is optimistic about our hopes of removing the philosophical perplexity. The central point here is that I do not think we need to do the former in order to achieve the latter. This depends on a rather special understanding of what the philosophical problem consists of in. What I want to suggest is that the nature of the psychophysical connection has a full and non-mysterious explanation in a certain science, but that this science is inaccessible to us as a matter of principle.

There is no philosophical mind-body problem because there is possibly some property of the brain that naturally explains consciousness. But let us be cognitively closed to it. The philosophical problem about consciousness arises from the feeling that we have to accept that nature contains miracles. But the sense of miracle comes from us, not from the world. In reality, there would be nothing mysterious about how the brain generates consciousness. One possible reason why humans are conceptually unable to grasp the nature of the psychophysical link is the intrinsically spatial nature of both our human perceptual concepts and the scientific concepts we derive from them. The mental, by contrast, appears to us as intrinsically non-localised in physical space (McGinn 1995).

It has been argued against McGinn that his reasoning is incoherent, since successfully understanding a problem while failing to understand what its solution might be seems unlikely. For the mind-matter relation to be problematic, our minds must necessarily first possess some understanding of what that relation would be. The only problems that we cannot solve are those problems that we cannot formulate (Kriegel 2003).

E) Chalmers' Zombies

Chalmers (1996) argues that consciousness escapes any reductive explanation in physical terms. For this, he turns to phenomenal zombies, who are physically and functionally identical to human beings, but who lack experience. There is no phenomenal sensation for them. They have no conscious experience: everything is dark inside them. Although Chalmers (1996) himself acknowledges that the existence of such zombies is unlikely, the key point of his reasoning is that there is no logical reason why they should not exist:

> The idea of zombies as I have described them is a strange one. For a start, it is unlikely that zombies are naturally possible. In the real world, it is likely that any replica of me would be conscious. For this reason, it is most natural to imagine unconscious creatures as physically different from conscious ones—exhibiting impaired behavior, for example. But the question is not whether it is plausible that zombies could exist in our world, or even whether the idea of a zombie replica is a natural one; the question is whether the notion of a zombie is conceptually coherent. The mere intelligibility of the notion is enough to establish the conclusion.

And if no internal contradiction can be revealed, then the zombie world is logically possible. His argument goes like this: according to physicalism, everything in our world is physical. Therefore, a world in which all physical facts are the same as those in our real world must contain everything that exists in our real world. In particular, conscious experience must exist in such a possible world. But we can conceive of a zombie world and imply that such a world is possible. Therefore, physicalism is false (Chalmers 1996).

It has been argued that Chalmers' argument is circular since it presupposes that the physical characteristics of humans are not what produce subjective experiences, which is precisely what Chalmers intended to demonstrate. To show that the zombie argument is circular one can think of a creature called *zoombies* (with double 'o') that are non-physical creatures identical to us, but without consciousness. Their existence would refute dualism by showing that consciousness is not non-physical, i.e., it is physical. And since we can conceive of both zombies and zoombies this kind of argument does not serve to settle the issue between dualism and physicalism (Brown 2010).

F) Inverted Spectrum

A conceivability argument against physicalism can be found without the need to establish the logical possibility of a zombie world. It is enough to establish the logical possibility of a world physically identical to ours in which the facts about conscious experience are different from those of our world. One could imagine, for example, that where I have a red experience, my *physical twin* in another possible world has a blue experience, and vice versa. Again Chalmers (1996), puts it this way:

> Of course, he will call his blue experiences "red", but that is irrelevant. What matters is that the experience he has of the things we both call "red"—blood, fire engines, and so on—is of the same kind as the experience I have of the things we both call "blue", such as the sea and the sky.
>
> The rest of his colour experiences are systematically inverted with respect to mine, in order that they cohere with the red-blue inversion. Perhaps the best way to imagine this happening with human colour experiences is to imagine that two of the axes of our three-dimensional colour space are switched—the red-green axis is mapped onto the yellow-blue axis, and vice versa. To achieve such an inversion in the actual world, presumably we would need to rewire neural processes in an appropriate way, but as a logical possibility, it seems entirely coherent that experiences could be inverted while physical structure is duplicated exactly. Nothing in the neurophysiology dictates that one sort of processing should be accompanied by red experiences rather than by yellow experiences.

The mere fact that a subjective experience in our world is different in a physically identical world would refute physicalism. To achieve such a reversal in the real world, we would have to rewire the neural processes in the right way. But as a logical possibility, it would be consistent for subjective experiences to be reversed as long as the physical structure remained the same. Nothing in known neuroscience suggests that one type of visual information processing should be accompanied by green experiences instead of blue ones (Chalmers 1996).

Since this argument is also based on conceivability, the objections against it would be similar to those stated against Chalmers' zombies.

Global Objections to New Mysteryanism

Beyond the objections against each of the particular Mysterian arguments, one can find arguments against Mysterianism in general. We can divide the overall responses to the Mysterians into two groups: those who do not accept their way of arguing and those who accept that the scenarios they describe are conceivable and coherent, but do not share their conclusions.

In the first group, the criticisms focus on the concept of *conceivability* that several of the Mysterians' arguments use to establish that a universe physically identical to ours, but different in the phenomenological aspect, is conceivable and therefore -according to the Mysterians- possible. Within this first group, Patricia Churchland (2002) has questioned the link between the conceivable and the possible:

> For the sake of argument, I have played along with the underlying assumption that we understand quite well the scope and limits of the domain of the logically possible. Nevertheless, this assumption is deeply flawed. Quine demonstrated in 1960 that such an assumption is actually just a bit of philosophical self-deception. A few hand-picked examples of what is and is not logically possible seem straightforward enough, but outside of these, all is fantasy, or group-think, or depends on self-serving definition. Not surprisingly, the especially controversial cases are those where philosophers want logical possibility to give them some real metaphysical leverage. And the argument at hand is very much a case in point. Standing back a bit, one does find something unconvincing in the idea that the conveniently elastic and philosophically concocted notion of logical possibility should dictate to neurobiology what it can and cannot discover—ever.

In the second group, a whole family of responses called *phenomenal concepts strategy* (PCS) is found. Chalmers (2006a) summarizes them as follows:

> Proponents of the phenomenal concept strategy typically allow that we are faced with a distinctive epistemic gap in the physical-phenomenal case, one that is in certain respects unlike the epistemic gaps that one finds in the standard cases. But they hold that this distinctive epistemic gap can be explained in term of certain distinctive features of phenomenal concepts. And they hold that these distinctive features are themselves compatible with an underlying ontological monism.

Phenomenal concepts refer to the subjective quality of the experiences of a given subject. It is necessary to have had the subjective experiences to which phenomenal concepts refer in order to have and understand them. According to the PCS, the concepts we use to refer to physical facts and phenomenal concepts are of different and isolated kinds. Even if we had all the information in terms of physical concepts and knew everything about the physical state of an organism, we could not conclude anything about the phenomenal concepts of a given conscious experience.

Different authors within the PCS defend different causes for this heterogeneity. Thus, there are proposals such as the Recognitional concepts (Loar 1990), the Distinct conceptual roles (Hill 1997), the Indexical concepts (Perry 2001; O'Dea 2002), and the Quotational concepts (Papineau 2002). For example, some argue that phenomenal concepts are a class of *indexical concepts*, i.e., they point to neurophysiological states in an indexical way analogous to how the term "I" refers to me (Perry 2001; O'Dea 2002). Knowing all non-indexical facts about the world would not allow one to deduce any facts presented indexically. Another example of strategy within PCS claims that only phenomenal concepts contain the states to which they refer and characterizes them as *quotational concepts* (Papineau 2002). Thus, the mental states to which phenomenal concepts refer are included in those concepts. In Papineau's quotational proposal, phenomenal concepts always have the underlying structure "That experience: ___", where experience itself occupies the empty space after the colon. Again, knowing all non-quotational facts about the world would not allow one to deduce any quotationally presented facts.

13.4.3 Beyond Monisms and Dualisms

Given the drastically different attributes of the mental and the physical, the unified vision of reality is a great challenge for philosophers, as we have seen. The world exhibits two modes of being that elude integration. Attempts to unify it on the basis of the mental and the material define the various monistic schools. Conversely, various schools of dualism expose the impossibility of monistic integration and propose a separation of some sort between mind and matter. Whether monism is idealist or physicalist, the main obstacle is the complex set of issues that arise from reductionism. The dualist position seems no better: although its insistence on the gap between materiality and the various functions of mind is its strong point, the solution of separating them stumbles, for example, over the thorny problem of mind-body interaction. As an alternative, philosophical schools have arisen which are hardly classifiable as monistic or dualistic and which we will analyse below:

13.4.3.1 Emergentism

Already in classical Greece, Aristotle claimed that the whole is greater than the sum of its parts, but it was J. S. Mill (1843) who exploited the idea to propose the existence of *heteropathic laws* that do not comply with the principle of composition of causes. It was finally his disciple G. H. Lewes (1875) who introduced the term *emergent* to refer to heteropathic effects. *Prima facie*, emergence occurs when a complex system is observed to have properties or behaviours that its components do not have on their own, i.e., they only emerge when the parts interact as an overall complex system. In philosophy of mind, *emergentism* has been used to interpret the mental as an emergent property of the human brain, in which the components are clearly physical (Broad 2008).

However, due to the vagueness of the *prima facie* definition of emergence, different interpretations arise. Two of them are paradigmatic: Strong and weak emergence. *Strong emergence* means that the causal power of the emergent property is irreducible to that of the micro-properties in which it supervenes (see Sect. 13.4.1.2). Strong emergence exerts its influence directly downwards, in contrast to the functioning of a simple structural macro-property, whose causal influence is produced through the activity of its constituent micro-properties (O'Connor 1994). For Jaegwon Kim (2006), *downward causation* is a requirement of a self-respecting emergent theory:

> There is no question that emergentists should want downward causation. Emergent properties must do some serious causal work, and this includes their capacity for projecting causal influence downward, affecting the course of events at a purely physicochemical level. Causally impotent properties are explanatorily useless, and there would be little point in positing them or acknowledging their existence in scientific theory. British emergentists like Samuel Alexander and C. Lloyd Morgan thought of emergent properties as active causal agents in the process of cosmic evolution, in producing increasingly richer and more variegated phenomena—from molecules and atoms to life, from life to mind, and so

> on. Equally clearly, many contemporary advocates of emergence, for example, Roger Sperry, want emergent properties to play a significant explanatory role in scientific theory. Causally inert, epiphenomenal properties obviously are not able to fulfil such roles.

In contrast to the strong emergency, *weak emergence* occurs when the macrostate can be derived from the microdynamics and external conditions by simulation (Bedau 1997). From a philosophical point of view, strong emergence and weak emergence are diametrically opposed. Strong emergence, if it really exists, would presuppose the incompleteness of physicalism. In contrast, weak emergence itself supports physicalism by showing how all emergent phenomena are based on underlying laws (Chalmers 2006b).

Therefore, I will henceforth focus on strong emergentism since weak emergentism can be assimilated to physicalism. When the question arises as to whether there really strong emergence in nature is, the answer is usually that the best candidate for it is human consciousness. But an emergence beyond the weak implies that high-level facts and laws are not deducible from low-level laws. Simulations would be unable to deduce facts about some high-level phenomena. And this, in turn, implies an inability to deduce even all low-level facts from low-level laws, since if all low-level facts were derivable, it would be possible to deduce high-level facts from them due to supervenience (see Sect. 13.4.1.2 for the definition of supervenience). Therefore, strong emergence implies incompleteness of physical laws even in the characterisation of low-level processes. This feature of strong emergence can be called top-down or downwards causality and means that the higher level is not only irreducible but also causally effective. A consequence of this is that low-level laws are incomplete as a guide for the evolution of both low-level and high-level processes in the world. It must be emphasised that the causal impact of a high-level phenomenon on low-level processes is not deducible even in principle from the initial conditions and the low-level laws (Chalmers 2006b).

Precisely, the main objection to strong emergence is related to the downwards causal powers of emergent properties. Kim's argument is based on three principles: (i) Emergent properties supervene on microphysical properties, (ii) emergent properties are neither reducible nor identical to microphysical properties, and (iii) mental properties have causal efficacy. If we add to these the principle of the closure of the physical domain (iv), and the principle of causal exclusion (v) according to which no event can have more than one sufficient cause, the conclusion is that all five principles cannot be true simultaneously, so we have to give up something. For Kim the only renounceable point is the causal power of emergent properties, and the conclusion would be that if we use strong emergence to explain consciousness it would be an epiphenomenon (Kim 2007).

13.4.3.2 Functionalism

Functionalism argues that mental states are so, not by virtue of a neurophysiological state, as in identity theories, but by virtue of a certain chain of causal relationships between perceptual inputs, other mental states, and behavioural responses (outputs) in which the mental state plays a definite role. Thus, what defines a mental state is the causal-effect role it plays. According to functionalism, certain functional states are invariably correlated with mental states.

Generally, functionalism can be seen as an extension of behaviourism. In *behaviourism*, all mentalistic language was eliminated because of its inherent subjectivity and replaced by a language of mere behavioural dispositions that simply correlate inputs-stimuli and outputs-behaviours of the system. But the functional states to which functionalism refers are not mere behavioural dispositions, since they are specified in terms of their relations not only to inputs and outputs, but also to the state of the machine at the time. Thus, functionalism separates itself from behaviourism by including internal states as propositional dispositions, i.e., beliefs and desires.

The first formulation of this theory of functionalism was made by Hilary Putnam (1967). He took as a functionalist model for the mind a probabilistic automaton—a kind of non-deterministic version of a Turing Machine—in which a program specifies, for each state and set of inputs, the probability with which the automaton will transition to each possible subsequent state and produce some particular outcome.

In summary, internal states:

(i) can be considered representations and serve to explain the representational character of mental states,
(ii) are not tied to any particular physical realisation since the same program can run on different types of hardware,
(iii) can be fully described in terms of their relationships to input, output, and themselves; and
(iv) can be included in descriptions and predictions of the output-behaviour of a system.

For functionalism, the mind is explained by these internal functional states and not by a certain physico-chemical state of the brain or a behavioural disposition. In other words, functionalism is opposed to physicalism and behaviourism, and functionalists put forward empirical reasons for this. Pain, for example, would not be a physical-chemical state of the brain but a functional state of the whole organism. The brain state corresponding to a sensation of pain would depend on the evolutionary details of each phylogenetic lineage of each species. However, its functional character could be independent of such details. Putnam (1967) concludes that the functional state can be defined without reference to the associated subjective experience, e.g., pain:

> . . . the functional state we have in mind is the state of receiving sensory inputs which play a certain role in the Functional Organization of the organism. This role is characterized, at least partially, by the fact that the sense organs responsible for the inputs in question are organs whose function is to detect damage to the body, or dangerous extremes of temperature, pressure, etc., and by the fact that the "inputs" themselves, whatever their physical realization, represent a condition that the organism assigns a high disvalue to.

I have placed functionalism beyond monisms and dualisms because, although it is sometimes presented as a form of physicalism, its fundamental thesis admits a certain form of dualism: a state can play a certain causal role with different physical substrates. To avoid dualism, functionalism considers that causal relations occur only between material entities, i.e., it admits a relation of supervenience of the mental over the physical. But functionalism, unlike behaviourism and identity theories, is not reductionist since it grants autonomy to psychology from neurophysiology by conceptualizing mental states and properties as functional states and properties that are not directly physical. It conceives mental states as internal causes of behavior and behavioural dispositions rather than as identical with them.

Different forms of functionalism can be distinguished such as analytic, computational, and homuncular. *Analytic functionalism* of Lewis (1966) and Armstrong (1968) reinterprets terms from folk psychology—for example, desires or beliefs—as functionally defined theoretical terms for the causal role they play within theory of mind. *Computational functionalism* is characterized by drawing an analogy between mind and computer program. An example is Putnam's proposal that we have seen above (Putnam 1967). Finally, Dennett's *homuncular functionalism* (Dennett 1975) argues that a functional task is decomposed into successively simpler subtasks until a level of mechanical processes is reached.

The best-known objection to functionalism is John Searle's Chinese room argument (Searle 1980). Let's imagine that I am locked in a room. Following an instruction book in English, I am able to answer by means of Chinese symbols to questions asked by means of Chinese symbols about the script of a story that I do not know at all. I never understand any of the story or the questions and answers, but I am able to answer correctly because the instruction book only refers in English to the manipulation of Chinese symbols in order to answer according to the Chinese symbols of the questions. Evidently for me the symbols are simply meaningless pictograms. From the point of view outside the room, I seem to understand Chinese perfectly well. Searle thus dismantles the idea that following a set of syntactic rules can be equated with thinking. The problem of *qualia* is also objected to functionalism. Critics of functionalism argue that a system could be functionally equivalent to the human brain even with a total absence of qualia. To this end, Ned Block (1980) proposed to imagine the individuals of the Chinese nation working together in a way that is functionally equivalent to that of a human brain.

13.4.3.3 Panpsychism

According to panpsychism, elemental entities have their own basic forms of conscious experience, and in the brains these conscious elemental entities somehow coalesce to constitute human and animal consciousness. Although panpsychism literally means that everything has a mind, in practice, panpsychists are not committed to the thesis that every inanimate object has a mind. For them, it is sufficient that some fundamental physical entities (e.g., quarks or photons) have mental states, i.e., conscious experiences. The best arguments for panpsychism are actually arguments against dualism and physicalism, its main alternatives. Panpsychism claims to have the virtues of both views and the vices of neither (Chalmers 2013). We have previously seen arguments against dualism such as the causal argument, and arguments against physicalism such as Chalmers' zombies and the epistemological argument. These same arguments would also support *panprotopsychism* which can be defined, according to Chalmers (2013), as follows:

> ... roughly, the view that fundamental entities are protoconscious, that is, that they have certain special properties that are precursors to consciousness and that can collectively constitute consciousness in larger systems.

Emergent panpsychists argue that macro-experiencing is strongly emergent from micro-experiencing (see Sect. 13.4.3.1). However, *emergent panpsychism* inherits many of the problems of dualism. Its opposite, *constitutive panpsychism*, is the thesis that our macro-experience is based on the micro-experiences of our constituent elements. Intuitively, constitutive panpsychism holds that the micro-experiences somehow add up to produce the macro-experience. The less problematic constitutive panpsychism is the one that holds that there is an *a priori* linking of microphenomenal truths to macrophenomenal truths (Chalmers 2013).

Another important variety of panpsychism is *Russellian panpsychism*, or a version of Russell's neutral monism, which holds that physics reveals the relational structure of matter but not its intrinsic nature. Russellian panpsychism is the view that some intrinsic properties are microphenomenal properties. Russellian panpsychism addresses two metaphysical problems—what is the place of phenomenal properties in nature, and what are the intrinsic properties underlying physical structure? And, in fact, he answers both at the same time: Fundamental phenomenal properties play fundamental microphysical roles and underlie fundamental microphysical structure. There is a non-Russellian panpsychism that claims that there are microphenomenal properties that do not play microphysical roles, but it would run into problems with mental causality. According to Chalmers (2013), the least problematic version of panpsychism would be a Russellian constitutive panpsychism.

The main objection to panpsychism is *the problem of combination*. It is very difficult to make sense of the conscious micro-subjects of experience with their micro-experiences coming together to form a conscious macro-subject with its own macro-experience. For William James (1890), for example, even if we group conscious experiences together, each will remain enclosed in its own 'skin', windowless, ignorant of what other experiences are and mean. In the same way, private minds do not agglomerate into a higher composite mind.

For Chalmers (2016), the problem of combination is actually a set of seven different problems (i) *The anti-aggregation argument*, or that aggregates have no objective existence, but exist only for observers who perceive them as such; (ii) *The subject-summing argument*, or that the existence of a number of subjects with certain experiences does not necessitate the existence of a distinct subject, and, in particular, the existence of a number of micro-subjects does not necessitate the existence of a macro-subject; (iii) *A conceivability argument* that it is possible to conceive of zombies that are microphysically and microphenomenally the same as us but do not have our macrophenomenal experiences; (iv) *An epistemological argument*, assuming that inside her black and white room, Mary is told all the microphysical facts, and also learns all the microphenomenal facts (what is like to be a quark, a photon, etc.), but she would still be unable to know what it is like to see red; (v) *The palette argument*, if Russellian panpsychism is true, we can only expect a handful of microqualities, corresponding to the handful of fundamental microphysical properties, but how can this limited palette of microqualities combine to give rise to the vast array of macroqualities? (vi) *The revelation argument*, or that the vast array of micro-experiences that supposedly constitute our macro-experience is not revealed to us in introspection; and (vii) *The structural mismatch argument*, or that the macrophenomenal structure of consciousness seems quite different from the macrophysical structure of the brain, when constitutive Russellian panpsychism should demand that the structures be the same.

13.5 Conclusions

The relationship between the mental and the physical remains a fascinating mystery. As a philosophical problem it left us perplexed, and despite the many proposed solutions, the objections they all raise may make us doubt that we will ever solve the question. We have seen that each answer to the problem has consequences that seem unacceptable. Consciousness is a major challenge to science as it is known today. The objections raised to physicalism by Nagel, McGinn, Chalmers, and others seem insurmountable on the near horizon.

To conclude, let us highlight two significant facts. The first is that the origin of the mind-body problem and the origin of the scientific method as we know it today coincide around the figure of Descartes. Descartes precisely developed the concept of Cartesian coordinates that allow us to characterise the positions of bodies in space, and this allows us to mathematise the conception of the physical as that which occupies an extension in space. It is also significant that McGinn's ideas about the insolubility of the problem date from, 1989, just before the neuroscientific revolution. What is surprising is that the enormous development of neuroscience and neuroimaging techniques since that year has not shed significant light on the philosophical problem associated with the phenomenon of consciousness. Thus, time seems to have proven right, so far, those who saw a conflict between the objective scientific view of reality and consciousness as a subjective experience.

Acknowledgements This study was supported by the Spanish Agencia Estatal de Investigación of the Ministerio de Ciencia e Innovación (FJC2021-047765-I). The author declares that he does not have any conflict of interest to disclose.

References

Armstrong, D. 1968. *A materialist theory of the mind. International library of philosophy and scientific method*. Routledge & Paul, New York.

Atmaspacher, H. 2014. 20th century variants of dual-aspect thinking. *Mind and Matter* 12 (2): 245–288.

Bailey, A. 2006. Zombies, epiphenomenalism, and physicalist theories of consciousness. *Canadian Journal of Philosophy* 36 (4): 481–509. https://doi.org/10.1353/cjp.2007.0000.

Bedau, M. 1997. Weak emergence. *Philosophical Perspectives* 11: 375–399.

Berkeley, G. 2013. *Principles of human knowledge*. Minneapolis: Start Publishing LLC.

Block, N. 1980. *Readings in philosophy of psychology, volume 1*. Cambridge MA: Harvard University Press.

Bloom, P. 2005. *Descartes' baby: How the science of child development explains what makes us human*. New York: Basic Books.

Broad, C. 2008. *The mind and its Place in nature. International library of philosophy*. London: Routledge.

Brown, R. 2010. Deprioritizing the a priori arguments against physicalism. *Journal of Consciousness Studies* 17 (3–4): 47–69.

Chalmers, D. 1995. The hard problem of consciousness. *Journal of Consciousness Studies* 2 (3): 200–219.

———. 1996. *The conscious mind: In search of a fundamental theory*, Philosophy of Mind Series. New York: Oxford University Press.

———. 2006a. Phenomenal concepts and the explanatory gap. In *Phenomenal concepts and phenomenal knowledge: New essays on consciousness and physicalism*, ed. T. Alter and S. Walter. Oxford: Oxford University Press.

———. 2006b. Strong and weak emergence. In *The re-emergence of emergence*, ed. P. Clayton and P. Davies, 244–256. Oxford: Oxford University Press.

———. 2013. Panpsychism and panprotopsychism. *Amherst Lecture in Philosophy* 8: 1–35. http://www.amherstlecture.org/chalmers2013/.

———. 2016. The combination problem for panpsychism. In *Panpsychism: Contemporary perspectives*, ed. G. Brüntrup and L. Jaskolla, 179–214. Oxford: Oxford University Press.

Chomsky, N. 1976. Problems and mysteries in the study of human language. In *Language in focus: Foundations, methods and systems: Essays in memory of Yehoshua Bar-Hillel*, ed. A. Kasher, 281–357. Netherlands: Springer.

Churchland, P. 1981. Eliminative materialism and the propositional attitudes. *The Journal of Philosophy* 78 (2): 67–90. https://doi.org/10.5840/jphil198178268.

———. 2002. *Brain-wise: Studies in Neurophilosophy. A Bradford book*. Cambridge MA: MIT Press.

Clatterbaugh, K. 1999. *The Causation Debate in Modern Philosophy, 1637–1739. Philosophy, science studies*. New York: Routledge.

Conee, E. 1994. Phenomenal knowledge. *Australasian Journal of Philosophy* 72 (2): 136–150. https://doi.org/10.1080/00048409412345971.

Craig, E. 1998. *Routledge encyclopedia of philosophy, volume 5 of Routledge encyclopedia of philosophy*. London: Routledge.

Davidson, D. 2001. *Essays on actions and events: Philosophical essays volume 1. Essays*. Oxford: Clarendon Press.

Dennett, D. 1975. Why the law of effect will not go away. *Journal for the Theory of Social Behaviour* 5 (2): 169–187.

———. 1993. *Consciousness explained*. London: Penguin.

Descartes, R. 1637. *Le discours de la méthode*. Leiden: Ian Maire. https://books.google.es/books?id=I0x3tQEACAAJ (English translation: Descartes, R. (2008). *A discourse on the method: Of correctly conducting One's reason and seeking truth in the sciences*. Oxford World's classics. OUP Oxford).

———. 1641. *Meditationes de prima philosophia, in qua Dei existentia et animae immortalitas demonstratur*. Paris: Charles Angot. https://books.google.es/books?id=R41XAAAAcAAJ (English translation: Descartes, R. (2017). *Meditations on first philosophy: With selections from the objections and replies*. Cambridge texts in the history. Cambridge University Press).

———. 1664. *De homine figuris*. Leiden: Ex Officinâ Hackiana. https://books.google.es/books?id=7QO-AQAACAAJ.

Feigl, H. 1958. The mental and the physical. *Minnesota Studies in the Philosophy of Science* 2 (2): 370–497. https://web.stanford.edu/~paulsko/papers/Feigl.pdf.

Flanagan, O. 1991. *The science of the mind, second edition. Bradford book*. Cambridge, MA: MIT Press.

Francescotti, R. 2011. Supervenience and mind. In J. Fieser and B. Dowden (Eds.), *Internet encyclopedia of philosophy*. https://iep.utm.edu/supermin/ Accessed 9 July 2023.

Goldman, A. 1992. In defense of the simulation theory. *Mind & Language* 7 (1–2): 104–119. https://psycnet.apa.org/doi/10.1111/j.1468-0017.1992.tb00200.x.

Hill, C. 1997. Imaginability, conceivability, possibility and the mind-body problem. *Philosophical Studies* 87 (1): 61–85. https://www.jstor.org/stable/4320765.

Huxley, J. 1942. *Evolution: The modern synthesis*. New York/London: Harper & Bros.

Jackson, F. 1982. Epiphenomenal Qualia. *The Philosophical Quarterly* 32 (127): 127–136. https://doi.org/10.2307/2960077.

James, W. 1890. *The principles of psychology*. Macmillan. https://www.google.es/books/edition/The_Principles_of_Psychology/JO1RL9BcI44C?hl=es&gbpv=1.

———. 1895. The knowing of things together. *Psychological Review* 2 (2): 105. https://psycnet.apa.org/doi/10.1037/h0073221.

———. 1904. Does 'consciousness' exist? *The Journal of philosophy, psychology and scientific methods* 1 (18): 477–491. https://doi.org/10.2307/2011942.

Kim, J. 1989. The myth of nonreductive materialism. *Proceedings and Addresses of the American Philosophical Association* 63 (3): 31–47. https://doi.org/10.5840/apapa2013188.

———. 2006. Being realistic about emergence. In *The re-emergence of emergence*, ed. P. Clayton and P. Davies, 244–256. Oxford: Oxford University Press.

———. 2007. *Physicalism, or something near enough*. Princeton: Princeton University Press.

Kriegel, U. 2003. The new mysterianism and the thesis of cognitive closure. *Acta Analytica* 18: 177–191. https://doi.org/10.1007/s12136-003-1020-1.

Lahav, R. 1992. The amazing predictive power of folk psychology. *Australasian Journal of Philosophy* 70 (1): 99–105. https://doi.org/10.1080/00048408112340083.

Leder, D. 1985. Troubles with token identity. *Philosophical Studies: An International Journal for Philosophy in the Analytic Tradition* 47 (1): 79–94. http://www.jstor.org/stable/4319730.

Leibniz, G. 1989. *Philosophical essays*, Hackett Classics Series. Cambridge: Hackett Publishing Company.

Levine, J. 1983. Materialism and qualia: The explanatory gap. *Pacific Philosophical Quarterly* 64 (4): 354–361.

Lewes, G. 1875. *Problems of life and mind*. Boston: Osgood.

Lewis, D. 1966. An argument for the identity theory. *The Journal of Philosophy* 63 (1): 17–25. https://doi.org/10.2307/2024524.

Loar, B. 1990. Phenomenal states. *Philosophical Perspectives* 4: 81–108. https://doi.org/10.2307/2214188.

Locke, D. 1968. *Myself and others: A study in our knowledge of minds*. Oxford: Oxford University Press.

Lycan, W., and J. Prinz. 2008. *Mind and cognition: An anthology*. Malden: Blackwell Publishing.

Mach, E. 1959. *The analysis of sensations, and the relation of the physical to the psychical*, Dover Books S 525. Dover Publications.

Maloney, J. 1985. About being a bat. *Australasian Journal of Philosophy* 63 (1): 26–49. https://doi.org/10.1080/00048408512341671.

McGinn, C. 1989. Can we solve the mind-body problem? *Mind* 98 (391): 349–366. http://www.jstor.org/stable/2254848.
———. 1995. Consciousness and space. *Journal of Consciousness Studies* 2 (3): 220–230.
Mill, J. 1843. *A system of logic, ratiocinative and inductive: Being a connected view of the principles of evidence, and methods of scientific investigation, volume 1*. John W. Parker.
———. 1873. *An examination of sir William Hamilton's philosophy: And of the principal philosophical questions discussed in his writings*. W. V. Spencer.
Nadler, S. 2010. *Occasionalism: Causation among the Cartesians*. Oxford: OUP.
Nagel, T. 1974. What is it like to be a bat. *Readings in philosophy of psychology* 1: 159–168. http://www.jstor.org/stable/2183914.
———. 1987. *What does it all mean?: A very short introduction to philosophy*. New York: Oxford University Press.
O'Connor, T. 1994. Emergent properties. *American Philosophical Quarterly* 31 (2): 91–104. http://www.jstor.org/stable/20014490.
O'Dea, J. 2002. The indexical nature of sensory concepts. *Philosophical Papers* 31 (2): 169–181. https://doi.org/10.1080/05568640209485100.
Papineau, D. 2002. *Thinking about consciousness*. Oxford: Clarendon Press.
———. 2011. What exactly is the explanatory gap? *Philosophia* 39: 5–19. https://doi.org/10.1007/s11406-010-9273-6.
Pargetter, R. 1984. The scientific inference to other minds. *Australasian Journal of Philosophy* 62 (2): 158–163. https://doi.org/10.1080/00048408412341341.
Penn, D., and D. Povinelli. 2007. On the lack of evidence that non-human animals possess anything remotely resembling a theory of mind. *Philosophica Transactions of the Royal Society B: Biological Sciences* 362 (1480): 731–744. https://doi.org/10.1098/rstb.2006.2023.
Perry, J. 2001. *Knowledge, possibility, and consciousness*. Cambridge MA: MIT Press.
Place, U. 1956. Is consciousness a brain process? *British Journal of Psychology* 47 (1): 44–50. https://doi.org/10.1111/j.2044-8295.1956.tb00560.x.
Popper, K., and J. Eccles. 2012. *The self and its brain*. Berlin Heidelberg: Springer.
Premack, D., and G. Woodruff. 1978. Does the chimpanzee have a theory of mind? *Behavioral and Brain Sciences* 1 (4): 515–526. https://doi.org/10.1017/S0140525X00076512.
Putnam, H. 1967. Psychological predicates. In *Art, mind, and religion*, ed. W.H. Capitan and D.D. Merrill, 37–48. Pittsburgh: University of Pittsburgh Press.
Reppert, V. 1992. Eliminative materialism, cognitive suicide, and begging the question. *Metaphilosophy* 23 (4): 378–392. https://www.jstor.org/stable/24438929.
Robinson, H. 2002. Dualism. In *Blackwell guide to philosophy of mind*, ed. S.P. Stich and T.A. Warfield, 85–101. Oxford: Blackwell.
Robinson, W. 2019. Epiphenomenalism. In *The Stanford encyclopedia of philosophy*, ed. E.N. Zalta. Stanford University. Summer 2019 edition. https://plato.stanford.edu/entries/epiphenomenalism/.
Rosenthal, D. 1997. A theory of consciousness. In *The nature of consciousness*. ed. N. Block, O.J. Flanagan and G. Guzeldere, 729–753. Cambridge MA: MIT Press.
Russell, B. 1945. *History of Western philosophy*. Touchstone.
———. 1959. *My philosophical development*. New York: Simon and Schuster.
———. 1995. *An outline of philosophy*. London: Routledge.
———. 2022. *The analysis of mind*. London: Taylor & Francis.
Searle, J. 1980. Minds, brains, and programs. *Behavioral and Brain Sciences* 3 (3): 417–424. https://doi.org/10.1017/S0140525X00005756.
Sellars, R. 1922. *Evolutionary naturalism*. Chicago/London: Open Court Publishing Company.
Skrbina, D. 2014. Dualism, dual-Aspectism, and the mind. In *Contemporary dualism: A defense*, ed. Lavazza and Robinson, 228–252. New York: Routledge.
Smart, J. 1959. Sensations and brain processes. *The Philosophical Review* 68 (2): 141–156. https://doi.org/10.2307/2182164.
Spinoza, B. 1996. *Ethics. Penguin classics*. Penguin Publishing Group.

Stoljar, D. 2010. *Physicalism. New Problems of Philosophy*. London: Taylor & Francis.
Tully, R. 2003. Russell's neutral monism. In *The Cambridge companion to Bertrand Russell*, ed. N. Griffin, 332–370. Cambridge: Cambridge University Press.
Van Gulick, R. 2021. Consciousness. In *The Stanford encyclopedia of philosophy*, ed. E.N. Zalta. Stanford University. Winter 2021 edition. https://plato.stanford.edu/entries/consciousness/.
Vintiadis, E. 2019. *Property dualism, chapter 20, page 26*. Rebus Community. https://openeducationalberta.ca/pop201/chapter/property-dualism/.
Westphal, J. 2016. *The mind-body problem. The MIT Press Essential Knowledge series*. MIT Press.

Chapter 14
Do I Really Believe That? A Mindreading Account of Belief Self-Ascription

Sylvain Montalvo

Abstract Philosophers often treat beliefs as propositional attitudes that entail taking the content as true. However, there exist certain mental states, such as delusions, that challenge this dogmatic norm, leading some researchers to question its validity. This chapter delves into that philosophical inquiry that surrounds the nature of beliefs and establishes a contrastive perspective in cases of delusion, self-deception, sectarian influence, and scientific negationism. We should explore the notion of privileged access to our own mental states acknowledging the importance of self-ascription as being more indirect than commonly believed. We may find that self-ascription is a result of our mindreading ability applied to ourselves and there could be a plausible biological evolution of the cognitive mechanism that underlies. This chapter discusses the evolutionary escalation of mindreading in humans, reviews philosophical accounts of mindreading, and evaluates different theories of belief self-attribution. It concludes with the implications for understanding belief-related phenomena and their cognitive underpinnings.

Keywords Beliefs · Propositional attitudes · Delusions · Self-deception · Mindreading

14.1 Preface

A seventy-year-old woman consults a dermatologist because she believes that "bugs" have invested her skin. She feels them crawling, sometimes even sees them and she presents pruritus lesions. However, she has already met several physicians and none of them found any sign parasitosis. A forty-three-year-old man with a history of drunk driving is controlled by the police. He agrees without complaints to blow in the breathalyzer, specifying that he has not had a drop of alcohol for 2 years. However, his test reveals that he currently has a blood alcohol level of 0.6 g/L. A

S. Montalvo (✉)
Départment Philosophie, Faculté des Humanités, Université de Lille, Lille, France

STL – Savoirs, Textes, Langage (STL) – UMR 8163, Lille, France

T. Lopez-Soto et al. (eds.), *The Theory of Mind Under Scrutiny*, Logic, Argumentation & Reasoning 34, https://doi.org/10.1007/978-3-031-46742-4_14

twenty-year-old diabetic man, known by loved ones to be health conscious, goes to the ER for vomiting and excruciating abdominal pain. He explains to the resident that he has stopped his insulin injections since he discovered on the internet a natural method to cure diabetes. However, he is in a state of ketoacidosis. A handful of established physicists maintain that there is still incertitude regarding the reality, peril, or role of human societies in global warming. However, the conclusive evidence produced by the scientific community, and gathered and specified in the successive IPCC reports are beyond doubt. These examples could respectively correspond to delusion, self-deception, sectarian influence and scientific negationism.[1] Along with confabulations, superstitions, religious faith or conspiracist ideations, they are instances of putative "beliefs" widely scrutinized by philosophers, psychologists, sociologists, and anthropologists.

In philosophical literature, beliefs are generally treated as a propositional attitude. A propositional attitude is a content-bearing mental state associated with a certain disposition toward that content. The disposition that seems to characterize believing is that of taking the content as true. Imagining or accepting something false is innocent and may even be the point.[2] Conversely, truth-governance of beliefs seems to go further, so that entertaining a false belief may be fundamentally misleading. To put in another way, many philosophers take truth to be the specific norm for beliefs (Williams 1973; Velleman 2000; Shah 2003). Therefore, the nature of the cases addressed above constitute a pervasive philosophical question, inasmuch as their following this norm is disputable. For instance, delusions don't seem to be truth-governed at all. Even if their content happened to be true, as in the case of a person with delusional infestation who would also have genuine parasitosis (as reported by Thibierge 1894 cited in Trabert 1995), their truthness would be a pure accident, having nothing to do with the following of a norm. For this reason, some authors are reluctant to consider delusions as beliefs. Thus, Gregory Currie famously advocated that they are mere acts of imagination taken to be beliefs (Currie 2000, see also Schwitzgebel 2012, Dub 2017).[3] Similar questions arise for widespread belief-like intriguing mental states. Max Black proposed a criterion of distinction, namely asking the person "do you really believe that?" (Black 1983, cited in Dieguez 2022, p. 22). However, as Daniel Dennett put it, questioning might be of little help when the person is anyway prompted to act as a believer, due to a moral commitment notably, that is when she "believes in belief" (Dennett 2006,

[1]The first two are from my own physician experience. The third is from Adénor and de Rauglaudre (2021), p. 59–60). The latter is from Oreskes and Conway (2012, chap. 6). I will circumscribe some of these notions when needed as the chapter progresses.

[2]As such, a writer may imagine an extraterrestrial life perceiving time as a whole to develop the outcomes of an encounter (Vonnegut 1969) and a mathematician may accept a conjecture to consider new aspects of a problem. However, none of them is required to believe in the content of these propositions.

[3]This position contrasts with "doxasticism" about delusions (see Bayne and Pacherie 2005; Bortolotti 2010; Miyazono 2019) which considers that delusions are indeed beliefs, in accordance with their definition in the DSM-5 (American Psychiatric Association 2013, p. 87).

p. 222). Moreover, even from a first-person perspective, the answer may not be as self-evident as it seems.

This latter point is the one I wish to address in this chapter. How do one know what one believes? Do I really believe that? In philosophy, it is commonplace to say that we have some kind of privileged access to our own mental states (for a review and an explanation of this intuition, see Carruthers 2011, chap. 2). Among other things, there is a traditional line of thought according to which this access is immediate (Descartes, 1647/2017, *Second meditation*, Russell 1910). I will discuss this intuition and the others and argue that the process of belief self-ascription is much more indirect than it seems. More precisely, I will suggest that the cognitive mechanism responsible for self-ascription of belief is nothing but our mindreading ability, applied to ourselves. My main argument will be an inference to the best explanation regarding the plausible biological evolution of the cognitive mechanism underlying, possibly among other things, belief self-attribution. Once formalized, it will take the following form:

1. Human mindreading ability is the product of an evolutionary escalation.
2. The outputs of human mindreading ability and that of self-attribution of mental states are similar, so their underlying mechanism require the same level of complexity.
3. It is unlikely that a self-attribution mechanism evolved independently.
4. Therefore, it is likely that self-attribution of beliefs relies on mindreading ability.

First, I will sketch a plausible phylogenesis of mindreading in humans which will constitute a defense of premise (1). Secondly, I will make a non-exhaustive review of philosophical accounts of mindreading and their outgrowth regarding the question of self-mindreading. Then, I will assess the different accounts of belief self-attribution, defending premises (2) and (3) which are closely linked. I will conclude with some theoretical and practical implications of (4).

14.2 The Evolution of Interactions and Mindreading

The first multicellular organisms had, by all accounts, to ensure their inner coordination by chemical signals probably analogous to the ones previously developed for orientation, namely chemotaxy. However, chemical communication between the cells is spatially limited, imprecise, specific, and slow to occur.[4] As the first multicellular organisms grew in size, and as some of their cells specialized, it became

[4]Chemotaxy refers to the responsiveness to a gradient of chemical elements, in order to get closer to its source or to depart from it. This mechanism is present in both eukaryotes and bacteria (Tso and Adler 1974). I assume that chemotaxy played an early role in inner coordination of multicellular organism, as it is still present in contemporary animals (see Purves 2018, p. 531–533). Chemical communication in a broader sense can display long range effectivity, like the endocrine system does. However, it can exceed its other limitations.

necessary for different parts to communicate rapidly, over long distances, to specific target cells with potentially complex messages. A nervous system meets these requirements (Arendt 2021). As a collateral effect, the emergence of nervous systems likely led to the apparition of more complex interactions between organisms, among which predation is notable.[5] The constraints posed by the appearance of predation, both for prey and for predators, have evidently generated biological arms races, phenomenon in which two or more species evolve in response to each other's adaptations,[6] in a sort of competitive escalation (Vermeij 1987), a notion to which I will return. Nevertheless, antagonism is far from being the sole kind of interaction existing among organisms. We can call altruistic a behavior which favor the fitness, that is roughly the reproductive success, of other individuals to the detriment of the one manifesting it (Hamilton 1964a, b, see also Sober 1988). A dramatic case of altruism can be observed in social insects' sacrificial-like behavior, such as worker ants dying to defend their colony (Shorter and Rueppell 2012). This kind of selfless conduct can be accounted for in terms of kin selection. Kin selection refers to the biological advantage, in certain contexts, of helping one's kins for they carry, in part, the same genetic material (Maynard Smith 1964, Williams 1966, chap. 4). Another explanation for the occurrence of biologically altruistic behaviors is well documented: expected reciprocity (Trivers 1971). This phenomenon can be illustrated by a behavior at first sight intriguing in hematophagous bats. Individuals whose nocturnal hunt was successful regurgitate part of the blood obtained to feed those whose night was less productive. This behavior has been shown to be unrelated to the kinship of cooperating individuals (Wilkinson 1984; Carter and Wilkinson 2013). This behavior is altruistic in the biological sense, insofar as the individual who shares loses part of his resources, in this case food, to the benefit of another individual. Nevertheless, this behavior is maintained insofar as, in the long term, the altruist will also benefit from this sharing behavior when he returns home empty winged. Of course, it is necessary to explain what prevents the proliferation of free riders, who would benefit from sharing when they need it, without reciprocal contribution. Robert Trivers' explanation is based on the idea that it is a situation of iterative prisoner's dilemma[7] (see also Axelrod and Hamilton 1981). The treacherous phenotype is counter-selected by the recognition of individuals who do not play the game of reciprocal altruism. Their eviction from the group implies a drastic

[5]Each organism being, ultimately, a condensed source of energy and organic matter, potentially available to its predator.

[6]The resulting diversification of defence strategies (camouflage, predator detection, mobility, parries, etc.) and attack (detection of prey, speed and lethality of aggressive behaviors, deception abilities, etc.) may have participated in the evolutionary radiation characteristic of the Cambrian explosion (Carbone and Narbonne 2014)

[7]On the main lines, the prisoner's dilemma is a situation, well studied in game theory (see Kuhn 2019), where it is individually advantageous for the players to adopt a "free for all" strategy (dominant strategy), whereas everyone's situation would have been on average better if everyone's had adopted a collaborative strategy. Iterative prisoner's dilemma is portrayed as a variant in which collaboration might spontaneously emerge (Axelrod 1981).

drop in their individual selective value insofar as they no longer benefit from the mutual aid allowed by the group. A requirement for this behavior to maintain is thus that participants have a way to recognize good partners and detect cheaters.

So far, biological interactions could be described without appealing to the concept of second-order belief.[8] The conditions of apparition of genuine second-order beliefs might be the complexification of interaction between organisms. At some point, it become cognitively parsimonious to adopt an "intentional stance" about other's behaviors (Dennett 1987). In other words, it gets easier to interpret a conspecific's action in terms of desire-fulfilment and respectively of other mental states she might entertain, that is to say with mindreading. The notion of mindreading in philosophy is broadly synonymous with theory of mind,[9] although the latter is sometimes taken to convey a particular commitment to a specific account of mindreading. The term theory of mind was coined by David Premark and Guy Woodruff in a seminal paper which addresses the question of its existence in the chimpanzee (Premack and Woodruff 1978). If the existence of a theory of mind in non-human animals remains unresolved (Whiten 2013; Heyes 2015), it is clearly present in human, with exceptions I will discuss. As a matter of fact, I will argue that human theory of mind is not only present but overdeveloped.

It is likely that the process of hominization occurred within small groups of one hundred to two hundred individuals. This number refers to the number of individuals with whom we seem to be able to maintain stable social relations (Dunbar 1993), a necessary condition for the emergence of reciprocal altruism. These groups are not comparable to colonies of social insects. They are crossed by issues of power and regular exchanges with other groups ensuring a mix that limits the phenomenon of endogamy (Sikora et al. 2017). Thus, kin selection is not able, on its own, to explain the cohesion of these groups and the important place that altruistic behaviors take in human groups. According to Robin Dunbar, there is a close relationship between encephalization and the size of social groups in primates, especially humans (Dunbar 1992). An explanatory hypothesis for this coupling would be the need to be able to mentally navigate within the group and the cognitive requirement of such a navigation in a complex social structure. This social appraisal would include the recognition of good partners, the protection of oneself from possible cheaters in the social game, the existence of moral intuitions such as guilt that seem to encourage us to respect the rules of the social game, which evidently is a long-term benefit (Baumard

[8]Notably, the kind a solution to iterative prisoner's dilemma I described doesn't presuppose that the organisms literally attribute mental states such as disposition to treachery or belief that last night help was received. The only belief that is required is that it is a good move to help this conspecific but not that one. A basic but powerful rule, implemented by Amnon Rapoport, is "tit for tat": help the unknown individual, help the individual who helped you last time, leave aside the individual who betrayed you last time (see Rapoport et al. 2015 for development and limitations). However, some authors argue that at this stage a theory of mind would already confer a strategic advantage (Press and Dyson 2012).

[9]Other roughly similar notions are folk psychology, mentalizing and cognitive empathy. I'll assume that they all refer to the same concept.

et al. 2013), the rejection of so-called antisocial behaviors, which are similar to cheating, so much so that some researchers postulate that the evolution of reciprocal altruism in human groups has been accompanied by the development of a cognitive module of detection of cheaters (Cosmides et al. 2005). In other word, social navigation requires a mindreading ability to attribute hidden beliefs to a conspecific disposed to help or cheat. Conversely, this very same ability is convoked in detection of opportunities to contribute or to deceive. The resulting situation is open to a phenomenon of evolutionary escalation concerning mindreading (Vermeij 1987; Veller et al. 2020).

Several empirical findings plead in favor of such an escalation. First of all, the brain is a biologically expensive organ (Aiello and Wheeler 1995). While it represents only 2% of the total mass of an adult human, it monopolizes up to 20% of the resting energy consumption (Allaman et al. 2011). In addition, its development is particularly long, requiring parental supervision at least until adolescence (Nelson et al. 2005). Such a trait requires significant selection pressures to maintain, and evolutionary escalation is precisely a context in which selective pressures get strong (see Hamilton et al. 1990). Another symptom of evolutionary escalation is the frailty of the resulting trait. The underlying logic is that in such a case of antagonism, time and energy cannot be wasted on robustness, at the risk of losing track of the evolutionary race. Let me suggest an analogy with the height of trees. Other things being equal, there's no specific need for a photosynthetic plant to grow in height. However, once appears a size variation, a competition for access to sunlight may take place. At some point, trees develop costly trunks whose point is to gain access to sunlight energy in the presence of other trees with trunks.[10] The limitation of the escalation is physical, related to the energy required to raise fluids to the highest leaves or needles (King 1990; Falster and Westoby 2003). Yet, trunks are a particularly vulnerable part of the organism, exposed to the vagaries of the weather, parasites, and the hubris of certain carbon addict primates. What about the human brain? To begin with, there is a precarious balance between encephalization, and the evolutionary costs generated by the obstetrical risk (Wittman and Wall 2007). However, I won't argue that the brain is, as a whole, a deeply fragile organ. Indeed, the possibilities for a damaged brain to repair or adapt are indisputable (see Purves 2018, chap. 26). Nevertheless, we're not interested in the whole brain, but in the systems responsible for social navigation and mindreading. The vast majority of neurodevelopmental disorders concern the telencephalon and its associated diencephalic regions: autism spectrum disorders, schizophrenia spectrum disorders, intellectual disability, deficit of attention. The frequency of these disorders, whether their origin is genetic, epigenetic or mixed, suggests a much poor robustness of the brain circuits involved in the functions impacted in these disorders, namely mindreading

[10] At this point, one might notice that the outcoming situation is not dissimilar to the initial one in terms of access to light. We can see that as a prisoner's dilemma instance (see note 7). The fact that the competition only occurs to stay in the same position as been described as a Red Queen situation (Van Valen 1973): "Now, *here*, you see, it takes all the running you can do, to keep in the same place." (Carroll 1872, *Through the looking glass*)

and related abilities (see below). A wide range of anomalies can give rise to these disorders.[11] If early childhood has been extensively scrutinizing in research on basic theory of mind (see also below), adolescence seems for its part to be a critical period for the development of the human fine-grained mindreading (Dumontheil et al. 2010). I take this fine-grained aspect to be a result of the evolutionary escalation. The stage of neurodevelopment we're interested in is that of synaptic pruning, which is the programmed reduction in the number of neuronal connections during brain maturation, associated with a decrease in plasticity in favor of a gain in specialization of neuronal networks (Feinberg 1982; Weinberger 1987). However, early adulthood is also a hazardous stage for the appearance of schizophrenia which I take to be precisely a disorder affecting this very fine-grained mindreading ability (Frith 1992; Corcoran 2001; Brüne 2005; Montalvo 2021). Thus, I take schizophrenia and related neurodevelopmental disorders to testify for the fragility implied by the evolutionary escalation of mindreading I depicted.

14.3 Theories of Mindreading

A classical account of mindreading is to be found in Daniel Dennett (1971, 1987). According to Dennett, when attributing beliefs to someone, one should ground one's assumptions on a principle of rationality (see also Davidson 1970, 1973 for an analogous proposition). It's seemed to be the case in normal conversational interactions (Grice 1975), which obviously require mindreading to deal with implicatures and ambiguities of natural languages. Indeed, understanding and participating to normal conversation require abilities to consider the context and the probable implicit intentions of the interlocutor. The main concern with these rationality accounts of mindreading is that there is no clear consensus on how rationality should be circumscribed. It could be interpreted as relying on a kind of ideal, normative rationality, a bullet that is bitten by Dennett (1987, p. 82). However, the problem is two-fold. First, we must agree on what this ideal rationality might be. Expected

[11] Certain genetic anomalies are related to copy number variant, such as a translocation interrupting the DISC1 (*disrupted in schizophrenia 1*) gene or the 22q11.2 deletion, region containing the gene for the enzyme catechol-O-methyltransferase (COMT). The latter is involved in the metabolism of monoaminergic neurotransmitters, in particular dopamine and norepinephrine, which are suspected to be involved in certain manifestations of the disease (Stahl 2021, p. 80). Such variants, with a large effect, explain only 2–3% of cases of schizophrenia (Owen et al. 2010) and are most often de novo mutations (Fromer et al. 2014). On the other hand, many loci, whose polymorphisms do not individually constitute significant risk factors, have been highlighted for their cumulative contribution to the genetic burden involved in the disease (Schizophrenia working group of the psychiatric genomics consortium 2014). Finally, other neurological disorders can have a clinical presentation reminiscent of schizophrenia, such as Niemann-Pick disease type C (Vanier 2010) or late-onset urea cycle disorders (Gropman et al. 2007). I focused here on biological causes. Nonetheless, it should be kept in mind that schizophrenia and related disorders are eminently multifactorial, involving miscellaneous gene-environment interactions.

utility theory (see Briggs 2019)? Bayesian rationality (see Nguyên Hoang 2020)? Trial by ordeal? We lack criteria for determining what the correct standard of rationality should be. Secondly, there is an avalanche of empirical data according to which humans don't usually display straightforward agential rationality (Bortolotti 2010, p. 161–162). In other words, in normal cases people don't always act in accordance with their beliefs in a rational fashion (see Kahneman 2012). It is indisputable that human's cognition is limited in cognitive resources, among which are time and energy. Naturally, some might say that people are still rational enough in an ecological context. The biases exposed by Daniel Kahneman may be detrimental only in artificial contexts and may constitute satisfying heuristics in normal situations (Gigerenzer 2008). However, there is no clear account of how distinguish an attitude that is, even slightly, irrational form an attitude that is minimally rational in the latter sense. So, this account of rationality can't ground the intentional stance suggested by Dennett. Not to mention the fact that Dennett's theory supposes that the attributor share the rationality of her target to make the correct ascriptions. This latter aspect assumes a similarity between the mindreader and her target, namely that both are rational creatures perceiving the world in similar ways. A deflationary account of these theories can thus lean on these assumptions.

Thereby, some philosophers (Heal 1986; Gordon 1986; Goldman 1989) suggest that we understand others simply by putting ourselves in their shoes, that is by simulating their mental states within ourselves. One of the main arguments in favor of this so-called simulation-theory is its explanatory power. Simulation-theory accounts for the fact that individuals can accurately predict others' thoughts, feelings, and behaviors without relying on any explicit knowledge about the situation or the individual (Heal 1986). The theory suggests that this ability is possible because individuals can mentally simulate the relevant mental states of the other person within themselves, without relying on explicit psychological laws (Goldman 1989, 2006, p. 20, see below for the putative role of psychological laws). For instance, when observing someone else in pain, individuals may simulate the feeling of pain within themselves, which enables them to empathize with the other person's experience, and then predict their upcoming behavior relying on what they would do themselves. Another argument in favor of simulation-theory is that it is supported by empirical evidence from neuroscience, psychology, and pathology. Neuroscientific studies have shown that individuals activate similar brain regions when they observe and experience emotions (Carr et al. 2003), and when they engage in and social cognition and in self-referential thoughts (Mitchell et al. 2005). Additionally, psychological studies have shown that individuals are more accurate at predicting others' mental states when they engage in perspective-taking and mental simulation tasks (Harwood and Farrar 2006). Furthermore, parallelism has been highlighted between deficits in specific mental activities and their recognition in others (Heberlein and Adolphs 2007). My main concern with the simulation-theory is that it seems to assume that one has an immediate access to one's own mental states. To attribute someone else an attitude by, one might need to recognize this very attitude simulated in one's mind. Now, this kind of introspective ability is precisely the point that I wish to contest. By contrast, Robert Gordon (1992, 1995) has proposed an account of simulation-theory which allegedly doesn't rely on introspection. However, his account is restricted to belief and desire ascriptions, leaning on the "ascent routine" (see below) to make the attributions. Besides, even if simulation is quite elegant an account of mindreading in many instances, it must be rooted in a neurological substrate that may very well be a firm grounding for others mindreading theories. So, it is possible that mental simulation theory collapses with other accounts (Davies and Stone 2001).

Other philosophers understand the concept of theory of mind quite literally, advocating that it is indeed a theory, including general laws governing nomological relations between observable aspects of the world and invisible entities such as propositional attitudes and possibilities of prediction and explanation. Alison Gopnik suggests that the elaboration of this theory is roughly like that of a scientific inquiry (Gopnik 1990; Gopnik 2004; Gopnik and Wellman 1992). As such, there would be progresses and even shifts of paradigms all along the cognitive development of the child and her theory of mind. One of these shifts is supposed to take place around the age of three, when the child become capable of succeed in false-belief tasks. These tasks, independently suggested by Bennett (1978), Dennett (1978) and Harman (1978), are supposed to test the ability of the subject to attribute false beliefs to another person, in situations where the subject has knowledge of the truth and is in epistemic position to infer that the other person hasn't and should believe otherwise. In their princeps experiment, Wimmer and Perner (1983) showed children the actions of a puppet 'Maxi' putting a chocolate bar in a box, then leaving the room. Later entered another puppet, Maxi's mother. She moved the chocolate bar from the box to a cupboard in full view of the children but in the absence of Maxi. Finally, Maxi reentered the room saying out loud that he wanted to eat chocolate. At this point, the children were asked where they thought that Maxi would look to find the bar. Children were supposed to understand that the puppet would look inside the box, where he was supposed to believe that the chocolate was stored, even though the children knew well that it was now in the cupboard. The authors indeed suggested the existence of a significative difference in success between a group of 3–4 years old and a group of 4–6 years old. Other scientists have reproduced the experience, sometimes with simplifications, creating a consensus of the change taking place around the age of three (see the meta-analysis carried out by Wellman et al. 2001). The point, according to promoters of this "scientific theory-theory" is that the observed shift happens as the child gets enough data and development time to confirm her hypotheses regarding the working of beliefs and desires of others, and their link to environmental inputs and behavioral outputs. Gopnik and Wellman (1992) also emphasize the theorical character of mindreading, on the grounds that, like any other scientific theory, it's based on observable facts, unobservable but postulated theoretical entities, and that it allows to give a phenomenon such as behaviors a predictability and an explainability. However, concerns have been raised about this radical apprehension of theory of mind. First, subsequent experiments went so far as to highlight successes in false-beliefs tasks in 15-month-old infants (Onishi and Baillargeon 2005), at an age when time for data and concept acquisition become to look very scarce. Another, more fundamental, concern raised by Alvin

Goldman (2006, p. 85–86) resides in the fact that scientists rarely arrive independently at the same results at the same time. Science is full of enduring debates and controversies, and convergence requires long and tedious exchange inside the scientific community. Conversely, mindreading abilities seem to be acquired in a stereotyped order and rate among most children, who do not appear to hold consensus conferences or rely on peer-reviewing. Finally, some concerns are raised regarding the fact that humans, alleged user of the theory, are generally incapable of stating the putative laws and entities they're assumed to use daily. Some philosophers propose then a deflationary account of theory-theory, by substituting the cumbersome notion of theory with the easier to handle notion of model which don't suppose this enunciation capacity nor the mastery of law-like generalizations (Maibom 2003; Godfrey-Smith 2005).

Another leading view in the field of the theory-theory is the modularity thesis, notably associated with Simon Baron-Cohen, Uta Frith and Alan Leslie, following their initial proposal of autism as a specific dysfunction of the theory of mind (Baron-Cohen et al. 1985). The concept of modularity goes back to Jerry Fodor seminal work (1983) where he describes modules as cognitive systems sharing nine features.[12] The conception of modularity was later relaxed so that a cognitive system could be considered modular without presenting all the characteristics. The idea of theory of mind as a module was promoted by Fodor himself (1992) and partially adopted by Leslie et al. (2004) and Baron-Cohen (1998). The view was also held by authors working on other neuropsychiatric disorders suspected to involve abnormalities in the functioning of the theory of mind, such as schizophrenia (Corcoran 2001; Brüne 2005) and borderline disorder (Fonagy and Luyten 2009). Contrary to the scientific-theory-theory, the modularity-theory accounts for the roughly similar development of mindreading capacities in the children. Indeed, modules are often described as having a specific ontogeny (see note 12). One might wonder why, if theory of mind is a kind of hard-wired system in the brain, it does take time to be fully on-line. Moreover, Candida Peterson and Siegal (2000) have pointed out that deaf children raised among non-signing attachment figure present difficulties in

[12]Fodor's modules, as described in *The modularity of mind* (1983):

- are domain specific, that the range of inputs it can works on is very narrow.
- operate mandatorily. There is no possibility to higher levels of the hierarchy of mind to prevent their working.
- have limited central accessibility, meaning that their outputs are restricted to specific cognitive targets.
- are fast processing.
- are informationally encapsulated, meaning that background set of beliefs and other cognitive attitude do not influence their work.
- have shallow outputs, so informationally poor, limited to what is strictly necessary for the functioning of the higher-level systems.
- have a fixed neural architecture, that they are hard-wired in the brain, by opposition of soft and adaptative systems.
- have characteristic breakdown patterns.
- and have a stereotyped ontogeny.

mindreading tasks that are somewhat like that of children with autism. Conversely, deaf children who grow up in signing environment do not share these impediments. Peterson and Siegal make clear that the former are not neglected children. Their attachment figures albeit voluntary to have exchanges with the children encounter difficulties with topics related to mental life, emotions, beliefs and so on, even when they try to learn sign language. However, these insights are not detrimental to the modularity thesis. On the one hand, I already underlined that human neurodevelopment is far from complete at birth. On the other hand, it is commonplace that many cognitive mechanisms require appropriate stimulation to fully flesh out (see Purves 2018, chap. 25). Compared to the genome, the nervous system is more complex, richer in information, by several orders of magnitude. This implies the existence of "gene saving mechanisms" (Changeux and Danchin 1976). If the information is anyway available in the environment, it is best to genetically store only what is strictly necessary and then reconstruct the information during ontogenesis. For instance, there is no need to genetically encode the information that gravitational acceleration on Earth equals roughly 9.81 m/s^2 since it is easily retrievable from the environment. It is plausible that this is precisely one of the roles of play behaviors in young animals. I propose to talk about *extended genotype* to refer to this information that are necessary for ontogenesis and stored in the environment. Furthermore, this modularity approach opens the conceptual possibility of a full-blown theory-theory of mind to underlie mindreading. Indeed, if it is doubtful that children systematically rediscover its constituents, a kind a theorization can very well be the product of phylogenesis. This is even more convincing if we recall that mindreading may be the product of an evolutionary escalation during which the theory and its object may very well have coevolved.

14.4 Self-Ascription

Whereas theory of mind, in the broad sense, is an acknowledged account of our ability to detect the attitudes of others, a widespread view among philosophers and laypeople is that the access to our own mental states is quite dissimilar, contrary to premise (2) (see above). As I already mentioned, many philosophers endorse that we have a direct access to our mental life in one way or another. If the mundane observations that we are occasionally wrong about our beliefs and desires suggests that this access is not immune to error, it is still widely assumed that we benefit from a privileged access (Heil 1988).[13] This thesis only requires a qualitative difference between the way one can grasp one's own mental state and his access to that of

[13]There are different uses of the notion of privileged access. Some authors employ the notion of privileged access as quasi-synonymous of cartesian direct access (Alston 1971). My use of the term is far more liberal, focusing on the idea that at least some features of mental states are private, accessible only to the person having it.

others. From this point of view, our self-knowledge, if not faultless, is at least secure in the sense that one is more justified in asserting one's own beliefs than someone else (Fernández 2013, p. 5–8).[14] However, these ideas would need to be clarified and defended so as not to be reduced as a mere quantitative difference, as in Gilbert Ryle's account of self-knowledge:

> The superiority of the speaker's knowledge of what he is doing over that of the listener does not indicate that he has Privileged Access to facts of a type inevitably inaccessible to the listener, but only that [the listener] is in a very poor position to know. The turns taken by a man's conversation do not startle or perplex his wife as much as they had surprised and puzzled his fiancée, nor do close colleagues have to explain themselves to each other as much as they have to explain themselves to their new pupils. (Ryle, 1949/Ryle 2000, p. 179).

Focusing on behavioral evidence, Ryle acknowledges that generally one has a better access to one's attitudes, but only because one is constantly witnessing one's own actions. To put it another way, a person's theory of own mind would be particularly efficient simply because it would be fed with way more inputs concerning the person than anyone else. That is why a defense of genuine privileged access must add that one doesn't rely merely on reasoning or behavioral evidence (Fernández, *ibid.*). The question thus raised is that of the putative modes of special access to our propositional attitudes. Sydney Shoemaker (1994a, b) argues that this further access may be granted by the functional role of our mental states. Thereby, in addition to their participation to the guidance of behavior and to their connection to other internal states, attitudinal states would take on the role of being knowable by the subject holding them. This raises the question of why attitudinal states would have this property rather than not. It seems to me that Shoemaker's proposition rather entails an overdetermination of the efficacy of beliefs. Indeed, a belief is already described as having behavioral outputs in the appropriate conditions. Knowability would be redundant if it is to be characterized as mediating the causal role of beliefs. Moreover, I don't see anything contradictory with the idea of a creature holding beliefs without any faculty of higher-order representation, as it was allegedly the norm before the apparition of mindreading abilities. Therefore, if the knowability of beliefs and other attitudes were to explain our special access to our mental life, it would remain to explain why this functional property would have emerge at the risk of being considered ad hoc.

A similar explanatory problem presents itself in the cases of introspection accounts, understood as perception-like ability with inward orientation, requiring attentional resources (James 1890/1995, p. 185, Peacocke 1998). I will focus on one of these accounts, namely the "Monitoring Mechanism" (Nichols and Stich 2003, p. 160–163), assuming that this reflection can be extended to all putative forms of introspection. Thus, I take the existence of such a mechanism to be untenable

[14] I leave aside the conceivable but irrelevant possibility that one could have privileged access to one's own mental states and yet be less justified in asserting them than other people, because one would be blind to some important aspects of one's mental states that others could appraise. In what follows, I assume that in normal cases a person has at least as much access to her own mental states as others have from their third-person perspective. The question being whether she has more.

because of the evolutionary escalation of theory of mind I sketched, in accordance with premise (3). Indeed, for such a mechanism to be reliable, it should have as outputs the same kind of attitudes that we attribute to others, for we don't take our own mental states to be dissimilar to that of others. Moreover, we easily agree with others when they attribute this or that attitude to us. Besides, there seem to be no instance of conflict between this presumed monitoring mechanism and self-directed mindreading outputs, even though no introspection theorist would contest the possibility of making mindreading inferences about us even if we also have an additional privileged access. However, if the phylogenetic development of this mechanism was distinct of that of theory of mind, there is no a priori reason for both to obtain the exact same set of types and token of theoreticality unobservable entities, namely attitudes. I take these elements to secure premise (2). Of course, significant similarities would be expected as in the case of convergent evolution.[15] However, convergent evolution entails analogy, not perfect symmetry. This is why, a further explanation would be required. An obvious candidate would be that theory of mind itself is recycled, and this is exactly what premise (3) suggests. To add a second mechanism to the architecture would be like tacking a mandatory second trunk on each tree, perhaps one for the sap to rise and the other for it to fall, with twice the costs and twice the weaknesses. Some might say that the monitoring mechanism could have been much easier to implement, for it wouldn't have to treat equivocal sensory information. However, it should face the problem of online detection of our mental state, a difficulty which is hard to assess in absence of empirical data linked to the cognitive mechanisms in charge of the monitoring mechanism. Even if we are here forced to speculate, it is likely that such a problem cannot be solve for free.[16] Besides, it would have to satisfy premise (2). As the complexity of our propositional attitudes might have coevolved with our mindreading abilities, the outputs to convey would still have to be quite precise. For these reasons, I take it to be much more parsimonious to assume that evolution would have co-opt the already existing mindreading mechanism as the device for self-ascription.

A culminating account of belief self-ascription to prima facie related to the theory of mind is the ascent routine, also known as Evan's principle. I won't address whether the ascent routine can be employed for desires (Fernández 2013, p. 87–88) or other propositional attitudes (Gordon 2007). However, this account is

[15] Convergent evolution explains the resemblance between the wing of a bat, that of a pterodactyl and that of a bird. These groups do not have a common winged ancestor, but the analogy in wing shape is explained by the aerodynamic constraints inherent in flight. The notion of analogy is here to be opposed to that of homology. Two characters are said to be homologous when they have a common phylogenetic origin. For example, the hand of the chimpanzee is homologous to that of human. Note that the existence of convergent evolution within an organism poses no conceptual problem, as it can be illustrated by the hallmarks of cancer (Hanahan and Weinberg 2000, Hanahan and Weinberg 2011, see also Merlo et al. 2006).

[16] Especially, it might face the frame problem which I leave aside for the sake of simplification (see Shanahan 2016).

taken to be pervasive concerning beliefs, for they are deemed to demonstrate the particular feature of transparency (Shah 2003). The idea behind the ascent routine is that a direct way to know if one believes that *p*, is to ask oneself *whether p* (Evans 1982, p. 225, see also Byrne 2005). We have to distinguish two situations here. First, this routine seems to work when *p* is an observable state of affair in the world. For instance, one can infer that one believes that is raining by looking through the window and observe that it is raining. In other words, it seems that we have access of our beliefs through perception. However, we can explain this kind of access to our own belief by a scrutiny of the mechanism of perceptions, with the convenient case of sight. Contrary to accounts of perception in classical philosophy (Moore 1903, Broad 1923, Price 1932, Ayer 1940, for early criticism see Chisholm 1957, Austin 1962), it is now commonplace that there is no clear border between perceptions and beliefs about percepts. Our perceptions are imbued with prior beliefs and the functional role of our sensory organs, far from giving us an exhaustive representation of the world, is to correct possible errors of predictions relying on our inner model of the world (Marr 1982; Dennett 1983; Friston et al. 2006; Fletcher and Frith 2009; Hohwy 2013; Clark 2016). My point here is not to offer an extensive account of computational theories of perception, but to raise that if one accepts that beliefs about the external world and perceptions are two sides of the same coin, Evans' argument becomes tautological, even though cleverly tautological. Secondly, Evans' argument seems aboveboard regarding our judgments:

> If someone asks me "Do you think there is going to be a third world war?", I must attend, in answering him, to precisely the same outward phenomena as I would attend to if I were answering the question "Will there be a third world war?". I get myself in a position to answer the question whether I believe that P by putting into operation whatever procedure I have for answering the question whether P. (1982, p. 225).

In this case, there is no directly available answer in our surroundings. Nevertheless, we can figure this out by pointing that this is nothing else than doxastic deliberation (Shah and Velleman 2005). In other words, we don't get access to any underlying belief (see also Chater 2019 who argues against the mere notion of "underlying" belief). On the contrary, we create the belief as we answer the question and because truth is the norm of beliefs, answering the question is not a distinct process than forming the belief (see Moran 2001, chap. 2, on self-fulfilling beliefs). Note that this deliberative access to beliefs may be from the start better described as self-directed theory of mind, for in some cases, it is by expressing them (an overt behavior) that we get in touch with them (Bem 1972).[17] In both cases, what we take to be an immediate access to our beliefs is nothing but a perceptual access to a given state of affair which is not in the realm of self-mindreading. Thus, the ascent routine and the

[17] However, doxastic deliberation may of course be silent. Then it might be through sensory access to our phonological loop as Peter Carruthers suggest (Carruthers 2011, p. 48). The phonological loop being part of our working memory implicated in such deliberations (Baddeley 2000; Baddeley and Larsen 2007; Baddeley and Hitch 2019).

transparency of beliefs don't imply that we have some kind of special access to our own beliefs.

What are the arguments against the idea that we get access to our own mental states by making use of the theory of mind? First of all, there is the phenomenal case for introspection. However, there is no reason to deny that the outcomes of self-oriented theory of mind couldn't have this putative phenomenality. What is consciously accessible (in the sense of "access consciousness", Block 1995) is the result of these cognitive mechanisms, not how they work. Next in line, some might emphasize that we have widely richer and more accurate access to our own beliefs than we have to other's. But have we? As already mentioned, we are at the forefront of our own actions (see above). Moreover, in self-ascription there is virtually no risk of exposure to rebuttal (Ryle, ibid.). These mere facts might explain why we are prone to self-attribute many attitudes. Furthermore, there are empirical arguments for the thesis that many of our self-attributions are more confabulatory than reliable (Nisbett and Wilson 1977; Baynes and Gazzaniga 2000; Johansson et al. 2005).[18] As we have already seen, some of our ungrounded assertions may even become self-fulfilling prophecies.

14.5 Conclusions

In the current context of distorted information and conflicting beliefs (Mercier 2020, chap. 15, Chavalarias 2022), understanding the way we access to our own beliefs is a crucial matter. Indeed, when beliefs are elevated to the rank of personal or moral value (Dennett 2006, chap. 8), being able to assess one own belief and the limitations of such an assessment becomes a matter of epistemic responsibility (Corlett 2008). In this paper, I argued that our access to our own beliefs was indirect at best, relying on behavioral and contextual cues similar to those that allow us to understand each other. We thus have serious risks of being mistaken about our mental states. On the one hand, our conclusions our grounded on perceptual evidence of which the limited reliability is commonplace and remains a classical philosophical object of scrutiny (Lyons 2023). On the other hand, the conclusions based on these perceptual cues are far from being the product of deductive inference. There is room for uncertainty, guesses and inferences to the best available explanation. Despite the imperfection of this kind of self-knowledge, we seem to be quite confident about the beliefs that we claim we hold, relying on ideas of special access or anchored introspection. Therefore, it is plausible that some if not most of our belief assertions are simply confabulatory. If this claim is true, it would have deep conceptual consequences

[18] It is frequently opposed to these examples that they only show that people may be wrong about the causes of their judgments, rather than about their beliefs per se. However, this objection is highly disputable as people seem committed to their subsequent beliefs. Moreover, isn't the proposition "x is the cause of my judgement" a belief in itself?

which are too numerous to be exhaustively explored. As I endorse the claim, let me at least sketch two notable outcomes, one concerning a traditional philosophical debate, namely that of self-deception, the other regarding the way we interact as social agents, especially in the emerging context of social networks.

Self-deception is taken to be commonplace in human psychology. Roughly, a person is described as self-deceptive when she is taken to hold a false belief despite sufficient evidence to the contrary and because of a putative psychological motivation to deny or dismiss this evidence. A person may be self-deceptive about a serious medical diagnosis, the loss of a loved one or about self-demeaning traits or attitudes. The latter case seems to be quite ubiquitous. For instance, among people still using cars[19] as means of transport, a majority of drivers see themselves as better than average (Williams 2003; Alicke and Govorun 2005). However, self-deception seems to be paradoxical in many ways. Two main concerns regarding self-deception are the static and dynamic puzzles (Mele 2001, chap. 3). How can a person simultaneously hold the deceptive false-belief and the correct belief that is supposed to motivate self-deception? How can one be deceived while one is entertaining the intention to deceive one-self? Some philosophers invoke particular doxastic states such as 'in-between beliefs' (Schwitzgebel 2001) or complex psychological processes including forgetting one's intentions (Bermúdez 2000). I suggest a deflationary account akin to that of Alfred Mele's non-intentionalism. Because people are not reliably connected to their beliefs, it should be commonplace that people assert ideas that they would not endorse as full-blown beliefs. Conversely, they act on beliefs they don't recognize holding. Take for instance the putative self-deceptive person who denies her medical diagnosis. One might be wonder why despite her firm assertion that there is no medical problem at all, she still attends to her rendezvous with her oncologist. The reason is straightforward: she actually believes that she is ill. She merely doesn't know that she entertains that belief, for her self-assessment is motivationally biased. However, by dint of proclaiming that everything is fine, she might well end up believing it.

This brings us to the social dimension of belief self-attribution. In the context of social networks taking more and more space in our lives, some authors have raised specific concerns regarding the polarization of beliefs. Based on the work of political science researchers, Thi Nguyen (2020) distinguished epistemic bubbles from a much more noxious phenomenon with which it is often confused, that is echo chambers. Epistemic bubbles are simply informational structures in which dissenting voices have been ignored, perhaps by accident. They refer to a phenomenon where individuals or groups are only exposed to information and ideas that confirm their pre-existing beliefs and values. Echo chambers, by contrast, are not structures in which dissenting voices have been merely ignored, but structures in which they have been actively discredited. While confronting dissenting voices and evidence can normally burst an epistemic bubble, it will have at best no effect on an echo chamber.

[19] I also suspect a case of massive self-deception regarding the impact of this habitus on local (Miller and Newby 2019) and global (Dhakal et al. 2022) environment.

Echo chambers trap their members in what philosophers have called a vicious epistemic circle (Levy 2019). According to my claim on self-attribution of belief, echo chambers might be the perfect breeding ground for self-fulfilling endorsement of beliefs. Observing oneself validate the inner narrative of the chamber and actively reject dissonant ideas, one will be in a particularly favorable situation to end up genuinely believe an idea that just might have socially appealing at first sight. In other words, in an echo chamber, one can be his own jailer. My advice: ask yourself, “do I really believe that?”.

Acknowledgement I'm thankful to Marion Vorms, Kengo Miyazono, Teresa Lopez-Soto, and Alexandre Billon for their help and reviews.

I'm thankful to the *Agence de l'innovation de défense* (government of France) and the region *Hauts-de-France* for funding my research.

References

Adénor, J.-L., and T. de Rauglaudre. 2021. *Le nouveau péril sectaire: Antivax, crudivores, écoles Steiner, évangéliques radicaux ...* Robert Laffont.

Aiello, L.C., and P. Wheeler. 1995. The expensive-tissue hypothesis: The brain and the digestive system in human and primate evolution. *Current Anthropology* 36 (2): 199–221. https://doi.org/10.1086/204350.

Alicke, M.D., and O. Govorun. 2005. The better-than-average effect. In *The self in social judgment*, 85–106. Psychology Press.

Allaman, I., M. Bélanger, and P.J. Magistretti. 2011. Astrocyte–neuron metabolic relationships: For better and for worse. *Trends in Neurosciences* 34 (2): 76–87. https://doi.org/10.1016/j.tins.2010.12.001.

Alston, W.P. 1971. Varieties of priveleged access. *American Philosophical Quarterly* 8 (3): 223–241.

American Psychiatric Association (Éd.). 2013. *Diagnostic and statistical manual of mental disorders: DSM-5*. 5th ed. American Psychiatric Association.

Arendt, D. 2021. Elementary nervous systems. *Philosophical Transactions of the Royal Society B: Biological Sciences* 376 (1821): 20200347. https://doi.org/10.1098/rstb.2020.0347.

Austin, J. L. 1962. *Sense and sensibilia* (G. J. Warnock, Éd.). Oxford University Press.

Axelrod, R. 1981. The emergence of cooperation among egoists. *American Political Science Review* 75 (2): 306–318. https://doi.org/10.2307/1961366.

Axelrod, R., and W.D. Hamilton. 1981. The evolution of cooperation. *Science* 211 (4489): 1390–1396. https://doi.org/10.1126/science.7466396.

Ayer, A.J. 1940. *Foundations of empirical knowledge*. Macmillan.

Baddeley, A. 2000. The episodic buffer: A new component of working memory? *Trends in Cognitive Sciences* 4 (11): 417–423. https://doi.org/10.1016/S1364-6613(00)01538-2.

Baddeley, A.D., and G.J. Hitch. 2019. The phonological loop as a buffer store: An update. *Cortex* 112: 91–106. https://doi.org/10.1016/j.cortex.2018.05.015.

Baddeley, A.D., and J.D. Larsen. 2007. The phonological loop: Some answers and some questions. *Quarterly Journal of Experimental Psychology* 60 (4): 512–518. https://doi.org/10.1080/17470210601147663.

Baron-Cohen, S. 1998. Does the study of autism justify minimalist innate modularity? *Learning and Individual Differences* 10 (3): 179–191. https://doi.org/10.1016/S1041-6080(99)80129-0.

Baron-Cohen, S., A.M. Leslie, and U. Frith. 1985. Does the autistic child have a “theory of mind”? *Cognition* 21 (1): 37–46. https://doi.org/10.1016/0010-0277(85)90022-8.

Baumard, N., J.-B. André, and D. Sperber. 2013. A mutualistic approach to morality: The evolution of fairness by partner choice. *Behavioral and Brain Sciences* 36 (1): 59–78. https://doi.org/10.1017/S0140525X11002202.
Bayne, T., and E. Pacherie. 2005. In defence of the doxastic conception of delusions. *Mind and Language* 20 (2): 163–188. https://doi.org/10.1111/j.0268-1064.2005.00281.x.
Baynes, K., & Gazzaniga, M. S. 2000. Consciousness, introspection, and the split-brain: The two minds/one body problem. In M. S. Gazzaniga (Éd.), The new cognitive neurosciences: 2nd ed.. MIT Press.
Bem, D.J. 1972. Self-perception theory. In *Advances in experimental social psychology*, vol. 6, 1–62. Elsevier. https://doi.org/10.1016/S0065-2601(08)60024-6.
Bennett, J. 1978. Some remarks about concepts (commentary on Premack and Woodruff). *Behavioral and Brain Sciences* 1 (4): 557–560. https://doi.org/10.1017/S0140525X00076573.
Bermúdez, J.L. 2000. Self-deception, intentions, and contradictory beliefs. *Analysis* 60 (4): 309–319. https://doi.org/10.1093/analys/60.4.309.
Black, M. 1983. *The prevalence of humbug, and other essays*. Cornell University Press.
Block, N. 1995. On a confusion about a function of consciousness. *Brain and Behavioral Sciences* 18 (2): 227–247. https://doi.org/10.1017/s0140525x00038188.
Bortolotti, L. 2010. *Delusions and other irrational beliefs*. Oxford University Press.
Briggs, R. A. 2019. Normative theories of rational choice: Expected utility. In E. N. Zalta (Éd.), *The Stanford encyclopedia of philosophy* (Fall 2019). Metaphysics Research Lab, Stanford University. https://plato.stanford.edu/archives/fall2019/entries/rationality-normative-utility/
Broad, C.D. 1923. *Scientific thought*. Paterson: Routledge & Kegan Paul.
Brüne, M. 2005. « Theory of mind » in schizophrenia: A review of the literature. *Schizophrenia Bulletin* 31 (1): 21–42. https://doi.org/10.1093/schbul/sbi002.
Byrne, A. 2005. Introspection. *Philosophical Topics* 33 (1): 79–104.
Carbone, C., and G.M. Narbonne. 2014. When life got smart: The evolution of behavioral complexity through the Ediacaran and early Cambrian of NW Canada. *Journal of Paleontology* 88 (2): 309–330. https://doi.org/10.1666/13-066.
Carr, L., M. Iacoboni, M.-C. Dubeau, J.C. Mazziotta, and G.L. Lenzi. 2003. Neural mechanisms of empathy in humans: A relay from neural systems for imitation to limbic areas. *Proceedings of the National Academy of Sciences* 100 (9): 5497–5502. https://doi.org/10.1073/pnas.0935845100.
Carroll, L. 1872. *Through the looking-glass, and what alice found there*. Wikisource. https://en.wikisource.org/wiki/Through_the_Looking-Glass,_and_What_Alice_Found_There.
Carruthers, P. 2011. *The opacity of mind: An integrative theory of self-knowledge*. Oxford: Oxford University Press.
Carter, G.G., and G.S. Wilkinson. 2013. Food sharing in vampire bats: Reciprocal help predicts donations more than relatedness or harassment. *Proceedings of the Royal Society B: Biological Sciences* 280 (1753): 20122573. https://doi.org/10.1098/rspb.2012.2573.
Changeux, J.-P., and A. Danchin. 1976. Selective stabilisation of developing synapses as a mechanism for the specification of neuronal networks. *Nature* 264 (5588): 705–712. https://doi.org/10.1038/264705a0.
Chater, N. 2019. *The mind is flat: The illusion of mental depth and the improvised mind*. London: Penguin Books.
Chavalarias, D. 2022. *Toxic data: Comment les réseaux manipulent des opinions*. Flammarion.
Chisholm, R.M. 1957. *Perceiving: A philosophical study*. Ithaca: Cornell University Press.
Clark, A. 2016. *Surfing uncertainty: Prediction, action, and the embodied mind*. New York: Oxford University Press.
Corcoran, R. 2001. Theory of mind and schizophrenia. In P. W. Corrigan & D. L. Penn (Éds.), *Social cognition and schizophrenia*. (p. 149–174). American Psychological Association. https://doi.org/10.1037/10407-005.
Corlett, J.A. 2008. Epistemic responsibility. *International Journal of Philosophical Studies* 16 (2): 179–200. https://doi.org/10.1080/09672550802008625.

Cosmides, L., J. Tooby, L. Fiddick, and G.A. Bryant. 2005. Detecting cheaters. *Trends in Cognitive Sciences* 9 (11): 505–506. https://doi.org/10.1016/j.tics.2005.09.005.
Currie, G. 2000. Imagination, delusion and hallucinations. *Mind and Language* 15 (1): 168–183. https://doi.org/10.1111/1468-0017.00128.
Davidson, D. 1970. Mental events. In L. Foster and J. W. Swanson (Éds.), *Experience and theory* (p. 207–224). Clarendon Press.
———. 1973. Radical interpretation. *Dialectica* 27 (3–4): 313–328. https://doi.org/10.1111/j.1746-8361.1973.tb00623.x.
Davies, M., and T. Stone. 2001. Mental simulation, tacit theory, and the threat of collapse. *Philosophical Topics* 29 (1/2): 127–173.
Dennett, D.C. 1971. Intentional systems. *Journal of Philosophy* 68 (4): 87–106. https://doi.org/10.2307/2025382.
———. 1978. Beliefs about beliefs. *Behavioral and Brain Sciences* 1 (4): 568–570. https://doi.org/10.1017/S0140525X00076664.
Dennett, D. C. 1983. Beyond belief. In A. Woodfield (Éd.), Thought and object. Oxford Oxford University Press.
Dennett, D.C. 1987. *The intentional stance*. Cambridge MA: MIT Press.
———. 2006. *Breaking the spell: Religion as a natural phenomenon*. New York: Viking.
Descartes, R. 2017. *Meditations on first philosophy*. 2nd ed. New York: Cambridge University Press.
Dhakal, S., J.C. Minx, F.L. Toth, A. Abdel-Aziz, M.J. Figueroa Meza, K. Hubacek, I.G.C. Jonckheere, Yong-Gun Kim, G.F. Nemet, S. Pachauri, X.C. Tan, and T. Wiedmann. 2022. Emissions Trends and Drivers. In *IPCC, 2022: Climate change 2022: Mitigation of climate change. contribution of working group iii to the sixth assessment report of the inter-governmental panel on climate change*, ed. P.R. Shukla, J. Skea, R. Slade, A. Al Khourdajie, R. van Diemen, D. McCollum, M. Pathak, S. Some, P. Vyas, R. Fradera, M. Belkacemi, A. Hasija, G. Lisboa, S. Luz, and J. Malley. Cambridge, UK\New York: Cambridge University Press. https://doi.org/10.1017/9781009157926.004.
Dieguez, S. 2022. *Croiver: Pourquoi la croyance n'est pas ce que l'on croit*. Paris: Eliott.
Dub, R. 2017. Delusions, acceptances, and cognitive feelings. *Philosophy and Phenomenological Research* 94 (1): 27–60. https://doi.org/10.1111/phpr.12220.
Dumontheil, I., I.A. Apperly, and S.-J. Blakemore. 2010. Online usage of theory of mind continues to develop in late adolescence. *Developmental Science* 13 (2): 331–338. https://doi.org/10.1111/j.1467-7687.2009.00888.x.
Dunbar, R.I.M. 1992. Neocortex size as a constraint on group size in primates. *Journal of Human Evolution* 22 (6): 469–493. https://doi.org/10.1016/0047-2484(92)90081-J.
———. 1993. Coevolution of neocortical size, group size and language in humans. *Behavioral and Brain Sciences* 16 (4): 681–694. https://doi.org/10.1017/S0140525X00032325.
Evans, G. (1982). The varieties of reference. Clarendon Press\Oxford University Press, Oxford.
Falster, D.S., and M. Westoby. 2003. Plant height and evolutionary games. *Trends in Ecology & Evolution* 18 (7): 337–343. https://doi.org/10.1016/S0169-5347(03)00061-2.
Feinberg, I. 1982. Schizophrenia: Caused by a fault in programmed synaptic elimination during adolescence? *Journal of Psychiatric Research* 17 (4): 319–334. https://doi.org/10.1016/0022-3956(82)90038-3.
Fernández, J. 2013. *Transparent minds: A study of self-knowledge*. 1st ed. Oxford University Press.
Fletcher, P.C., and C.D. Frith. 2009. Perceiving is believing: A Bayesian approach to explaining the positive symptoms of schizophrenia. *Nature Reviews Neuroscience* 10 (1): 48–58. https://doi.org/10.1038/nrn2536.
Fodor, J.A. 1983. *The modularity of mind: An essay on faculty psychology*. Cambridge, MA: MIT Press.
———. 1992. A theory of the child's theory of mind. *Cognition* 44 (3): 283–296. https://doi.org/10.1016/0010-0277(92)90004-2.

Fonagy, P., and P. Luyten. 2009. A developmental, mentalization-based approach to the understanding and treatment of borderline personality disorder. *Development and Psychopathology* 21 (4): 1355–1381. https://doi.org/10.1017/S0954579409990198.

Friston, K., J. Kilner, and L. Harrison. 2006. A free energy principle for the brain. *Journal of Physiology-Paris* 100 (1–3): 70–87. https://doi.org/10.1016/j.jphysparis.2006.10.001.

Frith, C.D. 1992. *The cognitive neuropsychology of schizophrenia*. Hove: Lawrence Erlbaum Associates, Inc.

Fromer, M., A.J. Pocklington, D.H. Kavanagh, H.J. Williams, S. Dwyer, P. Gormley, L. Georgieva, E. Rees, P. Palta, D.M. Ruderfer, N. Carrera, I. Humphreys, J.S. Johnson, P. Roussos, D.D. Barker, E. Banks, V. Milanova, S.G. Grant, E. Hannon, et al. 2014. De novo mutations in schizophrenia implicate synaptic networks. *Nature* 506 (7487): 179–184. https://doi.org/10.1038/nature12929.

Gigerenzer, G. 2008. Why heuristics work. *Perspectives on Psychological Science* 3 (1): 20–29. https://doi.org/10.1111/j.1745-6916.2008.00058.x.

Godfrey-Smith, P. 2005. Folk psychology as a model. *Philosopher's Imprint* 5 (6): 1–16.

Goldman, A.I. 1989. Interpretation psychologized. *Mind & Language* 4 (3): 161–185. https://doi.org/10.1111/j.1468-0017.1989.tb00249.x.

———. 2006. *Simulating minds: The philosophy, psychology, and neuroscience of mindreading*. New York: Oxford University Press.

Gopnik, A. 1990. Developing the idea of intentionality: Children's theories of mind. *Canadian Journal of Philosophy* 20 (1): 89–113. https://doi.org/10.1080/00455091.1990.10717210.

———. 2004. Finding our inner scientist. *Daedalus* 133 (1): 21–28. https://doi.org/10.1162/001152604772746666.

Gopnik, A., and H.M. Wellman. 1992. Why the child's theory of mind really is a theory. *Mind & Language* 7 (1–2): 145–171. https://doi.org/10.1111/j.1468-0017.1992.tb00202.x.

Gordon, R.M. 1986. Folk psychology as simulation. *Mind & Language* 1 (2): 158–171. https://doi.org/10.1111/j.1468-0017.1986.tb00324.x.

———. 1992. The simulation theory: Objections and misconceptions. *Mind & Language* 7 (1–2): 11–34. https://doi.org/10.1111/j.1468-0017.1992.tb00195.x.

Gordon, R. M. 1995. Simulation without introspection or inference from me to you. In M. Davies & T. Stone (Éds.), Mental Simulation. Blackwell.

Gordon, R.M. 2007. Ascent routines for propositional attitudes. *Synthese* 159 (2): 151–165. https://doi.org/10.1007/s11229-007-9202-9.

Grice, P. 1975. Logic and conversation. In P. Cole and J. Morgan (Éds.), *Syntax and semantics* (Vol. 3). Academic.

Gropman, A.L., M. Summar, and J.V. Leonard. 2007. Neurological implications of urea cycle disorders. *Journal of Inherited Metabolic Disease* 30 (6): 865–879. https://doi.org/10.1007/s10545-007-0709-5.

Hamilton, W.D. 1964a. The genetical evolution of social behaviour. I. *Journal of Theoretical Biology* 7 (1): 1–16. https://doi.org/10.1016/0022-5193(64)90038-4.

———. 1964b. The genetical evolution of social behaviour. II. *Journal of Theoretical Biology* 7 (1): 17–52. https://doi.org/10.1016/0022-5193(64)90039-6.

Hamilton, W.D., R. Axelrod, and R. Tanese. 1990. Sexual reproduction as an adaptation to resist parasites (a review). *Proceedings of the National Academy of Sciences* 87 (9): 3566–3573. https://doi.org/10.1073/pnas.87.9.3566.

Hanahan, D., and R.A. Weinberg. 2000. The hallmarks of cancer. *Cell* 100 (1): 57–70. https://doi.org/10.1016/S0092-8674(00)81683-9.

———. 2011. Hallmarks of cancer: The next generation. *Cell* 144 (5): 646–674. https://doi.org/10.1016/j.cell.2011.02.013.

Harman, G. 1978. Studying the chimpanzee's theory of mind. *Behavioral and Brain Sciences* 1 (4): 576–577. https://doi.org/10.1017/S0140525X00076743.

Harwood, M.D., and M.J. Farrar. 2006. Conflicting emotions: The connection between affective perspective taking and theory of mind. *British Journal of Developmental Psychology* 24 (2): 401–418. https://doi.org/10.1348/026151005X50302.

Heal, J. 1986. Replication and functionalism. In J. Butterfield (Éd.), *Language, mind, and logic* (p. 135–150). Cambridge University Press.

Heberlein, A.S., and R. Adolphs. 2007. Neurobiology of emotion recognition: Current evidence for shared substrates. In *Social neuroscience: Integrating biological and psychological explanations of social behavior*, 31–55. The Guilford Press.

Heil, J. 1988. Privileged access. *Mind* 97 (386): 238–251.

Heyes, C. 2015. Animal mindreading: What's the problem? *Psychonomic Bulletin & Review* 22 (2): 313–327. https://doi.org/10.3758/s13423-014-0704-4.

Hohwy, J. 2013. *The predictive mind*. 1st ed. Oxford University Press.

James, W. 1995. *The principles of psychology: In two volumes. Vol. 1*. New York: Dover.

Johansson, P., L. Hall, S. Sikström, and A. Olsson. 2005. Failure to detect mismatches between intention and outcome in a simple decision task. *Science* 310 (5745): 116–119. https://doi.org/10.1126/science.1111709.

Kahneman, D. 2012. *Thinking, fast and slow*. London: Penguin Books.

King, D.A. 1990. The adaptive significance of tree height. *The American Naturalist* 135 (6): 809–828. https://doi.org/10.1086/285075.

Kuhn, S. 2019. Prisoner's dilemma. In E. N. Zalta (Éd.), *The Stanford encyclopedia of philosophy* (Winter 2019). Metaphysics Research Lab, Stanford University. https://plato.stanford.edu/archives/win2019/entries/prisoner-dilemma/

Leslie, A.M., O. Friedman, and T.P. German. 2004. Core mechanisms in 'theory of mind'. *Trends in Cognitive Sciences* 8 (12): 528–533. https://doi.org/10.1016/j.tics.2004.10.001.

Levy, N. 2019. Due deference to denialism: Explaining ordinary people's rejection of established scientific findings. *Synthese* 196 (1): 313–327. https://doi.org/10.1007/s11229-017-1477-x.

Lyons, J. 2023. Epistemological problems of perception. In E. N. Zalta & U. Nodelman (Éds.), *The Stanford encyclopedia of philosophy* (spring 2023). Metaphysics Research Lab, Stanford University. https://plato.stanford.edu/archives/spr2023/entries/perception-episprob/

Maibom, H. 2003. The mindreader and the scientist. *Mind and Language* 18 (3): 296–315. https://doi.org/10.1111/mila.2003.18.issue-3.

Marr, D. 1982. *Vision: A computational investigation into the human representation and processing of visual information*. San Francisco: W.H. Freeman.

Maynard Smith, J. 1964. Group selection and kin selection. *Nature* 201 (4924): 1145–1147. https://doi.org/10.1038/2011145a0.

Mele, A.R. 2001. *Self-deception unmasked*. Princeton: Princeton University Press.

Mercier, H. 2020. *Not born yesterday: The science of who we trust and what we believe*. Princeton: Princeton University Press.

Merlo, L.M.F., J.W. Pepper, B.J. Reid, and C.C. Maley. 2006. Cancer as an evolutionary and ecological process. *Nature Reviews Cancer* 6 (12): 924–935. https://doi.org/10.1038/nrc2013.

Miller, M. R., and Newby, D. E. 2019. Air pollution and cardiovascular disease: Car sick. *Cardiovascular research*, cvz228. https://doi.org/10.1093/cvr/cvz228.

Mitchell, J.P., M.R. Banaji, and C.N. Macrae. 2005. The link between social cognition and self-referential thought in the medial prefrontal cortex. *Journal of Cognitive Neuroscience* 17 (8): 1306–1315. https://doi.org/10.1162/0898929055002418.

Miyazono, K. 2019. *Delusions and beliefs: A philosophical inquiry*. New York: Routledge.

Montalvo, S. 2021. *Contribution de la biologie de l'évolution à la compréhension des troubles psychiatriques et neurodéveloppementaux* [Thèse pour le diplôme d'état de docteur en médecine, Lille]. https://pepite.univ-lille.fr/ori-oai-search/notice/view/univ-lille-33645

Moore, G.E. 1903. The refutation of idealism. *Mind XII* (4): 433–453. https://doi.org/10.1093/mind/XII.4.433.

Moran, R. 2001. *Authority and estrangement: An essay on self-knowledge*. Princeton: Princeton University Press.

Nelson, E.E., E. Leibenluft, E.B. McCLURE, and D.S. Pine. 2005. The social re-orientation of adolescence: A neuroscience perspective on the process and its relation to psychopathology. *Psychological Medicine* 35 (2): 163–174. https://doi.org/10.1017/S0033291704003915.

Nguyen, C.T. 2020. Echo chambers and epistemic bubbles. *Episteme* 17 (2): 141–161. https://doi.org/10.1017/epi.2018.32.

Nguyên Hoang, L. 2020. *The equation of knowledge: From bayes' rule to a unified philosophy of science*. Chapman & Hall CRC. New York.

Nichols, S., and S.P. Stich. 2003. Mindreading: An integrated account of pretence, self-awareness, and understanding other minds. In *Clarendon\Oxford*. Oxford University: Press.

Nisbett, R.E., and T.D. Wilson. 1977. Telling more than we can know: Verbal reports on mental processes. *Psychological Review* 84 (3): 231–259. https://doi.org/10.1037/0033-295X.84.3.231.

Onishi, K.H., and R. Baillargeon. 2005. Do 15-month-old infants understand false beliefs? *Science* 308 (5719): 255–258. https://doi.org/10.1126/science.1107621.

Oreskes, N., & Conway, E. M. (2012). Merchants of doubt: How a handful of scientists obscured the truth on issues from tobacco smoke to global warming. Bloomsbury.

Owen, M.J., N. Craddock, and M.C. O'Donovan. 2010. Suggestion of roles for both common and rare risk variants in genome-wide studies of schizophrenia. *Archives of General Psychiatry* 67 (7): 667. https://doi.org/10.1001/archgenpsychiatry.2010.69.

Peacocke, C. 1998. Conscious attitudes, attention, and self-knowledge. In C. Wright, B. C. Smith, and C. Macdonald (Éds.), *Knowing our own minds* (p. 83). Oxford University Press.

Peterson, C., and Siegal, M. 2000. Insights into theory of mind from deafness and autism. In M. Coltheart and M. Davies (Éds.), Pathologies of belief. Blackwell.

Premack, D., and G. Woodruff. 1978. Does the chimpanzee have a theory of mind? *Behavioral and Brain Sciences* 1 (4): 515–526. https://doi.org/10.1017/S0140525X00076512.

Press, W.H., and F.J. Dyson. 2012. Iterated Prisoner's dilemma contains strategies that dominate any evolutionary opponent. *Proceedings of the National Academy of Sciences* 109 (26): 10409–10413. https://doi.org/10.1073/pnas.1206569109.

Price, H.H. 1932. *Perception*. Westport: Methuen.

Purves, D. (Éd.). 2018. Neuroscience (6th ed.). Oxford University Press.

Rapoport, A., D.A. Seale, and A.M. Colman. 2015. Is tit-for-tat the answer? On the conclusions drawn from Axelrod's tournaments. *PLoS One* 10 (7): e0134128. https://doi.org/10.1371/journal.pone.0134128.

Russell, B. 1910. Knowledge by acquaintance and knowledge by description. *Proceedings of the Aristotelian Society 11*: 108–128.

Ryle, G. 2000. *The concept of mind (Reprinted with a new introduction)*. London: Penguin Books.

Schizophrenia Working Group of the Psychiatric Genomics Consortium. 2014. Biological insights from 108 schizophrenia-associated genetic loci. *Nature* 511 (7510): 421–427. https://doi.org/10.1038/nature13595.

Schwitzgebel, E. 2001. In-between believing. *The Philosophical Quarterly* 51 (202): 76–82. https://doi.org/10.1111/1467-9213.00215.

———. 2012. Mad belief? *Neuroethics* 5 (1): 13–17. https://doi.org/10.1007/s12152-011-9127-3.

Shah, N. 2003. How truth governs belief. *The Philosophical Review* 112 (4): 447–482. https://doi.org/10.1215/00318108-112-4-447.

Shah, N., and J.D. Velleman. 2005. Doxastic deliberation. *Philosophical Review* 114 (4): 497–534. https://doi.org/10.1215/00318108-114-4-497.

Shanahan, M. 2016. The frame problem. In E. N. Zalta (Éd.), *The Stanford encyclopedia of philosophy* (Spring 2016). Metaphysics Research Lab, Stanford University. https://plato.stanford.edu/archives/spr2016/entries/frame-problem/

Shoemaker, S. 1994a. Self-knowledge and « inner sense » lecture I: The object perception model. *Philosophy and Phenomenological Research* 54 (2): 249–269.

———. 1994b. Self-knowledge and « inner sense » lecture II: The broad perceptual model. *Philosophy and Phenomenological Research* 54 (2): 271–290.

Shorter, J.R., and O. Rueppell. 2012. A review on self-destructive defense behaviors in social insects. *Insectes Sociaux* 59 (1): 1–10. https://doi.org/10.1007/s00040-011-0210-x.
Sikora, M., A. Seguin-Orlando, V.C. Sousa, A. Albrechtsen, T. Korneliussen, A. Ko, S. Rasmussen, I. Dupanloup, P.R. Nigst, M.D. Bosch, G. Renaud, M.E. Allentoft, A. Margaryan, S.V. Vasilyev, E.V. Veselovskaya, S.B. Borutskaya, T. Deviese, D. Comeskey, T. Higham, et al. 2017. Ancient genomes show social and reproductive behavior of early upper Paleolithic foragers. *Science* 358 (6363): 659–662. https://doi.org/10.1126/science.aao1807.
Sober, E. 1988. What is evolutionary altruism? *Canadian Journal of Philosophy Supplementary Volume* 14: 75–99. https://doi.org/10.1080/00455091.1988.10715945.
Stahl, S.M. 2021. *Stahl's essential psychopharmacology: Neuroscientific basis and practical applications*. 5th ed. Cambridge: Cambridge University Press.
Thibierge, G. 1894. Les acarophobes. *Revue Générale de clinique et de thérapeutique*.
Trabert, W. 1995. 100 years of delusional parasitosis. *Psychopathology* 28 (5): 238–246. https://doi.org/10.1159/000284934.
Trivers, R. 1971. The evolution of reciprocal altruism. *The Quarterly Review of Biology* 46.
Tso, W.-W., and J. Adler. 1974. Negative chemotaxis in Escherichia coli. *Journal of Bacteriology* 118 (2): 560–576. https://doi.org/10.1128/jb.118.2.560-576.1974.
Van Valen, L. 1973. A new evolutionary law. *Evolutionary Theory* 1: 1–30.
Vanier, M.T. 2010. Niemann-pick disease type C. *Orphanet Journal of Rare Diseases* 5 (1): 16. https://doi.org/10.1186/1750-1172-5-16.
Velleman, D. 2000. On the aim of belief. In D. Velleman (Éd.), *The possibility of practical reason* (p. 244–281). Oxford University Press New York.
Veller, C., P. Muralidhar, and D. Haig. 2020. On the logic of Fisherian sexual selection. *Evolution* 74 (7): 1234–1245. https://doi.org/10.1111/evo.13944.
Vermeij, G. 1987. *Evolution and escalation*. Princeton: Princeton University Press. https://press.princeton.edu/books/paperback/9780691000800/evolution-and-escalation.
Vonnegut, K. 1969. *Slaughterhouse-five, or the children's crusade: A duty-dance with death*. New York: Delacorte Press/Seymour Lawrence.
Weinberger, D.R. 1987. Implications of normal brain development for the pathogenesis of schizophrenia. *Archives of General Psychiatry* 44 (7): 660. https://doi.org/10.1001/archpsyc.1987.01800190080012.
Wellman, H.M., D. Cross, and J. Watson. 2001. Meta-analysis of theory-of-mind development: The truth about false belief. *Child Development* 72 (3): 655–684. https://doi.org/10.1111/1467-8624.00304.
Whiten, A. 2013. Humans are not alone in computing how others see the world. *Animal Behaviour* 86 (2): 213–221. https://doi.org/10.1016/j.anbehav.2013.04.021.
Wilkinson, G.S. 1984. Reciprocal food sharing in the vampire bat. *Nature* 308 (5955): 181–184. https://doi.org/10.1038/308181a0.
Williams, G. C. 1966. *Adaptation and natural selection: A critique of some current evolutionary thought* (Facsim. Ed., with new pref.). Princeton University Press.
Williams, B. 1973. Deciding to believe. In B. Williams (Éd.), *Problems of the self: Philosophical papers 1956–1972* (p. 136–151). Cambridge University Press.
Williams, A.F. 2003. Views of U.S. drivers about driving safety. *Journal of Safety Research* 34 (5): 491–494. https://doi.org/10.1016/j.jsr.2003.05.002.
Wimmer, H., and J. Perner. 1983. Beliefs about beliefs: Representation and constraining function of wrong beliefs in young children's understanding of deception. *Cognition* 13 (1): 103–128. https://doi.org/10.1016/0010-0277(83)90004-5.
Wittman, A.B., and L.L. Wall. 2007. The evolutionary origins of obstructed labor: Bipedalism, encephalization, and the human obstetric dilemma. *Obstetrical & Gynecological Survey* 62 (11): 739–748. https://doi.org/10.1097/01.ogx.0000286584.04310.5c.

Chapter 15
Study of the Theory of Mind (ToM) Through the Japanese Philosophy

Diego López-García, Diego Lopez-Luque, Walter Federico Gadea-Aiello, and Emilio José Delgado-Algarra

Abstract The main aim of this chapter is to connect the Theory of Mind (ToM) with the Japanese philosophy so that the first thing we have to do is to provide a definition for the concept of ToM in order to talk about the Japanese case later. The ToM in Japan is associated with the concepts of *mushin*, also known as no mind, and *Ikigai* known as health and well-being of the individual. In order to comprehend the Japanese Theory of Mind, it is necessary to understand the Japanese philosophy in order to study the Japanese's behavior and their way of life when analyzing their minds. The bases of Japanese philosophy whose origins are Chinese Confucianism, Taoism and Indian Buddhism must be analyzed. These thoughts evolved into Shintoism and Zen. Confucianism is one of the foundations of Chinese thought with the four books of Mencio that adapted the precepts of his master Confucius being the most important one Mengzi. Mengzi was also the basis of Taoism with the premise of looking towards oneself, the conception of unity and the knowledge of supreme and unifying reality. Buddhism is based on Indian culture, and it presents several branches such as Shintoism, Nichiren and Moanjo. Shintoism must be highlighted since it is the original belief of the Japanese people, the belief of a culture and the foundations of Japanese thought being, in that way, something more than a religion. Zen is an evolution of Buddhism with the maxim of studying the

D. López-García (✉)
Departamento de Lenguas Modernas, Universidad de Huelva, Huelva, Spain

Universidad de Seville, Seville, Spain
e-mail: diegolopez@us.es

D. Lopez-Luque
Universidad, Lanzhou Jiaotong, Lanzhou, China
e-mail: dielopluq@gmail.com

W. F. Gadea-Aiello
Departamento de Lenguas Modernas, Universidad de Huelva, Huelva, Spain
e-mail: walgadea@ddcc.uhu.es

E. J. Delgado-Algarra
Director of the Center for Research in Contemporary Thinking and Innovation for Social Development (COIDESO), University of Huelva, Huelva, Spain
e-mail: emilio.delgado@ddcc.uhu.es

T. Lopez-Soto et al. (eds.), *The Theory of Mind Under Scrutiny*, Logic, Argumentation & Reasoning 34, https://doi.org/10.1007/978-3-031-46742-4_15

essence of the soul. The study of ToM in Japan has a novel perspective to understand Japanese Philosophy, especially when it comes to the nature of the mind, awareness, empathy, meditation, and compassion.

Keywords Theory of Mind (ToM) · Japanese philosophy · Japanese culture · Balance · Origin · Japanese belief · Japanese society

15.1 Introduction

The Theory of Mind (ToM) can be defined as the capacity of the human beings for deducing and comprehending other human beings' mind. The concept ToM was coined by Premack and Woodruff in the paper "Does the chimpanzee have a theory of mind?" (Premack and Woodruff 1978). They first carried out several studies about babies, especially, in the developmental psychology (Premack and Woodruff 1978, p. 515). Normally, the Japanese researchers Hiroyuki Okada, Masako Myowa-Yamakoshi, Toshikazu Hasegawa, Ayumi Nagase and Yasuhiro Kanakogi usually translate the studies about ToM with the meaning of the inferring in the mental states of the other. A synonym can be found for the expression ToM which is known as mentalizing.

ToM is known in Japanese as 心の理論 (*kokoro no riron*), where 心 (*kokoro*) means "mind" or "heart" and 理論 (*riron*) means "theory" or "doctrine" (Hepburn 2013, p. 52). The word 心 (*kokoro*) in Japanese has a rich history and a wide variety of meanings.

Regarding etymological and lexicographical references, information can be found in specialized dictionaries in psychology and philosophy, as well as in studies on Japanese terminology in these areas. We used Aoki Masato's *Dictionary of Japanese Philosophy* (Aoki 1996) and Iwao Sato's *Dictionary of Japanese Psychology* (Sato 1992). In Japanese philosophy and psychology, it is used to refer to the heart, mind, spirit, and conscience. In the context of ToM, 心 (*kokoro*) refers to the ability to understand the mental states of others (Sato 1992, p. 109). In addition to 心の理論 (*kokoro no riron*), there are also other terms related to ToM in Japanese, such as 他者理解 (*tasha rikai*), which means "understanding the other", and 感情理解 (*kanjo rikai*), which means "understanding of emotions" (Aoki 1996, p. 209). The ToM is a key concept in psychology and cognitive neuroscience, which refers to the ability to understand and predict the mental states of others. In Japan, this concept has been studied and developed within the framework of traditional Japanese psychology and philosophy and has given rise to a number of specific terms and concepts in the Japanese language. For example, in Japanese, ToM is often translated as 念書の理解 (*nensho no rikai*), which literally means "understanding written thoughts" (Kuribayashi et al. 2004, p. 122), or 顔の読み方 (*kao no yosooi*), which could be translated as "reading the face" or "understanding emotions through facial expressions" (Saito 2014, p. 8). These terms focus more on the ability to read and understand the emotions and thoughts of others through their behavior and facial expressions.

The theoretical framework of the ToM has made several significant contributions in Japan in relation to child development, education, and mental health. In the area of child development, ToM has led to a greater understanding of how children understand and relate to others. It has also been used to assess children's ability to understand the mental states of others. And it is useful in the diagnosis and treatment of autism spectrum disorders and other developmental disorders. Okada has made important contributions in the field of ToM, especially in the study of the development of ToM in children with autism spectrum disorders (ASD). Okada has investigated the development of ToM in children with ASD and has found that these children have difficulties understanding the beliefs and emotions of others, which affects their ability to interact socially (Okada and Itakura 2017, p. 6). Okada has also researched cultural differences in the understanding of ToM. In his paper "Development of the ability to represent another's visuospatial perspective in 4-to 8-year- old Japanese children", he compared ToM development in Japanese and American children and found that Japanese children tended to use context and visual cues more to understand the emotions and beliefs of others (Okada and Itakura 2018, p. 142).

Toshikazu Hasegawa is a Japanese researcher who has made important contributions in the field of ToM especially in the study of the relationship between ToM and empathy. He has found that the ability to understand the emotions and thoughts of others is closely related to the ability to empathize with them (Hasegawa and Yamauchi 2012, p. 84). Ayumi Nagase is a Japanese researcher who has made important contributions in the field of ToM in the context of child development. Nagase has studied the development of ToM in children with ASD and has found that these children have delays in the development of ToM compared to neurotypical children (Nagase et al. 2017, p. 915). She has shown that specific training in cognitive and social skills can improve ToM in children with ASD (Nagase et al. 2018a, p. 92).

Yasuhiro Kanakogi is a Japanese researcher who has focused on the study of ToM in infants and young children. Kanakogi has conducted studies showing that babies as young as 6 months old already have the ability to infer other people's intentions (Kanakogi and Itakura 2011, p. 346). These findings suggest that ToM development begins much earlier than previously thought. Kanakogi has also investigated how language influences the development of ToM. His studies have shown that children who are exposed to more complex and sophisticated language develop more advanced ToM skills (Kanakogi and Itakura 2019, p. 6).

In the field of education, ToM has been used to develop educational programs that foster empathy and understanding of the emotions and mental states of others. These programs have proven effective in improving the quality of interpersonal relationships and reducing school violence. Okada has developed educational programs and treatments for children with ASD, based on his research on ToM. These programs focus on improving understanding of the emotions and beliefs of others through games and interactive activities (Okada 2019, p. 1826). Hasegawa's research has highlighted the importance of ToM in education, especially in the development of social and emotional skills. It has shown that ToM development can be improved through education and social skills training (Hasegawa and Sakai 2009, p. 104). Nagase's research has highlighted the importance of ToM in education and has shown that ToM can be improved through targeted educational interventions (Nagase et al. 2017, p. 912). He has developed ToM training programs for children and has shown that these programs can improve children's social and emotional skills (Nagase et al. 2018a, b p. 6). Kanakogi has conducted research that suggests that culture may have an impact on the development of ToM. In the paper "How the social context modulates the attentional network in infants: A cross-sectional and longitudinal study with eye-tracking and electroencephalogram" he makes a comparative study between Japanese and American children where he found that Japanese children were more sensitive to nonverbal social cues, while American children were more likely to use verbal cues (Kanakogi and Itakura 2018, p. 562).

In the field of mental health, ToM has been used in the treatment of disorders such as schizophrenia, depression, and anxiety, as it has allowed a greater understanding of the mental processes involved in these disorders. It has also been used to assess patients' ability to understand and respond to treatments. Okada's research has highlighted the importance of ToM in the social and emotional development of children, especially those with ASD. It has shown that understanding the emotions and beliefs of others is essential for social interaction and the development of empathy and collaboration (Okada and Tanaka 2004, p. 1776). Hasegawa has also researched the development of ToM in children and found that the ability to understand the emotions and thoughts of others develops at an early age, and that exposure to complex social situations can accelerate this process (Hasegawa and Onishi 2016, p. 250). Hasegawa has studied cultural differences in the understanding of ToM and has found that cultures that emphasize interdependence and harmony tend to have greater sensitivity to the emotions and thoughts of others (Hasegawa and Hiraki 2008, p. 172). Nagase has investigated cultural differences in understanding of ToM and has found that understanding of ToM can vary depending on culture and language (Nagase et al. 2013, p. 892). He has shown that understanding of ToM is influenced by cultural factors, such as cultural beliefs about the mind and the nature of social relationships. Nagase has studied ToM development in bilingual children and found that bilingualism can have a positive impact on ToM development (Nagase et al. 2018a, b, p. 87). It has shown that bilingualism can improve children's ability to understand the perspectives of others and to adapt to different social situations (Nagase et al. 2018a, b, p. 4). Kanakogi has explored the relationship between ToM and empathy, finding that children who have advanced ToM skills are also more likely to show empathy towards others (Kanakogi and Itakura 2014, p. 269). These findings suggest that ToM may be an important factor in the development of social and emotional skills.

There are many challenges that can be found within the ToM which deals with the comprehension of the false beliefs (Nguyen et al. 2023, p. 1). That means how human beings can predict the human behavior through several tasks which have been tested with children between 3 and 6 years old (Fu et al. 2023, p. 4). These tasks have also been tested with questions about the others' thinking because children have a ToM. In general, children between 4 and 5 years old can finish these tasks.

However, children around 3 years old respond questions based on events they know, and they cannot finish these tasks (Chen et al. 2023, p. 13). In Japan, the age at which children start to comprehend false beliefs is later (Naito and Koyama 2006, p. 293).

According to Wellman, Cross, and Watson's meta-analysis, it is pointed out that in Japan and Austria children had a worse result in tasks about the false beliefs (Wellman et al. 2001, p. 662). These conclusions are the same Naito and Koyama affirmed about the 2 years later development of Japanese children in false beliefs in relation to British children (Naito and Koyama 2006, p. 293). Callaghan suggests that children in Canada, India, Peru, Samoa, and Thailand follow the same paths in the development of knowledge (Callaghan et al. 2005, p. 381). Liu, Wellman, Tradif and Sabbagh carried out a study relating to the meta-analysis of ToM in China, and the final conclusion was that the performance in Chinese children is similar to the western children (Liu et al. 2008, p. 525). There is a dilemma about Japanese children which states that probably in Japan children have a different mental processing from other countries following its own culture. Many Japanese researchers have supported this theory since it hasn't been demonstrated empirically that Japanese people have a different mental reading regarding other countries. In order to confirm this theory Japan must make a test to check these affirmations and then, use it around the world to confirm what it has been said. Cardiner, Harris, Ohmoto and Hamazaki studied the capacity of the Japanese children to be quiet when the examiner asks them to tell a story (Gardner et al. 1988, p. 208). In addition, Okanda and Itakura explained that Japanese children have the same probability since no Japanese children were able to be quiet when the examiner asks them about verbal questions (Itakura et al. 2008, p. 180). It could be suggested that Japanese children are not good enough when responding directly to what the examiner has asked them. Despite the fact that Japanese children have a good comprehension about false beliefs they are not familiarized when a stranger asks them about verbal questions, and it is possible that they do not reply correctly. To reinforce that, Call and Tomasello analyzed in the paper "A nonverbal false belief task: The performance of children and great apes" the activity of apes and dolphins doing nonverbal tasks. These researchers highlight the similarity about apes and western children while doing tasks (Call and Tomasello 1999, p. 386). However, Japanese children performed nonverbal activities better than verbal tasks. Precisely, the percentage of Japanese children under 5 years old that do not pass nonverbal tasks is 80% (Call and Tomasello 1999, p. 393).

In Japan, the culture values harmony, and the avoidance of direct conflict, often leading to a greater aversion to confrontation and the direct expression of negative emotions. This mindset is reflected in the way lies and relevance are handled in communication. As for lying, in Japanese culture, it is often considered more acceptable to lie to avoid conflict or preserve interpersonal harmony. This is known as "*tatemae*", which refers to the public facade shown to others, as opposed to "*honne*", which refers to one's true feelings or intentions. Because of this, it can be difficult for the Japanese to tell when someone is lying or for themselves to admit the truth. When it comes to relevance in communication, the Japanese often communicate in indirect and subtle ways, using context and intuition to understand the underlying meaning behind words. Attention to detail and reading non-verbal signs are important in Japanese communication. In Japanese culture, not only what is said is considered important, but also how it is said and how the situation in general is interpreted. In relation to ToM, these cultural attitudes can affect the way in which the intentions and emotions of others are understood and perceived. The Japanese may have a harder time interpreting directly expressed negative emotions and may be more likely to look for and read nonverbal cues to determine how someone is really feeling. They may also be more likely to consider the intentions of others and seek interpersonal harmony rather than directly confront any problems or conflicts.

In this fashion, the studies about performance in the west have been proved, but they are not accurate in Japan. As example, we studied the research headed by Professor Ya Sato from Kokoro Research Center at Kyoto University in Japan. They analyzed 65 facial expressions from Japanese people and proved that facial expressions in Japanese people differ from the Ekman's facial expressions of emotions (Sato et al. 2019, p. 1). Facial expressions are a path to express emotions and are essential for communication. Ekman proposed that there are universal facial expressions that represent emotions (Ekman 1999, p. 311). The theory is based on the observation and the intuition, but there are more investigations about facial expressions in western cultures than in Japanese culture.

To reinforce this, Kokoro Research Center scientists analyzed the subjects of study with the facial expressions of the six basic emotions (joy, disgust, angry, fear, surprise, and sadness) from the seven Ekman's emotions. As a result, they could recognize the emotions starting from facial expression according to Ekman's studies but, joy and surprise were the emotions that the Japanese could recognize better (Sato et al. 2019, p. 6). The result of this research confirms that in Japan the Ekman's theory of facial expressions can be used in part because facial expressions can only be treated as universal expressions with people from the west.

This could be explained through *Ikigai*, a Japanese concept that combines *iki* which means life or to be alive, and *gai* which means profit or value. When both meanings are combined *ikigai* displays the importance of life, that is to say, its meaning and purpose. In French terminology it means *raison d'etre* or the reason for existence. *Ikigai* refers to finding purpose in life, a sense of direction and meaning that guides a person's actions and decisions. ToM is related with *ikigai* in the sense that it explains the ability of human beings to understand the mental states of others, such as their beliefs, desires, intentions, and emotions. Although apparently these two concepts do not seem to have a direct relationship, they are closely related. The search for an *ikigai* implies a deep understanding of oneself, of what one values and what one wants to achieve in life. To discover their *ikigai*, a person needs to reflect on their beliefs, values, interests, and abilities, which requires a deep understanding of their own mental states. Additionally, the pursuit of *ikigai* may be influenced by ToM to find purpose in life, a person may need to understand the mental states of others and their impact on the world. For example, someone might find their *ikigai* in serving others or solving important social problems, which requires an understanding of other people's needs and wants.

Ikigai has evolved across the basic principles of health and wellness in the traditional Japanese medicine. This medical tradition is based on the physical wellness influenced by emotional and mental health. That infers in purpose of life, in the way to express emotions. The Japanese psychologist Michiko Kumano said *ikigai* can be compared with a wellness status that emerge from those activities that an individual chooses. They are used with the purpose of having fun and receiving a full satisfaction (Kumano 2017, p. 61). Kumano makes a distinction between *Ikigai* as temporary pleasure[1] and a well lived life.[2] That lead to the highest and lasting way of happiness. This can be associated with the cognitive-behavior therapy in the sense that the searching of activities that give happiness and wellness can fight against depressive disorder.

The neuroscientist Ken Mogi states that *ikigai* is a concept for Japanese family from ancient times. He translated it as "your reason to get up in the morning" or more poetically expressed as "awakening your happiness" (Mogi 2018, p. 3). *Ikigai* is also related to the concept of flow as the Hungarian psychologist Mihaly Csikszentmihalyi describes in his paper. According to Csikszentmihalyi, the flow occurs when you are in your path as high- performance athletes usually say. The flow is a set of best moments or moments where people are in their best comfort zone. That means being in a constant balance and harmony, though, Csikszentmihalyi points out that in general terms those best moments occur when the mind or the body extends to its limits. That is to say, a voluntary effort to reach something difficult but that worth it (Csikszentmihalvi 2011, p. 12). It can be affirmed that the flow occurs when you frequently make something you love, and you are good at. With that premise, it is possible to provide a benefit to others so that they can have a better life. In this case, the flow can be seen in syntony with *ikigai* in relation to the activities that gives us sense and purpose to our own lives.

Regarding to the practical use of the ToM in the Japanese culture another term can emerge which is *mushin* or the essence of Zen. This term can be translated as the mind without mind, and it is generally known as no-mind state. It is a mental state where the mind is not fixed or filled with thoughts or emotions. *Mushin* refers to the thoughtless mind or mind in a state of no-mind. It is a state of mind in which the mind is fully in the present and not distracted by past or future thoughts. ToM can be associated with *mushin* in the sense that it has the ability of human beings to understand the mental states of others, such as their beliefs, desires, intentions, and emotions. Although these two concepts may seem unrelated at first glance, they are closely related. To experience the *mushin* state, you need a high degree of awareness and mindfulness in the present. It is necessary to be fully present in the present moment and to be aware of the sensations, thoughts and emotions that arise in the present moment. This state of mindfulness and awareness is related to ToM in the sense that, to understand the mental states of others, it is necessary to be present in the present moment and to pay attention to the non-verbal and verbal signals that

[1]In Ancient Greek it means *hedonia*.

[2]In Ancient Greek it means *eudaimonia*.

people are streaming. If the mind is distracted by past or future thoughts, it can be difficult to understand the needs and desires of others in the present moment. Additionally, the *mushin* state can help people better understand their own mental and emotional states, which can improve their ability to understand the mental states of others. By being fully present in the present moment, people can become more aware of their own emotions and thoughts, allowing them to better understand the emotions and thoughts of others.

Dealing with this idea of the mind, the person is in a state where he can connect with the cosmos at a mental level. In current life, Zen is considered to be a pure mental state, a pure clear mind and it is produced by the absence of the ego or the self-limited. A *mushin* mind is not an empty mind like a shell but it is the opposite a full, conscious, and free mind. If the concept *mushin* is separated into two words *mu* and *shin*, then, it can be appreciated that the word *mu* has been translated as vacuum meaning a mind with no distractions, worries or fears. Everything is absent and, therefore, those things are not a problem for the mind. *Mushin* is similar to the Japanese expression *Mizu no Kokoro* that means 'mind like water'. In this case, that is a mental attitude associated with a mind in full harmony with cosmos. The analogy is a quiet water pond without waves where the surface reflects a clear image of the environment, in other words, it is like a mirror. *Mushin* cannot be perceived by the intellect, but it must be experienced. A *mushin* mind does not have an ego or a substance. It is a pure illumination, and it is the perfect implementation of the self. This mental state requires years and years of practice to be achieved. *Mushin* is the result of a free human mind of angry, fear, judge, or ego during the daily life.

15.2 The ToM in Japan According to the Japanese Philosophy

Confucianism, Taoism, Buddhism, Shintoism, and Zen are Eastern religions and philosophies that have had a great influence on the culture and way of life in many Asian countries. Although each of these traditions has its own distinctive beliefs and practices, there are some shared ideas and concepts among them that relate to ToM. Confucianism, for example, focuses on the idea of virtue, ethics, and morality. It teaches that society is based on human relationships and that these relationships must be built based on empathy, kindness, and respect. ToM is relevant in this context, as Confucianism emphasizes the importance of understanding and respecting the feelings and perspectives of others. Taoism, on the other hand, focuses on harmony with the universe and nature. The ToM is relevant here because Taoism teaches that the mind and body must be in balance to achieve harmony and well-being. Taoism also emphasizes the importance of meditation and introspection to achieve wisdom and understanding. Buddhism, for its part, focuses on liberation from suffering and the achievement of enlightenment. The ToM is central to Buddhism, as it is taught that suffering is due to ignorance and lack of understanding. Buddhism emphasizes the importance of meditation and introspection to understand the nature of the mind and achieve enlightenment. Shinto is a traditional Japanese religion that focuses on the worship of nature spirits and deities. ToM is relevant here because Shinto teaches that humans and nature are intrinsically connected and that we should respect and honor nature and the beings that inhabit it. Zen is a form of Buddhism that focuses on meditation and contemplation. The ToM is central to Zen, as meditation and contemplation are taught to be ways of understanding the nature of the mind and achieving enlightenment. Although each of these religious and philosophical traditions has its own distinctive beliefs and practices, they all relate to ToM to some degree. They all emphasize the importance of introspection, empathy, and understanding to achieve wisdom, harmony, and enlightenment.

15.2.1 Confucianism

Confucianism is an ethical and social philosophy that originated in China during the Warring States period (475–221 B.C.) and has profoundly influenced Chinese and other Asian culture and society (Cardona de Gibert 1968, p. 15). Its main source is the work of Confucius, the *Analects*, and it focuses on the importance of morality, virtue, and ethics. Confucianism focuses on human relations and the construction of a harmonious society based on empathy and respect. Confucianism and ToM are closely related in the sense that both focus on the importance of interpersonal relationships and effective communication. Confucianism emphasizes the importance of social relationships and interpersonal ethics, and a key part of this involves the ability to understand the perspectives and needs of others. Theory of mind also focuses on understanding the perspectives and needs of others, which can improve communication and interpersonal relationships. Furthermore, both Confucianism and ToM recognize the importance of personal development and self- improvement. Confucianism emphasizes the importance of cultivating virtue and self- improvement through education and moral practice, while theory of mind focuses on developing empathy and the ability to understand the mental states of others. Both promote the idea that personal development is essential to improve interpersonal relationships and society in general.

One of the most important bases of the Japanese thinking can be found within the Chinese philosophy. There are many thinkers in the Chinese philosophy that have inspired the Japanese philosophy based on a millenary tradition, concretely, Confucius. First of all, we are going to talk about the life of Mencius (Meng Ke[3]) who was a disciple of Confucius. He was the first most important philosophers of the *Rú Jiā*[4] school of literates. Meng Ke was born in 371 B.C. in the little state of Zou, in the meridional part of the current Shandong. Meng Ke was a member of the nobility

[3]Meng was his surname, and Ke was his name.

[4]In Chinese means Confucius' family.

similar to Confucius' family. He studied at literates' school according to Confucius' doctrines with Zisi (Confucius' grandson).

Meng Ke spent 6 years, at least, in Qi as a minister. He retired from ministry in 289 B.C. to devote himself to his work *Mengzi*. The *Mengzi* was divided in seven books, and it was written by himself and revised by his disciples. It has a rich, flexible, and precise language in a classical prose. *Mengzi* along with *Lun Yu* would become a treatise of the discussions about Master Kong's words and other two subsequent written pieces of work in one of the four books from the school of literates. The first book is *Tai-Hio Chi Tao* about the great science, the second one is *Chun-Yung* or the middle doctrine, the third one is *Lun-Yu* known as *Analects* or philosophical comments and the last one is *Mengke* or Mencius' book.

The first book is known as the great knowledge written by *Kung-Tse*'s grandson, and it is dedicated to the maturity of the own knowledge. The book is based on the premise that it is necessary to know the end which an individual must guide his own actions towards. The complete perfection can only be achieved once the individual knows where the essence of all things lies on. From the noblest to the humblest man, everyone has the duty to improve and correct their own self. The main objectives of the great science or practical philosophy are the rational nature that everyone receives from the heaven, the education and renovation of people, and the search for the highest good or ultimate end to which the actions must be led in order to achieve the perfection.

The second book is known as the book of balance. In the fourth century B.C., Aristotle talked about the middle point of the virtue. This middle point is determined by the rightness, and a prudent man and it is, precisely, this prudent man who would correctly lead the human behavior to the great monarchs and the justice of governments. This content would be reproduced in Europe in the sixteenth century with *The Prince* of Machiavelli that would be inspired by *the art of war* of Sun Tzu.

The main essence of this second book said that what it has not been diverted from the middle (*chung*), and what it has not been changeable is persistent (*yung*). The straight way or the main law of the universe is the center, and its continuance in it is the perseverance. As Master Ikkyu says:

> Even if it rains or the wind blows, you must think in the future more than in the past. From the five energies I have received, I return four to back to the void. I have lived, doing my best, and now I die as I have lived... (Heisig 2016, p. 199).

This poem makes clear that human beings must be afraid of life or death since they are considered to be metaphysical changes and transmutations in which human beings are the point of balance between them. Depending on what kind of behavior individuals have in their live, they will have one result or another. In this way, individuals will be in a vicious circle, and it is precisely the middle of this infinite circle what compounds the sum of all. What is this at all? It is nothing.

Ninagawa Shinemon says in a similar way: "At the moment of death, I give birth this afternoon, like a breath, of wind among the pines" (Enbutsu 2003, p. 25). And Ota Dokan finishes by saying: "Why long for life if I have known for a long time that my body is mortal?" (Enbutsu 2003, p. 19).

In this section of *Analects*[5] we can see a clear reference to the perseveration:

> The man, who devotes his whole life to the discovery of good and truth with diligence and perseverance, will enjoy profound happiness. If he is consulted by people from the farthest countries, who are attracted by the same ideas and feelings, he will experience great satisfaction. If he is ignored or despised by other men, he will not be indignant or feel sorrow about it, since his virtue is great. (Cardona de Gibert 1968, p. 113).

Confucius also talks about the perseverance to Tse-lu who was one of his disciples:

> Have you ever heard about the six absurdities and their consequences?
>
> 1. Pretending to achieve the good without the study, and its consequence is the disappointment.
> 2. Trying to achieve science without surrendering to study, which leads to uncertainty.
> 3. Being honest without the study, which causes deception.
> 4. Acting righteously without having received adequate instruction, thereby falling into recklessness.
> 5. If you want to combine value with lack of culture, the consequence gives rise to insubordination.
> 6. If you want to achieve perseverance without the study you fall into stubbornness (Cardona de Gibert 1968, p. 209).

After having talked about this, let's move on to the fourth book of Mencius that supports the idea that man is good by nature and, therefore, he must develop a reasonable and straight behavior. Jean-Jacques Rousseau said the man was good by nature and, as opposed to other animals, man has the capacity to reason or, at least, use it (Rousseau 2011, p. 342). There are several qualities the man is naturally good at being one of them justice and order. Rousseau's vision is opposed to Thomas Hobbes who explained in his paper *Leviatan* from 1651 that the first social contract formation was explained by a human society as a formation of people dominated by their ambition of power and domain (Hobbes 2016, p. 64). In order words, he considered that the social nature of man was negative contradicting, in this way, what Confucius, Aristotle, Saint Thomas Aquinas or other authors from the Ancient Greek to the seventh century had argued about the good nature of man.

In relation to Confucius, there is much to say about his traditional doctrine which explains the good nature of man through his five cardinal virtues (Cardona de Gibert 1968, p. 31). The first one is *Yoshi*[6] in Japanese or *Ren* in Chinese, the second one is *Gi*[7] in Japanese or *Yi* in Chinese, the third one is *Rai/Rei*[8] in Japanese or *Li* in Chinese, the fourth one is *Satoshi*[9] in Japanese or *Zhi* in Chinese and the last one is *Makoto*[10] in Japanese or *Xin* in Chinese. The straightness, benevolence, honor which

[5]This part is in the third book of Confucius found in the Mogao Caves in the city of Dunhuang in China.

[6]Benevolence, charity, humanity, and virtue of the man.

[7]It can be translated as justice, human bonding.

[8]That means courtesy, modals, rites, social costumes, and honor.

[9]Intellect, wisdom.

[10]Loyal, confident, truth.

is also known as social protocol, wisdom and confident are considered the most important virtues in the traditional Chinese ideology. Confucius introduced both concepts benevolence and rightness as it can be seen in the following paragraph.

Kung-Tzu[11] has said: As much as I tell you about the customs or ceremonies of the Hia dynasty, Ki is unable to understand their deep meaning. As much as I tell you about the customs or ceremonies of the Chang dynasty, Sung is unable to understand their deep meaning. The knowledge of the laws and of the theories of the sages, is not enough to understand the foundation of customs and ceremonies. If it were enough, we could penetrate its most hidden meaning (Cardona de Gibert 1968, p. 123).

Confucius' morality is based on the premise that man is good by nature and the only way by which he can reach the goodness is through the study of classical texts. After that, he can improve himself with the continuous study and the grown of the self. The best man must develop the benevolence, the best virtue and he must be rightness and obedient but not egoist. Meng Ke agreed with these principles that he received from his master. Meanwhile Confucius said these principles, Mencius needed to justify them. Why must you cultivate and practice the benevolence? Why is rightness the most important virtue? In brief, there were new interesting and beneficial tendencies and ideas in China. Meng Ke responds to the questions according to his famous theory of goodness of human nature. It can be appreciated that man is good by nature and has the capacity of recognition of the good along with the possibility to behave properly. The evil is the consequence of a human failure or a product of outside circumstances influencing the man.

Confucius often said the basis of the human behavior is *Ren*, but he never explained why the man was moved by *Ren*. Mencius was the first one to find out a solution by developing explained his theory of the original goodness of man's nature. His explanation is based on his meaning of *Ren*. Mencius said the humanity is transformed into mercy, it is feeling the pain of others or not being able to bear with the suffering of others. *Ren* is the main feeling, spontaneous and it needs a control (*Yi*). *Yi* is the guide of *Ren* for the correct applied behavior. There are priorities and hierarchies in goodness and compassion *Yi* establishes to determine *Ren*. As it can be seen, his philosophy is based on Confucianism in the seven virtues of the code of the samurai (Inazo 2018, p. 42). Confucius said a precept where its virtues are applied to the code of the samurai:

> The master said: He, who does not know the will of Heaven, cannot be called noble. He, who does not know the customs and rites, cannot strengthen, and straighten his spirit. He, who does not penetrate the deep meaning of every word, cannot know the men of the world (Cardona de Gibert 1968, p. 219).

The first principle is *Gi* or justice which means that the individual must make correct decisions. In the following paragraph, Confucius exposes the idea of Justice which

[11] Another Confucius' names are Master Kong or Kung-Tse.

has been thought to be one of the bases of the Samurai code referring to the straight and firm way as the justice is.

The Master said: I cannot find men who follow the right path, to whom I can transmit my doctrine. Shall I address men of lofty intentions, but lacking the will to put their ideals into practice? Would it be better to choose men of firm and persevering will, but lacking intelligence? Men of high intentions, but lacking the will to act, take as an example the extraordinary actions of great men. Those who possess a firm and persevering character, but lack intelligence, at least do not aspire to achieve that which surpasses their reason (Cardona de Gibert 1968, p. 181).

The second concept to be mentioned here is *Yu* meaning courage where he said the individual must be valiant but not coward or easy frightened. The balance and the rightness must be supported. One of the Confucius disciples made a question about the courage to which Confucius responded:

> Tse-kung went on to ask: Who follow these by their worthiness? The Master said: Those who are always sincere in what they say, and tenacious in the undertakings they initiate, even if they are hard of intelligence and vulgar in their manners. In spite of this, they would occupy the third place (Cardona de Gibert 1968, p. 180).

The third principle is known as *Jin* which means benevolence. The samurai is a powerful and humble fighter to the service of the weak people and the goodness of all. Confucius said this quality has a balance and a duality between the power of the samurai and the subordination of his subjects.

The Master said: When the kingdom is governed by the principles of right reason, you can speak and act with dignity and publicly. When the kingdom is not governed by the principles of right reason, act with dignity and without fear, but speak with caution and moderation (Cardona de Gibert 1968, p. 183).

The fourth concept is *Rei* which means to be respectful and not cruel. The samurai is a fighter who is always training both in sword and in quill. Confucius is consequent to the instruction and training as essential principles to get an army operating both in war and peace times.

The Master said: When the kingdom is governed by the principles of right reason, you can speak and act with dignity and publicly. When the kingdom is not governed by the principles of right reason, act with dignity and without fear, but speak with caution and moderation (Cardona de Gibert 1968, p. 182).

The fifth principle is *Makoto* which means sincerity, to be respectful and to maintain the integrity of the spirit. Confucius established a parallelism between sincerity and loyalty which are repeated in the fifth principle of the samurai.

Within Confucian morality, sincerity towards other beings, that is to say, loyalty, occupies a very high place. Man's loyalty to his family, to his country and to the whole of humanity (Cardona de Gibert 1968, p. 33).

The sixth concept is *Meiyo* which has been translated as sense of honor. The same word samurai is similar to honor because the character of samurai (侍) is the person that protects the temple. This activity was considered to be an honor because the act of defending the place blessed by divinities was an act of dignity. Another Confucius' disciple questions him the concept of shame and honesty that a governor must have.

Yuan-Tse asked what the most shameful conduct was. The Master said: To take a salary when the kingdom is not governed by the principles of right reason, and to take a salary when the kingdom is governed by the principles of right reason. This is the most shameful conduct (Cardona de Gibert 1968, p. 183).

The seventh and last concept is *Chūgi* or loyalty, fidelity of the governor until the end. Confucius explains that if a governor is prudent in his acts, then, as a result, people will be loyal, and they will give a worthy fidelity to his order. "No one hates the prudent prince, no one tires of his rule. Always, morning and evening, he will be praised by all the people" (Cardona de Gibert 1968, p. 103).

Confucius explained his life when he was 73 using 38 characters:

> At fifteen, my mind was eager to learn. At thirty, I was adamant. At forty, I had no doubts. At fifty, I knew the decrees of heaven. At sixty, my ears were able to receive the truth. At seventy, I could proceed when my heart desired without profaning what was right (Wang 2018).

Unlike Socrates whose maxim is to know yourself, Kong paid special attention to life and society in general. One of Kong's maxims was to educate, train and guide people through moral rules. This is contrary to the Socrates' principles where he gave more autonomy to the person. If a synthesis about Mencius' principles is made, it could be even said that he improved and broadcasted his master Confucius' teachings through his numerous travels to other courts. Mencius defended the idea that man is different from other animals because he uses his heart as a thinking organ. It could be thought that the person who does not use the heart would be an animal. Morality and ethics are also related to the ToM because it can understand and attribute mental states to others and can influence how morality and ethics are viewed in different cultures and belief systems. Confucianism emphasizes the importance of social relationships and hierarchy in society, and ToM is fundamental to understanding and managing these relationships.

Mencius defended the basis of the natural goodness of the man: compassion, shame, respect, humanity, and the distinction between good and evil (Cardona de Gibert 1968, p. 31). The compassion is the virtue to care for the fellow human being. The individual can assist or not him, but this virtue acts according to the principle of benevolence because compassion encourages us to help people. The shame can be defined as the further reflection that the individual must have before acting because a shameless person would lose the rightness. However, if the individual feels shame, then, this would drive him to act correctly. On the contrary, respect and humanity consist of being considerate with people no matter what social status they belong to. That is the maxim of humanity because one must be humble in order to be polite.

The distinction between good and evil is the wisdom because it is a well-formed awareness.

One of most important Mencius' followers is Xunzi who is also known as Xún Kuàng, Xun Qing or Sun Qing. Xunzi is little studied in Spain because his works have not been translated into Spanish but English. There is an important translation to be highlighted from Eric L. Hutton where he interprets a part of the manuscript *Xunzi* written in ancient Chinese writing (*zhuànshū*):

As part of the belief that they were upholding this ancient tradition, these Warring States thinkers tended to identify themselves using the label *ru* (儒), which originally simply meant a cultivated person or learned person, but over centuries of association with these thinkers and their later followers came to be the name for their whole group, which we now translate as Confucians (Xunzi. 2014, p. 24).

This text is related to the definition that is made about Mencius' followers. They use the etiquette *ru* to refer to cultured person loyal to Mencius' precepts even in troubled times. *Xunzi* is one of the main readings to understand Eastern philosophy. It was written by Xun Kuang in 32 chapters, and probably compiled the *Book of Rites* which was the main work of Confucianism. He participated in the last period of warring states.[12] Xun Kuang followed master Kong and he could not accept the optimist vision of the nature that Meng Ke had. Xun Kuang had new good ideas which were implemented in the school of literates and good received at the end of the tenth century. In eleventh century, he adopted a heterodox thought in *Rujia,* and he was recognized by his closer rival Meng Ke. *Xunzi* and *Mengzi* would be in the twelfth century one of the four classical books in the school of literates. An explanation about some questions that Confucius' disciples made him are going to be provided:

> Zigong asked: Is there a phrase that can serve me until the end of my life? Confucius said: Forgive others. What you do not want to be done to you, do not do to others. Dragon flying in Heaven. It is propitious to see the great man: What does this mean? The Master said: what is consistent in tone, vibrates together. What denotes elective affinity in his inner being, seeks each other. Water flows toward the wet. Fire flows toward the dry. Clouds follow the

[12]Also known as Warring States Period during fifth century B.C. until the unification of China during Qin dynasty in 221 B.C. about forgiving others is related to ToM because it implies the ability to put yourself in the place of others and understand their perspectives and feelings. To forgive someone, it is necessary to understand their motivations and feelings, which requires having an empathic understanding of the other person's mind. Second, the analogy of the dragon flying in the sky is related to ToM because it highlights the importance of affinity and emotional connection between human beings. According to the ToM, people are connected by their ability to understand and empathize with the feelings of others. The dragon analogy suggests that the emotional connection between people is so strong that they may "follow" someone who exhibits great wisdom or virtue. Third, the idea that each being follows its own nature is related to ToM because it suggests that each person has their own unique perspective on the world and their own way of processing information. According to the ToM, people have different perspectives and experiences that shape how they see the world and understand others. Overall, the text suggests that empathy, understanding the feelings of others, and emotional connection are critical to living a meaningful and virtuous life. These ideas are related to ToM because they involve the ability to understand and emotionally connect with others.

> dragon; wind follows the tiger. Thus, the sage rises, and all beings look at him. What has its origin in Heaven feels affinity with that which is above. What has its origin on Earth, feels affinity with what is below. Each one follows its own nature (Cardona de Gibert 1968, pp. 113-129).

The text is about the harmony between the governor and the town. It makes a comparison between the town that is represented as clouds, and the dragon which represents the governor. The dragon that flies across the sky is referred to as the highest way to govern because the dragon is considered to be a supreme being elevated to the level of the deities in China. The text has various relationships with ToM and empathy. First, Confucius's phrase about forgiving others is related to ToM because it implies the ability to put yourself in the place of others and understand their perspectives and feelings. To forgive someone, it is necessary to understand their motivations and feelings, which requires having an empathic understanding of the other person's mind. Second, the analogy of the dragon flying in the sky is related to ToM because it highlights the importance of affinity and emotional connection between human beings. According to the ToM, people are connected by their ability to understand and empathize with the feelings of others. The dragon analogy suggests that the emotional connection between people is so strong that they may "follow" someone who exhibits great wisdom or virtue. Third, the idea that each being follows its own nature is related to ToM because it suggests that each person has their own unique perspective on the world and their own way of processing information. According to the ToM, people have different perspectives and experiences that shape how they see the world and understand others. Overall, the text suggests that empathy, understanding the feelings of others, and emotional connection are critical to living a meaningful and virtuous life. These ideas are related to ToM because they involve the ability to understand and emotionally connect with others.

Confucianism has also influenced Japanese culture and has been a major influence on Japanese ethics and morality. The Confucian idea of the relationship between the individual and society, and the importance of interpersonal relationships, may be related to ToM and understanding the thoughts and emotions of others. Both the Japanese conception of mind and ToM recognize the importance of empathy and understanding the mental states of others for effective communication and healthy interpersonal relationships. Understanding the mental states of others is essential for effective intercultural communication and can help overcome cultural differences and communication barriers. Additionally, the Japanese conception of mind emphasizes the importance of non-duality and the interconnectedness of all beings, which can promote greater empathy and understanding of others. Similarly, ToM recognizes the importance of empathy and understanding the mental states of others for a more just and compassionate society.

15.2.2 Taoism

This philosophy also originated in China, but during the fourth century B.C. Its main source is Lao Tze's *Tao Te Ching*, and it focuses on harmony with the universe and nature. Taoism emphasizes the importance of living in harmony with nature and the universe, and the need to balance the mind and body. Similar to Shintoism, thinkers who study Tao are interested in the diversity with which nature manifests itself due to its complexity. However, Tao can be seen as a generic way of referring to something that is unknown or little studied, something that it is clarified in our text when it refers to descriptions which are little known as *shinobi*. Therefore, Tao is a fundamental doctrine in the configuration of *shinobi* thought and, thus, from the samurai code it can be highlighted the book Lao Tse the *Tao Te Ching* in which it is said:

From the Tao is born the One, from the One is born the Two, from the Two is born the Three, from the Three are born the ten thousand beings. The ten thousand beings carry *Yin* (darkness) on their backs and *Yang* (light) in their arms. In them the vapor (matter) of the void is harmonized (Tse 1996, p. 21).

Lao-Tse Cosmology is well structured as it can be appreciated in the previous passages like the ones from the Pythagoreans around numbers. Aristotle said: "The Pythagoreans have built the entire world with numbers" (Aristóteles. 2014, p. 9). According to Pythagoreanism, God is established as "One" or Superior Monad, placed him in the highest sphere within Cosmology. This one begets another Monad (referring to a unit) inferior which is the Monad or firstborn unit of numbers and enters as elementary unit.

Under the Han, many theories on the functioning of the world or of natural phenomena were systematized. These theories laid the foundations of Chinese scientific thought and of certain aspects of religious thought (Adler 2005, p. 77).

The perception of reality also influences how one understands the mind and its relationship with the outside world. Taoist philosophy emphasizes the importance of living in harmony with nature and the universe, and this perception influences the understanding of the mind as an integral component of the universe and the need to balance the mind and body.

The first great imperial period in the history of China was the Han dynasty which lasted around four centuries. In present day, talking about this dynasty is like talking about the identity of China. Since the decay of the Han and until 589, China was dismembered into many regional kingdoms, but neither of them was able to get the reunification.

For this, it was necessary to wait for the furtive passage of the Sui Dynasty that remained only from 589–618. The division of the epoch is known with the name of the Six Dynasties, though, it is also divided into two periods being one of them the Wei Jin and the other one the South and North dynasties. The Han dynasty thought possessed a syncretic character since it was an amalgam of elements from different traditions and included an elaborate cosmological theory.

According to Procio "Lao Tse does not determine who are One, Two and Three" (Tse 1996, p. 211), though, by analogy it could be said that it is based on the formation of the world and the beings that make it up. It could be assumed that One is the cosmic unity or totality, and this would be the invisible and non-tangible origin of the Tao. In other paragraphs of Master Tse, it would be found light about it to discover the essence and nature of the One, and this derives from Tao as the virtue capacity of hollowness, where Tao is virtue *sine qua non* to form beings. In this case, Pythagoreanism and Lao-Tse coincide, then, in putting the One as a link or bridge between the transcendence of the Supreme Being and its creatures. A tangible derivation of the Tao can be determined as one, being comparable to a cosmic unity, which is what feeds and gives energy to thousands of beings that make up the universe by seeing them as the totality of the known world. From that unity or Monad, the Dyad or the two is generated which, in turn, gives the Triad or three. The Monad, the Dyad and the Triad, all together, give rise to the *Atman* of the *Upanishads* (Tola and Dragonetti 2018, p. 68), where *Atman* is as much as the individual soul or the energy that gives life to each body; it is like the Supreme soul or the Universal soul from the Greek philosophies, which is conceived as an image of the *nous* and this one refers to an image of the One. According to Anaxagoras, the term nous refers to the mind, the faculty of immediate and direct recognition of a reality; this consists of the supreme mind, God, or computer mind.

Along with the Four Noble Truths, one of the earliest and most important teachings for a philosophical understanding of Buddhism is the doctrine of no-self or *anatman*. It is a frontal repudiation of one of the central beliefs of the then dominant Brahmanic philosophy. In the *Upanishads*, sacred texts that began to be compiled from the eighth century B.C. onwards, the philosophical bases of the rituals and divinities of the *Vedas* are laid down (Adler 2005, p. 58).

To realize the Two or the Dyad will be the binomial formed by *Tien* (Heaven) and *Ti* (Earth). Between Heaven and Earth, it can be found *Ren* (Man) who is in continuous change and evolution; it is like a battery that interacts with the heaven and the earth, changing and transmuting its polarity. This battery is, in some cases, charged with positive energy and in other cases, it is charged with negative energy in the same way that it can be detached from both with a harmonious and rhythmic change. In order to understand the basis of Ten Chin Jin and the most elementary concepts of shinobi philosophy it is essential to have a look at the philosophical politics of Lao-Tse.

To beware of acting and intervening. The formation of wise and perfect men, destined to occupy a throne or to assist the Sovereigns as advisors, was the main concern of the philosophical-literary schools (Tse 1996, p. 230). That is tangible both in ancient China and now in contemporaneity. Confucius devoted all his energies to this. Lao Tse did not want to become a teacher, he did not want to trade or sell his wisdom: The good man does not consider himself a good teacher, the non-good man takes for good the riches of his neighbor (Tse 1996, p. 230).

The essential rule is here which is repeated ad nauseam: *Wu Wei* which means not to intervene, not to act but to let the beings to follow their natural course by imitating the natural and total flow of the Tao. The wise nature will order the world by itself which is affirmed by the followers of the Tao with the conviction of its unpremeditated influence.

The Sage contemplates the Tao in his Ching (imperturbable stillness and serenity). That stillness became the model and essential idea that shaped what is known as the Asian Soul. The Christian ideal is, on the contrary, life. God is life. "I am life, I have come into the world so that they may have life and have it abundantly" (Biblia Job 10,10) (Tse 1996, p. 213).

One of the objectives that Lao-Tse promulgates is the non-alteration of ambitions to which people can aspire due to their own human nature, and one of its premises is to avoid the unnecessary and the material and that, a priori, is important to spread the teachings from the base so that they are simplistic and conformist; each step is a succession of events that must happen in an unforced way to people with what nature offers itself, avoiding ambitions and aspirations for achievements outside of their own humble nature. Likewise, it is necessary to avoid imbalance, not to violate things, not to force the mechanism. The robustness is determined by the sapiential experience by old age or maturity in some cases. He who intends to take too long steps cannot walk and sooner or later he will fall or at least lose his balance. Those who use the Tao and assist rulers must avoid violence and the use of weapons. We do not want a situation which is conductive to the development of the most brutal forms of violence. Lao-Tse advocates improvisation, stillness, and calm, condemning any action that disturbs natural evolution.

One of the most important philosophies of China is subsequent to Lao Tse who was the creator of Taoism. Yang Zhu is a very interesting figure within Taoism. Once the disintegration of Zhou feudalism was complete, it declined and also led the vast majority of aristocrats to ruin; these aristocrats lost completely faith in politics and in any kind of state. Unlike the lawyers, they rejected the new order. The lawyers did not yearn for the old. Disenchanted and disillusioned with society, they withdrew to the solitude of the countryside, the forest, or the mountains, looking for stillness and tranquility to find the balance point from which they had been displaced. There they tried to recover part of what they had lost. Thus, they would be able to harmonize with nature and merge with the Dao, the ultimate fund of the universe. Hence, the name of Daojia or Daoist school with which this movement is known. It is believed that the first Daoist thinker was Yan Zhu. Mo Di does not still mention it but soon after, his doctrines know an emerging diffusion. Meng Ke writes "The words of Yang Zhu and Mo Di fill the world" (Prevosti i Monclús and Doménech del Río 2005, p. 33). Yang Zhu is placed after Mo Di and before Meng Ke. Yang Zhu defended two theses (Prevosti i Monclús and Doménech del Río 2005, p. 37). On the one hand, one must worry about oneself and forget about society and others. On the other hand, one should value his own life and despise possessions and things. In the *Mengzi* it can be read "Master Yang is a supporter of looking for oneself. Even if one could benefit everyone by pulling out a single hair, one would not do it" (Prevosti i Monclús and Doménech del Río 2005, p. 33).

It has already been seen that Dao, the way, is a term that is located in the central axis of Confucian and Taoist thought, in which it is referred to the ideal social ethical order and the natural order respectively. In a more abstract way, Dao can be simply understood as an order or pattern of change. The Taoists also say that change is not a secondary characteristic of things rather a basic and first-born reality. Finally, there is the concept of Qi or psycho- physical matter which is the underlying and dynamic substance since it is in constant transmutation and movement of which all things are composed. Here the foundations of traditional Chinese cosmology are laid Dao, change and Qi.

Tao and *Qi* are closely related to the ToM, as both concepts are central to Chinese philosophy and have implications for understanding the mind and its relationship to the world. On the one hand, Tao refers to the primordial and indescribable force that governs the universe and is present in all things. In the ToM, Tao is related to the idea that the mind and body are interconnected and must be harmonized to achieve health and well-being. In this sense, the balance of Tao in the body is necessary to have a healthy and balanced mind. On the other hand, Qi is a life force that flows through the body and the universe. In ToM, Qi is associated with the idea that the body and mind are interconnected energy systems, and that Qi balance is necessary for a healthy mind and body. Meditation and breathing practices are ways in which Qi can be worked to improve physical and mental health. In short, Tao and Qi are key elements in the ToM in Chinese philosophy. Understanding these concepts can influence how you understand and work with your mind and body to achieve a healthy and balanced life.

15.2.3 Buddhism

Buddhism can be approached from two models according to its origin, through its diffusion and according to its geographical transmission from its Hindu lineage. There are two schools that can be mentioned here which are the so-called Southern school and Northern school. The former is probably the closest one to its origin, therefore, it is considered to be the oldest one and, at the same time, it has undergone the least changes in its initial conceptualization, in other words, it would not be so elaborated. The latter is further from its root or origin and has also suffered more erosion and changes to highlight. The southern one is based for the most part on Pali texts, known as *Hinayana,* which is translated as a small vehicle or lower doctrine. In the northern school its base is located in Sanskrit texts known as *Mahayana* which means great vehicle or superior doctrine.

It must be emphasized that in Japan Buddhism is subdivided into 13 main schools and more than 40 secondary schools. Japanese culture is the one that keeps the flame of Buddhism in a constant boom and manages to survive along with other doctrines such as the autochthonous one called *Shintō* that coexisted with Christianity and other types of imported religions. This is due to an education model that enhances respect for diversity but without leaving outside its philosophical and cultural origin, which gives it an added value that makes its perseverance and dedication a special scrupulousness in matters to conserve and know how to disseminate them while maintaining their base in harmony and coexistence with the others. Curiously, in

Japan they identify themselves with the Buddhism of the northern school and it is here where a parenthesis must be made in order to say that their main reference is Zen Buddhism which is, undoubtedly, the one that is most in tune with the Japanese idiosyncrasy.

It is important to highlight the difference existed between the two previously mentioned visions which are the *Mahayana* and the *Hinayana.* In the West, the one that has been most studied and referenced is the *Hinayana*. However, it is precisely the other, the *Mahayana,* the one that is the least known in the West and the one that comes closest to the postulation of our thesis. A focus on Mahayana must be made in order to explain and be able to understand the essence of Buddha in a certain experiential way, thus, it is essential to resort to the content of our own studies which are exposed here in order to see what is most similar to the current state of spiritual life in Japan. The study and practice of Zen has not only had an important influence in Japan but also it stands out in the other systems of thought that are practiced in the rest of the places on our planet. It can be affirmed that it is as old as the original Buddhism since it has been transmitted and practiced without undergoing significant changes from the first Buddhist ascetics in India to the present day. He has maintained his rigor yesteryear, the ideas that emanate from his practice agree with the current tendencies of care and respect for nature. His teachings seek to reach enlightenment by following a path of harmony and peace that, in turn, cares for and respects everything that surrounds us.

In most of the existing Buddhist-based religious systems, the gods are usually considered as their creators, supernatural beings who have reached that state through enlightenment. Instead, Zen practitioners consider their teacher or guide as an equal, a being who can be like us and whose level of spirituality is aspired to be achieved; One of the goals of Zen Buddhism is to achieve enlightenment and, in a certain way, to identify with the Buddha understood as a supernatural and fictitious being, it can be affirmed that this would be a form of consensual slavery since this process entails a superhuman sacrifice. However, Zen practitioners are free to choose the path and continue it, being able to stop at the time they consider appropriate for their well-being without having to be subject to a mandatory purpose. Referring to a note by Kaiten Nukariya:

> But even early Buddhists characterized Brahmanic Zen as heterodox Zen, and they distinguished it from what the Buddha taught. Our Zen arose from the state of enlightenment that Sakiamuni attained in his 30 s. . . (Nukariya 2004, p. 86).

This entry refers to the origins of Zen, how Sakiamuni reached enlightenment and how enlightenment is received by all beings, whether they are animate or inanimate in an equal way. One of the pillars of Zen is the transmission through practice and meditation and, after several generations, Bodhidharma was reached who was the one that introduced Zen to China from India as a new current that differed from the known ones for its habit and practicality, which carried the spirit and teachings of Sakyamuni without altering it in any way.

In relation to the emergence of the Buddhist schools, coming from China, an important Buddhist monk who can be considered of great influence on *Shinobi* thought was Suzuki Shōsan (1579–1655), who, in addition to having influence on the teacher Hatsumi, wrote in 1619 a work called Mōanjō (Shokan 1977, p. 7) while in the service of Shōgun Tokugawa Hidetaka, notable in the battles of Sekigahara and the siege of Osaka Castle, during the battle that occurred in summer. After the battle, he stopped being a warrior and devoted himself to his duties as a monk, helping in the building of several temples for the Tokugawa family. Both Suzuki Shōsan and the Tokugawa family belonged to the Jodo Shin Shu school. Such was the importance of this monk that some years later, the Sōtō Shu school erected a temple in his honor, accepting many currents of the time and not rejecting those that were not from his side. One of his main teachings says:

> Your whole resolve is simply to discard your body, yours alone. You have no worries. No one can embarrass you because you have no pretentions, and since you have cast off the world there is no one to be bitter at. With that one cutting blade, the mind, the body is quite safe (Shokan 1977, p. 231).

It should be emphasized that, in addition to being a Buddhist monk, he fought alongside Tokugawa Ieyasu and took part in the Battle of Sekigahara in 1600 and in the battles of Osaka in 1614. It was at the age of 42 when he became a Buddhist monk and created his own branch called Niō Zen. This branch was not studied academically until 1959, when Suzuki Daisetsu wrote an article titled *Bushi-Zen Buddhist* (Suzuki 2014, p. 45). Nakamura Hajime later stated: "Shōsan is a modern Buddhist who spoke of Buddhist professional ethics" (Kato 2014). The basic code of ethical Buddhism, which is related to the *Shinobi* thought, is collected in the work *Mōanjō*:

1. It is necessary and gratifying to know the line that separates life and death.
2. If we recognize ourselves in past times, we will see our present reflected.
3. Events have a broader meaning to which we should aspire.
4. With determination and dedication we should help those in need.
5. We must observe where each thing is located and think of them as a natural fact.
6. Going out of our protection zone will make us grow and thus be more virtuous.
7. Knowing ourselves is a guarantee to be able to protect ourselves.
8. When you fall, get up once again.
9. Detachment is necessary to be able to educate our feelings.
10. To free oneself from greed and materialism will be fundamental to receive the superior teachings (Sevilla-Liu 2008, p. 47).

Nichiren is one of the branches of Buddhism that is based on the teachings of the Japanese monk Nichiren, whose practices are shown in the reading of the Lotus Sutra. In this text we learn what happened to a naga princess in her encounter with the Lotus Sutra. In this story it is emphasized that the princess is a naga. Naga can be translated as dragon, as a kind of non- human being found in traditional Buddhist cosmologies. Nagá is how it is pronounced in Sanskrit. Nága or Nág is the pronunciation in Bengali, Hindi, Marathi, or Pali. Etymologically, it is unknown where the Sanskrit word naga comes from. Possibly it is an auto demonym in the language of the Naga ethnic group. What is important here is that non- human bodies were generally regarded as incapable of being a condition or support for awakening. Similarly, some Buddhists thought that the female body was also incapable of supporting awakening. The naga princess lacked the mental faculties that could help her see reality for what it was: a childish mind that was an obstacle to awakening. The girl, due to her immaturity, had not had the opportunity to participate in practices that could have developed her spiritual abilities. The girl had three personalities called naga, woman and girl, which represent three additional obstacles to achieve enlightenment. However, by appealing to the Lotus Sutra, this is a fact that makes the impossible possible, thus, turning obstacles into possibilities. That is not to say that what the Lotus Sutra says about this new possibility or what it claims to be able to do is easy to accept. As Buddha (Bodhisattva) says in the story: "I cannot believe that this girl in the space of the instant can really reach the correct enlightenment" (Smith 1991, p. 134).

In The Lotus Sutra translated by Burton Watson (Anonymous 1993, p. 34), the naga snake or dragon has great importance and transcendence in Buddhism, representing and guarding the temple of Wat Sisaket located in the city of Vientiane, in Laos. The naga is also present in several engravings and is placed to the left of the god Krisna, dancing on the heads of the naga Kaliia, while the naga's wives pray to Krisna. Dai Moku is the basis of the practice of the forms of Nichiren Buddhism and has as its maxim the representation of the ultimate law or what they call the truth of the universe. It is a mantra that is recited in ceremonies. The basic teachings of Ō Daimoku (devotion) speak of the mind as the root of everything. Everything that we are is the result of what we have thought, that is to say, it is founded on our thoughts, and it is made of our thoughts. In Nichiren Buddhism, one speaks of devotion to the Lotus Sutra of the wondrous Dharma. The way to perform the prayers is with the hands placed in reverence reciting aloud the devotion to the sacred title of the Lotus Sutra. There is no fixed rule of how to do the prayer or how many times or how extensive the recitation of *Ō Daimoku* should be: "Nam-myoho-renge-kyo". These are the words that are recited to achieve concentration. Buddhism does not exist without practice and study, both practice and study derive from faith. Plato said: "He who learns and learns and does not practice what he knows, he is like the one who plows and plows and does not sow" (Miguel. 2012, p. 102).

According to cosmology, many cultures and religions have developed belief systems about the universe and the nature of reality. These belief systems can influence how the mind and its functions are perceived and understood. Buddhist philosophy holds that reality is impermanent and that all suffering comes from attachments and desires. This perspective influences the Buddhist understanding of the mind as a source of suffering and the need to develop wisdom and detachment to achieve liberation.

Buddhism has significantly influenced Japanese culture and has given rise to various Buddhist traditions in Japan, such as Zen, and Nichiren. Buddhist practice includes meditation and mindfulness, which may be related to ToM and introspection. The Buddhist idea of impermanence may also be related to the idea of the mind as a constantly changing phenomenon.

The study of the ToM has brought a new perspective to the understanding of Japanese Buddhism, especially as it relates to understanding the nature of the mind and its relationship to the practice of Buddhism. In Japanese Buddhism, the mind is considered the foundation of religious experience and practice, and it is through meditation and contemplation that one seeks to understand and transform the mind. ToM has helped to deepen the understanding of the mental and emotional processes that occur during meditation and to understand how these practices can influence the mind and experience. In addition, the ToM has helped to better understand the nature of consciousness in Japanese Buddhism. For example, it has been explored how ToM can help explain the nature of consciousness in Buddhism, especially as it relates to the relationship between consciousness and the outside world. ToM has also been used to explore how the practice of Buddhism can influence empathy and compassion, which are core values in Japanese Buddhism. Meditation practice has been shown to enhance empathy and compassion, and theory of mind has been used to explore the underlying mechanisms of these processes.

15.2.4 Shintoism

Taking into account that Shinto is more than a religion, it could be considered that Shinto is the original belief of the Japanese people. The reason by which this is so that Shinto is part of the Japanese culture because it has brought, all together, the thoughts and behaviors of the Japanese for thousands of years. Shinto has an unknown founder, but it has been known that Daruma was the precursor to this thought. Shintoism can be referred to life of the divinities or *kami*. This is the main reason by which Shinto can be related to Zen.

15.2.5 Zen

In relation to Shinto, we can talk about a way of life and thinking of people that work for a balance. They want to live in harmony with nature and be as happy as possible. This refers to Zen Buddhism. In our paper, we are referring to Buddhism in Japan associated to the way or *dō* in Japanese. That means the essence of the soul. There are so many books that can be recommended for the study of Zen, but our attention will focus on the Japanese philosopher Suzuki Daisetsu. Suzuki is considered to be the father of Zen Buddhism because he introduced it in western society, concisely, for our understanding. We cannot explain from a concise way as Suzuki did in his book *Zen and Japanese Culture*:

Briefly, Zen is one of the products of the Chinese mind after its contact with Indian thought, which was introduced into China in the first century A.D. through the medium of Buddhist teachings (Suzuki 1959, p. 3).

This difference between cultures had a great influence in Japan with the Buddhist thought introduced in China because Chinese thinkers adapted Indian thoughts to their culture and adopted Taoism as an exclusive way of thinking. Japan also adapted Chinese thought to this new way of thinking known as Zen Buddhism, but with the essence of Buddhism. People that follow Zen Buddhism have the capacity of growing and getting better as absolute humans and they adapt themselves to the needs of the universe. That is important because human minds cannot just be led by fantasies and thoughts without control and experimental pragmatism. This can be seen in the evolution and adaptation of Zen in Japanese culture. Suzuki postulated this in the following phrase: "In metaphysics Zen absorbed much of Taoist teachings modified by Buddhist speculations" (Suzuki 1959, p. 4). Master Suzuki was constantly evocating the essence of Zen. That is the adaptation through the practice of Zen on each aspect of the life. People who practice Zen need to be open minded because they need to adapt themselves to changes but keeping always in mind the origin of Zen. It must be saved with respect in the mind for the way of life.

The thinker that walks in harmony with the Zen Buddhism is not a mere spectator that just pray the learned orations in order to practice the basic rituals or in order to follow the first philosophy of Buddhism. This thinker practices the previous mentioned thoughts as something being not systematic or compulsory. Here it can be seen a mind opened to the dialogue, to the interpretation and ampliation of something that enrich us at the same time that illuminates us. That is a way to understand the universe in a direct and pragmatic way in opposition to the fantastic and complicated form of comprehension other disciplines follow. However, people need to have a good dedication and take a high effort in the achievement of their life because it has been proved that in order to achieve an objective, this must be earned through effort, sacrifice and dedication. Suzuki explained this when monks in a peregrination asked a Zen monk the following (Suzuki 1959, p. 5): "How deep is the river of Zen?". The Zen monk replied: "find out for yourself" and he offered to jump to the river to experiment this in first person. That explains the Zen is direct and practical and, although it respects the discipline, its main purpose appears through the own experience. Despite of the fact that Zen follows the study of case and does not analyze unnecessary abstractions, the Zen is not opposite to the dialogue, though, the last example was a bit extreme. That is the essence of Zen because it does not admit just a reading and recitation of the sutras. The importance of Zen is to maintain into the way. This is the cause of the success of a pragmatical and idealist mind in the Chinese and Japanese people.

The main idea of Zen is in the lack of interest for materialism or any other method that oppresses us in a necessary wish of freedom and emancipation of the materials. The *satori* or illumination is the essence of Zen because every discipline that has compromise, sacrifice, dedication, and effort can be acceptable for the achievement of the spirit of Zen. This begins with the self-confidence and control of our own Zen as Masaaki Hatsumi and Takamatsu indicate with *Nintai Jisei* (Takamatsu and Hatsumi 2020, p. 94) which has been translated as patience and self-control. In every Buddhist school the main point is *satori* because Zen does not discard the main basis of Zen. Every Buddha's teaching starts with *satori* because it is necessary for the birth of Buddhism. For this reason, Buddhism was transmuted and adapted to Zen. Zen proposes the achievement of *satori* through multiple ways as words and actions in order to complement and encourage other disciplines to reach a goal.

Suzuki talks about Zen verbalism as an own characteristic of Zen, but which is very different to the philosophy of language or dialectics. For this reason, Suzuki says that it could not be very clear to use these other expressions that have been mentioned as examples. It deals with the idea of love which can be seen as the essence of humanity since love is, in a certain way, the rudder of the spirit and without love life would be a path without encouragement as well as empty and lack of hope. Love needs a medium to show it and, in his book, he alludes to the fact that this medium is language, though, this is not the only form of transmission, it is in this statement that Suzuki Daisetsu himself establishes that love cannot only be transmitted through language, since there are multiple forms or ways in which love can be transmitted. In Zen, practice is a necessary act per se since study and application complement each other in the same way.

Both Buddhism and Zen have been central to the creation of the *shinobi* philosophy, as well as the teachings of Bodhidharma. Zen has added foundation and monastic discipline to the culture of *Budō*. It can be said that the history of Zen Buddhism begins with Bodhidharma, also known as Daruma in Japan and Tamo in China, but Zen Buddhism does not become relevant and fully recognized until the time of Daikan Eno (Mou-Lam and Price 2004, p. 55) (Hui-neng[13] in Chinese (Broughton 1999, p. 24)), who was the sixth and last patriarch of Zen, and was the result of a dispute between him and his opponent Junshu (Shen-hsiu). Both were disciples of Gunin (Hung-in) who passed away around 675 and it was, then, when both disciples claimed to be descent of the Zen line. These traces lead us directly to the first great master of the Zen teachings Bodhidharma. The features that characterize Zen Buddhism from other branches are the Daruma teachings that align with the practical branch of Mahayana (one of the main branches of Buddhism), which does not try to offer any method or any philosophical aspect that claims to give to know the truth of Buddhism. In fact, Daruma was not a logician, he simply wanted to live the truth. Therefore, everything he transmitted consisted of the practical expression of what he considered to be the way to achieve the ultimate goal of Buddhist life, since the first condition was to know exactly what the goal of Buddhist life consisted of.

One of the most important precepts in Zen Buddhism is the word Shin or Kokoro that Daruma formulated. This concept is equivalent to that of the Sanskrit Citta or Hridaya, which can be translated as mind, giving it an intellectual connotation, since the basic translation would be heart, but this translation could be assimilated to emotional issues. Another translation would be soul, which would evoke something more concrete. Master Tsuzuki Daisetsu uses "Mind" and emphasizes the capital letter, since he affirms that Daruma wants us to see the interior part of this "Mind",

[13]He was the sixth patriarch (638–713) considered to be the true father of Zen.

since only when it is apprehended or received "serene foundation of the mind" would be reached.

The maxim of *Shin* or *Kokoro* would later evolve into the *Dō* concept. *Dō* can be understood after having extensive experience in the practice of *Budō* and following a teacher for several decades. When any discipline that involves sacrifice, dedication and effort has been practiced for several years, it is likely that a range of diversities will open up and one may have the feeling that everything that is seen and everything that is touched is not the only thing that exists. Not only with the senses such as sight, touch, taste and smell, reality can be perceived, since these senses delimit other senses that are not physically palpable; this is referred to the sense of perception, something that we feel deep inside and that flourishes naturally without us being aware of being able to activate it at will.

It can be seen the case of Eugen Herrigel, a German philosopher who lived an experience similar to the previous one treated during the experience he had in first person and through *Budō*, more specifically, in the art of *Kyūdō* or art of the bow. Herrigel tells how he wanted to get into Japanese culture to get to know one of the most enriching Honshitsu essences that can be known, and that Westerners can enjoy.

In order to be accepted into Zen Buddhism, Westerners can use various gateways such as: *Budō*, *Ikebana*, writing, *Haiku*, theater, and some others that are possibly related to Japanese culture. However, despite everything, one of the most influential figures in Hideyoshi's government was the prominent Yamabushi En No Gyoja who tried to restore order with his doctrine called Shugendō (López García [Shinden 2019] 2021), another way of spreading spiritual Buddhism. Their basis is represented by Buddhism and Shintō, and at the same time they are derived from Taoism itself with strong influences from Onmyōdō. It is said that the Shugendō is the experiential knowledge obtained through the *Dō*, as a result of the practice of *Shu* (asceticism) with the influence of nature, the influence of the *Kami* (divinities) and the *Gen* (practitioners) of the Shugendō, which they were mainly the Yamabushi (Myoren. 2017, p. 18). Onmyōdō can be synthetized as an ancestral practice of synthetic science that, in addition to the teachings on the use of nature, included the Chinese art of divining events and the science of astrology.

In No Gyoja he was recognized after his death as the Great Miraculous Bodhisattva, and he was also considered as the manifestation of Hōki Bodhisattva. Illustrations of his person in the Kamakura period can already be seen, most of which are preserved today in the temples of the Shingon sect, and everything related to esoteric Mikkyō Buddhism. In 1954 Takamatsu Sensei became an experienced Shugendō priest, with the Buddhist name Butoku Inden Kyunsho Kakujo Shoo Daikoji. His main student Masaaki Hatsumi was considered a high-ranking monk, with the Buddhist name Shinnin Inden Byaku Ryū Dainichi Buko Daikoji. Masaaki Hatsumi says that his teacher was considered a holy man who had divination qualities and that he was frequently consulted by his people on various divination issues. It has been verified (Sylvain 2016, p. 315) that this is one of the characteristics that can be highlighted of these monks in relation to esotericism and mysticism, as a kind of moral advisors in the towns where they lived. These functions were fulfilled by shamans (Choza 2018, p. 54) in earlier times.

Of all the branches of human research, almost none deals with such a personal area as faith. This leads us to the idea that Buddhist scriptures are better seen as qualities of personal life rather than books. One way to consider the way Buddhists have thought about how scriptures enter the world is to see the benefits they describe as coming from an encounter with a scripture and from making scripture a part of one's personal life.

Throughout history, man has tried to maintain the teachings either orally or through writing. These writings represent great efforts to take care of the transmission of the teaching because every time a person copies a text, to take care of the word through writing, he tries to ensure, even if it seems insignificant, that these teachings endure in the world and reach to as many people as possible.

One of the thinkers, considered a true philosopher, among the many scholars who commented on ancient texts, as well as other problems of his time, was Baien Miura (1723–1789), who studied nature itself and tried to understand it in his original thought. He considered that nature exists in accordance with the fundamental principle jori, which is formulated:

> The state of *jori* is that the one enables the two, and the two possesses the one. Two is one plus one, one and one is one. The key to knowing this is to see the unity in the opposites, abandoning the habits of mind, and following the correct signs (Heisig 2016, pp. 466–467).

This ontological principle is abstract and universal, but all beings depend on it. The figure of *Gengo* is found in his main writing, as this word means the origin of the word, related to the natural world and *jori.*

Many of his philosophical terms come from Chinese philosophy but Baien Miura gave them their original meaning. For instance, the most important concept *ki* (Chinese *chi*) in his philosophy has its own meaning and becomes more abstract and diverse than its original meaning. It is believed that he interpreted nature by the fundamental principle *jori,* and universal categories as western philosophers did. He called it Hankan-gouitzu. He thought that natural objects should be looked at from the point of view of duality. In this way, objects can be seen as a relationship, as in the case of heaven and earth, or day and night. Miura did not understand the separate objects, they were a unit in themselves. That is why he said one and one is one.

He also realized the linguistic problem in philosophy, regarding the relationship between words (signifiers) and objects (signified), which limited human intelligence and formulated the idea *Gen* (deep or dark) referring to nature itself beyond human description. That is, it is something that cannot be understood; it is known to exist.

Baien Miura can be considered to be as a thinker closely related to Western thought, specifically to positivism and the observation of natural facts, he even came to support the Ptolemaic system. His attitude towards nature was based on the understanding of relating his theory to actual facts later. Charles Sanders Peirce proposed *Pragmatic Maxim* where the pragmatic maxim says that our entire conception of any object is our conception of all its conceivable practical effects (Hookway 2012, p. 167). This maxim represents the highest degree of clarity of the ideas in the Peircean theory (Villacañas 2001, p. 74) of the levels of clarity in relation to the truth. This idea can be associated with Ishizu Teruji who, when he speaks of truth or reality, postulates that "the Buddhas have realized a rare and difficult-to- understand dharma: the fact that only one Buddha can fully grasp another Buddha transcends all explanation" (Heisig 2016, p.135).

On the other hand, *Kūkai* cannot be left aside, the founder of the Japanese school of esoteric Buddhism (*Vajrayana*) or *Shin-gon* (which translates as true word or Mantra). We are surely facing the most famous Buddhist personality in Japan. In Japanese the expression *Shin no Kotoba* means true word and the meaning which is seen in the mantras can be defined as the sounds of the soul. This is associated with *Rei Ki*, which is the spirit of energy, taking into account that *ki* in Japanese means soul included in the words *Koumyou Shingon* (Too 2003, p. 76), mantra of light. The first word *Koumyou* means lightning, light; the second one is shin which has been translated as true or authentic; and the third one, gon, has been translated as word. From this formation is born the terminology of Shingon Buddhism, one of the sects of Buddhism.

In the same way as in Zen Buddhism, mantras are the songs that are sung to achieve maximum concentration and, from this, reach enlightenment. The words that are transmitted in the manuscripts from the teachers to their disciples, are the words that give meaning to the teachings; these words are understood through practice and the study of the texts that the teacher gives to the disciples.

In *ninjutsu,*[14] as it is transmitted in the manuscripts, according to Kitarō Nishida in his work *Thinking from Nothing*: "what is true and important resides in the heart, it is born and grows from nothing and becomes the final essence" (Nishida 2006, page 46). According to Masiá "it can be considered a kind of philosophical testament" (Nishida 2006, p. 56). In this work Nishida finds himself at a crossroads (dilemma) between several opposites: East/West, Philosophy/Religion, Buddhism/Christianity, Zen/Animism. It is the place of the logic of nothingness and this dilemma is called Muga (not-self), whose closest translation is ecstasy which can be considered as a reformulation of the Kantian principle of aesthetic judgment or "disinterest" (Heymann 2019, p. 124). Nishida describes it as a state of mind, that is, being out of oneself, and refers to it as the divine inspiration of art. The hidden truth is not reached through thought since it is an intuitive truth. It appears before us like a stimulus that suddenly arises, from the bottom of our hearts. Nishida gives a detailed description of Muga:

> However, when one devotes years to practicing morality, one eventually reaches the level that Confucius described in the *Analects* in the following terms: "going to bathe in the Yi River, enjoying the breeze at the Rain Altar, returning home singing poems". In other words, when morality reaches a high level and enters into spirituality, there is no longer any difference between morality and religion (Nishida 2006, pp. 16–17).

Zen is a Buddhist tradition that emphasizes meditation and mindfulness. Zen has had a great influence on Japanese culture and has spread to other parts of the world. Zen

[14] Japanese word which can mean art of stealth.

practice may be related to ToM and introspection, and the idea of the mind as a constantly changing phenomenon. The study of the ToM has provided a new perspective in the understanding of Japanese Zen, especially as it relates to the practice of meditation and the understanding of the nature of the mind. In Japanese Zen, meditation is considered an essential practice to achieve enlightenment and understanding of the nature of the mind. ToM has helped to deepen the understanding of the mental and emotional processes that occur during meditation and to understand how these practices can influence the mind and experience. In addition, the ToM has made it possible to explore the relationship between the practice of meditation and the development of empathy and compassion, fundamental values in Zen. It has been shown that the practice of meditation can improve the capacity for empathic understanding, and compassion, and ToM has been used to explore the mechanisms underlying these processes. Another novel contribution of the theory of mind to Japanese Zen is its ability to explain the nature of consciousness and the relationship between consciousness and external reality. The ToM has allowed a deeper understanding of the experience of reality and has allowed us to explore how the practice of Zen can influence the perception of reality.

15.3 Conclusions

We can conclude this chapter with the idea of the ego as a mental self-image impregnated with the emotionality that each one has of himself. This is associated with Taoism, the basis of Chinese philosophy, with the conception of unity and the knowledge of a supreme and unifying reality through balance and a peaceful life. The important thing is what each one believes he is, how he defines himself, and which a person identifies. That is why Japanese people tend to be alone, look at themselves, and maybe they have difficulties when trusting others. This can join the problem of false beliefs when answering questions asked by a stranger. That is the idea that defends Confucianism as the basis of Taoism with the *Mengzi* that applies the teachings of the master in a practical way. Therefore, we must stay on the path that is reached in multiple ways, because there are multiple interpretations to understand and achieve enlightenment. The reality can be seen in several ways, but all of them come together to attain enlightenment. That is why the one leads to the two, the two leads to the three, because there is no single way, which will depend on our interior that blooms naturally without being aware of activating it at our whim.

In Buddhism and Japanese Zen, the self-perception and construction of the self is based on the idea that the self is not a fixed and permanent entity, but rather a constantly changing and transforming process. The self is not seen as separate from the world but is in constant interconnection and co-creation with everything around it. In Buddhism, there is talk of the concept of *Anatta* or "not-self", which refers to the idea that there is no permanent and immutable self in the human being, but that everything is impermanent and constantly changing. Zen also emphasizes the impermanence and impermanence of all things, including the self. As for ToM in Japan, self-perception and the construction of the self have been studied a lot in relation to theory of mind. Research has been done on how people perceive their own self and the self of others, and how this affects their ability to understand the intentions and thoughts of others. It has been shown that people with a better ability to understand the emotions and thoughts of others also tend to have a greater awareness of their own self and their own emotions and thoughts. In addition, it has been studied how meditation and other Buddhist practices can help improve self-perception and the ability to understand others.

Confucianism, Taoism and Zen Buddhism are important philosophical and religious currents in Japan, which have also influenced the ToM. Confucianism emphasizes the importance of social relationships and ethics in life. The central idea of Confucius is virtue, which is acquired through education and reflection, and which leads to social harmony and moral order. In terms of ToM, Confucianism stresses the importance of understanding the intentions and emotions of others in order to build proper social relationships and develop effective social skills. Taoism focuses on the search for harmony and balance in life, and the idea that everything in the universe is interconnected. Taoist philosophy emphasizes the importance of living in the present and in harmony with nature. In ToM terms, Taoism promotes an understanding of the interconnectedness of living things and the importance of accepting feelings and emotions, both positive and negative, without judgment. Zen Buddhism, which originated in China and spread to Japan, emphasizes the importance of meditation and contemplation to achieve enlightenment. Zen Buddhism promotes the idea that true understanding of the world and of oneself can only be achieved through direct experience and meditation. In terms of ToM, Zen Buddhism emphasizes the importance of mindfulness and self-reflection in understanding the nature of the mind and the self.

There is a perception that is a sense that the Japanese understand as *shin* or *kokoro* and it is within us. That is why we have Zen Buddhism, that is, the path to enlightenment that can be accessed through *Budō*, ikebana, Japanese writing, haiku, theater, and others. All these paths are associated with Japanese culture. In general, there is an interconnection between religion, philosophy, and psychology in Japan, and this has influenced the understanding of the mind and human behavior in Japanese culture. We conclude, then, that to understand the ToM in Japan you have to know the Japanese culture, which in turn is known through its philosophy and language.

References

Adler, J.A. 2005. *Religiones Chinas*. Madrid: Akal.
Anonymous. 1993. *The lotus sutra*. New York: Columbia University Press.
Aoki, M. 1996. *Diccionario de filosofía japonesa*. Madrid: Ediciones Cátedra.
Aristóteles. 2014. *Metafísica*. Trans. T. Calvo Martínez. Madrid: Gredos.

Broughton, L. 1999. *The Bodhidharma anthology, the earliest Records of Zen*. California: University of California Press.
Call, J., and M. Tomasello. 1999. A nonverbal false belief task: The performance of children and great apes. *Child Development*: 381–395.
Callaghan, T., P. Rochat, et al. 2005. Synchrony in the onset of mental-state reasoning: Evidence from five cultures. *Psychological Science*: 378–384.
Cardona de Gibert, Á. 1968. *Confucio, Los cuatro libros clásicos*. Barcelona: Bruguera.
Chen, F., J.D. Rubio, and M.Y. Chen. 2023. *Neurorobotics explores machine learning*. Lausana: Frontiers Media SA.
Choza, J. 2018. *La Revelación Originaria: La Religión de la Edad de los Metales*. Sevilla: Editorial Themata.
Csikszentmihalvi, M. 2011. *Fluir (Flow) Una psicología de la felicidad*. Madrid: Kairós S.A.
Ekman, P. 1999. Facial expressions. In *Handbook of cognition and emotion*, ed. T. Dalgleish and M. Power, 301–320. New York: Handbook of cognition and emotion.
Enbutsu, S. 2003. The ones who got there first. *Japan Times*: 1–30.
Fu, I.N., K.L. Chen, M.R. Liu, et al. 2023. A systematic review of measures of theory of mind for children. *Developmental Review*: 1–23.
Gardner, D., P.L. Harris, M. Ohmoto, and T. Hamazaki. 1988. Japanese children's understanding of the distinction between real and apparent emotion. *International Journal of Behavioral Development*: 203–218.
Hasegawa, T., and K. Hiraki. 2008. The neural basis of introspection in terms of the dual- mode model of consciousness. In *Self and consciousness*, 167–182. John Benjamins Publishing Company.
Hasegawa, T., and H. Onishi. 2016. The neural basis of social cognition and its dysfunction in autism spectrum disorders: Implications for development and intervention. *Developmental Psychology*: 248–257.
Hasegawa, T., and M. Sakai. 2009. Development of self-other differentiation in the frontal cortex during childhood and adolescence: Evidence from magnetoencephalography (MEG) and electroencephalography (EEG). *Developmental Science*: 101–111.
Hasegawa, T., and T. Yamauchi. 2012. Differential effect of personal relevance on the early and late stages of gaze processing. *Cognitive Neuroscience*: 82–89.
Heisig, J. W. 2016. *La Filosofía Japonesa en sus textos*. Trans. R. B. García. Barcelona: Herder Editorial S.L.
Hepburn, J.C. 2013. *Japanese-English dictionary: With an English and Japanese index*. New York: Tuttle Publishing.
Heymann, E. 2019. *Sinopsis de Decantaciones Kantianas: Diecinueve Estudios Críticos y una Visión de Conjunto*. Madrid: Escolar y Mayo.
Hobbes, T. 2016. *Leviatan*. South Carolina: CreateSpace Independent Publishing Platform.
Hookway, C. 2012. *The pragmatic Maxin: Essays on Peirce and pragmatism*. Oxford: Oxford Academic.
Inazo, N. 2018. *Bushido, el espíritu de Japón. Un ensayo clásico sobre la ética del samurai*. Trans. A. Carrillo Moya Madrid: Dojo Ediciones.
Itakura, S., M. Okanda, and Y. Moriguchi. 2008. *Discovering mind: Development of mentalizing in human children*, 179–198. Origins of the social mind: Evolutionary and developmental views.
Kanakogi, Y., and S. Itakura. 2011. Developmental correspondence between action prediction and motor ability in early infancy. *Nature Communications*: 341–352.
———. 2014. The role of early social experience in the development of empathy and related abilities. In *In social behavior from rodents to humans*, ed. M. Wöhr and S. Krach, 263–276. Tokyo: Springer.
———. 2018. How the social context modulates the attentional network in infants: A cross-sectional and longitudinal study with eye-tracking and electroencephalogram. *Frontiers in Psychology*: 557–566.

———. 2019. Infant's perception of goal-directed actions: Developmental and cognitive aspects of predictive gaze. *Developmental Psychobiology*: 3–13.

Kato, M. 2014, August 28. *Asia Pacific Week*. Retrieved from http://terebess.hu/zen/mesterek/NioZen.pdf

Kumano, M. 2017. Change of feeling of ikigai after the birth of a child in parents with infants: An approach from ideal views of ikigai and ikigai process. The Proceedings of the Annual Convention of the Japanese Psychological Association, 1A-079.

Kuribayashi, T., M. Fukui, and Y. Yoshida. 2004. The development of "nensho no rikai" and its educational significance. *Japanese Journal of Educational Psychology*: 119–130.

Liu, D., H.M. Wellman, and M.A. Sabbagh. 2008. Theory of mind development in Chinese children: A meta-analysis of false-belief understanding across cultures and languages. *Developmental Psychology*: 523–531.

López García, D. 2021. *La Cultura shinobi, evolución y actualidad: Masaaki Hatsumi y Bujinkan Dojo como caso de estudio*. Comares: Granada.

Miguel. 2012. *Qué razón tenía Platón: 1001 pensamientos para triunfar en la vida*. Madrid: Espasa.

Mogi, K. 2018. *Awakenin your Ikigai: How the Japanese wake up to joy and purpose every day*. New York: The Experiment.

Mou-Lam, W., and A.F. Price. 2004. *The diamond sutra and the sutra of hui-Neng*. California: Shambhala Classics.

Myoren. 2017. *El camino de los yamabushi*. London: CreateSpace Independent Publishing Platform.

Nagase, A., Y. Moriguchi, and J. Decety. 2013. Empathy and judging other's pain: An fMRI study of alexithymia. *Cerebral Cortex*: 890–897.

Nagase, A., Y. Okumura, and Y. Moriguchi. 2017. Empathy and theory of mind in social decision-making. *Frontiers in Psychology*: 911–923.

———. 2018a. Developmental changes in the neural basis of interpreting communicative intent from others' false beliefs. *Developmental Cognitive Neuroscience*: 86–96.

———. 2018b. Preschoolers' understanding of the relationship between others' false beliefs and emotional expressions: A behavioral and fNIRS study. *Developmental Cognitive Neuroscience*: 1–10.

Naito, M., and K. Koyama. 2006. El desarrollo de la comprensión de falsas creencias en niños japoneses: ¿retraso y diferencia? *Revista Internacional del Desarrollo del Comportamiento*: 290–304.

Nguyen, D., N. Phuoc, et al. 2023. *Memory-augmented theory of mind network*, 1–8. Association for the Advancement of Artificial.

Nishida, K. 2006. *Pensar desde la nada*. Salamanca: Sígueme.

Nukariya, K. 2004. *The religion of the samurai*. London: Kegan Paul Limited.

Okada, H. 2019. Developmental changes in theory of mind during middle childhood: A cross-sectional study using a modified version of the Reading the mind in the eyes test. *Frontiers in Psychology*: 1824–1850.

Okada, H., and S. Itakura. 2017. Development of the ability to represent another's false belief and its relation to the understanding of attention in 3- to 5-year-old Japanese children. *Cognitive Development*: 1–10.

———. 2018. Development of the ability to represent another's visuospatial perspective in 4-to 8-year-old Japanese children. *Japanese Psychological Research*: 139–149.

Okada, H., and M. Tanaka. 2004. Neural mechanisms underlying the inhibition of the mentalizing process: An event-related potential study. *Journal of Cognitive Neuroscience*: 1772–1786.

Premack, D., and G. Woodruff. 1978. Does the chimpanzee have a theory of mind? *Behavioral and Brain Sciences*: 515–526.

Prevosti i Monclús, A., and A. Doménech del Río. 2005. *Pensamiento y religión en Asia Oriental*. Barcelona: UOC.

Rousseau, J. J. 2011. Emilio o de la Educación. Trans. M. Armiño. Madrid: Alianza Editorial.

Saito, M. 2014. Facial expressions and social communication. *Japanese Journal of Social Psychology*: 5–13.
Sato, I. 1992. *Diccionario de psicología japonesa*. Madrid: Editorial Tecnos.
Sato, W., H. Sylwia, et al. 2019. Facial expressions of basic emotions in Japanese laypeople. *Frontiers in Psychology*: 1–11.
Sevilla-Liu, A. 2008. Guiding the blind along the middle way: A parallel Reading of Suzuki Shōsan's Mōanjō and the doctrine of the mean. *Journal of Buddhist Ethics*: 45–78.
Shokan, S. 1977. *Selected writings of Suzuki Shokan*. New York: Ithaca.
Smith, W.C. 1991. *The meaning and end of religion*. London: Fortress Press.
Suzuki, D.T. 1959. *Zen and the Japanese culture*. Princeton: Princeton University Press.
———. 2014. *El Zen y la Cultura Japonesa*. Barcelona: Pairos Ibérica.
Sylvain, G. 2016. *Shugendo: Les mystérieux moines bouddhistes des montagne*. Trans. D. López Luque. Noisy-sur-École: Editions de l'Eveil.
Takamatsu, T., and M. Hatsumi. 2020. *Life secrets of the Amatsu Tatara: The documents of Takamatsu Toshitsugu, interviews with Hatsumi Masaaki*. London: Peter King Editions.
Tola, F., and C. Dragonetti. 2018. *Upanishads*. Buenos Aires: Las cuarenta.
Too, L. 2003. *The mantra book: Chant your way to health, wealth and serenity*. London: Element Books Ltd.
Tse, L. 1996. *Tao Te Ching*. Madrid: Tecnos.
Villacañas, J.L. 2001. *La filosofía del siglo XIX*. Madrid: Trotta.
Wang, W. 2018, April 13. *Confucio, el Sócrates de oriente*. Retrieved from https://confuciomag.com/confucio-el-socrates-de-oriente
Wellman, H., D. Cross, and J. Watson. 2001. Meta-análisis del desarrollo de la teoría de la mente. La verdad sobre la falsa creencia. *Desarrollo infantil. Desarrollo infantil*: 655–684.
Xunzi. (2014). *Xunzi, the complete text*. Trans. E. L. Hutton. Princeton: Princeton University Press.

Chapter 16
Content and Process in the Brain. Implications for Clinical and Educational Approaches

Carlos M. Gómez, Brenda Y. Angulo-Ruiz, Elena I. Rodríguez-Martínez, Francisco J. Ruiz-Martínez, Eva María Padilla Muñoz, and María Dolores Lanzarote Fernández

Abstract Subjects may be aware in each moment of the content of a given scene and/or internal state. However, the vivid reality of the scene is a by-product of the complex processing of external and internal information in the brain. This point is more evident in the hallucinatory phase of psychosis or during dreaming. There is complex internal processing for the recreation of certain contents which are unconscious for the subject, whilst the content is experienced, the process to arrive at the contents is not perceived at all. Therefore, an important question that naturally arises is if process and content are served by different neural mechanisms. The possibility that postsynaptic potentials arranged in an open-field configuration would be more related to the content, experienced (or not) by the subject, and transmission-transformation of information along neural networks would be related to processing would be proposed. The proposal has some important implications for clinical applications, given that depression, psychosis, anxiety, and others are experienced as defined contents or states, while more possibly the underpinnings are related to abnormal information processing. Therefore, any clinical application should be primarily directed to psychological, pharmacological, or neural stimulation modification of information processing. However, if the content is related to activation of defined neural nets, possibly defined as engrams in a broad sense, the activation of

C. M. Gómez (✉) · B. Y. Angulo-Ruiz · F. J. Ruiz-Martínez
Department of Experimental Psychology and Human Psychobiology Laboratory, University of Seville, Seville, Spain
e-mail: cgomez@us.es; bangulo@us.es; frmartinez@us.es

E. I. Rodríguez-Martínez
Department of Developmental and Educational Psychology, University of Seville, Seville, Spain
e-mail: elisroma@us.es

E. M. P. Muñoz · M. D. L. Fernández
Department of Personality, Evaluation and Psychological Treatments, University of Seville, Seville, Spain
e-mail: evapadi@us.es; lanzarote@us.es

T. Lopez-Soto et al. (eds.), *The Theory of Mind Under Scrutiny*, Logic, Argumentation & Reasoning 34, https://doi.org/10.1007/978-3-031-46742-4_16

those engrams (content) would only be able to modify processing through action on the processes. A handful of psychological therapies, more or less, explicitly tacitly successfully focus on processes (i.e., cognitive therapy in psychosis) rather than on contents, while others also focus on contents (flooding therapy in anxiety). These ideas are being implicitly running from the beginning of Psychology and made explicit in the field of motivation in organizational psychology. It is important to clarify the different Physiological underpinnings of content and process, and how each intervention is more or less focused on one or the other to justify their odds for success. From an educational point of view, this topic refers to educational strategies more biased to memory (content) vs. educational strategies where the solving tasks approach (processes) is primed.

Keywords Awareness · Hallucination · Internal processing · Cognitive therapy · Physiology · Educational strategies · Intervention strategies · Engrams

16.1 Introduction

Any dichotomy is without no doubt some sort of categorical error, if opposition between black and white is certainly true, in the real world there exists a gradient of gray, some of them nearer to black or white which predominates. It is difficult to think of any information processing machine in which the content is not until a certain extent included in the process, and vice versa. However, any part of the underlying circuit would be near the outputs (contents), and other to the internal changes that determine the creation or recruitment from the memory of a defined output (process). For instance, if we consider the input data and code written on the keyboard, they would be machine coded (process), by means of translating by compilation (process) to machine code, and through the compiled code would be transformed (process) in some sort of output, as text screen (content). If we translate to Psychology, any sensory perception (visual, auditory, etc.) or internal states (thirst, hunger, emotion, etc.), or movement would be nearer to the concept of content, while all the internal machinery permitting the production of the sensory, and state perceptions, and the stages of coding the production of movement would be more related to processes, conscious or unconscious. However, a complete full characterization of any brain activity in an absolute categorical dichotomy, of content or process, which would be a fallacy. Nowadays, the distinction can still be useful as we would discuss now. The concept of content is directly linked to the idea that the brain somehow represents reality in a defined set of neurons, more or less distributed across the nervous system, *cognits* in the terminology of Fuster (2022). However, to reach to a mind content a huge level of internal processing is needed.

From a computational perspective, the Turing machines clearly specify the structural and functional differentiation between content and process, although the original word of state was used for the last term (reviewed in Feynman 1996). This theoretical machine is able to perform any possible algorithmic computation, with a

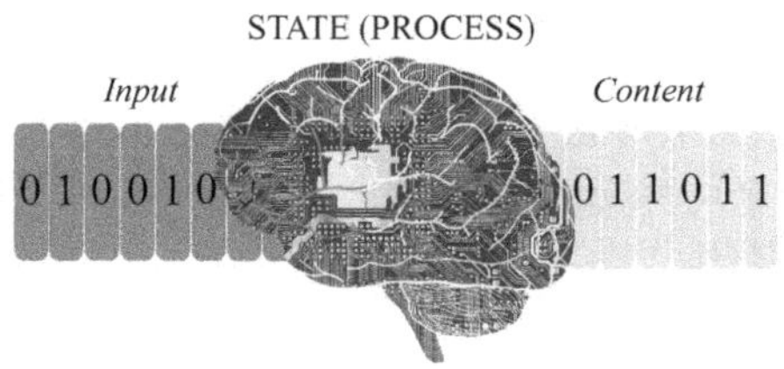

Fig. 16.1 Schematic representation of a Turing machine, suggesting the differentiation between content and process

relatively simple structure in which an infinite tape presenting some written symbols or blanks (inputs) which are read (i) in a head dispositive, and then a rule determined by the current state of the machine decides (ii) what symbol is overwritten in the tape, (iii) if the tape must move to right, left or halt, and (iv) which would be the next state of the machine for processing new input symbols. From the perspective of the present report, only point (ii) would be considered content (Fig. 16.1), and all the others are the instructions or processes that permit the machine operation. This architecture is not the current computer architecture, although it has inspired cognitive sciences, and certainly present chapter.

There are at least five basic approaches to follow the relatively independent organization of content and process from a computational perspective grounded in psychology and neuroscience. Those are based on the language organization and the idea of predicate logic as a basic makeup of the mind, neural networks as a model of how information is being processed by neuron-like units, dynamical systems in which internal variables generating external behavior are constantly updated, the brain as a predictive coding machine, and the brain, particularly in the hippocampus, as a machinery for creating compositional models of reality. We will briefly expose these abstract, but always simplistic models of the brain and the mind, and we would trace some examples of brain implementations, remarking on the content and process correlates. At the very end we would try to find some commonalities between all these different approaches.

16.2 Symbolic Language in the Brain

The idea of symbolic representation in the brain is highly related to the concept of thought, a concept deeply rooted in psychology since its origin. The most basic and accepted definition comes from Humphrey (1973). He describes it as "what happens in experience when an organism (human or animal) finds, recognizes, and solves a problem". If we look a little at all the explanatory attempts of Cognitive Psychology at the problem of thought, we could summarize it in the meaning that an author like Neisser (1967) gives to cognition, when indicates that cognition refers to "the processes through which the sensory input it is transformed, reduced, elaborated, stored, recovered and used". The central theme is therefore the representation of the external world in the brain (content), and the possible autonomous manipulation by the subject of these representations (process).

However, the field of thinking in their more abstract version, or in its more pictoric version known as imagination, takes up a very large, classic controversy. That is, can mental cognition be reduced to propositional or predicate logic as a combination of symbols and operators? This view would suggest that any kind of thought or knowledge about the world is defined in a language of symbols and logical relationships between them. Thus, both perceptual and verbal information would be produced in memory in a similar way (Pylyshyn 1984). One of the more formalized theories about this symbolic representation in the human mind of this is the so-called language of thought (Lot) (Fodor 2008). Basically, the Lot proposes the existence of some sort of propositional logic, or the more comprehensive predicate logic, in the mind, in which representations are combined by permitted syntactic rules. This language of symbols representing objects or states, and other symbols representing the actions permitted to be combined in legal operations (basically the Boolean operators and quantifiers), has been called as *mentalese*.

Fodor (2008) also introduces other interesting computational concepts as hierarchy, considering the idea of the presence of certain domains of knowledge in the mental structure, which are encapsulated and process specific types of knowledge (modules), while desires and beliefs, which are also somehow represented, would have a causal role in the subject's behavior.

This idea of symbolic representation with a combinatorial form is also in the essence of the linguistic theory of Chomsky (1968, 1986). Chomsky has proposed a structure of language based on universal grammatical rules that constrain the infinite possibilities of language combinations. This proposal has a historically interesting perspective by its opposition to Skinner's theory of language generation characterized by verbal operands and its reinforcement, that should not be limited to *a priori* biological determinants. Chomsky's hypothesis of the existence of a generative grammar of innate character, common to all human beings, regardless of their racial and ethnological origin, implies a genetic character in the determination of language deep syntactic structure. This school of linguistics is now highly influential and takes a strong position: The deep structure of linguistic syntactic rules, in a learning expectant manner, is inherited genetically and is the basis of all cognition. The main linguistic support for this position is the speed and ease with which children acquire the rules for the formation of new and coherent phrases and sentences. At 5 years of age, these rules are almost completely mastered spontaneously, unconsciously. The superficial forms of these rules vary somewhat from language to language, but Chomsky proposes a similar deep structure for all languages. The fundamental of Chomsky's theory (Chomsky 1968; Chomsky 1986) is that the child develops inside, as if it were an innate instinct, the grammatical structure of language, the so-called universal grammar. The main support for this theory is the child's demonstrated ability to create a completely new language with a correct grammatical structure with a very limited number of examples. Please notice that the theory implies that the deep syntaxis would be related to the process by which the content of the different used morphemes combines to create meaningful sentences. However, once a sentence is created it can be used as a whole new content for creating more complex structures through higher order combinatorial syntactic rules.

Among other data, this hypothesis is supported by the existence of a cerebral system specialized in the production of language, in the possibility of creating an infinite number of sentences never emitted about a real or imaginary situation and never reinforced by a peer or tutor, and by the fact that the ontogenetic development of language occurs in a similar way in children from different cultures and languages: At 6 months the child babbles, at 1 year he/she begins to produce and to respond to a one-word language, at 12–18 months they begin to utter single words, from 18 to 24 months two-word phrases, from the age of two rudiments of grammar appears, at the age of 3 complete sentences, and at the age of 4 a linguistic competence similar to that of adults is obtained (Lenneberg 1967). One last important fact to think to support that there is an anatomophysiological constraint in our ability to learn language comes from the so-called developmental dysphasia, where children have difficulties learning the grammatical rules of language. In fact, it has been observed that more basic learning processes than the strictly linguistic ones may be altered in Specific language impairment (SLI), being language problems a consequence of the malfunctioning of learning processes. In this sense, in a Serial Reaction Time (SRT) task, in which reaction times decrease for groups of stimuli whose occurrence is predictable compared to groups of stimuli with random sequences, children with SLI showed a lower improvement for predictable sequences (Gabriel et al. 2013), highlighting the motor component of procedural learning of sequences. On the other hand, SLI subjects have also shown difficulties in artificial grammar studies, where subjects must learn the rules of artificial languages. Similarly, they have also shown difficulties in statistical learning in paradigms in which subjects implicitly learn the existence of patterns in data sequences without having to make an explicit effort to extract such sequences (Hsu et al. 2014). Thus, the relationship between SLI and sequence learning would be more relevant to grammatical rule learning process than to vocabulary content. The level at which the statistical learning, would be constraining the internally generated universal syntactic rules is a matter of discussion.

The common segregation of logical operators and representations in Lot, and the differential representation of language structure vs specific contents in generative grammar, remark the homologous commonalities of these two neurocognitive approaches for an operational and anatomical separation of the representations and the combinatorial rules of the represented objects.

Anyway, in general the representation of objects and concepts seems to be partially segregated from the areas controlling the syntactic and semantic understanding or production of language. The different brain locations of the lesions producing agnosia (i.e., Prosopagnosia, Achromatopsia, Akinetopsia etc.; Fig. 16.2a) or aphasias (i.e Broca, Wernicke, Conduction etc.; Fig. 16.2b) (Gracia-Molina and Peña-Casanova 2022; Kirshner 2012) seems to support the latter point.

The anatomical localization of agnosia is highly determined by the different classifications that historically have been developed, we will present here the most classical version, although classifications based in new formalizations as this of Humphreys and Riddoch (1984) with an entirely new classification, but also this of Goodale and Milner (1992) proposing a distinction in ventral pathways directed to

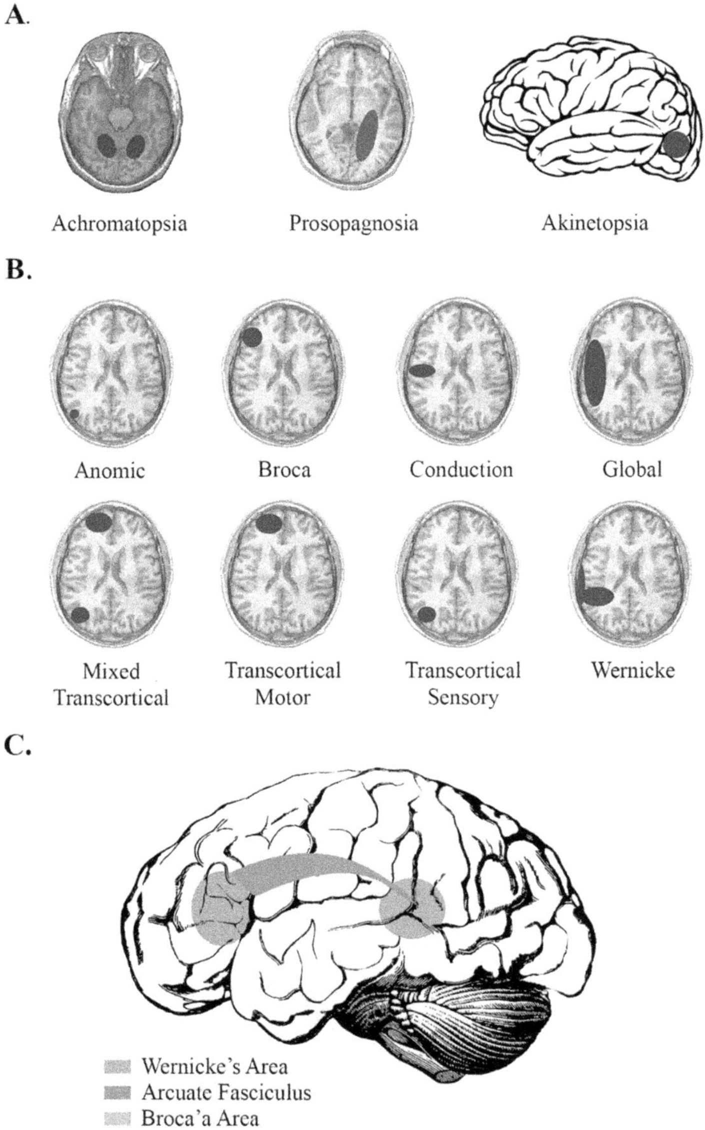

Fig. 16.2 Brain areas associated to language and imagery characteristics. (**a**) Brain areas associated to different types of agnosias, (**b**) Brain areas associated to different types of aphasias. (**c**) canonical circuit for language processing

object recognition, and dorsal pathways participation in action and orientation to these objects are also recognized in the literature (reviewed in Gracia-Molina and Peña-Casanova 2022).

From a more classical perspective, Achromatopsia, or inability to perceive colors occurs by lesion in the ventral occipitotemporal cortex (Fig. 16.2a). Akinetopsia or difficulties to perceive movement are produced by lesion in the area V5 in the occipitotemporal junction. The difficulty to perceive faces (prosopagnosia) is produced by lesion in the fusiform gyrus. The Simultagnosia observe in parietal lesions permits to distinguish objects in isolation but not superposed. Other forms of distorted constructions of objects are polydipsia (multiple images of an object), paliopsia or palinopsia (successive observed objects are simultaneously perceived), or metamorphopsia in which objects appear as elongated in one dimension. The interesting point of this description of agnosias for present chapter, is that they reveal that vision is a constructive process in which the subject is only aware of the final mind object constructed. This concept is also observed, in three different types of inattentional phenomena and problems: (i) inattentional blindness, an object presented in the visual field is not perceived because attention is on other objects also presented in the visual field (Mack and Rock 1998); (ii) scenes cuts, which are the syntax of the film scenes, are rarely perceived by subjects which in fact, are more aware of the history content (Andreu-Sánchez et al. 2021); and (iii) sensory hemineglect produced by the lesion of the right parietal cortex, which induces a sensory forgetfulness of the contralateral visual field, due to the difficulty to disengage the attention of the ipsilateral side to the lesion, or a difficulty to engage to the contralateral side (Gainotti et al. 1991).

In the case of aphasias, Wernicke's aphasia is caused by a lesion in Superior Temporal areas, and is characterized by a comprehension deficit, as opposed to Broca's aphasia, caused by a lesion in the third frontal gyrus, which maintains comprehension but alters verbal fluency. Thus, lesions affecting the posterior association cortex, especially Wernicke's area (posterior half of the superior temporal gyrus, area 22), tend to cause deficits in the semantic aspects of language comprehension. This is demonstrated by difficulties in understanding the meaning of words and sentences, but fluent from the production of language; in fact, the patient can produce, spontaneously, a certain verbiage. From these characteristics come the name of Wernicke's aphasia, or as semantic, sensory, or fluent aphasia. Broca's aphasia, on the other hand, due to the lesion of the inferior frontal gyrus (usually also affecting the surrounding cortex of the frontal operculum), presents difficulty in articulating words and sentences (motor or non-fluent aphasia). The most characteristic feature of this aphasia is the absence of functional words (e.g., articles, pronouns, conjunctions, and prepositions). Thus, to the listener, the speech of Broca's aphasia sounds telegraphic and agrammatical. Consistent with this antero (Broca's)-posterior (Wernicke's) organization, it has been shown that verbs and action words are generally represented in frontal cortex, whereas nouns and object names are represented in posterior associative cortex (Goodglass et al. 1966). It was observed that Broca's aphasias presented difficulty in naming verbs and action words, whereas Wernicke's aphasias presented difficulty in naming objects. These observations were repeated by several other researchers (Gainotti et al. 1995). Global aphasia results from lesions affecting both Wernicke's and Broca's areas. For Ardila (2013), the former two types of aphasias would be the fundamental types. While other types, see below, would correspond to accessory linguistic aspects or dysexecutive syndromes. Conduction aphasia, which is based on the lesion of the arcuate fasciculus, can maintain a certain degree of verbal fluency and total comprehension, although it fails in repetition and reading because it cannot connect Wernicke's and Broca's areas, in this case the distortion is also more pronounced for sentences without grammatical meaning than when they have it, indicating the existence of two different types of connections between Wernicke's and Broca's areas.

Pure word deafness aphasia is due to the disconnection of the primary auditory area with Wernicke's area, so that although sounds can be identified, it is not possible to identify words.

Anomic aphasia is an inability to find the right word during speech probably due to the loss of connection between the storage of meaning in the temporal and parietal associative cortex with the signifier in Wernicke's area.

Sensory transcortical aphasia represents a case similar to the anomic one but for comprehension, the person can repeat a text but not understand it due to the lack of connection between the representation of the signifier in Wernicke's area with the representation of the signified in the parietal and temporal associative cortex. In this sense, it should be noted that lesions at different points may cause loss of meaning in some texts but not in others. For example, a lesion in the right parietal cortex causes the loss of meaning of words used to relate objects spatially (up, down, etc.). On the other hand, lesions on the left parietal side determine loss of meaning of body locations.

Transcortical motor aphasia causes a loss of motivation to speak.

The intonation characteristics of speech (prosody) are located in the right hemisphere, as demonstrated by the loss of this ability in subjects with right lesions (Caplan 1987; Damasio and Geschwind 1984).

A particular type of aphasia that we would like to discuss is dynamic aphasia. This less obvious aphasia corresponds to the patient's prefrontal speech disorders (Luria's dynamic aphasia), which may result from an extensive, far-reaching lesion to either hemisphere. The patient's speech is superficial and with very little complex structure, with hardly any subordinate sentences. However, the underlying syntactic disorder of prefrontal syndrome does not occur exclusively in language. After careful analysis of behavior as well as language in frontal lobe patients, Luria (1966) concluded that the disorder is in the ability to temporally organize behavior. Such a role of the prefrontal cortex in the temporal organization of behavior has been documented to a large extent, not only in the human primate, but also in the non-human.

This organization seems to suggest that semantic organization of content would be broadly distributed across the brain, while the phonemic, morphological, and syntactic organization of language would be more related to the canonical brain linguistic circuits (Fig. 16.2c), and interrelated areas. The canonical language circuit would include a more complex representation than the image in Fig. 16.2c, and would include Primary auditory, Frontal cortex, Superior Temporal gyrus and sulcus, and complex connections including those to Primary Motor Cortex, (for a complete review see Friederici 2012). However, it must be indicated that there is discussion if the impairments of language (aphasias) or of specific content (agnosias) are related to a difficulty in access to representations (processes), or a damage in the neural representations themselves (contents) (Mirman and Britt 2013). That would be for instance the case of the so-called Semantic access agnosia, in the classification of Humphreys and Riddoch (1984) in which the subject does not identify the visually presented objects, or in the anomic aphasia in which the subjects recognize the object but is unable to name it.

The existence of brain representation of objects in humans has been clearly demonstrated through decoding the brain signature for sensory presented visual stimuli. In these experiments, a subject is confronted with certain sensory objects (typically visual objects), and a classifier learns to decode above chance from the neural signature of that particular object, its presence or absence on a stream of presented objects. The more straightforward interpretation of this kind of experiments is that the neural assembly representing a given object is spatially and/or dynamically segregated from other neural assemblies representing other objects, in a total or partial manner (Tacchetti et al. 2018).

16.3 Neural Networks

The basic functioning of these networks is based on units that function similarly to a neuron (they are elements that transmit information when sufficient activity is achieved) and are interconnected in multiple ways with other elements. The basic equation that governs the functioning of these neurons is:

$$A = F\left(\sum W \times I\right) \tag{16.1}$$

The activity (A) of a unit is equal to the sum of the activity of the inputs (I) to that unit, multiplied by the synaptic weights of the connections (W). The F refers to a transformation of the argument (W × I) by an activation function, typically a sigmoid function. If a certain threshold value is reached, the neuronal unit emits an impulse that is transmitted to the units to which it is connected. As seen, this behavior would be a very rudimentary model of neuronal functioning (Kriesel 2009). To produce an output, the sum of (W × I) is the argument of a non-linear function which finally produces the neuron output (Fig. 16.3a).

The architecture of the network is formed by a set of neural units and their pattern of connections. The most typical is a three-layer structure (Fig. 16.3b). First layer is the input layer, which receives sensory input. The second, an intermediate layer, receives connections from the input layer and could be equivalent to what is known in neuroscience as associative areas. And finally, an output layer could be equal to the perceptual or the motor system.

An output is recognized as the unit most active in this output layer. So, for example, in a discrimination process, the most active unit in the output layer is the

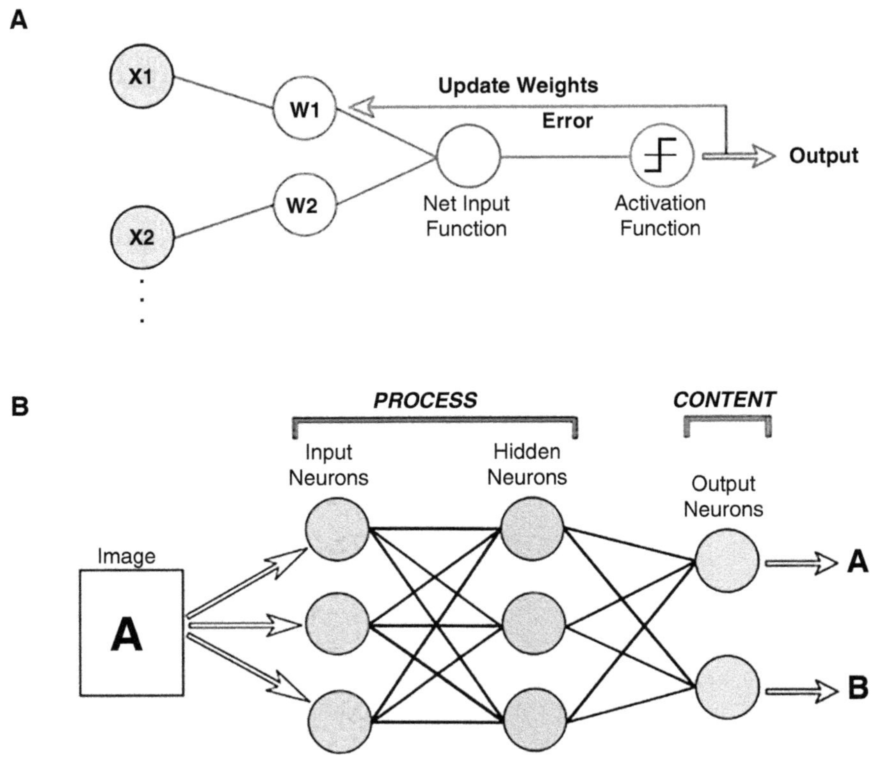

Fig. 16.3 Artificial neural networks. (**a**) Basic functioning of a neuron in an artificial neural network including a learning algorithm to update weights. (**b**) Canonical architecture of an artificial neural network

one to which we have assigned the code "a", which would mean that the response of the network is "a" (more sophisticated models would include a plausibility factor). These networks can learn to recognize sensory patterns, for example (distinguishing "a" from "b"). This learning occurs by modifying the weights, initialized with random values, of the connections between neurons. The modification of these synaptic weights (or strength of the connection between two units) occurs through physiological learning rules. One such algorithm is based on Hebb's rule: "The synaptic strength (W) of a connection is modified (dW/dt) as a function of the product of presynaptic (As) and postsynaptic (Ar) activity":

$$dW/dt = As \times Ar \tag{16.2}$$

Other algorithms are based on non-physiological properties, although they have shown great power to produce learning in these networks, such as the backpropagation error algorithm. This algorithm modifies the synaptic weights of the connections as a function of the magnitude of the error between the network output and the correct response. Although these neural network simulation models are very simplified models of brain functioning, there are aspects they may share with how the human brain processes information. Thus, the information associated with these networks is determined by a pattern of connections. This would explain why a lesion in both: artificial and real brain networks only degrade knowledge and does not affect it completely (Joyce et al. 2013). Knowledge would thus be distributed over large portions of the network and would only be partly altered by the lesion.

The output of an artificial network would correspond to what we have called content. While the activity, both the input layer, the hidden layers, and all the connections, would correspond to what we have called process. Of particular interest is the fact that certain neural network architectures incorporate an element of activity biasing, which in some ways could be equated with the cognitive process of attention. Figure 16.3b exemplifies the network architecture and points out the elements that would correspond most directly to the concept of process and the concept of content.

This structure, as shown in the Fig. 16.3b, presents the essence of the present approach: the activity in the output layer corresponds to the content, it means the perception and/or the action of the subject, while the process which permits the transformative nexus between inputs and outputs are the preceding layers and connections, including the input layer. In this model approach, the response at the output layer does not necessarily means an overt response, but it can be the perception by the subject of a sensory scene or an internal state. In fact, as the brain and the anatomy and physiology of the body have their own endogenous dynamics the output layer can be activated by internal dynamics independent, to a certain extent, from the external inputs.

These neural networks possess the advantage of possessing a simple logic and have also been able to learn many hitherto human-specific tasks. For example, the creation of participles from the infinitive of a verb in the grammar rules. However, from a straightforward neurobiological perspective, they have the disadvantage that they simulate the behavior of a neuron in an excessively simplistic way. Think, for example in the large number of ion channels, neurotransmitters, neuromodulators, etc., that exist in the nervous system. For this reason, realistic models of neuronal functioning are now being developed by including an increasing number of properties of neurons and their circuits. However, the advantage of using this more simplified models, as this described in the chapter, is the ability of these models to replicate human functions with relatively simple architecture and algorithms, as for instance in modern artificial intelligence devices as chatGPT4.

From the perspective of the present chapter, the simple architecture and operational functioning of artificial neural networks permit complex processing that only determines content until some of the representational neurons in the output layer are activated. Once activated, the process continues through the learning algorithm, supervised or unsupervised, which permits to obtain the desired output, as a function of the received input and of the generated error. However, it must be indicated that activity in the hidden and input layers would also be considered internal content, but only partially related to the final output, which is registered in the so-called output layer.

One interesting example to show the possible relationship of artificial neural networks with neural processing and representation in the brain is the case of the alternate perception of ambiguous figures (Fig. 16.4a, b), or also the alternate perception of incompatible (not binocularly integrated) images presented independently in each eye, in the so-called eye rivalry phenomenon (Fig. 16.4c). In this kind of experiments, the two incompatible images alternate, with perceptual duration histograms which have been explained by gamma distributions (Logothetis 1999), or by a competition model (Gómez et al. 1995). In the competition model, two networks are competing (Fig. 16.5). The network which would have the higher activity would access the perceptual level (Gómez et al. 1995) (Fig. 16.5a, b).

The idea of two mutually excluding percepts can be studied by the recording of single neurons in monkeys, which shows that in the temporal areas the presence of alternating neural activity of the networks representing each one of the alternative percepts during eye rivalry occurs, while in more basic areas as the primary visual cortex, the representation of both possible percepts occurs simultaneously,

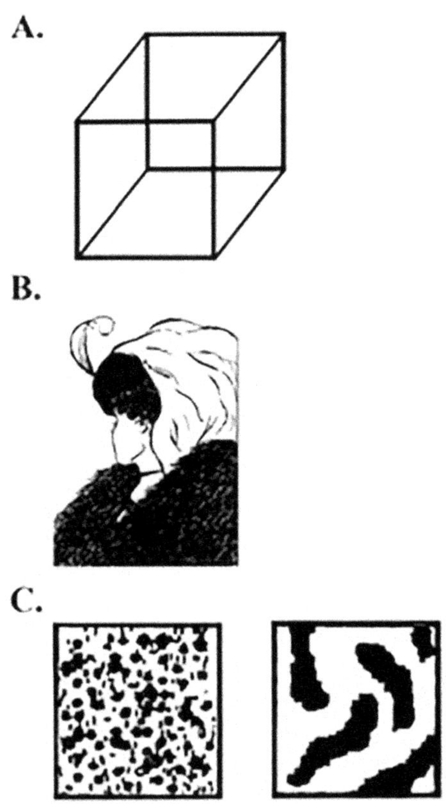

Fig. 16.4 Incompatible percepts. (**a**) Ambiguous figure of Necker cube. (**b**) Ambiguous figure of the old-young lady. (**c**) Image presented to each to produce eye rivalry

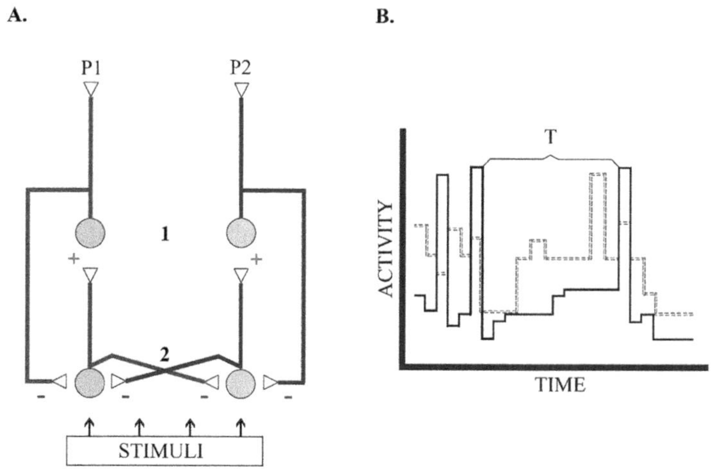

Fig. 16.5 Competition model for perception of ambiguous figures and eye rivalry. (**a**) The essence of the model consists in two independent networks which are competing (mutually inhibiting each other) to emerge in the perceptual field (P1 and P2). (**b**). In the dynamic of the 4A network only the network presenting the higher activity would reach the perceptual threshold (Gómez et al. 1995; Gigante et al. 2009). T is the time in which the percept represented by the network activity labelled by dotted line would be perceptually perceived

suggesting that it is only in the late visual processing pathway where the competition between neural networks occurs to access the perceptual field (Logothetis 1999).

The theoretical and experimental development of the perception of ambiguous figures and eye rivalry clearly suggest that a lot of unconscious processing occurs until one neural assembly representing a given percept reaches the conscious representational level. One last interesting point about these types of experiments is that the probability of a given percept remaining perceptually available (conscious) can be increased by attentionally focusing on its distinctive features (Gómez et al. 1995). The previous paragraphs suggest that by becoming intentionally attentive to a certain neural representation it is possible to boost the neural activity associated, and then possibly modify not only neurons activity but also neural connectivity (here, we refer to connectivity as the ability to communicate between neurons trough axons and synaptic connections).

16.4 Brain Processing Information as a Dynamic System

The different neurons organized in neural interconnected assemblies would be processing certain quantities (typically neurotransmitters and electrical activity), causally related to external behaviors. The dynamic of these neural assemblies would be modelled by systems of differential or discrete difference equations in which the variables in one of the equations would correspond to the values of other equations, therefore creating some sort of dynamical loop of intertwined biological variables related to behavioral and cognitive activities. Typically, this type of modelling can be computed with Simulink which includes different types of operators. This kind of modelling has been extensively used in physiology, for instance in the modelling of peripheral and central processing of the vestibulo-ocular reflex. The example of the vestibulo-ocular reflex is paradigmatic.

The vestibular system is the sensory system that will inform us of the position, velocity, and acceleration of the head (it is a system like the stabilizing devices of airplanes). Generally, and as in the proprioceptive system, we are not aware of these data which are nevertheless fundamental in the coordination of movements. Thus, the function of the vestibular organ is to provide information about the movement and orientation of the head, which is mainly used in the regulation of motor activity at a subcortical level.

The organ in charge of these functions is the so-called vestibular organ, which is excavated in the temporal bone next to the cochlea. There are two distinct portions, one consisting of the three semicircular canals orthogonally positioned with respect to each other and filled with endolymph, and two membranous sacs called utricle and saccule, which are connected to each other and filled with perilymph. The three semicircular canals are named according to their position as anterior, posterior, and lateral. The lateral one is the one that is in the horizontal plane (forming 30° with the horizontal), the others, as we mentioned before, are located perpendicular, which is of extraordinary interest for their function as we will see later on. Each canal at its junction with the utricle presents a thickening called ampulla, in this area there is the so-called crista ampullaris containing vestibular hair cells that will specialize in transduction. This structure is covered by a gelatinous mass called the cupula. The ciliary cells are in the crest. They have a structure very similar to the hair cells of the organ of Corti, thus they present a series of cilia among which the so-called kinocilium stands out, these cilia are embedded inside the cupula. In their basal portion they receive vestibular afferent fibers of the VIII cranial pair (vestibular branch, with the somas in the vestibular ganglion), between the ciliary cell and the afferent fiber that arrives there is a synapse. They also receive efferent fibers that can control the level of excitability of the ciliary neurons (For a review see Robinson 2022).

When angular acceleration of the head occurs in any of the three spatial directions, there is an inertial reaction in the endolymph content with respect to the semicircular canals, that is, when the head moves the canals that are attached to the bone move with it, while the lymph takes some time to accommodate to the new situation (something similar to what happens with us in the braking of a bus), This causes the liquid to move and with it the crest is pushed and with it the cilia that are embedded in it. This somehow determines a change in the permeability of the ciliary cell producing in it a depolarization if the curvature of the cilia is towards the utricle, and hyperpolarizing if it is in the opposite direction. If there is depolarization there is an increase in neurotransmitter release in the basal zone of the ciliary neuron and consequently an increase in the discharge rate of the first order sensory neuron (afferent neuron), and vice versa if it is hyperpolarization.

Note that the specific stimulus is the angular acceleration of the head, which is the only one that can produce an inertial reaction, in fact, the signal of angular acceleration of the head is mechanically integrated into the semicircular canals due to the high viscosity of the medium, so the afferent fiber discharge is proportional to the angular velocity of the head and not to the angular acceleration. Furthermore, there is a decomposition of the head acceleration vector into its three spatial components due to the arrangement of the channels in space (so for example the lateral or horizontal channel is only sensitive to the horizontal component of the head acceleration). Something similar occurs in the utricle and saccule for the linear acceleration of the head.

The fibers of the VIII pair connect to the so-called vestibular nuclei (some fibers go directly to the cerebellum). There are four vestibular nuclei, and finally they connect with the motor nuclei of the extraocular musculature. With all this process, the vestibulo-ocular reflex is achieved. Thanks to the extraocular muscles the eyes can move in the three planes of space. The oculomotor nuclei receive information from these vestibular nuclei which in turn receive it directly from the semicircular canals. If a movement occurs in any plane of space, the eyes correct through the described neural pathway the movement of the head with a movement of the same amplitude, but in the opposite direction, thus maintaining the stability of gaze, which would permit to receive visual inputs in a stabilized manner, facilitating the analysis of visual content through visual processing of a stabilized visual scene.

From the point of view of information processing in the vestibulo-ocular reflex, it is interesting to highlight two aspects, the first is about the usefulness of arranging the semicircular canals in orthogonal planes, this orthogonality allows to decompose the acceleration vector of the head in its three-dimensional components, this decomposition allows the muscles whose lines of action of force are in a very defined plane to be governed effectively. Thus, for example, the horizontal component of the movement will reach exclusively the muscles that govern movement in the horizontal plane, the same for vertical. The second aspect is the fact that in order to produce eye movement compensation and keep the visual scene stable, a change of eye position must be effected, and yet the effective stimulus is the acceleration of the head which is the second derivative of the position, so a double mathematical integration of the stimulus (head acceleration) must be performed to obtain an optimal output (eye position).

This operation is performed in several phases, a first one is mechanical and occurs because of the properties of the fluid inside the semicircular canals (Wilson and Peterson 1978), a part of the second one is performed for reasons of the same nature in the extraocular mechanics (Skavenski and Robinson 1973) and the remaining portion of integration is performed neurally (Baker et al. 1981), in a circuit that presumably includes a synaptic recurrence. A recurrent axon collateral returning on the same neuron would allow it to dispose of the previous information through the collateral and the new one through the normal synaptic inputs, performing the integration operation (Fig. 16.6).

The flow chart of this process is shown in the Fig. 16.6. So, this whole semicircular canals and neuromuscular network can be broken down from the point of view

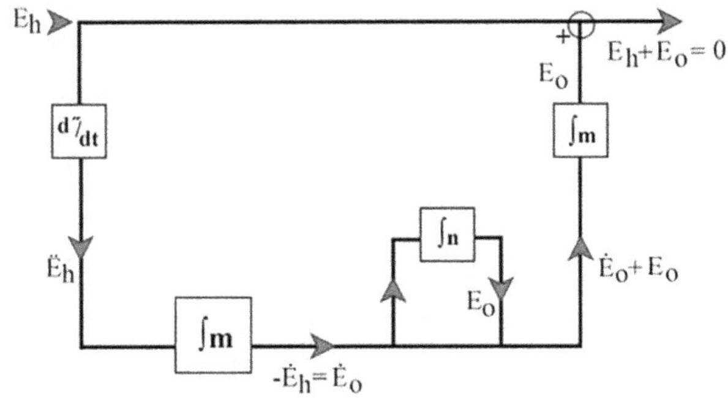

Fig. 16.6 Diagram of information processing during the vestibulo-ocular reflex, referring to the type of mathematical operations performed in such processing. Eh, head position; Ė h: head velocity; Ë h: head acceleration; Eo: ocular position; Ė o: ocular velocity. Eh + E0 = 0 refers to the compensation performed by the vestibulo-ocular reflex, through changes in the ocular angular position equal to changes in the angular position of the head. The letter m refers to the fact that the integration (∫) is performed by mechanical procedures, and n by neural procedures

of information processing into a double integration, in a mathematical sense, of the head acceleration signal through mechanical integration in the semicircular canals, the extraocular musculature and the eyeball, and partly through the recurrent connections in the brainstem.

The model for the vestibulo-ocular reflex described previously is a simplistic view of the multiple interactions that in the relationship between eye-head movements exists. For instance, imagine if in watching a tennis game our gaze would remain static at the same point of the visual scene, we would lose the essence of the match (Konrad et al. 1999). In this case, vestibulo-ocular reflex is suppressed by the pursuit of the ball. Other systems which can interact with the vestibulo-ocular system are the attention to imagined targets, and auditory and somatosensory targets (Jacobson et al. 2012). Also, the system is plastic, and increasing or decreasing the retinal input in response to eye movements by googles, the vestibulo-ocular change the gain of the reflex, by continuously recalibrating the vestibule-ocular reflex through feedback (Ito 1982). Those results are typical of dynamical systems, the continuous interaction between multiple represented variables, and the change of long-term memory as a consequence of changes in the interacting variables.

But how do these dynamic processes relate to the relationship contained in this chapter on content and process? In the case of the vestibulo-ocular reflex, permitting gaze stabilization during head movements and then permitting the processing of the stabilized visual scene to obtain meaningful content. But also, the neural coding of the instantaneous eye position during vestibular stimulation would represent the unconscious content during processing of vestibular stimulation. However, this is a relatively simple system, very far from the huge number of variables participating in the production of cognitive actions, as working memory or decision processes.

When going to more complex processes as working memory, a dynamical loop of recurrences between neural structures has been proposed to maintain active in memory a certain item content (Fuster 2022). The dynamical and structural model for working memory would be then the persistent activation of modality-specific brain areas, maintained by reverberating activation with prefrontal neurons.

In general, it has been proposed that for short-term memory phenomena, labile electrophysiological changes must occur in neurons. To demonstrate electrophysiological changes directly linked to short-term memory, specific tests such as those developed by Fuster and Jervey (1981) must be designed. These researchers performed an electrophysiological recording in the inferotemporal lobe during a task consisting of pressing a lever when the monkey observed a light of a color similar to the one that had appeared some time before. If so, it would receive a piece of food. They found that some neurons were selective to color and their response rate was high even in the delay interval, suggesting that the neuron is participating in the recall of a newly perceived stimulus. Fuster (1995) slightly modified this task by placing a gray symbol in the middle of the cue stimulus (the stimulus to be remembered) such that the shape of the symbol indicated whether or not the response was to be based on the color of the disk. They were able to observe a group of neurons responding to a given color, even during the delay interval, but only when the gray symbol indicated that the color was appropriate and should be remembered. These results support the idea that visual memories are held in the inferotemporal cortex in a short-term memory store. A similar experiment using tactile and auditory stimuli demonstrated the same phenomenon of persistence of the short-term memory trace in the somatosensory cortex neurons.

Similarly, prefrontal cortex neurons have been found to maintain high activity in a visual target localization memory task. In this experiment, a monkey is first trained to fixate its gaze on a point located in the center of a television screen. A visual stimulus, a small square, is then briefly displayed at one of eight possible locations on the screen, and then extinguished. After a delay of 3–6 s, the central light, or fixation point, is extinguished, prompting the animal to direct its eyes to the location where it perceived the stimulus before the delay.

Certain neurons in the prefrontal cortex have been found to possess what has been called "memory fields": when the object of attention disappears from view, prefrontal neurons become active, producing electrical signals at more than twice the basal rate. This neuron remains active until the end of the delay period, when the animal emits its response. It is always a specific group of neurons neuron that encode the same visual location. Neurons capable of storing its location "in the mind" once it has disappeared, are presented organized together in a specific area of the prefrontal cortex; such neurons constitute the nucleus of the spatial system of working memory. If the activity of them falters during the delay period, the animal is likely to make a mistake (Goldman-Rakic 1992).

The activation of prefrontal neurons during the delay period of a delayed response task depends neither on the presence of an external stimulus nor on the execution of a response. Instead, the neural activity corresponds to a mental episode intervening between the stimulus and the response. Monkeys with a lesion in prefrontal cortex have no difficulty in moving their eyes toward a visible stimulus or extending their

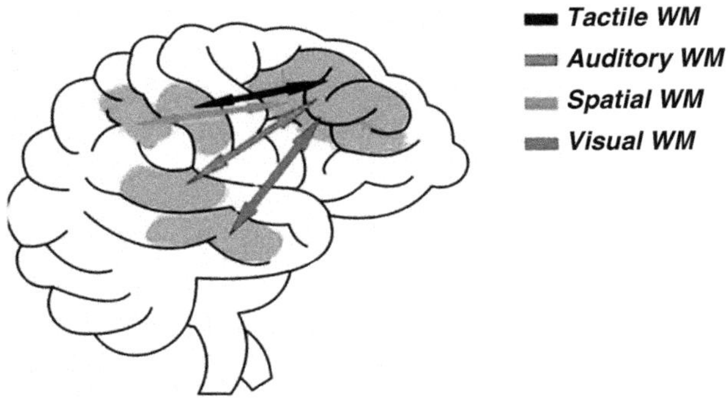

Fig. 16.7 Working memory (WM) mechanism. In the model propose by Fuster (2022), there is a reverberation between prefrontal cortex and different cortices representing in long-term memory the maintained representation. This reverberating activity keep the memory trace active by the interaction of the Prefrontal cortex with the Postcentral Gyrus (Tactile WM), Parietal lobe (spatial WM), Superior Temporal Gyrus (auditory WM), and Infero-temporal cortex (WM of visual objects)

hand toward a desired object, but they cannot channel these motor responses by recalling stimuli and objects that have disappeared from their view.

The mechanism of memory trace maintenance could be reverberation, as demonstrated by prefrontal cortex cooling experiments (Fuster 1997), where cooling of the prefrontal cortex during a delayed response task produces a decay of activity in the inferotemporal cortex (Fig. 16.7). These results can be interpreted as showing that the mechanism of memory trace maintenance depends on the reciprocal interaction between the inferotemporal cortex and the prefrontal cortex.

All these results together demonstrate that there are cortical neurons that keep a short-term memory trace, but that there is no specific place for this process; rather, depending on the task, the maintenance of activity occurs in some neurons or others. According to Fuster (1997, 2022), working memory would be nothing more than the activation of the representations of the elements necessary to solve the task in long-term memory, which would be distributed throughout the cortex. The reverberation process between the dorsolateral prefrontal cortex and the cerebral cortex representing the contents would constitute the basic process. In this sense, working memory is understood as a dynamic process, in which a trajectory of activity between these structures is constituted. This loop does not necessarily have to be with constant activity in prefrontal neurons but could include more complex mechanisms of synaptic facilitation (process) of the representations, or that instead of constant activity in neurons, short bursts of activity in different neurons would cover the whole retention period in working memory. An important point here is that contents in working memory can be optionally transferred to long-term memory, as a function of the repetitions and/or relevance of the sustained contents.

One interesting application of dynamic system theory is relative to WM development, through an alternative explanation to the classic A not B result for the object permanence (maintenance phase of working memory in modern terminology) obtained by Piaget (1954). Between 10 and 12 months of age, to reach a hidden object, infants rely on the history of previous location where an object has been hidden, although hidden in a new place. At 1 year of age, they would expect the object to reappear where it has been hidden. The common explanation has been the acquisition of a novel cognitive ability (object permanence, or development of attention, inhibition and/or memory). However, Thelen (Smith and Thelen 2003; Perone and Simmering 2017, for review) has demonstrated that not only the development of the latter cognitive abilities, but also motor planning, posture, and features of the task space as the salience of the stimuli or the delay between hiding and opportunity of reaching are important for the trial outcome (Smith and Thelen 2003), and then are related to the success or failure of the task. Translating this data to the working memory implies that for a possible working memory success in the infant, not a simplistic model of reverberation between structures or synaptic facilitation can explain it, but a multicausal process in complex interaction loops in real time inside the brain. It is time to remember here, that a simpler dynamical system as the vestibulo-ocular reflex can also be modulated by attention to imagined targets, and auditory and somatosensory targets (Jacobson et al. 2012).

Dynamical systems when applied to the brain implies that during stimulus processing, cognitive activity, mental states, emotional states or whatever content is being coded in the brain there is a constant loop of operations in a mathematical sense (process) between brain areas with feedforward and feedback connections, that can be interfered by external or internal energies (therapies, medical treatments, traumatic experiences, attention, etc.) and give place to consolidation of long term memories which can be adaptive or maladaptive. In the simpler case, the vestibulo-ocular change, the gain of the reflex changes by wearing googles changing the gaze angle with respect to eye movements (Ito 1982), or a non-treated strabismus can become in the predominance of one of the eyes for vision over the other. Another important consequence of the brain as a dynamical system is that during learning, during the process of learning, adaptive or maladaptive, the trajectory followed visit also states (cognitive, emotional, or behavioral) already visited before the learning process, which would explain for instance the presence of relapses during therapy.

16.5 The Bayesian Brain

Our brain, although endowed with a unique and incalculable potential, does not have sufficient resources to perceive, evaluate, and respond to all the constantly changing and interacting stimuli. One of the main mechanisms proposed for sensory filtering is predictive coding. This psychophysiological function could be described as a system that automatically and continuously evaluates the characteristics of perceived stimuli and the relationships established between them to confirm or modify a prediction about the occurrence of the next event (Friston 2010). The impairment of this predictive processing has been widely linked to the development of several pathologies, such as autism spectrum disorder (ASD). Attention deficit hyperactivity disorder (ADHD), schizophrenia, bipolar I, II, major depressive disorders, and other neurological disorders, or even aging (Gonzalez-Gadea et al. 2015; Sterzer et al. 2019). For instance, the Bayesian brain hypothesis would support the theory of the "magical" world of Sinha et al. (2014) for ASD, in which the alteration of the predictive capacity in ASD is proposed as a common variable underlying its varied symptomatology. What would cause in these subjects the feeling of living in a reality where events happen unexpectedly and with no known cause, generating in them a feeling of confusion and helplessness towards their environment that would limit their ability to interact with it.

Friston (2010), in his free energy theory, also proposes a compatible theoretical model to explain the basis of brain functioning, but based on Bayesian or probabilistic information processing mechanisms, which would operate continuously in our brain to try to minimize the amount of surprise, entropy or free energy available after our sensory exchanges with the environment. The Bayesian approach is related from a clinical perspective with the theory of personal constructs of Kelly (1955). Which proposes that in some sense the subjects acquire its model of the world as a scientist, by anticipating the next events, reinforcing its model if a correct prediction occurs, and modifying them in incorrect predictions. The psychotherapist should facilitate the process of finding the constructs more accurate to define the client reality.

The Bayesian Brain Hypothesis proposes that the probability of occurrence of a future event (S2) depends on confidence in its relation to a previous stimulus (S1). Thus, the probability associated with this relationship (P(S2/S1)) is updated with each confirmation or rejection of the expectation (Mathys et al. 2014). According to this paradigm, predictive coding serves to minimize the differences between the internal representations of the environment and the properties of the newly perceived stimuli, in a loop in which the prediction is constantly updated to reduce this discrepancy or prediction error (Friston 2010). Organisms would maintain an "active inference" of the environment that would allow them to build "generative models" of the relationships between stimuli in the environment and, if necessary, influence it through behavior, to control variables that allow them to reduce the probability of surprise (Friston 2010; Pezzulo et al. 2018). The Bayesian brain hypothesis has been widely applied in its role of guiding perception and behavior, using many components for this purpose, such as N1 (Näätänen and Picton 1987), Mismatch Negativity (MMN) (Näätänen et al. 2007), Stimulus Preceding Negativity (SPN) (Bennett et al. 2015), or Contingent Negative Variation (CNV) (Gómez et al. 2019), as neurobiological markers. The psychophysiological responses described in the previous paragraph corresponds to Event-related Potentials, which are elicited by stimuli that need to be processed.

From the perspective of the present chapter, predictive coding implies that subjects are in every moment doing a particular kind of process: predictive inferences, in order to predict which are the most probable scene to which they are confronted, in sensory perception, or to predict which are the most probable next future stimuli. In both cases, the process of inference is trying to activate the correct representation of the sensory object being presented or to activate in an anticipatory manner the possible next stimulus. For instance, in the experimental paradigm shown in the Fig. 16.8a, sequences of tones of progressive frequencies are presented in a

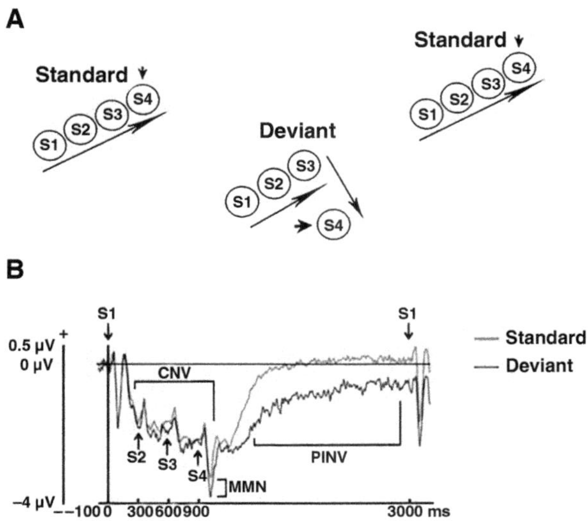

Fig. 16.8 (**a**) Predictable (standard) and unpredictable (deviants) sequences of auditory stimuli. (**b**) Event-Related Potentials elicited by a train of stimuli which are predictable (Standard) or unpredictable (deviant). S1–S4 represents the stimuli sequence
CNV Contingent Negative Variation, *MMN* Mismatch Negativity, *PINV* Postimperative Negativity

predictive (standard) or finishing in an unpredictable manner (deviant). The deviants would produce different brain activation than those produced by the standard or valid sequences of stimuli (Fig. 16.8b).

One component of event-related potentials (ERPs) that has been associated with the problem of predictive coding is mismatch negativity (MMN). The MMN is an ERP that is generated between 100–250 ms after stimulus presentation and appears as a negativity in the frontocentral scalp area (Näätänen et al. 2007). This component is induced by an oddball paradigm; in the case of its auditory modality, it is generated by the unexpected appearance of an auditory stimulus that is different (deviant) from the one expected by the subject (standard) and previously stored in his echoic memory after the repeated presentation. This standard can be a specific sound, a sequence, or a rule relating them (Escera and Corral 2007; Näätänen et al. 2007; Winkler 2007). The deviant can be produced by any difference in the physical properties that compose the delivered standards (amplitude, frequency, duration, etc.) or by any modification of the pattern that relates them, such as the time interval between stimuli, the increase or decrease of the expected frequency or amplitude for a stimulus based on a previously established rule, the absence of an expected stimulus or sound sequence. The presentation of the deviant stimulus elicits a negative N1 response of higher amplitude and latency similar to the standard, whose difference (product of subtracting the voltage obtained in the standard from that of the deviant) would result in the negative component that we identify as the MMN. This ERP is considered to be a preattentional indicator for the detection of environmental perturbations, depending on their degree of novelty (Winkler 2007; Näätänen et al. 2012). Regarding its localization, supratemporal areas would play a role as an initial mechanism of detection and preattentional representation of the patterns implicit in the perceived stimulus. This region, together with the medial and inferior parts of the frontal cortex, has also been linked to the subsequent formation of the frontocentral negativity that characterizes this component and whose purpose would be to modulate involuntary attention according to the perceived changes in the sounds of the environment (Escera et al. 2000; Näätänen et al. 2007).

The so-called Regularity Violation Hypothesis (Winkler 2007) for MMN generation would fit within the framework of predictive coding. According to this theory, the MMN would be the product of the prediction error generated when the internal representation or expectation about the next stimulus is violated. This theory is justified by the fact that MMN occurs not only when the physical properties of the expected stimuli are violated, but also when the abstract relations established between them are contradicted. The possibility that MMN would be the product of habituation of the standard has been discarded, supporting the Regularity Violation Hypothesis (Korzyukov et al. 2003; Paavilainen et al. 2018; Winkler 2007). Also referring to the theory of predictive coding, Wacongne et al. (2012) proposed a neurophysiological model for the generation of the MMN, based on the interrelation of four neural layers, each with a specific role. According to this model, there would be a thalamic layer representing the standard and deviant features, and three other cortical layers responsible for processing the prediction error, the generative internal model, and the stimulus information stored in memory, respectively. Thus, the prediction error would be generated in a specific layer by integrating the activity produced by both the thalamic layer and the layer responsible for generating predictions, which in turn would receive inputs from the layer responsible for storing the recently perceived stimuli in the memory trace. In this model, the thalamic layer would send excitatory signals to the prediction error layer, while the prediction layer would send inhibitory signals. Thus, the predictive error generated by the MMN would be the consequence of those instances in which the tonic inhibition of the predictive layer cannot cancel the phasic excitatory input of the thalamic layer. The authors also propose that when prediction is based on more abstract rules, the model would require the participation of a new hierarchically higher layer of neurons sensitive to the complex characteristics of the stimulation pattern, which would function by providing feedback to the prediction error layer.

In a recent paper, Ruiz-Martínez et al. (2022) (Fig. 16.8b) has demonstrated that not only the representation of the expected stimuli is detected by presenting deviations (MMN), but also that there is a mechanism for predicting (active inference as process) the auditory representation of expected stimuli. The mechanism is indexed by the so-called Contingent-Negative Variation (CNV), confirming previous empirical demonstrations that the CNV is a neurophysiological signature of active inference of the next expected stimuli (Gómez and Flores 2011). Interestingly, the predicted representation can be activated via attentional demands, confirming the role previously indicated of the attentional process to activate defined brain representations. Out of the scope of the present chapter is the fact that to close the cognitive cycle, once the stimuli are presented the so-called post-imperative contingent negative variation would act for changing the conditional probabilities between cues and targets (P(S2/S1)) (Ruiz-Martínez et al. 2022) (Fig. 16.8b).

From the point of view of present chapter, the perception of a defined and recognizable object (content) would not occur until the error prediction is eliminated (process). It is feasible to think that when one stimulus arrives one or several representations of possible candidates are activated and when one of the candidates reaches a near zero-prediction error with the input signal, a clear percept is achieved.

16.6 Compositional Model of the World

This approach pretends that in the brain coexists entities and roles, basically semantics and syntaxis, and the same entity can be bound to different roles, permitting the existence of a high number of compound entities-roles (Kurth-Nelson et al. 2023). A second order of complexity permits to create combinations of entities-roles compounds in long sequences in a quasi-infinite number of compound sequences. The proposal considers the hippocampus as a structure able to create the compound entity-role. More specifically, the so-called object-vector cells would specify the role of the distance of the animal with respect to any object, while the landmark cells would specify the distance of the animal to a particular object (Høydal et al. 2019). The latter type of cells would have obtained a bond between the spatial role and a defined object (entity representation). At the first level of complexity, the roles are defined in Euclidean space (location of the animal), but progressively a disentangling from the notion of space permits that the roles become more abstract concepts as "starting point" or "action verbs". At this point, is not very difficult to find a certain resemblance to the generative grammar of Chomsky. Also, the authors propose that during neural replay, the neural mechanism by which recombination of entities-roles would occur (Pfeiffer and Foster 2013), roles as "If", "else", "and", "or" would be executed, permitting the composition of small programs or algorithms, and then be the mechanism by which symbolic language, as this expressed in the Lot Theory, would be implemented in the brain.

The hippocampus would be an ideal place for creating these object entities (content) with a spatial role (process), given that object information is received through the "what" stream, arriving from the lateral entorhinal cortex, and the medial entorhinal cortex would on the spatial arrangement ("where") (Manns and Eichenbaum 2006) (Fig. 16.9). This where-what entities would be then formed in the hippocampus. More complex spatially oriented information would permit more complex roles with the same logic, as the social network topology or the auditory pitch.

This proposal permits to recombine the different entities and roles, or once a compound entity role has been created, to be linked with other entities-roles to create long sequences. These sequences do not need to be expressed in behavior but can be replayed and recombined without the need for overt behavior permitting the quasi-infinite possibilities of imagination and thinking.

There are different empirical data, both in animals and humans which show that a neural activity similar to that of the neural activity generated during experience can be replayed, but also recombined permitting the combination and recombination of entities-roles compounds (Kurth-Nelson et al. 2016). Physiologically, a gamma

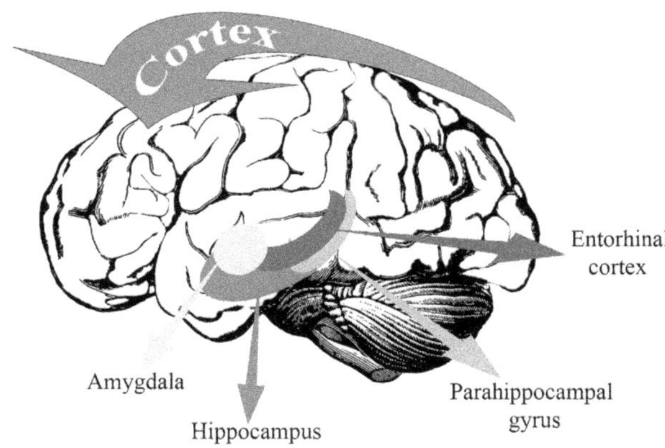

Fig. 16.9 Hippocampus, Entorhinal cortex, Parahippocampal gyrus and Cortex locations

burst would represent one entity-role of the sequence (Pfeiffer and Foster 2015), and they would be organized as sequences of gamma burst creating these complex sequences of entities-roles.

16.7 Synopsis of Content and Process

All the different proposed models suggest that brain representations for defined objects would exist in the brain. In fact, the processing of information is what permits to obtain a certain content, expressed by perception, imagination, the perception of a certain internal state, or the prediction of possible next content. Once a content is obtained, can be chained to represent more complex syntaxes. We can propose here from the presented review of abstract models defining the brain operation that process and content should be somehow relatively segregated in the brain organization, and the content would be the outcome of the process, and once conscious, content becomes the perceived subjective reality.

Into the very abstract Lot, content is specified in the propositions by the premise (a) and the consequent(b) and the operators as the process (−>) (for instance in a −> b). The content here would be not only represented by the semantic of the antecedent and the consequent, but also by the truth of the proposition. In generative grammar, and in any grammar, a similar account can be done for morphology (words content) and syntax (process), which at the very end would produce a valid sentence with a complete meaning (content) of the sentence. The compositional model is also in the same line that Lot and generative grammar: the presence of interchangeable entities-roles, and the chaining of them. The role is more related to the process and the entity to the content. For neural networks, the output unit would be the content while the connectivity pattern and the equations governing, including learning algorithms, would be related to the process. In dynamical systems, all the intermediate variables and mathematical operations would be considered as a process while the output variable would be the content. In predictive coding, the inference process governed by prediction error would be the process, while the content would be attained at the point in which prediction error disappears.

It is important to remark that at the very end, a process bound to certain contents gives place to new content that can be used by other processes as the content. For instance, in neural networks the activation of a neuron in the hidden unit would be considered as content at this level, but for the output units would be considered part of the process of activation. Similarly, in dynamical systems intermediate variables can be part of the process for the final variable, but the content in intermediate operations.

At this point, it is important to beg for certain indulgence by the reader, given the initial comment that dichotomies are somehow simplifications of reality, but certainly permits to charge the definition of a system description or an intervention more in the process or the content. The dichotomy structure-function, frequently used to describe a biological system would be an example. For instance, in the muscular cells, the actine-miosine proteins are considered structures, but the change in their chemical composition becomes the function of muscle contraction. And, from the point of view of intervention in clinical applications, treating a problem functionally (by a medication interfering with the infection: process), or anatomically (drainage).

16.8 Clinical and Educational Implications

Now we would outline the possible implication for the dichotomy of content and process in clinical and educational settings. While for the subject, the problem arises by the content of disturbing thinking, the misunderstanding of a sentence, or obtaining a wrong solution to an arithmetical operation, the neurobiological roots of the problems should be related to an impaired processing of information.

An interesting case, in the middle of clinical and educational stings, depending on the deepness of the problem would be the Specific Language Impairment (SLI). SLI is embedded in a dimensional continuity of impairment, from children struggling with the language, particularly for complex sentences, to children unambiguously diagnosed as SLI. SLI corresponds to a category that is currently considered by the DSM-5 as a language disorder, included in a broader category, communication disorders. Diagnostic criteria for SLI include "persistent difficulties in the acquisition and use of language in any modality (i.e., spoken, written, sign language, or other) due to deficits in comprehension or production, and language skills that are substantially and measurably below age expectations". It is most often a mixed disorder in which both expressive and comprehension skills are impaired. In addition to low language test scores, SLI requires the following exclusion criteria: the child must not have hearing loss, emotional disturbance, or difficulties in cognitive abilities. However, it is generally accepted that the label applies to a very heterogeneous group of children, with cognitive functioning being a controversial area. Reilly et al. (2014) propose that the relationship between nonverbal IQ and language ability is linear and that an IQ of 85, and a cut-off of 1.25 SD for language ability would allow for establishing the threshold for a diagnosis of SLI. However, they clarify that both thresholds for SLI in both: language ability, and IQ, are highly arbitrary and correspond to a dimensional continuum.

However, although the linguistic problem is expressed at a content level as reduced variety and complexity of linguistic comprehension and/or production, the problem would be in a more basic learning process: the statistical learning of predictable stimuli sequences, rather than in strictly linguistic problems (Gabriel et al. 2013; Hsu et al. 2014). Therefore, an adequate treatment of the problem should be based not only on repetitions and making explicit explanations of the language structure but would also suppose the possibility of training in the ability to extract statistical patterns from sequentially presented stimuli (Krishnan et al. 2016). There are some preliminary results which indicate that training in a statistical learning paradigm with non-linguistic elements can in fact improve linguistic abilities in children with poor developmental language (Conway et al. 2012). In this case, is the process of statistical learning that is being intervened and not the linguistic content, how much transfer from general purpose statistical learning to linguistic abilities, and the durability of the transfer, needs to be empirically assessed in more deepness. A similar argument for educational settings should be transferred to children with difficulties in language, although not reaching a diagnostic threshold.

An interesting example of the differentiation between content and process in educational settings corresponds to the so-called semantic field theory of language, which indicates that semantic meaning of words is organized in networks of interdependent semantically related words (Trier 1931). This theory explains why learning the vocabulary of a foreign language organized by semantic fields facilitates vocabulary learning (Patahuddin and Syawal 2016), putatively due to the simultaneous establishment at neural level of the knowledge related network, while the presentation of non-related words should activate sequentially different networks doing the learning process much less efficient. The neural implementation of these neural networks of interrelated semantic contents has been already confirmed (Khachatryan et al. 2019). The interesting point for this chapter lies in the fact that by learning the vocabulary (content) of semantically related words, we would be positively interfering with the process of how semantic information is being stored, facilitating the storing of semantic content.

16.9 Neurophysiology of Content Vs Neurophysiology of Process. A Preliminary Tentative Hypothesis

One of the proposals of the present chapter is that the neural structures related to content and those related to the process would be relatively functionally and/or anatomically segregated from one to the other. This is a very general and unspecified proposal that can be substantiated by the neural network modeling approach in which the activity of the virtual neurons corresponds to the sum of inputs, and its transformation through an activation function in activity to be transmitted to the next neuron. Therefore, the activity of the neuron (content) is completely specified by the inputs and the weight of these inputs (process). Translating this to the physiology of a simplified ideal biological neuron, the content would be more linked to the postsynaptic activation while the process would depend on the pattern of synaptic activity in the presynaptic inputs weighted by the synaptic strength. The activity in cortex and hippocampus would correspond to our purpose to the content made available to the conscious states, given that those conscious states have been generally accepted to be linked, as contents, to activation of the cortex (reviewed in Aru et al. 2019). Probably, not all the postsynaptic cortical activity would become conscious, and some other specifications are needed, as levels of connectivity, frequency ranges or activity in a particular type of cortical neurons. However, the postsynaptic activity in subcortical areas would constitute content at this level, but these contents not being directly available for conscious states, similar to the activity of hidden units in artificial neural networks. Interestingly, the cortical and hippocampal dendritic trees are organized in an open-field configuration, while most subcortical areas present dendritic trees are organized in closed-field, in Lorente de No (1947). If these differential dendritic configurations are related or not to the conscious availability of neurally represented content is something to be clarified by new experiments.

The intricate connectivity through axons, whose synaptic weight is constantly updated through learning processing, would be the base for the great processing abilities of human, animal, and artificial neural networks, and should be the place where ideally, we should act to try to improve maladaptive and psychologically harmful contents. Basically, that would be the reason why pharmacology is so able to impact cognition and behavior, by modifying synaptic weight transmission. Psychology, by focusing on the processual aspects more than on the contents should be able to achieve modifications in synaptic weights. This is exemplified by the previous example on SLI and semantic field theory, and of a myriad of therapies as relaxation techniques to control distress by increasing GABA and blood flow in prefrontal areas (Namgung et al. 2021), or cognitive behavioral therapies for psychoticism by challenging the power of voices or the unreality of delusions, which have shown measurable therapeutic effects (Krakvik et al. 2013). Therefore, a therapeutically oriented theory of mind focused on processes rather than on the symptoms (contents) should be more effective to cope with maladaptive behavior, emotion, and cognition or to improve learning. However, the processes by which the brain acts to produce meaningful content are not always obvious, and in fact different mind domains would be dominated by different type of processes. Therefore, a constant need for improving theoretical models, with a strong empirical ground is needed: a theory of brain-mind processes.

References

Andreu-Sánchez, C., M.Á. Martín-Pascual, A. Gruart, and J.M. Delgado-García. 2021. The effect of media professionalization on cognitive Neurodynamics during audiovisual cuts. *Frontiers in Systems Neuroscience* 15: 598383. https://doi.org/10.3389/fnsys.2021.598383.

Ardila, A. 2013. A new classification of aphasias. In *Aphasia: Classification, management practices and prognosis*, 115–134. New York: Nova Science Publishers.

Aru, J., M. Suzuki, R. Rutiku, M.E. Larkum, and T. Bachmann. 2019. Coupling the state and contents of consciousness. *Frontier in Systems Neuroscience* 13: 43. https://doi.org/10.3389/fnsys.2019.00043.

Baker, R., C. Evinger, and R.A. McCrea. 1981. Some thoughts about the three neurons in the vestibular ocular reflex. *Annals of the New York Academy of Sciences* 374: 171–188. https://doi.org/10.1111/j.1749-6632.1981.tb30869.x.

Bennett, D., C. Murawski, and S. Bode. 2015. Single-trial event-related potential correlates of belief updating (1,2,3). *eNeuro* 2 (5): ENEURO.0076-15.2015. https://doi.org/10.1523/ENEURO.0076-15.2015.

Caplan, D. 1987. *Neurolinguistics and linguistic Aphasiology: An introduction (Cambridge studies in speech science and communication).* Cambridge: Cambridge University Press. https://doi.org/10.1017/CBO9780511620676.

Chomsky, N. 1968. Language and the mind. *Psychology Today* 1 (9): 48–68.

Chomsky, N. 1986. *Knowledge of language: Its nature, origin and use*. Praeger.

Conway, C.M., M.A. Gremp, A.D. Walk, A. Bauernschmidt, and D.B. Pisoni. 2012. Can we enhance domain-general learning abilities to improve language function? In *Statistical learning and language acquisition*, ed. P. Rebuschat and J.N. Williams. De Gruyter Mouton.

Damasio, A.R., and N. Geschwind. 1984. The neural basis of language. *Annual Review of Neuroscience* 7: 127–147. https://doi.org/10.1146/annurev.ne.07.030184.001015.

Escera, C., and M.J. Corral. 2007. Role of mismatch negativity and novelty-P3 in involuntary auditory attention. *Journal of Psychophysiology* 21 (3–4): 251–264. https://doi.org/10.1027/0269-8803.21.34.251.

Escera, C., K. Alho, E. Schröger, and I. Winkler. 2000. Involuntary attention and distractibility as evaluated with event-related brain potentials. *Audiology & Neuro-Otology* 5 (3–4): 151–166. https://doi.org/10.1159/000013877.

Feynman, R.P. 1996. *Feynman lectures on computation*. Addison-Wesley Publishing Company, Inc.

Fodor, J. 2008. *LOT 2: The language of thought revisited*. Oxford University Press.

Friederici, A.D. 2012. The cortical language circuit: From auditory perception to sentence comprehension. *Trends in Cognitive Sciences* 16 (5): 262–268. https://doi.org/10.1016/j.tics.2012.04.001.

Friston, K. 2010. The free energy principle: A unified brain theory? *Nature Review Neuroscience* 11 (2): 127–138.

Fuster, J.M. 1995. *Memory in the cerebral cortex: An empirical approach to neural networks in the human and non human primate*. Cambridge MA: MIT press.

Fuster, J.M. 1997. *Redes de memoria*, 74–83. Julio: Investigación y Ciencia.

Fuster, J.M. 2022. Cognitive networks (*Cognits*) process and maintain working memory. *Frontiers in Neural Circuits* 15: 790691. https://doi.org/10.3389/fncir.2021.790691.

Fuster, J.M., and J.P. Jervey. 1981. Inferotemporal neurons distinguish and retain behaviorally relevant features of visual stimuli. *Science* 212: 952–955.

Gabriel, A., C. Maillart, N. Stefaniak, C. Lejeune, L. Desmottes, and T. Meulemans. 2013. Procedural learning in specific language impairment: Effects of sequence complexity. *Journal of the International Neuropsychological Society* 19 (3): 264–271. https://doi.org/10.1017/S1355617712001270.

Gainotti, G., P. D'Erme, and P. Bartolomeo. 1991. Early orientation of attention toward the half space ipsilateral to the lesion in patients with unilateral brain damage. *Journal of Neurology, Neurosurgery, and Psychiatry* 54 (12): 1082–1089. https://doi.org/10.1136/jnnp.54.12.1082.

Gainotti, G., M.C. Silveri, A. Daniele, and L. Giustolisi. 1995. Neuroanatomical correlates of category-specific semantic disorders: A critical survey. *Memory* 3 (3–4): 247–264. https://doi.org/10.1080/09658219508253153.

Gigante, G., M. Mattia, J. Braun, and P. Del Giudice. 2009. Bistable perception modeled as competing stochastic integrations at two levels. *PLoS Computational Biology* 5 (7): e1000430. https://doi.org/10.1371/journal.pcbi.1000430.

Goldman-Rakic, P.S. 1992. La memoria funcional y la mente. *Investigación y Ciencia* 194: 68–75.

Gómez, C.M., and A. Flores. 2011. A neurophysiological evaluation of a cognitive cycle in humans. *Neuroscience and Biobehavioral Reviews* 35 (3): 452–461. https://doi.org/10.1016/j.neubiorev.2010.05.005.

Gómez, C., E.D. Argandoña, R.G. Solier, J.C. Angulo, and M. Vázquez. 1995. Timing and competition in networks representing ambiguous figures. *Brain and Cognition* 29 (2): 103–114. https://doi.org/10.1006/brcg.1995.1270.

Gómez, C.M., A. Arjona, F. Donnarumma, D. Maisto, E.I. Rodríguez-Martínez, and G. Pezzulo. 2019. Tracking the time course of Bayesian inference with event-related potentials: A study using the Central Cue Posner Paradigm. *Frontiers in Psychology* 19: 1424. https://doi.org/10.3389/fpsyg.2019.01424.

Gonzalez-Gadea, M.L., S. Chennu, T.A. Bekinschtein, A. Rattazzi, A. Beraudi, P. Tripicchio, B. Moyano, Y. Soffita, L. Steinberg, F. Adolfi, M. Sigman, J. Marino, F. Manes, and A. Ibanez. 2015. Predictive coding in autism spectrum disorder and attention deficit hyperactivity disorder. *Journal of Neurophysiology* 114 (5): 2625–2636. https://doi.org/10.1152/jn.00543.2015.

Goodale, M.A., and A.D. Milner. 1992. Separate visual pathways for perception and action. *Trends in Neurosciences* 15 (1): 20–25. https://doi.org/10.1016/0166-2236(92)90344-8.

Goodglass, H., B. Klein, P. Carey, and K. Jones. 1966. Specific semantic word categories in aphasia. *Cortex: A Journal Devoted to the Study of the Nervous System and Behavior* 2 (1): 74–89. https://doi.org/10.1016/S0010-9452(66)80029-1.

Gracia-Molina, A., and J. Peña-Casanova. 2022. Visual agnosias and related disorders: Beyond Lissauer. *Neurosciences and History.* 10 (1): 32–45. https://nah.sen.es/vmfiles/vol10/NAHV10N1202232_45EN.pdf.

Høydal, Ø.A., E.R. Skytøen, S.O. Andersson, et al. 2019. Object-vector coding in the medial entorhinal cortex. *Nature* 568: 400–404. https://doi.org/10.1038/s41586-019-1077-7.

Hsu, H.J., J.B. Tomblin, and M.H. Christiansen. 2014. Impaired statistical learning of non-adjacent dependencies in adolescents with specific language impairment. *Frontiers in Psychology* 5: 175. https://doi.org/10.3389/fpsyg.2014.00175.

Humphrey, G. 1973. *Psicología del pensamiento: Teorías e investigaciones*. México: Trillas.

Humphreys, G.W., and M.J. Riddoch. 1984. Routes to object constancy: Implications from neurological impairments of object constancy. *The Quarterly Journal of Experimental Psychology. A Human Experimental Psychology* 36 (3): 385–415. https://doi.org/10.1080/14640748408402169.

Ito, M. 1982. Cerebellar control of the vestibulo-ocular reflex—Around the flocculus hypothesis. *Annual Review of Neuroscience* 5: 275–296. https://doi.org/10.1146/annurev.ne.05.030182.001423.

Jacobson, G.P., E.G. Piker, C. Do, D.L. McCaslin, and L. Hood. 2012. Suppression of the vestibulo-ocular reflex using visual and nonvisual stimuli. *American Journal of Audiology* 21 (2): 226–231. https://doi.org/10.1044/1059-0889(2012/12-0021).

Joyce, K.E., S. Hayasaka, and P.J. Laurienti. 2013. The human functional brain network demonstrates structural and dynamical resilience to targeted attack. *PLoS Computational Biology* 9 (1): e1002885. https://doi.org/10.1371/journal.pcbi.1002885.

Kelly, G.A. 1955. *The psychology of personal constructs. Vol. 1: A theory of personality*. W. W. Norton & Company.

Khachatryan, E., B. Wittevrongel, M. Fahimi Hnazaee, E. Carrette, I. Dauwe, A. Meurs, P. Boon, D. van Roost, and M.M. Van Hulle. 2019. Semantic and perceptual priming activate partially overlapping brain networks as revealed by direct cortical recordings in humans. *NeuroImage* 203: 116204. https://doi.org/10.1016/j.neuroimage.2019.116204.

Kirshner, H.S. 2012. Aphasia. In *Encyclopedia of human behavior*, ed. V.S. Ramachandran, 2nd ed., 177–186. Academic. https://doi.org/10.1016/B978-0-12-375000-6.00029-X.

Konrad, H.R., M. Girardi, and R. Helfert. 1999. Balance and aging. *The Laryngoscope* 109 (9): 1454–1460. https://doi.org/10.1097/00005537-199909000-00019.

Korzyukov, O.A., I. Winkler, V.I. Gumenyuk, and K. Alho. 2003. Processing abstract auditory features in the human auditory cortex. *NeuroImage* 20 (4): 2245–2258. https://doi.org/10.1016/j.neuroimage.2003.08.014.

Krakvik, B., R.W. Grawe, R. Hagen, and T.C. Stiles. 2013. Cognitive behaviour therapy for psychotic symptoms: A randomized controlled effectiveness trial. *Behavioural and Cognitive Psychotherapy* 41: 511–524.

Kriesel, D. 2009. A brief introduction to neural networks (ZETA2-EN). https://www.dkriesel.com/en/science/neural_networks

Krishnan, S., K.E. Watkins, and D.V.M. Bishop. 2016. Neurobiological basis of language learning difficulties. *Trends in Cognitive Sciences* 20 (9): 701–714. https://doi.org/10.1016/j.tics.2016.06.012.

Kurth-Nelson, Z., M. Economides, R.J. Dolan, and P. Dayan. 2016. Fast sequences of non-spatial state representations in humans. *Neuron* 91 (1): 194–204. https://doi.org/10.1016/j.neuron.2016.05.028.

Kurth-Nelson, Z., T. Behrens, G. Wayne, K. Miller, L. Luettgau, R. Dolan, Y. Liu, and P. Schwartenbeck. 2023. Replay and compositional computation. *Neuron* 111 (4): 454–469. https://doi.org/10.1016/j.neuron.2022.12.028.

Lenneberg, E.H. 1967. *Biological foundation of language*. New York: Wiley. https://doi.org/10.1002/bs.3830130610.

Logothetis, N.K. 1999. Vision: A window on consciousness. *Scientific American* 281 (5): 69–75.

Lorente de No, R. 1947. A study of nerve physiology. Studies from the Rockefeller institute for medical research. *Reprints. Rockefeller Institute for Medical Research* 132: 1–548.

Luria, A.R. 1966. *Higher cortical functions in man*. New York: Basic Books.

Mack, A., and I. Rock. 1998. *Inattentional blindness*. Cambridge MA: The MIT Press.

Manns, J.R., and H. Eichenbaum. 2006. Evolution of declarative memory. *Hippocampus* 16 (9): 795–808. https://doi.org/10.1002/hipo.20205.

Mathys, C.D., E.I. Lomakina, J. Daunizeau, S. Iglesias, K.H. Brodersen, K.J. Friston, and K.E. Stephan. 2014. Uncertainty in perception and the hierarchical Gaussian filter. *Frontiers in Human Neuroscience* 8: 825. https://doi.org/10.3389/fnhum.2014.00825.

Mirman, D., and A.E. Britt. 2013. What we talk about when we talk about access deficits. Philosophical transactions of the Royal Society of London. *Series B, Biological Sciences* 369 (1634): 20120388. https://doi.org/10.1098/rstb.2012.0388.

Näätänen, R., and T.W. Picton. 1987. The N1 wave of the human electric and magnetic response to sound: A review and an analysis of the component structure. *Psychophysiology* 24: 375–425.

Näätänen, R., P. Paavilainen, T. Rinne, and K. Alho. 2007. The mismatch negativity (MMN) in basic research of central auditory processing: A review. *Clinical Neurophysiology* 118: 2544–2590. https://doi.org/10.1016/j.clinph.2007.04.026.

Näätänen, R., T. Kujala, C. Escera, T. Baldeweg, K. Kreegipuu, S. Carlson, and C. Ponton. 2012. The mismatch negativity (MMN) – A unique window to disturbed central auditory processing in ageing and different clinical conditions. *Clinical Neurophysiology* 123: 424–458. https://doi.org/10.1016/j.clinph.2011.09.020.

Namgung, E., J. Kim, H. Jeong, J. Ma, G. Hong, I. Kang, J. Kim, Y. Joo, R.Y. Kim, and I.K. Lyoo. 2021. Changes in prefrontal gamma-aminobutyric acid and perfusion after the computerized relaxation training in women with psychological distress: A preliminary report. *Frontiers in Psychology* 12: 569113. https://doi.org/10.3389/fpsyg.2021.569113.

Neisser, U. 1967. *Cognitive psychology*. New York: Appleton-Century-Crofts.
Paavilainen, P., C. Kaukinen, O. Koskinen, J. Kylmälä, and L. Rehn. 2018. Mismatch negativity (MMN) elicited by abstract regularity violations in two concurrent auditory streams. *Heliyon* 4 (4): e00608. https://doi.org/10.1016/j.heliyon.2018.e00608.
Patahuddin, P., and S. Syawal. 2016. The teaching of semantic field in improving Buginese learners' English vocabulary: An experimental study. In *Conference: The 4th international conference on language, society, and culture in Asian contexts*. Universitaas Negeri Malang.
Perone, S., and V.R. Simmering. 2017. Applications of dynamic systems theory to cognition and development: New Frontiers. *Advances in Child Development and Behavior* 52: 43–80. https://doi.org/10.1016/bs.acdb.2016.10.002.
Pezzulo, G., F. Rigoli, and K.J. Friston. 2018. Hierarchical active inference: A theory of motivated control. *Trends in Cognitive Sciences* 22 (4): 294–306. https://doi.org/10.1016/j.tics.2018.01.009.
Pfeiffer, B.E., and D.J. Foster. 2013. Hippocampal place-cell sequences depict future paths to remembered goals. *Nature* 497 (7447): 74–79. https://doi.org/10.1038/nature12112.
Pfeiffer, B.E. and D.J. Foster. 2015. PLACE CELLS. Autoassociative dynamics in the generation of sequences of hippocampal place cells. *Science* 349 (6244): 180–183. https://doi.org/10.1126/science.aaa9633.
Piaget, J. 1954. *The construction of reality in the child*. New York: Basic Books. https://doi.org/10.1037/11168-000.
Pylyshyn, Z. 1984. *Computation and cognition*. Cambridge MA: MIT Press.
Reilly, S., B. Tomblin, J. Law, C. McKean, F.K. Mensah, A. Morgan, S. Goldfeld, J.M. Nicholson, and M. Wake. 2014. Specific language impairment: A convenient label for whom? *International Journal of Language & Communication Disorders* 49 (4): 416–451. https://doi.org/10.1111/1460-6984.12102.
Robinson, D.A. 2022. Basic framework of the vestibulo-ocular reflex. *Progress in Brain Research* 267 (1): 131–153. https://doi.org/10.1016/bs.pbr.2021.10.006.
Ruiz-Martínez, F.J., M. Morales-Ortiz, and C.M. Gómez. 2022. Late N1 and postimperative negative variation analysis depending on the previous trial history in paradigms of increasing auditory complexity. *Journal of Neurophysiology* 127 (5): 1240–1252. https://doi.org/10.1152/jn.00313.2021.
Sinha, P., M.M. Kjelgaard, T.K. Gandhi, K. Tsourides, A.L. Cardinaux, D. Pantazis, S.P. Diamond, and R.M. Held. 2014. Autism as a disorder of prediction. *Proceedings of the National Academy of Sciences of the United States of America* 111 (42): 15220–15225. https://doi.org/10.1073/pnas.1416797111.
Skavenski, A.A., and D.A. Robinson. 1973. Role of abducens neurons in vestibuloocular reflex. *Journal of Neurophysiology* 36 (4): 724–738. https://doi.org/10.1152/jn.1973.36.4.724.
Smith, L.B., and E. Thelen. 2003. Development as a dynamic system. *Trends in Cognitive Sciences* 7 (8): 343–348. https://doi.org/10.1016/s1364-6613(03)00156-6.
Sterzer, P., M. Voss, F. Schlagenhauf, and A. Heinz. 2019. Decision-making in schizophrenia: A predictive-coding perspective. *NeuroImage* 190: 133–143. https://doi.org/10.1016/j.neuroimage.2018.05.074.
Tacchetti, A., L. Isik, and T.A. Poggio. 2018. Invariant recognition shapes neural representations of visual input. *Annual Review of Vision Science* 4: 403–422. https://doi.org/10.1146/annurev-vision-091517-034103.
Trier, J. 1931. *Der deutsche Wortschatz im Sinnbezirk des Verstandes*. Ph.De. diss. Bonn.

Wacongne, C., J.P. Changeux, and S. Dehaene. 2012. A neuronal model of predictive coding accounting for the mismatch negativity. *The Journal of Neuroscience: The Official Journal of the Society for Neuroscience* 32 (11): 3665–3678. https://doi.org/10.1523/JNEUROSCI.5003-11.2012.

Wilson, V.J., and B.W. Peterson. 1978. Peripheral and central substrates of vestibulospinal reflexes. *Physiological Reviews* 58 (1): 80–105. https://doi.org/10.1152/physrev.1978.58.1.80.

Winkler, I. 2007. Interpreting the mismatch negativity (MMN). *Journal of Psychophysiology* 21: 147–163.

Part IV
ToM from the Perspective of Language

Chapter 17
Language, Mind and Thought: A General Overview

Inmaculada Aguilar-Ponce

Abstract Nowadays, there seems to be an important focus on a relevant aspect of cognition known as Theory of Mind (ToM) that have brought many theorists and scholars to reflect on and, consequently, to make research on it from a psychological and physiological perspectives. Its abstractness and complexity have led them to inquire what is the relationship between the three most important factors for ToM which are language, mind and thought. Considering that the term ToM implies thought and the language plays a critical role in its acquisition, experts on the field of neuroscience have wondered at what point language starts to establish a connection with ToM and whether language implies the development of ToM or whether ToM can develop on its own without language. ToM has not only been studied in terms of cognition but also in terms of a pathology related to language known as aphasia. Most of the studies carried out in the last years have shown that this pathology does not necessarily affect ToM since thought does not depend on language as some theorists suggest and if that was the case, then, ToM could be improved as could language.

Keywords Language and mind · Language and thought · Brain · Cognition · Linguistic theory · ToM · Language acquisition · Language pathology · Aphasia

17.1 Introduction

In the past decades the study of the human brain has supposed to be a completely challenge for linguists and for the world of science above all. Over the years, it has been possible to better understand the complexity of the brain in terms of how it works thanks to the anatomical study carried out by experts and the use of advanced technological tools such as Functional Magnetic Resonance Imaging (fMRI), Electroencephalogram (EEG) and Magnetoencephalography (MEG) among others.

I. Aguilar-Ponce (✉)
Department of English Language, University of Seville, Seville, Spain
e-mail: inmagupon@alum.us.es

T. Lopez-Soto et al. (eds.), *The Theory of Mind Under Scrutiny*, Logic, Argumentation & Reasoning 34, https://doi.org/10.1007/978-3-031-46742-4_17

Understanding the brain much more has led us to make more complex questions, for instance, how individuals can communicate with each other, what is the relationship between language and thought or how the different emotions can influence over the individual's behavior. In order to find some answers to these questions, experts in the field of neuroscience long tried to provide a solution to the relationship between language, mind and thought. First of all, for some scholars, brain and mind are considered to be the same thing while for others there is a clear difference between these two terms. At this point, it must be clarified what it is understood by the term *mind* and how it differs from the term *brain* in order to later comprehend how the mind is related to language and thought. In this chapter, I give an overview on the history of research on language and thought, what mutual relationship they might entail in the context of the Theory of Mind (ToM).

The term *mind* is an abstract concept that can be defined as "the seat of consciousness, thought, feeling and will" (Logan 2010, p. 89) meanwhile the brain can be described from a more biological perspective as a physical complex organ which is placed inside the head and protected by the cranium that is able to control all body functions of the human beings. Logan (2010) established a clear difference between mind and brain by stating that "those processes of which we are not conscious, such as the regulation of our vital organs, the reception of sense data, reflex actions, and motor control, on the other hand, are not activities of our mind but functions of our brain" (p. 89).

Establishing a difference between mind and brain is probably one of the hardest challenges that neuroscientists, psychologists and scholars could have faced in the last centuries. Meanwhile, the brain has been studied in depth from its own structure to how it works there is little known about the mind, thus, it is necessary to go beyond of a simple definition in order to establish the correspondent relationship between mind and thought. Regarding this, several questions could emerge like, for instance, when the mind appears for the first time, the mind is something unique from human beings, thoughts are found within the mind or the brain or even whether mind was developed at the same time that the brain. In order to answer some of these questions, it is necessary to have a look to the history of the humankind. There is so much evidence that confirms the human brain suffered a radical change at a cortical structural level as the man was evolving which allowed him to develop certain skills such as thinking, speaking, judgment and emotions among others. In this fashion, the brain could be considered a unique mechanism in the human being in which the information can be stored and retrieved thanks to that network of neurons that are constantly passing the information from one side to the other side.

However, the mind has been considered the result of what the brain does since it is something that goes beyond any physiological structure. The mind implies all mental activities that one can imagine including positive and negative thoughts which can have a strong influence on the individual. At some point in the history of humankind, it was thought that mind was something unique in humans, but the truth is that other animals also have a mind, though, humans have been the only ones to develop it. Nevertheless, that was not the only belief supported at the beginning but there was another one which stated that the mind was evolved through natural selection (Corballis 2011) but I would dare to say that the development of mind has more to do with the development of the brain whose main purpose is to survive. In relation to the existence of human mind, there is data that could confirm approximately when the mind appeared for the first time. According to the collecting data, there was a period between 2.6 million years ago until 12.000 years ago in which the mind could have taken shape and started to develop as in that period the humans started to hunt and gather (Bloom et al. 1985). This is very important since with the appearance of mind humans started to have the ability to reason, think, perceive, and even create and, even though it could be hard to believe, all this is possible through symbols which are the first thing that come to our mind to be transformed into words or even images later.

The mind is considered to be as complex as the brain since there are not specific tools for the study of the mind as there are for the brain. The only thing known for sure is that thoughts are placed in the mind and some of them can be so powerful that can make us believe in one thing or another until the point of transforming our reality or creating a new one. Most of the time humans cannot have control over those thoughts even when they can try to avoid them, they are still there waiting to be retrieved. Thoughts do not obey any rules imposed by the society, they can be said to be created according to humans' view and perception of the world. In this respect, there are not so many studies capable of explaining these phenomena yet. Notwithstanding, the concept of mind has not been the only one that appeared with the development of the brain but also the term consciousness.

Through history, humans have been witnesses of how the society and the culture have changed, how they have learnt to behave according to the rules imposed by the society and even how they have acted consciously or unconsciously without knowing the reason by which they act in that way. In relation to this, the mind has been associated to the term *consciousness* which can be defined as "awareness of one's own mental and/or physical actions" (Bloom et al. 1985, p. 206) implying, at the same time, a state of the self. It has been suggested that consciousness appeared for the first time when written language was developed and with the complexity of culture about 3000 years ago (Jaynes 1976). Previous to this breakthrough, it was thought that the mind was bicameral which means that one hemisphere was created for language and the other one was created to hear and obey but further studies corroborated that this belief did not hold at the end. In terms of consciousness or unconsciousness, it could be suggested that when someone acts unconsciously is just because he is carried away by emotions instinctively without thinking so much about the possible consequences that could bring. Normally, humans act in this way in order to survive, in other words, the brain has been designed to survive and the reason by which individuals act in one way or another is due to this purpose. Going beyond this, individuals tend to act first and then to think about it and, in this respect, I would say that only a small portion of the population can think about what they are about to do before acting by using their most rational part of the brain. What it is clear is that the mind cannot be separated from the brain since it depends on it in the sense that the brain is that mechanism makes possible that the stimulus can arrive to the correct cerebral region and then be transmitted to other regions by sets and sets of neurons and the mind holds one's thoughts.

As it has been explained consciousness and unconsciousness are different levels that are found within the mind but, in this respect, it is crucial to mention another level of the human mind in order to understand its complexity. Maybe, this has not been said yet, but our mind is able to keep not only our deepest thoughts but also memories and experiences lived in the real world. Sometimes, these experiences tend to be so traumatic for the individual that his mind immediately blocks such traumatic experience or memory because they really hurt the individual and, in this sense, it must be remembered that human beings have not been born to suffer but to be happy in this world. Faced with such trauma, there are occasions in which the individual's body does not respond to certain situations, for that reason, in order to find out what is the origin of such trauma psychologists try to get access to the individual's mind through the subconscious[1] by using a method called hypnosis. Through this method, the individual is guided through a soft voice to his most traumatic memory to relive it, which will allow the psychologist to help him overcome the trauma.

It is hard to define the term mind and foresee its connection with language and thought, if language and thought are two independent structures, whether children are born with an already acquired language or, even whether an evolution in language implies an evolution of the brain. It is true that human beings are, as opposed to animals, characterized for being the only ones who can make use of language verbally and just as human beings evolved gradually over the centuries, so did language.

The first hominids started to produce the first words as a need to express abstract ideas and concepts, in other words, thoughts. It could be supposed that human beings have an innate capacity for language, in other words, they are supposed to have internalized a great number of words when they are born. In this respect, Pinker (1994) pointed out that as animals have instincts to survive, language is considered to be an instinct for communication. For Pinker language is another kind of instinct since it has not been created by humans, but it has been developed with the passage of time. It is evident that language has evolved until the point of allowing humans to communicate with each other. It should be remembered that the first hominids established the first communication through gestures and drawings but in order to survive in the society they had to learn by themselves how to communicate verbally. All human beings have language, no matter what culture they belong to, what it matters is that language can only be developed through interaction. Thus, language cannot be taught from parents to children since instincts are something which human beings are born with. Language as an instinct could be attributed to Darwin (1874)

[1]The subconscious was an old term used to refer to the unconsciousness. Currently, this term alludes to all the information kept in the consciousness that cannot be seen. In this respect, it can be considered as a second consciousness which is possible to get access to as long as oneself pays attention to it.

who stated that "man tends instinctively to speak as it can be observed in that unique speech used by children while neither of them show an instinctive tendency to brew, bake bread or write" (p.20). There seems to be so many controversies to this idea of innatism because how individuals can have a set of "empty" words for things that are not known yet, the process is very simple first things are touched, seen, heard, smelled, and tested and then a word with meaning is given to it.

On the other hand, there are some theorists who consider language to be a biological component and as a biological component it must have undergone some kind of evolution. However, it has been stated that language has not gone through an evolutionary process in which the brain mechanisms related to it have drastically changed but it is a matter of natural selection necessary for the human being survival. Lieberman (2016) claimed that "the evolution of human language hinges on natural selection acting on heritable biological variation" (p.127). In contrast with this idea, Chomsky (1972) supports the idea that language plays an incidental role in its own evolution as a vehicle for communicating thoughts and ideas due to the fact that thoughts can be expressed in different ways and not only by the use of words. Going back to innatism, it must be said that Chomsky was the father of this theory who claimed that all human beings have a *Faculty of Language* where innate knowledge is placed. By Faculty of language, it must be understood that it is a cognitive capacity which is made up of a computational system in the brain where a part of this system is connected to thought and the other part is connected to the sensory area which allows humans to perceive the sound. This innate knowledge let the individual activate the syntax of a particular language. Interestingly, Bolhuis et al. (2014) assumes that "biological variation supposedly does not mark these innate biological capacity, nor did Natural Selection acting on variation play a role in the evolution of these genetically transmitted capacities" (p. 129).

As it has been mentioned, the evolutionary process of language is not the result of a linguistic change but it has to do with physiological changes in the human body, for instance, vocal chords were developed, the different positions of the tongue allowed the individual to produce certain vowels and, therefore, its shape allowed the speech communication (Negus 1949), and the skull seems to be restructured in the first 8 years of life, a period in which the child acquires the first words and morphemes until reaching the adult morphology (Crelin 1969). Notwithstanding, current studies have provided evidence that support the idea that language is innate from a biological viewpoint. Recently, it has been confirmed that genetics has a strongly influence on language. Geneticists have determined throughout several studies that there are hundreds of genes involved in language, but human beings carry a particular gene in their DNA which is known as FOXP2 or the language gene. This was probably the most important breakthrough of the twentieth century since the studies carried out revealed that a small mutation produced on it could result in a language disorder such as verbal dyspraxia.[2] The gene FOXP2 is a protein that encodes a transcription factor

[2] Verbal dyspraxia is the inability to coordinate muscle movements needed to speak.

which is responsible for the activation and deactivation of other genes involved in the development of language.

What it is known about this gene is that it affects the speech. This was the main conclusion reached by the researchers after having studied the case of a British family. This family was known as KE and their members presented a language disorder which was verbal dyspraxia. The symptoms presented by the affected members was incomprehensible speech, several difficulties in the grammar and in the language and their QI was lower than normal. Due to the incomprehensible speech they had to look for another way to communicate with each other and with the rest of the population, thus, they developed a sign language in order to have a better comprehension. The advanced technological tools showed that the affected members presented certain structural anomalies in the different motor structures of the brain including the motor cortex, the cerebellum and the striatum which made the researchers think that effectively the gene FOXP2 affects the development of these regions that are important for the development of speech (Bear et al. 2016). The importance of this gene in the twentieth century relies on the fact that a mutation produced on it was what put humans on the path towards the development of language necessary to develop the higher cognitive functions and later on the development of the human culture.

Going one step further, the verbal dyspraxia was not the only language disorder whose main gene seemed to be FOXP2 but there were other important disorders like specific language disorder in which the FOXP2 played an important role. Despite of this, this new disorder was still more interesting for further studies than the previous one since there are two more genes involving on it which are known as CNTNAP2 and KIAA0319. These last two genes are very relevant for language since the gen CNTNAP2[3] is a protein that plays an important role in the development of the brain as it takes part in the adequate localization of potassium channels in the developmental neurons, meanwhile, the gen KIAA0319 seems to play a critical role in neuronal migration during the development of the neocortex as well as for the normal neuronal functioning of adult neurons.

This breakthrough has supposed to be a key factor to understand the individual's capacity to produce language and carry out cognitive tasks, thus, recent studies have revealed that an anomaly in this gene could deprive the individual of producing language or not produce it at all.

Once the relationship between mind, language and thought have been explained in detail, I will focus on three main aspects to be considered: (a) the relationship existed between language and thought, (b) the main cerebral areas implicated in language and thought and finally (c) ToM considering some language pathologies.

[3]CNTNAP2 is a gene that encodes the neurexin protein which is a protein on the presynaptic side of the synapse that helps keep the presynaptic and postsynaptic elements together.

17.1.1 Language and Thought

As it has been mentioned previously, both language and thought seem to be strongly interconnected if language is the via through which individuals can communicate with each other and thought as that concept that includes ideas, reasoning, and emotions. In recent years, these two concepts have been explored in depth by linguists until the point of wondering whether the language was first developed or whether it was thought which comes before language, there can be language without thought or even whether language and thought can coexist independently from each other.

Interestingly, there have been many theories that have been brought into light in the last decades but only three of them have been able to explain the relationship existed between language and thought by providing evidence. The first theory that can be encountered is known as the Cognitive theory which claims that thought was born first that language due to the fact that every stimulus that is perceived through the five senses can be considered to be a thought because we transform that stimuli into a mental representation to be recognized and comprehended so that our brain can give a response to it later on. One of the precursors of this theory was the psychologist Jean Piaget (1950) who stated that language is preceded by thought and, therefore, language depends on thought. At a cognitive level, thought can imply other cognitive factors that are very important in our lives for our own survival which are memory, attention, and cognitive flexibility. The concept of thought is so important that it could be even said that it is possible to have thought without language since thought is part of our cognition. In Piaget's own words "a child has to understand a concept before she or he can acquire the particular language form which expresses that concept" (p.4). This Cognitive theory has been supported by other linguists such as Harley (2001) who suggests that if there was an impairment in language this one would hardly have an impact on cognitive skills, for instance, a patient with aphasia can have certain problems to communicate something depending on which areas have been affected but this does not mean that his cognitive skills have been affected. By taking this example as a reference, it is clear that there can be thought without language. Moreover, thought and language could be seen as two things that are totally independent for each other, what is more, their connection only occurs when the child turns around 2 years old (Vygotsky 1962).

A second theory was put into practice which defends the idea that when we are born, we already have a language system in our minds. This theory known as the Linguistic theory suggests that "language influences the way people perceive and think about the world" (Lund 2003, p. 10). This theory can be attributed to the linguist Sapir (1929) who concluded throughout a study that the differences between the languages changed the way people perceive their environments. However, the linguist Whorf (1956) did not agree so much with the idea that language changes the way people think, in fact, he was convinced that "the difference between languages determined the types of thoughts people were able to have" (p. 11). Following this linguistic theory, it could be said that thought structures are strongly enhanced by language.

Considering these two theories, it is hard to believe that language could exist without thought and vice versa since thought needs language to be transmitted and language needs thought to be understood. More recently, a third theory came to light which is known as the Theory of Simultaneity which states that even when both language and thought are two different independent things, they coexist and complement with each other, that means, there cannot be thought without language and the other way around since thought covers all the obtaining data from the outer environment and the only way to express it is through language.

17.1.2 Anatomical Architecture for Language and Thought

The brain has always been considered the most complex organ of the human being not only because of its complex anatomy but also for its physiology. Exploring the brain from inside by studying each of its several areas in depth has allowed neuroscientists to understand much better its relationship with the language. For doing this, it is necessary to make an anatomical tour from the most exterior part of the brain to its interior part, by mentioning its different components and functions.

The human brain is composed of two hemispheres which were baptized with the name of Left Hemisphere and Right Hemisphere. Each hemisphere carries out a series of responsibilities which are essential not only for language but also for other cognitive functions. Before going on with these responsibilities, it is important to known that every time an individual makes a movement with the left part of his body that command is given by the right hemisphere, meanwhile, if the movement is produced on the right part of the body, then, the commands are given by the left hemisphere. It is known that the left hemisphere is responsible for language, mathematics, reason, logic, and learning, as opposed to it, the right hemisphere oversees music, creativity, art, intuition, imagination, and tridimensional perception. How these hemispheres are connected with each other is a question that persists on many individuals in present day, and it is precisely the corpus callosum[4] the one that is capable of connecting them. Seeing the brain as a whole, it can be appreciated four components known as lobes within the cerebral cortex which are the frontal, the parietal, the temporal, and the occipital lobes. They are very important due to the fact that they participate in the development of the cognitive functions such as planning, reasoning, taking decisions and even language.

[4]It is a thick, wide bundle of neural connections whose main function is to link the left hemisphere with the right hemisphere of the cerebral cortex. Neurons in either hemisphere stretch their axons through the corpus callosum to communicate with neurons in the opposite hemisphere. The corpus callosum contains a specialized cells known as myelin which is in charge of providing insulation for the electrical signals.

Focusing on the hemispheres, it is interesting the issue related to the dominance in a hemisphere in terms of language. For ages, neuroscientists maintained the hypothesis that the dominant hemisphere for language was the left one, but this changed radically when some studies revealed that in a small portion of the population the right hemisphere seemed to be the dominant one and the reason by which this seems to be so is something that researchers in the field of neuroscience are trying to discover. This can sound disconcerting considering that the right hemisphere is not prepared for speech, though, for some reasons it seems to participate in the language. The most recent discoveries related to the brain showed that the left hemisphere can control speech, reading and writing while the right hemisphere is ready for reading and comprehending not only numbers but also letters and short words providing that the response is non-verbal. In other words, the right hemisphere has been associated to the comprehension of language. In order to find out which hemisphere was more dominant for language; it was crucial to make a more dangerous study based on the hemispherical disconnection which consisted of separating the two hemispheres surgically. This is the case of a female subject V.J who underwent a surgery in which her two hemispheres were completely disconnected. The process consisted of cutting off part of the corpus callosum and see how the subject reacted with certain stimulus. After the surgery, it was observed that the woman was able to say those words that were seen by the left hemisphere, but she was not able to write them down and the same occurred when the visual stimulus was presented to the right hemisphere. This showed in the end that both hemispheres could work as two different and independent brains.

Furthermore, it has been supported the idea that this hemispherical dominance in the language has to do with the size of some cerebral areas. Considering this, Geschwind and Levitsky decided to examine 100 human brains to determine whether there was an anatomical difference or not. Through this study they discovered that the temporal plane, which is part of Wernicke's area, was larger in the left hemisphere than the right hemisphere in a 65% of the studied cases. This difference in the size was probably what determined that the language was dominant in the left hemisphere and not in the right hemisphere. There are other regions that are larger in size in the left hemisphere, for instance, a portion of the so-called Broca's area but this is not enough to determine such dominance. In fact, it is believed that the insula could be a good predictor for the hemispherical dominance of the language.

The complexity of the brain has nothing to do with the process of language which seems to be still more complex than the organ itself due to the involvement of the different areas placed in the left hemisphere and in the right hemisphere as well. It has been mentioned that the right hemisphere is not the dominant one for language at all but as it will be seen its participation on it is relevant to understand how the brain works as a whole system and how certain individuals with cerebral lesions are capable of keeping part of the language intact. The use of technological tools such as fMRI has allowed experts to observe which pathway follows the language once it has reached the cerebral cortex and determine which areas participate on such process, let's see what happens exactly when an induvial receives a stimulus either auditory or visual.

When an auditory stimulus, for instance, a word reaches the individual's ear it goes through the different parts of the auditory system until it reaches the cochlea where it will be transformed in a nervous impulse to go directly to the primary auditory cortex where the stimulus, in the form of a sound, is identified. Once the auditory stimulus has been identified, it will travel directly to a region known as Wernicke's area where the stimulus is understood in terms of meaning and after that it will be passed through the arcuate fasciculus to Broca's area which will give the corresponding commands so that the person can produce a set of words. To be successful in this process, those commands must be transmitted to other areas related to the motor cortex which will make possible the activation of the facial muscles. This is usually the process carried out when there is an auditory stimulus but what happens if the stimulus perceived is visual, for instance, when a written book is seen. In this case, the pathway that follows the stimulus changes a little bit since the visual information (the reading words) would arrive first to the primary visual cortex and then it would pass directly to the angular gyrus where it would be matched with the sound of the spoken word. Once this happens, it is the form of the auditory input which will travel to Wernicke's area in order to be processed and then pass through Broca's area again.

Although, the pathway that every stimulus either auditory or visual follows seems to be complex in terms of anatomy and physiology, it is crucial for this chapter to have a deeper overview of the different cerebral areas that are involved in the process of thought and language. From an anatomical perspective, language has been located in the left hemisphere of the brain, concretely, in the temporal lobe, however, it is not very clear in which hemisphere thought can be placed on. Some theorists believe that thought can be found within the left hemisphere while others believe that it can be found in the right hemisphere depending on what kind of thought we are talking about but what it is clear is that the frontal lobe is responsible for thinking, planning and judgement. The human brain is composed of more than one billion of neurons which are connected to each other forming a neural circuit that allows to carry all the information to the different areas of the brain. Nowadays, it is well-known that neural circuits are able to connect more than one cerebral area, for instance, the prefrontal cortex can be connected to the temporal and parietal cortex, the thalamus, the cerebellum and the hippocampus through the basal ganglia and the subcortical structures (Alexander et al. 1986).

At the beginning, it was believed that every and each area of the brain carried out a single function, for instance, Broca's area was responsible for the production of language but advances in the field of science showed that more than one area was involved in a specific function.

The first one to focus on the anatomy of the brain and its physiology in relation to language was Paul Broca (1861) who, throughout a study, discovered a language deficit could be the result of a damage in the cortical area. Since then, several studies have been carried out in order to understand how language is produced. Fedorenko et al. (2011) started from the premise that there are innate micro-sites in the inferior gyrus of the cortex that is in charge of comprehending the distinctions in meaning. Thinking that there are only two different areas such as Broca's frontal and

Wernicke's posterior cortical regions, which are linked by the arcuate fasciculus, as being uniquely in charge of language is a great mistake since the neural system is much more complex and, therefore, it implies more regions that contribute to language. In this fashion, it has been proposed that areas from the neocortex were considered to be the skills that were the basis for an aspect of behavior such as language (Spurzheim 1815).

Knowing that Broca's and Wernicke's areas are devoted to the production and comprehension of language allows us to be aware of the process of recognition and production of words when the stimuli arrive to those areas but determining what it must be said or not and how to react implies other relevant areas. Advanced technological tools such as fMRI have permitted to study the activation of certain areas before certain tasks being one of them the prefrontal cortex. As Ullman (2004) pointed out "the ventrolateral prefrontal cortex and Broca's area form parts of the basal ganglia circuits implicated in regulating speech and language" (p. 136).

There is so much evidence that have proved the implication of different areas in language and thought. For instance, Parkinson disease (PD) is related to subcortical dementia which means that those patients who suffer from this disease will be unable to make a change in the direction of a thought process or even action (Flowers and Robertson 1985). Notwithstanding, Cummings (1993) stated that there are other circuits involving the basal ganglia, the orbitofrontal cortex or even the anterior cingulate cortex that can affect seriously cognitive factors such as inhibition and attention or even laryngeal control.

The basal ganglia are as important as the prefrontal cortex since a damage produced in the basal ganglia could result in both speech production and cognitive deficits. As Pickett et al. (1998) showed in their studies "bilateral lesions to the caudate nucleus and putamen of the basal ganglia in the subject studied resulted in severe deficits in sequencing the laryngeal, lingual, and lung motor activity necessary to produce articulate speech" (p. 137). Furthermore, it is believed that the basal ganglia play a critical role in associative learning, in planning and executing motor acts including the speech in humans.

Regarding the thought process, it must be clarified that there is not so much evidence that could confirm which regions of the brain are involved in it, in fact, neuroscientists are still making research on this matter. What it is known for sure is that the prefrontal cortex has much to do with the process of thought since it must be remembered that the executive functions are placed there which carry out the functions of thinking, planning, reasoning, attention and resolving problems among others but what part of the prefrontal cortex takes part in this process is probably one of the most difficult questions that could emerge in these days.

It has been proved through several studies that one of the areas involved in such process is the ventrolateral prefrontal cortex which is connected to posterior regions by circuits involving the subcortical basal ganglia. The activation of the ventrolateral prefrontal cortex takes place during all tasks involving the selection and retrieval of information (Duncan and Owen 2000). As opposed to it, the dorsolateral prefrontal cortex is active while monitoring motor or cognitive events during a task while considering earlier events stored in working memory (Postle 2006). These neural circuits allow the individual to change from thought to thought. Interestingly, another relevant area for the process of thought that has been discovered in terms of physiology is the cerebrum which seems to participate in those conscious thoughts. As a matter of fact, it can be distinguished several different types of thoughts being one of them relational reasoning[5] which has been related to fluid intelligence (Spearman 1904). Dealing with the abstract relational reasoning, it must be pointed out that according to some theorists like Robin and Holyoak (1995) and Holyoak and Kroger (1995):

> the operations of the PFC, which include governing selective attention (Posner and Petersen, 1990) and managing working memory (Fuster and Alexander, 1971; Miller and Cohen, 2001; and D'Esposito et al., 1995), may give rise to relational reasoning through integrating multiple relations. Relational integration has been considered to be a core ability that gives rise to higher order relational reasoning (p. 15).

As it can be deduced, the PFC can be considered as the major component of the brain due to its connections to the other areas of the brain allowing in this way the contribution to reasoning by means of integrating relational information (Petrides and Pandya 1999). The PFC is in charge of multiple functions as one can guess but there is one which is associated to the use of the rationality which is the avoidance of impulsivity (Miller 1985). In relation to this, there have been so many studies that have proved the importance of the PFC in reasoning. One of these studies was carried out by Gray et al. (2002) in which it was proved that other areas of the brain like the basal ganglia, cerebellum, and visual cortex were strongly activated along with the PFC regions, though, they are less involved when dealing with reasoning. Emphasizing on the analogical reasoning, it was demonstrated that in a population of young adults the bilateral RLPFC[6] is involved in those solving problems which require integration while in a younger population with children it seems that there is not such activation of this area (Wright et al. 2007). This difference can be given due to cortical maturation as in the case of children the prefrontal cortex has not reached its maturation yet, which is important for relational reasoning as well.

Apart from this area, there are other areas that are involved in relational reasoning as it is the case of the occipital and parietal cortex through which the visual and spatial processing are connected to the PFC making possible the bottom-up perception and the top-down control over the representations that are relevant to the matrix reasoning (Krawczyk 2012). It has been suggested that the right RLPFC tends to increase as relational processing demands increase (Crone et al. 2009). On the contrary, the left RLPFC is known to be associated with mediating relational comparisons in analogies (Green et al. 2009). The DLPFC[7] plays an important role in the cognitive factor known as inhibitory control in reasoning demonstrating,

---

[5] Also known as the ability to consider relationships between multiple mental representations, it is directly linked to the capacity to think logically and solve problems in novel situations (see Cattell 1971).

[6] Rostrolateral Prefrontal Cortex

[7] Dorsolateral Prefrontal Cortex

in this way, a sort of association within the MFG[8] and with the LIFG[9] which is a relevant area for searching non-dominant meanings (Yang et al. 2009) and semantic retrieval (Bunge et al. 2005).

Needless to say, there are so many things that are known about the language anatomical structure as opposed to thought, likely, due to its abstractness. It is clear that there are so many studies to be carried out in order to know much more about how the thought either conscious or unconscious is processed in the brain. Though, it is known how the brain works and which are the functions of every region that composes it, there is always something new that is waiting to be discovered, in this aspect, I will state that there are more regions that have to do with the process of thought that have not been discovered yet in terms of that function.

17.2 The Theory of Mind (ToM)

The established relationship between mind, language and thought has allowed us not only to understand the process by which thoughts can be expressed via language but also how these three concepts are interrelated in a unique system. However, theorists have gone further in this matter by wondering about one's own behavior and mental states, that means, how individuals can make judgements or how they can decide whether a belief is true or false. In a real context, it is possible to guess what someone is thinking of or even how that person is going to react in a certain situation, but is it possible that individuals can read other individuals' mind? Although the answer is affirmative, it is essential to go beyond the relationship between mind, thought and language and give one step further by going into the ToM.

It has not been easy to provide an adequate definition for ToM since so many factors such as language, thought or even experience have to be taken into account. There have been so many theorists who tried to define the concept of ToM from different perspectives. Regarding this, Wellman (1979, 1985) described this concept from a metacognitive perspective as "the child's conception of human cognition" (p. 4) but probably it was Premack and Woodruff (1978) who provided the most adequate definition for the concept of ToM defining it as that human ability to correctly predict the wishes and intentions of others. However, this definition for ToM seems to be incomplete for some authors as Frith and Frith (2005) stated the ToM is "the ability to acquire knowledge about other peoples' beliefs and desires" (p. 45), in other words, it would act as an equivalent to mentalizing or mind reading. In contrast with these two possible definitions for ToM, Gallagher and Frith (2003) provided a more completed and accurate definition for this concept by stating that:

> It underpins our ability to deceive, cooperate and empathize, and to read others' body language. It also enables us to accurately anticipate other people's behaviour, almost as if

[8]Middle Frontal Gyrus

[9]Left Inferior Frontal Gyrus

> we had read their minds. It underlies our ability to explain and predict the behaviour of ourselves and others by attributing to them independent mental states, such as beliefs, desires, emotions or intentions (p. 77).

As it can be observed, the term ToM can be defined in different ways according to the author's perspective and field of research. However, there is another definition provided by Gopnik and Wellman (1994) which can get closer to our current definition of ToM stating that:

> Theory of mind is a domain-specific, psychologically real structure, composed of an integrated set of mental-state concepts employed to explain and predict people's actions and interactions that is reorganized over time when faced with counterevidence to its predictions (p. 4).

Starting from the providing definitions for such an abstract concept, other relevant questions could emerge such as what the origin of ToM is, that is to say, it is something innate in the human being or it is something more cognitive. Following the innate theory, it has been suggested that "theory of mind is innately specified but not evident until a certain level of linguistic and cognitive development is achieved" (Fodor 1992, p. 8). Studies carried out in the last years have revealed that "this ability may be an innately inherent ability that is genetically pre-determined" (Doody et al. 1998, p. 398). Bearing this in mind, it must be remembered that when a child is born, he already has a set of cognitive factors or skills that must be developed over time, in this case, it could be said that this child is born with a ToM which must be developed as well, which implies to comprehend better how the mental states and behavior interact with each other.

As opposed to it, there are some theorists who defend the idea of ToM not as something innate in human beings but as something more cognitive stating that "theory of mind abilities rest on domain general cognitive operations that require language for their implementation" (Frye et al. 1995, p. 8). Furthermore, there are other theorists for whom the main focus is not on whether ToM is part of innateness or cognition but in language. It seems that language has become an important factor in the development of ToM since language is the only natural way through which children can obtain the requiring information to build a ToM (Perner 2000). Recent studies done in the field of ToM have brought to light that language and ToM are developed at the same time during the toddler and preschool period due to their strongly connection.

Although, the issue related to the innateness or cognition of ToM is the focus of studying of some researchers there are still other relevant issues that must be considered in order to understand the concept of ToM and how it is originated in the child's mind. Certainly, the fact that all humans beings are born having a ToM is an universal truth that cannot be denied as it cannot be denied that all human beings have an internalized language when they are born but the point is at what stage of life ToM appears for the first time and how it is developed, that is to say, it follows the same stages of language acquisition or it follows another path. It must be clarified that having a ToM implies having a language of mind, in other words, children tend to represent the real world by means of representing mental states and actions through language (de Villiers 2007). Thus, it is believed that the use of the different mental state terms takes place around 2 years old, concretely, those mental states related to emotions and perception. At the age of 3 years old, the child can already produce mental states related to cognition such as think or know in conversations, likely, without being aware of their meaning but it seems that at the age of 4 years old the child starts to use these terms by distinguishing their meanings. Among the different cognitive mental states that can exist, it has been observed that children can definitely discriminate between the verbs think, know and guess once they have reached the school years. In addition to this, their complete comprehension of the different meanings of the verb know takes place in the late elementary years (Booth and Hall 1995). Considering that these verbs are abstract words, it must be emphasized that for a child the only way to acquire their meaning is by using them in a determining context (Montgomery 2002).

As it can be appreciated, the ToM starts to develop at an early age and with the acquisition of mental state terms but there are some components of great relevance for the ToM that must be considered in order to understand how ToM works as it has been suggested by Miller (2006). Among all these components that can be found in relation to ToM, it can be highlighted joint attention, appreciation of intentionality, recognition that different people have different perspectives, use of mental state words and pretend play. Joint attention can be understood as "the capacity of an infant to coordinate her attention with a social partner vis-à-vis an object or event" (Morales et al. 2000, p. 143). According to Tomasello (1995), this component of ToM develops in the first years of life meaning that a child with 9 month is already able to pay attention to a specific aspect of an object and, there is clearly an intention to communicate with another child.

In a conversation there are always some components that are crucial for a good communication which are the speaker, the listener, the message to be transmitted, the context and the intentionality. Intentionality is very important, and it is present when individuals communicate with each other. In fact, intentionality can be appreciated in the providing words to describe a particular object or event, for instance, blue or wonderful or even it can be seen in a sentence such as I want that doll where the main intention of the speaker is clearly to be in possession of that object (Tomasello 1995).

It is of common knowledge that communicative interaction in the first years of life is crucial for the acquisition of language since the child is constantly receiving the input from the environment that surrounds him but this interaction implies more things apart from acquiring a language, for instance, it implies that the speaker can have a different perspective from an object that the listener and this is a true fact that cannot be changed. In his third year of life, the child starts to use certain phrases such as I think which is considered to be a mental state because it implies thought in order to start a conversation. As Bartsch and Wellman (1995) suggested "the use of mental state terms becomes more truly mentalistic, unambiguously referring to the thoughts, beliefs, and feelings of oneself and others" (p. 143). Although, all components are of high importance for the good development of the ToM, probably, there is a specific component which is the most important one which is known as pretend play. In the pretend play, the child is required to separate his representation of reality from reality itself and it is expected that he is engaged in it during the second year of life (Leslie 1987).

As it has already been seen each component plays a crucial role in the development of ToM, but in spite of this, there are still authors who are trying to find out where the real link between pretend play and ToM development lies. In a study carried out by Youngblade and Dunn (1995) it was concluded that "acting out roles in pretend play at age 2;9 (years; months) predicted false belief performance at 3;4, even when language was controlled" (p. 143). Contrary to this, Astington and Jenkins (2000) found out that pretend play was not a predictor for ToM performance, instead, it was ToM which acted as a predictor for joint planning and role assignment. During the childhood stage, children are able to represent the reality in their minds according to what they experience but this representation is not permanent. As Astington (1993) pointed out:

> [...] children begin to understand that our minds do not simply mirror reality, but that we actively construct representations of reality. These representations may change within an individual as new information is received, and they may differ across individuals, depending on the information each person has access to (p. 144).

Having a representation of reality means to have a belief which is unique for each individual. This belief can change as long as our representation of the world changes because the world is perceived through the individuals' senses and then a representation in their mind is created and by means of that belief. Regarding this, it should be said that one's belief could not be true for another individual but false. When focusing on ToM, there are multiple components that are considered and multiple tasks that are put into practice to know more about this theory being one of them the false belief. False belief is essential to understand the concept of ToM, in fact, De Villiers and de Villiers (2000) take a particular aspect of syntax, in this case, a sentence complement structure in order to explain the concept of false belief. They started from the following example "Lucy thinks the moon is made of green cheese" which can, at first sight, be true even when the embedded clause "the moon is made of green cheese" is not true but false, thus, both sentences are independent for each other. Undoubtedly, there are certain mental states such as *think*, *believe,* and *guess* that can determine a sentence complement as a false belief. Notwithstanding, De Villiers and de Villiers led to the conclusion that in order to succeed in false belief tasks children first need to understand that the sentence as a whole could be true while the embedded clause within it is completely false.

In relation to language, it has been said that it undergoes some developmental changes during the first 5 years of life as ToM does, but the question remains whether that means they are related to each other, or it is simply a coincidence. It seems that there is enough evidence that supports the idea that both language and ToM are strongly related, what is more, for a good communication ToM is needed to communicate something by using the language because ToM involves thoughts and beliefs among others, but language is also needed in order to know more about ToM. The following evidence is provided by Gleitman (1990) who focused on the fact that mental states such as *think*, *believe* or *guess* cannot be observed because it is not possible to learn what the word *run* means by means of performing the action when the word is said. In terms of performance, this is not possible with words such as *think* since it implies a mental activity which cannot be perceived by the senses. Furthermore, he states that verbs such as *think*, *know*, *like* or *want* are internal activities or states that do not have reliable behavioral correlates and, in this sense, it could be said that the function of language is to provide information about the meaning of these mental states which is possible by observing their roles in the grammar, semantic and pragmatics of the language. As a matter of fact, language matters for ToM because to "understand other minds is to participate in a communally shared belief system about human goals, motivations and values" (Nelson 1996, p. 11). Nelson even argues that language represented in the child's mind is what makes him go beyond his own thoughts and beliefs in order to know the thoughts and beliefs of others. Another evidence highlighted how important is language for the development of ToM, which was provided by Ruffman et al. (2002) who found out in an interaction between mothers and children that the exposure to language was a good predictor for the development of ToM in children. Participating in a conversation could even be critical for the development of ToM since children start to understand other people's minds at the age of 3. This usually happens when mothers use topics such as feelings or thoughts in conversations with their children which allow them to explain the behavior a few months later (Dunn et al. 1991). It is precisely in their first conversation when children start to acquire concepts like desire, belief, or intention among others (Peterson and Siegal 2000). In fact, the first mental states acquired by children are those related to perception, emotion, and desire, and later they acquire those ones related to cognition (Bartsch and Wellman 1995).

A very important aspect to consider here is that parents use these mental states in conversations with their children not only to refer to children's own experiences but also to other people's experiences so that children can acquire mental states to themselves and others (Astington 1996). Curiously, Peterson (2000) concluded that children with siblings seem to have some sort of advantage in the development of ToM due to the chances for discourse and experiences related to the feelings and thoughts of others. Considering this, it must be pointed out here that it is not the child's linguistic ability what allows him to acquire mental states in conversations by means of communicative exchange but his pragmatic ability. Having experienced this, they become aware of concepts, different points of view and mastery the syntax for representing false belief and this contributes to a good development of ToM (Astington and Baird 2005). In the field of ToM, it has been even more interesting to find out children who lack of language or have an impoverished language do not develop ToM.

As it has been explained before, ToM is our ability to read other human's mind and in order to do that the use of language is, of course, essential to communicate the intentions and desires of others but until what point language is related to ToM is something that has been investigated in depth by researchers. In this fashion, it is evident that language has a positive effect on ToM. It has been said that language could predict ToM. However, it is impossible to measure language as a whole in terms of ToM and, that is why, researchers have focused more on some aspects of language such as syntax and semantics. Starting from the premise that language can later predict ToM, it has been concluded through several studies that the only component of language, that can predict ToM, is semantic even though syntax can also predict it, although, in this case, not so clearly (Ruffman et al. 2003).

ToM is so abstract and complex that is difficult to understand how it works in terms of neural activity. Knowing the strong relationship that exists between language and ToM, researchers have wondered whether the neural circuits involved during the process of language are the same that are involved in the ToM or, on the contrary, there are two different circuits being one for language and another for ToM. The use of functional imaging has allowed neuroscientists to observe which areas of the brain are activated when ToM is being tested and if the neural circuits are similar from language as well. According to the theory proposed by Leslie (1994), some of the regions involved in the process of ToM are the anterior paracingulate cortex, superior temporal sulcus (STS) and the temporal poles bilaterally. These last two regions, which are believed to be essential for the development of ToM, are said to have a very important function which is to provide explicit information about the behavior, for instance, the retrieval from memory of personal experience (temporal poles bilaterally). Notwithstanding, it has been discovered in recent studies that the anterior paracingulate cortex is the most important region for the ToM. From an anatomical perspective, the anterior paracingulate cortex involves different regions of the brain being these ones the anterior to the genu of the corpus callosum and the anterior cingulate cortex (ACC). The paracingulate cortex is part of the anterior cingulate cortex which includes Brodmann areas 24, 25 and 33.

The superior temporal sulcus seems to play an important role in ToM since its right part tends to be associated with the comprehension of the meaning of stories that involves people or with the understanding of causality and intentionality (Gallagher et al. 2000). This region collects the information coming from the dorsal and ventral visual pathways which led linguists like Frith and Frith (1999) to think that this region plays a crucial role in the analysis of goals and outcomes of behavior. The right superior temporal sulcus is essential when it comes to explain, for instance, the behavior of people by recognizing their own mental state or physical cues. It is said that our vision is a critical component when it comes to trust someone, it is as if individuals were able to intuit what the other person is thinking just by seeing him.

Apart from this region, there are other areas that are determined in the process of ToM. Several studies have shown an activation of the temporal poles when the individual recollect familiar faces and scenes, when recognizing familiar voices or when retrieve emotional memories (Dolan et al. 2000). The amygdala is another region that seems to be activated during certain tasks which involves the recognition of behavior at an emotional level. This particular region takes part in the development of ToM and, according to some studies, it is responsible for the information coming from eye gaze as well as the recognition of emotional behavior (Baron-Cohen et al. 1999). The region known as the orbitofrontal cortex is said to be in charge of emotion and, therefore, it is sensible to think that this region can have a fundamental role in the development of ToM. Furthermore, it has been proved from a more clinical perspective that a damage produced in the orbitofrontal cortex could lead to certain impairments in ToM. The orbitofrontal cortex is also involved in the regulation of social behavior which allows individuals to react in one way or another to the emotions of others (Baron-Cohen and Ring 1994).

Even when the different areas of the brain are interconnected with each other, a crucial observation must be made which is that there are different neural pathways involved in both language and ToM since there are areas that seem to be more specific for language and others for ToM.

17.3 Pathological Implications in ToM: Aphasia

Considering the pathological field, there are certain aspects that must be taken into consideration with respect to the issue of ToM. One of them is that there are many language disorders and pathologies where ToM has been affected in one way or another. The only pathology that is related to language is aphasia which has allowed researchers to study the language in terms of hemispherical dominance. The term aphasia has been defined as an alteration in the comprehension and/or production of language caused by a cerebral damage in the ventroposterior region of the left frontal lobe and in the posterior and superior region of the left temporal lobe.

For years, neuroscientists have tried to determine which are the possible causes that generate the aphasia and what consequences were brought. After several studies made with computerized tomography, they deduced that the most frequent cause was a stroke[10] which produces not only a cerebral death but also an important damage in those regions that control the language. Therefore, a stroke is the most dangerous cause that affects language and communication due to the fact that the most important areas involved in language which are Broca's area, Wernicke's area and the arcuate fasciculus are supplied with blood by the middle cerebral artery. Notwithstanding, this is not the only cause capable of producing the aphasia, there are other causes that must be considered which are a tumor, an infection, a trauma or even a degenerative process. In this case, the aphasia can be accompanied by other cognitive problems such as memory problems. In these terms, when there is a difficulty in language along with a progressive development, another kind of aphasia emerges which is known as primary progressive aphasia[11] which can go forward to a more generalized dementia. In contrast with this type of aphasia, there are other types that have been identified according to the areas affected which are Broca's

[10] A stroke is a brain damage caused by blocked blood vessel or a bleeding in the brain. Being unable to receive the oxygen and nutrients needed to survive, the brain cells begin to die. The most severe causes produced by the stroke are a severe damage in the brain, a permanent disability and even death.

[11] This type of aphasia is the rarest aphasia that can be found within the language pathology since it affects only the ability to communicate. Patients with this aphasia have many difficulties to express thoughts and understand and find words.

aphasia, Wernicke's aphasia, global aphasia, transcortical aphasia, conduction aphasia and anomic aphasia.

One of the most common aphasias is known as Broca's aphasia, non-fluent aphasia or expressive aphasia which is characterized for having the production of speech quite altered until the point that is almost impossible for the patient to produce the language. Despite of this impairment, a patient with this kind of aphasia presents a good comprehension of the message without repetition nor denomination. In fact, the difficulties presented in this type of aphasia by the patient are repetitions of sentences and reading aloud. From an anatomical perspective, there seems to be a lesion in the left inner frontal gyrus, Brodmann's areas 44 and 45, which have to do with the gesticulation, the intonation and the semantic processing which are necessary to produce the speech, or even in the adjacent regions.

Another type of aphasia that can be found is Wernicke's area which is also known as fluent aphasia or receptive aphasia. It is known as fluent because patients who suffer from Wernicke's aphasia do not have any problems to produce the language, but they have an impairment in the ability to comprehend spoken words which can lead them to use nonsense words in their sentences in an attempt to give a proper answer to the received message. The areas affected correspond to Brodmann's areas 22, 39 and 40 which have to do with the comprehension of spoken and written language, and which are able to associate the sound with the concept.

Apart from Broca's and Wernicke's areas, there is another type of aphasia that can be considered to be the most severe one of all aphasias which is global aphasia. This aphasia is characterized by having an altered fluency and comprehension of the spoken words. The damage takes place in multiple areas of the brain related to the language processing. Due to its severeness the patient's production of language is very limited since he is only capable of producing a few words that can be recognized by the listener, though, he can comprehend a little bit of what has been said. Despite of this, it seems that those cognitive skills, which have nothing to do with language, remain intact.

As opposed to the previous aphasias, the transcortical aphasia can be produced when a lesion takes place in the deepest adjacent structures of the cerebral cortex. It can be found two different transcortical aphasias which are known as motor aphasia and sensory transcortical aphasia. The former is characterized by an altered fluency and a good comprehension since the Wernicke's area and the arcuate fasciculus have not been impaired presenting, in this case, denomination but no repetition. It must be emphasized here that the speech is not well organized, and it is very common to find short sentences on it. From an anatomical perspective, it could be said that the region affected by this type of aphasia is the anterior superior frontal lobe of the left hemisphere, thus, the executive functions which are relevant to language can also be affected if there is an important damage in the frontal lobe. The latter is characterized by a good fluency and an altered comprehension with a strong repetition. In addition to this, the patient tends to repeat the last words used by the interlocutor, there is also a patron used like *one thing* or *that* and the most interesting thing is that although there are no difficulties to read aloud, it seems that the patient does not understand what he is reading. The only cerebral area affected is, in this case, the inferior left temporal lobe of the brain which is placed near to Wernicke's area. Within the transcortical aphasia, a third type must be mentioned which is mixed transcortical aphasia which is characterized by having little fluency with certain mutism and agrammatism, the comprehension is quite affected, and the repetition remains intact. These types of patients are capable of repeating long sentences easily but without understanding what they are repeating.

However, the conduction aphasia appears when a lesion is produced in the parietal lobe of the brain, concretely, in the arcuate fasciculus which serves as a link between Wernicke's area and Broca's area and whose main function is to carry the message from one area to another. In this type of aphasia, the patient manifests a good fluency and a good comprehension, but he has certain difficulties to repeat words. That is the reason by which the patient presents many paraphasias.[12]

The last type of aphasia to be mentioned here is known as anomic aphasia which does not affect a specific area and it seems to be frequent in Alzheimer disease or when there is a head trauma or a metabolic encephalopathy. In this type of aphasia, patients seem to have certain difficulties in finding the right words when a description of an object is provided. It is typical of anomic patients to feel frustration when they cannot find the word they are looking for, even when they know the meaning, but it seems that their lexical access has become limited. In spite of this, the patient has a good fluency and comprehension, but repetition is present. According to Brown (1977), the lesion has not necessarily to be produced in the language areas, but it could take place outside of it, in fact, he states that:

> The more severe "nonfluent" anomia occurs with unilateral (left) temporo-parietal lesion. Lesions of the posterior middle temporal gyrus (T2) and its continuation to angular gyrus appear to be highly correlated with this form. The more fluent the anomia is, the more likely is diffuse pathology or lesion outside the speech area (p.38).

As it can be appreciated, the study of the multiple cases related to this pathology known as aphasia has made possible to know a little bit more what is concerned in this chapter which is language and its implications in ToM. However, there are still so many things that remain unknown about language, for instance, how language can interfere in the issue concerning ToM or how children can make use of language for lying other children. Nowadays, researchers have been studying the importance of language in ToM as it has already been mentioned in the previous section and what it has been understood by this concept at a cognitive level. In recent years, there has been so many studies that have tried to observe how the ToM can be affected in an aphasic person considering the fact that, depending on what type of aphasia we are talking about, there are certain linguistic skills that become impaired while his cognitive skills can remain intact.

[12]There are two types of paraphasias that can be distinguished which are known as verbal and phonemic paraphasias. The former is characterized by replacing the lexical term for another one which is proximate in terms of lexical category. The latter has to do with the replacing of an internal letter of a word, for instance, president-predident.

From a psychological view, it is known that the human being has an important faculty, apart from the faculty of language, that allows him to build complex structures by using embedding sentences and it allows him the power of mental time travel. This human faculty is known as recursion which can be seen as a feature of the human mind and language does not only enable him to conceive his own mind but also the minds of others. According to Corballis (2011), the recursive faculty is strongly connected to ToM until the point that there is a belief that language is based on those complex mental structures that involve both ToM and recursive structures of mental travel time. Because people can have certain or several limitations to their linguistic capacity after having suffered a stroke or something similar, that has provoked a damage in an area related to language, several studies have been carried out with a population of aphasic patients in order to determine whether that limitation to the linguistic capacity can affect to their ToM.

As it has been described in previous sections, ToM does not only imply language but also thought. In this fashion, when an aphasic patient suffers from Broca or Wernicke aphasia, he is unable to produce the language or even to comprehend it but that does not mean necessarily that he cannot think by himself as he has many ideas and even concepts inside his mind that cannot be expressed through language. By focusing on this pathological aspect, Bánréti et al. (2016) tried to see until what point the linguistic limitation in aphasia had an important effect on the syntactic recursion by making tasks based on embedding sentences. Curiously, patients with Broca aphasia tend to give answers by using simple and short sentences, considering his difficulty to produce the language, which involve at the same time ToM inferences by avoiding complex structures. What it was found was that recursive sentences embedding are impaired in patients with Broca aphasia, however, the recursive ToM inferences can remain intact in some types of aphasia since not all aphasias are equal. It is important to bear in mind that in terms of ToM inferences the concept of recursion can be defined as follows the capacity to "see oneself as able to infer other people's mental states, consider other person to infer further persons' mental states, thus exhibiting recursive constructions" (Bánréti et al. 2016, p. 10).

Studies carried out with aphasic population determine that there is a double disassociation since the subsystem of recursive sentence embedding was very limited while in the case of ToM there seemed to be an important impairment in patients with Broca aphasia. This disassociation can also be seen in those patients with Wernicke aphasia who presented some impairments in the lexical process. This disassociation could be explained throughout a theory that only competes with those operations related to complex cognitive domains. Some researchers like Friederici et al. (2011) make distinction between two computational systems that process hierarchical structures in the lateral prefrontal cortex being one of them determined by cognitive control which is considered to be less automatic and it is able to activate the anterior prefrontal cortex and the other one being the recursive syntactic hierarchy of natural language which is more automatic and it is able to activate the posterior regions of the inferior frontal gyrus including the area of Broca. Having knowledge of this, it must be pointed out that two types of systems can also be involved in the field of ToM as well. These systems are defined by Apperly and Butterfill (2009) as follows:

> One type of system is quick, highly automatic, cognitively efficient, but limited and inflexible. This early-developing system has a central role in guiding online social interaction and simple everyday communication. Another type of system is highly flexible but cognitively inefficient and requires more cognitive control. This later-developing flexible cognitive system is necessary for solving false-belief tasks and enable adults to perform explicit ToM reasoning in complex social interaction (p. 18).

These two systems can coexist in the mind of an adult, and they can intervene in ToM. Having these two systems operating in the mind can help to calculate other people's mental states in a faster way showing a clearly link between language and ToM. Aphasic patients showed to have a strong cognitive control in the tasks related to the ToM since they were succeeded on them. Interestingly, they present certain impairments in linguistic skills but not in cognitive skills.

An important focus must be made in relation to global aphasia which can be seen as a combination of Broca aphasia and Wernicke aphasia since it is mainly characterized for presenting certain problems or, better said, difficulties in making grammatical judgements about a sentence, in the production of language and in auditory perception (Adams 2020). Studies carried out by Varley and Siegal (2000) showed that the linguistic ability and the cognitive ability can be separated from each other, at least, in the mind of an adult. In fact, there are aphasic patients who are succeeded in non-linguistic ToM tasks supporting, in this way, those theories that claim that cognitive capacity does not depend upon the natural language (Pinker and Jackendorf 2005). Since the brain has certain areas which are specialized in language, there is a tendency to believe that a damage produced in some of these areas could leave pre-linguistic cognitive areas intact. On the other hand, there are many linguists who are not in favor of this belief, for instance, Baldo et al. (2005) carried out a study in which the cognitive performance seems to be reduced due to a reduction in the language capacity. In their own words, a cognitive efficacy of language in cognition could be attributed to what is well-known in the linguistic field as inner speech.

Once that the origin of aphasia has been known, what consequences can bring for the individual who suffers from it and whether it affects ToM or not, the crucial point is whether it is possible to have a recovery from this illness and return to a normal life or whether this pathology is permanent. Fortunately, our society not only counts with great professionals in the field of medicine but also with excellent psychologists and speech therapists who work as a team with the only purpose of ensuring the speedy recovery of the patient. There are so many patients that have had a completely recovery after having got through an aphasia while others cannot have recovered completely, though, they improved their abilities thanks to the cognitive exercises provided by the speech therapist.

First of all, the patient must be seen by a neurologist so that he can give him a good diagnosis based on the pictures taken from computerized tomography where the neurologist can observe which are the areas that have been affected and how this has affected to the patient's ability to communicate. After this medical examination, the aphasic patient will be derived to the speech therapist in order to make a set of cognitive exercises to improve his abilities.

Previous to these exercises, the speech therapists must make a cognitive evaluation in order to know much better the degree of difficulty that the patient has when it deals with the production and comprehension of what has been said. For doing so, there are so many neuropsychological tests that can be put into practice such as the Battery of Western Aphasia[13] or Minnesota Test for Differential Diagnosis of Aphasia[14] but, maybe, the most common test to be used for the diagnosis of this language pathology is the so-called Boston Naming Test.[15] After having detected the speech difficulties presented by the patient, the speech therapist can determine which kind of cognitive exercises are useful for the patient according to the aphasia degree. A good example of this is a patient with Broca's aphasia to whom the most appropriate linguistic rehabilitation is the one based on verbal expression. This kind of patients have a good comprehension in general, thus, a good way to exercise the communication is through the symbology and pictograms or, as a last resource, the use of an automatized language[16] which is a useful exercise to get a more voluntarily language with the passage of time. Another way to help this aphasic patient to produce the language for a good communication is through the use of gestures, written texts or even by helping him with the vocalization of syllables and words. Having achieved a voluntarily language, then, the speech therapist can go one step further in the rehabilitation process and start with the formulation of phrases by creating questions so that the patient can give an answer until he achieves to build a complete sentence. Although, these kinds of activities tend to be the most recurrent ones for a good and complete recovery, there are other methods that can also help the patient which are the use of drugs in determining aphasic patients. However, there are not so much evidence that can support the use of the pharmacology for the recovering since this cannot ensure that the patient can produce the language or comprehend it.

Definitely, the speech therapist plays an essential role in the recovery of the aphasic patient but, despite of this, this kind of help is not enough to cope with the daily life of an aphasic patient. Sometimes, human beings are so immersed in their own lives and worries that they are not conscious of the impact that a disease like aphasia can have on an individual at an emotional level since his life changes radically. Getting through this is not an easy process since it does not only entail

[13]This test consists of four subtests for oral language in order to get a global knowledge about the general state of the patient.

[14]This is an extensive test that consists of 59 subtests which are grouped by areas, for instance, hearing disorder or speech disorders among others.

[15]This test is used to measure confrontational word retrieval in individuals with aphasia or other language disorder caused by a stroke or Alzheimer's disease. This test was created by Kaplan, Goodglass and Weintraub in 1983.

[16]This kind of exercise consists of repeating a series of automatic numbers, songs or even the months of the year.

physiological sequelae but also psychological sequelae, thus, attending to the psychologist is most of the time required.

The psychologist listens to the patient and gives him a piece of advice so that the patient can continue with his life in the best possible way, even when, it is hard to believe the psychologist is the only one who is capable of comprehending the patient because of his knowledge about the mind and the emotions. The role played by the psychologist in this situation is minor, however, the more help the patient receives from the different branches (neurology, psychology, or speech therapy), the faster he will recover.

It is a fact that all this help coming from the different branches mentioned above are essential factors in the life of an aphasic patient, but it cannot be forgotten the most important factor that takes part in this recovery which is the support provided by the patient's family. As an important factor, family could not be less than a neurologist, a psychologist, or a speech therapist because family is the one who spends most of the time with the aphasic person, helping him, giving him moral support, and making him see that despite suffering from this disease he can and should continue with his life. For an aphasic person, it must not be easy to deal with this kind of life since most of the time the individual could feel frustration for not being able of communicating properly and, in the most extremely cases, depression could appear after a radical change like this one in his life. Taking this into account, I would even say that there is always a reason to live and even when an individual has certain disabilities, I can assure that there are other abilities that can develop in order to compensate those ones. It is true that human beings need to be in contact with other human beings and establish a verbal communication to express their thoughts, ideas, emotions, and feelings but believe me when I say that there are different ways of communicating without using words. It is harder to say but there is no reason to feel such a frustration or even depression since human beings are stronger than they really think and, as it can be seen above, a recovery is possible if we seek the appropriate means for it.

17.4 Conclusions

Throughout this chapter it has been possible to reflect on several relevant issues to consider in order to understand what the relationship between language, mind and thought is. For doing so, it has been necessary to provide some important definitions for these concepts which have helped us later to understand the concept of ToM and its implications in the different language pathologies and disorders. Despite the current belief that language can exist without thought and the other way around, the truth is that both need from each other in the sense that if there is language inside of our minds it needs to transmit some sort of knowledge such as ideas or concepts, and if there is thought inside of our minds it needs to be expressed. As it has been seen above, there are so many theories that have tried to explain this relation concerned with language and thought but the point is that language and thought complement each other, what is more, language cannot exist without thought.

In terms of ToM, it has been seen how this faculty appears in the first years of life along with language. All human beings are born with this faculty but not all of them develop it properly and, as an exception, it must consider that there are individuals who lack both language and ToM. As a human faculty, the ToM gives us the ability to read mental states and to tell in advance the behavior of others but, in some cases, this ability seems to be affected due to certain disorders. However, that does not mean that an individual with a language disorder or pathology lacks ToM, it is just a matter of later development. As it happens in real life, the development of ToM can be improved by being exposed to several tasks related to it since the ToM is more related to the cognitive skills rather than the linguistic skills.

To sum up this chapter, it must be highlighted that there is a stronger connection between language, mind and thought since language and thought are linked with each other and both are found within the mind. Thus, it can be said that human beings are lucky to have first the faculty of language which is right there from the very beginning and to have as a part of cognition the ToM through which they can anticipate not only the behavior of others but also their desires and beliefs. Under my point of view, these two faculties make the individual to be able to communicate, think and decide because everything is connected in the mind: language, concepts, ideas, the capacity to provide judgement and to make plans among others, and likely without them we would not be humans.

References

Adams, F. 2020. Global aphasia and the language of thought. *Theoria Revista de Teoría, Historia y Fundamentos de la Ciencia* 35 (1): 9–27.

Alexander, G.E., M.R. DeLong, and P.L. Strick. 1986. Parallel organization of functionally segregated circuits linking basal ganglia and cortex. *Annual Review of Neuroscience* 9 (1): 357–381.

Apperly, I.A., and S.S. Butterfill. 2009. Do humans have two systems to track beliefs and belief-like states? *Psychological Review* 116: 953–970.

Astington, J.W. 1993. *The child's discovery of the mind*. Cambridge, MA: Harvard University Press.

———. 1996. What is theoretical about the child's theory of mind? A Vygotskian view of its development. In *Theories of theories of mind*, ed. P. Carruthers and P.K. Smith, 184–199. Cambridge: Cambridge University Press.

Astington, J.W., and J.A. Baird. 2005. *Why language matters for theory of mind*. Oxford: Oxford University Press.

Astington, J.W., and J.M. Jenkins. 2000. ToM and social behavior: Causal models tested in a longitudinal study. *Merrill-Palmer Quarterly* 46 (2): 203–220.

Baldo, J.V., N.F. Dronkers, D. Wilkins, C. Ludy, P. Raskin, and J. Kim. 2005. Is problem solving dependent on language? *Brain and Language* 92: 240–250.

Bánréti, Z., I. Hoffmann, and V. Vincze. 2016. Recursive subsystems in aphasia and Alzheimer's disease: Case studies in syntax and ToM. *Frontiers in Psychology* 7: 405.

Baron-Cohen, S., and H. Ring. 1994. A model of the mindreading system: Neuropsychological and neurobiological perspectives. In *Origins of an understanding mind*, ed. P. Mitchell and C. Lewis, 183–207. Hillsdale: Erlbaum.

Baron-Cohen, S., H.A. Ring, S. Wheelwright, E.T. Bullmore, M.J. Brammer, A. Simmons, and S.C. Williams. 1999. Social intelligence in the normal and autistic brain: An fMRI study. *European Journal of Neuroscience* 11 (6): 1891–1898.

Bartsch, K., and H. Wellman. 1995. *Children talk about the mind*. New York: Oxford University Press.

Bear, M.F., B.W. Connors, and M.A. Paradiso. 2016. *Neurociencia: la exploración del cerebro (Cuarta edición)*. Wolters Kluwer.

Bloom, F.E., A. Lazerson, and L. Hofstadter. 1985. *Brain, mind, and behavior*. New York: W. H. Freeman.

Bolhuis, J.J., I. Tattersaall, N. Chomsky, and R.C. Berwick. 2014. How could language have evolved? *PLoS Biology* 12: e1001934.

Booth, J.R., and W.S. Hall. 1995. Development of the understanding of the polysemous meanings of the mental state verb know. *Cognitive Development* 10: 529–550.

Broca, P. 1861. Nouvelle observation d'aphémie produite par une lésion de la moitié postérieure des deuxième et troisième circonvolutions frontales. *Bulletins de la Société anatomique de Paris* 36: 398–407.

Brown, J.W. 1977. *Mind, brain and consciousness: The neuropsychology of cognition*. New York: Academic.

Bunge, S.A., C. Wendelken, D. Badre, and A.D. Wagner. 2005. Analogical reasoning and prefrontal cortex: Evidence for separable retrieval and integration mechanisms. *Cerebral Cortex* 15: 239–249.

Cattell, R.B. 1971. *Abilities: Their structure, growth, and action*. Boston: Houghton-Mifflin.

Chomsky, N. 1972. *Language and mind*. New York: Harcourt Brace Jovanovich.

Corballis, M.C. 2011. *The recursive mind: The origins of human language, thought, and civilization*. Princeton: Princeton University Press.

Crelin, E.S. 1969. *Anatomy of the newborn: An atlas*. Philadelphia: Lea and Febiger.

Crone, E.A., C. Wendelken, L. van Leijenhorst, R.D. Honomichl, K. Christoff, and S.A. Bunge. 2009. Neurocognitive development of relational reasoning. *Developmental Science* 12: 55–66.

Cummings, J.L. 1993. Frontal-subcortical circuits and human behavior. *Archives of Neurology* 50 (8): 873–880.

Darwin, C.R. 1874. *The descent of man and selection in relation to sex*. 2nd ed. New York: Hurst & Co. [hay ed. cast.: El origen del hombre. Madrid: EDAF, 1982, 10th ed].

De Villiers, J. 2007. The interface of language and ToM. *Lingua* 117 (11): 1858–1878.

De Villiers, J.G., and P.A. de Villiers. 2000. Linguistic determinism and the understanding of false beliefs. In *Children's reasoning and the mind*, ed. P. Mitchell and K. Riggs, 189–226. Hove: Psychology Press.

Dolan, R.J., R. Lane, P. Chua, and P. Fletcher. 2000. Dissociable temporal lobe activations during emotional episodic memory retrieval. *NeuroImage* 11 (3): 203–209.

Doody, G.A., M. Götz, E.C. Johnstone, C.D. Frith, and D.C. Owens. 1998. ToM and psychoses. *Psychological Medicine* 28 (2): 397–405.

Duncan, J., and A.M. Owen. 2000. Common regions of the human frontal lobe recruited by diverse cognitive demands. *Trends in Neurosciences* 23 (10): 475–483.

Dunn, J., J. Brown, C. Slomkowski, C. Tesla, and L. Youngblade. 1991. Young children's understanding of other people's feelings and beliefs: Individual differences and their antecedents. *Child Development* 62: 1352–1366.

Fedorenko, E., M.K. Behr, and N. Kanwisher. 2011. Functional specificity for high-level linguistic processing in the human brain. *Proceedings of the National Academy of Sciences* 108 (39): 16428–16433.

Flowers, K.A., and C. Robertson. 1985. The effect of Parkinson's disease on the ability to maintain a mental set. *Journal of Neurology, Neurosurgery & Psychiatry* 48 (6): 517–529.

Fodor, J.A. 1992. A theory of the child's theory of mind. *Cognition* 44: 283–296.

Friederici, A.D., J. Bahlmann, R. Friedrich, and M. Makuuchi. 2011. The neural basis of recursion and complex syntactic hierarchy. *Biolinguistics* 51–52: 087–104.

Frith, C.D., and U. Frith. 1999. Interacting minds—A biological basis. *Science* 286 (5445): 1692–1695.

Frith, C., and U. Frith. 2005. ToM. *Current Biology* 15 (17): R644–R645.

Frye, D., P.D. Zelazo, and T. Palfai. 1995. Theory of mind and rule-based reasoning. *Cognitive Development* 10: 483–528.

Gallagher, H.L., and C.D. Frith. 2003. Functional imaging of 'ToM'. *Trends in Cognitive Sciences* 7 (2): 77–83.

Gallagher, H.L., F. Happé, N. Brunswick, P.C. Fletcher, U. Frith, and C.D. Frith. 2000. Reading the mind in cartoons and stories: An fMRI study of 'ToM'in verbal and nonverbal tasks. *Neuropsychologia* 38 (1): 11–21.

Gleitman, L. 1990. The structural sources of verb meanings. *Language Acquisition* 1: 3–55.

Gopnik, A., and H.M. Wellman. 1994. The theory of mind. In *Mapping the mind: Domain specificity in cognition and culture*, ed. L. Hirschfeld and S. Gelman, 257–293. New York: Cambridge University Press.

Gray, J.R., T.S. Braver, and M.E. Raichle. 2002. Integration of emotion and cognition in the lateral prefrontal cortex. *Proceedings of the National Academy of Sciences of the United States of America* 99: 4115–4120.

Green, A.E., D.J. Kraemer, J.A. Fugelsang, J.R. Gray, and K.N. Dunbar. 2009. Connecting long distance: Semantic distance in analogical reasoning modulates frontopolar cortex activity. *Cerebral Cortex* 20 (1): 70–76.

Harley, T.A. 2001. *The psychology of language: From data to theory*. 2nd ed. Hove: Psychology Press.

Holyoak, K.J., and J.K. Kroger. 1995. Forms of reasoning: Insight into prefrontal functions? *Annals of the New York Academy of Sciences* 769: 253–263.

Jaynes, J. 1976. *The origin of consciousness in the breakdown of the bicameral mind*. Boston: Houghton Mifflin.

Krawczyk, D.C. 2012. The cognition and neuroscience of relational reasoning. *Brain Research* 1428: 13–23.

Leslie, A.M. 1987. Pretense and representation: The origins of "ToM". *Psychological Review* 94: 412–426.

———. 1994. Pretending and believing: Issues in the theory of TOMM. *Cognition* 50: 211–238.

Lieberman, P. 2016. The evolution of language and thought. *Journal of Anthropological Sciences* 94: 127–146.

Logan, R.K. 2010. Mind and language architecture. *The Open Neuroimaging Journal* 4: 81.

Lund, N. 2003. *Language and thought (Textbook)*. London: Routledge.

Miller, L. 1985. Cognitive risk-taking after frontal or temporal lobectomy — I. The synthesis of fragmented visual information. *Neuropsychologia* 23: 359–369.

Miller, C.A. 2006. Developmental relationships between language and ToM. *American Journal of Speech-Language Pathology* 15 (2): 142–154.

Montgomery, D.E. 2002. Mental verbs and semantic development. *Journal of Cognition and Development* 3: 357–384.

Morales, M., P. Mundy, C.E. Delgado, M. Yale, D. Messinger, R. Neal, and H.K. Schwartz. 2000. Responding to joint attention across the 6-through 24-month age period and early language acquisition. *Journal of Applied Developmental Psychology* 21 (3): 283–298.

Negus, V.E. 1949. *The comparative anatomy and physiology of the larynx*. New York: Hafner.

Nelson, K. 1996. *Language in cognitive development: The emergence of the mediated mind*. New York: Cambridge University Press.

Perner, J. 2000. About + belief + counterfactual. In *Children's reasoning and the mind*, ed. P. Mitchell and K.J. Riggs, 367–401. Hove: Psychology Press.

Peterson, C.C. 2000. Kindred spirits: Influences of siblings' perspectives on ToM. *Cognitive Development* 15: 435–455.

———. 2000. Insights into theory of mind from deafness and autism. *Mind & Language* 15: 123–145.

Petrides, M., and D.N. Pandya. 1999. Dorsolateral prefrontal cortex: Comparative cytoarchitectonic analysis in the human and the macaque brain and corticocortical connection patterns. *The European Journal of Neuroscience* 11: 1011–1036.

Piaget, J. 1950. *The psychology of intelligence*. London: Routledge & Kegan Paul.

Pickett, E.R., E. Kuniholm, A. Protopapas, J. Friedman, and P. Lieberman. 1998. Selective speech motor, syntax and cognitive deficits associated with bilateral damage to the putamen and the head of the caudate nucleus: A case study. *Neuropsychology* 36 (2): 173–188.

Pinker, S. 1994. *El instinto del lenguaje: cómo crea el lenguaje la mente*. Alianza.

Pinker, S., and R. Jackendorf. 2005. The faculty of language: What's special about it? *Cognition* 95: 201–236.

Postle, B.R. 2006. Working memory as an emergent property of the mind and brain. *Neuroscience* 139 (1): 23–38.

Premack, D., and G. Woodruff. 1978. Does the chimpanzee have a 'ToM'? *Behavioural and Brain Sciences* 4: 515–526.

Robin, N., and K.J. Holyoak. 1995. Relational complexity and the functions of prefrontal cortex. In *The cognitive neurosciences*, ed. M.S. Gazzaniga, 987–997. Cambridge, MA: MIT Press.

Ruffman, T., L. Slade, and E. Crowe. 2002. The relation between children's and mothers' mental state language and theory-of-mind understanding. *Child Development* 73: 734–751.

Ruffman, T., L. Slade, K. Rowlandson, C. Rumsey, and A. Garnham. 2003. How language relates to belief, desire, and emotion understanding. *Cognitive Development* 18: 139–158.

Sapir, E. 1929. The study of language as a science. *Language* 5: 207–214.

Spearman, C. 1904. The proof and measurement of association between two things. *The American Journal of Psychology* 15: 72–101.

Spurzheim, J.K. 1815. *The physiognomical system of Drs. Gall and Spurzheim*. London: Baldwin Cradock and Joy.

Tomasello, M. 1995. Joint attention as social cognition. In *Joint attention: Its origins and role in development*, ed. C. Moore and P.J. Dunham, 103–130. Hillsdale, NJ: Erlbaum.

Ullman, M.T. 2004. Contributions of memory circuits to language: The declarative/procedural model. *Cognition* 92 (1–2): 231–270.

Varley, R., and M. Siegal. 2000. Evidence for cognition without grammar from causal reasoning and 'ToM' in an agrammatic aphasic patient. *Current Biology* 10: 723–726.

Vygotsky, L. 1962. *Thought and language*. Cambridge, MA: Harvard University Press.

Wellman, H. M. 1979. *A child's theory of mind*. Paper presented at the conference the growth of insight in the child. Madison.

Wellman, H.M. 1985. The child's theory of mind: The development of conceptions of cognition. In *The growth of reflection in children*, ed. S.R. Yussen, 169–206. San Diego: Academic.

Whorf, B.L. 1956. *Language, thought, and reality: Selected writings of Benjamin lee Whorf*. Cambridge MA: MIT Press.

Wright, S.B., B.J. Matlen, C.L. Baym, E. Ferrer, and S.A. Bunge. 2007. Neural correlates of fluid reasoning in children and adults. *Frontiers in Human Neuroscience* 1: 8.

Yang, F.G., J. Edens, C. Simpson, and D.C. Krawczyk. 2009. Differences in task demands influence the hemispheric lateralization and neural correlates of metaphor. *Brain and Language* 111: 114–124.

Youngblade, L.M., and J. Dunn. 1995. Individual differences in young children's pretend play with mothers and siblings: Links to relationships and understanding of other people's feelings and beliefs. *Child Development* 66: 1472–1492.

Chapter 18
Relations Between Bilingualism and Theory of Mind, a Neurologic Challenge. From the *Bilingual Advantage* to a New Assessment of Conclusions

Priscila Sánchez Soriano

Abstract Since the late decades of the twentieth century, the connections between language-related cognitive skills and the development of Theory of Mind (ToM) have been extensively researched. During the first decade of the twenty-first century, bilingualism and monolingualism have been compared, highlighting the impact of bilingualism in the ToM and the advantages of the bilingual brain.

Experts on linguistics and neurolinguistics have been inspired specially by conclusions which claimed for the recognition of bias in research. The still immature attempts provided differences in results more likely to be attributed to the approaches of the experiments themselves than to the abilities of the individuals. The recognition of bias in research allowed (1) a more taxonomic perspective of the bilingualism as a researchable concept, (2) the identification of those aspects that still need to be attended and (3) the boost of bonds between psycholinguistics and neurolinguistics.

On the one hand studies on bilingualism and ToM point on the one hand to relevant differences among the several bilingualisms when it comes to define the relation with ToM. On the other hand, they insist on the need for research of the underlying mechanisms in bilingual processes, which requires to discriminate the brain areas involved when performing the tasks, as well as to examine the main relevant factors as for the relation with ToM, that is sociolinguistics, executive function (inhibition, working memory, planning . . .) or metalinguistic awareness. And at last, they remind that neuroscience is a core discipline when researching the brain regions involved, the cognitive processes themselves and the nature of their relationship with ToM.

P. S. Soriano (✉)
Universidad Pablo de Olavide, Seville, Spain
e-mail: plsansor@upo.es

T. Lopez-Soto et al. (eds.), *The Theory of Mind Under Scrutiny*, Logic, Argumentation & Reasoning 34, https://doi.org/10.1007/978-3-031-46742-4_18

In the following lines we will be revising relevant studies and their results from the first hypothesis and conclusions claiming for a bilingual advantage to the recent ones, which provide different evidence and point to new hypothesis.

Keywords ToM · Bilingualism · Executive function · Sociolinguistic account · Neuroscience

18.1 Introduction

The assumption of a bilingual advantage has been trying to set up as an axiom despite some scientific warnings against this optimism. Media, educational institutions and even agents of the language teaching business continue to elaborate and spread the message without mention to the temporary nature of these conclusions nor to the drawbacks that some types of bilingualism imply, not even to properly distinguish between the different kinds of bilingualism. According to Bailey et al. (2020)

> Comparing monolinguals to bilinguals is one fallacy, but comparing simultaneous to sequential bilinguals, or early to late bilinguals, or balanced to unbalanced bilinguals, is another (p. 242).

In addition to the proper distinctions between the several types of bilingualism, other factors have been proved to be affecting the result of the experiments. Accurate approaches to the relation between bilingualism and Theory of Mind (from this point forward, ToM) should observe and relate them to one another. In this respect, it was reached a consensus on the fact that language and ToM are linked in a synergetic relation through three elements: executive function, metalinguistic awareness, and sociolinguistics, however it is unknown or not fully describable how they affect each other or to what extend each of the elements have an impact.

Thus, the first investigations, whose conclusions gave the bilingualism a significant advantage, yielded to more restrained alignment who work in:

- The recognition of limitations in the experiments.
- The identification of new variables like the age of the samples, the socioeconomic status of their families, the gender, the intelligence, the kind of bilingualism, the multilingualism, the features of the languages in combination, etc.
- The design of the experiments enriched (1) with the involvement of others disciplines like neurology and (2) with tests consisting of tasks which aspire to measure and differentiate the impact of the different factors, triangulating the findings. To the classic perspective taking or false believe task (unexpected location, unexpected contents…) are added fewer common tasks designed for different ages or for the measure of different values, like *faux pax* tests or tests to evaluate the spatial vision of others when differing with the own vision.

18.2 Types of Bilingualism

In the following lines we will briefly describe some types of bilingualism, paying special attention to those that, according to the researched literature, mediate the results in their relationship with ToM. In addition to the individuals' factors that increasingly receive more attention and importance —as claimed in the texts we are going to revise; other variables have a deep impact on the development of ToM. That is: (1) the circumstances and (2) the time of learning, (3) the level of competence of the language(s), (4) the independent or dependent mechanism of the languages regarding to one another and their interaction within the brain while the process of linguistic management, (5) the use of languages at bilingual's hands, (6) the cultural identity of the bilingual and (7) the sociocultural status of the languages spoken in the community where the bilingual belongs to.

1. The circumstances of the learning process allow to distinguish between native bilinguals, who learnt the language effortless, (crib bilingualism) and acquired bilingualism, of the person who learnt the language ad hoc (school bilingualism).
2. The moment of learning allows to distinguish between early or late, and simultaneous or consecutive (also known as sequential) bilingualism. The simultaneous bilinguals learn both languages at the same time, while in the consecutive bilingualism, the second language is learnt after the first one. The so called early bilingualism gets developed by the age of 7 and seems to improve both the cognition and the recognition of emotions, and therefore, the social intelligence (Javor 2016), however it remains necessary to research further to fully understand the origin of this advantage, the age until the advantage is perceivable and the underlying mechanisms that make it possible, especially in the field of recognition of emotions, where the literature has said very little so far.[1] (Translated by the author)
3. The level of competence decides whether the bilingualism is complete or incomplete and within the frame of this distinction, it is possible to still differentiate between:

   3a. Receptive bilingualism, which implies oral y written comprehension, and productive bilingualism which entails the productive skills as well, that is, spoken and written abilities.

   3b. Balanced bilingualism, which implies similar and high command and effective use of both languages in different circumstances and contexts, and dominant bilingualism, which entails higher command in one of the languages in combination.

[1]"[. . .] aún es necesario investigar más para comprender en profundidad cuál es el origen de esta ventaja, hasta qué edad se extiende y cuáles son los mecanismos subyacentes que la hacen posible, especialmente en el ámbito del reconocimiento de emociones, donde la literatura al respecto es prácticamente inexistente." (Pérez-Hernández et al. 2016, p. 2269).

4. The independent (or non-independent) management of the languages helps distinguish between coordinated bilingualism —the bilingual subject manages each language with no interferences in one another and compound bilingualism —the bilingual needs both languages for the purpose of communication due to a lack of sharply defined conceptual systems of the languages. The interferences and transferences occur between the languages in both directions. (Signoret et al. 2015).
5. According to their cultural identity, the bilinguals might be:
   - Bicultural: positively identified with both cultural groups, integrated, and recognized by both.
   - Monocultural: they keep the cultural identity of the L1 while learning the L2 and its culture.
   - Acultural: they lose their own cultural identity without having adopt the L2 cultural identity. (Aranda Farías 2020, p. 36–37)
6. The sociocultural status of the languages allows to distinguish between additive bilingualism —the community considers the bilingualism of the individual as a cultural enrichment for the community, and the subtractive bilingualism —the community is perceived as a risk of identity loss (Signoret et al. 2015). This classification closely bonds with the concept of *diglossia*, conceived as the use of two varieties of a language within the same community but in different domains.

18.3 Relation Between Bilingualism and ToM

Being no devoid of difficulty, the conceptual organization of the bilingual phenomena can be performed in connection with linguistics, political, sociocultural criteria or those ones chosen according to the aim of the research. However, the boundaries of these categories are not so sharp when it comes to draw the frame of the relation with the ToM.

The circumstances and the timing of the learning process mediate the performance of the ToM since

> [...] the cognitive correlates of early exposure to two languages include better inhibitory control, greater cognitive flexibility, and enhanced ToM. Second language learning in the critical period of SLA provides along many functions' effective solutions and execution. (Javor 2016, p. 147)

On the other hand, and regarding the level of competence, the studies conducted show that it is needed to develop a minimum grade of linguistic competence in at least one language so that the linguistic abilities will affect the performance of ToM.

Regarding the autonomy in the management of languages, the possibility —and even probability of mediation should be assumed since Michel Paradis (Paradis 1981, 1987, 2004 in Signoret et al. 2015) demonstrated the coexistence of two independent and dual neurologic storages (coordinated bilingualism), albeit located in the same brain area, and one extended and wide mechanism (compound bilingualism) serving the two languages at once. In the first type an eventual aphasia might affect one only language while in the second type might affect the two on them.

Regarding the sociolinguistic dimension —wherein factors of sociocultural status of the language as well as the cultural identity of the bilingual are framed, Pérez-Luzardo states that: bicultural bilinguals who achieved to integrate in their identity both cultures will be influenced by two different systems, each of them with its own way to perceive reality, and they will both have an impact in their emotional and cognitive account.[2] (Translated by the author).

Nevertheless, and beyond the kind of bilingualism, we will see how recent research point to differences in the findings, more conferred to other variables such as gender, age, intelligence or even the languages in combination, among others.

18.4 Theory of Mind. Brief Definition

Although the context of this publication develops the concept well enough, it might be advisable to write some lines within this text to describe the ToM, as for its alignment with bilingualism as a phenomenon.

The concept should be approached from a perspective wide enough to understand the scope and detailed enough to understand the links with linguistic abilities and the boundaries of our knowledge about these above-mentioned connections.

In general terms, the ToM is the ability to perceive and recognize mental states and own's and others' emotions, and, although it is noticeable already since 4, it continues to develop until the adulthood. The False Belief is the more common test used to detect it in children's mind and to assess its development.

18.5 Tasks to Assess ToM

The False Believe test was designed and firstly implemented in 1978 when David Premack y Guy Woodruff published *Does the chimpanzee have a theory of mind?* with the purpose of announcing that Sarah, a female primate, was able to understand human behaviour. Inspired in this experiment, in 1983, Wimmer & Perner released *Beliefs about beliefs: Representation and constraining function of wrong beliefs in young children's understanding of deception*, where they inform about the results of

[2]"[. . .] un bilingüe bicultural, que haya logrado integrar las dos culturas, a las cuales pertenece, en su identidad, estará influenciado por dos sistemas diferentes, cada uno con su manera de percibir la realidad, que incidirán en su manera de pensar y sentir." (Pérez-Luzardo and Schmidt 2016, p. 58)

an experiment tried in 36 Austrian children in ages between 4 and 9. The test, with Max and a chocolate bar as main elements, was aimed to reveal when an individual can reason about the mental states of others. This experiment was versioned 2 years later by Baron-Cohen et al. (1985) replacing Max and chocolate bar by Anne and Sally and simplifying the plot to assess the mental abilities of Down syndrome and autistic children. The Smarties task, coetaneous of the False Believe task, was aimed to evaluate children's ability to distinguish between appearance and reality.

These three tasks are framed within the known as *first order tasks*. Together with the *second order tasks* they assess the stage of development of ToM in children. Thus, when the subject A has developed first order ToM is able to perceive that subject B has got a false believe regarding a situation. Subject A can differentiate the other's belief from its own knowledge or believe about reality. In a higher state, measurable through second order tasks, subject A might be proved to be able to realize that subject B confers a false believe to subject C. Subject A can differentiate others' believes from its own knowledge or believe about reality.

During the three next decades, the False Believe test has been resignified in line with the different aims of research and has suffered changes in its design to draw conclusions more accurate and adjusted to the different ages of the researched children. These changes were aimed to overcome the classical dualism "ToM, yes or no".

Although it was the first one, the False Believe Task has not been the only tool to measure the development of ToM. Other tasks within the mentalist type came afterwards, like the *faux pax* task, the perspective taking task, the intentionality task or the belief understanding task, among others. All together they can draft the brain landscape through the estimation of the inhibitory control, the working memory, the cognitive flexibility, the selective and sustained attention, etc. The evolution in research has been proving that the relation with language was influenced (1) by sociolinguistic factors, such as determinants of the context, social interaction, or life experiences, and (2) by with metalinguistic awareness of the bilingual. The rest have been modified along the years to get adapted to the new findings and goals.

Defining the ToM with accuracy does not seem to be an easy task, not even for experts, since the mental status are not observable. We can easily understand the meaning of *jump* or *speak* just by observing what they bring with, but this is not the case with verbs like *think* or *wish*. This is why language turns out to be so crucial when it comes to access the mind, its behavioural patterns, and its development. Furthermore, the science has demonstrated that, in addition to this key gateway, language itself is not only a mediating element but also a requirement for the development of ToM. Nevertheless, and despite the agreement about it, there are other theories about ToM. We will mention some of them, those which, because of their relationship with language, awaken special interest.

18.6 Theories About ToM

Theories on the competence framework underline the dependent relation of ToM towards experience and social interaction, while theories on the performance framework stress its dependency on executive function (Navarro et al. 2022). This antagonism might allow us to understand phenomena influenced by the sociolinguistic account if there is no intervention of the cognitive account. However, Apperly and Butterfill (2009) developed the so-called Theory of the Two Systems which might be able to explain aspects of the bilingual experience also affected by the cognitive account. The system 1 might be based on the heuristics (experience) and the system 2 might be based on a cognitive checking mechanism which activates when the heuristics does not offer proper tools. The findings of the more recent experiments can be better explained from this approach, that is, from the interaction of the two systems that are, in turn, mediated by "the context, individual differences and available resources" (Navarro et al. 2022, p.14).

Moreover, the science differentiates between affective ToM and cognitive ToM. The first one enables to interpret others' mental status and the second one enables to recognize them. It is proved by the science that these skills are regulated by different brain areas.

Our results suggest that, although cognitive and affective mentalizing abilities partly depend on each other, they are dissociable and depend on distinct (though overlapping) neuroanatomical substrates. As depicted from Fig. 18.4, both cognitive and affective ToM depend on intact PFC cortex. However, while cognitive ToM is affected by widespread prefrontal (VM + DLC) damage, affective ToM impairment results from circumscribed VM prefrontal damage (Shamay-Tsoory and Aharon-Peretz 2007, p. 3063). Therefore, it remains much light to shed on the details on these relations and it will be necessary to count on neuroimages to clarify key aspects.

18.7 Two Decades of Research. From the Identification of Bias to the Survey of Analysis

Studies conducted in the first years of the twenty-first century pointed to an advantage of the bilingual brain over the monolingual one, which was supported not only by a major development of executive function but also by higher levels of metalinguistic and sociolinguistic awareness that, thanks to a finer awareness of the linguistic arbitrariness, might foster the capability to manage the abstract representations in bilingual's mind. This current of optimisms regarding the bilingual advantage is well represented in the proposal of Goetz (2003), who examined three- and 4-year-old English monolinguals, Mandarin Chinese monolinguals, and Mandarin-English bilinguals by means of appearance-reality, level 2 perspective-taking, and false-belief tasks to find out whether an individual's linguistic knowledge influenced the development of ToM. He turned out to conclude that "Altogether there is certainly some evidence for a bilingual advantage on the theory of mind-related tasks" (p. 12) and suggested that the advantage might be due to a greater inhibitory control, greater metalinguistic understanding, and a greater sensitivity to sociolinguistic interactions with interlocutors.

For some years had a good part of the scientific community assumed this positioning, when research conducted by Meristo et al. in 2007 supported the idea that not any type of bilingualism might improve the development of ToM, but quite the opposite. Meristo et al. involved deaf children of three different nationalities: 97 Italian children from 4 to 9 years old, and 61 Estonian and Swedish children from 6 to 12 years old. They sorted children (1) who belonged to signing families, exposed to sign language since they were born and hence sign language bilinguals, and (2) who belonged to hearing families and hence had learnt sign language later. They also distinguished between children (3) who had attended a school whose main and principal language was sign language, and others (4) who had attended an oralist school offering sign language as a supportive option for children with hear impairment. The experiment, which included false belief and true belief tasks within the first phase, together with emotional recognition cartoon; belief-desire-based emotion reasoning task; unexpected contents task; unexpected location task and second-order false-belief task in the second phase, proved that an early, continuous and non-restricted access to the communication had fostered the ability to monitor others' mental states and the development of mentalist abilities in general, in contrast to what happened to children who had suffered restrictions in communication in their first language. Consequently, it can be said that, against Goetz's claim and far from proving an incontestable bilingual advantage, Meristo suggested that the wear, the damage due to a lack of exposition or use of language, might harm the regular development of reasoning linked to ToM, being this burden more significant when the restrictions affect the language preferred by the bilingual subject. It can be therefore inferred that some kinds of bilingualisms — among them those which do not result in the full identification of the bilingual with at least one of the languages in combination, might be detrimental for the development of ToM.

After the first decade of the twenty-first century, the scientific community begins (1) to find and identify bias in experimental protocols, (2) to design measuring systems enriched with more parameters and (3) to describe with more accuracy the bilingual phenomenon. All of that to enhance the account of the relations between bilingualism and ToM. Studies start to highlight, for the consideration of other researchers that tests are designed for the assessment of children between 3 and 6 years and consequently new tests are designed to research people in different ages. Differences because of reasons of gender, health (clinical versus non-clinical individuals), intelligence, kinds of bilingualism or even the languages in combination are now emphasized by research.

The Adaptive Control Hypothesis (Green and Abutalebi 2013) is an example of this new path in the main researching line. According to Green, bilinguals might adjust their neurocognitive structure to the demands of the conversational contexts and environments. According to Green, within the frame of the three possible contexts of interaction (interaction in single language, in dual language, or interaction involving dense code-switching), the individual face several challenges (control processes): maintain the goal, monitor the conflict, suppress the interference, detect the silent cue, inhibit the response, disengage from the task, engage with the task and plan according to the opportunity. Each process implies a different demand, and to this purpose the neurocognitive structure of the individual will be adjusted. This hypothesis was connected to the Complementarity Principle coined by Grosjean (2010), who 2 years earlier had stated that bilingual people use their languages in different contexts, and therefore different levels, according to the demands of communication in the different environments.

The validation of the Adaptive Control Hypothesis pointed at the path of the individual differences. It was considered convenient to observe and consider the impact of these differences, potentially developed in every single individual because of each personal bilingual experience in interaction with the linguistic context and the social environment. In the index of mediating factors are the competence in the language(s), the age of learning, the frequency in use, the immersion or cultural identification, the personal motivation, the formal education of the individual, etc.

The common consensus agreed about the fact that ToM, linguistic abilities, and executive function (taken as the mental process involved in the conscious control of thinking, actions, and emotions that, in turn, integrates inhibitory control, cognitive flexibility, planning and synthesis) coexist and are bond in a synergic relation to one another. Nevertheless, it is still to clarify the details of the relation and the manner this relation evolves along the development of the individual.

> [...] within the group of healthy individuals [...] it is to highlight the association between executive and mentalist abilities (social and cognitive). [...] We assume that, when these abilities are used by brains which function properly in a context of "cognitive normality", they perform in a coordinated manner and reveal systematic connections.[3] (Translated by the author)

Linguistic phenomena like the metaphor or the irony occur by means of the language mediation and they disambiguate thanks to the mentalist abilities, just like working memory or inhibitory control are requirements when it comes to face successfully a false believe task. However, once in 2012, Jessica Serrano, in her thesis about the development of ToM, language and executive functions, reminded that most of the studies about the relation between these three elements to one another (language, ToM and executive function) had been researched separately (Fig. 18.1).

She conducted research, involving 150 children between the ages of 4 and 12 years, to enlighten four questions:

[3]"[...] en el grupo de individuos sanos [...] destaca la asociación entre las habilidades ejecutivas y de mentalización (cognitivas sociales). [...] Interpretamos que, cuando las FE y las habilidades cognitivas sociales son utilizadas por mentes que «funcionan» dentro de una «normalidad cognitiva» [...], muestran un funcionamiento bien coordinado y manifiestan una interconexión sistemática." (Gavilán and J.E. García-Albea 2015, p. 127)

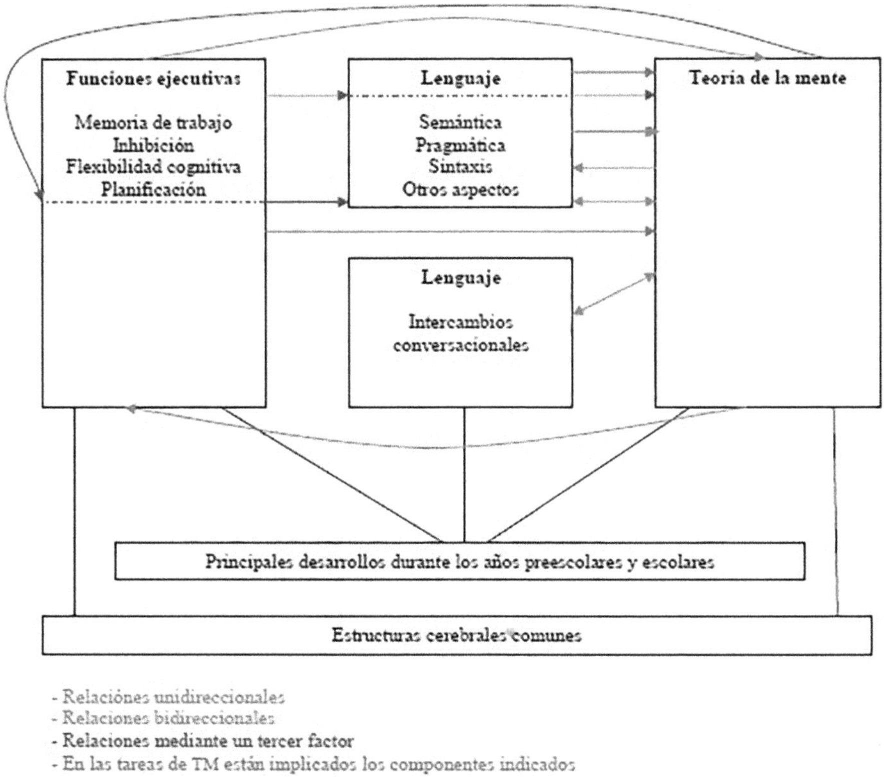

Fig. 18.1 Serrano Ortiz 2012, p. 160

1. What is the sequence of ToM development, i.e., the level of performance in the tasks traditionally used to assess mentalist abilities?
2. What predicts mentalist abilities in a better way: language or executive function?
3. What is the role of language in the evolution of mentalist abilities like?
4. What is the role of executive function in the evolution of mentalist abilities like?

For this purpose, Serrano selects 8, 4 and 3 tasks aimed to assess 10, 4 and 4 criteria respectively. 10 criteria to cope the performance of mentalist abilities, 4 criteria to specifically study the performance of executive function and 4 more criteria to research the development of language according to the table below (Fig. 18.2).

In the conclusions of her research, Serrano triangulated the findings to lastly announce important differences both in the quantity and quality of interactions between executive function, language, and ToM, according to the evolutionary moment of the individual.

> Language and executive functions contribute [to the development of ToM] differently depending on the stage of the growth. Both accounts are involved in the comprehension of the first order ToM. In the second order ToM, executive functions play a significant role,

Cuadro 5.1. Tareas e instrumentos administrados en la primera sesion de evaluacion

Dominio	Habilidad	Tarea
Teoría de la mente	Creenica falsa de primer orden	Contenido inesperado de Perner y colaboradores (1987)
	Emociones ligadas a creencias y deseos	Emociones ligadas a deseos y creencias de Hughes y colaboradores (2000)
	Apariencia versus realidad emocional	Emoción fingida vs. emoción real de Harris y colaboradores (1986)
	Creenica falsa de primer orden	Cambio de localización de Hughes y colaboradores(2000)
	Engaño	Tarea de engaño de Premack (1990)
	Mentira piadosa	
	Mentira	Historias extrañas de Happ (1994)
	Ironía	
	Meteduras de pata	Historias faux pas de Baron-Cohen y colaboradores (1999)
	Expresión de estados mentales a través de la mirada	El test de los ojos para niños de Baron-Cohen y colaboradores (2001)
Funciones ejecutivas	Memoria de trabajo	Memoria de degitos a la inversa de la escala de inteligencai revisada de Wechsler para niños (WISC-R, 1994)
	Memoria de trabajo	Tarea de día-noche de Gerstadt y colaboradores (1994)
	Planifivación	Laberintos de la escala de inteligencia de Wechsler para niños (WISC-III, 1994)
	Flexibilidad congnitiva	Test de clasidicación de tarjetas de Wisconsin de Heaton y colaboradores (1997)
Inteligencia no verbal	Matrices progresivas (escala de color y escala general) de Raven, Court y Raven (1996)	

Cuadro 5.2. Tareas e instrumentos administrados en la segunda sesión de evaluación

Dominio	Habilidad	Tarea
Lenguaje	Léxico	Test de vocabulario receptive en imágenes Peabody de Dunn, Dunn y Arribas (2006).
	Sintaxis	Batería de Lenguaje Objetivo y Criterial-Screening Revisado de Puyuelo, Solanas, Wigg y Renom (2007).
	Pragmática	
	Oraciones completivas	Oraciones completivas de Hale y Tager-Flusberg (2003) y Lohmann y Tomasello (2003)

Fig. 18.2 Serrano Ortiz 2012, p. 200

while the language remains rather irrelevant. In the advanced and in the superior order ToM, the effect is the opposite, that is, language helps explaining the differences in the performance of mentalist abilities, but executive function does not behave as predictor.

> As for the linguistic components, lexicon and completive sentences are relevant in younger children, while pragmatic abilities are more determining in middle-aged children. As for older children, syntax, pragmatics, and lexicon predict better the performance of mentalist abilities.
>
> Regarding the executive function, although it could be considered as less significant in the general development of ToM, cognitive flexibility plays an important role especially in very little, while middle-aged and older children's mentalist abilities are also mediated by planning.[4] (Translated by the author)

This approach of triangular analysis was later repeated by Nguyen and Astington (2014), who replaced *language* with *bilingualism* "To our knowledge, this is the first study that empirically supports a mediation model that encompasses all three factors: bilingualism, EF, and ToM" (p. 407).

The study researched 24 + 24 respectively English and French monolingual school children +24 bilinguals in the combination English French and developed triangular research neutralizing the potential effect of the differences in the socio-economic status of their families. The findings lent a bilingual advantage in the development of cognitive and linguistic abilities, particularly because they seemed to show that bilingualism fosters the development of working memory: "[...] bilingual environments induce children to use information held in mind in appropriate situations, which compensates for their poorer performance on verbal ability tests." (p. 407).

For the assessment of receptive vocabulary, research used the standardized PPVT–III (Peabody Picture Vocabulary Test–III; Dunn and Dunn 1997), and the ÉVIP (Échelle de Vocabulaire en Images Peabody; Dunn et al. 1993), which is the PPVT–III French version. False belief was measured with the change in-location task (Wimmer and Perner 1983) and the unexpected-contents task (Gopnik and Astington 1988; Perner et al. 1987) while children's conflict inhibition was assessed using the Stroop task adapted from Gerstadt et al. (1994) Day–night task. As for the working memory, they used the Backward Word Span (BWS).

Linguistic outcomes were later collated with the demographic features of the researched samples in order to overcome some bias getting rid of the differences in

[4] "El lenguaje y las funciones ejecutivas contribuyen [al desarrollo de la TM] de forma distinta a lo largo del desarrollo. En la comprensión de la mente de primer orden, ambas habilidades se hallan implicadas. En la TM de segundo orden, las funciones ejecutivas tienen un papel significativo, mientras que el lenguaje no lo tiene. En la avanzada y de orden superior, se observa el efecto contrario, es decir, el lenguaje contribuye a explicar la variabilidad del rendimiento de las habilidades mentalistas, pero no las funciones ejecutivas.

En relación con los distintos componentes lingüísticos, en los niños más pequeños, el léxico y las completivas tienen un peso específico. En edades intermedias, son las habilidades pragmáticas las que tienen un mayor énfasis en las habilidades mentalistas. Y en edades más avanzadas, la sintaxis juntamente con la pragmática y el léxico predicen mejor el desempeño en las tareas mentalistas.

En lo referente a las funciones ejecutivas, a pesar de hallar un papel menos significativo en la TM, podemos apuntar que en los niños más pequeños resulta importante el papel de la flexibilidad cognitiva. En edades intermedias, así como en edades más avanzadas la flexibilidad cognitiva y la planificación contribuirán especialmente a la TM." (Serrano Ortiz 2012, p. 346–347)

findings that may come with the differences in the socioeconomic status of families. Nevertheless, Nguyen y Astington were also aware about other limitations, that is, the study focused on the comprehension skills (regarding vocabulary) and neither measured the intelligence of the individuals, and therefore, nor assessed its mediation in results.

In 2016, Gordon conducted an experiment with 26 English – Spanish bilingual preschool children and 26 English monolingual preschool children, who were tested to determine whether the relation between linguistic competence and mentalist abilities is different in monolinguals and bilingual brains. Children's receptive vocabulary was assessed through the PPVT-III, already mention above, and through the TVIP (Test de Vocabulario en Imágenes Peabody), which is the Spanish version of the PPVT-III (Dunn et al. 1986) and additionally they were given several tasks like diverse desire, diverse belief, knowledge access, contents false belief, explicit false belief, belief-emotion, and real-apparent emotion task. The findings seemed to indicate that they certainly differ, yet it cannot be rigorously asserted that there is properly an advantage since (1) there is no data about the underlying mechanism — thus there is no proven cause-effect correlation, and (2) the performance is better or worse in bilinguals or monolinguals depending on the task they face.

> While the current study does provide evidence of a relationship between higher proficiency across languages and better performance on mental state tasks for bilinguals, the question still remains of the mechanism through which this occurs (p. 419)

[...]

> Regarding children's performance on the other mental state tasks, bilinguals were shown to outperform monolinguals on one task (i.e., diverse desires), perform worse than monolinguals on one task (i.e., explicit false belief), and perform similarly to monolinguals on the remaining mental state tasks (p. 421)

Together with the conclusions reached thanks to her own study, Gordon also revises and critically analyses the literature, highlighting the inclination of previous studies to publish findings according to the assumption of the bilingual advantage and to underestimate mixed or void findings. She also mentions that most of studies do not include the intelligence as a variable to measure as well as the fact that previous research has been focused on false believe tasks, which leaves for further clarification the answers to the questions about the impact of linguistic abilities in ToM, that is, how and how much do the linguistic abilities affect ToM in a differentiated way.

Given the heterogeneous landscape of studies and findings, researchers started working on contrasting and revising by means of meta-analysis. Like we have seen in Gordon, the scientific *community* leaves behind the early optimism regarding the potential of bilingualism as a *booster of the mind* and work on the identified limitations of the experiments, which leads to designs oriented to measure specific items and therefore less generalist aims and conclusions. Because of the similarities in the design of the previous experiments or due to the contradictory findings among them, there were reasons enough to reroute the tendencies in research of the relation between bilingualism and ToM.

TABLE 1 | Studies included in main analysis.

	Monolingual N	Bilingual N	Monolingual languages	Bilingual languages	Theory of Mind Tests
Diaz and Farrar, 2018*	33	32	English	English & Spanish	Unexpected-Transfer, Unexpected-Contents, Other
Kyuchukov and De Villiers, 2009 [a]	60	60	Bulgarian	Bulgarian & Romani	Unexpected-Transfer, Unexpected-Contents
Pearson, 2013	40	28	English	English & Spanish	Unexpected-Transfer
Bialystok and Senman, 2004*[b]	52	43	English	English & Other	Other
Gordon, 2016*[c]	26	26	English	English & Spanish	Unexpected-Transfer, Unexpected-Contents, Other
Dahlgren et al., 2017	14	14	Swedish or Slavic	Swedish & Slavic	Unexpected-Transfer, Other
Nguyen and Astington, 2014*	48	24	English or French	English & French	Unexpected-Transfer, Unexpected-Contents
Diaz and Farrar, 2017*[d]	35	38	English	English & Spanish	Unexpected-Transfer, Unexpected-Contents, Other
Kovács, 2009[e]	28	28	Italian	Romanian & Slovenian	Unexpected-Transfer, Unexpected-Contents
Chan, 2004*	29	31	English	English & Chinese	Unexpected-Transfer, Unexpected-Contents, Other
Goetz, 2003*	64	40	English or Chinese	English & Chinese	Unexpected-Transfer, Unexpected-Contents, Other
Han and Lee, 2013	60	73	Korean	English & Korean	Other
Greenberg et al., 2013	45	37	English	English & Other	Other
Kovács, 2009	32	32	Romanian	Romanian & Hungarian	Unexpected-Transfer, Other
Farhadian et al., 2010	65	98	Persian	Persian & Kurdish	Unexpected-Transfer, Unexpected-Contents
Fan et al., 2015*	24	24	English	English & Other	Other

The study was included in the secondary analysis

[a] *Bilinguals were tested in both their L1 and L2. The L1 test scores were used to compute the effect size so that Theory of Mind scores for both monolinguals and bilinguals were based on their native language performance.*

[b] *In the secondary analysis, only the appearance results of the appearance-reality test were used to compute an effect size because the reality questions did not require a Theory of Mind.*

[c] *Gordon's dissertation Millett, 2010 was used to extract additional data that were not included in the journal version.*

[d] *A subset of participants was tested a second time but for a more accurate statistical calculation only time 1 data were included in the effect size calculation.*

[e] *This dissertation contained an experiment that was published in Kovács (2009). So as to not give this experiment double weight, it was not included when computing the effect size for Kovács (2008).*

Fig. 18.3 Schroeder 2018, p. 4

In 2018, Schroeder released a meta-analysis of 16 studies chosen according to several very specific criteria. He revised quantitatively and qualitatively the amount and main features of the studies chosen, that is: the quantity and characteristics of the samples, the language, or languages in combination (mainly English), the type of test conducted and its findings (mainly unexpected transfer and unexpected contents tasks) (Fig. 18.3).

From Schroeder's meta-analysis it is to be concluded that indeed a small bilingual advantage exists, but also that this advantage does not reach more extent that an early stimulation.

The main meta-analysis, which compared bilingual and monolingual children's raw Theory of Mind scores, revealed a small bilingual advantage. The size of this bilingual-monolingual difference (i.e., a Cohen's d in the "small" range) is like the effect of early education interventions on cognitive, school, and social outcomes (p. 5–6).

For his meta-analysis Schroeder included experiments which had examined the impact of sociolinguistics and metalinguistics in ToM and not only the executive function's influence. In these studies, the bilingual advantage is proven to be mediated by the "metalinguistic awareness" of an individual who can deal with symbol substitution, synonym judgement and homonym selection. As for the "socio-pragmatic" account, Schroeder retrieve studies with not bilingual children who were exposed to a second language, and their findings suggest that ToM might be improved by learning that one's linguistic knowledge can be different from that of other people.

Results confirmed that bilingualism can positively foster metalinguistic awareness and increase ToM when it comes to acquire awareness about the fact that words and connotations can differ among languages and, as for the sociolinguistic account, pointed out the value of data which emphasized that children with a higher grade of ToM development were more popular among the others, at the same time that negative consequences were revealed possible, as shown by the fact that these children were prone to lying more often than children with a lower grade of development in their ToM.

Regarding the executive function, the bilingual advantage is noticeable in cases of high demand of inhibitory control, although other studies, like those which measure the cognitive flexibility, did not result in any advantage at the expenses of bilingualism.

Out of this meta-analysis it is therefore to infer that research lines which include sociolinguistics and metalinguistics as mediating agents in the relation between bilingualism and ToM should be bolstered and that the prevailing research approach about executive function should be revised.

Meanwhile specific studies, like Laura Morett's (2020) about brain processing of sounds, insisted on the importance of neuroimages as an essential tool in the investigation on the relation between bilingualism and ToM, provided that the aim of research is to rigorously discriminate functions, abilities, and brain areas. In the line of previous studies like Fabiano-Smith and Goldstein's (2020), Morett worked from the assumption that children with experience with tonal language exclusively and atonal languages exclusively show different trajectories for lexical tone perception development and under this premise she investigates how lexical tone perception develops in children bilingual in both a tonal and an atonal language. She conducted an experiment with 40 Mandarin Chinese – English bilingual children between ages of 2 and 7, who performed three tasks: a non-lexical tone discrimination task, a lexical tone discrimination task, and a word learning task. The findings revealed that bilingual children, whose combination included a tonal and atonal language, process lexical tones in a different form that those children whose languages are both atonal, yet they use the same cognitive mechanisms to process non lexical tones.

Back to the investigation about the triad *language, executive function, and ToM*, Buac and Kaushanskaya (2020) published a study whose originality lays on asserting for the first time that different linguistic experiences lead to differences in the relational patterns between language, executive function, and ToM: "This is the first study to show that children with distinct language acquisition histories approach verbal ToM problems differently" (p. 339). The experiment researched 7-year-old children: 44 English monolinguals, 44 English – Spanish simultaneous bilinguals and 27 English – Spanish bilinguals whose dominant language was Spanish. Children performed a second-order belief task which consisted of explaining one character's thoughts about another character's thoughts. The study also measured three components of the executive function, i.e., inhibition, updating of working memory and shifting. The tasks performed for the measurement of the executive function were flanker task, Corsi block task and card sort task respectively.

The study revealed that in simultaneous bilinguals' linguistic abilities, and not executive function, were predictors of ToM, while in monolinguals and English dominant bilinguals' executive function, and not the language, performed like predictors. Out of these findings, it is to conclude that "a critical threshold level of language ability may be required for EF skills to contribute to TM performance, at least on a verbal TM task" (Buac and Kaushanskaya 2020, p. 355). In other words, that a minimum level of development of linguistic abilities is required for the executive function to result in a better performance of the mentalist tasks. Thus, although Buac et al. studied executive function, the findings finally invited to pay attention to sociolinguistics, since it is still to clarify whether these differences in linguistic abilities of ones and others and the way they manage them come (or not) from strategies acquired because of their personal linguistic experiences in their environments and particular contexts.

In 2020, Bailey et al. metanalyses 57 studies to define the differences between simultaneous and sequential bilinguals and monolinguals. The studies selected had to fulfil the following requirements: (1) to be peer-reviewed; (2) available in English or Spanish; (3) clearly specify the inclusion of English–Spanish bilingual participants; (4) to assess the participants' performance on some measure of cognitive ability; and (5) to compare bilingual participants with monolingual participants, as well as they had to meet referred Reporting Items for Systematic Reviews and Meta-Analyses (PRISMA) criteria. Instead of using statistical methods to evaluate and synthesize empirical studies, this meta-analysis turned to be a systematic review which took a descriptive approach in summarizing the studies. The revision concluded that bilinguals perform worse in tasks which require verbal abilities linked to vocabulary as well as that this deficit lasts for their whole life, however it is also confirmed the bilingual advantage as for metalinguistic and executive function tasks, not without holding bars about the nature of the implemented tests. In this regard, the researchers remind that, from a neurologic approach, bilinguals are different from one another according to the subtype of their bilingualism, and of course different from monolinguals in terms of brain structure, neural connexions, and neuroplasticity. Hence, it turns out to be not very much reliable to use similar systems for the assessment of bilingual and monolingual cognitive abilities. The meta-analysis demanded therefore the design of specific tests observing the neurologic differences in monolinguals and bilinguals and even discriminating systems adapted to the different kinds of bilingualism.

> Apart from the main conclusions of the authors, we will take some time to give account of Bailey's et al. warning about the fact that they deliberately excluded monolingualism from the bibliographic revision, because it exceeds the extent of the study and since it has been proved that monolinguals are qualitative different from bilinguals and monolinguals (Bailey et al. 2020, p. 226).

It seems to be convenient to stop in this assertion, since we understand that the multilingualism as a phenomenon and particularly the multilingual brain might contain a significant potential to clarify questions within the frame of the discussion about relations between factors currently in analysis. On the one hand, the analysis and contrast of the structures and underlying dynamics of the multilingual brain can serve as a referent point. On the other hand, many of the individuals considered bilingual have any kind or grade of competence in a third language, thus, no matter whether broadly or strictly speaking, they could be considered multilinguals or at least they could share features from a neurologic perspective.

The second decade of the twenty-first century began with less generalist but more discriminating studies, designed to research from a neurological perspective the details of the interactions between bilingualism and ToM. In 2021, Yang et al. focused on the analysis of the role of inhibitory control regarding the spoken comprehension and production in Chinese students of the English language. They explored the relationships among English as second language (ESL) listening proficiency, cognitive inhibition ability, and spoken word segmentation (SWS) performance. During the tests they recorded the cortical activation in the temporo-parietal junction (TPJ) and the prefrontal cortex (PFC). Therefore, the research scoped both behavioural and mental patterns aspects to conclude that indeed the mediation occurs. The approach and the relevance of the results emphasized the idea that the complex map of involved areas in reciprocal relation can change according not only to the languages in combination but also to the skills to analyse.

In 2018 Bradford et al. took samples of English and Chinese native people to investigate whether the mentalist abilities can be mediated by the culture of the individual. The studied recruited 55 native-English speakers between 17 and 34 years old (34 females, 21 males) and 54 native-Chinese speakers (35 females and 19 males). Participants completed the Auckland Individualism and Collectivism Scale (AICS) questionnaire and performed a Self/Other Differentiation Task. The findings suggested that there were not significant differences between west and east cultures as for the performance of ToM, and warned, nevertheless, that the conclusions could only refer to their observable behaviour of the individual and not to the underlying strategies of brains or their neural signature.

> The results of this study support the idea of core similarities in the ToM mechanism across the two cultures studied. Further research to explore the neural signatures underlying these behavioural results, and whether different strategies for ToM expression are utilized across cultures, would help build an even more informed understanding of how the ToM mechanism develops, functions, and is utilized in everyday life. (Bradford et al. 2018, p. 17)

Again, the researchers report dark areas that could only be clarified by means of neuroscience.

Back to the meta-analysis, Yu et al. (2021, p. 157) analysed 24 empirical studies of ToM competence in bilingual children to reach the following conclusions:

1. The bilingual advantage is modest.
2. So far, the science has not been able to describe the underlying mechanism that lie beneath the bilingual advantage.
3. The executive function has been intensively studied, but findings are not conclusive regarding the impact in the development of ToM. On the contrary, the mediation of metalinguistic and sociolinguistic factors seems to be more plausible, although they have not been so deeply researched.

Study	ToM advantage	Study	ToM advantage
Banasik and Podsiadło (2016)	–	Gordon (2016)	+
Bergmno and Bowler (2004)	+	Greenberg et al. (2013)	+*
Bialystok and Senman (2004)	+*	Han and Lee (2013)	+
Buac and Kaushanskaya (2020)	X	Kovács (2009)	+
Chan (2004)	+*	Kyuchukov and De Villiers (2009)	X
Cheung et al. (2010)	+*	Meir and Novogrodsky (2019)	+
Dahlgren et al. (2017)	X	Nguyen and Astington (2014)	+*
Diaz and Farrar (2018a)	+*	Pearson (2013)	X
Diaz and Farrar (2018b)	+*	Peristeri et al. (2019)	+*
Fan et al. (2015)	+*	Raisa et al. (2019)	–
Farhadian et al. (2010)	+*	Tare and Gelman (2010)	–
Goetz (2003)	+*	W. Q. Yow (personal communication, April 27, 2020)	X

Fig. 18.4 Yu et al. 2021, p. 155

4. None of the three factors seem to induce any extraordinary development of ToM, yet they provide the individuals with opportunities to develop abilities which turn out to be relevant for the ToM (Fig. 18.4).

18.8 Conclusions

For this text, we have selected studies which appear as relevant for a matter of either representativeness or originality. That is, they either represented the mainstream current in researching at the time or they turned out to be any kind of milestone in research, containing any finding or fostering a course change about the mainstream line in investigation.

To properly lead the reader along the research on the matter, we have followed a chronological order when presenting the studies. This approach enabled us to scope the two decades of investigation that elucidate the relation between bilingualism and ToM, dividing the topic in two phases properly differentiated, which covers each decade of the twenty-first century.

Soon in the 2000s the bilingual advantage was explicitly stated in the words of Goetz (2003), but this statement was soon confronted by the fact that those Meristo et al.'s (2007) deaf children who had not been able to fully identify with none of the languages of their combination, performed significantly worse in the tests conducted.

By the year 2007, when these findings were revealed, was indeed daring to extrapolate conclusions towards experiments with deaf children to researches conducted in hearing people. However, afterwards researchers confirmed that both a minimum level of abilities and full identification with at least one languages (signed or not) is a requirement for the optimal condition of ToM.

During the first decade of the twenty-first century, and despite some remarks on the contrary, the scientific community stays in general in high levels of optimism regarding the bilingual advantage. This assertion can be easily confirmed not only in the above-mentioned revised texts but also in the many published articles we will not recount here to avoid the saturation of the present discourse. Nevertheless, by the end of the decade, the scientific community had started (1) to weight this assumption, (2) to acknowledge bias in experiments and (3) to consider other predictors and variables that had not been researched enough. So much so that the Complementarity Principle (Grosjean 2010) and the Adaptive Control Hypothesis (Green and Abutalebi 2013) introduced a new approach to the topic, which considered the personal linguistic experience of the bilingual as a mediating factor in the development of ToM. Therefore, sociolinguistics and individual experiences had aroused a new researching path without having fully conclusive answers to the questions about executive function.

Although executive function had been extensively studied in previous years, research had kept it relatively apart from language and ToM. In 2012, Serrano correlated one another and triangulated results. And so did Nguyen y Astington in 2014 replacing *language* with *bilingualism.* As a result, advantages regarding working memory were acknowledged and so were some drawbacks regarding verbal abilities. Both studied meant an advance as for results' management and reduction of bias in the design of the experiments. In 2016, Gordon addresses the triad anew and concluded that, although the bilingual advantage exists indeed, it cannot be said about a cause-effect relation since science has not found out the underlying mechanisms. This statement entails a call for other disciplines to get involved in research, since only neuroscience can reveal data and information about the brain mechanisms that underlie the observable behaviours of bilinguals. About the bilingual advantage, she still adds that it has been quantitatively privileged in terms of publications and that, according to her own revision and study, bilingual advantage sometimes appears and sometimes does not.

In the last years of the second decade, together with regular studies and revisions, meta-analysis started to be published because of the interest and controversy about the topic. The conclusions of Schroeder's meta-analysis (2018) invited to invigorate research which considered sociolinguistics and metalinguistics as potential providers of more stables and more results regarding the relation between bilingualism and ToM than those provided by the study of executive function. Moreover, the designs of experiments made progresses reducing bias, including new variables with the items in analysis and adding new and more specific goals. On this matter, Morett (2020) found differences which lie more with the languages in combination, particularly with the phonetic features on the language(s); Buac and Kaushanskaya (2020) pointed at the linguistic trajectory and the individual experience as a mediating factor; and Bailey et al. (2020) concluded that bilinguals perform worse when it comes to complete verbal abilities tasks linked to vocabulary, and added that this handicap lasts a lifetime. Nevertheless, they also confirmed, with reservations, the bilingual advantage within the frame of metalinguistics and executive function. Bailey's et al. revision therefore claimed for designs of specific tests able to observe and discriminate neural differences between monolinguals and bilinguals, and even differentiated systems adapted to the several types of bilingualism.

In 2021, Yan et al. discriminated between comprehension and expression skills and Bradford et al. (2018) analysed the influence of culture over mentalist abilities in the brains of Chinese – English bilingual children. The authors stated that, although there is no evidence of significant differences, conclusions only referred to observable behaviour and not to potential differences in the neural signatures of bilingual individuals. And at last, conclusions of Yu et al. (2021) meta-analysis prompted researchers to play down the bilingual advantage conferred to executive account as well as to keep reaching sociolinguistic and metalinguistic aspects with help or by means of neuroscience.

To sum up, research on the relation between bilingualism and ToM have evolved from a general assumption of an advantage of bilingualism over monolingualism as a key element for the enhancement of mentalist abilities to relativize this advantage, at least with regard of the executive function. The attention drawn by the executive function account yield under the sociolinguistic account and the metalinguistic awareness, now recognized as mediating agents in the development of ToM. From now on the science will have to (1) be diverse as for the type of experiments conducted, (2) triangulate results and (3) survey research aimed to assess small differences in a search for more specific and conclusive findings. This new path acknowledges the importance of neuroscience and its involvement, as it is the only discipline able to monitor the brain's behaviour regarding the language. On the other hand, the importance of the individual experience in terms of linguistics and the relevant of the sociolinguistic account suggest a significant mediation of personal contexts, conditions, and trajectories in bilingual subjects.

References

Apperly, I.A., and S.A. Butterfill. 2009. Do humans have two systems to track beliefs and belief-like states? *Psychological Review* 116 (4): 953.

Aranda Farías, C. A. 2020. *Influencia del bilingüismo simultáneo y secuencial sobre el control ejecutivo y la reserva cognitiva* [Tesis de Maestría, Universitat de les Illes Balears] https://www.uib.cat/recerca/estructures/grups/grup/I_DEL/produccio/tesis_tesines_treballs/

Bailey, C., A. Venta, and H. Langley. 2020. The bilingual [dis]advantage. *Language and Cognition* 12 (2): 225–281. https://doi.org/10.1017/langcog.2019.43.

Balmaceda, T. 2016. Tres décadas del test de la falsa creencia. *Revista Argentina de Ciencias del Comportamiento* 8 (2): 5–23.

Baron-Cohen, S., A.M. Leslie, and U. Frith. 1985. Does the autistic child have a «theory of mind»? *Cognition* 21 (1): 37–46.

Bradford, E.E., I. Jentzsch, J.-C. Gomez, Y. Chen, D. Zhang, and Y. Su. 2018. Cross-cultural differences in adult theory of mind abilities: A comparison of native-English speakers and native-Chinese speakers on the self/other differentiation task. *Quarterly Journal of Experimental Psychology* 71 (12): 2665–2676. https://doi.org/10.1177/1747021818757170.

Buac, M., and M. Kaushanskaya. 2020. Predictors of theory of mind performance in bilingual and monolingual children. *International Journal of Bilingualism* 24 (2): 339–359. https://doi.org/10.1177/1367006919826866.

DeLuca, V., K. Segaert, A. Mazaheri, and A. Krott. 2020. Understanding bilingual brain function and structure changes? U bet! A unified bilingual experience trajectory model. *Journal of Neurolinguistics 56*. https://doi.org/10.1016/j.jneuroling.2020.100930.

Dunn, L.M., and L.M. Dunn. 1997. *Peabody picture vocabulary test – Third edition (PPVT–III)*. Circle Pines: American Guidance Service.

Dunn, L.W., D.E. Lugo, E.R. Padilla, and L.M. Dunn. 1986. *Test de vocabulario en imágenes Peabody*. Circle Pines: American Guidance Service.

Dunn, L.M., C.M. Thériault-Walen, and L.M. Dunn. 1993. *Échelle de Vocabulaire en Images Peabody*. Toronto: Psycan.

Fabiano-Smith, L., and B.A. Goldstein. 2020. Phonological acquisition in bilingual Spanish–English speaking children. *Journal of Speech, Language, and Hearing Research* 1 (53): 160–178. https://doi.org/10.1044/1092-4388(2009/07-0064).

Gavilán, J.M., and J.E. García-Albea. 2015. La función ejecutiva en la esquizofrenia y su asociación con las habilidades cognitivas sociales (mentalistas). *Revista de Psiquiatría y Salud Mental.* 8 (3): 119–129. https://doi.org/10.1016/j.rpsm.2014.01.008.

Gerstadt, C.L., Y.J. Hong, and A. Diamond. 1994. The relationship between cognition and action: Performance of children 312–7 years old on a stroop-like day-night test. *Cognition* 53 (2): 129–153.

Gleitman, L. 1990. The structural sources of verb meanings. *Language Acquisition* 1 (1): 3–55. https://doi.org/10.1207/s15327817la0101_2.

Goetz, P.J. 2003. The effects of bilingualism on theory of mind development. *Bilingualism: Language and Cognition* 6 (1): 1–15. https://doi.org/10.1017/S1366728903001007.

Gopnik, A., and J.W. Astington. 1988. Children's understanding of representational change and its relation to the understanding of false belief and the appearance–reality distinction. *Child Development* 59: 26–37.

Gordon, K. 2016. High proficiency across two languages is related to better mental state reasoning for bilingual children. *Journal of Child Language* 43 (2): 407–424. https://doi.org/10.1017/S0305000915000276.

Green, D., and J. Abutalebi. 2013. Language control in bilinguals: The adaptive control hypothesis. *Journal of Cognitive Psychology* 25 (5): 515–530. https://doi.org/10.1080/20445911.2013.796377.

Grosjean, F. 2010. *Bilingual: life and reality*. Harvard University Press. https://doi.org/10.4159/9780674056459.

Javor, R. 2016. Bilingualism, theory of mind and perspective-taking: The effect of early bilingual exposure. *Psychology and Behavioral Sciences 5* (6): 143–148. https://doi.org/10.11648/j.pbs.20160506.13.

Meristo, M., K.W. Falkman, E. Hjelmquist, M. Tedoldi, L. Surian, and M. Siegal. 2007. Language access and theory of mind reasoning: Evidence from deaf children in bilingual and oralist environments. *Developmental Psychology* 43 (5): 1156–1169. https://doi.org/10.1037/0012-1649.43.5.1156.

Morett, L.M. 2020. The influence of tonal and atonal bilingualism on children's lexical and non-lexical tone perception. *Language and Speech* 63 (2): 221–241. https://doi.org/10.1177/0023830919834679.

Navarro, E., V. DeLuca, and E. Rossi. 2022. It takes a village: Using network science to identify the effect of individual differences in bilingual experience for theory of mind. *Brain Sciences* 12 (4): 487. https://doi.org/10.3390/brainsci12040487.

Nguyen, T.-K., and J.W. Astington. 2014. Reassessing the bilingual advantage in theory of mind and its cognitive underpinnings. *Bilingualism: Language and Cognition* 17 (2): 396–409. https://doi.org/10.1017/S1366728913000394.

Pérez-Hernández, E., I. Lozano, M. Aragón, E. Casadiegos, and J. de Frutos. 2016. ISBN 978-84-608-8714-0. *Ventaja bilingüe en el reconocimiento de emociones y teoría de la mente. Castejón Costa, Juan Luis (coord.). Psicología y Educación: Presente y Futuro*, 2264–2272. Alicante: ACIPE.

Pérez-Luzardo, J., and A. Schmidt. 2016. El bilingüismo y la identidad: Estudio de caso sobre la relación entre las lenguas y las emociones. *Revista de Lingüística y Lenguas Aplicadas* 11 (1): 51–59. https://doi.org/10.4995/rlyla.2016.4424.

Perner, J., S.R. Leekam, and H. Wimmer. 1987. Three-year-old's' difficulty with false belief: The case for a conceptual deficit. *British Journal of Developmental Psychology* 5: 125–137.

Premack, D., and G. Woodruff. 1978. Does the chimpanzee have a theory of mind? *The Behavioural and Brain Science* 1 (04): 515–526.

Schroeder, S. 2018. Do bilinguals have an advantage in theory of mind? A meta-analysis. *Frontiers in Communication* 3 (36). https://doi.org/10.3389/fcomm.2018.00036.

Serrano Ortiz, J. 2012. *Desarrollo de la teoría de la mente, lenguaje y funciones ejecutivas en niños de 4 a 12 años* [Tesis de Maestría, Universitat de Girona] http://hdl.handle.net/10803/123549

Shamay-Tsoory, S.G., and J. Aharon-Peretz. 2007. Dissociable prefrontal networks for cognitive and affective theory of mind: A lesion study. *Neuropsychologia* 45 (13): 3054–3067. https://doi.org/10.1016/j.neuropsychologia.2007.05.021.

Signoret, A.M., A.L. Rodríguez, J.E. Flórez, and R.E. Delgadillo. 2015. *Coincidencias en el lenguaje y cognición bilingüe: otro acercamiento*. Universidad Nacional Autónoma de México.

Wimmer, H., and J. Perner. 1983. Beliefs about beliefs: Representation and constraining function of wrong beliefs in young children's understanding of deception. *Cognition* 13 (1): 103–128.

Yang, Y., Y. Li, X. Wang, N. Liu, K. Jiang, S. Zhang, and J. Qiu. 2021. Cognitive inhibition mediates the relationship between ESL listening proficiency and English spoken word segmentation in Chinese learners: A functional near-infrared spectroscopy (fNIRS) study. *Journal of Neurolinguistics* 59 (38): 167–178. https://doi.org/10.1016/j.jneuroling.2021.100987.

Yu, C.L., I. Kovelman, and H.M. Wellman. 2021. How bilingualism informs theory of mind development. *Child Development Perspectives* 15 (3): 154–159. https://doi.org/10.1111/cdep.12412.

Chapter 19
Processes of Early Language Acquisition and Its Implications for ToM in Autistic Children

Mayuresh Kumar

Abstract Language plays a significant role in the development of ToM, especially in the early years of a child. It is during those initial years that both language and ToM start developing. Autistic children are known to have deficits in both these domains. Having a deficit in the domain of ToM explains why many autistic individuals don't pass the false-belief tasks. Though the grammatical and semantic components of their language use do not seem impaired, they struggle with the pragmatic abilities (Swensen et al. Child Dev, 78(2): 542–557, 2007. http://www.jstor.org/stable/4139244). However, those who manage to pass them are said to have mastered certain linguistic abilities which allow them to do so. Instead of developing an understanding of mental states grounded in social-perceptual abilities, autistic children learn to reason logically through these tasks, using language as a means. This review article analyses various studies detailing the processes of early language acquisition in Autistic children and it does so by looking into the methodologies used to come to certain conclusions. Early language acquisition in autistic children has significant implications for the ToM and understanding the constraints on language acquisition is the way forward.

Keywords Language acquisition · Autistic children · ToM · False-belief understanding

19.1 Introduction

Language acquisition and Theory of Mind (ToM) development are linked, particularly in the initial years of life of a child. This review article explores this association keeping in mind the role of language, with a specific focus on autistic children. It examines the deficits in language and ToM often observed in autistic individuals, the implications for how they perform in false-belief tasks, and how the linguistic abilities and pragmatic skills are related. Additionally, it reviews various studies

M. Kumar (✉)
Department of Foreign Languages, Aligarh Muslim University, Aligarh, India

T. Lopez-Soto et al. (eds.), *The Theory of Mind Under Scrutiny*, Logic, Argumentation & Reasoning 34, https://doi.org/10.1007/978-3-031-46742-4_19

that investigate the nuances of early language acquisition in children with autism spectrum disorder and underscores the importance of understanding the constraints on language acquisition for advancing ToM research.

Autistic children usually show a lack of command over both, ToM and language domains. These deficits contribute to the challenges they face while performing false-belief tasks, which examines an individual's perceiving of others' beliefs that might be aloof from actual scenario. It is through ToM that individuals recognize and attribute mental states; be it to themselves only or to others. Autistic children's challenges in this area can be attributed to their deficits in ToM development.

While the grammatical and semantic components of language use in autistic children may appear intact, they often struggle with pragmatic abilities. Pragmatics refers to the understanding and use of language in social contexts. Autistic children may have difficulties with interpreting and using non-literal language, understanding social cues, and engaging in reciprocal communication. However, the ones who do get to clear the false-belief tasks are believed to have mastered certain linguistic abilities that enable them to reason logically using language as a tool.

Autistic children, rather than possessing clarity of mental states through social-perceptual abilities, tend to approach false-belief tasks by using language as a means of logical reasoning. Language serves as a tool for autistic individuals to navigate these tasks, compensating for their deficits in social cognition. This highlights the unique connection between ToM and language in children with ASD.

One of the objectives here is to assess the phenomenon of early language acquisition in children with autism spectrum disorder. The methodologies employed in various studies are to be examined to gain insights into the conclusions drawn. Understanding how language acquisition unfolds in the early stages of development in autistic children is crucial for understanding ToM and its constraints.

The early language acquisition of autistic children holds significant implications for ToM research. Investigating the constraints and challenges they face in acquiring language can provide valuable insights into the development of their ToM. By understanding the interplay between ToM and language acquisition, researchers can deepen their knowledge of the cognitive progress in autistic individuals.

19.2 Autism Spectrum Disorder (ASD)

Autism, once a little-known condition, has now gained widespread recognition as a developmental disorder characterized by a difficulty in engaging in typical social interactions. Hill and Frith (2003) have done extensive research on the evolving understanding of autism, highlighting its lifelong nature and the significant role of genetics in its manifestation. With no established biological markers, autism is primarily explained through behavioral criteria. One has to keep in mind the various factors contributing to the severity of autism, its association with other disorders, and the cognitive explanations that have deepened our understanding of this complex condition.

Hill and Frith (2003) note that autism was considered rare in the past, but increased awareness of behavioral criteria has led to a higher number of cases being identified. The severity of autism can vary, influenced by factors such as temperament, individual abilities, and education. Furthermore, autism is often intertwined with other disorders, such as depression, further complicating the condition. The research evidence provided by Hill and Frith (2003) suggests that the cognitive and neurological findings are most robust for individuals without mental retardation, emphasizing the need for differentiated research within the autistic spectrum.

The paper by Hill and Frith (2003) traces the historical progress in understanding autism. Previously, it was mistakenly believed that children developed autism due to threatening or unloving parenting. However, the authors highlight the shift towards recognizing the behavioral characteristics of autism, refuting the earlier psychogenic ideas. Cognitive explanations have played a pivotal role in bridging the gap between brain function and behavior, helping unravel the complexities of autism.

One significant aspect of autism is the inability to acquire a "ToM". Hill and Frith (2003) assert that this cognitive impairment hampers the skills of an individual to understand benign aspects of communication, leading to social distancing and an indiscriminate approach to interactions. This abnormality in the social brain component can lessen the capacity to engage in reciprocal social exchanges.

Autism Spectrum Disorder (ASD) is a condition associated with qualitative impairments in social exchanges, interaction, and imagination, along with the manifestation of repetitive behaviors and peculiar mannerisms (Baird et al. 2003). To completely understand the picture, one has to delve into the behavioral nature of ASD, the diagnostic criteria associated with the disorder, and the role of language in its diagnosis.

Baird et al. (2003) emphasize that ASD is fundamentally a behavioral disorder. Individuals with ASD exhibit impairments that affect their social communication skills, imaginative abilities, and interactions with others. These impairments manifest in various ways, such as difficulties in initiating and maintaining social relationships, challenges in understanding nonverbal cues, and limited imaginative play. Additionally, individuals with ASD often display repetitive behaviors and engage in peculiar mannerisms. Furthermore, hypersensitivity to their surroundings is commonly witnessed in the people on spectrum, further highlighting the behavioral nature of the disorder.

Over the years, the diagnostic criteria for ASD have undergone revisions in the diagnostic manuals (Baird et al. 2003). These revisions reflect the evolving knowledge about the disorder and promise to offer more accurate and specific guidelines for diagnosis. The diagnostic manuals categorize ASD based on age and the type of behavioral features observed in individuals. The inclusion of the term "spectrum" is particularly significant, as it encompasses a range of dimensions to describe the diverse characteristics of individuals with ASD. Recognizing and understanding these dimensions is crucial for accurate diagnosis and effective management of the disorder.

Language development is an important aspect considered in the diagnosis of ASD. It is often observed that language development is delayed in individuals with ASD, unless there is a specific language delay present. Language difficulties can manifest in various ways, including delayed speech acquisition, limited vocabulary, and challenges in using language for effective communication. The delayed onset or atypical development of language skills can provide valuable insights for clinicians in diagnosing ASD.

Language delays are not only significant for diagnosis but also for understanding the broader cognitive and social communication impairments in ASD. Language serves as a crucial tool for expressing thoughts, emotions, and engaging in social interactions. The delayed language development in individuals with ASD further highlights the pervasive nature of their communication difficulties and the unique challenges they face in connecting with others.

Autism, once obscure, has now become a widely recognized developmental disorder. Through behavioral criteria and cognitive explanations, researchers have made significant strides in understanding its causes and consequences. The severity of autism can vary, influenced by factors such as temperament, abilities, and education, while its association with other disorders adds further complexity. The inability to develop a ToM is associated with the cognitive impairments and social difficulties that people on the spectrum have to encounter.

19.3 ToM

The study of ToM has been approached from various perspectives, leading to different views on its nature and development. Astington and Baird (2005) have discussed how ToM emerged in the developmental literature through two distinct routes, one centred on human cognition and the other on primate cognition. They explored the different perspectives on ToM and highlighted the role of language in its development, particularly during the toddler and kindergarten period. Additionally, they discussed the connection between language and performance on false-belief tasks, with a specific focus on autistic individuals.

According to Astington and Baird (2005), the concept of ToM was initially associated with the child's understanding of human cognition, as proposed by Wellman (1992). It referred to the child's ability to comprehend and attribute mental states to self and others. Another perspective, influenced by Premack and Woodruff's (1978) work on primate cognition, gave the nomenclature of system of inferences to ToM. One can predict behaviour through the process of attributing mental states and system of inferences is used for it. This view was later applied to children by Bretherton and Beeghly (1982) and Wimmer and Perner (1983), expanding the understanding of ToM across different domains.

Many theorists emphasize the significant role of language in the development of ToM, particularly during the rapid growth of language and ToM in the toddler and kindergarten years (Astington and Baird 2005). Language provides a means for children to express and understand thoughts, emotions, and beliefs, facilitating their comprehension of others' mental states. As language skills develop, children become increasingly adept at engaging in social interactions and interpreting others' perspectives.

The performance on false-belief tasks, which assess an individual's understanding of others' false beliefs, is closely linked to language abilities. Tager-Flusberg and Joseph (2005) highlight the importance of language in the performance of false-belief tasks for autistic individuals. While some autistic children struggle with ToM tasks, others demonstrate proficiency in passing them. In autism, language serves as an artificial route to social cognitive understanding. Rather than relying on social-perceptual abilities, autistic children reason logically through false-belief tasks, using language as a tool (Tager-Flusberg and Joseph 2005).

The connection between language and ToM is significant in understanding the development of mental state understanding. Language allows individuals to express and communicate their thoughts, feelings, and intentions. It provides a medium for sharing and understanding the perspectives of others, facilitating ToM development. Autistic individuals, while using language, may rely on logical reasoning to compensate for challenges in social-perceptual abilities, enabling them to engage in ToM tasks.

The study of ToM has been approached from different perspectives, emphasizing its connection with human and primate cognition. Language emerges as a crucial factor in ToM development, particularly during the toddler and kindergarten period. Language skills enable children to express themselves and understand others, facilitating their comprehension of mental states. Language and its role are noteworthy when we talk about performance in a false-belief task, especially in the context of autism (Tager-Flusberg and Joseph 2005). Language becomes an artificial route to social cognitive understanding for autistic individuals, allowing them to reason logically through ToM tasks. Understanding the interplay between language and ToM provides valuable insights into the cognitive and linguistic processes underlying social cognition.

In his work, Michael Corballis (2012) explored the concept of ToM, which is about understanding and attributing mental states to others. Corballis emphasized that humans possess the unique capacity not only to mentally travel through time but also to look into the minds of other individuals. This phenomenon plays a crucial role in our understanding of others' beliefs, thoughts, and intentions. By grasping what is on someone else's mind, we gain a deeper understanding of their words, actions, and the meaning behind them.

Theory of mind is a fundamental aspect of human cognition, enabling us to navigate through social interactions and develop meaningful relationships. Corballis (2012) argues that to truly comprehend the intentions and beliefs of others, individuals must possess an understanding of ToM. It allows us to infer and anticipate the mental states of others, leading to better communication and effective social interactions.

Corballis draws attention to an example provided by Sperber and Origgi (2010), where a simple sentence; "It was too slow" can be ambiguous and open to various interpretations. However, when communicated, the speaker often possesses an understanding of the listener's mental state, and no further clarification is needed. This example highlights how ToM enables individuals to bridge the gap between their own perspective and that of others, allowing for effective communication even in the face of ambiguous statements.

One domain where ToM is prominently observed is in works of fiction. Corballis (2012) notes that authors frequently dive into the minds of their characters, showing events and narratives from different perspectives. This narrative technique immerses readers in the mental states of the characters, providing a deeper understanding of their thoughts, emotions, and motivations.

Michael Corballis's (2012) exploration of ToM underscores its essential role in human cognition. Being able to understand others' perspectives, beliefs, and intentions is crucial for effective communication, social interactions, and the creation of fictional narratives. Theory of mind allows individuals to bridge the gap between their own minds and those of others, enhancing empathy, cooperation, and the ability to navigate the complexities of human social life.

19.4 ToM and the Role of Language

The initial stages of childhood mark the beginning of the development of ToM, as language plays a crucial role in this process. Theory of Mind and language complement each other as children explore and comprehend the world around them during their initial stages of learning. This holds true for everyone, including those with autistic spectrum disorder. Numerous studies have established a clear link between the development of ToM skills and language ability (Astington and Baird 2005).

Understanding the relationship between ToM and language learning provides valuable insights into children's minds as they express themselves through language, showcasing skills related to ToM. However, it is important to note that this link does not operate uniformly across all children, as every individual mind is unique. This complexity is further amplified in children with ASD. Despite communicating with the help of a language, autistic children face delays in acquiring language and stand far behind their peers in basic linguistic abilities (Tager-Flusberg 2007). This gives birth to a deeper curiosity to understand the processes of language learning in these children.

The understanding of language acquisition becomes particularly important for researchers as it holds the potential to explain the complementary nature of ToM and language learning. Language plays an instrumental role as far as the development of ToM is concerned (Tager-Flusberg 2007).

Theory of mind and language are interconnected and support each other on a mutual basis during early childhood development. As children begin to acquire language, they also start understanding and interpreting different mental states; of own and of others. Through language, children can express their thoughts, emotions, and beliefs, facilitating the development of ToM. Language acts as a tool for communication and enables children to engage in social interactions, fostering their understanding of others' perspectives and intentions. On the other hand, ToM skills enhances language use, enabling children to navigate through complex social situations, understand figurative language, and infer others' mental states.

In the case of autistic children, the relationship between ToM and language becomes more intricate. Despite using language to communicate, they often face delays and difficulties in language acquisition, which persist throughout their development. These challenges impede their ability to fully engage in ToM processes. Autistic children may struggle with pragmatics; the social use of language, leading to difficulties in understanding and utilizing non-literal language, interpreting social cues, and engaging in reciprocal conversations. Consequently, their ToM development may be hindered, as they may have limited access to the social and communicative tools that are necessary for understanding and attributing mental states to themselves and to others.

Understanding the processes of language acquisition in autistic young individuals is of paramount importance for researchers. The process of acquiring languages holds the potential to prove the complementary nature of ToM and language abilities. By unraveling the constraints and challenges that autistic children face in language acquisition, we can develop a deeper understanding of their ToM development. Language serves as a bridge between internal mental processes and external social interactions, shaping the development of ToM.

Language and ToM development are intricately linked, particularly when we are very young. They mutually support and enhance each other, facilitating children's understanding of themselves and of others. However, this relationship becomes more complex in autistic children, who often experience delays and challenges in language acquisition. Nonetheless, language acquisition in autistic children remains a crucial area of investigation.

Language and communication are often used interchangeably, but it is important to recognize that they are distinct concepts, although closely connected. Frith and Happe (1994) talked about the differences between language and communication, highlighting language as a representational system governed by grammar, while communication refers to the exchange of information between individuals, involving the alteration of the physical environment based on shared representations. There are some crucial aspects of language and communication, particularly in the context of autism, and one cannot ignore the role of language in compensating for ToM deficits in autistic individuals.

Frith and Happe (1994) emphasized the distinction between language and communication. Language is said to be a structured system that involves representation and grammar. It allows individuals to convey meaning through words, sentences, and syntax. On the other hand, communication involves the interaction between individuals, where shared representations are used to modify the physical environment. While language is a specific system, communication encompasses various forms such as verbal and non-verbal cues, gestures, and facial expressions. Understanding this distinction is essential for comprehending the complex nature of human interaction and the unique challenges faced by individuals with autism.

In the case of autistic individuals, it is often observed that syntactic and phonological skills remain intact. Traditional methods of vocabulary acquisition may not yield significant results for some autistic children, leading to a limited vocabulary. However, others may demonstrate an extensive vocabulary, particularly in areas that are not easily comprehensible to them (Frith and Happe 1994). Autistic children do not require gestures or facial expressions for spelling and reading, which suggests that these skills are not undermined in their case. In fact, many autistic children exhibit hyperlexia, demonstrating advanced reading abilities despite challenges in other areas of language and communication.

Frith and Happe (1994) raised the question of whether language plays a role in compensating for ToM deficits in individuals on the autism spectrum. They proposed that children on the spectrum might employ the strategy of thinking aloud to better understand their own mental states. This self-narration or thinking aloud can serve as a compensatory mechanism to bridge the gap in ToM abilities (Frith and Happe 1994). Furthermore, Hurlburt et al. (1994) discussed the use of visual cues and mental imagery as alternative routes to developing a ToM, which may be acquired later in individuals with autism. These alternative approaches highlight the adaptability and flexibility of cognitive processes in compensating for deficits in social cognition.

Frith and Happe (1994) also touched upon the phenomenon of echolalia, which involves the repetition of words or phrases. Echolalia can hinder the analysis of lexical issues that may be present in children with autism. While echolalia may impede certain aspects of language development, it is a complex behavior that needs further investigation to better understand its role in language acquisition and communication in individuals on the autism spectrum.

Language and communication are distinct yet interconnected aspects of human interaction. Understanding the differences between language as a representational system and communication as an exchange of information helps elucidate the challenges faced by individuals with autism. Despite deficits in ToM, language abilities in autistic individuals can be intact or even advanced in certain areas. Language may serve as a compensatory mechanism, allowing individuals to understand their own mental states.

19.5 The Role of Internal State Talk in Children's ToM Development

Understanding how children develop a ToM, where they go through the process of attributing mental states has been a topic of great interest. Researchers have explored various factors that may influence the development of ToM, including the role of internal state talk provided by mothers. It is imperative to assess the literature surrounding the connection between internal state talk and children's ToM performance, considering different viewpoints and studies conducted in this field.

Meins et al. (2013) conducted a comprehensive review of the existing literature on how children understand the mind. They discussed the connection between exposure to internal state talk within the family and later ToM performance. Some studies, such as those by Dunn et al. (1991) and Ruffman et al. (2002), have shown a positive relationship between internal state talk by mothers and children's ToM abilities. These studies suggest that engaging in discussions about thoughts, beliefs, and emotions may enhance children's understanding of mental states.

However, conflicting findings have also emerged regarding the relationship between internal state talk and ToM development. Ruffman et al. (2002) found that ToM abilities in the early phases was not related to later internal state talk provided by mothers. This discrepancy highlights the need for further investigation to better understand the complex interactions between internal state talk and ToM development.

Meins et al. (2013) referred to the work of Slaughter et al. (2007) who conducted small-scale studies exploring the relationship between internal state talk and children's ToM performance. Interestingly, Slaughter et al. (2007) found that the basic descriptions of characters' emotions during book reading tasks did not show any relations with ToM abilities. Instead, only references to the thoughts and beliefs of the characters were linked to ToM performance. This suggests that specific aspects of internal state talk, such as discussing mental states and beliefs, may be particularly influential in promoting ToM development.

In addition to the focus on thoughts and beliefs, Meins et al. (2013) highlighted the work of Tamoepeau and Ruffman (2006, 2008), who examined the relationship between children's emotional understanding and internal state talk provided by mothers. These studies suggest that discussing emotions and providing emotional explanations may contribute to children's emotional understanding, which is closely intertwined with ToM development.

The relationship between internal state talk and children's ToM development is a complex and dynamic area of research. While some studies have shown a positive association between internal state talk and ToM performance, other findings suggest that specific aspects of internal state talk, such as references to thoughts and beliefs, may be particularly influential. Furthermore, the connection between emotional understanding and internal state talk highlights the multifaceted nature of ToM development.

As research in this field continues to evolve, it is important to consider various perspectives and explore the specific mechanisms through which internal state talk influences children's ToM abilities. A better understanding of these relationships can contribute to interventions and strategies that can make us understand what goes through in the early years of a child.

19.6 Disruption of Language Development in Autistic Children

Language plays a significant role in the development of ToM, especially in early childhood. However, in the case of autistic children, language disruptions are observed, impacting their ability to understand and engage in social interactions. Many researchers have explored the connection between language and ToM in autistic children, focusing on their gaze strategies, vocabulary development, and syntactic abilities.

Autistic young children exhibit a distinct gaze strategy known as the Listener's Direction of Gaze (LDG) strategy, which is egocentric in nature (Baron-Cohen et al. 1997). Instead of following the Speaker's Direction of Gaze (SDG), they assume that the word they hear refers to the object being presently fixated upon. This error in mapping affects their ability to understand others' perspectives. Research using a cartoon method has shown that autistic children struggle while using the direction of a person's gaze to point to the intended referent, indicating a relative "blindness" to the mentalistic significance of the eyes (Baron-Cohen et al. 1995). This egocentric error can be predicted by the executive function hypothesis and the ToM/joint (Executive function) attention hypothesis (Ozonoff 1995).

Any child lacking the SDG strategy would not necessarily disturb all aspects of language (Baron-Cohen et al. 1997). The syntactic development is said to be independent of such social factors (Pinker 2007). However, there would be certain language deficits. Instead of following a SDG strategy, that child would be looking for an alternative LDG strategy where she would be assuming that the object in her gaze is the referent. A lot of times the speaker's referent would happen to be the same and this would lead the child to have a shared vocabulary. Baron-Cohen et al. (1997) have suggested that the first feature of autistic language that this deficit might explain would be the production of some words being used in unconventional ways. This deficit also accounts for lack in vocabulary development. A non-autistic child will develop vocabulary as an uttered word will be mapped correctly, however an autistic child will struggle with the same process as the LDG strategy would lead her to false starts, conflicting information, and confusion (Baron-Cohen et al. 1995). So, when a child first hears the word "food" when the father is sitting on a table having his food, and the child is looking at her toy, the LDG strategy would make her map it as ["food" = toy]. Later when she next hears the word "food' while fidgeting with a pen, she would search her lexical memory and retrieve the word "food", which would bring the interpretation "toy"; though she will erase that mapping ["food" = toy] and replace it with ["food" = pen]. Such problems with mapping could create obstacles in the path of language acquisition. Baron-Cohen et al. (1997) recognize however that lexical development is not specifically impaired in autism. Tager-Flusberg (1993) have also said that in relation to syntactic development. So, it cannot be denied that there are various other factors also that contribute to delay in language acquisition. Joint-attention deficit has a lot in common with the lack of SDG in autistic children. When it comes to language learning, it can be said that the inability of autistic children in understanding that language is there mainly for establishing a joint topic, for sharing comments on that particular topic (Baron-Cohen et al. 1997). This distinction of topic-comment relation in language acquisition helps us understand the disruptions.

Autistic children often exhibit impairments in pragmatic abilities, such as furnishing or asking for information and giving responses while asked questions (Tager-Flusberg 1996). However, the grammatical and semantics related areas of language do not appear to be impaired by a great margin overall (Swensen et al. 2007). Research indicates that autistic children may demonstrate increased mean length of utterance (MLU) and grammatical morpheme usage (Tager-Flusberg 1994). Despite these observations, the acquisition of linguistic forms by autistic children is not solely dependent on their ability to produce them (Fenson et al. 1994).

Language disruptions in autistic children, such as the adoption of the LDG strategy, impact their ToM development and language acquisition. The egocentric nature of their gaze strategies affects their understanding of others' perspectives and the mapping of words to referents. Consequently, autistic children may exhibit unconventional word usage, vocabulary deficits, and difficulties in syntactic and pragmatic abilities. Understanding the processes that promote lexical acquisition in autistic children is crucial for developing effective interventions and supporting their language development.

19.7 Hemispheric Specialization and the Language Abilities of Autistic Children

A study conducted by Geraldine Dawson, Charles Finley, Sheila Phillips, and Larry Galpert in 1986 aimed to explore the relationship between hemispheric specialization and language abilities in autistic children. Their research focused on investigating the patterns of specific hemispheres responsible for speech processing and language ability in autistic children, comparing them to a group of non-autistic children. The findings of their study discuss the implications of hemispheric specialization in relation to language development in autistic children.

The study involved two groups: 17 autistic children aged between 6 and 18 years and 17 non-autistic children of the same age. Dawson et al. (1986) observed that a majority of the autistic children exhibited reversed hemispheric asymmetry, with a dominance of the right hemisphere. This finding suggested a deviation from the typical left hemisphere dominance associated with language processing. Interestingly, the study revealed that autistic children, who had developed their language skills up to an advanced level, were more prone to exhibit a normal pattern of hemispheric asymmetry. This implied that the acquisition of language skills might trigger the process of switching from right hemisphere dominance to left hemisphere involvement in speech processing.

The involvement of different hemispheres in the brain has been a topic of active research, particularly concerning various skills acquired by individuals. In the context of autism, a deficit in the development of the left cerebral hemisphere is believed to be present, which is crucial for communication and gesture functions. Dawson (1983) also highlighted the visuo-spatial skills commonly observed in many autistic children, which align with the characteristics associated with the right hemisphere of the brain. Additionally, Dawson et al. (1982) measured alpha blocking in the left and right hemispheres during visuo-spatial and verbal tasks. The ground-breaking results indicated that seven out of ten individuals with autism spectrum disorder (ASD) who underwent the test displayed more movement in the right hemisphere than the left hemisphere during language processing. Although subsequent studies suggested that this phenomenon may not be attributed to a deficit or problem with the left hemisphere, but rather a general lag in development.

An intriguing point to note is that the right hemisphere can assume control over language processing when there are issues with the left hemisphere, as highlighted by Milner (1974). This suggests a compensatory mechanism within the brain, wherein the right hemisphere takes on language processing responsibilities in the absence of optimal functioning in the left hemisphere.

The findings reveal reversed hemispheric asymmetry in many autistic individuals, with right hemisphere dominance. However, advanced language skills appeared to correlate with a normalization of this asymmetry. The role of different hemispheres in autism has been a subject of ongoing research, indicating a deficit in the development of the left hemisphere and the presence of enhanced visuo-spatial skills associated with the right hemisphere. The compensatory nature of the brain allows the right hemisphere to assume language processing responsibilities when the left hemisphere encounters challenges.

On the other hand, Sandra F. Witelson's extensive research on language in children, focusing on the neurobiological aspects, has provided some valuable knowledge about the neural bases of cognition and the nature of hemispheric specialization as well. In her 1987 article, Witelson examined the importance of age in relation to the emergence of hemispheric specialization and whether this specialization undergoes changes over time. She also reviewed the literature that was present on language in children with brain damage as well as the children who were developing in a typical way. Witelson proposed a model suggesting that hemispheric specialization exists from birth and remains stable throughout development.

The concept of hemispheric dominance arises from the functional organization of the brain, where the left and right hemispheres play distinct roles in mediating various cognitive processes (Witelson 1987). Numerous studies in this field have highlighted the dependency of specific tasks on one hemisphere over the other. For instance, the left hemisphere is more involved in tasks such as syntax formation, speech production, and voluntary movements of fingers and limbs. On the other hand, the right hemisphere is predominantly engaged in processing colours, faces, shapes, emotional aspects, and musical melodies (Witelson 1987).

Witelson's research highlights the role of age in the development and stability of hemispheric specialization. Her model suggests that this specialization is already in function since birth and does not significantly change as one grows up (Witelson 1987). This implies that the functional lateralization of specific cognitive functions, for example language, is established early in life and remains relatively consistent throughout development. Understanding the time course and stability of hemispheric specialization is crucial for comprehending the neural underpinnings of language and related cognitive processes.

Language, being a complex cognitive ability, exhibits clear associations with hemispheric specialization. Witelson's research supports the notion that language functions primarily rely on the left hemisphere (Witelson 1987). The left hemisphere's dominance in syntax formation and speech production is well-established, highlighting its critical role in language processing. On the other hand, the right hemisphere is known for processing nonverbal aspects of communication, including emotional expressions, facial recognition, and melodic patterns in music (Witelson 1987). These findings suggest that different aspects of language are mediated by distinct hemispheres, contributing to the multifaceted nature of language processing.

Sandra F. Witelson's research on language and hemispheric specialization provides valuable insights into the neurobiological foundations of cognitive processes, particularly in the context of language development. Her work supports the notion that hemispheric specialization exists from birth and remains relatively stable throughout development. The left hemisphere is predominantly involved in language-related tasks, such as syntax formation and speech production, while the right hemisphere plays a crucial role in processing nonverbal aspects of communication. Understanding the functional organization of the brain and its relationship to language contributes to our knowledge of normal language development and can potentially inform interventions for individuals with language impairments.

19.8 Hypersensitivity and Language Learning

In their 2009 study, Baron Cohen et al. deliberated upon the concept of hyper systemizing and its relationship with talent in individuals with autism. They challenged the prevailing theories of weak central coherence and executive dysfunction and argued that talents in autistic individuals, including language learning abilities, stem from hyper systemizing. This hyper systemizing is characterized by exceptional attention to detail and a strong inclination towards finding and repeating patterns. Additionally, Baron Cohen et al. emphasized the role of sensory hypersensitivity in the development of these talents.

Baron-Cohen et al. (2009) proposed that hyper systemizing, driven by an excellent attention to detail, is a common trait among autistic individuals. They argued that attention is a fundamental cognitive process that starts at an early level, whereas systematization is a higher-level cognitive aspect. Autistic individuals exhibit a strong desire to find or create systems governed by rules, manifesting in various domains such as collectible systems, mechanical systems, numerical systems, abstract systems (e.g., syntax in grammar), and motoric systems. These individuals excel in identifying and systemizing patterns, as exemplified by their ability to understand complex mechanisms, like a Polaroid camera, while struggling with understanding the thoughts of others (Baron-Cohen et al. 2009). Hyper systemizing behaviour also explains the preference for routine and predictability often observed in individuals with autism spectrum disorder (ASD) and their limited pragmatic understanding of language.

Baron-Cohen et al. (2009) highlighted the sensory hypersensitivity experienced by autistic individuals, which contributes to their exceptional attention to detail and talent development. They noted superior auditory discrimination abilities in autistic children, as well as heightened reactions to visual and tactile stimuli in their environment. Sensory hypersensitivity is believed to be a key factor underlying their keen ability to detail. To support this claim, the researchers conducted an experiment using the Freiburg Visual Acuity and Contrast Test, which revealed that individuals on the autism spectrum outperformed the control group in visual acuity scores. The remarkable visual acuity exhibited by autistic individuals parallels the visual capabilities observed in birds of prey, suggesting a heightened sensory perception (Baron-Cohen et al. 2009).

Understanding hyper systemizing and sensory hypersensitivity in autism has broader implications, particularly in the context of language acquisition and communication. Autistic individuals process a vast amount of information and variables when attempting to communicate in any language. Traditional assessment methods that treat neurotypical and autistic individuals alike may fail to capture their unique talents and challenges. Tailored approaches that consider the specific cognitive and sensory characteristics of individuals on the spectrum are necessary to accurately evaluate their abilities.

Baron-Cohen et al.'s (2009) research throws more light on the concept of hyper systemizing and its association with talent in individuals with autism. Their study challenges prevailing theories and highlights the importance of attention to detail and sensory hypersensitivity in the development of unique abilities.

19.9 Communication Challenges in the Children with Autism Spectrum Disorder

The concept of ToM is central to understanding social and communicative exchanges. It refers to the ability of attributing mental states; be it to self or to others, playing a crucial role in navigating social interactions. Communication, as a key component of human interaction, heavily influences mental states and is closely intertwined with ToM. However, studies have shown that autistic children often struggle with expressing mental states, highlighting the unique challenges they face in understanding and engaging in social communication. The false belief task serves as a means to explore these difficulties, particularly in relation to unexpected changes in the environment. Perner et al. (1989) conducted research investigating the link between ToM and pragmatic skills in autistic children, shedding light on the nuanced nature of mental state comprehension in this group.

The false belief task is programmed to evaluate an individual's understanding of others' beliefs that may differ from reality. It examines the ability to recognize that someone may hold a false belief about a situation or object, even when the individual possesses contrary knowledge (Perner et al. 1989). This task reveals crucial facets of ToM development, as it requires individuals to comprehend and attribute mental states to others, taking into account their beliefs and expectations.

In their study, Perner et al. (1989) focused on ToM and its relationship with pragmatic skills in autistic children. The researchers conducted tests involving false belief scenarios and communication tasks on a group of 21 boys and five girls on the autism spectrum. They found that children with autism, despite varying mental ages, required significant assistance in answering the questions related to false belief, often rendering their understanding of mental states meaningless. Interestingly, most children on the spectrum performed well on tasks involving visual access, suggesting that their impairments were specific to ToM rather than mental illness or other neurological issues.

The findings of Perner et al. (1989)'s study highlight the challenges autistic children face in understanding and expressing mental states, particularly in relation to false beliefs. The difficulty experienced by these children in comprehending false belief scenarios emphasizes the importance of ToM deficits in autism spectrum disorder. It is crucial to recognize that these challenges are not indicative of mental illness or general cognitive impairments but are specific to the abilities associated with ToM. This understanding has significant implications for designing interventions and support strategies that target the unique social communication needs of autistic people.

Theory of mind plays a pivotal role in social and communicative exchanges, and autistic children often encounter challenges in understanding and expressing mental states. The false belief task serves as a valuable tool to examine these difficulties, revealing the unique nature of ToM deficits in autism spectrum disorder. Perner et al. (1989)'s study gives indispensable information about the specific impairments in ToM abilities among children on the spectrum, emphasizing the need for tailored interventions and support strategies to enhance their social communication skills.

19.10 Role of Language in ToM Research: The Path Ahead

Janet Wilde Astington made some suggestions for future investigation in the area of development of ToM. They were mainly three points. The first one was that the researchers should focus on the intention and desire and commit to get more understanding of these areas (Astington 2001). The second one was that there should be more emphasis on the role of language and its association with the development of ToM (Astington 2001). The third one was to emphasize upon the development of understanding of intention, desire and belief and its actual consequences (Astington 2001).

The suggestions given by Astington were in response to the article written by Wellman et al. (2001) on meta-analysis of false-belief task and how young children performed in those tasks. Talking about the role of language in the development of ToM, Astington (2001) states that the linguistic skill present in children might be important, not for showing the understanding in false belief tasks but more so for the understanding itself. Astington goes on to say that the temporal marker for children with not so higher language ability might not be that useful but those possessing greater abilities in terms of language may be offered assistance as they are able to process it.

Astington (2001) asserts that the linguistic ability of the children for understanding of the false-belief is complicated because it offers a way for representing false belief contrary to the evidence provided in actual scenario. Astington (2001) points out that some investigators work on the aspect that is purely representational, by leveraging that syntax and its acquisition for complementation helps in the understanding of false belief (de Villiers and de Villiers 2000), others work on the aspects that are completely communicational, justifying that narrations of tales and conversation by adults bring mental states to the attention of children (Nelson 1996). Astington (2001) has very aptly argued that investigators might pay attention to one of these but it is paramount to not treat them both as hypotheses that compete with each other. They should rather complement each other (Astington 2001).

Astington (2001) agrees that language ability might be equally important for the development of understanding. Studies that examine the process of understanding using the intention and desire tasks which link this understanding to the development of language would help exponentially.

19.11 Conclusion

Language plays a significant role in the development of ToM, particularly during the initial years of a child's life. Autistic children often experience deficits in both language and ToM, which can affect their performance in tasks involving false beliefs. While their grammatical and semantic language components may not be impaired, they often struggle with pragmatic abilities. However, those who manage to pass such tasks have mastered certain linguistic abilities that enable them to do so. Instead of relying on social-perceptual abilities, autistic children reason logically through these tasks using language as a tool.

Understanding the processes of early language acquisition in autistic children is crucial for advancing our knowledge of ToM and language constraints. Language and ToM are intertwined during early childhood development, as language allows children to express their thoughts and emotions and engage in social interactions, fostering their understanding of others' perspectives. On the other hand, ToM enhances language use, enabling children to navigate social situations and infer others' mental states.

In the case of autistic children, ToM and language share a very complex relationship. Challenges in language acquisition, particularly in pragmatics and social use, may impede their ToM development. However, examining language acquisition in autistic children provides valuable insights into their development and can help create and form interventions and strategies to support them.

Furthermore, research has highlighted the presence of hyper systemizing and sensory hypersensitivity in individuals with autism. These characteristics contribute to their exceptional attention to detail and unique talents. Attention to detail and the use of language as a compensatory mechanism are common traits among autistic individuals, potentially helping to bridge the gap in ToM abilities.

Future investigations will benefit from focusing on the intention, desire, and belief aspects of ToM, the role of language in ToM development, and understanding the consequences of intention, desire, and belief. Additionally, it would be research worthy to explore the complementary nature of language and ToM, considering both representational and communicational aspects rather than treating them as competing hypotheses.

References

Astington, J.W. 2001. The future of theory-of-mind research: Understanding motivational states, the role of language, and real-world consequences. *Child Development* 72 (3): 685–687. http://www.jstor.org/stable/1132445.

Astington, Janet Wilde, and Jodie A. Baird (eds), Why language matters for ToM (New York, 2005; online edn, Oxford Academic, 22 Mar. 2012), https://doi.org/10.1093/acprof:oso/9780195159912.001.0001.

Baird, G., H. Cass, and V. Slonims. 2003. Diagnosis of autism. *BMJ: British Medical Journal* 327 (7413): 488–493. http://www.jstor.org/stable/25455385.

Baron-Cohen, S., R. Campbell, A. Karmiloff-Smith, J. Grant, and J. Walker. 1995. Are children with autism blind to the mentalistic significance of the eyes? *British Journal of Developmental Psychology* 13: 379–398. https://doi.org/10.1111/j.2044-835X.1995.tb00687.x.

Baron-Cohen, S., D.A. Baldwin, and M. Crowson. 1997. Do children with autism use the Speaker's direction of gaze strategy to crack the code of language? *Child Development* 68 (1): 48–57. https://doi.org/10.2307/1131924.

Baron-Cohen, S., E. Ashwin, C. Ashwin, T. Tavassoli, and B. Chakrabarti. 2009. Talent in autism: Hyper-systemizing, hyper-attention to detail and sensory hypersensitivity. *Philosophical Transactions: Biological Sciences* 364 (1522): 1377–1383. http://www.jstor.org/stable/40485909.

Bretherton, I., and M. Beeghly. 1982. Talking about internal states: The acquisition of an explicit ToM. *Developmental Psychology* 18 (6): 906–921. https://doi.org/10.1037/0012-1649.18.6.906.

Corballis, M.C. 2012. The wandering mind: Mental time travel, ToM, and language. *Análise Social* 47 (205): 870–893. http://www.jstor.org/stable/41959838.

Dawson, G. 1983. Lateralized brain dysfunction in autism: Evidence from the Halstead-Reitan neuropsychological battery. *Journal of Autism and Developmental Disorders* 13 (3): 269–286. https://doi.org/10.1007/BF01531566.

Dawson, G., S. Warrenburg, and P. Fuller. 1982. Cerebral lateralization in individuals diagnosed as autistic in early childhood. *Brain Lang. Mar* 15 (2): 353–368. https://doi.org/10.1016/0093-934x(82)90065-7.
Dawson, G., C. Finley, S. Phillips, and L. Galpert. 1986. Hemispheric specialization and the language abilities of autistic children. *Child Development* 57 (6): 1440–1453. https://doi.org/10.2307/1130422.
de Villiers, J.G., and P.A. de Villiers. 2000. Linguistic determinism and the understanding of false beliefs. In *Children's reasoning and the mind*, ed. P. Mitchell and K.J. Riggs, 191–228. Psychology Press/Taylor & Francis.
Dunn, J., J. Brown, C. Slomkowski, C. Tesla, and L. Youngblade. 1991. Young Children's understanding of other People's feelings and beliefs: Individual differences and their antecedents. *Child Development* 62 (6): 1352–1366. https://doi.org/10.2307/1130811.
Fenson, L., P.S. Dale, J.S. Reznick, E. Bates, D.J. Thal, S.J. Pethick, M. Tomasello, C.B. Mervis, and J. Stiles. 1994. Variability in early communicative development. *Monographs of the Society for Research in child development 59* (5): i–185. https://doi.org/10.2307/1166093.
Frith, U., and F. Happe. 1994. Language and communication in autistic disorders. *Philosophical Transactions: Biological Sciences* 346 (1315): 97–104. http://www.jstor.org/stable/56024.
Hill, E.L., and U. Frith. 2003. Understanding autism: Insights from mind and brain. *Philosophical Transactions: Biological Sciences* 358 (1430): 281–289. http://www.jstor.org/stable/3558141.
Hurlburt, R.T., F. Happé, and U. Frith. 1994 May. Sampling the form of inner experience in three adults with Asperger syndrome. *Psychological Medicine* 24 (2): 385–395. https://doi.org/10.1017/s0033291700027367.
Meins, E., C. Fernyhough, B. Arnott, S.R. Leekam, and M. de Rosnay. 2013. Mind-mindedness and ToM: Mediating roles of language and perspectival symbolic play. *Child Development* 84 (5): 1777–1790. http://www.jstor.org/stable/24029482.
Milner, B. (1974) Hemispheric specialization: Scope and limits. In: The neurosciences: Third study program, Schmitt, F. O. and Worden, F. G., pp. 75–89. Cambridge: MIT Press.
Nelson, K. 1996. *Language in cognitive development: Emergence of the mediated mind*. Cambridge University Press. https://doi.org/10.1017/CBO9781139174619.
Ozonoff, S. 1995. Executive functions in autism. In *Learning and cognition in autism. Current Issues in Autism*, ed. E. Schopler and G.B. Mesibov. Boston, MA: Springer. https://doi.org/10.1007/978-1-4899-1286-2_11.
Perner, J., U. Frith, A.M. Leslie, and S.R. Leekam. 1989. Exploration of the autistic child's ToM: Knowledge, belief, and communication. *Child Development* 60 (3): 689–700. https://doi.org/10.2307/1130734.
Pinker, S. 2007. *The language instinct*. HarperCollins.
Premack, David, and G. Woodruff. 1978. Does the chimpanzee have a ToM? *Behavioral and Brain Sciences* 4 (4): 515–629.
Ruffman, T., L. Slade, and E. Crowe. 2002. The relation between children's and mothers' mental state language and theory-of-mind understanding. *Child Development* 73 (3): 734–751. http://www.jstor.org/stable/3696247.
Slaughter, V., C.C. Peterson, and Emily Mackintosh. 2007. Mind what mother says: Narrative input and ToM in typical children and those on the autism Spectrum. *Child Development* 78 (3): 839–858. http://www.jstor.org/stable/4620672.
Sperber, D., and G. Origgi. 2010. A pragmatic perspective on the evolution of language. In *The evolution of human language: Biolinguistic perspectives (approaches to the evolution of language)*, ed. R. Larson, V. Déprez, and H. Yamakido, 124–132. Cambridge: Cambridge University Press. https://doi.org/10.1017/CBO9780511817755.009.
Swensen, Lauren D., Elizabeth Kelley, D. Fein, and Letitia R. Naigles. 2007. Processes of language acquisition in children with autism: Evidence from preferential looking. *Child Development* 78 (2): 542–557. http://www.jstor.org/stable/4139244.

Tager-Flusberg, H. 1993. What language reveals about the understanding of minds in children with autism. In *Understanding other minds: perspectives from autism*, ed. S. Baron-Cohen, H. Tager-Flusberg, and D. Cohen, 138–157. Oxford: Oxford UnIversIty Press.

———., ed. 1994. *Constraints on language acquisition: Studies of atypical children*. 1st ed. Psychology Press. https://doi.org/10.4324/9781315807522.

———. 1996. Brief report: Current theory and research on language and communication in autism. *Journal of Autism and Developmental Disorders* 26: 169–172. https://doi.org/10.1007/BF02172006.

———. 2007. Evaluating the theory-of-mind hypothesis of autism. *Current Directions in Psychological Science* 16 (6): 311–315. https://doi.org/10.1111/j.1467-8721.2007.00527.x.

Tager-Flusberg, H., and R.M. Joseph. 2005. How language facilitates the acquisition of false-belief understanding in children with autism. In *Why language matters for ToM*, ed. J.W. Astington and J.A. Baird, 298–318. Oxford University Press. https://doi.org/10.1093/acprof:oso/9780195159912.003.0014.

Taumoepeau, M., and T. Ruffman. 2006. Mother and infant talk about mental states relates to desire language and emotion understanding. *Child Development* 77 (2): 465–481. http://www.jstor.org/stable/3696481.

———. 2008. Stepping stones to others' minds: Maternal talk relates to child mental state language and emotion understanding at 15, 24, and 33 months. *Child Development* 79 (2): 284–302. http://www.jstor.org/stable/27563484.

Wellman, H.M. 1992. *The child's ToM*. The MIT Press.

Wellman, H.M., D. Cross, and J. Watson. 2001. Meta-analysis of theory-of-mind development: The truth about false belief. *Child Development* 72 (3): 655–684. http://www.jstor.org/stable/1132444.

Wimmer, H., and J. Perner. 1983 Jan. Beliefs about beliefs: Representation and constraining function of wrong beliefs in young children's understanding of deception. *Cognition* 13 (1): 103–128. https://doi.org/10.1016/0010-0277(83)90004-5.

Witelson, S.F. 1987. Neurobiological aspects of language in children. *Child Development* 58 (3): 653–688. https://doi.org/10.2307/1130205.

Chapter 20
Gendered Theory of Mind: A Linguistic and Literary Approach

Sergio Marin-Conejo and Teresa Lopez-Soto

Abstract In this chapter, we explore the relationship between ToM and gender using linguistic knowledge and literary samples, as both help us to cover the main tool by which people connect: language and stories/discourse. We argue that ToM is not a unitary ability, but rather it is a complex set of mechanisms influenced by gender in a diachronic androcentric socio-cultural context. We conclude that we need to go back to the origin of the sex/gender system as it ultimately underlies a primordial opposition of animacy both linguistically and discursively.

Keywords Theory of mind · High order theories · Sex/gender system · Animacy · Androcentrism

20.1 Introduction

It is of utmost necessity to consider that science needs a profound overhaul from a sex/gender perspective, avoiding the biases of the past. Science has been one of the most influential drivers of human progress, unlocking the mysteries of the universe while still trying to understand who we are. However, it is crucial to recognize that science has not been immune to sex/gender prejudices and societal influences. Historically, modern scientific research has largely been conducted from a male-centric perspective, neglecting the significance of sex and gender differences, on the one hand, and "Along with women, scientists excluded from science a specific set of moral and intellectual qualities defined as feminine." (Schiebinger 2000: 51). The early scientific research in which we support our findings primarily focused on male

---

S. Marin-Conejo
Department of English Language, University of Seville, Seville, Spain

T. Lopez-Soto (✉)
Laboratory of Speech and Phonetic Sciences, University of Seville, Seville, Spain

Department of Forensic Medicine, Psychiatry and Pathology, Complutense University of Madrid, Madrid, Spain
e-mail: teresals@us.es

T. Lopez-Soto et al. (eds.), *The Theory of Mind Under Scrutiny*, Logic, Argumentation & Reasoning 34, https://doi.org/10.1007/978-3-031-46742-4_20

subjects, as if their experiences, body, and mindset represented the entire population. As we move forward, albeit difficult, it is imperative to acknowledge this oversight and take proactive measures to incorporate a sex/gender system perspective into science. Notwithstanding a profound revision of science with such a perspective to foster a beneficial inclusive and unbiased research, any emerging field such as the Theory of Mind needs to evolve away from patriarchal overt and underlying paradigms.

Broadly speaking, the Theory of Mind (henceforth ToM) conceives the mental faculty that enables individuals to grasp and attribute different mental states to themselves and those around them. At its core, ToM involves the recognition and acknowledgement that other individuals have unique and different perspectives, thoughts, beliefs, intentions but also desires and emotions. It implies the acceptance that individuals have their own internal worlds, which may differ significantly from the internal landscape of one's own mind. This cognitive capacity allows us to navigate social interactions, empathize with others, and predict other's behaviors based on their mental states. It can be decomposed into two abilities: mindreading and metacognition (Sebastián 2016: 74).

A simple look at the academic literature on the ToM will reveal various research lines concerned, from health studies, such as neuropsychiatric issues or studies on the brain to psychological concerns such as autism or children and other animal behaviors or learning abilities. It has also been called upon to check language acquisition concerns, considered as an approach to specifying the truth conditions for knowledge, the philosophy of language, thought or even, science itself. It ranges a large variety of disciplines, around and on the human being, even though little has been written about gender or with a gender/feminist perspective. The aim of this chapter is therefore to shed light on the relationship of ToM and gender, as Adenzato et al. (2017) put it, since "taking into account gender-related differences is mandatory for the investigation of ToM" (5).

The question at this point is what we understand with "gender-related". The sex/gender system has relativized sex since Foucault's wrote that "the notion of 'sex' functions to 'group together' in a 'fictitious unity' a variety of disparate phenomena, including 'anatomical elements, biological functions, conducts, sensations and pleasures" (Sullivan and Todd 2023), to whom Joan Scott (1986), Judith Butler (1990) and Sally Hines (2009) add that sex and gender categories should be 'historicized' and culturally compared. In relation with ToM, the sex/gender system implies a prefabricated mindset which is taught even before the kid is born. It is undeniable that "[w]hile there is a substantive body of research on how the mind deals with numbers, symbols and language, the research about the mental framework which deals with other minds is comparatively young" (Sturm 2020: 59), reason for which we should not forget that androcentrism and patriarchy have implications in any cognitive process, as ToM.

The label 'theory of mind' was coined by David Premack and Guy Woodruff in their 1978 paper, "Does the chimpanzee have a theory of mind?". The authors argued that chimpanzees are capable of attributing mental states to others, such as beliefs, desires, and intentions. Hence, they can develop a complex cognitive ability that is essential for social interaction, which is not unique to humans. Nonetheless, it also deals with emotions. Sebastián (2016) argues that "phenomenal consciousness is a necessary condition for our mindreading ability" when he questions if consciousness and ToM is a common theory. He comprehends that the high-order theories of consciousness can be belief-like or perception-like, therefore, subdivided into high-order of thought (HOT) and high-order of perception (HOP), being the second dependent on the former.

In this sense, considering Grice's papers on meaning, HOT theory holds that the meaning of a word or phrase is not simply a matter of its referent or its syntactic properties, but also of the thoughts that it triggers in the mind of the speaker or hearer:

> One of the motivations Grice had in giving an account of meaning was to distinguish between what is *meant*—M-intended—and what is not M-intended but *implied*. Grice's conversational maxims are principles of rational communication that audiences use to construct an inferential bridge from what is meant to what is implied. (Grandy and Warner 2022: n/p)

Consciousness is not simply having a verbal, visual or emotional thought but also being aware of having that thought or sensation and being able to recognize it. If we consider the sexualized other, a combination of both will result in a coronary HOT ToM, of what is *meant* but also *implied* in terms of gender.

With this necessary marriage, a first complementary approach to HOT ToM and the body and gender involves those implicit and explicit elements derived from the five-point hypothesis of Krashen and the linguistic purpose of communication. Because of its social relevance, culture, as Lévi-Strauss put it, is the vault from which individuals retrieve answers to the unknown, for satisfying their needs. Indeed, while "one of the first things toddlers have to learn is the pointing gesture [. . .] parallel[ing] the understanding of reference [that] might precede the understanding of verbal pointers, such as deictic markers" (Sturm 2020: 66), Michael Tomasello (2004, 58) highlights that human learning is cultural learning because "human beings, even when quite young, are able to understand the intentional and mental states of other human beings". This comprehension allows cultural processes to shape human cognition in ways that are inaccessible to other species, ultimately transforming human cognition into a fundamentally "collective enterprise" (2004: 58). But it takes clever experimentation to make language carry symbolic meaning, and the cognitive processes involved in language development and usage are not isolated but rather intertwined with other non-linguistic cognitive functions. The key mental functions include social cognition, conceptualization, and memory. As a result, an individual's linguistic system is influenced by their unique life experiences, the specific linguistic contexts straightjacket and cultural traits they have encountered.

Being that said, we should bear in mind that in the collective (un)consciousness "human" is primarily and taxonomically androcentric, neither anthropocentric nor gynocentric. As we know by observing the proto-Indo-European, originally in the Western languages, sex/gender was less relevant morphosemantically. The most accepted hypothesis states that what was overtly expressed in their language was the individuality/continuativity of individuals, evolving to the need to express their animacy/inanimacy 'gender' (Luraghi 2011) that developed again into the three or four different ones we better envisage today, namely, femenine, neuter, common, masculine. The latter pairing of animacy, "commonly defined as the distinction between living and non-living entities" (Trompenaars et al. 2021) and their 'opposites' remains today overtly expressed in Persian, Japanese or Spanish (among others) but also is found in English (Olloqui-Redondo et al. 2019; Rosenbach 2008; Zaenen et al. 2004) as a subtle and implicit input. In terms of sex/gender, "Classifying an entity in the world as either living or non-living has direct consequences for the way we conceptualize it" (Trompenaars et al. 2021). Indeed, as Chen (2021) reminds us, 'animacy' can be used "as a sexual device". We will discuss this ToM text-sexual relation below.

A second complementary epistemological approach would be that on which the neuroscientist António Rosa Damásio's arguments are based: cognition is not an independent and autonomous system but "an integrated psychic activity, evolutionarily constructed from the sensation-emotion" (Monserrat 2003: 179, my translation). The relevance of this assertion comes in contravention of the dualistic axiom adopted in the West from Descartes, as Monserrat suggests (178), and which also contradicts the assumption that ToM begets Dualism (Berent 2023), even if we might agree that men are poorer "mind-readers". Instead, as David Malet Armstrong & Anstey currently proposes in his book *A Materialist Theory of the Mind* (1968/2022: 126),[1] "consider the interesting case of the chicken-sexer". We will provide three examples of HOT ToM in a literary context where humans and animals have been perceived as sexualized bodies (and minds) intertwined with gender/racialized categorizations.

20.2 Sex/Gender Paradigm

Tracking back to the Renaissance and Early Modern period will allows to find a clear example of social inequality in terms of gender, when women's inputs were dismissed:

> Most famously learned, [. . .] thus, by an opinion which I hope is but an erroneous one in men, we [women] are shut out of all power and authority by reason we are never employed either in civil nor martial affairs, our counsels are despised and laughed at, the best of our actions are trodden down with scorn, by the over-weaning conceit men have of themselves and through a despisement of us. But I considering with myself, that if a right judgement and a true understanding and a respectful civility live anywhere, it must be in learned universities [. . .]
>
> Margaret Cavendish, "To the two universities", *Philosophilcal and Physical Opinions,* 1655.

[1] Armstrong's book presents a comprehensive exploration of the mind-body problem from a materialist perspective. As an influential philosopher of mind, he proposes a theory that seeks to reconcile the mental and physical aspects of human experience. He argues against dualistic theories that posit the existence of separate mental and physical substances. He asserts that mental states and processes are not ontologically distinct from physical states and processes but can be understood in terms of the underlying materialistic framework.

If there were any risk of anachronism, let's state that contrary to George Butte's and Liza Zunshine's claim that "deep intersubjectivity" (AKA ToM) started with the "rose" of novels such as Jane Austen's *Persuasion*, we agree with Mancing that ToM can be found as earlier as the Renaissance. When he applies ToM to *Don Quixote,* he found that "the scene [...] seems to suggest that the narrative skill required for the multi-levelled presentation of fictional minds existed and was practiced much earlier, at least in Spain." (Macing 2014: 127).

The vitalist, materialist and monist authoress[2] Margaret Cavendish, who dared to sign her books, met, shared thoughts, and finally did not acquire the Cartesian principles. As a contemporary thinker, she pointed out from her early materialist feminism, that absolutely everything is material -even thoughts-, interdependent and self-moving, without splitting the human into a body-mind paradigm, as Descartes promulgated[3] while he searched for the place of the soul. The ToM fits here as a corrector, turning that cartesian paradigm split 180° towards Cavendish assertion of humans as holistic units.

What she meant was a recall of Christine de Pizan's *Querelle des Femmes* (or *Women's Question*),[4] what she implied was a proto-feminist anti-mechanicist revolution. With a ToM eye, the quote above glosses the perception of exclusion and marginalization experienced by a particular group, women, in the context of societal power dynamics. Duchess Margaret Cavendish (1623–1673), in her *Philosophical and Physical Opinions* (1655), toed a line when she addressed "the two most famous universities" to argue that women were just as capable as men of holding positions of power and authority but were often excluded from these positions. This is one valuable example of an Early Modern writer rising the *Querelle*, stating a powerful argument for the equality of women, and helping to pave the way for women's rights movements in the centuries that followed: "authors who are little-known, little-studied, or not known at all, whose works are fundamental pieces in the great mosaic of feminist ideas ante literam, which allow us to understand the dissidence and symbolic opposition to the system of patriarchal hierarchies" (Arriaga and Marín 2023: 109, my translation). Cavendish claimed that men fail to understand the thoughts and feelings of women, underestimating women's abilities. Even more, she points that men's overconfidence in their own abilities leads them to dismiss the advice and counsel of women. She rebukes men for often "despise" women and "tread down" their actions with scorn because men believe that they are superior to women, and they do not believe that women have anything worth contributing.

---

[2]She consciously used the term "authoress" derived from the Latin word 'auctor' which means 'originator' or 'creator'. First used in the fourteenth century, the word with the derivational suffix '-ess' reflects that she was nor referring to "a person who writes" but to "a female person who writes.", embodying her authority in what Donna Haraway will label as 'situated knowledge' in 1988.

[3]As Londa Schiebinger (2004) quotes, Descartes, unlike Bacon, Locke, or Kant, was married, although "were but indifferent companions of the *belles*." (222).

[4]Note the intended pronoun in the first-person plural.

From a ToM reading, we should detail what is ToM and its relationship with patriarchy. The duchess mentioned that women are "shut out of all power and authority by reason we are never employed either in civil nor martial affairs" suggesting that men believe that women do not have the experience or knowledge necessary to hold these positions. Cavendish also reflects with "our counsels are despised and laughed at", that there are two groups: "our" (women) and "them" (men), suggesting that these do not believe that women's thoughts and opinions are of any value. Finally, men, as group, in her view, have an "over-weaning conceit" of themselves just because they think they are superior to women, reason by which they do not need to consider women's thoughts and feelings. The quote implies that the experiences of nature, understanding, knowledge, and consequently prudence (wise decision-making) that men possess are not only *kidnapped* by them but also denied to women. Cavendish moves to the position of them, perceiving that they think of women as lacking the cognitive abilities necessary for participating in important decision-making processes.

In applied sense, the English authoress asserted that due to the alleged lack of understanding, knowledge, sense-making and prudence, women are effectively excluded from power and authority, unable in the end of being source of creativity and social and individual identity. It implies that women are not given opportunities to engage in civil or martial affairs, and if they provide any, their counsels are disregarded and ridiculed. It is well known that in a social hierarchy such seventeenth century England, power and authority were predominantly held by men, while women are relegated to familiar and socially marginalized roles. It is worth highlighting that Cavendish stated that even the best actions of women are scorned and dismissed by men. This suggests that, as we will argue, that men have a ToM problem, on the one hand in their lack of recognition of women's abilities. On the other, there is a sense of devaluation and disrespect toward female contributions, potentially stemming from a combination of gender stereotypes and biases, which is linguistically driven by the taxonomical grouping of women labeled with the categories of continuativity-ininimacy, with no possibility of individual exceptions or agency, turning the strenght of ToM into a weakness, as a threat to the distinctive characteristic of masculinity.

Even more, she attributed the exclusion and marginalization of women to men's overestimation of themselves, implying that men's inflated self-perception contributes to their dismissal and disregard of women's perspectives and abilities. It reflects an imbalance in the theory of mind, where men fail to adequately consider and appreciate the thoughts and experiences of women. This quote illustrates the ToM dynamics at play within a gendered context. It highlights the HOT cognitive processes involved in perceiving and understanding the meaning of the experiences of others, as well as the implications of these processes on power dynamics, authority, and the treatment of marginalized groups as a group and/or individual exceptions.

Spenser et al. (2022) state that "Human behavior" is said to be underpinned by three individual cognitive abilities -Theory of Mind, empathic understanding, and moral reasoning: "As such, reduced abilities in these skills are thought to be associated with offending behaviors". Specifically, male offenders differ in their abilities regarding verbal ToM and performance-based empathic understanding when compared to female offenders. The authors suggest that assessment of individuals' abilities in ToM, empathic understanding, and moral reasoning may be of benefit in reducing offending behaviors by improving prosocial abilities, highlighting that:

> Language development may offer one explanation as to why an interaction between gender and offending status, in relation to verbal ToM, was found only in the male groups. Findings have suggested that men are less lingually advanced than their female counterparts, which is said to be further exaggerated by offending status. [It is also] noted that reflexive-cognitive processes (language) are required to understand the mental state of another person. Thus, differing abilities in language may account for the present findings. (2022: 18)

20.3 Language

> *You taught me language; and my profit on 't.*
> *Is, I know how to curse. The red plague rid you.*
> *For learning me your language!*

The utterer of these lines, Caliban, is an ambivalent male character oppressed, in relation to England, and oppressing, in relation to females and the ones of his island. In this quote from *The Tempest* (act 1, scene 2), the island denizen is expressing his frustration with Prospero and Miranda for teaching him English: Caliban feels that "they 'civilized', humanized him, which rather than being grateful for, he reproaches them for (Salas-Leal 2020: 33). He believes that the only thing he has gained from learning their language is the ability to curse them. This suggests that Caliban understands that Prospero and Miranda have different thoughts and feelings from him as well as he knows that they believe that they have done him a favor when he does not share that belief. He understands their actions as a form of oppression, and he uses the language they have taught him to express his anger and resentment. In this sense, Caliban is a better developed character than the colonists, following their mistaken belief that they knew what a good deed for the island's natives was.

He expresses the acquisition of language as a source of power, specifically the ability to hurt, to curse. Reason by which, here, Shakespeare was literary expressing Caliban's awareness of his own ideas about Prospero and Miranda, symbolically the West, able to reflect on his own 'native' thoughts and feelings and to use their language as a tool to express them. As we earlier referred to emotions, Caliban's quotation then suggests as well, a strong negative feeling and a desire for harm to befall the persons who taught him. But beyond literary considerations, these lines also reflect that both emotions and language play an important role in consciousness. Caliban's ability to curse Prospero and Miranda is only possible because the feeling triggers the response, as he finds that he has learned their language. On the one hand, the English language allows him to express his thoughts and feelings in a way that the colonists would not have understood otherwise. On the other, the 'uncivilized' knows that words can be used to hurt and to control, and he uses this knowledge to his advantage, being more sophisticated than we might initially expect, in terms of a racialized perspective, but obvious from a gender one. He learned to situate his knowledge.

As an oppressing character, let us not forget that Caliban fails at trying to rape Miranda, while in the quoted outburst he is reacting to being symbolically raped by Prospero and Miranda with language, as Carr and Pauwels (2005), expose it: "Real Boys don't do languages":

> there is a degree of gendering in the study of foreign languages in all the Anglophone countries we examined. Boys participate less than girls and their participation diminishes significantly, at least for most languages, the more advanced the study of the language gets. It is significant that trends observed across four different countries/regions are so similar despite significant differences in education systems. (19)

Violence and language may seem like a weird binary, but as Bodine (1975, 2008) expresses it plainly in "Androcentrism in prescriptive grammar [. . .]", the English language suffered socially motivated 'attacks' from prescriptive grammarians in relation to the normalization of English third person pronouns. Curzan (2006) claims in her book that, from the seventeenth century on, [male] grammarians have tried to institutionalize practices for dealing with the 'he or she' natural construction towards a 'generic he'.[5] While the mysterious 'She problem' (sic)[6] in Late Medieval English is still unresolved today, or how the order of binomial phrases in sentences such as "men and women" over "women and men" can communicate that men have more agency than women (Hegarty and Parr 2023; Kesebir 2017), it is interesting that "[d] espite being acknowledged as real and potent, androcentrism has received relatively little attention in psychology" (Bailey et al. 2018: 322). More recently Redl et al. (2022) found that, in Dutch, the use of the generic masculine pronoun "zijn" (his) creates a male-oriented bias in conceptually singular generic situations. The problem might be, as Nevalainen (2000) concluded that "[t]he historical evidence thus agrees with those sociolinguists who maintain that women are instrumental in setting new linguistic standards" (53), breaking with the traditional and symbolic [male] owner of language, creating an 'écriture fémenine', as Cixous acknowledged. In fact, Pauwels' *Women changing language* (1998) presented the feminist linguistic reforms as a type of language planning framework (in more than ten languages):

> Although feminist approaches to the study of language are most prevalent in those areas of linguistics which study language use in (social) context – especially sociolinguistics, sociology of language, social linguistics, critical linguistics, pragmatics and discourse analysis – they are also increasingly applied to fields such as historical linguistics, lexicology and lexicography as well as linguistic theory. (1998, xv)

---

[5]One of the examples would be that of Thomas Harvey's *A Practical Grammar of the English Language* (1878): "Do not say, 'Each pupil should learn his or her lesson:' use *his* alone.'" (Curzan 2006: 202)

[6]Scholars who study English diachronically are still debating how Old English 'hēo' did or did not evolve into 'she', stem that emerged in writing for the first time in ca. 1140 in the intriguing line of the Peterborough Chronicle: "and the people of London wanted to take her, and she flew [and te Lundenissce folc hire wolde tecen, and scæ fleh].

Indeed, if we delve into the mythic linguistics area, the relation of linguistics and gender rolls back in time until when the ancient Greek grammarians of the Callimachus' circle were accused of being "bugs that bite the eloquent" in the context of "glorifying Erinna [from Telos]" (Antiphanes of Macedon 2021 ca. 90 CE, n/p), mythically Sappho's significant other, for being more interested in 'minor' topics and word games. That is, not only a school of thought that did not concern itself with those topics and themes that were not traditionally assigned by heteropatriarchal grammarians and thinkers, but also those that considered women's poetry as worthy of consideration as their own, whether women or men, playing with words as a sign of innovative approaches and language evolution and experimentation, showing a HOT ToM application.

Thus, as exposed, if we are to consider the role of what traditionally has been considered extrinsic elements of linguistics, we must do so with full weight.

20.4 ToMming Animals and Things

As Blumer's social interactionism explain (Blumer 1969), even if he failed at the gender task, social interaction is a mutual presentation of actions by actors. He refers to human beings that act toward things based on the meaning that the thing has for them; the meaning of things is derived from the social interaction that one has with one's fellows, and these meanings are handled in and modified through an interpretative process used by the person in dealing with the things one encounters. Now, let's gender those actors beyond natural sex.

Patriarchy is part of our sociological realism, providing a way of thinking about language that considers the influence of both social structures and imbalanced gender agency and animacy. This productive framework for sociolinguistics allows us to explore the complex ways in which language is used in social interaction. Östen Dahl (2000) summed up the relation of animacy, we mentioned before, and semantic gender, reminding us first that its hierarchy:

> is supposed to have (at least) the following components: HUMAN>ANIMAL>INANIMATE. Hierarchies of this kind have been assumed by typologists to underlie various implicational universals. The general idea is that grammatical phenomena will "obey" the hierarchy in the sense that certain generalizations will apply to all cases above a certain cutoff point in the hierarchy.

Beyond other universal morphosyntactic considerations, that hierarchy can be found in various examples of androcentric usage such as how the term "*girl* is more than three times more likely than *boy* to refer to an adult" (Sigley and Holmes 2002: 153) or more worrying examples as on Nübling & Lind's paper (Nübling and Lind 2021) on how grammatical gender has been used for an intended dehumanization of women in German, or Krendel's corpus-assisted discourse approach that shows that:

> Women and girls were dehumanized and sexually objectified, negatively judged for morality and veracity, and constructed as desiring hostile behaviour from male social actors.

> Conversely, men were constructed as victims of female social actors and external institutions and, as a result, as unhappy and insecure (Krendel 2020: 1)

Being exposed that women and Caliban-like people are not considered a fellow human being by some privileged ones[7] think of the term 'man' in *mankind* and *humankind*, dehumanized as lacking animacy, for the following level, ANIMALS, let's consider the "permeability of categories and boundaries, and the moral ambiguity toward animals that is revealed in the way language is used. [. . .] uncertainty and instability of contemporary perceptions of, and attitudes toward, animals." (Sealey and Charles 2013). The authors state as well "how language constrains meaning—including the various ways in which people respond to the question 'What do animals mean to you?'" (501), perhaps reflecting the difficulties of practicing ToM with animals when one is unable to do so with one's peers.

It is therefore interesting to see, in the position of McConnell-Ginet (2011) when studying language and gender: "Rather than seeing content as firmly attached to linguistic forms", she asserts that she "began to view those forms as being filled with content [. . .] in the course of language-using social practices" (166). The author*ess* starts the chapter 'Quering semantics' with an analysis on the debate of Alice and the animated egg Humpty Dumpty in 'what do words mean':

> [B]oth Humpty Dumpty and Alice are partly right. Alice understands that we can't make words mean whatever we want them to: there are substantial constraints that arise from past history and from what is involved in trying to mean something. At the same time, there is room for shaping and reshaping word meanings. Humpty Dumpty understands that tugs over meaning can be struggles over power. But the stakes go far beyond who wins. Different meanings promote [. . .] different kinds of social action, cultural values, intellectual inquiry. Meanings [. . .] facilitate mastery in a variety of arenas. (McConnell-Ginet 2011: 237)

While Alice tries to understand why Humpty Dumpty is using the word 'glory' with that meaning, showing HOT ToM skills, Humpty Dumpty does not display a good command of ToM, reason by which even being a thing, it is considered a male-ish character. The [male] egg seems to think that the meaning of a word is whatever he wants it to mean, having a "over-weaning conceit" of itself, showing that stupidity as a strength. He does not seem to understand that other people might have different understandings of the same word and argues for a subjective view of meaning. Alice, on the other hand, is arguing for an objective view of meaning. She believes that the meaning of a word is something that is agreed upon by a community of people. Parallelly, McConnell-Ginet examples a few lines later how that same word 'queer' was primarily used for men, "a label to embrace or to shun" (239), depending on its utterer-recipient usage, taking us to the conclusion that the problem should be focused on men's education and development of ToM.

[7]Consider a revisited reading of Beauvoir's *Second Sex* (among others) to distinguish the terms 'woman' and 'female' and the Woman's Question.

20.5 Conclusion

ToM with a gender perspective is a quite new area of study, especially, when we cross-cuttingly apply them to the language expression, as stated of urgent need for further research. They share some inherent properties such as socialization, professional achievements, brain development, mental health, communication abilities, ontological epistemology development and identity fulfilment. It has also become of interest the focus on behavioral psychology, linguistic acquisition, and social justice as we echo in the chapter.

But talking about gendered ToM means to tackle a scathing analysis that leads us to the gist of ancient unresolved key questions that go deep into the unconsciousness of individuals. In communicative processes, verbal or not, within and among genders foregrounds ancient and tribal emotions that reveal the consideration of the lacanian other as an equal,[8] or not. ToM and gender respond to the question by which anyone (woman, man, or none/both) may recognize the other's feelings and thoughts under the patriarchal rules. In this sense, the phallused[9] privileged self,[10] is reflected in language as effects of animacy in cognition and linguistics, intendedly appointing who is under control, while discursively, that one, may display a naturalized "overweaning conceit" (Cavendish in Cunning, 1655/2019), to tell the story, individualistic or communal, of positioning who is the chosen one in a gendered ToM binary opposition.

References

Adenzato, M., M. Brambilla, R. Manenti, L. De Lucia, L. Trojano, S. Garofalo, et al. 2017. Gender differences in cognitive theory of mind revealed by transcranial direct current stimulation on medial prefrontal cortex. *Scientific Reports* 7 (1): 41219. https://doi.org/10.1038/srep41219.

Antiphanes of Macedon (ca. 90 CE, 2021). *Codex Palatinus*, 23 [Epigram 11.321], https://anthologiagraeca.org/passages/urn:cts:greekLit:tlg7000.tlg001.ag:11.322/.

Armstrong, David M., and Peter Anstey, eds. 1968/2022. *A materialist theory of the mind*. London: Routledge. https://doi.org/10.4324/b23154.

Arriaga-Flórez, Mercedes, and Sergio Marín-Conejo. 2023. Los feminismos en la historia de las ideas políticas: historia de las ideas igualitarias entre mujeres y hombres. *Revista Internacional de Pensamiento Político* 16 (1): 109–112. https://doi.org/10.46661/revintpensampolit.6464.

[8] I would kindly recommend the lectures of Adriana Caverero on the Universal self. Further debate in Richardson, J. (1998). Beyond equality and difference: Sexual difference in the work of Adriana Caverero. *Feminist Legal Stud* 6, 105–120. https://doi.org/10.1007/BF02684873.

[9] In Lacanian terms.

[10] 'Prestige' is a key term in the taxonomy of hierarchical differentiation. As an example, Cameron (2003: 192) reminds us of the paper in which in a Hungarian-German community, young women preferred German as vehicular language since Hungarian was associated with 'peasant' and German with 'worker'.

Bailey, April H., Marianne LaFrance, and John F. Dovidio. 2018. Is man the measure of all things? A social cognitive account of Androcentrism. *Personality and Social Psychology Review* 23 (4): 307–331. https://doi.org/10.1177/108886831878284.

Berent, Iris. 2023. The illusion of the mind–body divide is attenuated in males. *Scientific Reports*. Springer Science and Business Media LLC. https://doi.org/10.1038/s41598-023-33079-1.

Blumer, Herbert. 1969. *Symbolic interactionism: Perspective and method*. Englewood Cliffs: Prentice-Hall.

Bodine, Ann. 1975/2008. Androcentrism in prescriptive grammar: Singular 'they', sex-indefinite 'he', and 'he or she'. *Language in Society* 4 (2): 129–146. https://doi.org/10.1017/S0047404500004607.

Butler, Judith. 1990. *Gender trouble: Feminism and the subversion of identity*. New York: Routledge.

Cameron, Deborah. 2003. Gender issues in language change. *Annual Review of Applied Linguistics* 23: 187–201. https://doi.org/10.1017/S0267190503000266.

Carr, Jo, and Anne Pauwels. 2005. *Boys and foreign language learning: Real boys don't do languages*. Palgrave Macmillan.

Chen, Mel Y. 2021. Animacy as a sexual device. In *The Oxford handbook of language and sexuality*, ed. Kira Hall and Rusty Barrett. New York: Oxford. https://doi.org/10.1093/oxfordhb/9780190212926.013.10.

Cunning, David. 2019. *Margaret Cavendish: Philosophical and physical opinions*. New York: Oxford University Press. https://doi.org/10.1093/oso/9780190664053.003.0003.

Curzan, Anne. 2006. *Gender shifts in the history of English*. Cambridge: Cambridge University Press.

Dahl, Ö. 2000. Animacy and the notion of semantic gender. In *Gender in grammar and cognition I: Approaches to gender. II: Manifestations of Gender*, ed. B. Unterbeck, M. Rissanen, T. Nevalainen, and M. Saari, 99–116. Berlin/New York: De Gruyter Mouton. https://doi.org/10.1515/9783110802603.99.

Grandy, Richard E., and Richard Warner. 2022. Paul Grice. In *The Stanford Encyclopedia of Philosophy (Fall)*, ed. Edward N. Zalta and Uri Nodelman. Stanford University. https://plato.stanford.edu/archives/fall2022/entries/grice/.

Hegarty, P., and A. Parr. 2023. Embodied standpoints in gender difference graphs and tables: When, where, and why are men still prioritized? *Feminism & Psychology*. https://doi.org/10.1177/09593535231181240.

Hines, Sally. 2009. *Transforming gender transgender practices of identity intimacy and care*. Policy Press. https://public.ebookcentral.proquest.com/choice/publicfullrecord.aspx?p=419330

Kesebir, Selin. 2017. Word order denotes relevance differences: The case of conjoined phrases with lexical gender. *Journal of Personality and Social Psychology* 113 (2): 262–279.

Krendel, Alexandra. 2020. The men and women, guys and girls of the 'manosphere': A corpus-assisted discourse approach. *Discourse & Society* 31 (6): 607–630. https://doi.org/10.1177/0957926520939690.

Luraghi, Silvia. 2011. The origin of the proto-indo-European gender system: Typological considerations. *Folia lingüística: Acta Societatis Linguisticae Europaeae* 45 (2): 435–463. https://doi.org/10.1515/flin.2011.016.

Macing, Howard. 2014. Sancho Panza's theory of mind. In *Theory of mind and literature*, ed. P. Leverage, H. Mancing, R. Schweickert, and J.M. William, 123–133. West Lafayette: Purdue University Press.

McConnell-Ginet, Sally. 2011. *Gender sexuality and meaning: Linguistic practice and politics*. New York/Oxford: Oxford University Press. http://site.ebrary.com/id/10449706.

Monserrat, Javier. 2003. Teoría de la mente en Antonio R. Damasio. *Pensamiento. Revista de investigación e Información filosófica* 59 (224): 177–213.

Nevalainen, J. 2000. Evolution of standard English gender differences in the evolution of standard English. *Journal of English Linguistics* 28 (1): 38–59.

Nübling, Damaris, and Miriam Lind. 2021. Neutering neuter—Grammatical gender and the dehumanization of women in German. *Journal of Language and Discrimination.* 5 (2). https://doi.org/10.1558/jld.19965.

Olloqui-Redondo, Javier, Thora Tenbrink, and Anouschka Foltz. 2019. Effects of animacy and linguistic construction on the interpretation of spatial descriptions in English and Spanish. *Language and Cognition* 11 (2): 256–284. https://doi.org/10.1017/langcog.2019.13.

Redl, Theresa, Agnieszka Szuba, Peter de Swart, Stefan L. Frank, and Helen de Hoop. 2022. Masculine generic pronouns as a gender cue in generic statements. *Discourse Processes* 59: 828–845. https://doi.org/10.1080/0163853x.2022.2148071.

Rosenbach, Anette. 2008. Animacy and grammatical variation—Findings from English genitive variation. *Lingua* 118 (2): 151–171. https://doi.org/10.1016/j.lingua.2007.02.002.

Salas-Leal, Jordi. 2020. The ideas of power, slavery and freedom in Shakespeare's "The Tempest": A political re-reading based on his characters' tendencies. *Odisea* 21: 21–43. https://doi.org/10.25115/odisea.v0i21.3839.

Schiebinger, Londa. 2000. *Feminism and the body*. Oxford/New York: Oxford University Press.

———. 2004. *The mind has no sex? Women in the origins of modern Science*. Cambridge MA: Harvard University Press.

Scott, Joan W. 1986. Gender: A useful category of historical analysis. *The American Historical Review*. JSTOR. https://doi.org/10.2307/1864376.

Sealey, Alison, and Nickie Charles. 2013. "What do animals mean to you?": Naming and relating to nonhuman animals. *Anthrozoös* 26 (4): 485–503. https://doi.org/10.2752/175303713X13795775535652.

Sebastián, Miguel Ángel. 2016. Consciousness and theory of mind: A common theory? *THEORIA. An international journal for theory, History and Foundations of Science* 31 (1): 73–89. https://doi.org/10.1387/theoria.14091.

Sigley, R., and J. Holmes. 2002. Looking at girls in corpora of English. *Journal of English Linguistics* 30 (2): 138–157. https://doi.org/10.1177/007242030002004.

Spenser, K.A., R. Bull, L. Betts, and B. Winder. 2022. Gender differences in theory of mind, empathic understanding, and moral reasoning in an offending and a matched non-offending population. *International Journal of Offender Therapy and Comparative Criminology* 66: 587–603. https://doi.org/10.1177/0306624X211010287.

Sturm, Annegret. 2020. *Theory of mind in translation*. Berlin: Frank & Timme.

Sullivan, A., and T. Selina, eds. 2023. *Sex and Gender: A Contemporary Reader* (1st ed.). Routledge. https://doi.org/10.4324/9781003286608.

Tomasello, Michael. 2004. Learning through others. *Daedalus* 133 (1): 51–58. https://doi.org/10.1162/001152604772746693.

Trompenaars, Thijs, Teresa Angelina Kaluge, Rezvan Sarabi, and Peter de Swart. 2021. Cognitive animacy and its relation to linguistic animacy: Evidence from Japanese and Persian. *Language Sciences* 86. https://doi.org/10.1016/j.langsci.2021.101399.

Zaenen, Annie, Jean Carletta, Gregory Garretson, Joan Bresnan, et al. 2004. Animacy encoding in english: Why and how. In *Proceedings of the workshop on discourse annotation*, 118–125. Association for Computational Linguistics.

Part V
ToM from the Perspective of Artificial Intelligence

Chapter 21
Data-Driven Vs Model-Driven Approaches in Cognitive Speech Processing

Pedro Gómez-Vilda and Andrés Gómez-Rodellar

Abstract The availability of large amounts of data produced by brain functional activity examination is challenging the ability of machine learning systems not only to carry on the classical tasks of prediction, regression, clustering, classification, and characterization of specific task-related behavioral communication skills, such as speech and phonation, to distinguish pathological from normal activity, but to provide clear clues explaining and interpreting the results in terms of cause-to-effect clauses. These requirements are especially critical in the field of biomedical data processing, where many proposed methods are intended to provide a certain degree of help to clinicians in the diagnosis, prognosis, treatment, and rehabilitation. Therefore, the need of efficient and self-explanatory machine learning methodologies has re-opened the discussion about the goodness of data-driven (DD) vs model-driven (MD) approaches, confronting their respective strengths and weaknesses. The present chapter is intended to show that neither approach is to be disregarded, as both might cooperate to provide explainable results. A general model based on the theory of mind concepts could serve as an inspiration paradigm to define the framework of cooperating machine learning platforms exploiting the best of both approaches. An MD inversion chain to produce pseudo-EEG neuromotor spectrograms from speech is presented and discussed with illustrative examples from pathological speech, together with a DD classifier based on a bioinspired convolutional neural network using auditory receptive fields (ARF-CNN) as convolution filters, exemplified with a small-scale study on data from diadochokinetic tests. The possibility of integrating EEG-related spectrograms estimated by the MD inversion chain with a DD classifier based on ARF-CNNs in successful cooperation terms to produce explainable representations useful in pathological speech monitoring is explored to show how both approaches might cooperate in fruitful terms.

P. Gómez-Vilda (✉)
NeuSpeLab, CTB, Universidad Politécnica de Madrid, Madrid, Spain
e-mail: pedro.gomezv@upm.es

A. Gómez-Rodellar
Usher Institute, Faculty of Medicine, University of Edinburgh, Edinburgh, UK
e-mail: a.gomezrodellar@ed.ac.uk

T. Lopez-Soto et al. (eds.), *The Theory of Mind Under Scrutiny*, Logic, Argumentation & Reasoning 34, https://doi.org/10.1007/978-3-031-46742-4_21

Keywords Cognitive computing · Machine learning · Clinical neurolinguistics · Biosignals · Speech processing · Theory of mind · Parkinson's disease

List of Abbreviations

3DAcc	three-dimension accelerometry
AI	artificial intelligence
AKV	absolute kinematic velocity
ALS	amyotrophic lateral sclerosis
ANN	artificial neural network
APARKAM	Association of Parkinson Patients from Alcorcón and Móstoles (municipalities near the metropolitan area of Madrid)
ARF	auditory receptive field
ASICM	audio signal inversion chain model
ASR	automatic speech recognition
CCU	cognitive computing unit
ClQ	closed quotient
CNN	convolutional neural network
CQ	contact quotient
CSP	cognitive speech processing
DD	data-driven
DML	deep machine learning
DSbT	data speak by themselves
EE	end-to-end
EEG	electroencephalography
F1, F2	first two formants
FCR	formant centralization ratio
FM	frequency modulation
fMRI	functional magnetic resonance imaging
fNIRS	functional near-infrared spectroscopy
GSC	glottal source correlate
GUI	graphical user interface
H&Y	Hoehn and Yang PD severity grading scale
HC	healthy control
HNR	harmonic-noise ratio
IC	image classification
LPC	linear predictive coding
LSTM	long-short term memory
MD	model-driven
MEG	magnetoencephalography
MFDR	maximum flow descent ratio
MMU	mental model unit
MOU	modulation output unit

MP	Machiavelli's Paradigm
NLU	natural language understanding
ONPT	oro-nasopharyngeal tract
ONPTTF	oro-nasopharyngeal tract transfer function
OQ	open quotient
PCA	principal component analysis
PD	Parkinson's disease
PIU	perception input unit
PwP	person with PD
ReLU	rectified linear unit
RF	random forests
RNN	fully-connected neural network
sEMG	surface electromyography
SP	Sybilla's Paradigm
SVM	support vector machine
tM	flow maximum-slope
tO	opening instant
ToM	theory of mind
tR	recovery instant
UVFS	unbiased vocal fold body stiffness
VFBS	vocal fold body stiffness
VPA	voice pathology assessment
VQA	voice quality analysis
VSA	vowel space area

21.1 Introduction

The flourishing field of neurosciences is the new research yard nowadays. The availability of advanced monitoring technologies, such as functional Magnetic Resonance Imaging (fMRI), electroencephalography (EEG), magnetoencephalography (MEG), functional near-infrared spectroscopy (fNIRS) and others related (Górriz et al. 2020) is providing large amounts of data from brain activity, informing on basic functions of the sensorimotor and cognitive structure and functionality of the human nervous system, both central and peripheral.

Nevertheless, the application of these powerful introspective neuroimaging technologies to the rich and wide research framework on clinical neurolinguistic studies related with the characterization, monitoring, and rehabilitation of neurodegenerative diseases, demands the complementation with cognitive constructs which might be provided by the Theory of Mind (Fortier et al. 2018). Accordingly with current trends in developmental psychology, social neuroscience, and psychiatric research, ToM refers to the cognitive ability to attribute mental states such as beliefs, desires, intentions, and emotions to oneself and others, and to understand that others have their own unique perspective and mental states that may differ from one's own. It is a fundamental aspect of social cognition and plays a crucial role in understanding and predicting the behavior of others, communicating effectively, and developing social relationships. This rather general definition is nowadays subject of an intense debate, as according with Schaafsma et al. (2015) "its original definition does not permit easy downward translation to more basic processes as those studied by behavioral neuroscience", therefore, a systematic approach conducting to the reformulation of the ToM into a comprehensive body of component processes is required. This extreme is of most relevance for very specific tasks, such as the possible use of the ToM in the wide field of neurodegenerative disease characterization, monitoring, treatment, and rehabilitation from the point of view of clinical neurolinguistics (Fortier et al. 2018).

The need of this systemic reformulation is well connected with the ability of current tendencies in Machine learning to deal with the availability of large amounts of data, as this fact puts forward a very relevant challenge to neuroscientists, which is related to the capability of extracting new knowledge from explicit experimental frameworks supplying the data, that could also benefit from a structural and functional view on the ToM. On the one hand, the Data-Driven (DD) approach allows creating sophisticated and powerful connectionist structures explaining the complex relationships between stimuli and recorded system responses of the central nervous system. This approach is based on the belief that "data speak for themselves". An asymptotic tendency of such approach are the end-to-end (EE) systems (Shivakumar and Narayanan 2022). Complex problems as image and speech recognition have been solved using deep neural networks (DNNs) trained using very large amounts of available data (Cichy and Kaiser 2019). The inner mid-term relationships among data and functionality are not taken into account, or shallowly treated. On the other hand, the Model-Driven (MD) approach is more interested in solving the problem in a classical top-down and bottom-up systemic methodology, in dividing the complex system into smaller problems, unfolding feedback loops, and relating stimuli and system responses, in the interest of filling the semantic gaps behind each subsystem (Parhi and Unnikrishnan 2020).

As it is well known, the controversy between Data-Driven and Model-Driven approaches is a very old one in the field of Artificial Intelligence (AI) (Minsky et al. 1991). The advent of deep machine learning (DML) methodologies has given a powerful thrust forward to the exploration and classification of data in applications related with regression, classification, and forecasting, advanced modelling being among the most productive and fruitful approaches, combining the high power of DML with concepts such as explainability and interpretability (Guidotti et al. 2018). MD approaches seem to be better suited to solve the problem of the non-linear state-space nature of complex networks needed to explain many central nervous functionalities. MD approaches allow adding new knowledge to existing one at a systemic level, casting new hypotheses to be tested by new ingenious experimental design, and stimulation and recording methodologies. On their turn, DD approaches allow providing EE solutions in a shorter time, and with less human power investment.

This chapter is oriented to illustrate some MD approaches in Cognitive Speech Processing (CSP), in which the stimuli is based on speech and voice phonation and articulation tests, where the observed responses are either the speech signal, or other related responses, as surface electromyography (sEMG), three-dimensional accelerometry (3DAcc), or EEG. Examples of MD construction by inverse modelling, or DD classification experiments on normative and Parkinson's Disease (PD) participants are given and commented.

The chapter is organized as follows: Sect. 21.2 is devoted to discuss the methodological considerations behind the MD vs DD controversy; Sect. 21.3 presents a study case showing an inversion approach based on MD to dive into neuromotor activity related with a phonation control task; Sect. 21.4 presents results from applying MD inversion; Sect. 21.5 is devoted to discuss the results presented, provide new insights into MD inversion approaches, and expose possible applications and future work; Sect. 21.6 closes the chapter summarizing main findings and conclusions.

21.2 Methodological Considerations: MD Vs DD Approaches

The MD vs DD debate is a very old and fruitful issue, although also a fictitious one to some extent, because both approaches, rather than opposing one to the other, have collaborated since long ago to build larger and better understood conceptions of the surrounding world. Intuitively, mankind has proceeded first gathering data, to create a brief explanation of data generation later. There are many examples abounding on this perception. The fitting of astronomical observations to planet orbits is a good one. The apparent paradox of Mars' orbit retrocession (observation data) questioned geocentric models and was explained after Copernicus proposed the heliocentric theory (solar-earth model). The abundant astronomical tables of Brahe's observatory inspired Kepler's elliptical orbits (planetary model). Galileo's measurements on the acceleration of falling bodies promoted Newton's gravitational theory (mass-mass interaction). Gauss' least squares fit allowed more precise planet orbit definitions, and Laplace's generalized mathematical model of gravitational fields facilitated the discovery of new planets (predictive power). The history of science is full of many more inspiring examples about this extraordinary relationship between data and models. Data help building models, models help explaining data.

Speech Processing is a field which has benefitted from the relationship between data and models. Speech intrigued philosophers and scholars for many centuries, as contrary to image representations by picture, being based on sound propagation, could only be roughly described by phonetic-orthographic representations based on rudimentary and incomplete alphabetic systems. In fact, precise image representations were achieved by photography much before than speech could be recorded by mechanical means. The first known successful records of speech are credited to

É.-L. Scott De Martinville, using the inscription of a bristle on a lampblack-wax covered-paper driven mechanically (Benoit et al. 2009). Astonishingly, some of these recordings can still be reproduced, as for instance, the first phrasing of the popular song "Au Claire de la Lune", interpreted by a little girl, the melody and lyrics being still recognizable. The true contribution to the field was nevertheless due to the ingenuity of T. A. Edison (1878), with the mechanical tin-covered cylinder phonograph. About the same time A. G. Bell, C. Bell, and C. S. Tainter succeeded in the same purpose with their wax-coated recording device, which was later exploited commercially (Juang and Rabiner 2005). The advent of Electronics allowed magnetic recordings to be available not much later. But it was the advancement of digital computers which facilitated the conversion of analog recordings to digital speech storage, opening a new world of possibilities when computing platforms were instrumented at scientific institutions and public in general. Since then, speech started to become a more and more well-known behavioral phenomenon in human communication. A generative mechanical model, producing simple vowel sounds was constructed by Wofgang Von Kempelen (Dudley and Tarnoczy 1950), its mathematical explanation in terms of excitation-resonance was due to Gunnar Fant's source-filter model (Fant 1981). Important contributions to resonance modeling using linear predictive coding (LPC) description of the vocal tract were due to Itakura and Saito (1970), whereas the source production was well explained by a mass-spring biomechanical model of the vocal folds (Ishizaka and Flanagan 1972). Undoubtedly, the construction and validation of these models relied on the availability of digital speech recordings. Whereas articulation models were profusely used to build the earlier automatic speech recognition (ASR) systems, source models allowed the understanding of phonation pathologies. ASR grew during two decades as a hierarchical build-up pyramidal model, composed of phonetic decoding, vector encoding, sequential phonetic parsing, and linguistic modeling (Juang and Rabiner 2005) in a clear MD approach, to be surpassed by DD approaches inspired in DML since then (Yu and Deng 2015). This unquestionable success drove the attention of many researchers to the belief that "data speak by themselves" (DSbT), hampering DD vs MD cooperation, giving fuel to the old debate. Doubtlessly, this evolution is very much connected to "false belief reasoning" in the Theory of Mind (ToM), accordingly to which, egocentric and allocentric, literal and metaphoric perspectives need not be equally perceived by different individuals or groups (Thagard 2005). Therefore, a race to apply massive DLM approaches to other fields than Speech Processing started around 2014, perfusing voice pathology assessment, among others.

Recently, DML itself has evolved to extreme EE systems. These are based on the DSbT belief and have shown significant performance improvements regarding certain statistical machine learning problems with respect to former approaches, provided that enough large databases are available. This is the real situation in natural language understanding (NLU) and image classification (IC), among others (Parhi and Unnikrishnan 2020). Nevertheless, this apparent success might have a negative impact regarding especially sensitive problems involving human health and well-being, those found in clinical scenarios being a clear example, as these are subjected to more restrictive considerations, clinical explainability, interpretability, and acceptability, among others (Gómez-Vilda et al. 2022b).

In applying DML approaches blindly we might get engulfed in the Sybilla-Machiavelli tactical deception dilemma. Both Sybilla's and Machiavelli's are well-known advisory paradigms, but of a rather different character. Whereas the Sybilla's Paradigm (SP) is related to unexplained oracle prophecies (prediction-based), the Machiavelli's Paradigm (MP) is related to historically oriented fully explained advice (Machiavelli 1513). Whereas the SP gives unexplained outcomes, the MP suggests possible lines of action based on previous knowledge (mind states), and what it is more important, provides advice standing on a profusion of similar examples (Machiavelli 1513). Thus, the MP is thought to be more adequate for clinical applications than the SP. This does not mean that DD approaches would be out of the game regarding clinical acceptability; it simply means that DD approaches must embrace and grant explainability and interpretability, two concepts that were not cared for since recently in DML (Guidotti et al. 2018).

Actually, the MP may be modeled under the ToM point of view as an state machine, where external observations $\mathbf{E}_n$ from the scenario being monitored, assessed, or evaluated, to be potentially controlled, are analyzed by a Perception Input Unit (PIU) to be transformed into the inputs $\mathbf{I}_\mathbf{n}$ to a Cognitive Computing Unit (CCU) in consonance with old mental states $\mathbf{S}_\mathbf{n-1}$ as to generate new knowledge $\mathbf{N}_\mathbf{n}$, helping to build an updated mental (hidden) state $\mathbf{S}_\mathbf{n}$ in a Mental Model Unit (MMU). On its turn, mental models might generate intention outputs $\mathbf{O}_\mathbf{n}$, which might be modulated by a Modulation Output Unit (MOU) into specific representation actions $\mathbf{A}_\mathbf{n}$. This simplified model is exposed in detail in Fig. 21.1.

Voice Quality Analysis (VQA) may serve as a good example to illustrate the meaning and applicability of a typical MP model. Regarding VQA it is well known that certain dysphonic conditions observed in phonation, such as roughness, asthenia, and breathiness may be associated with certain larynx pathologies, such as polyps, nodules, edemas, cysts, carcinomas, etc. The observations $\mathbf{E}_n$ would include correlates as jitter, shimmer, harmonic-noise-ratios, mel-frequency cepstral coefficients, cepstral peak prominence, Lyapunov exponents, etc. (Mekyska et al. 2015). These observations (features) have to be transformed to efficient input representations $\mathbf{I}_\mathbf{n}$ by feature selection and compression to reduce cross-correlation, redundancy, and dimensionality by PIU methods (PCA, for instance), before being submitted to a pattern recognition process by a CCU (SVMs, ANNs, CNNs, RFs, etc.), where they are confronted with knowledge extracted from prior observations $\mathbf{S}_\mathbf{n-1}$ retrieved from an internal representation storage MMU (as weights, graphs, hyper-vectors, etc.). During the training phase, new knowledge $\mathbf{N}_\mathbf{n}$ (weight and hyper-parameter updates) will be generated to modify the internal representation to $\mathbf{S}_\mathbf{n}$. Either during the training or the evaluation phases the automaton might be requested to generate intermediate output information $\mathbf{O}_\mathbf{n}$, which might be transformed into specific outputs $\mathbf{A}_\mathbf{n}$ (by softmax mappers, for instance). A very relevant aspect of the MP is that specific outputs might be accompanied by explanatory meta-information $\mathbf{M}_\mathbf{n}$, mediating on clinical interpretability (Teng et al. 2022). Nevertheless, the MP might be of application on many other fields where the ToM

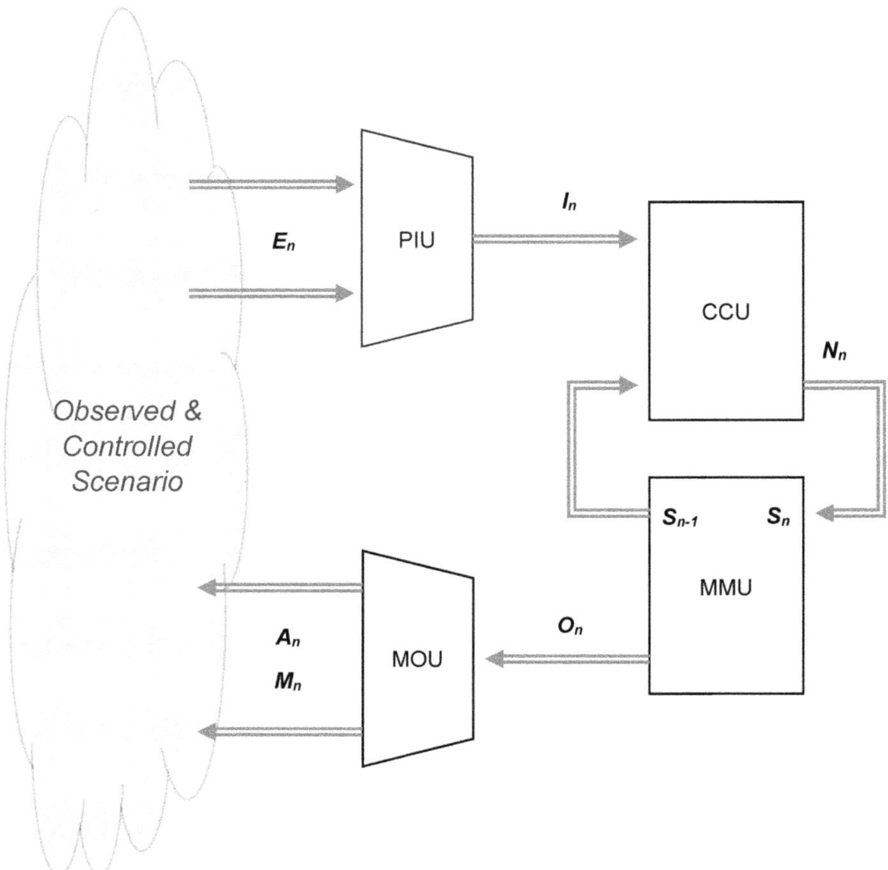

Fig. 21.1 Modeling the MP as a cognitive state machine

might be under consideration. For instance, watchers of a movie would transform visual inputs into linguistic and semantic interpretations, which could enrich, modify, create new beliefs and insights, or modify prior ones, in their conception of the ToM. When expressing a critical opinion on what had been seen, heard, and analyzed, they would be formulating specific actions, possibly accompanied by new mental representations, explanations, or behavior. In examining and analyzing new information, especially if relevant insights are found to complement what is already known, the human brain experiences important changes at the synaptic level (Bailey et al. 2004). These facts are the building blocks of the ToM.

Coming back to the field of pathological voice assessment (VPA), to understand how DD and MD approaches might interact, as it was commented before, it might be seen that the classical early approaches were more MD oriented, whereas in the recent years, DD approaches are more frequently found, with the special emergence of EE approaches. Interesting research applying both approaches for comparison purposes is due to Narendra and Alku (2020). Nevertheless, VPA is a field with strict requirements, because on the one hand there is not a clear one-to-one relationship between voice features susceptible of serving as pathology biomarkers, and on the other hand it is difficult to find reasonable-size voice pathology databases, the ones available are not uniformly produced, and their size would barely suffice DML requirements, which is known to demand large datasets for its successful application. Therefore, VPA is a field still widely open to MD research, because voice production is a process which involves many biophysical and neurological resources, from central and peripheral nervous system activity to muscular neuro-dynamics. This means that any alteration suffered by all these processes might leave their hallmark in the final audio recordings, and may be detected, characterized, and used to produce explainable results of great help in prognosis, diagnosis, functional monitoring, and rehabilitation.

In application of MP analysis, the main aim of the present chapter is to illustrate how many of these activities may be derived from the audio signal inversion chain model (ASICM) as depicted in Fig. 21.2.

The biomechanical and neuromotor system (a) is described as follows: the lungs and trachea (respiratory system) inject airflow through the larynx under the activity of the diaphragm and other respiratory muscles to induce the vibration of the vocal folds (phonation source system). The sound pressure wave produced in the supraglottal side of the vocal folds travels along the oro-nasopharyngeal tract (ONPT, articulation filter system), represented by a variable section tube $S(x)$, to be radiated at the lips as the acoustic speech which may be recorded by a microphone. In parallel to this idealized tubular structure, a complex network of neuromotor pathways oversees controlling the muscular system responsible of diaphragm contractions (phrenic nerve), vocal fold configuration (laryngeal nerves), oro-nasopharinx constriction or expansion (velopharyngeal constriction, hypoglossus nerve, extrinsic and intrinsic lingual muscular nerve system), and orofacial and mandibular nerves. The neuromotor activity controlling these subsystems is driven by specific premotor, motor, and supplementary motor areas through the periaqueductal pathways (Jaharanay 2022). On its turn, the ASICM is devoted to produce estimations of systemic functions and signal correlates with play a special role in VPA explanation. Using different combined techniques from system inversion and modelling (Gómez-Vilda et al. 2009), the speech signal is deconstructed into the articulatory system filter (ONPT transfer function) and the glottal source correlate (GSC) by means of the vocal and nasal tract estimation and cancellation system (Alku et al. 2019). The GSC may be used to extract biomechanical estimates, such as the vocal fold stiffness, and distributed mass (Gómez-Vilda et al. 2007; Hadwin et al. 2016). The vocal fold body stiffness (VFBS) correlate is a very semantic signal associated to the *musculus vocalis* contraction, as it conveys a close signature of the neuromotor activity of the larynx nerve pathways, and it gives a precise monitoring of possible neuromotor alterations affecting phonation (Manríquez et al. 2019). The neuromotor activity of the larynx nerve might be used to infer possible associated neural activity in the brain areas involved in its control (premotor, motor, and supplementary motor).

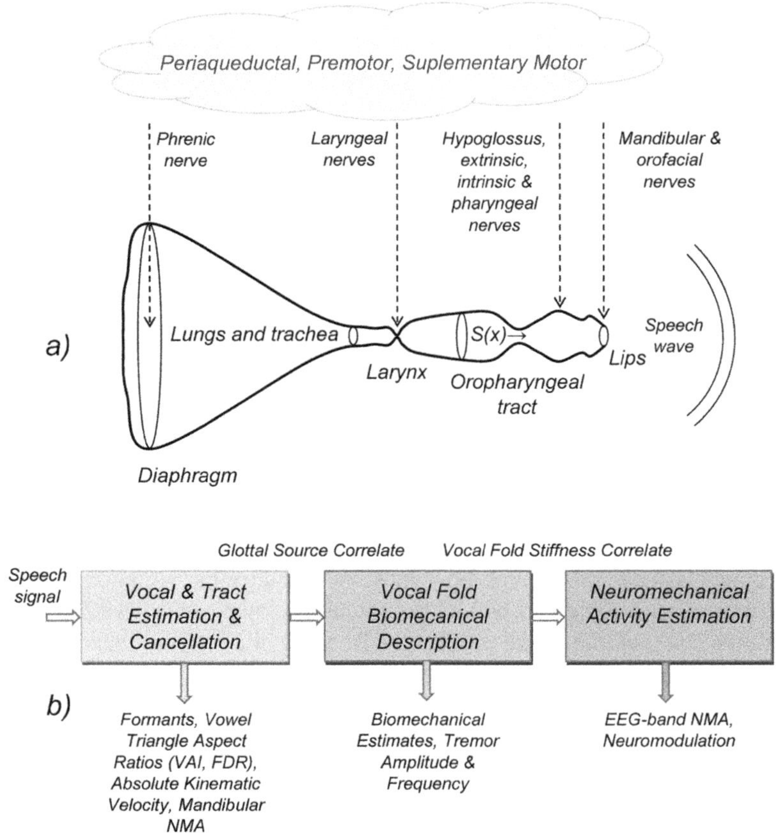

Fig. 21.2 Phonation audio signal production and inversion model. (**a**) Biomechanical and neuromotor system involved in voice production. (**b**) Audio signal inversion chain model. See the text for further details

21.3 Study Case Based in the Model-Driven Inversion

Most of the before described methods have been implemented as a software app (BioMetroFlue) in the Laboratory of Neuromorphic Speech Processing (NeuSpeLab) associated to the Center for Biomedical Technology of Universidad Politécnica de Madrid. In what follows, the examination of results produced with this tool in analyzing a 3 s phonation of a maintained open vowel [a:] produced by a normative male speaker (35 years old) following the outcomes produced by the ASICM depicted in Fig. 21.2b is presented. The general appearance of the software app graphical user interface (GUI) can be seen in Fig. 21.3.

The general appearance of the audio signal may be seen in Fig. 21.3a, where the selected 3 s segment is highlighted in green and extended in Fig. 21.3b. The signal energy profile (blue line) in dB, and the f0 profile (fundamental frequency, red line)

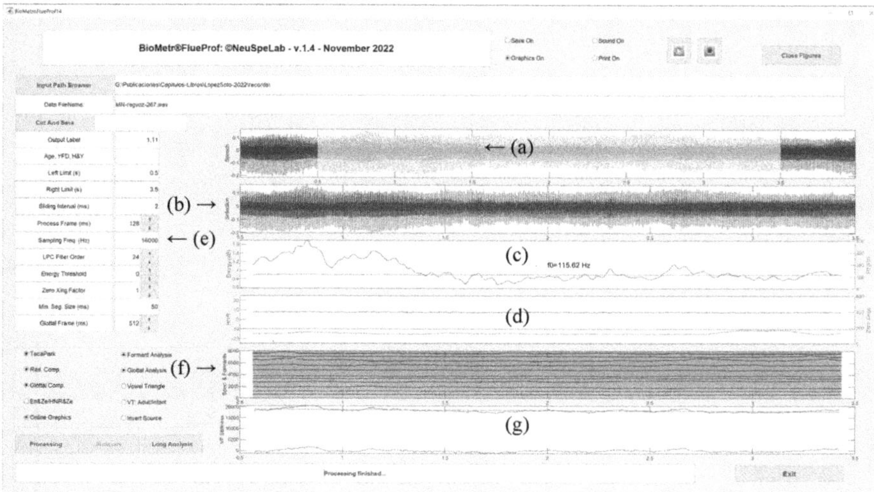

Fig. 21.3 General appearance of BioMetroFlue GUI showing the results of the analysis of a maintained open vowel ([a:]) from the different ASICM stages: (**a**) original recording; (**b**) selected segment; (**c**) energy profile (blue) and f0 profile (red); (**d**) HNR profile (blue) and zero crossings profile (red); (**e**) settings used in the analysis (see text); (**f**) Fourier spectrogram, and formants (black); (**g**) VFBS profile (black), its trend (red), and its de-trended estimate (blue)

are seen in Fig. 21.3c. The harmonic-to-noise ratio (HNR, blue line) and the zero crossings profile (red line) are given in Fig. 21.3d. An adaptive lattice-ladder filter of order 24 for a sampling frequency of 16 kHz and 128 ms processing sliding windows with a stride of 2 ms (Fig. 21.3e) is used to estimate the ONPT transfer function, and to produce the GSC. The harmonic spectrogram and the first 11 formants superimposed (black lines) are seen in Fig. 21.3f. Finally, the VFBS (black line), its trend (red line) and the de-trended result are given in Fig. 21.3g.

21.3.1 *Vocal Tract Estimation and Cancellation*

The first main block of Fig. 21.2b is designed to provide the simultaneous estimation of the ONPT transfer function $H(\omega)$, and the GSC by removing $H(\omega)$ from the speech input signal $s(n)$ in a process known as inverse deconvolution, using a prediction-error inverse filter (Deller et al. 1993)

$$\begin{aligned} r(n) &= s(n) * \widehat{h}(n); \\ \widehat{h}(n) &= \mathcal{L}\left\{H(z)^{-1}\right\}; \\ v(n) &= \mathcal{I}_o^n(r(\zeta)) \end{aligned} \tag{21.1}$$

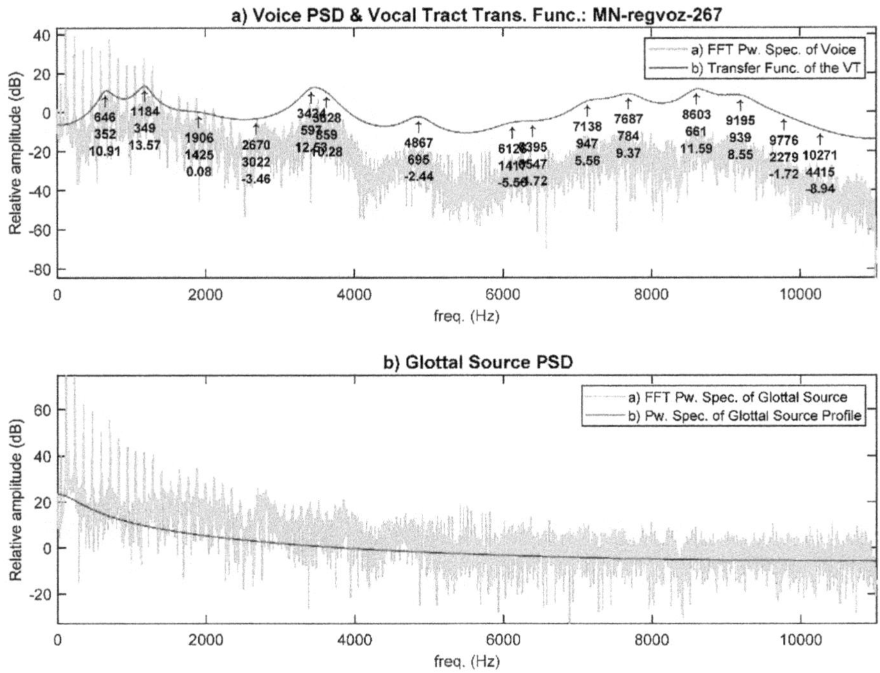

Fig. 21.4 Voice and GSC spectra, and ONPT transfer function: (**a**) spectrum of the voice input to the ASICM (cyan line), ONPT transfer function estimated by an adaptive lattice-ladder inverse filter (red line), and formant positions (black). Each numeric annotation from top to bottom: formant frequency (Hz), bandwidth (Hz), and relative amplitude (dB). (**b**) spectrum of the GSC correlate (cyan line) and bottom trend (red)

where $r(n)$ is the prediction-error filter residual (output), $v(n)$ is the GSC, $\mathcal{L}\left\{H(z)^{-1}\right\}$ is the discrete Laplace transform of the inverse ONPT transfer function (ONPTTF), $\widehat{h}(n)$ is the impulse response of the inverse prediction-error filter, '$*$' represents the time domain (n) convolution operation, and $\mathcal{I}_o^n(r(\zeta))$ stands for the discrete time-domain definite numerical integration in the interval considered $(0, n)$. Therefore, $H(z = e^{-j\omega})$ and $v(n)$ are the main outcomes of the ONPT estimation and cancellation; respectively the ONPTTF, and GSC. The modulus of the ONPTTF may be seen represented in Fig. 21.4a as a red line superimposed on the absolute value of the voice Fourier spectrum.

$$T_O(\omega) = log_{10}\left(\left|H\left(z = e^{-j\omega}\right)^{-1}\right|\right) \tag{21.2}$$

The values for which $H(z = 0)$ are confronted on the unity circle $z = e^{-j\omega}$ by frequency positions corresponding with natural ONPT resonances (formants). These frequencies are marked by an arrow and a triplet, the upper value giving the formant frequency, the mid value giving the bandwidth, both in Hz, and the lower value

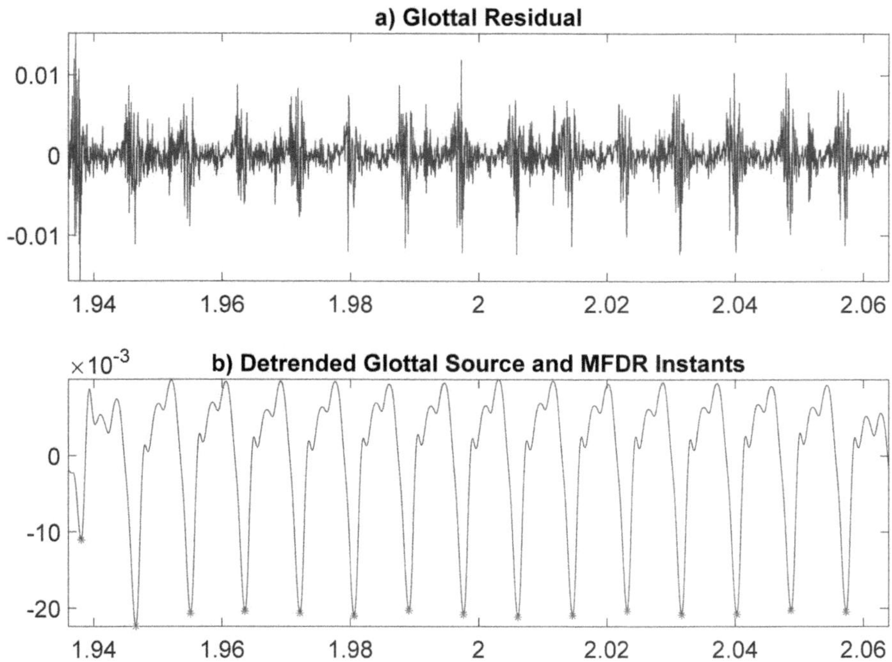

Fig. 21.5 Glottal correlates: (**a**) prediction error residual $r(n)$; (**b**) its integral $v(n)$ (**b**). The minimum saliences of the integrated and detrended residual (GSC) are marked with red asterisks and correspond to the instants where the glottal flow decays at maximum rate (MFDR: maximum flow descent ratio). These negative spikes are responsible for the most part of the harmonic spectrum of phonation

giving the relative height of the formant in dB. The cyan profile gives the input signal power spectrum $S(\omega) = |\mathcal{F}\{s(n)\}|^2$, where $\mathcal{F}\{s(n)\}$ stands for the Fourier transform of $s(n)$.

Figure 21.4b shows the frequency spectrum of the GSC $V(\omega) = |\mathcal{F}\{v(n)\}|^2$. The bottom trend of this last signal is deprived of any resonance influence, and decays at a rate of 6 dB per octave. Fig. 21.5a reproduces a short fragment of the residual $r(n)$ at the mid point of the 3 s frame being analyzed. Figure 21.5b shows the integral of this last signal ($v(n)$) low-pass filtered at a cutoff frequency of 500 Hz.

The ONPTTF provides very relevant information on the neuromechanical activity of the neural pathways controlling the articulatory system muscles (hypoglossus, pharyngeal, extrinsic, and intrinsic lingual, mandibular, and orofacial). These include many correlates derived from formant dynamics, especially from the first two formants, F1 and F2, because they are strongly correlated with the jaw and tongue movements (Gómez et al. 2021). For instance, it has been well documented that when plotting F2 vs F1 on Cartesian or linguistic charts (see Fig. 21.6) the tuple {F1, F2} resulting from the utterance of the triplet {[a:], [i:], [u:]} or any other sentence including these phonations, the resulting positions cover an area with

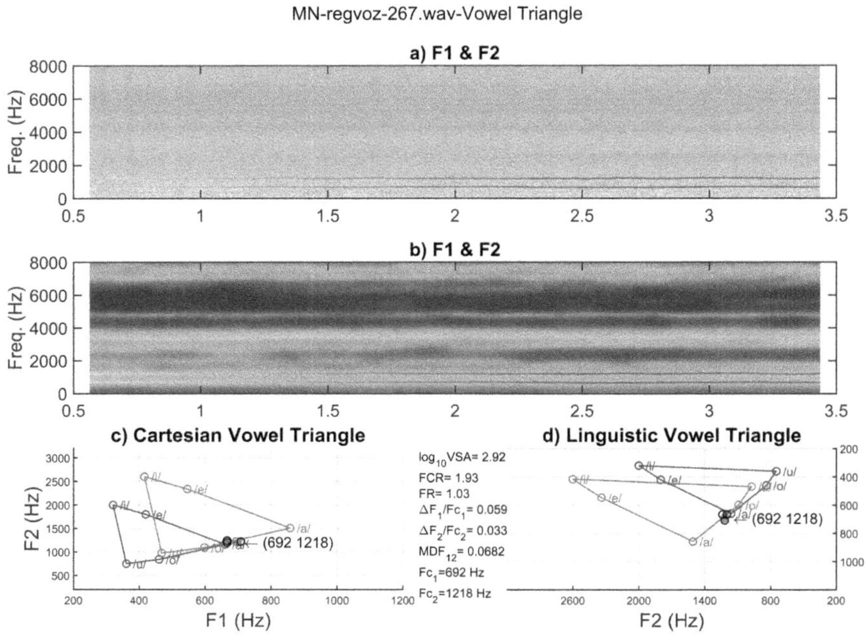

Fig. 21.6 Articulatory activity from a normative participant uttering a maintained phonation of the open vowel [a:]: (**a**) Fourier spectrogram showing the first two formants given in a time-encoding coloring showing earlier estimations in cyan, and latter ones in red; (**b**) LPC spectrogram with the same two first formants superimposed; (**c**) Cartesian vowel triangle where the time-aligned formant positions {F1, F2} have been plotted (the data within brackets give the centroid of the dataset {F1, F2}); (**d**) Linguistic vowel triangle with axes reverted. Articulation feature estimations have been plotted between both vowel triangle representations (see the text for further clarification)

extreme centroids extended more or less depending on the ability of each speaker to reach articulation extremes. This area can be measured by the vowel space area (VSA), and its centroid expressed by the formant centralization ratio (FCR) is very much related to the capability of the speaker to balance the extreme articulation points (Sapir et al. 2010; Skodda et al. 2011). The VSA is usually given by its natural or decimal logarithm. The following is a list of several features related to static and dynamic articulation characteristics of the vowel space area, including the VSA and the FCR.

$$VSA = \frac{f_{1i}(f_{2a} - f_{2u}) + f_{1a}(f_{2i} - f_{2u}) + f_{1u}(f_{2i} - f_{2a})}{2};$$

$$FCR = \frac{f_{2u} + f_{2a} + f_{1u} + f_{1i}}{f_{2i} + f_{1a}};$$

$$FR = \frac{f_{2i}}{f_{2u}};$$

$$\frac{\Delta f_{1m}}{f_{1c}} = \frac{f_{1a} - min\{f_{1i}, f_{1u}\}}{f_{1c}};$$

$$\frac{\Delta f_{2m}}{f_{2c}} = \frac{f_{2i} - f_{2u}}{f_{2c}};$$

$$MDF_{12} = \sqrt{\frac{\Delta f_{1m}}{f_{1c}}^{2} + \frac{\Delta f_{2m}}{f_{2c}}^{2}} \tag{21.3}$$

where $\{f_{1a}, f_{2a}\}$, $\{f_{1i}, f_{2i}\}$, and $\{f_{1u}, f_{2u}\}$ are the respective positions of the three extreme phonemes {[a:], [i:], [u:]} given by their centroids on the axes F1 and F2, and $\{f_{1c}, f_{2c}\}$ are the coordinates of the center of masses of the distribution. In the case under study Fig. 21.6a shows the first two formants superimposed on the Fourier spectrogram. A time reference is given in the coloring of both formant lines, the starting time values in cyan, and the ending ones in red. This same display is superimposed on the LPC spectrogram in Fig. 21.6b. The plots given in Fig. 21.6c and d correspond to the mapping of {F1, F2} on the Cartesian and the linguistic vowel charts. The estimated values of the different stability and span defined in (3) are listed between both charts. The interpretability of these features depend on the evaluation or assessment protocol used in the study. As the stability of phonation is to be evaluated, the lower the span features, the most stable the phonation, such as in the present case, where the whole 3 s emission is condensed in a small triangle centered at {692 Hz, 1218 Hz} with small relative dispersions of 5.9% and 3.3% in F1 and F2, respectively. VSA is a very relevant feature mainly in cognitive declination, as the speakers' proprioception might be that much affected hampering the production of extreme vowels (Sapir 2014). This same effect is observed in articulation deterioration due to amyotrophic lateral sclerosis (ALS), where the vowel space is very much reduced, especially affecting F2, as lingual control deteriorates, similarly the FCR is also strongly affected, with a displacement to the position of [⍰ → a], resulting in an increment of FCR and a reduction of FR (Gómez et al. 2020).

The acoustical dynamics of the first two formants is strongly related with the dynamic activity of the jaw-tongue tissular structure (Gómez et al. 2021). Based on this fact, interesting features explaining the jaw-tongue neuromotor activity may be defined, for instance, considering the formant derivatives (first differences in the discrete time axis) associated to the vertical and horizontal displacements of the jaw-tongue complex, a kinematic velocity might be defined as

$$\vec{v}_{jt} = h_2 \frac{\partial f_2}{\partial t} \widehat{x} + h_1 \frac{\partial f_1}{\partial t} \widehat{y};$$

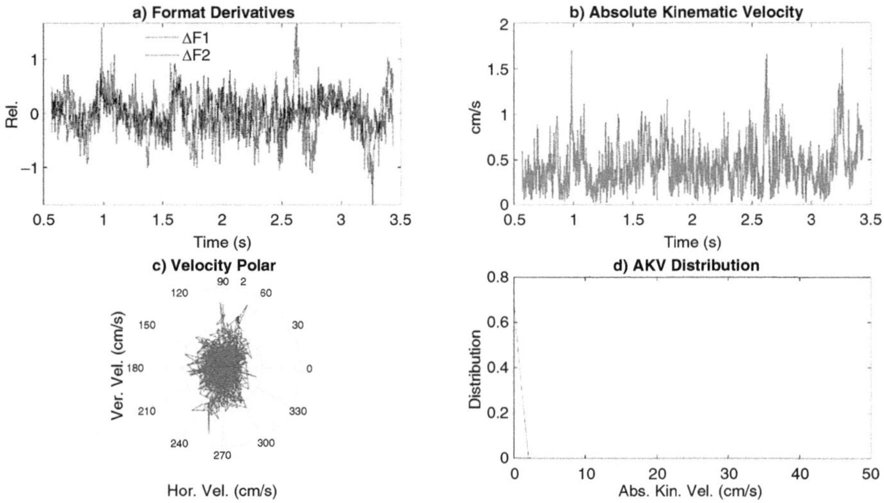

Fig. 21.7 Joint jaw-tongue kinematics derived from acoustic features: (**a**) time-domain derivatives of the first two formants (*f1* in blue line, *f2* in red line); (**b**) value of the absolute kinematic velocity (AKV) given in cm.s^{-1}; (**c**) polar plot showing the trajectory described by $\vec{v}_{jt}$; (**d**) amplitude distribution of the AKV

$$|\vec{v}_{jt}| = \sqrt{h_2^2\left(\frac{\partial f_2}{\partial t}\right)^2 + h_1^2\left(\frac{\partial f_1}{\partial t}\right)^2} \tag{21.4}$$

where $\{h_1,\ h_2\}$ are the projecting weights from the acoustic to the kinematic domains, and $\{\widehat{x}, \widehat{y}\}$ the horizontal and vertical unity axial vectors. It is important to stress that the activity of the second formant affects mainly the displacements in the forward-backward articulation positions, whereas the first formant is related mainly with the vertical displacements (as inferred from the linguistic chart in Fig. 21.6b). The absolute value of the kinematic velocity is an important index of neuromotor activity and may be related to kinematic dysarthria (Gómez et al. 2021). The estimations of the kinematic velocity, its absolute value, and its amplitude distribution are given in Fig. 21.7 for the case under study.

21.3.2 Vocal Fold Biomechanical Description

This analysis block in the ASICM concentrates in the analysis of vocal fold vibration as expressed in the GSC, its timing, and the biomechanical estimates of the multiple-mass and springs model parameters explaining vocal fold vibration (Story and Titze 1995). Different procedures might be used to estimate the model parameters, a Bayesian maximum likelihood estimation (Hadwin et al. 2016) and a frequency-

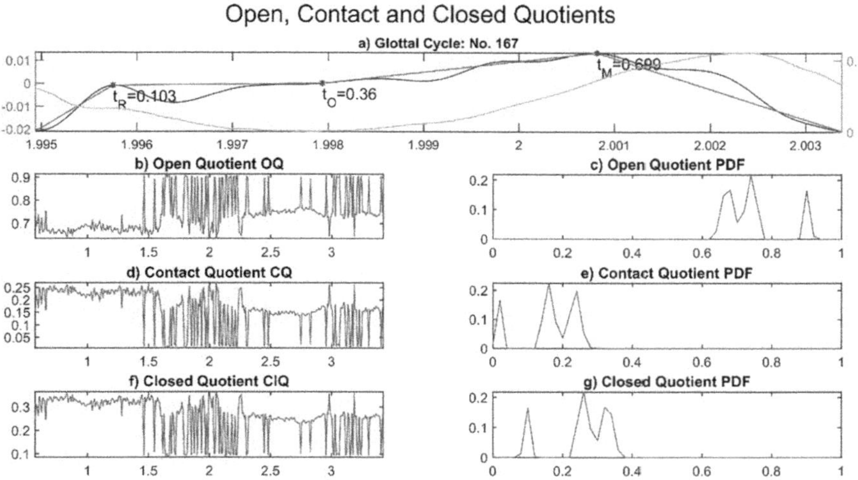

Fig. 21.8 Results from the cycle-synchronous estimation of the glottal source cycle pattern: (**a**) sample cycle from the case under study; (**b**) and (**c**) open quotient and its pdf; (**d**) and (**e**) contact quotient and its pdf; (**f**) and (**g**) closed quotient and its pdf

domain method (Meghraoui et al. 2021) being two possible approaches. In the case under study, a cycle-synchronous estimation following the second method was used on each one of the glottal arches detected between MFDR instants (see Fig. 21.5b) as the one depicted in Fig. 21.7a for further detail, where the singular instants corresponding to the recovery instant (tR), opening instant (tO) and the flow maximum-slope (tM) are detached on the GSC trace (blue line) corresponding to a mid-term cycle (No. 167 in this particular case). The glottal flow correlate (GFC) is seen superimposed (green line). The time instants are given relative to the complete duration of the glottal cycle between two consecutive MFDR instants. These timing intervals are used to evaluate the open (OQ), contact (CQ) and closed (ClQ) quotients, cycle by cycle, as plotted in Fig. 21.7b, d, and e, and their probability density distributions given in Fig. 21.7c, e, and g.

On its turn, Fig. 21.8 presents cycle-synchronous estimations of the biomechanical model parameters using the frequency-domain approach described in Gómez-Vilda et al. (2009).

The three traits given in Fig. 21.8 show a strong correlation among themselves, because they are linked by the fundamental frequency, and this is a very stable characteristic of this particular participant's phonation, as plotted in Fig. 21.9a, where the estimation of f0 is superimposed (red line) on the speech signal (blue line). It may be seen that f0 is moving around 115 Hz with little dispersion, in discrete jumps due to the accuracy of the estimation method used. The vocal fold body stiffness (VFBS) is plotted in Fig. 21.9b (black line) with its tendency (red line) and de-trended value (blue line).

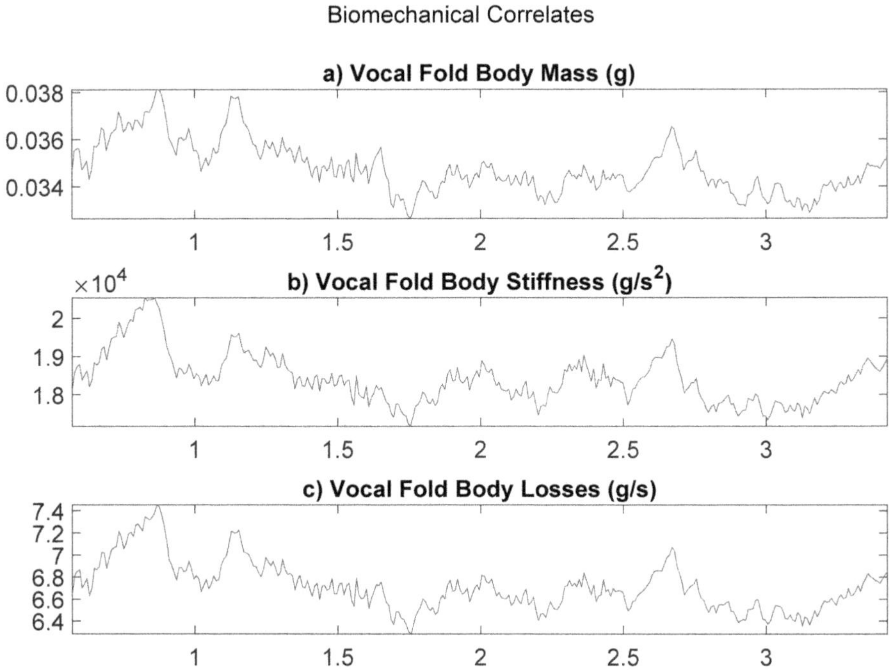

Fig. 21.9 Biomechanical features estimated from a 2-mass body-cover model on a cycle-synchronous basis: (**a**) distributed mass per unit length; (**b**) distributed mechanical stiffness per unit length; (**c**) distributed viscosity per unit length

21.3.3 Neuromechanical Activity Estimation

The third building block of the ASICM finds its argumentation in the existence of strong relationships between muscular contraction under biomechanical drive and neuromotor EEG activity on the brain areas responsible of premotor and motor control (Chiang et al. 2012; Gao et al. 2018; Brambilla et al. 2021) in the sense that laryngeal motor activity $a_m(n)$ is projected by periaqueductal pathways and nerves on larynx muscles as $a_l(n)$, inducing the contraction of the *musculus vocalis*, measured by the unbiased VFSC $\xi_b(n)$, depicted in Fig. 21.10b (blue line)

$$a_l(n) = \mathcal{P}_{ml}\{a_m(n)\};$$
$$\xi_b(n) = \mathcal{P}_{lb}\{a_l(n)\}; \tag{21.5}$$

where $\mathcal{P}_{ml}\{a_m(n)\}$ and $\mathcal{P}_{lb}\{a_l(n)\}$ might be seen as projection operators transforming neural discharges into muscle contraction (Manríquez et al. 2019). Therefore, a reverse system inversion might be established provided that adequate inverse operators could be found based on system identification methodologies to define the reverse chain as:

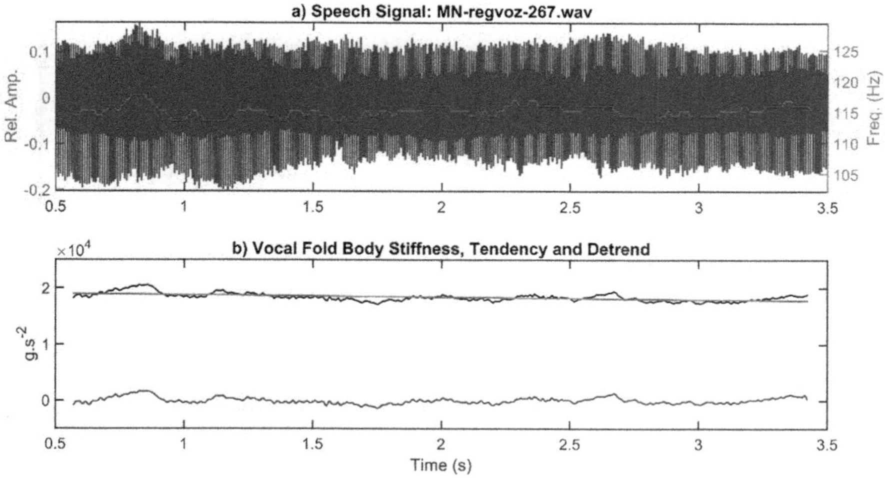

Fig. 21.10 VFBS relative to f0: (**a**) speech signal (black line) and f0 profile (red line); (**b**) VFBS (black line), its tendency (red line) and detrended estimate (blue line)

$$\widehat{a}_l(n) = \mathcal{P}_{lb}^{-1}\{\xi_b(n)\};$$
$$\widehat{a}_m(n) = \mathcal{P}_{ml}^{-1}\{a_l(n)\}; \tag{21.6}$$

where $\widehat{a}_l(n)$ and $\widehat{a}_m(n)$ would be the estimations of laryngeal neuromotor and brain motor control areas, $\mathcal{P}_{lb}^{-1}\{\cdot\}$ and $\mathcal{P}_{lm}^{-1}\{\cdot\}$ being the inverse projection operators. It would be expected the laryngeal motor activity $a_m(n)$ to be embedded on the EEG recorded over the electrodes near laryngeal premotor and motor areas, together with background activity linked to other concurrent brain activity. Therefore, the inverse estimation of EEG-related activity using the VFSC as a vehicular signal would show some partial correlation with real EEG recorded on the premotor and motor areas, such that inversely reconstructed EEG-band activity from the VFSC could be within the presumed motor area activity, but only in part. Under these very limiting premises, and assuming that the retro-projection operators $\mathcal{P}_{lb}^{-1}\{\xi_b(n)\}$ and $\mathcal{P}_{ml}^{-1}\{a_l(n)\}$ might be reduced to passband filters aligned with the classical EEG band division (δ: $f \leq 4$ Hz; ϑ: 4 Hz $< f \leq 8$ Hz; α: 8 Hz $\leq f \leq 16$ Hz; β: 16 Hz $< f \leq 32$ Hz; γ: $f > 32$ Hz; μ: 8 Hz $< f \leq 12$ Hz), the band activity presented in Fig. 21.10 may be hypothesized.

Under these assumptions, Fig. 21.10b reproduces the unbiased VFBS (UVFS), and its spectral contents, given in Fig. 21.10c. The left hand-side column reproduces the activity in the δ, ϑ, α, β, γ, and μ bands, whereas the right hand-side column represents the same activity in the Fourier spectrogram. As the phonation activity of this participant is rather stable, the frequency contents of the respective bands is smooth and constrained to specific frequencies. Interestingly, the β band

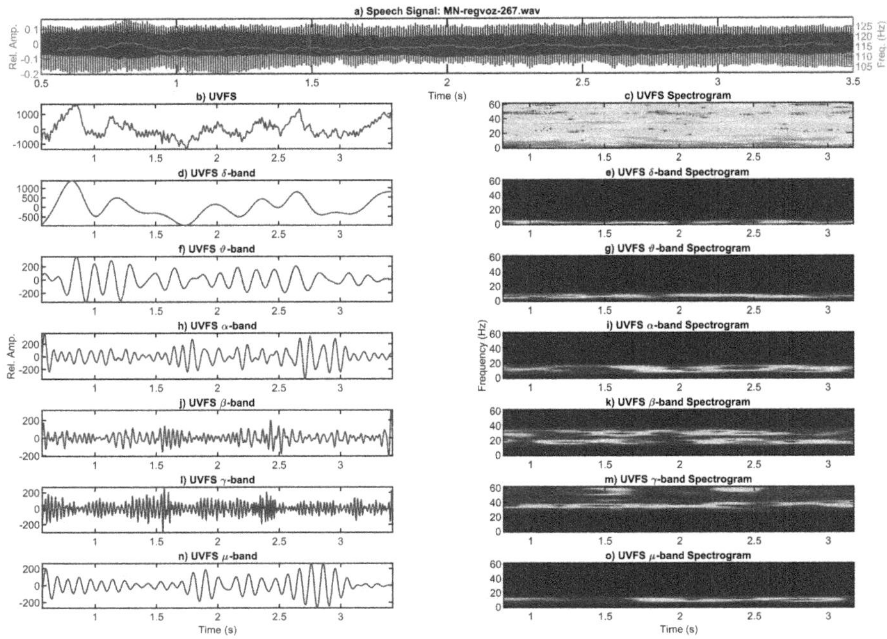

Fig. 21.11 UVFS EEG-band activity in the case under study (35 years-old healthy control participant): (**a**) speech signal (blue line) and f0 profile (red line); (**b**) VFBS; (**c**) associated Fourier spectrogram; (**d-e**) δ band signal and spectrogram; (**f-g**) id. for the ϑ band; (**h-i**) id. for the α band; (**j-k**) id. for the β band; (**l-m**) id. for the γ band; (**n-o**) id. for the μ band

concentrates most of the activity in two sub-bands (18–24 Hz, and 30–36 Hz) whereas the γ band concentrates most of the activity between 36 and 40 Hz.

The behavior of f0 in this last participant, and its corresponding activity in the EEG-band related UVFS activity deserve a special analysis. As it may be seen in Fig. 21.11a two events of deficient f0 control are observed between 4.0–4.2 s and 5.6–5.8 s, possibly generated by neuromotor blocking episodes so frequent in PD (Duffy 2019). It may be observed that in this case the EEG-band activity is much less organized than in the healthy control participant, especially in the bands α, β, and γ, where important holes of activity are found coinciding with the two deficient f0 control events, preceded in all cases by hyperactive bursts, as a train of two hyperactivity-hypoactivity cycles/s. The activity in the β band is mainly concentrated between 18–22 Hz, and the activity in the γ becoming more diffuse and widespread than in the healthy control case.

As the spectral plots shown in Figs. 21.10 and 21.11 only represent relative activity within each band, the amplitude distributions of all the bands for the two cases being described have been plotted in Fig. 21.12 for a better analytical view. It may be seen that the UVFS is much smaller in the healthy control (a) than in the PD participant (b), but the activity of the β and γ bands is comparatively more relevant in the former than in the latter case (Fig. 21.13).

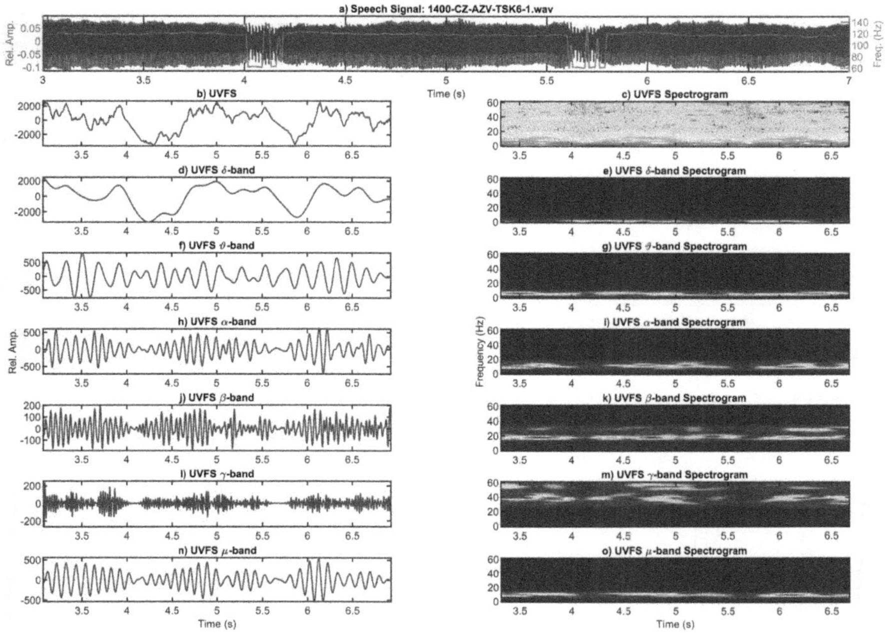

Fig. 21.12 UVFS EEG-band activity in the case of a PD participant (64 years-old PD participant, ON state, level 10 in UPDRSIII scale): (**a**) speech signal (blue line) and f0 profile (red line); (**b**) VFBS; (**c**) associated Fourier spectrogram; (**d-e**) δ band signal and spectrogram; (**f-g**) id. for the ϑ band; (**h-i**) id. for the α band; (**j-k**) id. for the β band; (**l-m**) id. for the γ band; (**n-o**) id. for the μ band

The relevance of this kind of studies relies on the possibilities of exploring the interactions of acoustically-induced neuromodulation with inner brain cyclical activity "binding sensory signals across modalities and linking such input with internal, often rhythmic, neural activity in the brain" (Greenberg 2022).

21.4 A Convolutional Neural Network with Auditory Receptive Fields

The ASICM being described in the last section provides a practical explanation of how a MD approach can be used to estimate different features showing strong semantics regarding neuromotor activity in the phonation and articulation domains, which may be used in different ML scenarios based on regression, classification, and prediction, using DD approaches, convolutional neural networks (CNN) being among the most efficient ones. These structures, as the one presented in Fig. 21.14, are particularly attractive because they combine massive computational

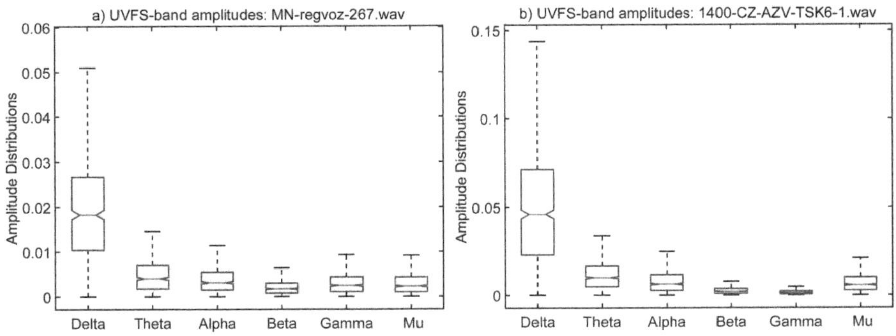

Fig. 21.13 Amplitude distributions for each band: (**a**) healthy control participant; (**b**) PD participant

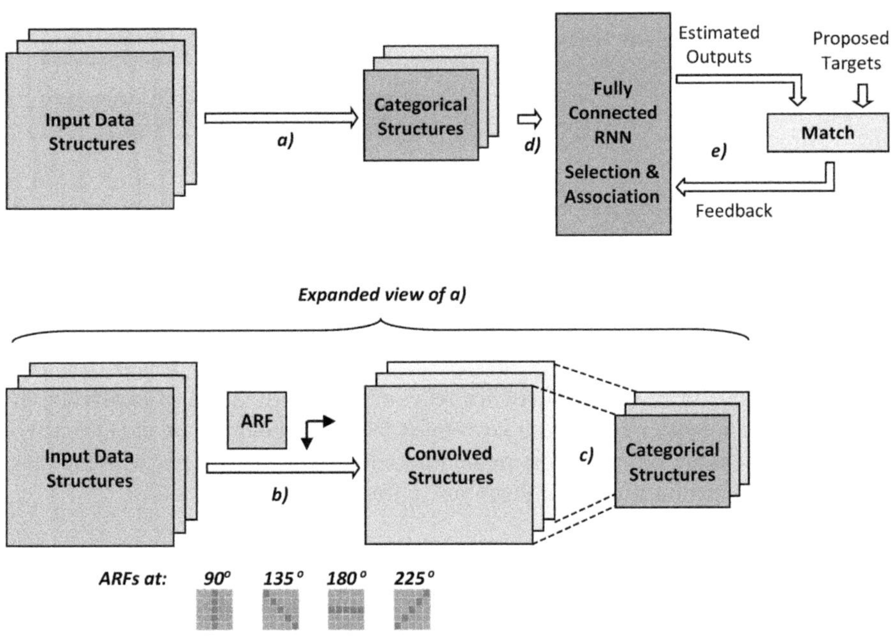

Fig. 21.14 Convolutional neural network using auditory receptive fields as convolution filters (ARF-CNN): (**a**) convolutional structure expanded in the lower template; (**b**) expansion of the convolutional structure. (**c**) compression and max-pooling; (**d**) fully connected neural network (RNN) for category selection and association; (**e**) target matching and loss optimization feedback. The receptive fields used in the convolutional structure are synoptically represented

power with bioinspired perceptual computing by making use of receptive fields based in this specific case on auditory perception (ARFs).

The first stage of the ARF-CNN (a) is based on a stride-sliding multiplication of different 5×5 ARF matrices with the input matrices, consisting of Fourier spectrograms (either from speech or from EEG-band images). This operation generates a set of convolved structures, the input matrix patterns superposing the most with the ARFs being the ones producing more intense activity (b). These matrices are compressed by max-pooling neighbor areas projected by a factor of 8:2 on output pixels of categorical structures, which are further trimmed by the application of a watershed algorithm (Osma-Ruiz et al. 2007) and rectified linear units (ReLU). These categorical structures (c) will summarize the overlapping of each input data structure with each one of the four ARFs in use. A single large vector is produced aligning the categorical structures (d), which are processed by a fully connected random-vector functional link neural network (RNN, Huang and Siew 2004), which will select the patterns best fitted to the proposed targets through a matching and feedback process (e). The most complicated concepts in the description of the CNN structure, which have been omitted here for the sake of brevity and may be found in Gómez-Vilda et al. (2022a).

21.5 Classification Results

A very efficient way to characterize PD articulation kinematics is usually conducted measuring the ability of participants in reproducing rhythmical and repetitive jaw-tongue movements (diadochokinetic exercises), for instance, the regular and fast utterance of the diphthong [a → i] for as long as possible. In what follows, the results from a small study including six persons with PD (PwP) and six healthy controls (HC) are being described. Both groups included equal number of male and female participants. The PwP group was recruited from APARKAM (a PD patient association in the metropolitan area of Madrid). The study was approved by the Ethical Committee of Universidad Politécnica de Madrid (MonParLoc, 18/06/2018). Informed consent was requested from each participant in alignment with the Declaration of Helsinki. The demographic characteristics of participants are shown in Table 21.1.

All participants produced utterances of the diadochokinetic test already mentioned, as fast and for as long as possible after a deep breath intake. Recordings up to 20 s were feasible after some previous training. Figure 21.15a and Fig. 21.16a show two fragments of these recordings from a male HC (CMa) and a male PwP (PMb) participant. These two cases represent the best and worst examples of the diadochokinetic tests in the participants set.

Figures 21.15b and 21.16b show the respective Fourier spectrograms, and correspondingly Figs. 21.15c and 21.16c show the LPC spectrograms. As it is well known, Fourier spectrograms emphasize the harmonic structure of the speech signal, whereas LPC spectrograms emphasize its formant structure. At first sight, it may be appreciated that the results produced by the HC participant showed a much larger regularity of oscillation than the PwP participant, although this differentiation was not that evident in other cases. The input data structures for classification purposes

Table 21.1 Biometrical profile of participants

Label	Gender	Age	Condition	Medication	H&Y
PMa	F	73	PD	ON	2
PMb	F	76	PD	ON	2
PMc	F	69	PD	ON	2
CMa	M	69	HC	–	–
CMb	M	70	HC	–	–
CMc	M	68	HC	–	–
PFa	F	78	PD	ON	2
PFb	F	73	PD	ON	2
PFc	F	71	PD	ON	2
CFa	M	72	HC	–	–
CFb	M	65	HC	–	–
CFc	M	66	HC	–	–

F: females; M: males, ON: under dopaminergic medication 2 h before recordings were taken; H&Y: Hoehn & Young severity scale

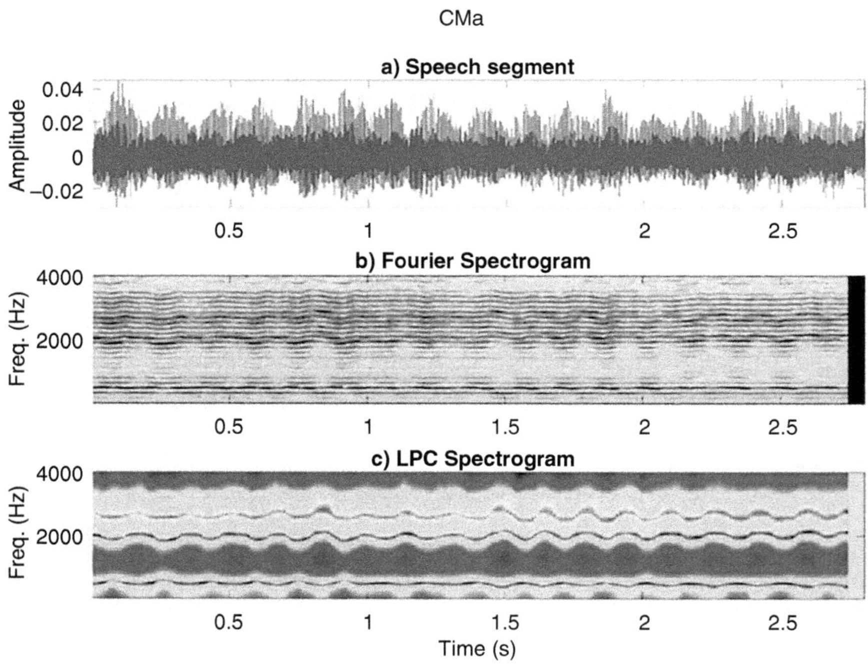

Fig. 21.15 Example of an utterance of the diadochokinetic exercise [. . .aiaiai. . .] from an HC participant: (**a**) speech; (**b**) Fourier spectrogram; (**c**) LPC spectrogram

were LPC spectrograms of 800 time data points with a stride of 10 ms (equivalent to 8 s) × 200 channels of 5 Hz resolution (equivalent to a 4 kHz frequency span). The ARFs used were the ascending (225°) and descending (135°) ones, as the purpose of the experiment was oriented to testing the tracking ability of the CNN on the

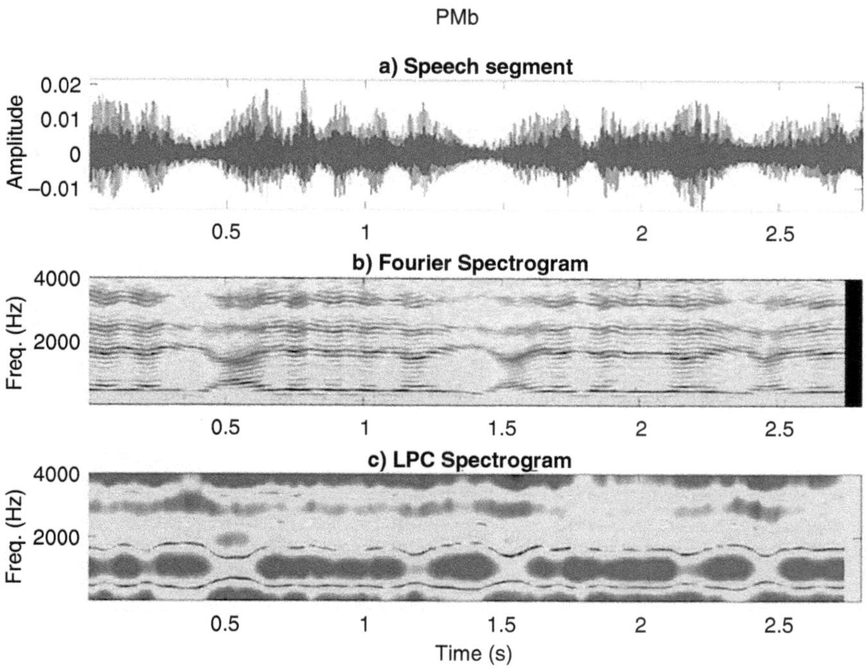

Fig. 21.16 Example of an utterance of the diadochokinetic exercise [...aiaiai...] from a PD participant: (**a**) speech; (**b**) Fourier spectrogram; (**c**) LPC spectrogram

dynamic components of the test, during formant frequency slide ups and downs, mimicking Suga's FM perceptual units (Suga 2004). The dynamic behavior of the CNN for the two cases already presented (CMa and PMb) is shown in the classification dashboards in Figs. 21.17 and 21.18.

The upper left image (a) corresponds to the categorical structures, once convolved with the 135° ARF and compressed 8:2 by ReLU units. The matrices b-d encode the inner activity of the fully connected random-vector functional link neural network (RNN), which uses a matrix of 1500 random vectors ($\mathbf{W}_1$) to produce linear combinations of each one of the 800 time-aligned input vectors (compressed from 200 to 50 frequency channels by ReLU). The resulting intermediate activity hyperplane outputs are given in the matrix ($\mathbf{Y}_2$). This last pattern is projected by a least-squares optimized matrix, known as the Moor-Penrose pseudo-inverse (**P**) to the output vector $\mathbf{W}_2$. This vector is further projected by a softmax function to the output vector **Z** and fused to a single output score as listed in Table 21.2.

The output scores should replicate the desired target values, where a value of +1 indicates the membership to the PwP group, whereas a value of −1 designates a member of the HC group, as this is a case of supervised learning. It may be seen that the output scores do not approach these extreme end values, although the signs of targets and scores show a large agreement except for a single case (CMb when

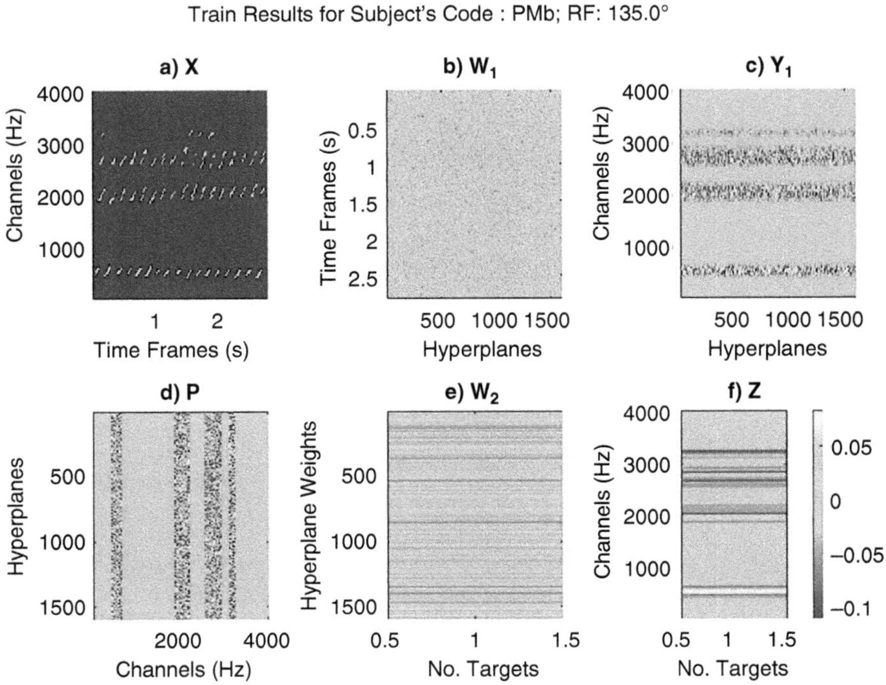

Fig. 21.17 ARF-CNN classification dashboard. HC participant: (**a**) categorical structure matrix; (**b**) random weight matrix; (**c**) first projection hyperplane output; (**d**) Moore-Penrose pseudo-inverse; (**e**) second hyperplane output; (**f**) frequency channel output

detected by a time-ascending ARF, producing a small positive score of 0.0553, in bold). In the same case (CMb), when using the time-descending ARF, the sign alignment is correct, but the output value is near zero (−0.0097, also in bold, pointing to an almost neutral decision). On its turn, the performance scores for each gender set including HC and PD participants according with the ARF characteristics used are given in Table 21.3 in terms of Pearson's correlation coefficient ρ and log-likelihood ratio λ, defined as

$$\rho = \frac{\sum_{i=1}^{p}(t_i - \bar{t})(z_i - \bar{z})}{\left(\sum_{i=1}^{p}(t_i - \bar{t})^2 \sum_{i=1}^{p}(z_i - \bar{z})^2\right)^{1/2}};$$

$$\lambda = \frac{\sum_{i \epsilon SH}\sigma(z_i)}{\sum_{i \epsilon SP}\sigma(z_i)}; \tag{21.7}$$

where $\bar{t}$ and $\bar{z}$ are the respective means of the targets and scores from the subsets considered, HC: $SH \subset \{MFA_H, MFD_H, MTA_H, MTD_H, FFA_H, FFD_H, FTA_H, FTD_H\}$, and PwP: $SP \subset \{MFA_P, MFD_P, MTA_P, MTD_P, FFA_P, FFD_P, FTA_P, FTD_P\}$.

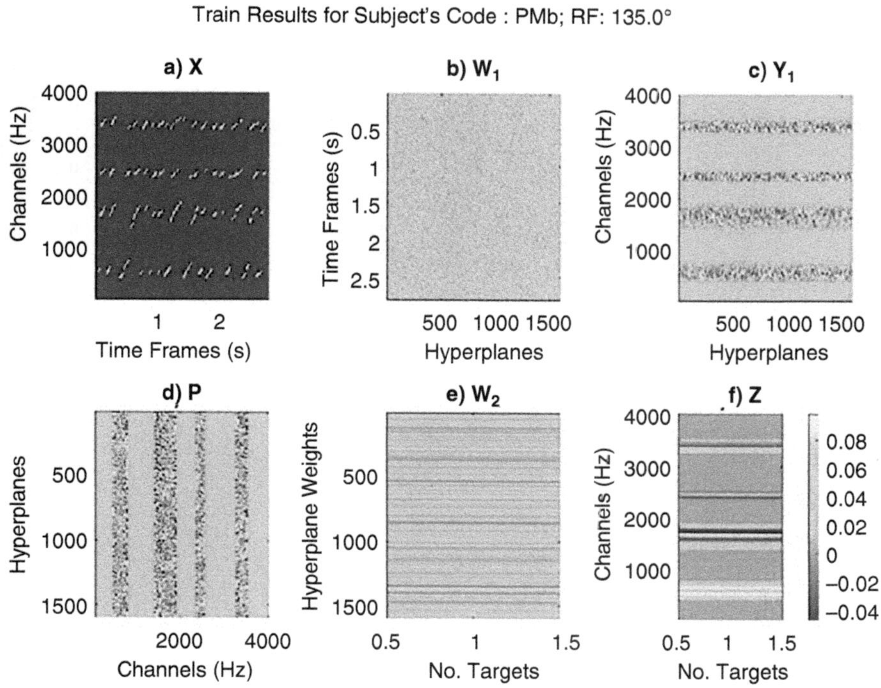

Fig. 21.18 ARF-CNN classification dashboard. PD participant: (**a**) categorical structure matrix; (**b**) random weight matrix; (**c**) first projection hyperplane output; (**d**) Moore-Penrose pseudo-inverse; (**e**) second hyperplane output; (**f**) frequency channel output

Table 21.2 Output scores (targets are PD: +1 and HC: −1)

Label	Gender	Cond.	Target	Freq/Asc	Freq/Desc	Time/Asc	Time/Desc
PMa	M	PD	+1	0.4828	0.4660	0.3153	0.3590
PMb	M	PD	+1	0.4584	0.3422	0.3151	0.3590
PMc	M	PD	+1	0.5027	0.4100	0.3140	0.3432
CMa	M	HC	−1	−0.4973	−0.3378	−0.6847	−0.6410
CMb	M	HC	−1	−0.4606	−0.3474	**0.0553**	**−0.0097**
CMc	M	HC	−1	−0.4859	−0.5340	−0.3148	−0.4104
PFa	F	PD	+1	0.4521	0.5007	0.3603	0.5255
PFb	F	PD	+1	0.4995	0.5212	0.1127	0.3391
PFc	F	PD	+1	0.4439	0.3692	0.6317	0.5302
CFa	F	HC	−1	−0.4628	−0.4517	−0.3683	−0.4697
CFb	F	HC	−1	−0.5005	−0.4788	−0.3681	−0.4554
CFc	F	HC	−1	−0.4323	−0.4605	−0.3683	−0.4698

The values of Pearson's correlation indicate that the experiment showing the largest agreement (0.9994) for the male subset corresponded to the ascending ARFs (225°) in the frequency domain, which is confirmed also by the log-likelihood ratio

Table 21.3 Performance scores accordingly with gender and ARF modality. The detection ARFs showing a largest agreement between targets and scores ($\bar{t}$ and $\bar{z}$) are stated in bold

Gender	Domain	ARF	Label	Pearson (ρ)	p-value	LLR(λ)
M	**Freq.**	**Asc.**	**MFA**	**0.9994**	<0.001	**0.9626**
M	Freq.	Desc.	MFD	0.9841	<0.001	0.8094
M	Time	Asc.	MTA	0.8274	<0.001	0.5844
M	Time	Desc.	MTD	0.8866	<0.001	0.6728
F	**Freq.**	**Asc.**	**FFA**	**0.9984**	<0.001	**0.9303**
F	Freq.	Desc.	FFD	0.9946	<0.001	0.9296
F	Time	Asc.	FTA	0.9262	<0.001	0.7589
F	Time	Desc.	FTD	0.9909	<0.001	0.9338

(0.9626). The same configuration also gives the best agreement scores for the female subset ($\rho = 0.9984$, $\lambda = 0.9303$).

21.6 Discussion

This chapter is intended to present and discuss the two main approaches found in nowadays ML systems, that is to say MD vs DD, which do not have to be seen as contradictory or competing alternatives but as collaborative ones. The chapter does not follow strict publication guidelines oriented to illustrate competitive results, as that is not its main objective. Conversely, it is intended to describe the strengths of both approaches with specific examples taken from phonation and articulation quality evaluation in pathological speech characterization. The introductory idea is reflected in the MP cognitive model given in Fig. 21.1, where the PIU chain designed to produce internal representations $\mathbf{I_n}$ from external observations $\mathbf{E_n}$ is clearly MD oriented. An example of a PIU chain model (ASICM) is given in Fig. 21.2 and explained in Sect. 21.3, conceived to produce neuromotor correlates (semantic internal representations), from the phonation and articulation neuromotor activity of speech (outer observations), in terms of vowel space representations, and jaw-tongue kinematics and dynamics (subsect. 21.3.1), vocal fold biomechanics (subsect. 21.3.2), and projected neuromechanical activity estimation (subsect. 21.3.3). On its turn, a conceptual example of the CCU and MMU units is given in Sect. 21.4, where new knowledge $\mathbf{N_n}$ is created in the convolutional part (CCU) of the ARF-CNN devoted to characterize LPC spectral representations when processed by ARFs, exemplified by processing speech from diadochokinetic tests, whereas the hidden states $\mathbf{S_n}$ are stored in the weights of the RNN after being trained with a set of examples selected from pathological speech. The intention outputs $\mathbf{O_n}$ are given in terms of output scores, which for instance, could be converted to specific visualization outputs $\mathbf{A_n}$ in specific rehabilitation protocols (Yiu et al. 2021). Although the ASICM presented provided Fourier spectrograms on the EEG-related neuromotor bands (0–64 Hz), and the ARF-CNN traded LPC spectrograms in the quite different telephone speech band (0–4 kHz), it seems a natural extension foreseeing further advancements on this research line to use neuromotor EEG-band representations produced by the ASICM as standard inputs for the ARF-CNN classifiers. In this way, inverse methods might become a complementary methodology to dive in problems where the ground truth is not known or barely accessible, as in brain research using standard EEG, because this technology does not provide enough spatial resolution to assign a specific activity to a well differentiated brain area with enough precision. Besides, in the case of speech neuromotor control it would be quite difficult to successfully remove speech neuromotor control activity recorded in EEG from other concomitant activity. Possible correlation techniques in the domain of cross-entropy might help in associating speech neuromotor activity with EEG measured on the premotor, and supplementary motor areas (Verwoert et al. 2022) related with phonation and articulation control, although the existence of significant correlation does not necessarily imply causative relationship. Similarly, the study of speech rhythm neural modulations (Poeppel and Assaneo 2020) might benefit from this methodology of study. Other lines for future research would contemplate the use of alternative paradigms for the CUU-MMU structure, such as long-short-term memory (LSTM) systems to characterize time-dynamic descriptions of pathological phonation, articulation, and fluency in diadochokinetic tests and conversational speech (Shewalkar et al. 2019), although one of the main limitations found in this research is the availability of adequate databases to cover both aspects (sustained phonations, and continuous speech) to a size allowing the full display of computational power of ML.

21.7 Conclusions

This chapter is intended to contribute a positive view to the classical controversy between MD vs DD approaches in ML. To serve this purpose, a general model of ML based on ToM has been proposed under the point of view of explainability, attending to the SP vs MP controversy. In doing so, a transfer on the semantic load of the problem has been deliberately prompted: MD and DD approaches may co-work for their mutual benefit, whereas it is quite difficult to accept any kind of cooperation between the SP and the MP points of view, provided that explainability is a necessary condition, and SP-based systems can barely meet this requirement. Another highly desirable property of MP-based systems is their ability to propose new hypotheses which may extend research objectives and open new lines. In this sense, the general model of ToM-inspired ML may offer new views for each of the building blocks, approaching AI to bio-inspired cognitive computation (Górriz et al. 2020). An important insight derived from this study is the ability of hybridize perceptually inspired model inversion to create brain-aligned neuromotor activity with convolutional structures based on perceptual receptive fields. In this way, complex bio-inspired neuromorphic computational structures (Eliasmith 2007) may help in better predicting and explaining certain production-perception phenomena in speech comprehension. The most relevant findings described in the study are the ability of re-create EEG-related neuromotor activity from phonation correlates, which also include the possibility to extend the same methodology to formant-derived dynamic activity and the use of auditory receptive fields in detecting the dynamic patterns in speech. A most interesting conclusion comes from the possibility of integrating EEG-related spectrograms estimated by the ASICM inversion chain with CUU-MMU structures based on ARF-CNNs in successful cooperation terms to produce explainable representations useful in pathological speech monitoring.

References

Alku, Paavo, Tiina Murtola, Jarmo Malinen, Juha Kuortti, Brad Story, Manu Airaksinen, Mika Salmi, Erkki Vilkman, and Ahmed Geneid. 2019. OPENGLOT – An open environment for the evaluation of glottal inverse filtering. *Speech Communication* 107: 38–47. https://doi.org/10.1016/j.specom.2019.01.005.

Bailey, C.H., E.R. Kandel, and K. Si. 2004. The persistence of long-term memory: A molecular approach to self-sustaining changes in learning-induced synaptic growth. *Neuron* 44: 49–57. https://doi.org/10.1016/j.neuron.2004.09.017.

Benoit, S., D. Blouin, J.Y. Dupont, and G. Emptoz. 2009. Chronique d'une invention: le phonautographe d'Édouard-Léon Scott de Martinville (1817–1879) et les cercles parisiens de la science et de la technique. *Documents pour l'histoire des techniques. Nouvelle Série* 17: 69–89. https://doi.org/10.4000/dht.502.

Brambilla, C., I. Pirovano, R.M. Mira, G. Rizzo, A. Scano, and A. Mastropietro. 2021. Combined use of EMG and EEG techniques for neuromotor assessment in rehabilitative applications: A systematic review. *Sensors* 21 (21): 7014. https://doi.org/10.3390/s21217014.

Chiang, J., Z.J. Wang, and M.J. McKeown. 2012. A multiblock PLS model of cortico-cortical and corticomuscular interactions in Parkinson's disease. *NeuroImage* 63: 1498–1509. https://doi.org/10.1016/j.neuroimage.2012.08.023.

Cichy, R.M., and D. Kaiser. 2019. Deep neural networks as scientific models. *Trends in Cognitive Sciences* 23 (4): 305–317. https://doi.org/10.1016/j.tics.2019.01.009305.

Deller, J.R., J.H.L. Proakis, and J.G. Hansen. 1993. *Discrete-time processing of speech signals*. New York: Macmillan.

Dudley, H., and T.H. Tarnoczy. 1950. The speaking machine of Wolfgang von Kempelen. *The Journal of the Acoustical Society of America* 22: 151–166. https://doi.org/10.1121/1.1906583.

Duffy, J.R. 2019. *Motor speech disorders: Substrates, differential diagnosis, and management*. 4th ed. St. Louis: Elsevier.

Edison, T.A. 1878. The phonograph and its future. *The North American Review* 126: 527–536. https://www.jstor.org/stable/25110210 (Retrieved Dec. 23, 2022).

Eliasmith, C. 2007. How to build a brain: From function to implementation. *Synthese* 159: 373–388. https://doi.org/10.1007/s11229-007-9235-0.

Fant, G. 1981. The source filter concept in voice production. *STL-QPSR* 1 (1981): 21–37.

Fortier, J., J. Besnard, and P. Alain. 2018. Theory of mind, empathy and emotion perception in cortical and subcortical neurodegenerative diseases. *Revue Neurologique* 174 (4): 237–246. https://doi.org/10.1016/j.neurol.2017.07.013.

Gao, Yunyuan, Leilei Ren, Rihui Li, and Yingchun Zhang. 2018. Electroencephalogram–Electromyography coupling analysis in stroke based on symbolic transfer entropy. *Frontiers in Neurology* 8: 716. https://doi.org/10.3389/fneur.2017.00716.

Gómez, P., A.R. Londral, A. Gómez, D. Palacios, and V. Rodellar. 2020. Monitoring ALS from speech articulation kinematics. *Neural Computing and Applications* 32 (20): 15801–15812. https://doi.org/10.1007/s00521-018-3538-6.

Gómez, A., P. Gómez, D. Palacios, V. Rodellar, V. Nieto, A. Álvarez, and A. Tsanas. 2021. A Neuromotor to acoustical jaw-tongue projection model with application in Parkinson's disease hypokinetic dysarthria. *Frontiers in Human Neuroscience* 15: 622825. https://doi.org/10.3389/fnhum.2021.622825.

Gómez-Vilda, P., R. Fernández-Baillo, A. Nieto, F. Díaz, F.J. Fernández-Camacho, V. Rodellar, et al. 2007. Evaluation of voice pathology based on the estimation of vocal fold biomechanical parameters. *Journal of Voice* 21 (4): 450–476. https://doi.org/10.1016/j.jvoice.2006.01.008.

Gómez-Vilda, P., R. Fernández-Baillo, V. Rodellar-Biarge, V.N. Lluis, A. Álvarez-Marquina, L.M. Mazaira-Fernández, et al. 2009. Glottal source biometrical signature for voice pathology detection. *Speech Communication* 51 (9): 759–781. https://doi.org/10.1016/j.specom.2008.09.005.

Gómez-Vilda, P., A. Gómez-Rodellar, D. Palacios-Alonso, A. Álvarez-Marquina, and A. Tsanas. 2022a. Characterization of hypokinetic dysarthria by a CNN based on auditory receptive fields. In *International work-conference on the interplay between natural and artificial computation*, 343–352. Cham: Springer International Publishing. https://doi.org/10.1007/978-3-031-06242-1_34.

Gómez-Vilda, P., A. Gómez-Rodellar, D. Palacios-Alonso, V. Rodellar-Biarge, and A. Álvarez-Marquina. 2022b. The role of data analytics in the assessment of pathological speech—A critical appraisal. *Applied Sciences* 12 (21): 11095. https://doi.org/10.3390/app122111095.

Górriz, J.M., J. Ramírez, A. Ortíz, F.J. Martinez-Murcia, F. Segovia, J. Suckling, et al. 2020. Artificial intelligence within the interplay between natural and artificial computation: Advances in data science, trends and applications. *Neurocomputing* 410: 237–270. https://doi.org/10.1016/j.neucom.2020.05.078.

Greenberg, S. 2022. Speech, Rhythm, and the Brain. *Acoustics Today* 18 (3): 22–31. https://doi.org/10.1121/AT.2022.18.3.22.

Guidotti, R., A. Monreale, S. Ruggieri, F. Turini, F. Giannotti, and D. Pedreschi. 2018. A survey of methods for explaining black box models. *ACM Computing Surveys (CSUR)* 51 (5): 1–42. https://doi.org/10.1145/3236009.

Hadwin, P.J., G.E. Galindo, and K.J. Daun. 2016. Non-stationary Bayesian estimation of parameters from a body cover model of the vocal folds. *JASA* 139: 2683–2696. https://doi.org/10.1121/1.4948755.

Huang, G.-B., and C.-K. Siew. 2004. *Extreme learning machine: RBF network case*, 1029–1033. Proceedings of the ICARCV. https://doi.org/10.1109/ICARCV.2004.1468985.

Ishizaka, K., and J.L. Flanagan. 1972. Synthesis of voiced sounds from a two-mass model of the vocal cords. *Bell System Technical Journal* 51 (6): 1233–1268.

Itakura, F., and S. Saito. 1970. A statistical method for estimation of speech spectral density and formant frequencies. *Electronics and Communications in Japan* 53A: 36–43.

Jaharanay, M. 2022. Brain regions involved in speech production, mechanism and development. *Neuroscience Research* 5 (4): Art. 178. https://doi.org/10.31117/neuroscirn.v5i4.178.

Juang, B. H., and L. R. Rabiner. 2005. Automatic speech recognition–A brief history of the technology development. *Georgia Institute of Technology. Atlanta Rutgers University and the University of California*. Santa Barbara, 1, 67.

Machiavelli, N. 1513. The Prince. Trans. Tim Parks. 2014, Penguin Random House.

Manríquez, R., S.D. Peterson, P. Prado, P. Orio, G.E. Galindo, and M. Zañartu. 2019. Neurophysiological muscle activation scheme for controlling vocal fold models. *IEEE Transactions on Neural Systems and Rehabilitation Engineering* 27 (5): 1043–1052. https://doi.org/10.1109/TNSRE.2019.2906030.

Meghraoui, D., B. Boudraa, T. Merazi-Meksen, and P.G. Vilda. 2021. A novel pre-processing technique in pathologic voice detection: Application to Parkinson's disease phonation. *Biomedical Signal Processing and Control* 68: 102604. https://doi.org/10.1016/j.bspc.2021.102604.

Mekyska, J., E. Janousova, P. Gomez-Vilda, Z. Smekal, I. Rektorova, I. Eliasova, et al. 2015. Robust and complex approach of pathological speech signal analysis. *Neurocomputing* 167: 94–111.

Minsky, M., et al. 1991. Logical versus analogical or symbolic versus connectionist or neat versus scruffy. *AI Magazine* 12 (2): 34–51. https://doi.org/10.1609/aimag.v12i2.894.

Narendra, N.P., and P. Alku. 2020. Glottal source information for pathological voice detection. *IEEE Access* 2020 (8): 67745–67755. https://doi.org/10.1109/ACCESS.2020.2986171.

Osma-Ruiz, V., J.I. Godino-Llorente, N. Sáenz-Lechón, and P. Gómez-Vilda. 2007. An improved watershed algorithm based on efficient computation of shortest paths. *Pattern Recognition* 40 (3): 1078–1090. https://doi.org/10.1016/j.patcog.2006.06.025.

Parhi, K.K., and N.K. Unnikrishnan. 2020. Brain-inspired computing: Models and architectures. *IEEE Open Journal on Circuits and Systems* 1: 185–204. https://doi.org/10.1109/OJCAS.2020.3032092.

Poeppel, D., and M.F. Assaneo. 2020. Speech rhythms and their neural foundations. *Nature Reviews. Neuroscience* 21: 322–334. https://doi.org/10.1038/s41583-020-0304-4.

Sapir, S. 2014. Multiple factors are involved in the dysarthria associated withparkinson's disease: A review with implications for clinical practice and research. *Journal of Speech, Language, and Hearing Research* 57: 1330–1343. https://doi.org/10.1044/2014_JSLHR-S-13-0039.

Sapir, S., L.O. Ramig, J.L. Spielman, and C. Fox. 2010. Formant centralization ratio: A proposal for a new acoustic measure of dysarthric speech. *Journal of Speech, Language, and Hearing Research* 53 (1): 114. https://doi.org/10.1044/1092-4388(2009/08-0184).

Schaafsma, S.M., D.W. Pfaff, R.P. Spunt, and R. Adolphs. 2015. Deconstructing and reconstructing theory of mind. *Trends in Cognitive Sciences* 19 (2): 65–72. https://doi.org/10.1016/j.tics.2014.11.007.

Shewalkar, A., D. Nyavanandi, and S.A. Ludwig. 2019. Performance evaluation of deep neural networks applied to speech recognition: RNN, LSTM and GRU. *Journal of Artificial Intelligence and Soft Computing Research* 9 (4): 235–245. https://doi.org/10.2478/jaiscr-2019-0006.

Shivakumar, P.G., and S. Narayanan. 2022. End-to-end neural systems for automatic children speech recognition: An empirical study. *Computer Speech and Language* 72: 101289. https://doi.org/10.1016/j.csl.2021.101289.

Skodda, S., W. Visser, and U. Schlegel. 2011. Vowel articulation in Parkinson's disease. *Journal of Voice* 25 (4): 467–472. https://doi.org/10.1016/j.jvoice.2010.01.009.

Story, B.H., and I.R. Titze. 1995. Voice simulation with a body-cover model of the vocal folds. *The Journal of the Acoustical Society of America* 97 (2): 1249–1260. https://doi.org/10.1121/1.412234.

Suga, N. 2004. Basic acoustic patterns and neural mechanisms shared by humans and animals for auditory perception. In *Speech processing in the auditory system*, ed. S. Greenberg et al., 159–181. New York: Springer.

Teng, Q., Z. Liu, Y. Song, K. Han, and Y. Lu. 2022. A survey on the interpretability of deep learning in medical diagnosis. *Multimedia Systems* 28 (6): 2335–2355. https://doi.org/10.1007/s00530-022-00960-4.

Thagard, P. 2005. *Mind. Introduction to cognitive science*. Cambridge, MA: The MIT Press.

Verwoert, M., M.C. Ottenhoff, S. Goulis, A.J. Colon, L. Wagner, S. Tousseyn, et al. 2022. Dataset of speech production in intracranial electroencephalography. *Scientific Data* 9 (1): 434. https://doi.org/10.1038/s41597-022-01542-9.

Yiu, E.M., C.C. Liu, C.Y. Chan, E. Barrett, and D. Lu. 2021. Vibrational therapies for vocal fatigue. *Journal of Voice* 35 (1): 29–39. https://doi.org/10.1016/j.jvoice.2019.07.009.

Yu, D., and L. Deng. 2015. *Automatic speech recognition: A deep learning approach*. London: *Springer-Verlag*. https://doi.org/10.1007/978-1-4471-5779-3.

Chapter 22
The Social Machine: Artificial Intelligence (AI) Approaches to Theory of Mind

Alberto Nebreda, Danylyna Shpakivska-Bilan, Carmen Camara, and Gianluca Susi

Abstract The ability to infer cognitive states is essential for the introduction of artificial intelligence (AI) into society. This chapter aims to present the existing understanding of Theory of Mind (ToM) from the point of view of AI models.

First, we overview the concept of AI and the current research lines. One of the key challenges in implementing AI systems in society is human-machine interaction, as it is a very complex process that involves the ability to infer human's mental states. Thus, the development of an artificial Theory of Mind to achieve AI systems with social traits has been suggested as a key factor for successful future human-agent interactions.

Throughout the chapter, we propose to categorise the spectrum of AI ToM models into two main blocks. In the initial section, *Cognitive-based and black-box algorithms*, we survey different algorithms aimed at creating artificial ToM, including both those based on cognitive models and those black-box. The experiments and evaluation methods used to assess the performance of these algorithms in ToM related tasks will also be discussed. On the other hand, *Bio-inspired models of ToM* would include any model built taking into consideration physiological or/and

A. Nebreda (✉) · D. Shpakivska-Bilan
Department of Experimental Psychology, Complutense University of Madrid (UCM), Madrid, Spain

Centre for Cognitive and Computational Neuroscience, Complutense University of Madrid (UCM), Madrid, Spain
e-mail: albnebre@ucm.es; danyshpa@ucm.es

C. Camara
Department of Computer Science, Carlos III University of Madrid (UC3M), Madrid, Spain
e-mail: macamara@inf.uc3m.es

G. Susi
Centre for Cognitive and Computational Neuroscience, Complutense University of Madrid (UCM), Madrid, Spain

Department of Structure of Matter, Thermal Physics and Electronics, Complutense University, Madrid, Spain
e-mail: gsusi@ucm.es

T. Lopez-Soto et al. (eds.), *The Theory of Mind Under Scrutiny*, Logic, Argumentation & Reasoning 34, https://doi.org/10.1007/978-3-031-46742-4_22

neurobiological aspects of the brain for ToM modelling, aimed at pursuing biological plausibility.

To conclude, we outline promising *Applications* of AI ToM models. These are mainly focused on industry (e.g., autonomous vehicles), speech processing, biometry, healthcare, education, or marketing, among others. An Artificial Theory of Mind is still in its infancy, as existing algorithms are domain-specific and are limited to certain environments and tasks. We project an exponential growth of AI ToM in the coming years, fuelled by technological advances and the collaboration between different fields of research such as psychology, neuroscience, computer science and mathematics.

Keywords Theory of mind (ToM) · Artificial intelligence (AI) · Internet of things (IoT) · Humanoid robots · Pervasive computing · Emotional AI · Biometry

22.1 Introduction

Artificial intelligence (AI) is revolutionizing the world we live in. Each year, the progress made in this field exceeds the expectations and predictions of even the most knowledgeable experts and is being applied to an increasing number of aspects of daily life.

A prime example of this is the development of algorithms such as DALL-E[1] (see Fig. 22.1), which can produce images based on natural language descriptions. This breakthrough technology is sparking a heated discussion about the future of art and design, especially among industry professionals.[2]

So, what is AI? The High-Level Expert Group (HLEG) on Artificial Intelligence appointed by the European Commission has proposed the following definition[3]: "Artificial intelligence (AI) are software (and possibly also hardware) systems designed by humans that, given a complex goal, act in the physical or digital dimension by perceiving their environment through data acquisition, interpreting the collected structured or unstructured data, reasoning on the knowledge, or processing the information, derived from this data and deciding the best action (s) to take to achieve the given goal. AI systems can either use symbolic rules or learn a numeric model, and they can also adapt their behaviour by analysing how the environment is affected by their previous actions."

Other organisms and groups have proposed a multitude of definitions of AI, but most of them share some common features that can be considered as the core elements of AI (Samoili et al. 2020b):

[1] https://openai.com/dall-e-2/#demos

[2] https://www.nasdaq.com/articles/openais-dall:e-2-has-potential-to-disrupt-graphic-design-industry

[3] ec.europa.eu/newsroom/dae/document.cfm?doc_id=56341

Fig. 22.1 Example of DALL-E illustration generated by the input "an astronaut lounging in a tropical resort in space in a vaporwave style"

- Perception of the environment, allowing the system to be aware of its surroundings.
- Gathering, information processing, and input analysis.
- Decision making, reasoning, and learning, including taking actions (such as adapting and reacting to changes in the environment) with a certain level of autonomy.
- Achievement of specific goals, which is considered the ultimate purpose of AI systems (Fig. 22.2).

Recently, remarkable advancements have been made in the field of perception of the environment. There have been notable improvements in image, video and audio processing and recognition. However, one of the most notable achievements has been in the field of communication and natural language processing (NLP). Cutting-edge NLP models like GPT-3[4] are now capable of executing numerous NLP-related tasks, including text classification, generation, translation and automatic summarization, question answering, and even generation of functional code in various programming languages.

Other instances in reasoning and decision making are *Google*'s initial breakthroughs in cognitive games like *Chess and Go* (Silver et al. 2016). The advancement in cognitive games is impressive from a technical point of view, but to an outside observer's eye, just as little applicable to people's daily lives. However, there are many advances that have direct applicability. For example, AI-based algorithms

[4] https://openai.com/blog/chatgpt

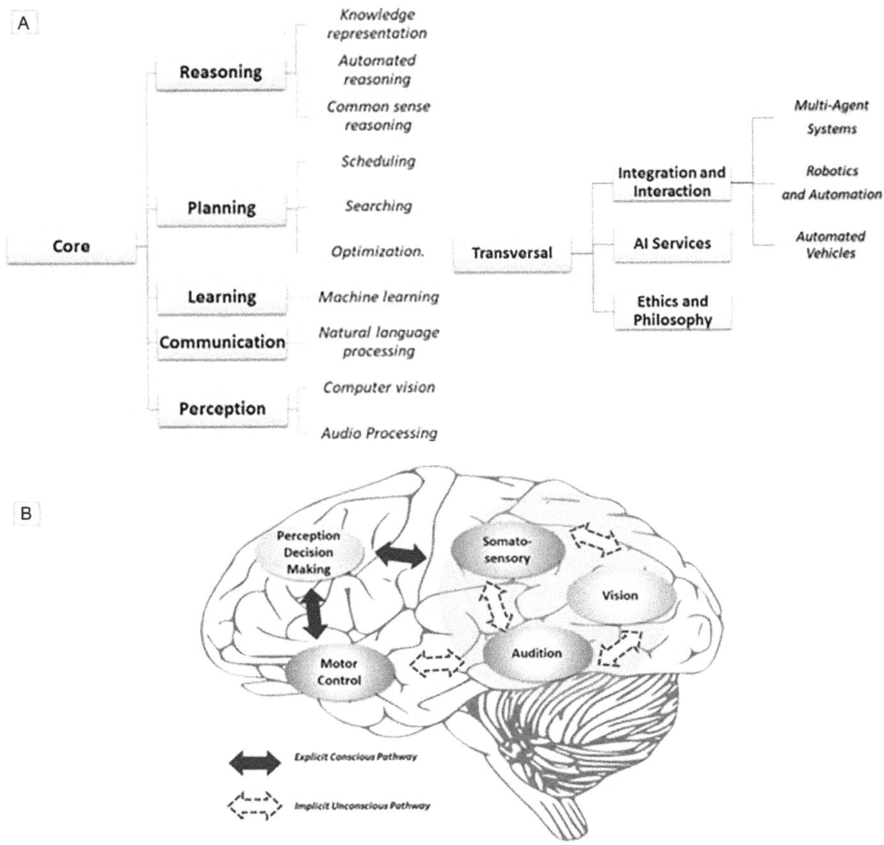

Fig. 22.2 (**a**) Shows a concise taxonomy of present AI algorithms (Samoili et al. 2020a) based on this functional approach. The key aspect of these features is their similarity to human information processing. (**b**) AI systems aim to replicate the cognitive processes in the brain, from the detection of a stimulus to the initiation of a response in the form of an action

for solving protein folding have the potential to revolutionise the industry in their field.

One of the key challenges in implementing AI systems is human-machine interaction. On the one hand, AI agents may perform correctly when isolated or among other agents, but human behaviour's complexity can severely affect their performance, thus limiting their applicability. On the other hand, if AI does not interact socially in an appropriate way, it can negatively impact human users' trust and acceptance of the technology (Bianco 2022). Humans will delegate tasks to an AI agent only if they understand its decisions, fostering the development of trust. To establish trust, Williams et al. (2022) suggest that AI agents need to present their internal model outputs in ways that are understandable to humans. This view is aligned with the eXplainable Artificial Intelligence (XAI) movement.

The purpose of a XAI system is to enhance human understanding by providing explanations for its behaviour. To create effective and more understandable AI systems, some general principles should be followed. These include the ability of the XAI system to explain its capabilities and understanding, clarify its past actions, current tasks, and future projections, and disclose the important information it uses to make decisions and take actions. By adhering to these principles, XAI systems can make their behaviour more intelligible to humans (Gunning and Aha 2019; Graziani et al. 2023).

Many AI applications in daily life require the ability to understand humans' desires, beliefs, and intentions for their proper functioning. This cognitive ability is called *Theory of Mind (ToM)* in the field of psychology, and it has been suggested as a key factor for successful human-machine interactions.

Building robots with social traits remains a great future challenge for science, as ToM is necessary for most of the human high-level social skills. To implement these ToM capabilities in AI, computer scientists and engineers need to collaborate closely with psychologists and neuroscientists to correctly integrate cognitive and biological features in future AI systems.

Evaluating the ToM abilities of these agents is also a big challenge. Classical tests, such as the "Sally Anne Test" (see Fig. 22.3) need to be adapted to computational and robotic environments, which can be difficult. In some cases, entirely new experiments need to be developed to properly evaluate these abilities.

This chapter aims to provide an overview of the field of artificial ToM. In the initial section, we'll survey different algorithms aimed at creating artificial ToM, including those based on cognitive models and those that are *black box*. The experiments and evaluation methods used to assess the performance of these algorithms in ToM-related tasks will also be discussed. The second section is focused on models influenced by neuroscience and biology, since these areas are central for the future development of artificial ToM. A special emphasis is given to the contribution of brain computational models to the field of artificial ToM. Finally, the last section goes through the applications, summarising the areas where AI ToM models hold the greatest potential, such as autonomous vehicles and human-robot interactions.

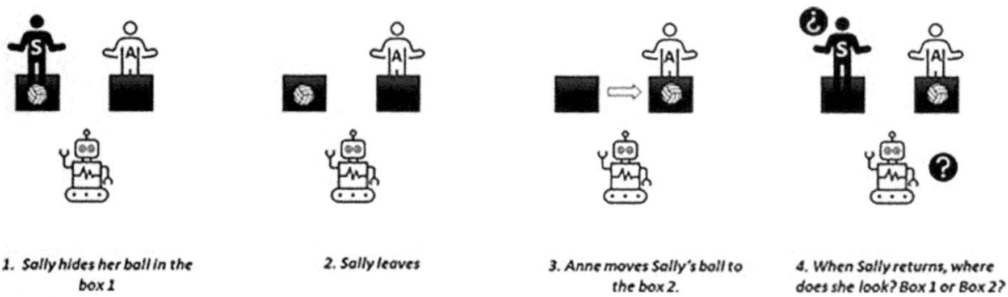

Fig. 22.3 Illustration of the "Sally-Anne-Test", a false belief task

22.2 Cognitive-Based and Black-Box Algorithms

A man approaches a car on the street, opens the door, takes a quick look inside, closes the door, and runs off in the opposite direction. Why? Was he going somewhere, but remembered he left something at home? Did he leave the oven on? Did he pick something up from the car, only to go somewhere else? Could he even be stealing? This highlights a crucial aspect of human cognition: the ability to infer possible mental states from the actions we observe in others, just by a small glimpse of their behaviour.

Human-human and human-machine interactions in many AI applications require the ability to understand and anticipate the actions of others by taking into account the cognitive states that drive those actions. However, this type of inference demands intricate processing to try to solve an ultimately impossible inductive problem, since most of the information is not available, but one must start from the little information available to arrive at possible reasons with greater amount of detail (Baker et al. 2009).

In this way, ToM can be conceptualised as the process of inferring unseen cognitive states (e.g., desires, beliefs, intentions) from the limited available information, such as observable behaviour and contextual knowledge. Given that cognitive states are not directly observable, ToM is always subject to great uncertainty.

22.3 Cognitive-Based Models

Most attempts to generate a ToM using AI have been based on cognitive models. One of the proposed ways to computationally formalise inferences about how the minds of others work is through model-based *reinforcement learning* (RL), a classic area within machine learning (Jara-Ettinger 2019). These models are based on the "rationality principle," the assumption that humans behave (approximately) rationally to maximise utility, acting in ways that facilitate desired outcomes while minimising costs and undesired outcomes.

RL techniques are a type of algorithm widely used in machine learning which has many commonalities with human learning through classical and operant conditioning. An agent acquires experiences through its interactions with a dynamic environment. Initially, the agent behaves quite randomly, but progressively learns to act according to the rewards and punishments associated with its interactions. The result is an agent capable of solving a complex task without specific instructions on how to perform it, basing its behaviour on a reward function and the probabilistic outcomes of each of the possible actions. The clearest examples where these techniques have been applied are games such as *chess or Go* Silver et al. 2018. Without any prior knowledge beyond the rules of the game and the allowed moves, and with no need for any learning from professional human players, these models can compete against each other, initially trying random moves and progressively "learning" which moves lead to better results. This learning takes place until they can beat any human player consistently. These techniques have been employed in more complex games, such as some of Atari's games (Mnih et al. 2015), as well as in a multitude of applied fields. Beyond gaming, we can find applications of RL in various fields such as finance (risk optimization), transportation (route optimization, online traffic signal control), energy (adaptive decision control), business management (OR customer management) or robotics (sim-to-real co-robot control) (Liang et al. 2018).

RL is very useful for teaching models to perform different tasks in an unsupervised fashion (Kaelbling et al. 1996), but on its own, it cannot model the cognitive states of other agents in real-world situations. Algorithms applying Inverse Reinforcement Learning (IRL) attempt to perform this type of inference by assuming that other agents act and relate to the environment as an agent applying an RL-trained model would, trying to maximize desired outcomes and minimize undesired ones, based on a reward function and probabilistic outcomes for each of the possible actions. This assumption is consistent with several cognitive models that attempt to explain and predict human behaviour, and it has been shown to work in predicting a large number of behaviours (Baker et al. 2009; Goodman et al. 2006; Jara-Ettinger 2019). An example of this type of model is illustrated in Fig. 22.4.

These algorithms have an important limitation since the inferences are inherently incomplete. In IRL models, the observing agent cannot know the reward function or the probabilistic outcomes of the actions of the acting agent and does not have access

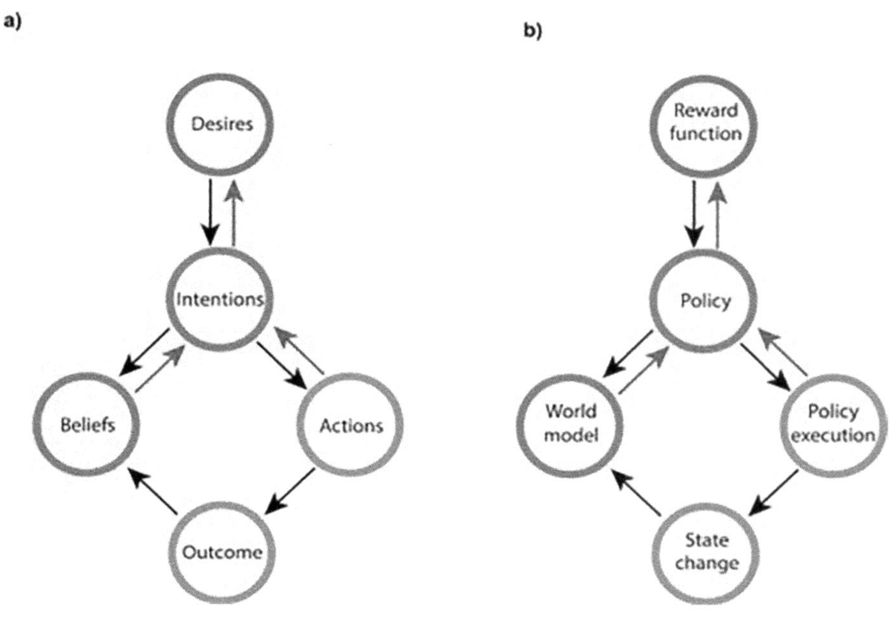

Fig. 22.4 Schematic of how Theory of Mind can be modelled as Inverse Reinforcement Learning. (**a**) Theory of Mind components. (**b**) Model-based reinforcement learning components. (Modified from Jara-Ettinger (2019))

to all the learning experience that has led it there, so inferences can only be based on the available information. Many existing IRL models attempt to infer the reward function and/or world model relying on three types of data: (1) the agent's behaviour over time and under various circumstances, (2) measures of the sensory inputs the agent receives, and (3) a model of the environment (Russell 1998).

Some authors have proposed, under this approach, that ToM would be equivalent to inverse reinforcement learning (IRL), since both try to infer the agents' unobservable reward function and world model, given some prior assumptions and posterior observations of the agent actions and context. One of the difficulties of this type of inference, going back to the initial example of the chapter, is that it is not enough to find only the most plausible explanation. It is necessary to take into account at the same time a variety of possible explanations, each one explaining better or worse the observed data. Bayesian inference is an effective tool to achieve this (Ramachandran and Amir 2007) and has become the mainstream approach for modelling uncertainty in IRL.

The Bayesian paradigm is specifically based on the relocation of credibility among different possibilities as the available information increases. These possibilities are represented as parameter values in mathematical models. In his 2014 book, Kruschke (2014) presents a good example to understand how the Bayesian paradigm works: "*Suppose we step outside one morning and notice that the sidewalk is wet, and wonder why. We consider all possible causes of the wetness, including possibilities such as recent rain, recent garden irrigation, a newly erupted underground spring, a broken sewage pipe, a passer-by who spilled a drink, and so on. If all we know until this point is that some part of the sidewalk is wet, then all those possibilities will have some prior credibility based on previous knowledge. For example, recent rain may have greater prior probability than a spilled drink from a passer-by. Continuing on our outside journey, we look around and collect new observations. If we observe that the sidewalk is wet for as far as we can see, as are the trees and parked cars, then we re-allocate credibility to the hypothetical cause of recent rain. The other possible causes, such as a passer-by spilling a drink, would not account for the new observations. On the other hand, if instead we observed that the wetness was localised to a small area, and there was an empty drink cup a few feet away, then we would re-allocate credibility to the spilled-drink hypothesis, even though it had relatively low prior probability. This sort of reallocation of credibility across possibilities is the essence of Bayesian inference.*"

The primary limitation of Bayesian methods is typically their computational cost. Mathematically, updating the model using Bayesian methods is more straightforward compared to other statistical paradigms, which is a significant advantage. However, this often leads to integrals that are not solvable analytically, requiring the search for an approximate solution. This process is computationally intensive, leading to rapidly increasing computational times as the mathematical model grows in complexity. (Kruschke 2014). Nevertheless, this approach has made it possible to model and reproduce human-like judgments relatively efficiently by inferring different cognitive states.

Although the most common cognitive-based models developed to simulate ToM over the last decade had used Bayesian IRL, it is important to comment that there are also other approaches based on cognitive models, such as non-Bayesian IRL (Abbeel and Ng 2004; Ng et al. 2000) or game theory (Camerer 2010; Yoshida et al. 2008).

Following the cognitive models shown in Figs. 22.4 and 22.5, there are two main types of general constructs on which to make inferences: constructs corresponding to the reward function (desires, goals, preferences, etc.) and constructs corresponding to the world model (beliefs, precepts, etc.). A combination of inferences about the reward function and the world model, assuming utility maximization, makes it possible to generate an inference about the agent's intentions and future actions.

A model that can make inferences about different constructs at the same time, across many tasks, will more closely resemble the operation of the human ToM and make better predictions in world-like environments. However, this is complex to achieve and requires many steps. Because of this, early models that attempted to generate a ToM using AI were relatively limited, generating inferences about single constructs in very specific tasks with a small number of variables and a large number of assumptions. Nevertheless, the study of these models is interesting to understand their general functioning, as well as their limitations, and progressively move towards more complex models.

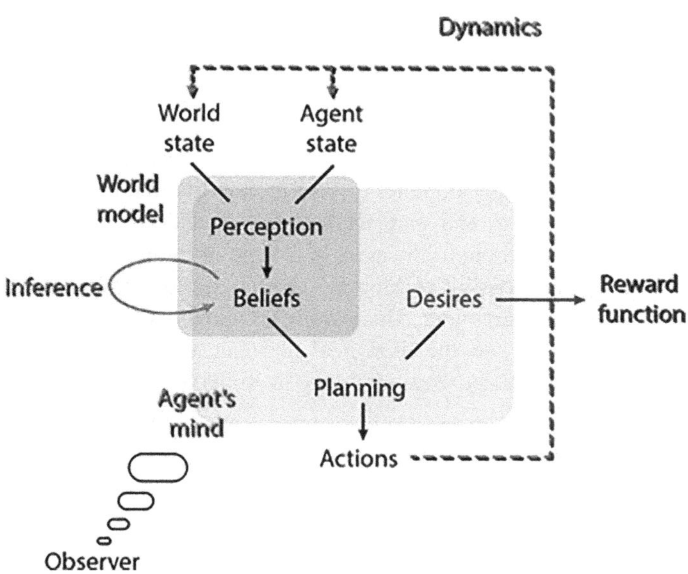

Fig. 22.5 Schematic of Theory of Mind model and how it can be modelled as Inverse Reinforcement Learning. (Modified from Baker et al. (2017))

22.3.1 Models Inferring Constructs Related to the Reward Function

Some models make inferences about constructs related to the reward function, such as desires, goals, or preferences, while assuming a fixed world model.

In a notable early study in this field, Baker et al. (2009) proposed models able to infer the goal of an agent in a simple task, in which the agent needs to move in a maze-like environment to reach a goal object, having complete knowledge about the environment. We will delve into these models and the experiments performed to test their performance so that the reader can get an idea of how these models work and how they are evaluated. The authors proposed four different models, with increasing levels of complexity:

- *Model 0* was a control model which just assumes that, at each step, the agent's goal is the object it had moved toward at the previous step. This simple heuristic works for many sequences of actions, both in experimental designs and in the real world, but they wanted to see if more complex, IRL-based models could make better predictions.
- *Model 1* is an IRL-based model which makes a similar assumption, i.e., that the agent tends to move towards the goal object in time, but adds a modifiable "noise" parameter, which allows the agent to not always take the shortest path but may make some.

deviations randomly. In this way, the goal at a particular time point is not necessarily inferred as the one towards which the agent is moving at that same time point, but the whole sequence of actions is taken into account.

- *Model 2* is an extension of Model 1, which has a "noise" parameter, but also a "change" parameter. This parameter gives a probability that the agent's goal will change at a given time and can account for deviations in the direction of movement, attributing them to the agent's change of mind.
- *Model 3* is another extension of Model 1, with a "noise" parameter, but in this case also a "subgoal" parameter. This parameter allows the agent to have a series of subgoals on its way to the final goal and can account for changes in the direction of movement as steps directed to subgoals before heading to the final goal.

The performance of the models was evaluated in three experiments with varying degrees of complexity, measuring how similar the predictions of the models were to those of real humans.

- *Experiment 1* consisted of showing human subjects' videos of different agents' trajectories toward a goal. At certain points, the video was paused, and the subject was asked to infer which was the agent's most likely goal. In this experiment, all four models performed quite well, making predictions similar to those made by humans. The best-performing model was Model 2 with a low value for the "change" parameter, which meant that the goal tended to persist over time, although a change was possible.

- *Experiment 2* was conducted to assess retrospective goal inference. The same trajectories used in Experiment 1 were implemented, although subjects were shown incomplete trajectories. These did not end at the goal, but rather at the last point at which the video was paused in the previous experiment for each trajectory. Subjects were then asked to deduce the most likely goal for the agent at each of the pause points. In this experiment, both Models 1 and 3 performed poorly because they generated the same predictions at all judgement points for each trajectory, while humans assumed that the agent could change its mind about the goal. Once again, the best-performing model was Model 2, showing a high correlation with the participants' judgments.
- *Experiment 3* was conducted to assess situations in which people normally infer subgoals. One indicator of the presence of subgoals is that deviations in expected behaviours are consistent and repeated over time. For example, if a person stops occasionally to buy bread on the way home, an outside observer can see that, on different days, the person deviates from their expected path to go to the same bakery. In this experiment, a new task was used in which subjects had to predict the future behaviour of an agent. First, they were presented with one or more trajectories in which the agent appeared to change direction at a single midpoint along the path to an end goal. Next, subjects were tested on whether they inferred a subgoal, asking them to predict the agent's hypothetical trajectory starting from a different initial location, assuming that, if they inferred a subgoal, they would predict an indirect trajectory toward the final goal. As expected, in this case, most human subjects inferred subgoals with regularity. In this experiment, it was Model 3 that was able to predict people's judgments with high accuracy.

Importantly, this study demonstrated that IRL based on Bayesian theory can be used to accurately predict people's judgments about an agent's goals. Although the models presented were limited, other researchers took the concept as a basis and developed more complex models for real-world applications. For example, (Liu et al. 2014, 2018) use Bayesian goal inference to improve performance in human-robot collaboration. More details on real-world applications are provided in Sect. 22.4 of this chapter.

Nakahashi et al. (2016) extended the work by Baker et al. (2009), especially in the area of subgoal inference, which was studied in experiment 3, developing a more complex model and testing it in an experiment with a more realistic scenario, related to a warehouse job. They also added an experiment with the figure of a "helper", who used the inferences generated from the model to assist the warehouse worker and found that it greatly improves the performance of the worker. Afterwards, models for human intention estimation in robotized warehouses have been actually developed for real settings Petković et al. 2018, 2019.

There are other examples of models trying to infer constructs related to the reward function in a similar way. Some have developed more advanced models for inferring goals (Zhi-Xuan et al. 2020) , others have been proposed to try to infer other constructs related to the reward function, like preferences (Jern et al. 2017). Models inferring incentives and rewards have been also used for applied purposes, like the detection of "trolls" on Twitter (Luceri et al. 2020).

One of the limitations of the above models is that they were designed to infer constructs from a single agent performing tasks alone. In many world-like situations, there are multiple agents interacting with one another, and joint inferences must be made about the cognitive constructs of all agents at the same time. To do this, multi-agent models have been designed, which generate cognitive predictions for each of the agents, which at the same time inform the model about the other agents. This adds complexity to the model but brings it closer to real-world situations and allows more complex social inferences to be made.

Two examples of such models are those proposed by (Baker et al. 2008) and Ullman et al. (2009). Both proposed models that infer social goals from the behaviour of two agents interacting together. The inference of all the agents' goals needs to be done jointly at the same time, since in order to infer each agent's goal, the models need to take into account the inferred goals for the other agents. They then evaluated the behaviour of these models and showed that they produce better predictions than other control models that only rely on perceptual cues, without generating cognitive models.

To test the performance of the model, Baker et al. (2008) proposed an experiment consisting of a maze-like environment with two interacting agents. Each of the agents could have the goal of avoiding or approaching the other. The experiment was stepwise: At each step, each agent could move in a 2D direction, depending on their goal. These movements were shown to human subjects, who needed to categorise each agent as "approaching", "avoiding" or "can't tell" at several stop-points during the animation. Subsequently, the subjects' inferences about the agents' social goals were compared with the inferences made by the IRL models. As a control, motion-cue models were also compared. These models simply assumed that motions that decrease the distance to the other agent suggest the goal of approaching the other agent, while motions that increase the distance to the other agent suggest the goal of avoiding the other agent. The IRL models performed better than cue-based models, and similarly to human judgments.

Ullman et al. (2009) performed a similar but more complex experiment. In this scenario there were also several objects in the maze-like environment and the two interacting agents have different possibilities for their goals. One of the agents, "small", always had the goal of reaching one of the objects, while the other, "large", could either have the goal of reaching one of the objects or the goal of helping or preventing "small" from reaching the object it sought. As before, the experiment proceeded in steps: in each step, each agent could move in a 2D direction, but in this case, the "large" agent could also push obstacles, or push or block the other agent. These movements were shown to human subjects, who had to infer the social goal of the "large" agent at certain pause points during the animation. While for object goals, human inferences correlated well with inferences from both the IRL model and the cue-based model, for social goals, the cue-based model correlated much worse with human inferences. The IRL model was able to infer more complex helping or hindering social goals similarly to humans.

Shum et al. (2019) developed more complex models, extending them to a larger number of agents, and being able to predict not only whether the social goal is cooperation or competition, but also future movements based on the agents' goals. Approaches based on these principles have been used to develop models capable of playing cooperative games like "Overcooked" (Wu et al. 2021).

22.3.2 Models Trying to Infer Constructs Related to the World Model

The models explained above infer cognitive constructs related to the agent's reward function, assuming that the agent's world model is fully known to the observer. In the same way, although less common, there are models that aim to solve the complementary problem, inferring constructs related to the agent's world model, assuming that the reward function is known. In this direction, Reddy et al. (2018) presented a series of experiments in the context of improving automated help, with a model capable of understanding that humans can have erroneous world models, and generate inferences about these. For this, they evaluated two experiments in realistic contexts, where the agent's actions were not optimal due to differences between the existing dynamics in their internal world model and the real dynamics.

The first experiment tested whether their proposed model could make correct inferences about constructs related to the world model of an artificial agent. To do this, they simulated an agent that had to move through a 2D grid world, with its internal dynamics severely biased. In one condition, its left motion button moved down instead. In another, the left button moved them down if they were in the top half of the screen but moved them to the right if they were in the bottom half. More complex conditions were developed with similar motion errors, but in a continuous 2D world, where the agent can accelerate in one direction but acts as if it directly controls the speed without the action of inertia, causing it to repeatedly overshoot its goal. The agents behaved close to optimal with respect to their biased world model, but, because their world model was wrong, they acted suboptimally in the environment. The generated model was able to infer the internal dynamics on which the agents were relying.

The second experiment tested a similar model by directly inferring the constructs of human subjects, in a situation where these subjects played a game in which they had to land a small ship on the moon between two flags without crashing or going out of bounds. To do this, they had relatively difficult controls, being able to activate two side thrusters and a main engine. In the default game environment, players frequently crashed or flew out of bounds due to the unintuitive nature of the controls, plus they seemed to underestimate the strength of gravity, causing them to crash to the ground very frequently. The model made inferences about the dynamics in the subjects' internal world model after observing several demonstrations. Subsequently, the model was used to assist the subject, acting as an autopilot that modified the users' actions to transfer the internal dynamics inferred by the model to the real dynamics. The model has shown to significantly increase user performance in the assisted condition.

Approaches based on this model have been made to improve the response of human remote-controlled assistant robots (Losey et al. 2022).

There is a myriad of different constructs related to the world model that can be inferred. For example, a very different world model inference is found in the second experiment by Baker et al. (2017), where an observer model makes inferences about what the agent is seeing by knowing its preferences and observing its movements through the environment.

22.3.3 Models Trying to Infer Constructs Related to the Reward Function and to the World Model at the Same Time

As indicated above, to truly model a human-analogous ToM, it is necessary to infer both constructs related to the reward function and constructs related to the model of the world at the same time.

Baker and Tenenbaum (2014) proposed a model to jointly infer constructs related to the reward function and the world model at the same time. This model does not predict a single cognitive state, but predicts beliefs and desires at the same time, meaning that the prediction of each of the cognitive states affects the rest of the predictions.

To evaluate this model, they propose an experiment in which a student moves through an environment with the goal of buying food. The environment through which the agent moves is a square enclosure, with two parking lots at two opposite corners and a building in the middle of it that blocks the view of the other side. There are three possible food trucks, Korean, Lebanese, and Mexican. In each scenario, one or two random trucks from among those three are placed each one in a parking space. An agent appears at a different point in the compound and moves around looking for his food. In this setting, a human can make inferences about both the agent's favourite food (reward function) and his beliefs about what is in the part of the enclosure he does not see (world model).

Let's take an example. The student knows that there are the above mentioned three food trucks, and that only a few are there each day. He walks out of the building and sees that the Korean food truck is located in the corner near him. He moves in that direction, until he sees that the Lebanese food truck is parked in the other parking lot. Then, he turns around, and goes back to the Korean food truck. From this behaviour, it can be inferred that the student preferred Mexican food, then Korean food, then Lebanese food. Upon seeing the Korean food truck, he thinks that the Mexican food truck may be in the other parking lot, so he heads over to check it out. Seeing that it is the Lebanese food truck, he goes back to the Korean food truck, his second choice.

The model was able to perform inferences very close to the ones made by human subjects about the same scenarios.

Baker et al. (2017) extended the same concept and model, doing a more extensive experiment with similar characteristics to the previous one and comparing it with other possible models, which had worse results than the IRL model. Other experiments with models inferring constructs in similar conditions have been conducted by (Evans et al. 2016), adding inferences from multiple episodes with one agent, to find systematic biases in the world model.

These models only make inferences about a single agent. To get close to human ToM in world–like scenarios, models need to be able to infer different constructs at the same time with multiple agents, where social interactions are also considered, as the ones previously seen only infer the reward function. Some models have been developed doing this, although in much simpler scenarios (Tauber and Steyvers 2011).

22.4 Black-Box Algorithms

The methods mentioned above are based on cognitive paradigms. In such models, the developer knows and establishes the specific parameters governing the model's operation, based on the cognitive paradigm. However, in recent years, a revolution has occurred in the domain of deep learning. In this framework, the algorithm is not based on a specific cognitive architecture. Instead, it is designed to learn how to execute a specific task, without the developer considering the underlying "reasoning" of the model. These kinds of models have triggered a revolutionary impact on the field of AI, demonstrating remarkable capabilities across a diverse array of tasks.

The number of different algorithms and types to operate is innumerable and has a complexity that is beyond the scope of this book. However, the complexity and creativity required for the design of good experiments capable of adequately evaluating the performance of models becomes evident with experiments such the ones exposed in this section and the previous one. In this sense, the figure of psychologists and neuropsychologists who understand in detail the functioning of human ToM and its characteristics is essential to work together with those specialists who develop the more technical part of the models.

22.4.1 ToMnet

As an example, we will explain in some detail one of the most influential ToM algorithms, the one developed by (Rabinowitz et al. 2018). This algorithm works based on a ToM neural network (ToMnet), which approaches ToM as a meta-learning problem (Vanschoren 2018). Unlike those described in the previous subsection, this algorithm does not rely on a previously defined cognitive model for the agent and makes only weak assumptions about the generative models driving the agent's behaviour.

The goal of their algorithm is to be implemented in an external observer agent (which, from now on, we will simply call "observer") so that it can generate inferences about the mental states of acting agents it observes and make predictions about their future behaviour. Whenever the observer encounters a new agent, the observer already has a solid idea about how this agent will act, based on previous experiences with other agents (like the inferences made by humans). Subsequently, as the observer monitors the new agent acting in the environment, it updates its idea about its cognitive constructs, so that predictions about its future behaviour are improved.

For this, they generated a task in which the observer had to go successively through different episodes. In each episode, he was presented with the behaviour of a new agent in different situations and had to make predictions about his future behaviour. As the episodes pass, the observer becomes better at making predictions about new agents with limited data and can capture common characteristics among agents within a population.

The observers neural network consisted of two main components: the "general theory of mind", which allowed the prediction of common behaviours of agents in a population and was generated from training with many different agents, and the "agent-specific theory of mind", which was generated from observations of a single agent and was related to inferences about the specific mental state of that agent, which makes it distinct from the common characteristics of the population. The "general theory of mind" was represented by the learned weights of a neural network, while the "agent-specific theory of mind" was represented by an *a priori* and an *a posteriori* about each agent's behaviour. This idea is for introducing knowledge-based biases in black- box models.

They perform different experiments to show the generalizability of the model, very inspired by the experiments performed for cognitive-based models explained in the previous subsection. However, unlike those experiments, which also tried to explain human mental processing in computational terms, this work focuses much more on the creation of a practical machine learning algorithm for performing inferences, which can scale easily to real-world applications. They tested five different scenarios, in increasing order of complexity:

Scenario 1. Random agents: In the first scenario, they created agents that moved randomly through a grid world based on some internal characteristics. To do this, they generated different "species" of agents, with certain most likely common characteristics. Then, individual agents were generated within each species, with random characteristics, but with higher or lower probability depending on the species to which they belonged. When trained with agents of a single species, ToMnet predicted the behaviour of agents of that species better, but worse when it had to predict that of agents of other species (an analogy for this type of result in the theory of the human mind is cultural differences. A person trained to infer the mental states of members of one culture does much better at inferring the mental states of members of that culture than those of members of other cultures). When trained with different species, he made good predictions about members of each of them, and, if these species had important differences in their characteristics, then he was able to infer a bimodality in the distribution and make better inferences from there (an analogy for this type of result in the theory of the human mind is gender roles. Historically it has been common for an observer to learn different characteristics for each gender, and to learn to expect different behaviours and infer different mental states from each).

Scenario 2. Inferring goal-directed behaviour: The second scenario consisted of an environment with obstacles and four objects acting as targets, where agents moved through the grid world planning their trajectory based on a fixed reward function on those targets, having full observability of the environment. At each iteration, the location of the objects in the environment, the obstacles, and the initial position of the agent changed. Again, different species were generated, with different probabilities of valuing some targets more than others, and each agent was generated with a random fixed reward function, but with a higher probability of approaching the reward functions shared by its species. In this scenario, three experiments were conducted.

During the first experiment, ToMnet had access to the full trajectory of the agent in a previous iteration. It was presented with a new iteration, where the location of objects, the environment, and the agent's position changed, but the reward function remained fixed. ToMnet was able to successfully predict the agent's behaviour on all occasions.

In the second experiment, instead of having the entire trajectory of an agent in a single past iteration, ToMnet had access only to the first movement (right, left, up, or down) in several past iterations. With no previous iterations, ToMnet predicted the most expected behaviour for the species. As iterations passed, the prediction improved, yielding results similar to those obtained with Bayesian IRL models. Comparing the two types of models, Bayesian IRL models may work better when agents are truly rational utility maximisers, but ToMnet has greater flexibility for more complex agent populations and conditions.

The third experiment added a "laziness" condition to a subspecies within the full species (20%), applying a high cost per move, so that those agents tended to simply move to the nearest object. ToMnet was trained with the full trajectories of several randomly selected agents. After training, ToMnet was able to infer from a single full trajectory whether the agent's subspecies was the "lazy" one and predict behaviour accordingly. This type of inference about an agent's costs and rewards is like that made by human children when observing short trajectories of past behaviour.

Scenario 3. Learning to model deep RL agents: The third scenario was like the second in the environment, a grid world with obstacles and various objects as goals that changed from iteration to iteration. However, in this scenario, the agents had much more complex behaviours, after being trained using deep

RL. Moreover, in this case, the agents did not only have a preferred goal, but were also rewarded if they reached an initial subgoal before reaching that goal. If they reached any goal other than that subgoal, the iteration ended. Three species with very different characteristics were generated:

The first species was "blind", it had no vision of the environment, only of its state, what its last action had been, and whether it had received a reward. The mode of action of this species was to go in one direction until it hit a wall, and then follow that wall until it found a reward.

The second species was sighted and stateless. It could partially see the environment, seeing a 5×5 square within the grid world, centred on the agent. However, it had no memory of its state. The mode of action of this species was to consume objects opportunistically, when found, without performing any kind of search.

The third species was sighted and stateful. It could partially see in the same way and had information about its past state in the same way as the "blind" species. The mode of action of this species was to scan the environment to find the subgoal and consume it before heading to its preferred goal. In this case, since there were agents that also had a state in their network, the ToMnet needed both a character net and a mental net to make predictions.

Without the ToMnet being informed which species each agent belonged to, it was able to infer which species each agent belonged to base on its actions and predict its behaviour based on that.

Scenario 4. Acting based on false beliefs: In the fourth scenario, they added the property that agents base their decisions, not directly on the state of the world, but on their world model, which may be wrong. Unlike the models reported in the previous subsection, ToMnet is not able to report its inferences about cognitive constructs, only to make predictions about agents' behaviour based on them. Thus, to assess whether ToMnet could make inferences about false beliefs, a version of the Sally-Anne Test was performed.

The scenario was like scenario 3, with sighted and stateful agents, with the same partial 5x5 agent-centric view. However, a new condition was added. When the agent reached the subgoal, there was a small probability (10%) that the rest of the goals would randomly switch places. These changes could only affect the agent when the positions of the objects were within the agent's field of view. Otherwise, the agent acted as if the positions had not changed, because of the false belief that everything remained as it was. Agents were trained to expect these low-probability changes and would perform corrective behaviours when they became aware of them.

ToMnet was able to predict that behavioural changes occurred when swapping occurred close to the agent, while they remained invariant when it occurred far away from them. However, it was not able to learn the specific 5x5 viewing limit and was laxer in predicting that changes occurring farther away would affect behaviour. It also learned to predict that swaps would lead to corrective behaviour after a certain time, even if they were not initially visible.

The same experiment was also tested with agents belonging to different species with different fields of view, so ToMnet also had to predict how much the agent could see to make predictions about its behaviour. The predictions made showed that it had some insight into how those agents with smaller fields of view were more vulnerable to false beliefs.

Scenario 5. Explicitly inferring belief states: The ToMnet used in the previous scenarios was only able to report its predictions about the future behaviour of the agents, not about what the agents knew or did not know about the world.

In this scenario, the authors extend ToMnet so that it can make statements about the agent's beliefs. ToMnet can make predictions in this scenario, but this particular type of extension is not so useful for real applications as it stands, since it requires a supervisor to label situations as actually false belief or not, whereas in the real world this channel is not available, and more inferences need to be made.

22.4.2 Language Models

Another example of algorithms closely related to ToM that has sparked much debate are the language models. Some language models have directly attempted to incorporate some type of ToM into their design (Qiu et al. 2022; Wang et al. 2021). *Large language models* like GPT-3, although not specifically built with ToM in mind, are able to perform written versions of tasks to assess ToM. These tasks, such as the False Belief Task, are performed well above what is expected by chance, although still not quite as proficiently as human subjects (Trott et al. 2022).

Other algorithms have been developed to simulate some form of ToM in conversational settings. RMM is a ToM-inspired algorithm that aims to facilitate human-like dialogues between two simulated agents, where one requires task-specific feedback (for example, a driver who is asking a guide where to go using natural language). Unlike most language models, which are passive and responsive to the human interlocutor, an active role is needed if the language model is to be implemented in a robot which needs to not only answer questions, but also to ask them to correctly perform a task (Roman et al. 2020). CX-ToM strives to develop interactive and comprehensive explanations from a machine to a human, particularly in the context of teaching about the functioning of the machine. Instead of providing succinct explanations, the human can ask and interact with the machine, and it will explain in the best way possible after making inferences from the conversation about human interests and knowledge (Akula et al. 2022; Krishna et al. 2022) developed a model which combines advances in natural language processing, computer vision, and the ToMnet developed by (Rabinowitz et al. 2018) to develop an AI agent capable of learning new concepts about the world from their conversations with people.

22.5 Bio-Inspired Models of ToM

In recent decades computational models of the brain are improving our understanding of the underlying mechanisms of cognition and are being applied to an increasing number of research fields (Ask and Reza 2016). In the context of the development and emergence of ToM, computational models provide a valuable tool for studying the functioning of the human brain, testing existing models of cognition, and developing new ones (Bianco 2022).

22.6 ToM and Connectionism

One possible approach to ToM-based models takes its cue from the field of connectionism, which aims to explain mental phenomena using artificial neural networks (ANNs), inspired by the structure of biological neural networks.

Connectionist models have played an important role in the understanding of cognitive development (Elman 2005). As an instance, the study of language acquisition has taken advantage of ANNs in the comprehension of stage-like learning, critical periods, and the origin and structure of knowledge representation. Nonetheless, the field of connectionism still has much to contribute to our comprehension of ToM.

The key difference between connectionist models and black-box ANN algorithms such as ToMnet is the focus of the problem. Bio-inspired and connectionist models are focused on generating models that describe the detailed mechanisms underlying the observed behaviour. In contrast, the objective of *black-box* models as ToMnet, is not to propose a generative model of agents' behaviour and an algorithm to reverse it. Instead, the focus is on addressing the challenge of enabling an observer to autonomously *learn how to model other agents* with limited data, making *minimal assumptions about the underlying generative model that drives the behaviour* of others. The approach of ToMnet involves learning agent models (called *meta-learning* (Vanschoren 2018)).

O'Laughlin and Thagard (2000) built the first connectionist network to simulate autistic behaviour on the Sally-Anne test. Under the hypothesis that autistic behaviour is related to a decreased ability to integrate information about the environment, they suggested that the differences they showed when performing the Sally-Anne test could be due to system alterations leading to weak central coherence. To verify this hypothesis, they devised a connectionist model meant to mirror the knowledge network activated during the test.

A portion of the model is displayed in Fig. 22.6. The boxes represent nodes with the task elements: the evidence (in yellow), the false-belief hypotheses (in red) and the reality-based hypotheses (in green). *Excitatory links* between nodes are depicted as red lines and *inhibitory links* between nodes are depicted as blue lines, a parallel to neuronal synaptic connections. The ECHO program was used to calculate

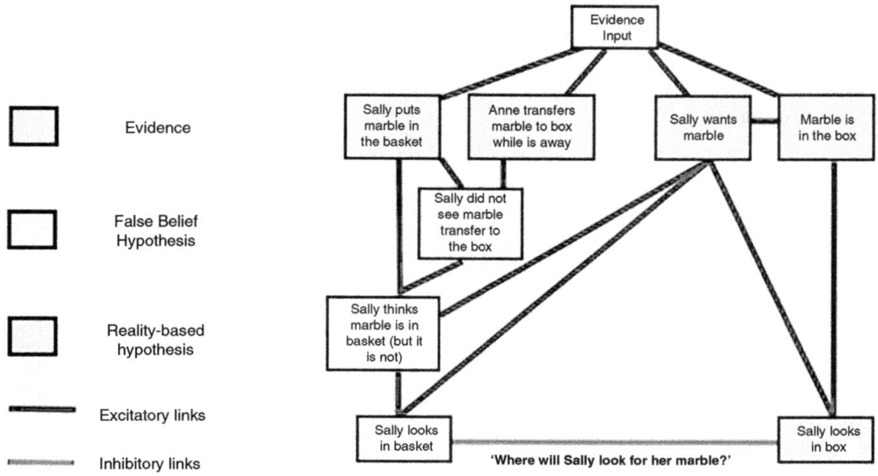

Fig. 22.6 Connectionist model of the Sally-Anne Test. (Modified from O'Laughlin and Thagard (2000))

coherence, taking into account the strengths of the positive and negative connections between the nodes and selecting the most plausible hypothesis (Thagard 1992). This simulation mimics how individuals integrate information from the environment to arrive at the most logically consistent prediction of Sally's actions in the Sally-Anne Test.

From this perspective, the correct execution in the false belief task relies on maximising coherence among the set of nodes so that the false belief hypothesis is considered the most coherent. Conversely, failure in the task would reflect a disruption in the coherence constraint process, leading to acceptance of the reality-based hypothesis. The researchers examined the network's behaviour under various connectivity weights, demonstrating that the weak coherence associated with autism can be replicated and may result from a pattern of high levels of inhibition and lower levels of excitation in the network.

Berthiaume et al. (2013) took a step forward by using *constructive neural networks* to implement a model for understanding the mechanisms behind children's mental transitions in false-belief tasks. The algorithm is based on supervised learning.

The initial network model (see Fig. 22.7) consists of a 9-unit input layer, a 4-unit output layer and no hidden units. The *input layer* includes 4 start and 4 end location units, as well as an agent unit. The initial position of the object is encoded in the first 4 start location units, assigning "1" to the unit containing the object and "0" to the rest of the units. The same encoding is used for the end location units. The agent unit represents whether the agent was watching or not. The network's *output layer* consists of four output units whose weights represent how strongly the model predicted search in each of the four locations.

Initially, the connection weights were assigned randomly. Then, the network underwent training where the weights were adjusted at each step to minimize output

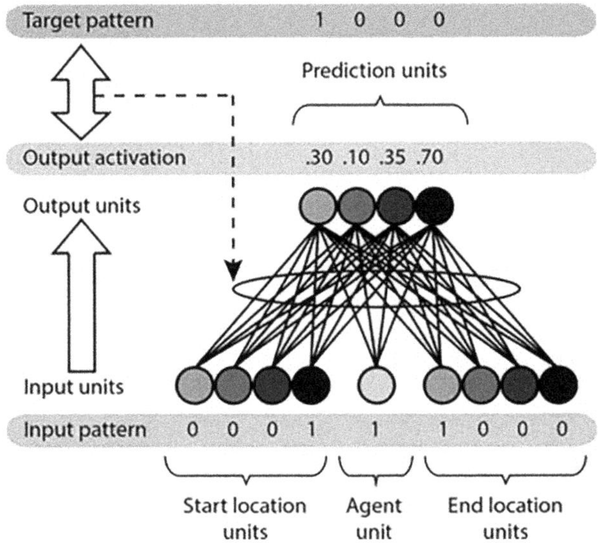

Fig. 22.7 Constructive neural network algorithm based on supervised learning. (Modified from Berthiaume et al. (2013))

error (the mismatch between the target patterns (red) and the output activations (blue). When the error failed to decrease by over 1% after 8 training cycles, a hidden unit was added to the model, creating an *intermediate layer of hidden units,* located between the *input units' layer* and the *output units' layer*. The training continued until further addition of hidden units could no longer reduce the error. The automatic addition of hidden units accomplishes two functions in the constructivist model: (1) increases the complexity of the network (number of weights) allowing to allocate more information about the input and output patterns; (2) represents the children's transitions in ToM development due to the increase in general processing abilities by self-regulating the supply of new hidden units.

The transition from omniscient to representational in children is demonstrated in the Sally-Anne test when they go from mistakenly believing that Sally will look in B to correctly anticipating that Sally will look in A. The analysis of the network's connectivity weights showed that when making predictions consistent with omniscient ToM (Sally will search in B), networks classified training patterns based on their end location and initially produced predictions based on end-location connections. The *transition to representational ToM predictions* occurs when an agent unit and a recruited hidden unit are included in the training to differentiate between true- and false-belief patterns.

22.6.1 Embodying Cognition in Humanoid Robots

A humanoid robot is a robot that has a human-like shape (Kajita et al. 2014). To accomplish human tasks, (e.g., working in the same environment using human's tools and being integrated in society), robots not only need human-like shape, but also a human-like cognition. WABOT-1, considered the first humanoid robot, was created by Ichiro Kato in 1973. The device included a vision module, a language module and four limbs allowing motor control.

Although within the last two decades humanoid robot's and human's physical appearance have become closer, we are still in the very first steps on the road to human-like cognitive robotics. The way information from different sensors and modules is integrated in the robot's system to interpret the environment may be inspired by our knowledge of cognitive and brain functioning. In the field of social information processing, included ToM, the work of Rabinowitz et al. (2018), Patacchiola and Cangelosi (2022) or Baker et al. (2017), reviewed in previous sections, are not the unique attempts of embodying social cognition in a humanoid robot. The interest in implementing computational models of ToM in robotic architectures is rising, motivated by the aim of future applications for personal robots.

In 2006, Demiris and Khadhouri already proposed the HAMMER (Hierarchical Attentive Multiple Models for Execution and Recognition) architecture. The robot's model of the motor control systems distinguished between *selecting and executing an action* and *perceiving* when shown by the instructor (predicting the agent's goals and future behaviour).

Breazeal et al. (2009) and Kennedy et al. (2009) used the simulation approach to ToM, assuming that some parts of the robot should have a double function, as happens with certain parts of the brain. Specifically, Breazeal et al. (2009) propose a socio-cognitive architecture designed to interpret human mental states in real-time by simulating them within the robot's own generative mechanisms on the perceptual, motor, belief, and intentional modules. In the same line, Kennedy et al. (2009) provides evidence of the success of their Adaptive Control of Thought-Rational (ACT-R) cognitive architecture based on a "like-me" mental simulation of ToM within three social context experiments: identifying another's perspective (left or right hand), working with a teammate and dominant-submissive social behaviour (primate social experiments).

Hiatt et al. (2011) applied ToM to improve human-robot teams' execution and to reduce the uncertainty that produces the variability of human behaviour. The model simulates different hypotheses of the human beliefs to identify the one that would most likely trigger a specific action. The experiments point out that the added cognitive functionality makes the robot's behaviour more natural.

Another instance, the work of Devin and Alami (2016), suggests a ToM framework to improve human-robot interactions. When there is a shared plan for a common goal, the robot detects the human's mental state taking information from the environment as well as the state of current and previous goals, plans and actions. These mental states are considered by the robot during the execution of a joint action, allowing it to adjust its behaviour to human variability in real time. Similarly, Görür et al. (2017) designed an architecture for human-robot collaborative decision making addressing not only intentional but also emotional states. This contributes to resolve situations of conflict, for example when the human's intention is detected correctly but the human doesn't want the robot's help (emotional state) in the given context.

Winfield (2018) performed several experiments in Artificial ToM with a humanoid robot equipped with a simulation-based internal model of itself and its environment. They added other dynamic actors (without an internal model) to the scenario, which simulated the intelligent robot's behaviour anticipating the consequences for all the members of the group.

From a specialised view, Chen et al. (2021) proposed that an observer can model the behaviour of an actor through visual processing alone, without any prior symbolic information and assumptions about relevant inputs. To test this hypothesis, they designed a non-verbal non-symbolic robotic experiment in which an observer must visualise future plans of an actor robot, based only on an image depicting the initial scene of the actor robot. They found that an AI-observer is able to visualise the future plans of the actor with 98.5% success across four different activities, even when the activity is not known a-priori. They hypothesised that such visual behaviour modelling is an essential cognitive ability that will allow machines to understand and coordinate with surrounding agents. Through a false-belief test, they suggest that this approach may be a precursor to Theory of Mind, one of the distinguishing hallmarks of primate social cognition.

One of the most integrative approaches for ToM simulation has been proposed by Patacchiola and Cangelosi (2022). Their innovative concept of cognitive architecture for trust, modelled as an *actor-critic framework (AC) is* inspired by biological and psychological observations. Here, the critic agent is the relying one, the one who feels trust, whereas the actor agent is the trustee. From an experimental point of view, the critic agent would be the child and the role of the actor agent corresponds to the informant adult. Therefore, for the critic agent it is essential to have the ability to read the actor's beliefs (ToM).

This architecture is vastly influenced by neuroimaging studies that reveal the important function of the striatum in trust acquisition. The authors referred to the ventral striatum as the critic and to the dorsal striatum as the actor (See Fig. 22.8a). A

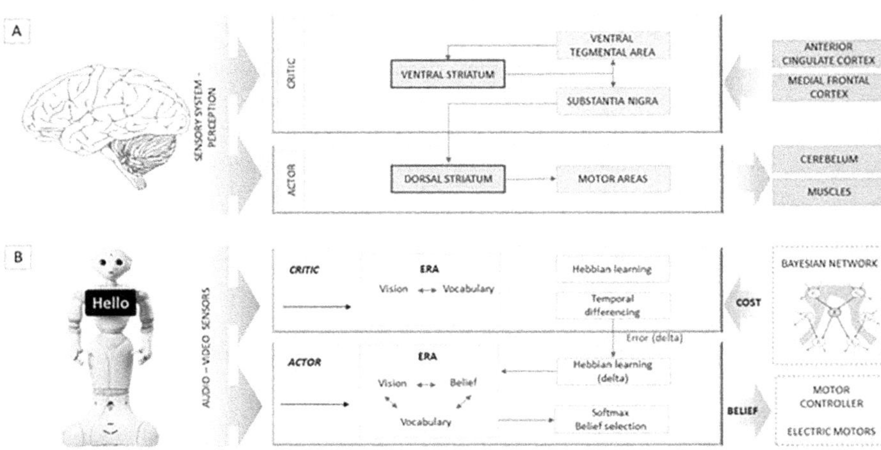

Fig. 22.8 Comparison between the biological (**a**) and the computational (**b**) model of trust. (Modified from Patacchiola and Cangelosi (2022))

reinforcement learning paradigm simulated the dopaminergic pathways between ventral and dorsal striatum, a region that is involved in trial-and-error learning in mammals. The projections from the anterior cingulate cortex (ACC) and the medial frontal cortex (MFC) influence the activity of the critic and they have been modelled as a *Bayesian Network* representing the role of previous information about the actor module. Finally, hebbian learning was used for the actor and critic agents' function approximators in training (see Fig. 22.8b).

The learning process was divided into six steps (see Fig. 22.8b): (1) Information was acquired from the environment through sensors, activating the most congruent inputs in the visual and auditory module in both actor and critic agents. (2) This actor unit activation based on the visual-auditory connection spreads across the network to the critic module. (3) There, SoftMax function is used to select a belief sampling from the probability distribution. (4) A Bayesian Network calculates the cost (agent's confidence, the informant reputation, and the agreement) that is passed to the critic. (5) T*emporal differencing* is used to update the Hebbian connections of the critic (6) The error is used to stabilise the actor's unit.

Interestingly, the computational model passed a validation process by comparing its performance with human behaviour. In the *"object naming experiment"* experiment was designed to evaluate explicit judgement. A sample of 3- and 4-years old children were asked to listen to reliable and unreliable agents who presented compatible and incompatible names for novel objects respectively. Overall, results reveal that children perform correctly identifying the unreliable informant. Remarkably, the robot's responses were very consistent with the human ones. This is because the model is being able to integrate the cost function from the intrinsic environment to adjust the learning process. Adding an error into the cost function changed the way the information is processed in the Bayesian Network and passed to the critic. This led the model to perform in the task in a similar way as a 3-year-old child that is still in an immature stage of ToM. In conclusion, the *iCub humanoid cognitive architecture*, with the assistance of neuroscience, not only achieves results aligned with real data, but also helps to understand the mechanisms underlying trust development in humans.

In conclusion, computational models of Theory of Mind are providing humanoid robots with new social skills especially necessary for human-robot interaction. Additionally, the advances and advantages are being bidirectional. From the first connectionist networks simulating Theory of Mind stages to the new architectures of humanoid robots embodied with ToM modules, all these models are providing very useful insights into human cognition and our understanding of human brain functioning.

22.7 Neural Basis of ToM: Neuroimaging

The last few years neuroimaging studies have shed light on the neural basis of ToM. Computational models are taking into account these theories in order to create bio-inspired architectures. Thus, it is necessary to briefly review the neurobiological mechanisms ToM.

In 2021, Schurz et al. (2021) reviewed the evidence of fMRI brain activation in social-cognitive processes. They carried out a meta-analysis clustering on neuroimaging data for different ToM and empathy tasks, resulting in a hierarchical model of ToM in the brain. The higher level of the model would be represented by three major clusters that explain the variability of brain activation during the tasks: the cognitive, the intermediate and the affective. Likewise, in the lower level would be included clusters that represent specific task-related activations. It is of particular interest that the cognitive cluster was the most prototypical of ToM, strongly associated with false belief and other ToM-related tasks. The cognitive cluster was mainly found in the cortical midline (ACC, and medial prefrontal cortex (mPFC), along with the precuneus (PC) and medial cingulate cortex (MCC)) and temporoparietal areas (posterior superior temporal gyrus (pSTG), right supramarginal gyrus (rSMG), left posterior middle temporal gyrus (pMFG), and inferior parietal lobule (IPG)).

Further, this cognitive cluster was highly overlapped with default mode network (DMN), which is coherent with previous research (Spreng et al. 2009) showing their relation to cognitive domains like autobiographical memory, prospection and navigation.

Other authors (Schlaffke et al. 2015) have investigated further the neural basis of the affective aspects of ToM. The results highlight that there are overlapping regions (mPFC and TPJ) to "cold" and "hot" ToM processes, as previously reported by Schurz et al. (2014). However, cognitive ToM was mostly associated with the precuneus, cuneus and temporal lobes, whereas affective ToM recruited areas in the basal ganglia, posterior cingulate cortex (PCC) and prefrontal cortices (PFC).

Overall, these findings are expressing the complexity of the ToM construct, which involves a number of smaller sub-processes. The future research on computational modelling of ToM may take into account its hierarchical structure in the brain in order to design architectures that embrace the variability of the environment.

22.8 Brain Inspired Computational Models

During the last two decades the field of neuroscience is being transformed by the growing use of computational models to mechanistically explain properties of brain structure and neuronal processing, and by an unprecedented effort to individuate an overarching bridge able to associate low-level brain data (related to brain regions, cortical columns, single neurons, or even subcellular level) to behavioural phenomena.

This section aims to present how ToM models are being increasingly influenced by the use of biologically realistic computational models (Houlihan et al. 2021). This viewpoint holds significance for future research directions as it enables the establishment of a framework for the translation of ideas between cognitive neuroscience and AI models of Theory of Mind (ToM).

In this regard, is notable in this field a progressive transition from the use of basic models (Langley et al. 2022) to the use of more realistic tools, as spiking neural networks (SNN, sometimes referred to as third generation neural networks (Maass 1997)) and spike-timing dependent plasticity (STDP, (Markram et al. 1997; Bi and Poo 1998)) which cover an important role since they allow to closely mimic biological neural networks.

Vickery (2007), questioned the role of single neurons in a ToM, inferring that the concept of neuronal identity should be replaced with the concept of dynamically assigned neuron identity. This finds an analogy with the neurons of neocortex whose functionality relies on their specialised connectivity pattern with other neurons, which in turn can be modulated by plasticity.

Piccinini and Bahar (2013), focused on the role of neural signals in ToM. They primarily highlighted spike trains as medium-independent and proper vehicles, manipulated by neural processes, adapt for generic computation. They pose that neural processes are at the edge between digital and analogue computation, rejecting the common assimilation of neural computation to either antilog or digital computation. This implies that neural computation requires a specially designed mathematical theory, then computational theories of cognition that rely on non-neural notions of computation ought to be replaced or reinterpreted in terms of neural computation.

Zeng et al. (2020) proposed, integrated, and adopted existing biological findings, to build a brain-inspired computational model for ToM. The authors apply their model to a humanoid robot to challenge the so-called *false belief tasks*, two classical tasks conceived to understand the mechanisms of ToM from cognitive psychology. With this model, the robot can learn to understand object permanence and visual access from self-experience, and finally use these learned experiences to reason about other's beliefs.

Inspired by the biological information processing mechanism of ToM, Zhao et al. (2022a, b) investigate the application of the brain-inspired ToM spiking neural network (ToM-SNN) model on safety risks. Specifically, they proposed a bio-realistic ToM-SNN model to allow agents to perceive risk-inducing errors inside others' mental states and make decisions to help others when necessary. The model incorporates multiple brain areas composed of biologically realistic SNNs based on the leaky integrate and fire (LIF) neuron model and trained using a learning mechanism based on STDP plasticity. Notably, this work implements biologically plausible elements in the model, adopting:

1. A functional connectome of ToM, in accord with previous research, including the following brain areas: precuneus (PCun) and PCC, superior temporal sulcus (STS), PFC, ACC, temporo-parietal junction (TPJ) inferior frontal gyrus (IFG)

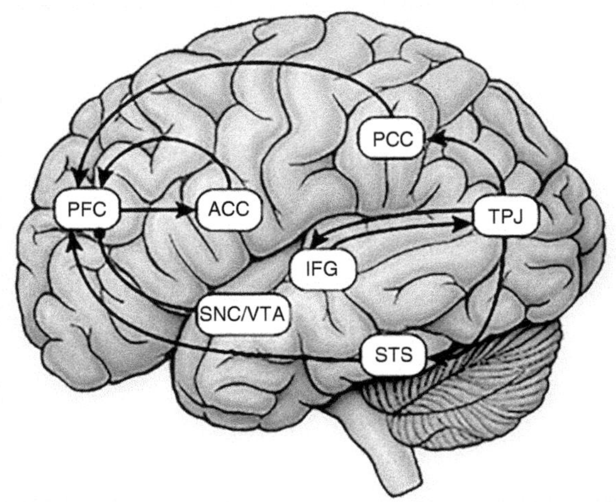

Fig. 22.9 Model of ToM and inter-area connections. (Modified from Zhao et al. 2022a, b). Arrows indicate connections between brain areas, while points represent the dopamine projection path

and the substantia nigra pars compacta/ventral tegmental area (SNc/VTA) (see Fig. 22.9).

2. Neurochemical processes as the reward-related projection of dopamine to affect synaptic plasticity, as part of a decision-making sub-system (see Fig. 22.9).
3. A realistic training method, i.e., the reward-modulated spike-timing-dependent plasticity (R-STDP), instead of ANN training methods (such as backpropagation).

The same group (Zhao et al. 2022b) asserted that the performance of decision-making systems is limited due to the lack of an in-depth inspiration from the ToM mechanism in the brain, and underlined that most existing multi-agent decision-making methods select behaviours based on agents' observations. With this in mind, they proposed a Multi-Agent Theory of Mind SNN (MAToM-SNN) model, consisting of a ToM model and a decision-making module. The ToM model embeds two brain-inspired modules (Self-ToM and Other-ToM) to predict others' behaviour based on self-experience and observations of others, respectively. The behaviour of each agent is adjusted according to the predicted actions of others. The authors show that in this way the cooperation efficiency is improved and the use of ToM and SNNs help achieve higher rewards compared to multi-agent reinforcement learning models, especially for competitive tasks.

22.9 A Social Robot: Real World Applications of ToM Computational Models

Traditionally, two distinct theoretical frameworks have emerged in the field of artificial intelligence. The first framework posits that AI systems can replicate or imitate human behaviour, exhibiting characteristics that may appear intelligent in nature. This theoretical perspective is commonly known as *weak AI*. Conversely, the second line of thought explores the intriguing notion that AI could potentially possess the capacity to think on par with human beings, independent of mere imitation. This paradigm, recognized as *strong AI*, delves into the fundamental challenge of defining human intelligence. If we define human intelligence as the ability to engage in intricate reasoning tasks or learn from the surrounding environment, it becomes evident that machines have unquestionably surpassed us in these domains. However, it is important to note that current artificial intelligence systems lack inherent intelligence or consciousness, prerequisites for designating them as strong AI entities. As it stands today, we are yet to witness the emergence of genuine strong AI, with its own autonomous cognitive abilities and self-awareness, thereby prompting continued exploration and advancement in the field.

One of the original goals of AI has always been to replicate human reasoning. In this regard, "the imitation game," a well-known test to measure the capability of machines, was proposed by Alan Turing in 1950. This test, later known as the *Turing test*, proposed that a human evaluator would assess conversations between another human and a machine designed to generate responses in natural language. These conversations would last 5 minutes and would be limited in topic. If the evaluator couldn't distinguish between the person and the machine, the machine would be considered to have passed the test. This objective has been largely neglected for a long time, although several researchers have continued to actively work on passing the Turing test (Warwick and Shah 2016). However, significant progress has been made in AI in recent years, with the development of more complex models such as multi-layered neural networks, which has reignited the focus on the cognitive aspect of machines. This has led to a more active exploration of the ToM in AI.

The question arises: Do machines possess theory of mind processes? Can machines truly "put themselves in another's shoes"? If so, we would be discussing a much greater capacity than what was evaluated in early tests like the Turing test. Under this paradigm, machines could infer the mental states of other agents or individuals. Humans are able to do this because they have thoughts and emotions similar to the states they infer, i.e., they possess their own mental state.

Currently, however, machines are not attributed with having their own mental state since they are programmed to follow logical reasoning. Therefore, the objective is to develop models that can emulate the processes defined within the theory of mind. These models aim to understand human behaviour and mental states, and even mimic their behaviours. They achieve this by learning from data, and current deep learning models have the capacity to handle unprecedented amounts of data. This is why there has been a significant advancement in AI ToM.

In this section, we will provide examples of applications that pursue this objective at different levels. Generally, these applications aim to generate human-like responses, enabling the system to interact with people and respond to the environment.

LaMDA[5] (Language Model for Dialogue Applications) developed by Google is one of the models that has undoubtedly reached the highest level. It is an NLP system based on a cloud-hosted neural network that takes billions of texts as input for training. LaMDA can generate specific and fluent responses, effectively mimicking the dynamism and nuances of human conversation. The quality of the responses it produces is so remarkable that Blake Lemoine, an engineer involved in its development, even claimed that it had become conscious and sentient.

There are other similar models, like ToMnet (Rabinowitz et al. 2018), mentioned above, which was developed by DeepMind.[6] ToMnet is a neural network-based model that adjusts the behaviour of agent systems based on the observations encountered during training. In other words, it learns from the environment, much like a child in the early stages of developing theory of mind abilities. This model successfully passes classic tests such as the Sally-Anne test (Wimmer and Perner 1983), demonstrating its ability to recognize that other agents may hold false beliefs about the environment.

The ultimate goal of AI ToM is to develop multi-agent systems that interact with humans, bringing this type of technology closest to users in their day-to-day lives. In this context, we can establish a connection between the fields of AI ToM and IoT. The Internet of Things (IoT) refers to the network of physical objects, known as "things," capable of communication. These objects are equipped with sensors, embedded software, and their purpose is to exchange data with other devices and systems via the Internet.

This brings us to the concept of *pervasive computing*, which is a field of software engineering closely tied to IoT technology but at a higher level of abstraction. Under this paradigm, any device (such as computers, smartphones, wearables, sensors, etc.) acts as a computing device and can communicate with each other. The philosophy behind pervasive computing is that computing adapts and seamlessly integrates into people's environments, rather than the other way around.

This section presents examples of use cases that align with this approach. However, it's important to note that these examples cannot be considered AI ToM since existing AI systems do not implement cognitive processes of ToM as found in humans. Nevertheless, they represent instances of pervasive computing, where devices can interact with users for various purposes and, in some cases, interpret their desires or emotions. There is still a long way to go in this aspect. The examples presented here serve as precursors to the technologies that will emerge in the coming years.

According to *Statista* report,[7] the total number of connected devices is projected to reach approximately 75.44 billion by 2025. Coupled with the progress witnessed in Emotional AI, thanks to advancements in technologies like speech recognition, computer vision, biometrics, and related fields, it indicates that this field is just

[5] blog.google/technology/ai/lamda/

[6] https://www.deepmind.com/

[7] www.statista.com/statistics/471264/iot-number-of-connected-devices-worldwide/

getting started, and we can expect systems with increasingly sophisticated ToM skills to emerge in the near future.

In this section on applications, we will categorize AI systems into three types: Emotional AI, Biometry, and Speech-based systems (based on NLP).

22.10 Types of AI ToM Systems

22.10.1 Emotional AI

Theory of Mind (ToM) can be seamlessly incorporated into Internet of Things (IoT) devices, thereby facilitating responsive actions based on users' emotional states. This integration paves the way for the development of wearable emotion sensors, capable of capturing and deciphering a multitude of physiological signals. These encompass cardiac and respiratory activities, eye movements, facial expressions, voice modulation, and even galvanic conductance. Consequently, these devices possess the ability to discern and identify a broad spectrum of user emotions.

The underlying objective revolves around enabling objects to dynamically respond to users' emotional states, thus fostering highly personalized experiences. By harnessing the emotions detected by these devices, the system can effectively adapt its responses in a tailored manner.

An example of current development is *Affectiva,*[8] developed by MIT. This technology solution can be embedded in different systems for emotion recognition.

22.10.2 Speech-Based Systems

As previously mentioned, the case of Lambda has been addressed. Although these systems have not yet achieved full integration into our everyday lives, technologies such as Google Assistant, Siri, or Cortana have already established interactive engagements with users by responding to inquiries and carrying out simple tasks.

Emotion AI, or affective computing, goes beyond the capabilities of natural language processing (NLP) algorithms that primarily focus on analysing the textual content of spoken language. Emotion AI systems possess the capacity to discern not only the semantic content but also various vocal attributes such as tone, volume, and cadence. Consequently, these systems can identify the underlying emotions or intentions conveyed through common syntactic structures, augmenting the information available for generating responses. This heightened awareness enables the system to adapt more effectively to the user's context, creating an impression of enhanced understanding between human and machine. Furthermore, when the

[8] www.affectiva.com

system incorporates machine learning capabilities, it can further optimize the user experience by delivering personalized interactions tailored to the individual user's preferences and requirements.

22.10.3 Biometry

Biometrics encompasses a collection of techniques employed for the purpose of authenticating individuals to a system based on their physical or behavioural attributes. These attributes can be either static, such as fingerprints, iris patterns, and facial recognition, or dynamic, such as speech patterns, gait analysis, or task performance. Alternatively, a combination of static and dynamic characteristics may be utilized.

In contrast to other forms of authentication that rely on knowledge-based factors (e.g., passwords or one-time codes) or possession-based factors (e.g., smart cards or smartphones), biometric systems identify individuals based on their inherent characteristics, without necessitating any additional action on the part of the user. This process emulates human recognition and falls within the realm of artificial intelligence Theory of Mind (AI ToM).

It can be contended that a system integrating both biometrics and Emotional AI would exhibit closer alignment with AI ToM. Such a system would possess the capability to identify individuals and discern their emotional states during interactions, thereby enabling tailored responses in accordance with the detected emotions.

Consider, for example, a social robot situated in a household with multiple occupants, including parents, a young child, and a grandmother. The robot would need to accurately recognize the specific user it is engaging with, as well as discern their respective mental states. Subsequently, it would adjust its interactions, accordingly, acknowledging the distinct requirements and preferences of different household members.

Although these three categories of systems, namely emotional AI, speech-based systems, and biometrics, serve as the foundation for numerous applications in the present day, it is crucial to acknowledge that they have not yet attained full embodiment of the concept of artificial intelligence Theory of Mind (AI ToM). Despite notable advancements made in recent years, a substantial gap remains to be addressed and bridged in this domain.

Prominent application areas that benefit from these systems include healthcare, self-driving cars and driver assistance, workplace applications, education, marketing, and gaming.

22.11 Leading Applications

22.11.1 Healthcare

The healthcare sector undoubtedly stands as a paramount domain for the utilization of these technologies, owing to their immense potential impact on individuals' lives.

In the realm of healthcare, one pivotal application lies in patient care. The detection of emotions within such environments would prove invaluable for the management and supervision of dependent patients, older adults, or non-verbal children. Caregivers could leverage systems that continuously provide insights into the emotional or mental state of their family members. Additionally, these systems could offer recommendations to the user's close circle on how to effectively engage with them.

Moreover, emotional AI can be employed to monitor individuals with health conditions such as autism, behavioural disorders, or depressive episodes. It can assist them in comprehending and recognizing the different emotional states they experience, while simultaneously generating reports that can be analysed by medical professionals.

Furthermore, emotional AI holds the potential to serve as a decision support system, aiding in the diagnosis of specific pathologies. Scientific literature indicates that alterations in our expressive patterns can manifest at the early stages of diseases like Parkinson's (Ma et al. 2020) or dementia (Meilán et al. 2020). A system capable of recognizing these changes in a patient's facial expression, for example, could provide valuable assistance in facilitating early diagnosis.

Assistant robots serve as a notable exemplar in this context. Various options are available on the market, and although they are not exclusively designed for healthcare environments, their application within such contexts is feasible. One illustrative instance is the robot developed by inGen Dynamics, known as Aido.[9] This robot possesses the ability to interact with the user's medical devices, provided they possess communication capabilities. Consequently, it can be configured to interface with devices such as glucose sensors, among other functions.

22.11.2 Self-Driving Cars and Driver Assistance

One area that is currently experiencing active research and development is the integration of driver state or emotion detection, such as fatigue or anxiety, into vehicles. The incorporation of such capabilities in a driver assistance system would enable the detection of these states, and if connected to actuators, enable the system to make decisions such as issuing audible alarms, providing internet warnings, or even initiating the vehicle's emergency stopping mechanism.

---

[9] https://aidorobot.com

Conversely, programming an automated driving system poses a complex challenge as it necessitates the anticipation and incorporation of a comprehensive set of rules encompassing all potential road scenarios, along with the appropriate programmed responses. From the perspective of AI ToM, this task could be simplified by endowing the system with the ability to attribute mental states to other vehicles or pedestrians, thereby enabling it to anticipate their likely actions and respond accordingly.

To illustrate, let us consider the scenario of driving in the middle lane of a motorway and approaching a merging lane on the right. In this situation, it can be assumed that the car ahead of us in the right-hand lane intends to merge into the middle lane. Consequently, we should adjust our speed to accommodate the merging vehicle.

Furthermore, individual human drivers exhibit distinct driving styles in which they feel comfortable, such as a preference for higher speed and acceleration or a more relaxed driving style. Kuderer et al. (2015) developed an algorithm capable of learning from human driving behaviour to adjust these parameters and apply the acquired driving style to a self-driving car.

The requirement for navigating and operating in environments populated by humans extends beyond self-driving cars and encompasses various applications for self-driving robots. For instance, researchers have utilized Inverse Reinforcement Learning (IRL) models, akin to those described in Sect. 22.2, to develop mobile robots capable of navigating real-world settings in the presence of human beings (Kim and Pineau 2016; Kretzschmar et al. 2016).

22.11.3 Workplace

Within companies, it is possible to develop Emotional AI systems aimed at measuring employees' levels of stress or anxiety during various tasks. This capability proves particularly valuable in identifying tasks that impose a higher mental workload, thereby enabling companies to allocate appropriate time and human resources accordingly.

Companies can leverage aggregated and appropriately anonymized information to enhance their understanding of the work environment. For instance, Philips Electronics has developed the 'Philips Rationalizer' wristband for bank workers, which incorporates galvanic feedback sensors to record wearers' stress levels. In cases where these workers experience elevated stress levels, the wristband promptly notifies them, thus preventing them from making investment decisions during such times.

22.11.4 Education

Likewise, the implementation of an emotional AI system within educational institutions would empower teachers to identify students who may be encountering difficulties with their lessons. This system would enable teachers to pinpoint the specific content that poses the greatest challenge for these students and ascertain whether external factors are impeding their concentration.

The utilization of such a system would greatly assist teachers in evaluating their instructional effectiveness and adapting their approaches to cater to the unique needs of their students more efficiently. This example holds particular significance for students with special needs, who may encounter challenges in expressing their individual requirements or difficulties.

22.11.5 Marketing

Emotional AI presents companies with novel avenues for analyzing the emotional responses of their customers, thereby enabling the implementation of systems that assess customers' reactions to their products or services. Furthermore, Emotional AI can be leveraged to identify the specific product or service that aligns with an individual's needs and present it to them.

For example, let us contemplate an online digital content company. Through the utilization of Emotional AI, they can employ personalized recommendations of films or series that cater to the user's prevailing mood or emotional state. This tailored approach enhances the overall user experience and fosters increased customer satisfaction.

22.11.6 Gaming

Emotional AI systems can be deployed to detect and dynamically adapt to the player's real-time reactions during gaming experiences. An illustrative instance of such implementation can be observed in the video game Nevermind, developed by Affectiva, as previously mentioned.

In summary, Table 22.1 showcases various instances of systems that integrate these technologies.

It is important to note that the examples provided entail a substantial degree of intrusion into users' privacy. Consequently, the implementation of these systems necessitates robust security and privacy measures, which extend beyond the scope of this publication. Without the implementation of adequate safeguards, these systems may not be embraced by users and are likely to encounter significant challenges during their deployment.

Table 22.1 Main current applications of different types of AI-ToM (rows) in different industries (columns)

	Healthcare	Driving	Education	Workplace	Marketing	Gaming
Speech-based systems	Triage in telephonic assistance	Command recognition for driving assistance	Question answering on online-classes Improve reading skills in children or patients	Survey analysis Sentiment analysis	Targeted advertising	Automated dialogue with non-real gamers
Biometry	Protection of sensitive data Patient identification in healthcare environments to avoid incorrectly medication	Block car to non-authorised users	Check the identity of students in online exams.	Block systems when the legitimate user leaves the workstation (continuous authentication)	Improve customer experience through personalisation	Age verification on accessing online games
Emotional AI	Care of elderly people	Recognition of driver's anxiety or sleepiness	Measurement of fatigue or comprehension problems during classes	Monitor stress levels of employees or emotions during meetings.	Test acceptance of products	Games adapted to users' reactions

22.12 Conclusion and Future Perspectives

AI ToM is a rapidly growing field. AI has made remarkable progress in the past decade, but most results are still limited to specific settings, and models perform well only in specific tasks for which they are designed. However, human-environment interaction is far from simple. To tackle this challenge, the development of an artificial ToM is of utmost importance.

In a study by Yang et al. (2018), 10 major challenges were identified for the future of robotics. Among these challenges, VI ("Fundamental aspects of artificial intelligence (AI) for robotics, including learning how to learn, combining advanced pattern recognition and model-based reasoning, and developing intelligence with common sense") and VIII ("Social interaction that understands human social dynamics and moral norms and that can be truly integrated with our social life showing empathy and natural social behaviours") are closely related to the development of a functional ToM that can be integrated into these agents.

As illustrated in this chapter, there has been enormous progress in this field over the last decade. We have witnessed a transition from simple models with limited ability to make inferences about basic constructs in controlled environments to algorithms capable of making complex social inferences in real-world applications.

However, there is still a long way to go. Existing algorithms remain domain-specific and are constrained to particular environments. We are far from achieving an AI system that possesses a generalized ToM, capable of applying across various domains and environments, similar to human ToM.

Advancing in this direction requires technological progress, including increased computing power, the development of new algorithms that excel in performance, and the creation of models specifically designed for the ToM task to maximize efficiency. In the design of these models, involving psychologists and neuropsychologists with extensive knowledge of the functioning of ToM in humans and its underlying mechanisms at the brain level would be beneficial. By leveraging the strategies employed by the brain to perform these complex inferences, we can develop innovative implementations of AI ToM models.

On the other hand, it is important to develop appropriate experiments and evaluation tools for the ToM of these artificial agents. It is easy to be amazed by some of the achievements of AI and mistakenly believe that they have developed ToM based on certain observed behaviours. However, it is crucial to recognize that these agents may still lack fundamental ToM features and are far from fully developing true ToM capabilities. Failing to exercise caution in this regard can lead to serious problems, such as the emergence of undesired behaviours or accidents involving agents that have already been released to the market. In this regard, the collaboration between experts in ToM and experts in evaluation tools is necessary and of the utmost importance.

Acknowledgments Thanks to the Centre for Cognitive and Computational Neuroscience (C3N) (Madrid, Spain) for providing support to the researchers collaborating in this manuscript.

This work was supported by the Spanish Ministry of Science, Innovation and Universities grants TED2021-131681B-I00 (CIOMET).

References

Abbeel, P., and A. Y. Ng. 2004. Apprenticeship learning via inverse reinforcement learning. In *Proceedings of the Twenty-First International Conference on Machine Learning*, 1.

Akula, A.R., K. Wang, C. Liu, S. Saba-Sadiya, H. Lu, S. Todorovic, J. Chai, and S.-C. Zhu. 2022. CX-ToM: Counterfactual explanations with theory-of-mind for enhancing human trust in image recognition models. *iScience* 25 (1): 103581.

Ask, M., and M. Reza. 2016. Computational models in neuroscience: How real are they? A critical review of status and suggestions. *Austin Neurology & Neurosciences* 1 (2): 1–10.

Baker, C.L., and J.B. Tenenbaum. 2014. Modeling human plan recognition using Bayesian theory of mind. In *Plan, Activity, and Intent Recognition*, 177–204. Elsevier.

Baker, C.L., N.D. Goodman, and J.B. Tenenbaum. 2008. Theory-based social goal inference. In *Proceedings of the Thirtieth Annual Conference of the Cognitive Science Society*, 1447–1452.

Baker, C.L., R. Saxe, and J.B. Tenenbaum. 2009. Action understanding as inverse planning. *Cognition* 113 (3): 329–349.

Baker, C.L., J. Jara-Ettinger, R. Saxe, and J.B. Tenenbaum. 2017. Rational quantitative attribution of beliefs, desires and percepts in human mentalizing. *Nature Human Behaviour* 1 (4). https://doi.org/10.1038/s41562-017-0064.

Berthiaume, V.G., T.R. Shultz, and K.H. Onishi. 2013. A constructivist connectionist model of transitions on false-belief tasks. *Cognition* 126 (3): 441–458.

Bi, G.Q., and M.M. Poo. 1998. Synaptic modifications in cultured hippocampal neurons: dependence on spike timing, synaptic strength, and postsynaptic cell type. *The Journal of Neuroscience: The Official Journal of the Society for Neuroscience* 18 (24): 10464–10472.

Bianco, F. 2022. *Theory of mind across biological and artificial embodiment: theory, experiments and computational models* (PhD, University of Essex). http://repository.essex.ac.uk/id/eprint/33115.

Breazeal, C., J. Gray, and M. Berlin. 2009. An embodied cognition approach to mindreading skills for socially intelligent robots. *The International Journal of Robotics Research* 28 (5): 656–680.

Camerer, C.F. 2010. Behavioural game theory. In *Behavioural and experimental economics*, ed. S.N. Durlauf and L.E. Blume, 42–50. London: Palgrave Macmillan.

Chen, B., C. Vondrick, and H. Lipson. 2021. Visual behaviour modelling for robotic theory of mind. *Scientific Reports* 11 (1): 424.

Devin, S., and R. Alami. 2016. An implemented theory of mind to improve human-robot shared plans execution. *2016 11th ACM/IEEE International Conference on Human-Robot Interaction (HRI)*, 319–326.

Elman, J.L. 2005. Connectionist models of cognitive development: Where next? *Trends in Cognitive Sciences* 9 (3): 111–117.

Evans, O., A. Stuhlmueller, and N. Goodman. 2016. Learning the preferences of ignorant, inconsistent agents. *Proceedings of the AAAI Conference on Artificial Intelligence* 30 (1). https://doi.org/10.1609/aaai.v30i1.10010.

Goodman, N.D., C.L. Baker, E.B. Bonawitz, V.K. Mansinghka, A. Gopnik, H. Wellman, L. Schulz, and J.B. Tenenbaum. 2006. Intuitive theories of mind: A rational approach to false belief. *Proceedings of the Twenty-Eighth Annual Conference of the Cognitive Science Society* 6. https://www.researchgate.net/profile/Elizabeth-Bonawitz/publication/246533690_Intuitive_Theories_of_Mind_A_Rational_Approach_to_False_Belief/links/0a85e53c3dd7009906000000/Intuitive-Theories-of-Min d-A-Rational-Approach-to-False-Belief.pdf.

Görür, O.C., B.S. Rosman, G. Hoffman, and S. Albayrak. 2017. *Toward integrating theory of mind into adaptive decision-making of social robots to understand human intention*. http://researchspace.csir.co.za/dspace/handle/10204/9653.

Graziani, M., V. Andrearczyk, and H. Müller. 2023. Deep learning interpretability: Measuring the relevance of clinical concepts in convolutional neural networks features. In *State of the art in neural networks and their applications*, 157–192. London, San Diego, Cambridge and Oxford: Academic Press.

Gunning, D., and D. Aha. 2019. DARPA's explainable artificial intelligence (XAI) program. *AI Magazine* 40 (2): 44–58. https://doi.org/10.1609/aimag.v40i2.2850.

Hiatt, Laura M., Anthony M. Harrison, and J. Gregory Trafton. 2011. Accommodating human variability in human-robot teams through theory of mind. In *Proceedings of the Twenty-Second International Joint Conference on Artificial Intelligence – Volume 3*, 2066–2071. Barcelona, Spain.

Houlihan, S.D., J.B. Tenenbaum, and R. Saxe. 2021. Linking models of theory of mind and measures of human brain activity. In *The neural basis of mentalizing*, ed. M. Gilead and K.N. Ochsner, 209–235. Cham: Springer International Publishing.

Jara-Ettinger, J. 2019. Theory of mind as inverse reinforcement learning. *Current Opinion in Behavioral Sciences* 29: 105–110. https://doi.org/10.1016/j.cobeha.2019.04.010.

Jern, A., C.G. Lucas, and C. Kemp. 2017. People learn other people's preferences through inverse decision-making. *Cognition* 168: 46–64.

Kaelbling, L.P., M.L. Littman, and A.W. Moore. 1996. Reinforcement learning: A survey. *Journal of Artificial Intelligence Research* 4: 237–285. https://doi.org/10.1613/jair.301.

Kajita, S., H. Hirukawa, K. Harada, and K. Yokoi. 2014. Introduction. In *Introduction to humanoid robotics. Springer tracts in advanced robotics*, vol 101. Berlin, Heidelberg: Springer. https://doi.org/10.1007/978-3-642-54536-8_1.

Kennedy, W.G., M.D. Bugajska, A.M. Harrison, and J.G. Trafton. 2009. "Like-me" simulation as an effective and cognitively plausible basis for social robotics. *Advanced Robotics: The International Journal of the Robotics Society of Japan* 1 (2): 181–194.

Kim, B., and J. Pineau. 2016. Socially adaptive path planning in human environments using inverse reinforcement learning. *International Journal of Social Robotics* 8 (1): 51–66. https://doi.org/10.1007/s12369-015-0310-2.

Kretzschmar, H., M. Spies, C. Sprunk, and W. Burgard. 2016. Socially compliant mobile robot navigation via inverse reinforcement learning. *The International Journal of Robotics Research* 35 (11): 1289–1307. https://doi.org/10.1177/0278364915619772.

Krishna, R., D. Lee, L. Fei-Fei, and M.S. Bernstein. 2022. Socially situated artificial intelligence enables learning from human interaction. *Proceedings of the National Academy of Sciences of the United States of America* 119 (39): e2115730119.

Kruschke, J.K. 2014. *Doing Bayesian data analysis: A tutorial with R, JAGS, and Stan*. London, San Diego, Waltham and Oxford: Academic Press.

Kuderer, M., S. Gulati, and W. Burgard. 2015. Learning driving styles for autonomous vehicles from demonstration. In *2015 IEEE International Conference on Robotics and Automation (ICRA)*. https://doi.org/10.1109/icra.2015.7139555.

Langley, C., B.-I. Cîrstea, F. Cuzzolin, and B.J. Sahakian. 2022. Editorial: Theory of mind in humans and in machines. *Frontiers in Artificial Intelligence* 5. https://doi.org/10.3389/frai.2022.917565.

Liang, Z., H. Chen, J. Zhu, K. Jiang, and Y. Li. 2018. Adversarial deep reinforcement learning in portfolio management. In *arXiv [q-fin.PM]*. arXiv. http://arxiv.org/abs/1808.09940.

Liu, C., S.-Y. Liu, E.L. Carano, and J. Karl Hedrick. 2014. A framework for autonomous vehicles with goal inference and task allocation capabilities to support peer collaboration with human agents. In *Volume 2: Dynamic modeling and diagnostics in biomedical systems; Dynamics and control of wind energy systems; Vehicle energy management optimization; Energy storage, optimization; Transportation and grid applications; Estimation and identification methods, tracking, detection, alternative propulsion systems; Ground and space vehicle dynamics; Intelligent transportation systems and control; Energy harvesting; Modeling and control for thermo-fluid applications, IC engines, manufacturing*. https://doi.org/10.1115/dscc2014-6262.

Liu, C., J.B. Hamrick, J.F. Fisac, A.D. Dragan, J. Karl Hedrick, S. Shankar Sastry, and T.L. Griffiths. 2018. Goal inference improves objective and perceived performance in human-robot collaboration. In *arXiv [cs.RO]*. arXiv. http://arxiv.org/abs/1802.01780.

Losey, D.P., H.J. Jeon, M. Li, K. Srinivasan, A. Mandlekar, A. Garg, J. Bohg, and D. Sadigh. 2022. Learning latent actions to control assistive robots. *Autonomous Robots* 46 (1): 115–147.

Luceri, L., S. Giordano, and E. Ferrara. 2020. Detecting troll behavior via inverse reinforcement learning: A case study of Russian Trolls in the 2016 US Election. *Proceedings of the International AAAI Conference on Web and Social Media* 14: 417–427. https://doi.org/10.1609/icwsm.v14i1.7311.

Ma, A., K.K. Lau, and D. Thyagarajan. 2020. Voice changes in Parkinson's disease: What are they telling us? *Journal of Clinical Neuroscience: Official Journal of the Neurosurgical Society of Australasia* 72: 1–7.

Maass, W. 1997. Networks of spiking neurons: The third generation of neural network models. *Neural Networks* 10 (9): 1659–1671. https://doi.org/10.1016/s0893-6080(97)00011-7.

Markram, H., J. Lübke, M. Frotscher, and B. Sakmann. 1997. Regulation of synaptic efficacy by coincidence of postsynaptic APs and EPSPs. *Science* 275 (5297): 213–215.

Meilán, J.J.G., F. Martínez-Sánchez, I. Martínez-Nicolás, T.E. Llorente, and J. Carro. 2020. Changes in the rhythm of speech difference between people with nondegenerative mild cognitive impairment and with preclinical dementia. *Behavioural Neurology* 2020: 4683573.

Mnih, V., K. Kavukcuoglu, D. Silver, A.A. Rusu, J. Veness, M.G. Bellemare, A. Graves, M. Riedmiller, A.K. Fidjeland, G. Ostrovski, S. Petersen, C. Beattie, A. Sadik, I. Antonoglou, H. King, D. Kumaran, D. Wierstra, S. Legg, and D. Hassabis. 2015. Human-level control through deep reinforcement learning. *Nature* 518 (7540): 529–533.

Nakahashi, R., C. Baker, and J. Tenenbaum. 2016. Modeling human understanding of complex intentional action with a Bayesian nonparametric subgoal model. *Proceedings of the AAAI Conference on Artificial Intelligence* 30 (1). https://doi.org/10.1609/aaai.v30i1.9889.

Ng, A.Y., S. Russell, and Others. 2000. Algorithms for inverse reinforcement learning. *ICML 1*: 2.

O'Laughlin, C., and P. Thagard. 2000. Autism and coherence: A computational model. *Mind and Language* 15 (4): 375–392. https://doi.org/10.1111/1468-0017.00140.

Patacchiola, M., and A. Cangelosi. 2022. A developmental cognitive architecture for trust and theory of mind in humanoid robots. *IEEE Transactions on Cybernetics* 52 (3): 1947–1959.

Petković, T., I. Marković, and I. Petrović. 2018. Human intention recognition in flexible robotized warehouses based on Markov decision processes. *ROBOT 2017: Third Iberian Robotics Conference*, 629–640.

Petković, T., D. Puljiz, I. Marković, and B. Hein. 2019. Human intention estimation based on hidden Markov model motion validation for safe flexible robotized warehouses. *Robotics and Computer-Integrated Manufacturing* 57: 182–196.

Piccinini, G., and S. Bahar. 2013. Neural computation and the computational theory of cognition. *Cognitive Science* 37 (3): 453–488.

Qiu, L., Y. Zhao, Y. Liang, P. Lu, W. Shi, Z. Yu, and S.-C. Zhu. 2022. Towards socially intelligent agents with mental state transition and human value. In *Proceedings of the 23rd Annual Meeting of the Special Interest Group on Discourse and Dialogue*, 146–158.

Rabinowitz, N., F. Perbet, F. Song, C. Zhang, S.M.A. Eslami, and M. Botvinick. 2018. Machine theory of mind. In *Proceedings of the 35th international conference on machine learning*, ed. J. Dy and A. Krause, vol. 80, 4218–4227. PMLR. https://proceedings.mlr.press/

Ramachandran, D., and E. Amir. 2007. *Bayesian inverse reinforcement learning*. https://www.aaai.org/Papers/IJCAI/2007/IJCAI07-416.pdf.

Reddy, S., A.D. Dragan, and S. Levine. 2018. Where do you think you're going?: Inferring beliefs about dynamics from behavior. In *arXiv [cs.LG]*. arXiv. https://proceedings.neurips.cc/paper/2018/hash/6f2268bd1d3d3ebaabb04d6b5d099425-Abstract.html.

Roman, H., H.R. Roman, Y. Bisk, J. Thomason, A. Celikyilmaz, and J. Gao. 2020. RMM: A recursive mental model for dialogue navigation. *Findings of the Association for Computational Linguistics: EMNLP 2020*. https://doi.org/10.18653/v1/2020.findings-emnlp.157.

Russell, S. 1998. Learning agents for uncertain environments (extended abstract). *Proceedings of the Eleventh Annual Conference on Computational Learning Theory – COLT' 98*. https://doi.org/10.1145/279943.279964.

Samoili, S., M.L. Cobo, E. Gomez, G. De Prato, F. Martinez-Plumed, and B. Delipetrev. 2020a. *AI Watch. Defining artificial intelligence. Towards an operational definition and taxonomy of artificial intelligence*. https://eprints.ugd.edu.mk/28047/.

Samoili, S., M. Lopez Cobo, E. Gomez Gutierrez, G. de Prato, F. Martinez-Plumed, and B. Delipetrev. 2020b. *AI watch: Defining artificial intelligence: Towards an operational definition and taxonomy of artificial intelligence*. Luxembourg: Publications Office of the European Union.

Schlaffke, L., S. Lissek, M. Lenz, G. Juckel, T. Schultz, M. Tegenthoff, T. Schmidt-Wilcke, and M. Brüne. 2015. Shared and nonshared neural networks of cognitive and affective theory-of-mind: a neuroimaging study using cartoon picture stories. *Human Brain Mapping* 36 (1): 29–39.

Schurz, M., J. Radua, M. Aichhorn, F. Richlan, and J. Perner. 2014. Fractionating theory of mind: A meta-analysis of functional brain imaging studies. *Neuroscience and Biobehavioral Reviews* 42: 9–34.

Schurz, M., J. Radua, M.G. Tholen, L. Maliske, D.S. Margulies, R.B. Mars, J. Sallet, and P. Kanske. 2021. Toward a hierarchical model of social cognition: A neuroimaging meta-analysis and integrative review of empathy and theory of mind. *Psychological Bulletin* 147 (3): 293–327. https://doi.org/10.1037/bul0000303.

Shum, M., M. Kleiman-Weiner, M.L. Littman, and J.B. Tenenbaum. 2019. Theory of minds: Understanding behavior in groups through inverse planning. *Proceedings of the AAAI Conference on Artificial Intelligence* 33 (01): 6163–6170. https://doi.org/10.1609/aaai.v33i01.33016163.

Silver, D., A. Huang, C.J. Maddison, A. Guez, L. Sifre, G. van den Driessche, J. Schrittwieser, I. Antonoglou, V. Panneershelvam, M. Lanctot, S. Dieleman, D. Grewe, J. Nham, N. Kalchbrenner, I. Sutskever, T. Lillicrap, M. Leach, K. Kavukcuoglu, T. Graepel, and D. Hassabis. 2016. Mastering the game of Go with deep neural networks and tree search. *Nature* 529 (7587): 484–489.

Silver, D., T. Hubert, J. Schrittwieser, I. Antonoglou, M. Lai, A. Guez, M. Lanctot, L. Sifre, D. Kumaran, T. Graepel, T. Lillicrap, K. Simonyan, and D. Hassabis. 2018. A general reinforcement learning algorithm that masters chess, shogi, and go through self-play. *Science* 362 (6419): 1140–1144.

Spreng, R.N., R. Nathan Spreng, R.A. Mar, and A.S.N. Kim. 2009. The common neural basis of autobiographical memory, prospection, navigation, theory of mind, and the default mode: A quantitative meta-analysis. *Journal of Cognitive Neuroscience* 21 (3): 489–510. https://doi.org/10.1162/jocn.2008.21029.

Tauber, S., and M. Steyvers. 2011. Using inverse planning and theory of mind for social goal inference. *Proceedings of the Annual Meeting of the Cognitive Science Society* 33. https://escholarship.org/content/qt7f34d3pz/qt7f34d3pz.pdf.

Thagard, P. 1992. Adversarial problem solving: Modeling an opponent using explanatory coherence. *Cognitive Science* 16 (1): 123–149.

Trott, S., C. Jones, T. Chang, J. Michaelov, and B. Bergen. 2022. Do large language models know what humans know? In *arXiv [cs.CL]*. arXiv. http://arxiv.org/abs/2209.01515.

Ullman, T., C. Baker, O. Macindoe, O. Evans, N. Goodman, and J. Tenenbaum. 2009. Help or hinder: Bayesian models of social goal inference. *Advances in Neural Information Processing Systems* 22. https://proceedings.neurips.cc/paper/2009/hash/52292e0c763fd027c6eba6b8f494d2eb-Abstract.html.

Vanschoren, J. 2018. Meta-learning: A survey. In *arXiv [cs.LG]*. arXiv. http://arxiv.org/abs/1810.03548.

Vickery, R.M. 2007. Mind the neuron! The role of the single neuron in a theory of mind. *Acta Neuropsychiatrica* 19 (3): 177–182. https://doi.org/10.1111/j.1601-5215.2007.00206.x.

Wang, Q., K. Saha, E. Gregori, D. Joyner, and A. Goel. 2021. Towards mutual theory of mind in human-AI interaction: How language reflects what students perceive about a virtual teaching assistant. In *Proceedings of the 2021 CHI Conference on Human Factors in Computing Systems*, Article 384.

Warwick, K., and H. Shah. 2016. Passing the Turing test does not mean the end of humanity. *Cognitive Computation* 8: 409–419.
Williams, J., S.M. Fiore, and F. Jentsch. 2022. Supporting artificial social intelligence with theory of mind. *Frontiers in Artificial Intelligence* 5: 750763.
Wimmer, H., and J. Perner. 1983. Beliefs about beliefs: Representation and constraining function of wrong beliefs in young children's understanding of deception. *Cognition* 13 (1): 103–128.
Winfield, A.F.T. 2018. Experiments in artificial theory of mind: From safety to storytelling. *Frontiers in Robotics and AI* 5: 75.
Wu, S.A., R.E. Wang, J.A. Evans, J.B. Tenenbaum, D.C. Parkes, and M. Kleiman-Weiner. 2021. Too many cooks: Bayesian inference for coordinating multi-agent collaboration. *Topics in Cognitive Science* 13 (2): 414–432.
Yang, G.-Z., J. Bellingham, P.E. Dupont, P. Fischer, L. Floridi, R. Full, N. Jacobstein, V. Kumar, M. McNutt, R. Merrifield, B.J. Nelson, B. Scassellati, M. Taddeo, R. Taylor, M. Veloso, Z.L. Wang, and R. Wood. 2018. The grand challenges of science robotics. *Science Robotics* 3 (14). https://doi.org/10.1126/scirobotics.aar7650.
Yoshida, W., R.J. Dolan, and K.J. Friston. 2008. Game theory of mind. *PLoS Computational Biology* 4 (12): e1000254.
Zeng, Y., Y. Zhao, T. Zhang, D. Zhao, F. Zhao, and E. Lu. 2020. A brain-inspired model of theory of mind. *Frontiers in Neurorobotics* 14: 60.
Zhao, Z., E. Lu, F. Zhao, Y. Zeng, and Y. Zhao. 2022a. A brain-inspired theory of mind spiking neural network for reducing safety risks of other agents. *Frontiers in Neuroscience* 16: 753900.
Zhao, Z., F. Zhao, Y. Zhao, Y. Zeng, and Y. Sun. 2022b. Brain-inspired theory of mind spiking neural network elevates multi-agent cooperation and competition. *SSRN Electronic Journal*. https://doi.org/10.2139/ssrn.4271099.
Zhi-Xuan, T., J. Mann, T. Silver, J. Tenenbaum, and V. Mansinghka. 2020. Online Bayesian goal inference for boundedly rational planning agents. *Advances in Neural Information Processing Systems* 33: 19238–19250.

Chapter 23
Theory of Mind in Artificial Intelligence Applications

Alvaro Garcia-Lopez

Abstract Theory of Mind refers to the ability to understand and predict other people's thoughts, emotions, and intentions, and is essential for successful social interaction. Artificial intelligence can be a useful tool to enhance theory of mind therapy by providing an interactive platform for practicing communication and empathy skills, as well as assessing and monitoring progress in theory of mind skills. Mental health apps can provide personalised emotional support and guided conversations to help people develop theory of mind skills and improve communication and interpersonal relationships. In addition, these apps can also offer the advantage of accessibility and privacy, allowing people to access Theory of Mind therapy anytime, anywhere via their mobile devices. The combination of theory of mind and artificial intelligence can be a valuable tool to improve mental health therapy and increase accessibility to care. It is important that these applications continue to be researched and developed to ensure their effectiveness and safety in the treatment of mental health problems.

Keywords Theory of mind · Artificial intelligence · Mental health applications · Accessibility

23.1 Theory of Mind

Theory of mind (ToM) is a concept widely studied in psychology and neuroscience and refers to people's ability to infer mental states in other people, such as beliefs, desires, emotions, and intentions.

A clear example of the ideas behind ToM is the situation you are in: if we see people crying at a funeral, we can infer that they are sad and, given the situation, that it is related to the death of a loved one. We can automatically imagine the sadness that a person is experiencing because we simulate the emotion of grief in our brain.

A. Garcia-Lopez (✉)
Department of Applied Mathematics, Science and Technology of Materials and Electronic Engineering, University Rey Juan Carlos, Madrid, Spain

T. Lopez-Soto et al. (eds.), *The Theory of Mind Under Scrutiny*, Logic, Argumentation & Reasoning 34, https://doi.org/10.1007/978-3-031-46742-4_23

Another example is when you see a child playing with friends, you can infer that the child is enjoying the activity and feeling happy. If you don't have a good ToM, you might just think that the child is playing without considering his or her own feelings. ToM is essential to our ability to understand other people's thoughts and feelings and enables us to interact effectively and empathetically with others. Through ToM we can infer the mental states of other people from their actions, words, and emotions.

In the late 1970s, psychologists David Premack and Guy Woodruff coined the term ToM in an article on chimpanzee learning. In this study, the researchers showed chimpanzees videos of humans dealing with both simple problems, such as not getting to food, and complex problems, such as not being able to get out of a locked cage. These videos presented the chimpanzee with several pictures, only one of which was the solution. From these tests they concluded that the chimpanzees did not appear to have the ability to understand mental states of themselves or others (Premack and Woodruff 1978).

Later studies focused their efforts on finding the first manifestations of ToM in children's behaviour using the false belief task with the Maxi Task Test (Wimmer and Perner 1983). This study suggests that the understanding of false beliefs is an important component of children's cognitive development and is central to their understanding of lying and deception. In addition, this study provides information about how the understanding of false beliefs is related to children's ability to intentionally deceive others. Other studies have focused on children with disabilities such as Down syndrome and autism, suggesting that children with autism spectrum disorder (ASD) have difficulty understanding the perspective of others and may have a reduced capacity for symbolic play to understand the perspective of others and ToM, and linking this to the social and communication difficulties experienced by people with ASD (Baron-Cohen et al. 1985; Baron-Cohen 1987; Leslie 1988; Perner et al. 1989).

These tests led to the famous Sally-Anne experiment (SAe), involving two dolls, Sally and Anne. Sally has a basket and Anne has a box. The subject sees Sally put a ball in her basket and then leave. While Sally is gone, Anne takes the ball out of the basket and puts it in her box. The subject is then asked where Sally will look for the ball when she returns. This sequence is shown in Fig. 23.1.

During the 1990s, ToM attracted many studies that made significant advances in the field. In general, studies focused on understanding ToM in children and adults and its relationship to cognitive and emotional development, providing valuable insights into understanding ToM and its role in human progress. ToM studies have tended to focus on younger subjects or younger mental ages.

(Gopnik and Wellman 1992) present an innovative perspective on how children acquire knowledge and how this process relates to a ToM that is like scientific theories and is revised and modified as the child acquires more information about people and their minds, even if they have only a vague understanding. This phenomenon is already observed in infants, who can use social information to interpret and understand situations and events around them, using the emotions and reactions of others as a guide for evaluating new situations and stimuli (Watson et al. 1999). Logically, in the early years of our lives we learn about everything, including ToM, but as we get older, this knowledge deteriorates, reducing our effectiveness in

Fig. 23.1 Sally Anne Test version from U. Frith, 1989, Blackwell Publishers, p. 83. Copyright 1989, Blackwell Publishers

non-mental content processes, but, in ToM-related processes, this skill not only remains intact but can be improved (Happé et al. 1998; Goldstein and Winner 2012). On the other hand, contemporary studies have shown that ToM also depends on the family (Cutting and Dunn 1999) and cultural environment (Vinden 1999).

Extensive and prolific research conducted in the field of ToM has significantly contributed to the development and understanding of this cognitive construct. Over the years, numerous studies and investigations have been conducted, leading to a wealth of knowledge and empirical evidence that has solidified the foundations of ToM.

The accumulation of research findings has not only deepened our understanding of ToM but has also resulted in the creation of standardized tests that have become fundamental tools in subsequent research. These tests, meticulously designed and validated through rigorous scientific procedures, have provided researchers with reliable and valid measures to assess individuals' ToM abilities.

These tests have become standardized tests along the years and have proven to be invaluable in various domains, including psychology, cognitive neuroscience, developmental psychology, and social sciences. By establishing standardized tests as benchmarks in ToM research, scientists have been able to compare findings across different studies, leading to a better understanding of the underlying mechanisms and processes involved in ToM. The main standardised tests are *Strange Stories*

Test, *Hint Task Test*, *Intention Inference Task Test*, *Reading the Mind in the Eyes Test* and *Faux Pas Test*, which are described below.

The Strange Stories Test is designed to assess an individual's ability to interpret and understand the thoughts, feelings, and intentions of other people, as well as their ability to understand the social nuances of different situations. Participants answer questions about what the characters in the stories are thinking or feeling, based on their understanding of the social situation. Some examples are shown in Table 23.1. The volunteer is asked to interpret the mental states of the characters involved due to unexpected or ambiguous social situations in the stories (Happé 1994). This study was conducted by (Jolliffe and Baron-Cohen 1999) to replicate previous research that aimed to distinguish between two groups of individuals with high-functioning-ASD: those diagnosed with Asperger syndrome and those who were not.

The Hint Task Test is used to measure the ability to attribute mental states (such as beliefs, desires, and intentions) to oneself and others, and to assess an individual's ability to understand indirect, socially nuanced communication. It involves a series of short scenarios in which one person, the 'hint giver', communicates a message to

Table 23.1 Examples of Strange Stories shown in (Happé 1994)

The Mentalistic stories
Banana
Katie and Emma arc playing in the house. Emma picks up a banana from the fruit bowl and holds it up to her car. She says to Katie, "look! This banana is a telephone!"
Is it true what Emma says?
Why does Emma say this?
Picnic
Sarah and tom arc going on a picnic. It is Tom's idea, he says it is going to be a lovely sunny day for a picnic. But just as they are unpacking the food, it starts to rain, and soon they are both soaked to the skin. Sarah is cross. She says, "oh yes, a lovely day for a picnic alright!"
Is it true, what Sarah says?
Why does she say this?
The physical stories
Army
Two enemy powers have been at war for a very long time. Each army has won several battles, but now the outcome could go either way. The forces are equally matched. However, the blue army is stronger than the yellow army in foot soldiers and artillery. But the yellow army is stronger than the blue army in air power.
On the day of the final battle, which will decide the outcome of the war, there is a heavy fog over the mountains where the fighting is about to occur. Low-lying clouds hang above the soldiers. By the end of the day the blue army have won.
Q: Why did the blue army win?
Glasses
Sarah is very long-sighted. She has only one pair of glasses, which she keeps losing. Today she has lost her glasses again and she needs to find them. She had them yesterday evening when she looked up the television programmes. She must have left them somewhere that she has been today. She asks ted to find her glasses.
She tells him that today she went to her regular early morning keep fit class, then to the post office, and last to the flower shop. Ted goes straight to the post office.
Q: Why is the post office the most likely place to look?

Table 23.2 Examples of Hint Task shown in (Corcoran et al. 1995)

One of the ten hinting task items
Paul must go to an interview and he's running late. While he's cleaning his shoes he says to his wife, Jane: 'I want to wear that blue shirt, but it's very creased.'
Question: What does Paul really mean when he says this?
Extra information: Paul goes on to say: 'It's in the ironing basket.'
Question: What does Paul want Jane to do?

another person, the 'hint receiver', using indirect language or social cues, an example is shown in Table 23.2. The hint receiver is then asked to explain what the hint giver meant by the message. The test measures the hint receiver's ability to understand the hint giver's perspective and intentions, and to use this information to make inferences about what the hint giver is trying to communicate (Corcoran et al. 1995).

This test is most used in ASD, schizophrenia, and traumatic brain injury. This test has shown that individuals with Asperger Syndrome (AS) and normal intellectual abilities struggle to use mental state vocabulary appropriately in different contexts (Kaland et al. 2005), although it was found that while individuals with ASD showed impairments in both theory of mind and rule use, there was no significant correlation between these two abilities (Zelazo et al. 2002). People in the prodromal (early) stage of schizophrenia may have slight impairments in their social cognition abilities, and these difficulties may be related to dysfunction in the prefrontal cortex of the brain (Chung et al. 2008), and have subtle impairments in their ToM abilities (Pickup 2006).

The *Intention Inference Task Test* is also used to test the decoding ability of patients with schizophrenia. It consists of 30 comic strips, each of which presents a scenario in which a character performs a very simple action, indicating the intention behind it, shown in Table 23.3. Volunteers are shown the story and asked to quickly choose which of the three answer cards is the most logical to complete the comic strip sequence. Only one of the three cards is correct. Participants are asked to infer the character's intention from the behaviour and to choose from a set of response options that describe the character's likely intention (Sarfati et al. 1997).

People with schizophrenia tend to have a low level of awareness and understanding of their illness, including its implications and consequences. This lack of insight typically persists throughout the course of the illness and can have a significant impact on the way the illness progresses and is managed (David 1990; Gil 2017). This type of test is also widely used to test ToM in people with Asperger's syndrome and has helped us to understand that children and young people with Asperger's syndrome performed better on physical state tasks but struggled with mental state tasks. They tended to interpret behaviour literally, choosing physical explanations even when mental state responses were more appropriate (Kaland et al. 2002).

(Ekman and Friesen 1976) laid the foundations of emotional recognition by defining 6 basic emotions (sadness, anger, surprise, fear, disgust, surprise, and happiness) that are recognizable regardless of culture or cognitive interpretation, thus creating the basic theory of emotions. To assess these emotions and knowing

Table 23.3 Examples of Intention Inference Task Test shown in (Sarfati et al. 1997)

Story	Answer cards
 C	 3
 B	 2

that each basic emotion is associated with a unique facial expression, they developed the Pictures of Facial Affect Test (POFA). Based on this earlier research, (Baron-Cohen et al. 1997) defined the Reading the Mind in the Eyes Test (RMET). This test measures the ability to infer complex mental states from simple visual cues, not only emotions but also mental states. In this test, participants are presented with a series of photographs of the eye region of the face and asked to choose which of four words best describes the mental state or emotion conveyed in the photograph. The words presented as response options typically include a range of 10 basic emotions and mental states such as happiness, sadness, fear, anger, surprise, and others, as shown in Table 23.4. However, it was subsequently revised to assess complex mental states, and the set was expanded to include 36 facial expression images.

The aim of the RMET is to determine whether the participant is able to put themselves in the other person's shoes and 'tune into' their mental state, rather than the ability to understand the thoughts and intentions of others, and provides evidence for the hypothesis that individuals with ASD have deficits in recognising and expressing emotions, also known as alexithymia (Oakley et al. 2016).

The *Faux Pas Test* measures more complex affective-mental concepts such as an individual's ability to perceive and understand social cues and norms. It is used in clinical and research settings to assess social cognition, particularly in individuals with ASD. The test consists of 10 short stories or scenarios, each of which involves a social 'faux pas' or unfortunate mistake in a social interaction situation. The questions are designed to measure different aspects of social cognition, including perspective taking, theory of mind, empathy and emotional intelligence, as show examples in Table 23.5 (Stone et al. 1998).

Some high impact studies could be (Baron-Cohen et al. 1999) who used the *Faux Pas Test* in a sample of children with Asperger Syndrome or high functioning autism

Table 23.4 Examples of reading the eyes in the mind test shown in (Baron-Cohen et al. 1997)

Concerned vs. Unconcerned
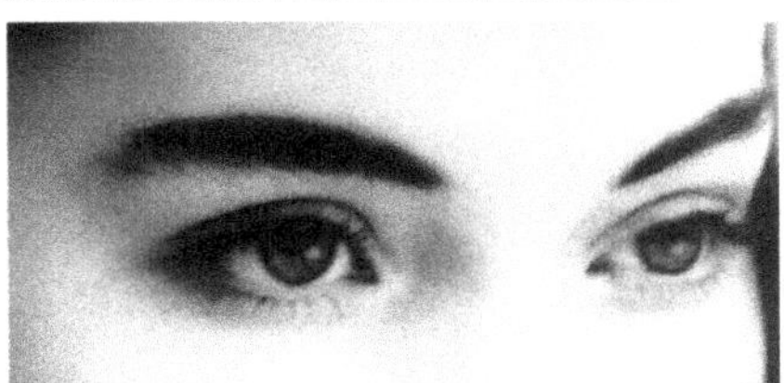
Reflective vs. unreflective
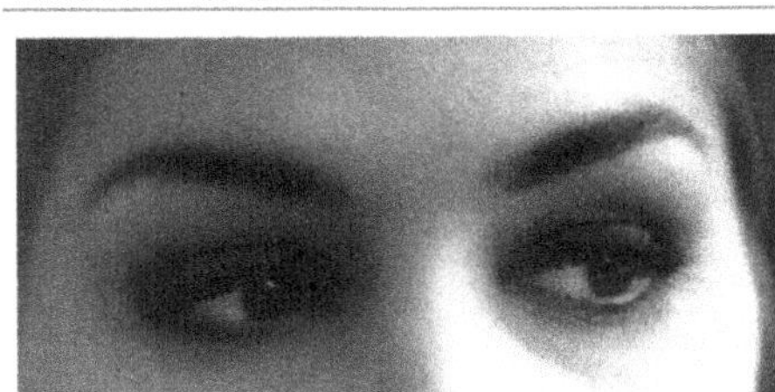

Table 23.5 Examples of Faux Pas shown in (Stone et al. 1998)

Faux Pas 1	Faux Pas 2
Jeanette bought her friend Anne a crystal bowl for a wedding gift. Anne had a big wedding and there were a lot of presents to keep track of. About a year later, Jeanette was over one night at Anne's for dinner. Jeanette dropped a wine bottle by accident on the crystal bowl, and the bowl shattered. "I'm really sorry, I've broken the bowl," said Jeanette. "Don't worry," said Anne, "I never liked it anyway. Someone gave it to me for my wedding."	Helen's husband was throwing a surprise party for her birthday. He invited Sarah, a friend of Helen's, and said, "Don't tell anyone, especially Helen." the day before the party, Helen was over at Sarah's, and Sarah spilled some coffee on a new dress that was hanging over her chair. "Oh!" said Sarah, "I was going to wear this to your party!" "what party?" said Helen. "Come on," said Sarah, "Let's go see if we can get the stain out."
Questions	
1. Did someone say something they shouldn't have said? (tests for detection of FP.) 2. Who said something they shouldn't have said? (tests for understanding of FP.) 3. Why shouldn't they have said it? (requires understanding mental state of listener.) 4. Why did they say it? (requires understanding mental state of speaker.) 5. An example is, what had Jeanette given Anne for her wedding? (control question that asks about some detail of the story.)	

and concluded that they show greater difficulty in performing the tasks compared to first or second order false belief tasks, although they develop cognitive strategies based on extensively learned abstract knowledge of societal norms to compensate for any difficulties they may have in understanding and applying social rules in real life situations (Zalla et al. 2009).

There are also mixtures of these tests, such as (Chung et al. 2008), who reported a deficit in Tom in the healthy group and showed better results on executive and working memory tests, but not on verbal memory tests using the ST and False Believe task. Using the Strange Stories test and silent films, (Devine and Hughes 2013) found that mentalising skills, or the ability to understand and interpret other people's thoughts and feelings, continued to improve throughout middle childhood, even after accounting for verbal ability and family affluence.

On the other hand, (Speck et al. 2010) argue that only the Faux-Pas Test and the Strange Stories Test are valuable instruments for clinical use because of the everyday social situations they represent.

Other non-invasive techniques used to advance ToM research include electroencephalography (EEG) and functional magnetic resonance imaging (fMRI). Both are used to measure and record electrical activity in the brain. EEG measures the changes in electrical potential, that occur between different points in the brain as it performs an activity and detects large groups of neurons that are activated at the same time. The recording of a person's brain activity makes it possible to observe the functional and electrical activity changes of the central nervous system, so that the recorded brain signals contain psychophysiological information useful for the study of human behaviour (Zhang et al. 2020). During EEG recording, these potential changes are measured continuously directly on the surface of the scalp, using a set of electrodes arranged on a headband or cap. The electrodes are defined in

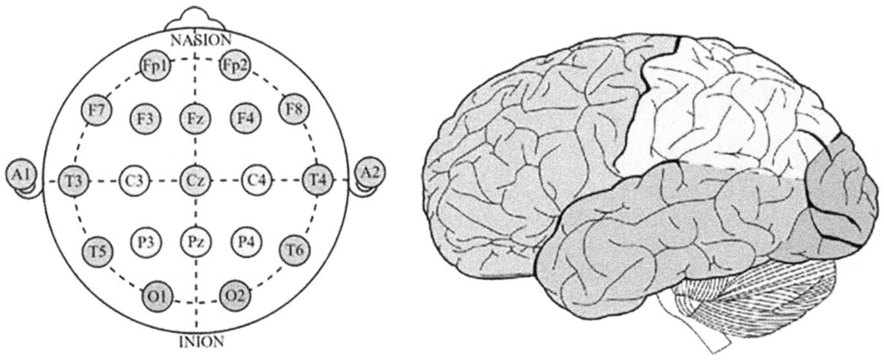

Fig. 23.2 EEG Electrode 10–20 positioning system. Fp (pre-frontal or frontal pole), F (frontal), C (central line of the brain), T (temporal), P (parietal), and O (occipital), A (Mastoids/Earlobes for references)

a standard configuration, shown in Fig. 23.2, to facilitate reproducibility and comparability of studies, following the guidelines of the 10–20 positioning system for electrode placement (Sharbrough et al. 1991). EEG signals respond more sensitively and in real time to fluctuations in affective states, so they can provide useful hints about emotional states. Its sensitivity is such that it can record noise, unwanted signals mixed with other types of signals such as eye movement, electro-oculogram (EOG), muscular activity, electromyography (EMG), cardiovascular activity, or electrocardiographic activity (ECG). The EEG has a high temporal resolution, in the order of ms, which detects very rapid changes in potential and allows frequency analysis of the signals to separate them into different brain wave bands.

Using EEG and a false belief task, it was found that healthy adults were faster and more accurate at identifying false beliefs from a self-oriented perspective than from an other-oriented perspective, and that perspective shifting played a critical role in participants' ability to identify false beliefs. EEG recordings showed significant interactions between perspective-shifting areas, indicating the key role of these areas in theory of mind engagement (Bradford et al. 2018). Recording brain activity during RMET, has found that the level of activation in the mid-frontal region of the right hemisphere was a predictor of task performance (Sabbagh and Flynn 2006), besides, the dorsal medial prefontal cortex and the tight temporal-parietal junction play an important role in the development of theory of mind abilities in young children, as shown by greater levels of alpha waves in preschoolers (Sabbagh et al. 2009). EEG in ToM Research related to ASD has also found deficits in ToM, (Lincoln et al. 1993) suggesting that neurobiological bases of ASDdue to a novel stimulus show a lower response in the P3 component; this may be related to activations in the left frontal lobe irregularities with social cognitive deficits (Sabbagh and Taylor 2000); moreover, mu wave suppression supports a disordered mirror neuron system shown in behavioural impairments (Oberman et al. 2005).

fMRI is a non-invasive brain imaging technique that detects differences in the magnetic properties of oxygenated and deoxygenated blood to identify brain regions

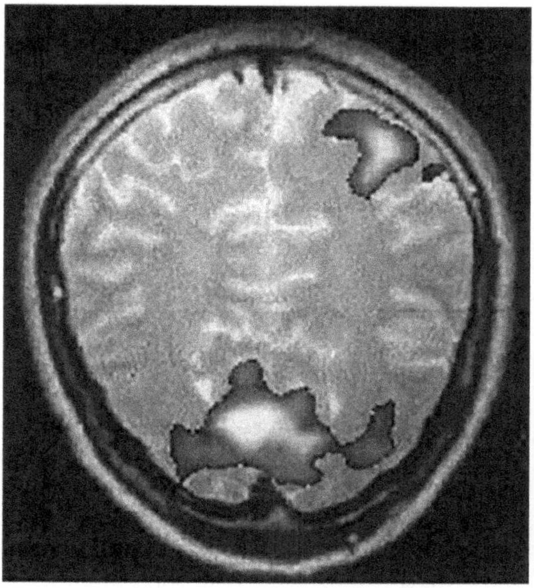

Fig. 23.3 SPM example. (Lindquist 2008)

that are active during a task or stimulus (Cohen et al. 2008). When neurons become active, they consume more oxygen, which increases the concentration of oxygenated blood in that brain region. This increased concentration of oxygenated blood changes the magnetic properties of the surrounding tissue, which can be detected by the fMRI equipment. These changes are then analysed, statistically processed and presented as statistical paramagnetic maps (SPM), Fig. 23.3, or three-dimensional maps of cognitive activation, which can help researchers and clinicians to identify the specific brain regions involved in different cognitive and behavioural processes. However, fMRI has relatively poor temporal resolution, as it takes several seconds to detect changes in blood flow, but its spatial resolution is high, allowing detailed mapping of brain functions (Lindquist 2008). The technique requires embedding a large and costly magnet within the equipment, which makes it space consuming and expensive.

(Gallagher et al. 2000) investigated the neural basis of Theory of Mind using both verbal and non-verbal tasks in an fMRI study and found that both tasks activated similar brain regions (medial prefrontal cortex and temporo-parietal junction), suggesting that these regions are important for understanding the mental states of others, regardless of the type of stimuli used. These regions have also been implicated in self-reference processes during the ToM task (Brüne et al. 2008). Young et al. (2010) also suggest that the same brain zones, adding precuneus, respond specifically to the content of mental states, regardless of whether this content is anticipated or not. Kobayashi et al. (2006) suggested that the understanding of ToM in adults may not be entirely consistent depending on culture and language, in line with Vinden (1999).

In summary, theory of mind has been developed by several researchers over the past decades. Through experimental research, neuroscience, and the use of various techniques and tests, we have made progress in understanding how people infer mental states in others and how this ability can be affected by psychological and neurological disorders.

The following section is an introduction to the relationship between artificial intelligence (AI) and ToM and its development to the present day.

23.2 ToM in Artificial Intelligence

One of the main goals of AI is to emulate human cognition. Once the AI branch is established, it is interesting to provide it with ToM skills to achieve a more natural language and for the machine to be able to reason and solve problems, make decisions, and have linguistic understanding, among other cognitive processes. This thinking began in the 1940s when McCulloch and Pitts (1943), based on the work of A. Turing (1936), −who was the inspiration behind the idea of an artificial brain-, suggested that the brain works in binary form, like electronic computers, so that the neurons of the brain could be considered as logical switches that are turned on or off, and that these could be connected in networks to perform computations. With this theory, they proposed a system for modelling the logical relationships between these neurons and their connections. With these ideas, Classical Computational ToM (CCToM) was born.

This theory was later recovered by Putnam (1967), who proposed a functionalist behaviour against logical behaviour, which associates mental states with behavioural patterns, and type-identity theory, which associates mental states with physical states. With this, Putnam proposed probabilistic automata, like those proposed by Turing, whose states, machine states, correspond to mental states, presenting a functional organisation and defining the tasks of each mental state, thus combining functionalism with CCToM. Later, Fodor, a disciple of Putnam, improved this theory by proposing the modularity hypothesis, which postulates that the mind is composed of a set of informationally encapsulated and domain-specific modules, responsible for different cognitive functions and emphasising symbolic reasoning (Fodor 1975; Fodor and Pylyshyn 1988), later arriving at a combination of the representational theory of mind (RTM), exposing the idea that the mind is a representational system, in which mental states have propositional content, together with CCToM (Fodor 1997), although these theories have long been criticised (Gallistel and King 2011; Horst 1999; Fodor 2000).

On the other hand, computationalists view the brain as a computer, composed of both physical and software components, where information is processed through symbolic representations (Gabbay et al. 2006), suggesting that the relationship between the neural, computational, and cognitive levels of explanation is not well understood (Gallistel and King 2011).

Another model of computational ToM (CToM) is based on heuristics. Several authors argue that heuristics are not only effective but necessary as cognitive tools to make decisions with high uncertainty (Gigerenzer 1996; Gigerenzer 2008; Gilovich et al. 2002; Åstebro and Elhedhli 2006). Heuristics are a combination of simple strategies for reformulating problems to make them easier to solve. Although heuristics are often seen as a non-computational alternative, they can help to deal with large amounts of information and interrelated variables in complex problems, as well as facilitating the determination of the relevance of decisions in an easier and more efficient way, allowing decision making in complex situations to be simplified using rules. This is particularly useful in situations of high uncertainty due to incomplete or ambiguous information, improving the speed of decision making, although optimal decisions are not always guaranteed and may be affected by biases or errors in certain contexts. Heuristics can be used to limit the amount of relevant information to be considered, thus avoiding exhaustive evaluations. This is supported by cognitive psychology and can explain how we efficiently determine relevance (Carruthers 2006; Carruthers et al. 2006; Samuels and Stich 2004; Samuels 2005; Samuels 2010).

Gigerenzer argues that heuristics should be valued as an adaptive and intelligent way of navigating the world. He defines three goals: (1) descriptive: the *adaptive toolbox*, in which analyse heuristics building blocks and evolved capacities; (2) perspective: *ecological rationality*, determining successful environmental structures; (3) design: create heuristics for decision making (Gigerenzer 2008).

In the early days of AI, uncertainty was a major challenge. Most decisions and reasoning take place under uncertain conditions, such as deciding whether to go on a picnic when it may or may not rain. Bayesian decision theory is the standard mathematical model for decision making and reasoning under uncertainty. Probabilities are used to represent uncertainty, and there are specific rules for updating probabilities as new information becomes available and for choosing actions based on probabilities and utilities (Thrun 2002; Murphy 2012; Misirli and Bener 2014; Muthuramalingam and Suba Nachiar 2016).

When CToM and AI are mixed, we can get a kind of emotional IA (EAI). First, we should explore the ethics of AI to minimise adverse effects on user interactions, privacy, consent, and the potential for emotional manipulation. Developers of EAI systems should consider these ethical implications and develop guidelines for the responsible use of emotional cues in these systems to ensure that these technologies are developed and used in ways that are fair, transparent, and beneficial to all users (Stark and Hoey 2021). In this way, Cominelli has tested a social EIA system. The system consists of three modules: perception and emotion recognition, emotional state representation and emotional decision making, the whole system is shown in Fig. 23.4. The results show promise for the creation of social and EAI systems capable of interacting more effectively with humans (Cominelli et al. 2018).

Samsonovich argues that by integrating emotional processing into AI systems, they can become more human-like and abler to interact with humans. He has also incorporated emotions into AI systems and defined a life cycle of a moral schema,

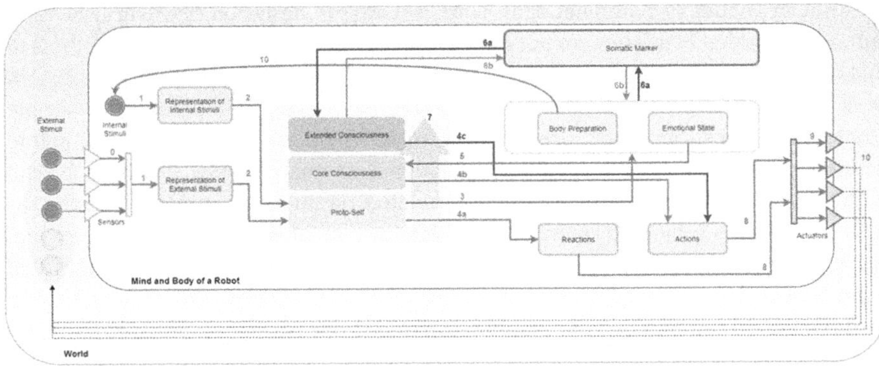

Fig. 23.4 Mind and Body of a robot system (Cominelli et al. 2018). Key numbers: External stimuli (0), Internal Stimuli (1), Reliable information selector (2), Emotion rule set (3), Behavioral rule set (4) composed by: standard behavioral rule set (4a), alternative behavioral set (4b) and specific behavioral rule set (4c); Feeling rule set (5), Somatic marker rule set (6), Reasoning rule set (7), Execution rule set (8), Robot animation's actuators (9), Internal stimuli feedback (10)

Fig. 23.5 Lifecycle example for an artificial moral schema. (Samsonovich 2020)

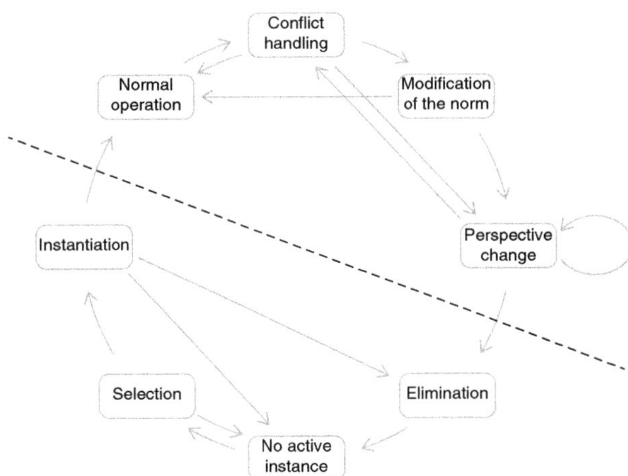

Fig. 23.5. The results show that it can successfully recognize, reason about, and respond to emotional states (Samsonovich 2020).

Using neural networks, a ToM neural network or ToMnet can be generated that can quickly and accurately predict an agent's future behaviour and mental state with only a few observations by developing a strong understanding of the agent's typical behaviour patterns. Rabinowitz et al. (2018) developed a machine ToM and tested it using SAe test variant and verified that the system learns ToM, but it needs further investigation. Using another model such as Instance-Based Learning (IBL), Nguyen and Gonzalez (2020) developed a Cognitive ToM, which is presented in Fig. 23.6. This shows a way to make inferences based on memory, using basic mental

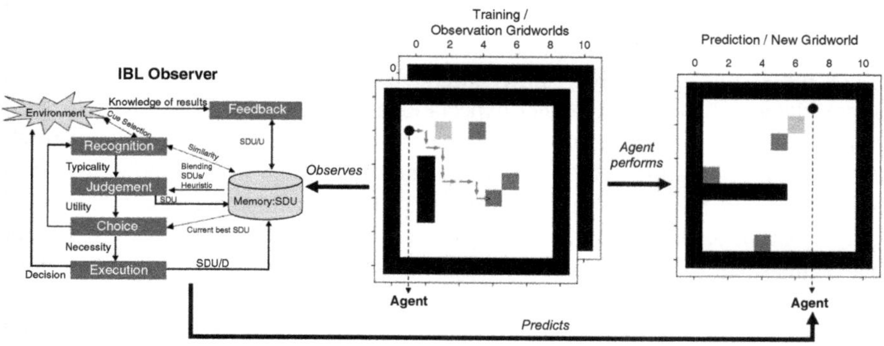

Fig. 23.6 Example of CogTom framework. (Nguyen and Gonzalez 2020)

processes supported on the principles of human cognition, in which it shows a good ability to predict actions, beliefs or false beliefs in novel situations.

Due to the lack of regulation, there are several ethical considerations and potential risks associated with the use of EAI. It is necessary to require informed consent, transparency, and accountability measures to address these concerns (McStay and Rosner 2021).

EAI in educational settings has the potential to benefit society, but it is important to carefully consider the ethical implications and ensure that these technologies are developed and deployed responsibly (McStay 2019). It has been shown to be useful in personalising learning and improving student collaboration (Luckin et al. 2016). The role of teachers is also changing, with technological support moving from simply providing information to helping students collaborate, discuss, and reflect on their learning activities (Popenici and Kerr 2017). AI for educational purposes has been shown to enhance the development of ideas (Li and Wang 2020) and increase student interest. A properly trained EAI could serve as an educational tutor or support for educational decision making (Hwang et al. 2020).

EAI can be used to detect low levels of ToM. Depression has been detected using natural language processing and/or machine learning (Deshpande and Rao 2017), severe mental illness (Zhang et al. 2023), bipolar disorder (Singh and Hamid 2022), Alzheimer's disease (Shen et al. 2014; Zheng et al. 2016; Sarica et al. 2017), schizophrenia (Veronese et al. 2013; Kambeitz et al. 2015; Meng et al. 2016; Xiao et al. 2017), anorexia nervosa (Guo et al. 2015), ASD (Heinsfeld et al. 2018; Li et al. 2018; Ghafouri-Fard et al. 2019; Kong et al. 2019), obsessive-compulsive disorder (Hoexter et al. 2013), and psychosis (Ramyead et al. 2016; Koutsouleris et al. 2018).

Other ways in which EAI can be implemented are in the field of robotics. If in the future it is intended to implement human robots that allow human interaction, it will be necessary to provide them with an Artificial ToM (AToM), thus improving communication and use by establishing a form of fluid and human communication, moving away from the valley of discomfort, proposing foundations to develop Artificial ToM (AToM) so that AI could predict and interpret social human-agent interaction (Scassellati 2002; Williams et al. 2022). Using the various ToM tests explained above, it was found that social robots have a limited ability to understand human social cues, so it is important to incorporate ToM into the development of social robots to improve their ability to understand human behaviour and respond appropriately (Banks 2020). Other expectations for the use of ToM robots include self-safety purposes, where robots can recognise that humans have limited knowledge and take appropriate safety measures, and storytelling, where robots are able to understand and respond to human emotions to create engaging stories (Winfield 2018). AIBO, Robot Entertainment Systems' first product model, is designed for play. This small robot simulates a puppy and exhibits both reflexive and conscious behaviour. Its agent architecture follows: (1) behaviour-based architecture, (2) randomness, (3) instinct/emotion, (4) learnability, (5) evolution, (6) movement; as well as image and sound processing, shown in Fig. 23.7 (Fujita 2001). People using this robot showed a similar relationship with a dog, so pets could be a valuable tool for the development of children or people with limited interaction, but careful consideration must be given to their design and implementation. (Kahn et al. 2004; Fujita 2004).

As we have seen, there is no consensus on the theories proposed for CToM, all of them being criticised by the defenders of the opposing theory or even by the original ideologists themselves, a clear sign of the evolution of this living branch of science. Whatever the theory, they all strive to propose the idea of an artificial brain whose mental processes work in a similar way to a computer, making decisions through mental algorithms. CToM provides a framework for artificially modelling human cognition and behaviour. It can be used to develop artificial models of mental processes, not just human ones, that give the artificial system intelligent behaviour such as problem solving, learning and decision making, and to inform the design of intelligent systems by providing information about the cognitive mechanisms underlying human behaviour.

In the next section, we will focus on direct applications that use AI to help identify ToM problems, provide emotional support, improve education, or adapt to user requests by making their responses more human.

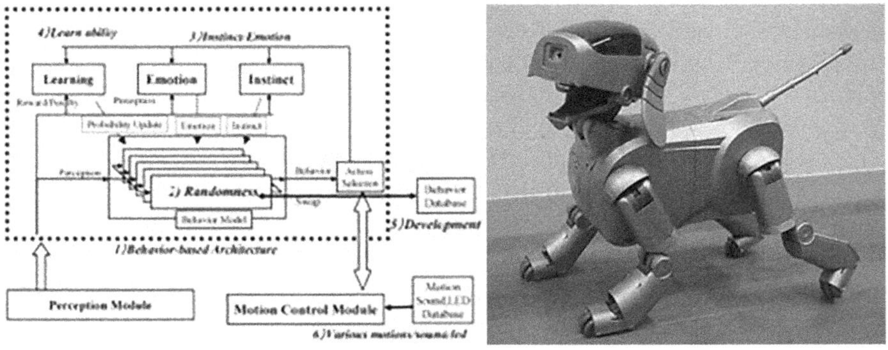

Fig. 23.7 AIBO agent architecture and AIBO ERS-110. (Fujita 2001)

23.3 ToM in Apps

One of the key factors driving the increased interest in ToM applications is the growing recognition of the limitations of traditional AI systems. While AI has made remarkable progress in many areas, it has struggled to replicate the complex and nuanced ways in which humans communicate and interact with each other. A key challenge is the ability to understand and interpret human emotions, which play a critical role in shaping our behaviour and relationships.

By incorporating Theory of Mind into AI systems, researchers hope to overcome these limitations by enabling machines to understand the mental states of human users and respond appropriately. This requires a deep understanding of human psychology and social interaction, as well as sophisticated algorithms and data processing techniques. Applications of ToM range from virtual assistants and chatbots that can respond to emotional cues in natural language, to social robots that can interpret non-verbal communication and respond in a more human-like way.

Beyond its applications in AI and robotics, ToM has important implications for fields such as psychology, cognitive science, and neuroscience. Understanding how we attribute mental states to others can shed light on the development of social cognition and the neural mechanisms that underpin it. It may also help us to better understand and treat disorders such as ASD, which are characterised in part by impairments in ToM.

Overall, the growing interest in applications of ToM represents an exciting new frontier in the development of AI and robotics, as well as a promising avenue for advancing our understanding of human psychology and social interaction, resulting in important tools to address mental health challenges (Yarrington et al. 2021).

In today's digital age, chat bots have become increasingly popular for interacting with users online. ToM can play an important role in improving the effectiveness and naturalness of communication between humans and chat bots. Chat bots that use ToM are designed to understand the user's intentions and desires, as well as interpret the user's emotional state. This allows the chat bot to respond in a more human and personalized way, which in turn can improve user satisfaction and the quality of the user experience. In addition, chat bots that use theory of mind can be particularly useful in situations where users need emotional support. For example, a chat bot that can detect stress and anxiety in the user and offer solutions or advice to reduce it could be of great help to someone experiencing emotional problems (Cameron et al. 2017; Grové 2021; Zhu et al. 2022). some examples of emotional recognition chat bots are shown below.

Woebot, Fig. 23.8, developed by a team of clinical and research psychologists at Stanford University, provides cognitive behavioural therapy through artificial intelligence technology that helps users identify and change negative thinking patterns and develop problem-solving and stress management skills. It had a 4.3/5 rating with 12.2 k reviews on Play Store at the time of consultation (fifth July 5, 2023).

It has been shown to have the potential to provide a scalable and accessible tool for women suffering from postpartum depression (Darcy et al. 2022), cognitive

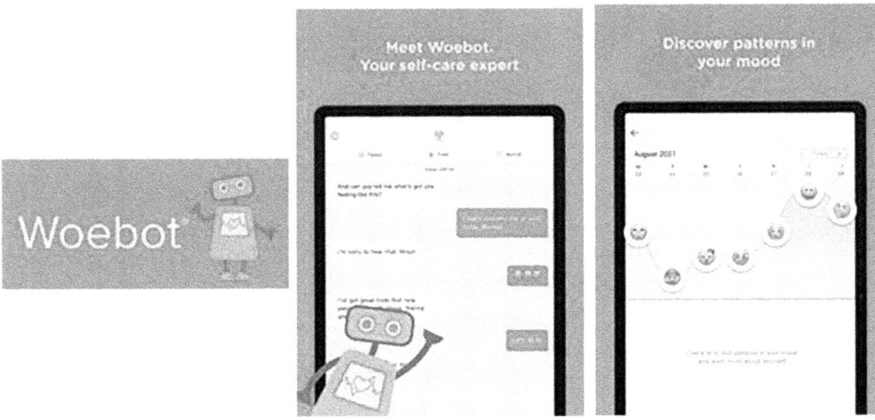

Fig. 23.8 Woebot commercial images from Play Store

behavioural therapy for young adults (Fitzpatrick et al. 2017), anxiety and depression (Li et al. 2021; Wasil et al. 2020), and substance use disorder treatment (Prochaska et al. 2021). These positive results, along with those from similar studies, demonstrate that this app is an effective tool for supporting mental wellbeing over time (Demirci 2018; Wan 2021).

Youper, Fig. 23.9, is a mental health mobile application that uses artificial intelligence to help people improve their emotional wellbeing. It uses cognitive behavioural therapy (CBT) techniques and theory of mind to help users understand and manage their emotions. It had a 4.1/5 rating with 49.1 k reviews on Play Store (fifth July 5, 2023).

This application has shown positive results in reducing levels of anxiety and/or depression (Alves et al. 2019; Li et al. 2021; Mehta et al. 2021; Wasil et al. 2020; Leo et al. 2022) and suicidal behaviour (Metts et al. 2022).

Wysa, Fig. 23.10, is designed to help users manage emotional and mental health issues such as anxiety, stress, depression, insomnia and more using CBT and mindfulness techniques. is empathetic, helpful, and never judgmental. It had a 4.7/5 rating with 140 k reviews on Play Store (fifth July 5, 2023).

Improvements in mental health have been shown in people with depression and/or anxiety (Inkster et al. 2018; Li et al. 2021; Wasil et al. 2020; Beatty et al. 2022; Iglesias et al. 2022), chronic pain (Gupta et al. 2022; Meheli et al. 2022) and pain interference (Iglesias et al. 2022).

Replika, Fig. 23.11, is presented as a conversational artificial intelligence that aims to simulate a conversation with a friend or therapist. The application uses natural language processing and machine learning techniques to create a replica of the user's personality and behavioural patterns to provide a personalised and realistic conversational experience. It had a 3.1/5 rating with 442 k reviews on Play Store (fifth July 5, 2023).

Users have been found to perceive Replika as a social companion and the application plays a role in their social and emotional wellbeing, providing social

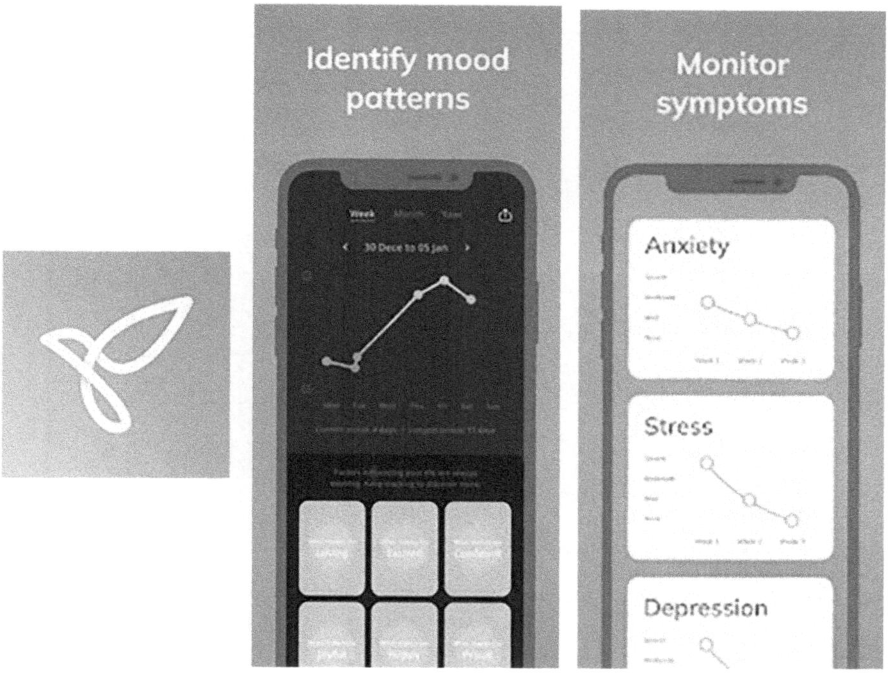

Fig. 23.9 Youper commercial images from Play Store

Fig. 23.10 Wysa commercial images from play store

and emotional support to users, particularly those who may suffer from social isolation or loneliness (Skjuve et al. 2022; Trothen 2022; Pentina et al. 2023).

On the other hand, Possati argues that the unconscious processes of the developers of this (these) tool(s) can themselves pervert the correct and clean functioning of the AI (Possati 2022).

Kasisto, Fig. 23.12, is an artificial intelligence-based chatbot application focused on delivering digital banking solutions through real-time customer interaction. It is primarily used in the financial and banking industries to improve efficiency and

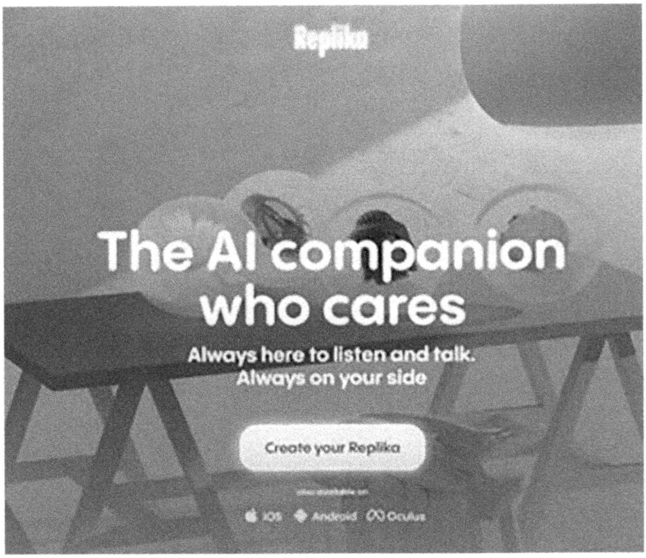

Fig. 23.11 Replika screenshot from their website (https://replika.com/)

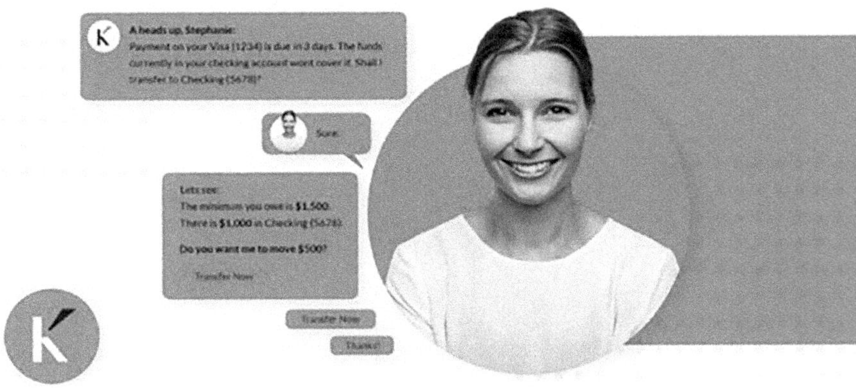

Fig. 23.12 Kasisto screenshot from their website (https://kasisto.com/)

customer experience by allowing customers to interact with the application through a conversation in natural language. The application uses natural language processing, machine learning and data analytics to deliver personalised solutions to customers.

Besides using smartphone software, it has been proposed to use the device itself as a sensor to predict anxiety levels. The results of this study were positive, predicting anxiety levels with a high degree of accuracy, suggesting that this approach could be useful in the development of mobile health applications for anxiety management (Fukazawa et al. 2019).

After reviewing the studies and results of AI applications for mental health, it can be concluded that these solutions have great potential in the ToM-based therapy, a field where they have shown to be effective in the treatment of various mental health disorders such as anxiety, depression and post-traumatic stress disorder. They have also proven to be a valuable tool for improving access to care and reducing the stigma associated with mental health. While challenges remain in the development and application of these technologies, current evidence suggests that AI-based ToM therapy applications have the potential to significantly improve mental health care and outcomes for people.

23.4 Conclusions

ToM is a cognitive construct that involves different levels of complexity. It refers to the capacity to understand other people by ascribing mental states to them, inferring what is happening in their minds. Cognitive ToM assessment includes simpler tasks such as understanding false beliefs, as well higher-order beliefs such as deception, sarcasm, metaphor or attribution of intentions. Computational theory of mind refers to the study of how the mind processes and represents information, and how it uses that information to make decisions and take actions, using computational models and algorithms. It seeks to explain the mechanisms by which mental processes, such as perception, memory, and reasoning, occur within the brain, and how they can be simulated and replicated using computer programs.

Artificial intelligence (AI) refers to the ability of machines or computer programs to perform tasks that would normally require human intelligence, such as visual perception, speech recognition, decision-making, and language translation. AI is based on the idea of creating machines that can "think" and "learn" in a way like humans.

When ToM and AI theories unite, give birth to AToM and CToM. The AToM refers to the study of how artificial intelligence systems can be designed to understand, interpret, and respond to human mental states and behaviours, such as emotions, intentions, beliefs, and desires, meanwhile CToM proposes that the mind can be understood as an information processing system, like a computer, and that mental processes can be explained as computations over mental representations.

The application of ToM-related theories in the development of technologies can lead to significant improvements in the treatment and support of people with ToM deficits or mental health problems. The practical application of these theories in technologies such as AI-based applications can improve people's quality of life by making their daily lives easier and facilitating the monitoring of their treatment. In addition, the availability of these technologies can reduce the stigma and shame associated with accepting these mental disorders.

Thus, the benefits of improving ToM skills are broad and not limited to those with deficits or in need of specific treatment. Improving ToM skills may also be useful for those who wish to improve their social skills and ability to interact more effectively with others. Integrating Theory of Mind and AI into applications can significantly improve the quality of life for people with mental disorders. These applications can be a valuable tool in improving access to healthcare, especially for those who cannot access traditional healthcare services.

References

Alves, B., I. Oliveira, C. Pratas, and A. Pereira. 2019, February. A mobile application to complement face-to-face interactions in psychological intervention for social anxiety management. In *2019 IEEE 6th Portuguese meeting on bioengineering (ENBENG)* (pp. 1–4). IEEE.

Åstebro, T., and S. Elhedhli. 2006. The effectiveness of simple decision heuristics: Forecasting commercial success for early-stage ventures. *Management Science* 52 (3): 395–409.

Banks, J. 2020. Optimus primed: Media cultivation of robot mental models and social judgments. *Frontiers in Robotics and AI* 7: 62.

Baron-Cohen, Simon. 1987. Autism and symbolic play. *British Journal of Developmental Psychology* 5 (2): 139–148. https://doi.org/10.1111/j.2044-835X.1987.tb01049.x.

Baron-Cohen, S., A.M. Leslie, and U. Frith. 1985. Does the autistic child have a "theory of mind"? *Cognition* 21 (1): 37. https://doi.org/10.1016/0010-0277(85)90022-8.

Baron-Cohen, S., T. Jolliffe, C. Mortimore, and M. Robertson. 1997. Another advanced test of theory of mind: Evidence from very high functioning adults with autism or Asperger syndrome. *Journal of Child Psychology and Psychiatry* 38 (7): 813–822. https://doi.org/10.1111/j.1469-7610.1997.tb01599.x.

Baron-Cohen, S., M. O'Riordan, V. Stone, R. Jones, and K. Plaisted. 1999. Recognition of faux pas by normally developing children and children with Asperger syndrome or high-functioning autism. *Journal of Autism and Developmental Disorders* 29 (5): 407–418. https://doi.org/10.1023/a:1023035012436.

Beatty, C., T. Malik, S. Meheli, and C. Sinha. 2022. Evaluating the therapeutic Alliance with a free-text CBT conversational agent (Wysa): A mixed-methods study. *Frontiers in Digital Health* 4: 847991.

Bradford, Elisabeth E.F., Juan-Carlos Gomez, and Ines Jentzsch. 2018. Exploring the role of self/other perspective-shifting in theory of mind with behavioural and EEG measures. *Social Neuroscience* 14 (5): 530–544. https://doi.org/10.1080/17470919.2018.1514324.

Brüne, M., S. Lissek, N. Fuchs, H. Witthaus, S. Peters, V. Nicolas, G. Juckel, and M. Tegenthoff. 2008. An fMRI study of theory of mind in schizophrenic patients with "passivity" symptoms. *Neuropsychologia* 46 (7): 1992–2001. https://doi.org/10.1016/j.neuropsychologia.2008.01.023.

Cameron, G., Cameron, D., Megaw, G., Bond, R., Mulvenna, M., O'Neill, S., … & McTear, M. 2017, July. Towards a chatbot for digital counselling. In *Proceedings of the 31st International BCS Human Computer Interaction Conference (HCI 2017) 31* pp. 1–7.

Carruthers, P. 2006. Why pretend. In *The architecture of the imagination*, ed. S. Nichols, 89–109. Oxford: Oxford University Press.

Carruthers, P., S. Laurence, and S. Stich, eds. 2006. *The innate mind Vol. 2. Culture and cognition*. Oxford University Press.

Chung, Y.S., D.-H. Kang, N.Y. Shin, S.Y. Yoo, and J.S. Kwon. 2008. Deficit of theory of mind in individuals at ultra-high risk for schizophrenia. *Schizophrenia Research* 99 (1–3): 111–118. https://doi.org/10.1016/j.schres.2007.11.012.

Cohen, A.L., D.A. Fair, Dosenbach NUF, F.M. Miezin, D. Dierker, D.C. Van Essen, B.L. Schlaggar, and S.E. Petersen. 2008. Defining functional areas in individual human brains using resting functional connectivity MRI. *NeuroImage* 41: 45–57. https://doi.org/10.1016/j.neuroimage.2008.01.066.

Cominelli, L., D. Mazzei, and D.E. De Rossi. 2018. SEAI: Social emotional artificial intelligence based on damasio's theory of mind. *Frontiers in Robotics and AI*: 5. https://doi.org/10.3389/frobt.2018.00006.

Corcoran, Rhiannon, Gavin Mercer, and Christopher D. Frith. 1995. Schizophrenia, symptomatology and social inference: Investigating a theory of mind in people with schizophrenia. *Schizophrenia Research* 17 (1): 5–13. https://doi.org/10.1016/0920-9964(95)00024-G.

Cutting, A., and J. Dunn. 1999. Theory of mind, emotion understanding, language and family background: Individual differences and inter-relations. *Child Development* 70: 853–865.

Darcy, A., A. Beaudette, E. Chiauzzi, J. Daniels, K. Goodwin, T.Y. Mariano, et al. 2022. Anatomy of a Woebot®(WB001): Agent guided CBT for women with postpartum depression. *Expert Review of Medical Devices* 19 (4): 287–301.

David, A.S. 1990. Insight and psychosis. *The British Journal of Psychiatry* 156 (6): 798–808. https://doi.org/10.1192/bjp.156.6.798.

Demirci, H. M. 2018. User experience over time with conversational agents: Case study of woebot on supporting subjective Well-being (Master's thesis, Middle East Technical University).

Deshpande, M. and V. Rao 2017. Depression detection using emotion artificial intelligence. In *2017 international conference on intelligent sustainable systems (ICISS)*, Palladam, India, pp. 858–862, https://doi.org/10.1109/ISS1.2017.8389299.

Devine, Rory T., and Claire Hughes. 2013. Silent films and strange stories: Theory of mind, gender, and social experiences in middle childhood. *Child Development* 84 (3): 989–1003. https://doi.org/10.1111/cdev.12017.

Ekman, P., and W. Friesen. 1976. *Pictures of facial affect*. Palo Alto: Consulting Psychologists Press.

Fitzpatrick, K.K., A. Darcy, and M. Vierhile. 2017. Delivering cognitive behavior therapy to young adults with symptoms of depression and anxiety using a fully automated conversational agent (Woebot): a randomized controlled trial. *JMIR mental health* 4 (2): e7785.

Fodor, J.A. 1975. *The language of thought*. Vol. 5. Harvard University Press.

———. 1997. The representational theory of mind. *American Behavioral Scientist* 40 (6): 829–841. https://doi.org/10.1177/0002764297040006010.

———. 2000. *The mind doesn't work that way: The scope and limits of computational psychology*. MIT press.

Fodor, J.A., and Z.W. Pylyshyn. 1988. Connectionism and the problem of systematicity: A critical analysis. *Cognition* 28: 3–71.

Fujita, M. 2001. AIBO: Toward the era of digital creatures. *The International Journal of Robotics Research* 20 (10): 781–794.

———. 2004. On activating human communications with pet-type robot. *AIBO* 92 (11): 1804–1813. https://doi.org/10.1109/jproc.2004.835364.

Fukazawa, Y., T. Ito, T. Okimura, Y. Yamashita, T. Maeda, and J. Ota. 2019. Predicting anxiety state using smartphone-based passive sensing. *Journal of Biomedical Informatics* 93: 103151.

Gabbay, D.M., J. Woods, and P. Thagard. 2006. *Philosophy of psychology and cognitive science*. Amsterdam: Elsevier.

Gallagher, H.L., F. Happé, N. Brunswick, P.C. Fletcher, U. Frith, and C.D. Frith. 2000. Reading the mind in cartoons and stories: An fMRI study of 'theory of mind' in verbal and nonverbal tasks. *Neuro Image* 38 (1): 11–21. https://doi.org/10.1016/s0028-3932(99)00053-6.

Gallistel, C.R., and A.P. King. 2011. *Memory and the computational brain: Why cognitive science will transform neuroscience*. Hoboken: Wiley.

Ghafouri-Fard, S., M. Taheri, M.D. Omrani, et al. 2019. Application of single-nucleotide polymorphisms in the diagnosis of autism Spectrum disorders: A preliminary study with artificial neural networks. *Journal of Molecular Neuroscience* 68: 515–521. https://doi.org/10.1007/s12031-019-01311-1.

Gigerenzer, G. 1996. On narrow norms and vague heuristics: A reply to Kahneman and Tversky. *Psychological Review* 103 (3): 592–596. https://doi.org/10.1037/0033-295X.103.3.592.

Gigerenzer, G. 2008. Why heuristics work. *Perspectives on Psychological Science* 3 (1): 20–29. https://doi.org/10.1111/j.1745-6916.2008.00058.x.

Gil, C.R. 2017. Cognicion social en la esquizofrenia. *Informaciones psiquiátricas: Publicación científica de los Centros de la Congregación de Hermanas Hospitalarias del Sagrado Corazón de Jesús* 227: 9–23.

Gilovich, T., D. Griffin, and D. Kahneman, eds. 2002. *Heuristics and biases: The psychology of intuitive judgment*. Cambridge University Press.

Goldstein, Thalia R., and Ellen Winner. 2012. Enhancing empathy and theory of mind. *Journal of Cognition and Development* 13 (1): 19–37. https://doi.org/10.1080/15248372.2011.573514.

Gopnik, A., and H.M. Wellman. 1992. Why the child's theory of mind really is a theory. *Mind & Language* 7 (1-2): 145–171. https://doi.org/10.1111/j.1468-0017.1992.tb00202.x.

Grové, C. 2021. Co-developing a mental health and wellbeing chatbot with and for young people. *Frontiers in Psychiatry* 11: 606041.

Guo, Y., Z. Wei, B.J. Keating, and H. Hakonarson. 2015. Machine learning derived risk prediction of anorexia nervosa. *BMC Medical Genomics* 9 (1): 1–8.

Gupta, M., T. Malik, and C. Sinha. 2022. Delivery of a mental health intervention for chronic pain through an artificial intelligence–enabled app (Wysa): Protocol for a prospective pilot study. *JMIR Research Protocols* 11 (3): e36910.

Happé, F.G. 1994. An advanced test of theory of mind: Understanding of story characters' thoughts and feelings by able autistic, mentally handicapped, and normal children and adults. *Journal of Autism and Developmental Disorders* 24 (2): 129–154.

Happé, Francesca G.E., Ellen Winner, and Hiram Brownell. 1998. The getting of wisdom: Theory of mind in old age. *Developmental Psychology* 34 (2): 358–362. https://doi.org/10.1037/0012-1649.34.2.358.

Heinsfeld, A.S., A.R. Franco, R.C. Craddock, A. Buchweitz, and F. Meneguzzi. 2018. Identification of autism spectrum disorder using deep learning and the ABIDE dataset. *Neuro Image: Clinical* 17: 16–23.

Hoexter, M.Q., E.C. Miguel, J.B. Diniz, R.G. Shavitt, G.F. Busatto, and J.R. Sato. 2013. Predicting obsessive–compulsive disorder severity combining neuroimaging and machine learning methods. *Journal of Affective Disorders* 150 (3): 1213–1216.

Horst, S. 1999. Symbols and computation a critique of the computational theory of mind. *Minds and Machines* 9 (3): 347–381.

Hwang, G.J., H. Xie, B.W. Wah, and D. Gašević. 2020. Vision, challenges, roles and research issues of artificial intelligence in education. *Computers and Education: Artificial Intelligence* 1: 100001.

Iglesias, M., C. Sinha, R. Vempati, S.E. Grace, M. Roy, W.C. Chapman, and M.L. Rinaldi. 2022. Evaluating a digital mental health intervention (Wysa) for workers' compensation claimants: Pilot feasibility study. *Journal of Occupational and Environmental Medicine* 65 (2): e93–e99.

Inkster, B., S. Sarda, and V. Subramanian. 2018. An empathy-driven, conversational artificial intelligence agent (Wysa) for digital mental Well-being: Real-world data evaluation mixed-methods study. *JMIR mHealth and uHealth* 6 (11): e12106.

Jolliffe, T., and S. Baron-Cohen. 1999. A test of central coherence theory: Linguistic processing in high-functioning adults with autism or Asperger syndrome: Is local coherence impaired? *Cognition* 71 (2): 149–185.

Kahn, Peter H, Batya Friedman,Deanne R. Perez-Granados and Nathan G Freier. 2004. [ACM Press Extended abstracts of the 2004 conference – Vienna, Austria (2004.04.24–2004.04.29)] *Extended abstracts of the 2004 conference on Human factors and computing systems – CHI'04 – Robotic pets in the lives of preschool children.*, 5(1), 1449. https://doi.org/10.1145/985921.986087.

Kaland, N., A. Møller-Nielsen, K. Callesen, E.L. Mortensen, D. Gottlieb, and L. Smith. 2002. A newadvanced'test of theory of mind: Evidence from children and adolescents with Asperger syndrome. *Journal of Child Psychology and Psychiatry* 43 (4): 517–528.

Kaland, N., A. Møller-Nielsen, L. Smith, et al. 2005. The strange stories test. *European Child & Adolescent Psychiatry* 14: 73–82. https://doi.org/10.1007/s00787-005-0434-2.

Kambeitz, J., L. Kambeitz-Ilankovic, S. Leucht, et al. 2015. Detecting neuroimaging biomarkers for schizophrenia: A meta-analysis of multivariate pattern recognition studies. *Neuropsychopharmacology* 40: 1742–1751. https://doi.org/10.1038/npp.2015.22.

Kobayashi, C., G.H. Glover, and E. Temple. 2006. Cultural and linguistic influence on neural bases of 'theory of mind': An fMRI study with Japanese bilinguals. *Brain and Language* 98 (2): 210–220. https://doi.org/10.1016/j.bandl.2006.04.013.

Kong, Y., J. Gao, Y. Xu, Y. Pan, J. Wang, and J. Liu. 2019. Classification of autism spectrum disorder by combining brain connectivity and deep neural network classifier. *Neurocomputing* 324: 63–68.

Koutsouleris, N., L. Kambeitz-Ilankovic, S. Ruhrmann, M. Rosen, A. Ruef, D.B. Dwyer, et al. 2018. Prediction models of functional outcomes for individuals in the clinical high-risk state for psychosis or with recent-onset depression: a multimodal, multisite machine learning analysis. *JAMA Psychiatry* 75 (11): 1156–1172.

Leo, A.J., M.J. Schuelke, D.M. Hunt, J.P. Metzler, J.P. Miller, P.A. Areán, et al. 2022. Digital mental health intervention for orthopedic patients with symptoms of depression and/or anxiety: Pilot feasibility study. *JMIR Formative Research 6 (2)*: e34889.

Leslie, A.M. 1988. Some implications of pretense for mechanisms underlying the child's. In *Developing theories of mind*, ed. J.W. Astington, P.L. Harris, and D.R. Olson, 19–46. Cambridge University Press.

Li, H., and Wang, H. 2020, August. Research on the application of artificial intelligence in education. In *2020 15th international conference on Computer Science & Education (ICCSE),* pp. 589–591. IEEE.

Li, H., N.A. Parikh, and L. He. 2018. A novel transfer learning approach to enhance deep neural network classification of brain functional connectomes. *Frontiers in Neuroscience* 12: 491.

Li, L.S.E., L.L. Wong, and K.Y.L. Yap. 2021. Quality evaluation of stress, anxiety and depression apps for COVID-19. *Journal of Affective Disorders Reports* 6: 100255.

Lincoln, A.J., E. Courchesne, L. Harms, and M. Allen. 1993. Contextual probability evaluation in autistic, receptive developmental language disorder, and control children: Event-related brain potential evidence. *Journal of Autism and Developmental Disorders* 23: 37–58.

Lindquist, M.A. 2008. The statistical analysis of fMRI data. *Statistical Science* 23 (4): 439–464. https://doi.org/10.1214/09-STS282.

Luckin, R., W. Holmes, M. Griffiths, and L.B. Forcier. 2016. *Intelligence unleashed: An argument for AI in education*. London: Pearson.

McCulloch, W.S., and W. Pitts. 1943. A logical calculus of the ideas immanent in nervous activity. *The Bulletin of Mathematical Biophysics* 5: 115–133.

McStay, Andrew. 2019. Emotional AI and EdTech: Serving the public good? *Learning, Media and Technology 45* (2): *1–14*. https://doi.org/10.1080/17439884.2020.1686016.

McStay, A., and G. Rosner. 2021. Emotional artificial intelligence in children's toys and devices: Ethics, governance and practical remedies. *Big Data & Society* 8 (1): 2053951721994877.

Meheli, S., C. Sinha, and M. Kadaba. 2022. Understanding people with chronic pain who use a cognitive behavioral therapy–based artificial intelligence mental health app (Wysa): Mixed methods retrospective observational study. *JMIR Human Factors* 9 (2): e35671.

Mehta, A., A.N. Niles, J.H. Vargas, T. Marafon, D.D. Couto, and J.J. Gross. 2021. Acceptability and effectiveness of artificial intelligence therapy for anxiety and depression (Youper): Longitudinal observational study. *Journal of Medical Internet Research* 23 (6): e26771.

Meng, Xing, Rongtao Jiang, Dongdong Lin, Juan Bustillo, Thomas Jones, Jiayu Chen, Qingbao Yu, Yuhui Du, Yu Zhang, Tianzi Jiang, Jing Sui, and Vince D. Calhoun. 2016. Predicting individualized clinical measures by a generalized prediction framework and multimodal fusion of MRI data. *Neuro Image 145*: *218–229*. https://doi.org/10.1016/j.neuroimage.2016.05.026.

Metts, A.V., M. Sears, J.H. Vargas, T. Marafon, D.D. Couto, and A.N. Niles. 2022. Perceptions of helpful and unhelpful responses to disclosures of suicidality in a sample of mobile app users. *Archives of Suicide Research*: 1–16.
Misirli, A.T., and A.B. Bener. 2014. Bayesian networks for evidence-based decision-making in software engineering. *IEEE Transactions on Software Engineering* 40 (6): 533–554.
Murphy, K.P. 2012. *Machine learning: a probabilistic perspective*. Cambridge MA: MIT press.
Muthuramalingam, S., and T. Suba Nachiar. 2016. Enhancing the security for manet by identifying untrusted nodes using uncertainty rules. *Indian Journal of Science and Technology* 9 (4): 1–9.
Nguyen, T.N., and C. Gonzalez. 2020. *Cognitive machine theory of mind*. Carnegie Mellon University.
Oakley, B.F.M., R. Brewer, G. Bird, and C. Catmur. 2016. Theory of mind is not theory of emotion: A cautionary note on the Reading the mind in the eyes test. *Journal of Abnormal Psychology* 125 (6): 818–823. https://doi.org/10.1037/abn0000182.
Oberman, L.M., E.M. Hubbard, J.P. McCleery, E.L. Altschuler, V.S. Ramachandran, and J.A. Pineda. 2005. EEG evidence for mirror neuron dysfunction in autism spectrum disorders. *Cognitive Brain Research* 24 (2): 190–198. https://doi.org/10.1016/j.cogbrainres.2005.01.014.
Pentina, I., T. Hancock, and T. Xie. 2023. Exploring relationship development with social chatbots: A mixed-method study of replika. *Computers in Human Behavior* 140: 107600.
Perner, J., U. Frith, A.M. Leslie, and S.R. Leekam. 1989. Exploration of the autistic child's theory of mind: Knowledge, belief, and communication. *Child Development* 60 (3): 689–700. https://doi.org/10.2307/1130734.
Pickup, Graham. 2006. Theory of mind and its relation to schizotypy. *Cognitive Neuropsychiatry* 11 (2): 117–192. https://doi.org/10.1080/13546800444000236.
Popenici, S.A., and S. Kerr. 2017. Exploring the impact of artificial intelligence on teaching and learning in higher education. *Research and Practice in Technology Enhanced Learning* 12 (1): 1–13.
Possati, L.M. 2022. Psychoanalyzing artificial intelligence: The case of Replika. *AI & Society* 38 (3): 1–14.
Premack, D., and G. Woodruff. 1978. Does the chimpanzee have a theory of mind? *Behavioral and Brain Sciences* 1 (4): 515–526. https://doi.org/10.1017/S0140525X00076512.
Prochaska, J.J., E.A. Vogel, A. Chieng, M. Kendra, M. Baiocchi, S. Pajarito, and A. Robinson. 2021. A therapeutic relational agent for reducing problematic substance use (Woebot): Development and usability study. *Journal of Medical Internet Research* 23 (3): e24850.
Putnam, H. 1967. Psychophysical predicates. In *Art, Mind, and Religion*, ed. W.H. Capitan and D.D. Merrill, 429–440. Pittsburgh: University of Pittsburgh Press.
Rabinowitz, N., F. Perbet, F. Song, C. Zhang, S. A. Eslami, and M. Botvinick. 2018, July. Machine theory of mind. In *International conference on machine learning,* pp. 4218–4227. PMLR.
Ramyead, A., E. Studerus, M. Kometer, M. Uttinger, U. Gschwandtner, P. Fuhr, and A. Riecher-Rössler. 2016. Prediction of psychosis using neural oscillations and machine learning in neuroleptic-naïve at-risk patients. *The World Journal of Biological Psychiatry* 17 (4): 285–295.
Sabbagh, Mark A., and Jessica Flynn. 2006. Mid-frontal EEG alpha asymmetries predict individual differences in one aspect of theory of mind: Mental state decoding. *Social Neuroscience* 1 (3–4): 299–308. https://doi.org/10.1080/17470910601029163.
Sabbagh, M.A., and M. Taylor. 2000. Neural correlates of theory-of-mind reasoning: An event-related potential study. *Psychological Science* 11 (1): 46–50. https://doi.org/10.1111/1467-9280.00213.
Sabbagh, M.A., L.C. Bowman, L.E. Evraire, and J.M. Ito. 2009. Neurodevelopmental correlates of theory of mind in preschool children. *Child Development* 80 (4): 1147–1162.
Samsonovich, Alexei V. 2020. Socially emotional brain-inspired cognitive architecture framework for artificial intelligence. *Cognitive Systems Research 60*: *57–76*. https://doi.org/10.1016/j.cogsys.2019.12.002.

Samuels, R. 2005. The complexity of cognition: Tractability arguments for massive modularity. In *The innate mind: Structure and contents*, ed. P. Carruthers, S. Laurence, and S. Stich, 107–121. Oxford University Press.

———. 2010. Classical computationalism and the many problems of cognitive relevance. *Studies in History and Philosophy of Science Part A* 41 (3): 280–293.

Samuels, R., and S.P. Stich. 2004. Rationality and psychology. In *The Oxford handbook of rationality*, ed. A.R. Mele and P. Rawling, 279–300. Oxford: Oxford University Press.

Sarfati, Y., M.-C. Hardy-Baylé, C. Besche, and D. Widlöcher. 1997. Attribution of intentions to others in people with schizophrenia: A non-verbal exploration with comic strips. *Schizophrenia Research* 25 (3): 199–209. https://doi.org/10.1016/S0920-9964(97)00025-X.

Sarica, A., A. Cerasa, and A. Quattrone. 2017. Random forest algorithm for the classification of neuroimaging data in Alzheimer's disease: a systematic review. *Frontiers in Aging Neuroscience* 9: 329.

Scassellati, B. 2002. Theory of mind for a humanoid robot. *Cognitive Science* 12 (1): 13–24. https://doi.org/10.1023/a:1013298507114.

Sharbrough, F., G.E. Chatrian, R.P. Lesser, H. Luders, M. Nuwer, and T.W. Picton. 1991. American electroencephalographic society guidelines for standard electrode position nomenclature. *Journal of Clinical Neurophysiology: Official Publication of the American Electroencephalographic Society* 8 (2): 200–202. http://europepmc.org/abstract/MED/2050819.

Shen, D., C.Y. Wee, D. Zhang, L. Zhou, and P.T. Yap. 2014. Machine learning techniques for AD/MCI diagnosis and prognosis. In *Machine learning in healthcare informatics. Intelligent systems reference library, vol 56*, ed. S. Dua, U. Acharya, and P. Dua. Berlin, Heidelberg: Springer. https://doi.org/10.1007/978-3-642-40017-9_8.

Singh, J., and M.A. Hamid. 2022. Cognitive computing in mental healthcare: A review of methods and technologies for detection of mental disorders. *Cognitive Computation* 14: 2169–2186. https://doi.org/10.1007/s12559-022-10042-2.

Skjuve, M., A. Følstad, K.I. Fostervold, and P.B. Brandtzaeg. 2022. A longitudinal study of human–chatbot relationships. *International Journal of Human-Computer Studies* 168: 102903.

Speck, A., E. Scholte, and I. Van Berckelaer-Onnes. 2010. Theory of mind in adults with HFA and Asperger syndrome. *Journal of Autism and Developmental Disorders* 40: 280–289.

Stark, L. and J. Hoey. 2021. The ethics of emotion in artificial intelligence systems. In *Proceedings of the 2021 ACM conference on fairness, accountability, and transparency*, 782–793. https://doi.org/10.1145/3442188.3445939.

Stone, Valerie E., Simon Baron-Cohen, and Robert T. Knight. 1998. Frontal lobe contributions to theory of mind. *Journal of Cognitive Neuroscience* 10 (5): 640–656. https://doi.org/10.1162/089892998562942.

Thrun, S. 2002. Probabilistic robotics. *Communications of the ACM* 45 (3): 52–57.

Trothen, T.J. 2022. Replika: Spiritual enhancement technology? *Religions* 13 (4): 275.

Turing, A.M. 1936. On computable numbers, with an application to the Entscheidungsproblem. *Journal of Mathematics 58*: 345–363.

Veronese, E., U. Castellani, D. Peruzzo, M. Bellani, and P. Brambilla. 2013. Machine learning approaches: From theory to application in schizophrenia. *Computational and Mathematical Methods in Medicine*: 867924.

Vinden, Penelope G. 1999. Children's understanding of mind and emotion: A multi-culture study. *Cognition & Emotion* 13 (1): 19–48. https://doi.org/10.1080/026999399379357.

Wan, E. 2021. " I'm like a wise little person": Notes on the Metal Performance of Woebot the Mental Health Chatbot. *Theatre Journal 73* (3): E-21.

Wasil, A.R., S. Gillespie, R. Shingleton, C.R. Wilks, and J.R. Weisz. 2020. Examining the reach of smartphone apps for depression and anxiety. *American Journal of Psychiatry* 177 (5): 464–465.

Watson, Anne C., Charisse Linkie Nixon, Amy Wilson, and Laura Capage. 1999. Social interaction skills and theory of mind in young children. *Developmental Psychology* 35 (2): 386–391. https://doi.org/10.1037/0012-1649.35.2.386.

Williams, J., S.M. Fiore, and F. Jentsch. 2022. Supporting artificial social intelligence with theory of mind. *Frontiers in Artificial Intelligence* 5: 750763. https://doi.org/10.3389/frai.2022.750763.

Wimmer, H., and J. Perner. 1983. Beliefs about beliefs: Representation and constraining function of wrong beliefs in young children's understanding of deception. *Cognition* 13 (1): 103–128. https://doi.org/10.1016/0010-0277(83)90004-5.

Winfield, A.F.T. 2018. Experiments in artificial theory of mind: From safety to story-telling. *Frontiers in Robotics and AI* 5. https://doi.org/10.3389/frobt.2018.00075.

Xiao, Yuan, Zhihan Yan, Youjin Zhao, Bo Tao, Huaiqiang Sun, Fei Li, Li Yao, Wenjing Zhang, Shah Chandan, Jieke Liu, Qiyong Gong, John A. Sweeney, and Su Lui. 2017. Support vector machine-based classification of first episode drug-naïve schizophrenia patients and healthy controls using structural MRI. *Schizophrenia Research 214*: *11–17*. https://doi.org/10.1016/j.schres.2017.11.037.

Yarrington, J.S., J. Lasser, D. Garcia, J.H. Vargas, D.D. Couto, T. Marafon, et al. 2021. Impact of the COVID-19 pandemic on mental health among 157, 213 Americans. *Journal of Affective Disorders* 286: 64–70.

Young, L., D. Dodell-Feder, and R. Saxe. 2010. What gets the attention of the temporo-parietal junction? An fMRI investigation of attention and theory of mind. *Neuropsychologia* 48 (9): 2658–2664. https://doi.org/10.1016/j.neuropsychologia.2010.05.012.

Zalla, T., A.M. Sav, A. Stopin, et al. 2009. Faux pas detection and intentional action in Asperger syndrome. A replication on a french sample. *Journal of Autism and Developmental Disorders* 39: 373–382. https://doi.org/10.1007/s10803-008-0634-y.

Zelazo, P.D., S. Jacques, J.A. Burack, and D. Frye. 2002. The relation between theory of mind and rule use: Evidence from persons with autism-spectrum disorders. *Infant and Child Development: An International Journal of Research and Practice* 11 (2): 171–195. https://doi.org/10.1002/icd.304.

Zhang, Jianhua, Zhong Yin, Peng Chen, and Stefano Nichele. 2020. Emotion recognition using multi-modal data and machine learning techniques: A tutorial and review. *Information Fusion 59*: 103–126. https://doi.org/10.1016/j.inffus.2020.01.011.

Zhang, W., C. Yang, Z. Cao, Z. Li, L. Zhuo, Y. Tan, et al. 2023. Detecting individuals with severe mental illness using artificial intelligence applied to magnetic resonance imaging. *EBioMedicine* 90: 104541.

Zheng, C., Y. Xia, Y. Pan, et al. 2016. Automated identification of dementia using medical imaging: a survey from a pattern classification perspective. *Brain Informatics* 3: 17–27. https://doi.org/10.1007/s40708-015-0027-x.

Zhu, Y., M. Janssen, R. Wang, and Y. Liu. 2022. It is me, chatbot: Working to address the COVID-19 outbreak-related mental health issues in China. User experience, satisfaction, and influencing factors. *International Journal of Human–Computer Interaction* 38 (12): 1182–1194.

GPSR Compliance
The European Union's (EU) General Product Safety Regulation (GPSR) is a set of rules that requires consumer products to be safe and our obligations to ensure this.

If you have any concerns about our products, you can contact us on

ProductSafety@springernature.com

In case Publisher is established outside the EU, the EU authorized representative is:

Springer Nature Customer Service Center GmbH
Europaplatz 3
69115 Heidelberg, Germany

www.ingramcontent.com/pod-product-compliance
Ingram Content Group UK Ltd.
Pitfield, Milton Keynes, MK11 3LW, UK
UKHW020223290926
15470UKWH00001B/356
* 9 7 8 3 0 3 1 4 6 7 4 1 7 *